Extensional Tectonics: Regional-scale Processes

Part I

AF704479

WITHDRAWN

Geological Society Key Issues in Earth Sciences
Society Book Editors
A. J. FLEET (CHIEF EDITOR)
P. DOYLE
F. J. GREGORY
J. S. GRIFFITHS
A. J. HARTLEY
R. E. HOLDSWORTH
A. C. MORTON
N. S. ROBINS
M. S. STOKER
J. P. TURNER

KEY ISSUES IN EARTH SCIENCES

Extensional Tectonics: Regional-scale Processes

Part I

COMPILED BY

R.E. HOLDSWORTH
University of Durham, UK

J.P. TURNER
University of Birmingham, UK

2002
Published by
The Geological Society
London

QE
511.4
.E983
2002

THE GEOLOGICAL SOCIETY

The Geological Society of London (GSL) was founded in 1807. It is the oldest national geological society in the world and the largest in Europe. It was incorporated under Royal Charter in 1825 and is Registered Charity 210161.

The Society is the UK national learned and professional society for geology with a worldwide Fellowship (FGS) of 9000. The Society has the power to confer Chartered status on suitably qualified Fellows, and about 2000 of the Fellowship carry the title (CGeol). Chartered Geologists may also obtain the equivalent European title, European Geologist (EurGeol). One fifth of the Society's fellowship resides outside the UK. To find out more about the Society, log on to *www.geolsoc.org.uk.*

The Geological Society Publishing House (Bath, UK) produces the Society's international journals and books, and acts as European distributor for selected publications of the American Association of Petroleum Geologists (AAPG), the American Geological Institute (AGI), the Indonesian Petroleum Association (IPA), the Geological Society of America (GSA), the Society for Sedimentary Geology (SEPM) and the Geologists' Association (GA). Joint marketing agreements ensure that GSL Fellows may purchase these societies' publications at a discount. The Society's online bookshop (accessible from *www.geolsoc.org.uk*) offers secure book purchasing with your credit or debit card.

To find out about joining the Society and benefiting from substantial discounts on publications of GSL and other societies worldwide, consult *www.geolsoc.org.uk,* or contact the Fellowship Department at: The Geological Society, Burlington House, Piccadilly, London W1J 0BG: Tel. +44 (0)20 7434 9944; Fax +44 (0)20 7439 8975; Email: *enquiries@geolsoc.org.uk.*

For information about the Society's meetings, consult *Events* on *www.geolsoc.org.uk.* To find out more about the Society's Corporate Affiliates Scheme, write to *enquiries@geolsoc.org.uk.*

Published by The Geological Society from:
The Geological Society Publishing House
Unit 7, Brassmill Enterprise Centre
Brassmill Lane
Bath BA1 3JN, UK
(*Orders:* Tel. +44 (0)1225 445046
Fax +44 (0)1225 442836)
Online bookshop: http://bookshop.geolsoc.org.uk

The publishers make no representation, express or implied, with regard to the accuracy of the information contained in this book and cannot accept any legal responsibility for any errors or omissions that may be made.

© The Geological Society of London 2002 All rights reserved. No reproduction, copy or transmission of this publication may be made without written permission. No paragraph of this publication may be reproduced, copied or transmitted save with the provisions of the Copyright Licensing Agency, 90 Tottenham Court Road, London W1P 9HE. Users registered with the Copyright Clearance Center, 27 Congress Street, Salem, MA 01970, USA: the item-fee code for this publication is 0305-8719/01$15.00.

British Library Cataloguing in Publication Data
A catalogue record for this book is available from the British Library.

ISBN 1-86239-114-9

Typeset by Alden Bookset, UK
Printed by The Alden Press, Oxford, UK.

Distributors

USA
AAPG Bookstore
PO Box 979
Tulsa
OK 74101-0979
USA
Orders: Tel. +1 918 584-2555
Fax +1 918 560-2652
E-mail *bookstore@aapg.org*

India
Affiliated East-West Press PVT Ltd
G-1/16 Ansari Road, Daryaganj,
New Delhi 110 002
India
Orders: Tel. +91 11 327-9113
Fax +91 11 326-0538
E-mail *affiliat@nda.vsnl.net.in*

Japan
Kanda Book Trading Co.
Cityhouse Tama 204
Tsurumaki 1-3-10
Tama-shi
Tokyo 206-0034
Japan
Orders: Tel. +81 (0)423 57-7650
Fax +81 (0)423 57-7651

Contents

Extensional tectonics: regional-scale processes

J. P. TURNER[1] & R. E. HOLDSWORTH[2]
[1]*University of Birmingham, School of Earth Sciences, Birmingham B15 2TT, UK (e-mail: j.p.turner@bham.ac.uk)*
[2]*Reactivation Research Group, University of Durham, Dept of Geological Sciences, Durham DH1 3LE, UK (e-mail: r.e.holdsworth@durham.ac.uk)*

Introduction

Aims of this book and background

The wealth of research concerning applications of structural geology and tectonics to problems in petroleum geoscience has provided a rich vein of scientific publishing for the Geological Society of London over the last twenty years or so. The research on which this prolific output was based has been facilitated by:

(i) the openness of the petroleum industry in sharing its commercially valuable subsurface datasets and ideas with the research community, and

(ii) the enthusiasm with which structural geologists embraced the shift from deeper crustal, 'ductile' deformation processes toward problems in brittle faulting and fracturing. Scientific collaboration between the petroleum industry and the structural geological research community has been tremendously successful through the 1980s and 90s. In the UK, a large number of collaborative research programmes and groups, funded jointly by industry and government agencies, have used structural geology and tectonics to tackle a number of major issues in petroleum geoscience, e.g. British Institutes Reflection Profiling Syndicate (BIRPS), UK DTI-LINK scheme, NERC Ocean Margins Programmes, Fault Analysis Group, Rock Deformation Research Group, Fault Dynamics Research Group.

The principal objective of the Geological Society's new *Key Issues in Earth Sciences* series is to provide a coherent overview of a current issue in geoscience using a mixture of ground-breaking papers and synoptic reviews that hitherto have been widely scattered among its many publications. As such, it aims to provide a readily accessible introduction to current issues in geoscience for students, HE teachers and professionals. To convey an impression of the history of development of ideas contained herein, papers are mostly reproduced in the order in which they were published. This volume on regional-scale processes is one of a two-part contribution to the *Key Issues in Earth Sciences* series on the general theme of extensional tectonic processes and their implications for petroleum geoscience. Thus, an accompanying part on faulting and fault-related processes (Holdsworth & Turner 2002) addresses all aspects of extensional faulting, with an emphasis on practical approaches to improved fault interpretation from seismic data.

This introductory paper summarises both the context of the present volume and the contents of each paper included here. Papers reproduced in the current volume are indicated in bold, whilst others referred to in the text are in the reference list as normal, including those from the companion part of this volume.

Role of structural geological research in petroleum geoscience

These have been exciting times for structural geologists working in petroleum geoscience, whether in academia, consultancy or oil companies. Fundamental advances in extensional tectonics on which the Geological Society has published include:

- The application to a diversity of sedimentary basins worldwide of McKenzie's (1978) ideas on the mechanical, thermal and stratigraphic consequences for the lithosphere of rifting (e.g. **Dewey 1982**; **Kusznir & Park 1987**; **Barr 1987**; **Kusznir *et al.* 1995**).

- Appreciation of the link between sedimentation and deformation and the development of models of interacting faulting and sedimentation (e.g. **Leeder & Gawthorpe 1987**; **Leeder *et al.* 1995**; Roberts & Jackson 1991).

- The widespread recognition of fault reactivation in a variety of geodynamic settings and development of the concepts of sedimentary basin inversion (e.g. **Williams *et al.* 1989**; **Hayward & Graham 1989**; **Sibson 1995**; Eisenstadt & Withjack 1995; **Stewart *et al.* 1996**; Holdsworth *et al.* 1997).

- Elucidation of order in complex natural fracture systems and the realization that faults and fractures grow and interconnect systematically according to simple deterministic rules (e.g. Walsh & Watterson 1991; Childs *et al.* 1995;

From: HOLDSWORTH, R.E. & TURNER, J.P. (compilers) *Extensional Tectonics: Regional-scale Processes.* The Geological Society, Key Issues in Earth Sciences, **2(1)**, 1–9.
1476-315X/02/$15.00 © The Geological Society of London 2002.

Walsh *et al.* 1996; Fisher & Knipe 1998; Ottesen Ellevset *et al.* 1998; Gibson 1998).

A natural consequence of the sharing of subsurface data between the petroleum industry and research communities has been widespread access to state-of-the-art datasets and the use of leading edge interpretational software in higher education. Virtually every Earth Sciences degree programme includes courses that offer the opportunity to work with seismic images of producing oil and gas fields and many students have had first-hand experience of the sometimes spectacular capabilities of 3-D reflection seismic data. Whilst many of the papers reproduced in this volume are based at least partly on data collected from traditional field studies, seismic interpretation has become a new kind of virtual fieldwork. Students enjoy making geological interpretations of seismic data and they benefit from the constraints it allows them to place on the scale of geological processes in a way that outcrop geology rarely permits.

Has all this research led to the discovery and production of more oil and gas? Undoubtedly the answer is yes, particularly with the increased maturity of major petroleum provinces and prioritization of the search for the so-called subtle trap. But the precise impact of geological research on exploration and production is difficult to quantify. Early in the exploration and production history of major petroliferous basins like the North Sea, the impact of research was probably minimal because most discoveries were made by drilling structural culminations, or 'highs', and more or less hoping for the best! However, the application of increasingly sophisticated geological modelling in combination with high-resolution reflection seismic data, especially 3-D seismic, has meant that subtle trapping configurations such as sub-salt structures, stratigraphic pinchouts, hanging wall closures and even fractured basement reservoirs are much more common drilling targets today than twenty years ago.

In addition to the tangible achievements of structural geological research in petroleum geoscience, it is worth remembering also its important contribution toward raising the general level of intellectual curiosity among petroleum geoscientists, with attendant benefits for professional efficacy and overall quality of life at work. For versatile scientists with a general interest in the structure and evolution of our planet, commerce and technology, geoscientific work in the petroleum industry is still the most stimulating of careers.

An example of the commercial impact of structural geological research: the Brent Province, northern North Sea

An illustration of the role of structural geology in evolving exploration and production strategies is provided by the North Viking Graben of the northern North Sea. Here, fields like Brent, Ninian, Statfjord and Snorre have produced billions of barrels of oil trapped in Middle Jurassic sandstone reservoirs in prominent tilted normal fault footwalls. Many of these 'supergiant' fields (e.g. Brent total reserves: 2.12 million barrels oil, 4400 billion cubic feet gas) were discovered in the early 1970s with the aid of single-fold reflection seismic data whose quality barely compares with the high-resolution data obtained from modern seismic surveys. But even on these old data, the Brent Field is visible as a prominent structural 'high' draped unconformably by a blanket of Cretaceous and Tertiary post-rift deposits.

Twenty five years on, the reservoirs are depressurized and the fields are long past their peak production. In order to realise the full value of the infrastructure that supports them, the commercial incentive for new geological thinking is enormous if field life is to be maximized. To a large extent, renaissance in the exploration for new reserves in the Brent Province has been stimulated by the ideas of fault scarp degradation coupled with the wealth of detailed research on prediction of the sealing capacity of seismically imaged faulting.

Fault scarp degradation describes the processes that occur when the footwall regions of normal faults in rift basins are uplifted, leading to their gravitational collapse into the basin. This collapse is facilitated by swarms of low-angle, curviplanar detachments that have the effect of emplacing significant volumes of Middle Jurassic reservoir into the otherwise unprospective hanging walls of the faults. A simple enough idea, but one whose commercial potential was not appreciated until the following breakthroughs:

(i) Reverse and forward flexural-isostatic modelling methods that quantified the magnitude of the uplift, often elevating the Jurassic footwalls above palaeo-sea level (e.g. Kusznir *et al.* 1991; Roberts *et al.* 1998).

(ii) Routine use of high-resolution 3-D seismic data that allowed seismic intepreters to delineate the fault-bounded slices of footwall reservoir material, now located beneath the denuded footwall crests (e.g. Underhill *et al.* 1997).

(iii) Development of methodologies that place quantifiable constraints on the sealing capacity of geological faulting, thereby reducing the commercial risk on these often small and highly exotic, fault-defined traps (e.g. Ottesen Ellevset *et al.* 1998; Fisher & Knipe 1998).

Context of the papers in the present volume

Mechanical and thermal consequences of extensional basin formation

All of the research presented in the papers reproduced in this volume has taken place against the backdrop of the plate tectonics paradigm (McKenzie & Parker 1967). Plate tectonics is an efficient, dominantly convective method for the Earth to lose its heat by the generation, cooling, subsidence and subduction of an outermost conducting layer segmented into a mosaic of plates (Oxburgh 1974). Given the three principal driving mechanisms for sedimentary basin subsidence (and uplift) – thermal effects, addition/removal of load and lithospheric thickness changes – plate tectonics has had wide implications for every aspect of sedimentary basin research. The paper of **Dewey** serves to reinterpret much of our understanding of the extensional basins of the British Isles in the context of the then youthful ideas embodied in plate tectonics.

The uniform lithospheric stretching model of McKenzie (1978) provides a unifying theory for the geological history of most extensional basins and Dewey devotes considerable space to examining its full implications. It describes an initial syn-rift phase of instantaneous, depth-independent extension (β = length of extended lithosphere/length of unextended lithosphere) and fault-controlled subsidence in which the base of the lithosphere becomes elevated. Subsequent cooling and thickening of the lithospheric mantle leads to a near-exponential phase of regional, post-rift (thermal) subsidence. Uniform stretching continues to serve as the basis for our explanation of a diverse range of geological phenomena in extensional basins. These include:

- Tripartite subdivision of sedimentary fills into pre-rift, syn-rift and post-rift sequences. Pre-rift and syn-rift sequences generally comprise continental and paralic facies cut by rift-related normal faults. Syn-rift successions usually exhibit pronounced thickening of stratigraphic units into the hanging walls of normal faults, and they are typically overlain by a blanket of marine post-rift sediment often several kilometres thick.
- A locally angular unconformity separating the post-rift succession from underlying syn-rift strata, with uplift and erosion concentrated at the crests of the footwalls of the major normal faults.
- Progressive onlap of the post-rift succession onto the cratonic margins of the underlying rift creating an overall 'Steer's Head' geometry in which post-rift sediments are thickest above the locus of rifting, becoming thinner with distance from the rift.
- Sustained post-rift sedimentation (>100 Ma) in response to regional 'sagging' of the lithosphere during which subsidence rate and heat flow decline steadily.

Dewey places considerable emphasis on the importance of the ratio of crust : lithosphere thickness and, therefore, the effect on average lithosphere density of uniform stretching. Given a 31 km-thick crust and 125 km-thick lithosphere, uniform stretching of the lithosphere will result in an increase in bulk lithosphere density leading to surface subsidence and the formation of a sedimentary basin. In stable, cratonic settings, however, where the crust/lithosphere ratio may be less than the critical 0.14 (due to the presence of a thick lithospheric mantle), uniform stretching will result in surface uplift. Conversely, island arcs are characterized by anomalously thin mantle lithosphere meaning that small amounts of stretching produce large magnitudes of subsidence.

Barr examines quantitatively the relation between lithosphere stretching, footwall uplift and hanging wall subsidence, processes that exert primary controls over the architecture and sedimentary infilling of modern and ancient extensional basins. For an instantaneously stretched, isostatically compensated basin, he defines the level of no vertical motion beneath which the lithosphere rises, and above which it subsides. Its depth will depend on the crust/lithosphere thickness and, for sediment-filled basins in normal lithosphere, it lies at depths of 7–8 km. In this scheme, footwall uplift is a consequence of vertical movement of the sole to a system of detached normal faults that, in general, will be localized within the brittle-ductile transition zone, beneath the level of no vertical motion. Barr's model successfully predicts the magnitude of footwall uplift – which depends primarily on initial fault spacing – and the dynamics of the uplift, and it shows good agreement with basins in the North Sea, Armorican margin and the Aegean Sea.

A significant departure from the uniform stretching model appears to be typified by the metamorphic core complexes of the locally hyper-extended Basin & Range Province of the western USA described by **Hamilton**. Here, mylonitized, domiform extensional detachments juxtapose amphibolite facies metamorphosed Proterozoic rocks in their footwalls with highly rotated Oligo-Holocene syn-rift siliciclastic sediment and basalt in their hanging walls. On average, the detachments accommodate $\beta = 2.0$ and characterize the upper crust to depths of *c.* 20 km. The detachment faults may originate at much steeper, 'Andersonian' angles, attaining progressively gentler inclinations as extension proceeds, according to the so-called domino mechanism of normal faulting. One consequence of stretching and thinning a volume of crust containing 'rigid' faults is that the fault planes will rotate to progressively gentler inclinations with increasing thinning. This is often envisioned in terms of an arrangement of initially vertical dominoes rotating within a horizontal shear couple. Ultimately, the upper crust becomes completely attenuated leading to the isostatic rise of the footwalls to the detachments and exhumation of their deeper, mylonitized segments. Descriptions of metamorphic core complexes such as those of Hamilton are important for two reasons:

(1) They are now recognized throughout the world and seismic examples have been described recently from the prospective Norwegian margin of the NE Atlantic (Lundin & Dore 1997).

(2) They exemplify non-uniform stretching of the upper and middle crust whereby the locus of fault-driven subsi-

dence of the uppermost crust is physically offset from the locus of thinning and isostatic uplift.

If it can be shown that the mode of extension exemplified by Hamilton's metamorphic core complexes takes place on the scale of the lithosphere, the implications of depth-dependent stretching for mechanisms of sedimentary basin formation are profound. In this scenario, McKenzie's (1978) uniform stretching model occupies one end of a spectrum of modes of lithosphere extension. The other extreme is represented by the non-uniform stretching model of Wernicke (1985) in which the lithosphere extends by simple shear accommodated by a trans-crustal detachment separating syn- and post-rift basins by tens of kilometres.

Lithospheric strength in the continents and its influence on the broad structural architecture and geological history of extensional sedimentary basins is addressed by **Kusznir & Park**. Mathematical models are employed to examine the large-scale rheological behaviour of the continental lithosphere and its evolution with time. Rheology describes the relation between the forces that cause deformation (stress) and the deformation that occurs (strain). Two broad classes of rheological behaviour are recognized in rocks: 'brittle' and 'ductile', the former describing materials which do not yield until their ultimate strength is exceeded and they fail along discrete surfaces, whereas ductile materials accumulate permament (non-elastic) strain by distributed flow before failure. Rheology is at the very heart of structural geology and tectonics because it links grain-scale deformation mechanisms (e.g. dislocation creep) to lithosphere-scale tectonics.

Kusznir & Park's models are calibrated by experimentally derived data concerning the deformation behaviour of minerals thought to be representative of the different parts of the lithosphere. Lithospheric strength is shown to be strongly dependent on geothermal gradient, the length of time since the application of the tectonic force and, critically, the thickness of relatively weak quartzo-feldspathic crust. Kusznir & Park demonstrate that whether the lithosphere undergoes strain hardening (progressive strengthening with time) or softening (progressive weakening with time) during extension is largely a trade-off between two factors: (i) the strengthening it experiences due to the replacement of relatively weak crust with strong mantle lithosphere, versus (ii) the weakening effect of increasing its geothermal gradient. The rate of stretching appears to be critical in determining which of these processes dominates. If it is slow (strain rates $<5 \times 10^{-15}$ sec^{-1}, or $\beta = 1.1$ in 634 Ma), the replacement of crust by mantle will dominate because heat loss to the surface will keep pace with extension (i.e. maintenance of thermal equilibrium throughout stretching) whereas rapid extension is dominated by increased geothermal gradient, leading to strain softening.

Kusznir & Park present scenarios in which rapid extension localizes deformation into narrow basins underlain by strain-softened lithosphere with high β values. Conversely, slow extension promotes stretching of the rift flanks and the development of broader basins with modest β values. These models are important for two reasons:

(1) They are predictive and the results can be compared with observational data collected from basins around the world. For example, the models predict that surface heat flows in excess of 65 mWm^{-2} are required for significant extension to occur, a value that seems to agree well with observed heat flow measurements from active rifts.

(2) Intra-crustal detachments, the precursors to non-uniform modes of lithospheric stretching, are shown to be a natural consequence of the marked strength discontinuities that develop at major compositional boundaries such as the Moho.

An emerging theme thus far is that extending lithosphere is a highly dynamic entity. Consequently, it is inherently difficult to reconstruct the mode of lithospheric stretching from ancient extensional basins because, as stretching progresses, the mode of extension may change drastically thereby masking the evidence for earlier processes. As Dewey points out, as β progresses beyond 2.0, mafic magmas will be injected into the lithosphere in rapidly increasing volume leading to significant changes in its density and rheology and, consequently, progressive departure from uniform stretching. Therefore, in reconstructing extensional tectonic processes, it is essential to examine a wide variety of basins at a range of stages in their development – from embryonic rifts to fully fledged rifted continental (passive) margins.

Morley provides a comprehensive account of young rift basins drawing mainly on examples from the Gulf of Suez and East Africa. He concentrates on the architecture and kinematic evolution of the fault systems that define them and his data, derived chiefly from examination of commercial seismic profiles and outcrops/satellite images, serve to illustrate the critical influence of pre-existing fabric on fault geometry. Unlike the mechanically isotropic layers in most mathematical simulations, the crust is highly anisotropic, criss-crossed by networks of fractures which, in a favourably oriented stress field, will undergo reactivation to form larger faults. This anisotropy commonly leads to highly irregular rift architecture in which rifts are divided into segments tens to hundreds of kilometres long.

Rifted continental margins are the scars of the break-up of continents and they represent the ultimate product of lithospheric extension. The Galicia margin (NW Iberia) is an excellent rifted margin at which to study the processes of lithospheric extension because it is a starved margin, covered only by a thin and discontinuous sedimentary veneer. Consequently, it is much easier to study the syn- and pre-rift basement here compared to more typical extensional basins which are buried beneath three kilometres of post-rift sediment. **Boillot *et al.*** reproduce spectacular seismic images of the Galicia margin, clearly showing highly rotated blocks of continental basement bounded by curviplanar normal faults detaching along a common level, the so called S-reflector. The significance of this undulating sole detachment is puzzling; it appears to separate crust from mantle, with the complication that the mantle rocks have been serpentinized during syn-rift hydrothermal activity. Given the significance of low-angle detachments in depth-dependent stretching, although Boillot *et al.* are careful not to imply as much, there is the intriguing possibility that the S-reflector is a fossilized trans-crustal detachment that defines the base of almost completely attenuated, stretched continental crust.

Stratigraphic consequences of extensional tectonics

Investigation of the inter-relations between hitherto unconnected sub-disciplines within geology, and indeed the sciences in general, has been one of the most fruitful recent developments and today inter-disciplinarity is a top priority among the research funding agencies. The leap in understanding of sedimentary basin dynamics has provided sedimentologists with an important addition to their interpretational toolkit – tectonics now joins eustasy and climate as one of the most significant allocyclic, or extrabasinal, controls on the sedimentary rock record and isolating their various signatures is a major research theme.

The models of **Leeder & Gawthorpe** of sedimentary facies development in half-graben basins in a range of depositional settings – continental, coastal, deep marine and carbonate ramp – provided striking new insights to tectonic controls over sedimentation. Their models are based on a large database collated from mainly active extensional basins where observations by sedimentologists and geomorphologists emphasize the primary influence that tectonics exerts upon geomorphology and sediment transfer mechanisms. Recurring themes in Leeder & Gawthorpe's models include the role of footwall uplift and erosion as the major sediment source, and the importance of en-echelon and transversely oriented faults in enabling sediment transfer from the uplifting footwall into the hanging wall basin.

Tectono-sedimentary interactions in ancient half-graben basins are also widely recognized and, in the North Sea, appreciation of the importance of relay ramps in controlling the entry of reservoir-quality sands into the basin has been a particularly important advance. However, because active extensional basins enable more or less direct measurement of the rates of processes, they continue to provide fresh insights into interpretation of the rock record. This point is highlighted by **Leeder, Seger & Stark** in their detailed description of the active Megara basin of Greece. They demonstrate spectacular differences in the rate of drainage basin growth and form between adjacent drainage basins along the Psatha-Skinos fault system. Given the uniformity of footwall uplift between these drainage basins (*c.* 350 m/Ma), the chief control on drainage basin development is shown to be the footwall lithological substrate. Whilst it was not possible to quantitatively compare sediment yields from the different drainage basins, it is clear that those parts of the uplifting footwall composed of poorly consolidated Plio-Pleistocene sediment erode considerably faster than drainage basins composed of limestone and peridotite.

Mathematical forward models of rifting can reproduce most thermo-mechanical and stratigraphic features of extensional basins. The methodology of **Kusznir, Roberts & Morley** has achieved particular success in a wide range of settings. Taking the uniform lithospheric stretching model as a starting point, they incorporate the effects of upper crustal faulting and lower crust/upper mantle plastic stretching, lithosphere thermal perturbations and re-equilibration, sediment loading and flexural isostasy to simulate the syn- and post-rift evolution of extensional basins. Unlike uniform stretching, which assumes Airy isostasy and distributed uniform ductile deformation of the continental lithosphere, Kusznir *et al.* present a geologically more realistic model, taking account of the finite flexural strength of the lithosphere and the vertical partitioning of strain between upper crustal brittle faulting (simple shear) and deeper, plastic deformation (pure shear). The models use β values computed directly from the reverse modelling of post-rift subsidence, with lithosphere flexural strength (effective elastic thickness: T_e) derived iteratively by matching observed and computed stratigraphic geometries and facies patterns. Thus, by using known palaeobathymetric markers like coals and shoreface sands to calibrate the models, they have provided petroleum geologists with a powerful tool for simulating rift architecture and water depth during key stages of reservoir deposition.

Uplift and erosion of sedimentary basins and the role of overpressure

Uplift describes vertical motion relative to a fixed datum like the geoid. It usually leads directly to surface erosion with subsequent amplification of the initial uplift due to isostatic rebound of the crust in response to unloading. Exhumation describes this positive feedback between uplift and erosion during which formerly deeply buried rocks are exposed at the surface. On the UK continental shelf, its magnitude, driving mechanisms and implications for petroleum geology are profound and have been intensively studied.

The main processes driving the exhumation of sedimentary basins are thermal-isostatic effects related to continental extension, orogenic crustal shortening (basin inversion, e.g. **Williams *et al.* 1989**, **Hayward & Graham 1989**) and epeirogeny. The latter group of processes encompasses the uplift of broad regions of continental interiors driven by plumes (e.g. **Nadin *et al.* 1995**), magmatic underplating (e.g. **Brodie & White 1995**), mantle delamination (e.g. Platt & England 1993), post-glacial isostasy (e.g. Lambeck 1991), and intra-plate stress (e.g. Cloetingh 1988). Due to the problem in constraining an absolute datum, very few studies are able to measure ancient uplift (e.g. Abbott *et al.* 1997). Furthermore, in cases such as burial history modelling of petroleum source rocks, exhumation is actually a far more useful parameter to constrain because it leads directly to cooling and lithostatic pressure release, with attendant implications for petroleum generation. In the absence of erosion, uplift on its own results in neither cooling nor pressure decrease.

Intuitively, exhumation is bad news for petroleum explorers:

- Sealing horizons are removed and/or their effectiveness is severely reduced.
- Faults are often reactivated causing them to become conduits for hydrocarbon leakage to the surface.
- Potentially attractive source rocks may be found to have exceeded their generative capacity due to earlier, deeper burial and heating.
- Potentially attractive reservoirs may, likewise, be over-compacted and downgraded.
- Lithostatic pressure reduction during exhumation causes oil accumulations to exsolve gas, pushing down the oil-water contact and spilling oil.
- Regional tilting during uplift results in changes to trap configurations, possibly destroying subtle one-way dip closures.

In spite of this, many prospective southern and western UK sedimentary basins have undergone severe late Mesozoic and Cenozoic exhumation and, in the Wessex basin of southern England, Europe's largest onshore oilfield at BP's Wytch Farm is located in the footwall to a major compressionally reactivated fault zone.

The reversal of slip on formerly extensional fault systems is termed basin inversion and, ultimately, it leads to folding, thrusting and expulsion of the syn-rift fill. Because of the characteristic change in displacement on normal faults in rift basins, from zero at their tips to maxima at their centres, basin inversion will often generate shortening in the cover sequence before the net extension at basement level has been cancelled. Illustrative examples of basin inversion from a range of geodynamic settings – rift basins, thrust belts and strike-slip fault zones – are presented by **Hayward & Graham**. They include discussion of the so called shortcut fault, a slice of basement horst subsequently incorporated in the hanging wall of a low-angle thrust that develops in the footwall of an unfavourably steeply inclined normal fault during its reactivation. Shortcut faults result in inliers of basement bounded on one side by a normal fault and on the other by a thrust. They represent one of several empirical criteria whose successful recognition can lead to diagnosis of basin inversion in areas of minimal seismic or well data.

Williams, Powell & Cooper define the null point on compressionally reactivated faults, the point of apparently zero fault displacement, above which displacements are net reverse, whilst below the null point they are normal. As an elegant explanation for apparently bizarre structural geometries, such as faults that change from normal to reverse along strike, seamless vertical transitions from hanging wall rollover to hanging wall anticline, upthrown hanging wall depositional 'thicks', and compressional folds in the hanging walls of normal faults, the null point concept is an essential tool for seismic interpreters working in inverted basins.

Virtually none of the deformation structures discussed in these papers form in the absence of fluid overpressure. Defined simply as pore fluid pressures greater than would be calculated from a hydrostatic gradient (*c.* 10 MPa/km), overpressure exercises profound influence over the mechanics of deformation through the reduction it brings about in effective stress (the difference between applied normal stress and pore fluid pressure). Thus, in Hubbert & Rubey's famous experiment, best repeated many times, an emptied beer can straight from the freezer, when inverted on a gently inclined smooth surface, will not slide down it until the air trapped inside has warmed and expanded sufficiently to reduce effective stress at the interface between the beer can and the inclined surface, at which point it will slide.

As well as its importance in structural geology, overpressure also inhibits siliciclastic compaction and retards vitrinite reflectance, the primary tool for assessing thermal maturity of petroleum source rocks. In terms of deformation of sedimentary rocks, overpressure is perhaps most usefully understood in terms of the increasing fraction of lithostatic pressure it is capable of supporting (i.e. the pressure exerted by the weight of the rock overburden). Hubbert & Rubey (1959) defined their λ factor to describe magnitude of overpressure as a fraction of lithostatic pressure. A λ of zero signifies hydrostatically pressured fluids while $\lambda = 1.0$ is not sustainable since it implies that fluid pressure exceeds rock strength. Despite this, the existence of hydrothermal breccias – triaxial networks of closely spaced, contemporaneous vein systems that resemble 3-D jigsaw puzzles – and vertical mineral fibres in horizontal veins attest to brief moments when fluid pressure must have attained lithostatic pressure, at least locally. On a similar theme, *in-situ* fluid pressures equivalent to $\lambda = 0.97$ have been inferred from the toes of actively deforming accretionary wedges in Barbados and the Makran (Davis *et al.* 1983).

Perhaps more than in any other setting, overpressures are potentially large during basin inversion. In what **Sibson** coined the 'sponge effect', basin inversion can be likened to wringing out a sodden sponge, liberating unknown volumes of water previously trapped in sedimentary pore spaces. The 200 km-long Cretaceous Mother Lode vein system of California provides a graphic example of the volumes involved. Given the low solubility of quartz in water, more than 10^6 m^3 of fault-hosted vein fill per kilometre strike length suggests that something like 10^9–10^{10} m^3 (6.3–63 billion barrels!) of aqueous fluid flushed through the vein system per kilometre strike length.

Sibson takes the undoubted importance of fluid overpressures during basin inversion to develop his model of fault valve behaviour. It describes cycles of

(i) overpressure build-up and progressive reduction in effective stress across a pre-existing fault,

(ii) fault reactivation, rupture and seal breach, and

(iii) sudden and rapid diminution of fluid pressure and fault seal, followed by the next cycle of overpressure build-up.

Fault valve behaviour accounts for the wealth of field evidence for the episodic nature of fault reactivation. It also provides an explanation for the selectivity of basin inversion in which only a few, or even one, of an existing set of comparable normal faults are reactivated (e.g. Abbotsbury-Purbeck Fault, southern England).

Basin inversion is promoted by large differential stress, low frictional strength and/or processes that serve to reduce effective frictional strength (e.g. overpressure), and the presence of weak layers like salt along which the shortening cover can detach itself from basement. In the inverted basins of the UK, salt is omnipresent and **Stewart *et al.*** provide a useful review of the influence of salt layers on cover faulting in UK basins. Rock salt confounds virtually every conventional assumption structural geologists like to make in sedimentary basins:

- Halokinetic deformation is highly three-dimensional (i.e. non-plane strain).
- Buoyancy-driven mobilization means that salt bodies change their shape and thickness drastically with progressive subsidence and sedimentation.
- Even at shallow depth, low temperature and high strain rate, salt deformation is generally plastic.
- Because of strong velocity contrasts and hence ray refraction at the boundaries between steep-sided salt bodies and encasing sediment, the precise geometry of salt bodies is the proverbial nightmare to interpret from migrated seismic profiles.

Stewart *et al.* present a large number of seismic-based cross-sections through inverted UK salt basins that exemplify the curious properties of salt and its role in vertically partitioning deformation between the sub-salt and supra-salt successions. Their paper serves as an excellent reminder of the pitfalls of making 'one-line' seismic interpretations in salt basins.

Late Mesozoic–Cenozoic exhumation was responsible for the erosion of up to three kilometres of Upper Jurassic and Cretaceous section from the western UK, with the exhumation bullseye centred on the East Irish Sea basin (Lewis *et al.* 1992; Ware & Turner 2002). From the point of view of understanding the likely mechanisms driving the uplift, it is important to appreciate the transition that exists in the UK between its southern basins, where the effects of far-field Alpine-driven basin inversion were dominant, to the inner and contiguous margins of the NE Atlantic where the influence of the Iceland Plume is evident. The Iceland Plume is a mantle upwelling that was active during the Palaeogene and which affected a 1000 km-wide region of the North Atlantic continents through (i) extensive volcanism (White & McKenzie 1989) and (ii) igneous underplating of the continental crust conjectured beneath a wide region extending far beyond the area of extrusive volcanism.

The link between exhumation and underplating is examined by **Brodie & White** who contend that, outwith the Wessex basin, shortening resulting from basin inversion generates insufficient crustal thickening to account for the observed exhumation, and therefore tectonically driven uplift. Whilst the inherent difficulties in quantifying the magnitude of shortening in inverted basins mean that basin inversion cannot be rejected, they present strong arguments for the existence of substantial quantity of underplated melt having permanently thickened and uplifted large parts of the western UK continental shelf. Evidence supporting this hypothesis includes:

- Absence of thermally subsided post-rift successions.
- The existence of only small free air gravity anomalies indicating that flexural effects arising from the opening of the North Atlantic Ocean are negligible.
- The results of inverse modelling of rare earth concentrations in extruded igneous rocks analyzed from the Tertiary Igneous Province which suggest that between 2–5 km of melt was added to the base of the crust during the early Tertiary.
- The widespread distribution of early Tertiary igneous rocks, from major flood basalts in the Tertiary Igneous Province to minor intrusive bodies like the Lundy Island granite of the Bristol Channel, is consistent with the presence of a large region of underplated igneous rock.

Given that the exhumation maximum is located in the East Irish Sea basin, well south of the Tertary Igneous Province, it seems unlikely that igneous underplating is the sole driving mechanism behind western UK exhumation. In fact, the regional pattern of exhumation is probably a product of the combination of northward declining inversion-related shortening and southward declining underplating. To date, direct evidence of several kilometres of underplated melt beneath the Irish Sea basins has been absent and over the coming years it is likely that deep crustal velocity information from wide-angle data will make an important contribution to this fascinating problem.

As well as the isostatic conseqences of thickening the crust with relatively buoyant basalt (buoyant, that is, compared to aesthenosphere density), plumes also generate transient flexural uplift in the adjacent lithosphere, beyond the limit of the hot plume material. This mechanism of transient uplift is investigated by **Nadin, Kusznir & Toth** who show that the wavelength and magnitude of the uplift is dependent on the buoyancy of the plume (i.e. magnitude of surface uplift at the plume head) and the viscosity of the lithosphere. In sedimentary basins like the North Viking Graben, on which their study focuses, plume-generated uplift exerts a relatively short-term buffer on the long-term pattern of post-rift subsidence. Nadin *et al.* use flexural-isostatic modelling to quantify how much the observed post-rift subsidence departs from the theoretical subsidence curve predicted from uniform stretching. The timing of the transient uplift in the northern North Sea, and the subsequent period of accelerated Eocene subsidence during the recovery of the anomaly, coincides with the deposition of many of the major Palaeocene basin floor fans in the North Sea. This provides a powerful illustration of the primary role of regional tectonics in controlling palaeogeography and sedimentary facies distribution.

To a great extent, the onshore outcrop pattern of British Isles geology reflects directly the effects of differential erosion due to the Iceland Plume. Thus, rocks become progressively younger with distance from the Atlantic margins, ranging from the Precambrian and Caledonian terrains of the Scottish Highalnds and western Ireland, to the Mesozoic–Cenozoic basins of southern Britain. Traditional geological training is founded on observing and interpreting evidence of Earth history preserved in the rock record. Perhaps the greatest challenge in investigating geological uplift and erosion is that such studies are often attempting to make reconstructions of what is missing from the rock record. Whilst the palaeobiology community has long grappled with conceptual difficulties posed by the inherent incompleteness of the palaeontological record, problems of 'missing' evidence are relatively new in other branches of geoscience. The problem is particularly acute in structural geology – in the absence of direct dating of syn-kinematic features (e.g. Ar–Ar dating of muscovite-bearing fault rocks), it is only possible to bracket the age of a structure between the youngest rocks it effects and the oldest rocks that post-date it. Thus, we might assume implicitly that major faults traversing ancient orogenic basement rocks formed during the orogenesis, even though in fact they may have originated much later.

Roberts & Holdsworth present an instructive case study from the Scottish Highlands in which they attribute many of the major faults affecting Precambrian and Caledonian basement to Mesozoic extensional basin formation. Whilst the evidence they cite to support their ideas is circumstantial, their map linking onshore and offshore geological relationships is powerfully persuasive. Roberts & Holdsworths' hypothesis challenges the widely held assumption that these Mesozoic faults owe their origin to early Palaeozoic orogenesis. More generally, their model represents an excellent example of kinematic coherence – the idea that fault systems accommodate smooth variations in strain across a wide range of microscopic to crustal scales – and it serves to remind us of the inherent pitfalls in making interpretations of the age and origin of structures on the basis of their trend.

Acknowledgements

We thank Tim Needham and Alan Roberts for their reviews of this article and numerous colleagues for suggestions as to which papers we should include.

References

ABBOTT, L.D. *et al.* 1997. Measurement of tectonic surface uplift rate in a young collisional mountain belt. *Nature,* **385,** 501–507.

CHILDS, C., WATTERSON, J. & WALSH, J.J. 1995. Fault overlap zones within developing normal fault systems. *Journal of the Geological Society, London,* **152,** 535–549 [also published in Holdsworth & Turner 2002].

CLOETINGH, S. 1988. Intraplate stresses: A new element in basin analysis. *In:* KLEINSPEHN, K.L. & PAOLA, C. (eds.) *New Perspectives in Basin Analysis.* Springer-Verlag, New York, 205– 230.

DAVIS, D., SUPPE, J. & DAHLEN, F.A. 1983. Mechanics of fold- and thrust belts and accretionary wedges. *Journal of Geophysical Research,* **88,** 1153–1172.

EISENSTADT, G. & WITHJACK, M.O. 1995. Estimating inversion: results from clay models. *In:* BUCHANAN, J.G. & BUCHANAN, P.G. (eds) *Basin Inversion.* Geological Society, London, Special Publications, **88,** 119–136 [also published in Holdsworth & Turner 2002].

FISHER, Q.J. & KNIPE, R.J. 1998. Fault sealing processes in siliciclastic sediments. *In:* JONES, G., FISHER, Q.J. & KNIPE, R.J. (eds) *Faulting, Fault Sealing and Fluid Flow in Hydrocarbon Reservoirs.* Geological Society, London, Special Publications, **147,** 117–134 [also published in Holdsworth & Turner 2002].

GIBSON, R.G. 1998. Physical character and fluid flow properties of sandstone-derived faults. *In:* COWARD, M.P., DALTABAN, T.S. & JOHNSON, H. (eds) *Structural Geology in Reservoir Characterisation.* Geological Society, London, Special Publications, **127,** 83–98. [also published in Holdsworth & Turner 2002]

HOLDSWORTH, R.E. & TURNER, J.P. 2002. *Extensional tectonics: faulting and fault-related processes.* Geological Society, London, Key Issues, in Earth Sciences, **2(2)**.

HOLDSWORTH, R.E., BUTLER, C.A. & ROBERTS, A.M. 1997. The recognition of reactivation during continental deformation. *Journal of the Geological Society, London,* **154,** 73–78.

HUBBERT, M.K. & RUBEY, W.W. 1959. Role of fluid pressure in the mechanics of overthrust faulting. *Geological Society of America Bulletin,* **70,** 115–166.

KUSZNIR, N.J., MARSDEN, G. & EGAN, S.S. 1991. A flexural cantilever simple-shear/pure-shear model of continental lithosphere extension: application to the Jeanne d'Arc Basin and Viking Graben. *In:* ROBERTS, A.M., YIELDING, G. & FREEMAN, B. (eds). *The Geometry of Normal Faults.* Geological Society, London, Special Publication, **56,** 41–60.

LAMBECK, K. 1991. Glacial rebound and sea level change in the British Isles. *Terra Nova,* **3,** 379–389.

LEWIS, C.L.E., GREEN, F., CARTER, A. & HURFORD, A.J., 1992, Elevated K/T palaeotemperatures throughout Northwest England: three kilometres of Tertiary erosion? *Earth and Planetary Science Letters,* **112,** 131–145.

LUNDIN, E.R. & DORE, A.G. 1997. A tectonic model for the Norwegian passive margin with implications for the NE Atlantic: Early Cretaceous to break-up. *Journal of the Geological Society, London,* **154,** 545–550.

MCKENZIE, D.P. 1978. Some remarks on the development of sedimentary basins. *Earth and Planetary Science Letters,* **40,** 25–32.

MCKENZIE, D.P. & PARKER, D.L. 1967. The North Pacific: an example of tectonics on a sphere. *Nature, London,* **216,** 1276–1280.

OTTESEN ELLEVSET, S., KNIPE, R.J., OLSEN, T.S., FISHER, Q.J. & JONES, G. 1998. Fault controlled communication in the Sleipner Vest Field, Norwegian Continental Shelf: detailed, quantitative input for reservoir simulation and well planning. *In:* JONES, G., FISHER, Q.J. & KNIPE, R.J. (eds) *Faulting, Fault Sealing and Fluid Flow in Hydrocarbon Reservoirs.* Geological Society, London, Special Publications, **147,** 283–297 [also published in Holdsworth & Turner 2002].

OXBURGH, E.R. 1974. The Plain Man's Guide to Plate Tectonics. *Proceedings of the Geologists Association,* **85,** 299–357.

PLATT, J.P. & ENGLAND, P.C. 1993. Convective removal of lithosphere beneath mountain belts: thermal and mechanical consequences. *American Journal of Science,* **293,** 307–336.

ROBERTS, S. & JACKSON, J.A. 1991. Active normal faulting in Central Greece: an overview. *In:* ROBERTS, A.M., YIELDING, G. & FREEMAN, B. (eds). *The Geometry of Normal Faults.* Geological Society, London, Special Publication, **56,** 125–142 [also published in Holdsworth & Turner 2002].

ROBERTS, A.M., KUSZNIR, N.J., YIELDING, G & STYLES, P. 1998. 2-D flexural backstripping of extensional basins: the need for a sideways glance. *Petroleum Geoscience,* **4,** 327–338.

UNDERHILL, J. R., SAWYER, M. J., HODGSON, P., SHALLCROSS, M.D. & GAWTHORPE, R.L. 1997. Implications of fault scarp degradation for Brent Group prospectivity, Ninian Field, Northern North Sea. *American Association of Petroleum Geologists Bulletin,* **81,** 295–311.

WALSH, J. J. & WATTERSON, J. 1991. Geometric and kinematic coherence and scale effects in normal fault systems. *In:* ROBERTS, A.M., YIELDING, G. & FREEMAN, B. (eds). *The Geometry of Normal Faults.* Geological Society, London, Special Publication, **56,** 193–203 [also published in Holdsworth & Turner 2002].

WALSH, J.J., WATTERSON, J., CHILDS, C. & NICOL, A. 1996. Ductile strain effects in the analysis of seismic intepretation of normal fault systems. *In:* BUCHANAN, P.G. & NIEUWLAND, D.A. (eds) *Modern Developments in Structural Interpretation, Validation and Modelling.* Geological Society, London, Special Publications, **99,** 27–40 [also published in Holdsworth & Turner 2002].

WARE, P.D. & TURNER, J.P. (2002). Magnitude and driving mechanisms of Tertiary exhumation of the East Irish Sea basin. *In:* DORE, A.G., CARTWRIGHT, J.A., STOKER, M.S., TURNER J.P. & WHITE, N.J. (eds). *Exhumation of the Circum-Atlantic Margins.* Geological Society Special Publications, In press.

WERNICKE, B.P. 1985. Uniform-sense normal simple shear of the continental lithosphere. *Canadian Journal of Earth Sciences,* **22,** 108–125.

WHITE, R.S. & MCKENZIE, D.P. 1989. Magmatism at rift zones: the generation of volcanic continental margins and flood basalts. *Journal of Geophysical Research,* **94,** 7685–7729.

Mechanical and thermal consequences of extensional basin formation

Plate tectonics and the evolution of the British Isles

John F. Dewey

Thirty-fifth William Smith Lecture

SUMMARY: This paper is an essay on the rationalism and difficulties of the geological corollaries of plate tectonics, with illustrations particularly from the geological history of the British Isles. There is special emphasis on the mechanics of continental lithosphere in extension, when basins are formed and in convergence whereby continental lithosphere thinned in extension is shortened and restacked to form orogenic belts. An upper lithosphere stress-concentrating elastic lid suffers extremely complex inhomogeneous deformations above a simpler, viscoelastic lower lithosphere whose deformation drives strain and displacement in the elastic lid. The very nature of plate tectonics as relative motion among a multi-plate mosaic of plates provides a rational semi-quantitative framework for large-scale geological complexity in space and time. The Caledonian and Hercynian history of the British Isles may be explained semi-quantitatively in terms of plate boundary evolution but it is not possible to define unequivocal solutions. The rationalism of plate tectonics is exhibited on a smaller, more direct field-scale by many of the complexities of ophiolite complexes, such as Ballantrae (Scotland), that can be explained as the effects of the evolution of ridge/ridge transforms and by diachronous deformation and sedimentation in subduction accretion prisms such as the Southern Uplands. Deformational and sedimentary patterns in NW Europe from Triassic times to the present day resulted from a complex interaction of extensional events related to diachronous Atlantic opening and compressional events related to sequential Alpine shortening phases. Those widespread events are typical of intracontinental plate boundary systems and create great difficulties for the geologist wishing to decipher relative plate motion from geological sequences, because huge areas, much larger than the British Isles, must be analysed and synthesized before much sense can be extracted. Further difficulties arise from the palinspastic and preservation problems of convergent boundaries; orogenic shortening telescopes and partially subducts facies belts. Most orogenic belts were assembled as a collage of displaced and rotated terrain elements that were subsequently flattened, elongated and straightened to an apparent but deceiving simplicity. Therefore, while plate tectonics provides a rational geological framework against which, for example, British basin evolution can be tested and explained, we cannot yet use the geological effects of plate boundary evolution to reconstruct relative plate motion quantitatively.

From Holdsworth, R.E. & Turner, J.P. (compilers) *Extensional Tectonics: Regional-scale Processes*. The Geological Society, Key Issues in Earth Sciences, **2 (1)**, 13–54.
1476-315X/02/$15.00

© The Geological Society of London 2002

Frist published in "*Journal of the Geological Society*" **139**, 371–412.

1. Introduction

It is impossible to write a rational, integrated and ordered account that quantitatively or even qualitatively explains the geological history of the British Isles, or anywhere, in a plate tectonic framework. The difficulties stem from many sources, as explained below, but arise principally from our present ignorance of how to relate the complex response of the continental crust to the generally simpler relative motions among the larger plates of the global plate mosaic. Although there are now many subtle facets of lithospheric evolution and mechanics that we are beginning to understand with considerable depth, we still do not understand properly how to build integrated and convincing plate tectonic models from geology. 14 years of plate tectonics has not yet allowed us instantly to explain geology but, rather, has enabled us to look at geology in new ways that have allowed us to move from a history of *ad hoc* model building into a more productive phase of rationalism. We can no longer generate special hypotheses to suit, but must operate from a basis of lithospheric mechanics, which, although sweeping aside irrationalism in the earth sciences and providing a rational frame of reference for explanation and model building, also makes life extremely difficult, as we shall see, for the geologist concerned with explaining the geological history of a particular region. However, in spite of the difficulties, we are now entering a very exciting phase in the earth sciences, in which, provided that we ask questions that have tractable solutions, we are rapidly approaching the position where not only the broader aspects of plate boundary evolution may be understood, but in which many of the fine subtle details of stratigraphy, igneous and metamorphic petrology and structural geology may become understandable in terms of the physics of the lithosphere. However, as Smith (1976) warned us, a particularly important caveat is that, since the advent of plate tectonics, it has become somewhat *de rigeur* to adopt dogmatic positions that gloss over the difficulties, so that new dogmas have replaced the old; for example, that an ultramafic sliver or, worse, a slice of basalt or gabbro, signifies the position of a vanished ocean. The response of rocks in the upper lithosphere to relative plate motions is exceedingly complicated; over-simple generalizations can obscure significant complexities and variations. For example, attempts, including my own, to explain the history of the early crust in simple plate boundary terms seem, in retrospect, to ignore fundamental differences between the Archaean and the modern world.

In another context, over-simplistic notions of ophiolites needing to have deep water sediment caps to be regarded as oceanic crustal remnants obscures the interesting and significant occurrences of oceanic crust at or above sea-level, for example, in fracture zones (Bonatti *et al.* 1977) above its normal age-depth. A useful perspective on the geological corollaries of plate tectonics is, perhaps, that we should seek a delicate balance between boldness in recasting our thinking of time/space relationships in geology, tempered with extreme caution in realizing the complexities of those corollaries.

This paper does not seek to provide a sequential plate tectonic-based palaeogeographic and palaeotectonic evolution of the British Isles from the Scourian 'event' to the Goodwin Sands. This, probably, will never be possible and, now, would be absurd. Instead, an attempt is made to develop some of the better-understood principles of plate tectonics, illustrated with examples drawn from the British Isles. The scheme adopted is as follows. First, some basic ideas of plate tectonics are outlined. Secondly, some of the powerful rationalism of the geological corollaries of plate tectonics is flavoured with the difficulties. Thirdly, the evolution of basins is discussed at some length, followed by basin inversion. Then follow sections on the Caledonides, Hercynides, and the Mesozoic/Cenozoic evolution of foreland NW Europe and, lastly, conclusions emphasizing the secular evolution of the continental crust. The number of figures is large but reflects the writer's preference for graphic presentation. The reference list is relatively short and does not properly reflect the effort that has produced the data and ideas upon which this paper is based; to include every source would require a bibliography of many hundreds of pages. I have relied extensively upon compendia, symposium volumes and textbooks such as Anderton *et al.* (1979), Harris *et al.* (1979) and Woodland (1975) and have included in the references only those papers that cannot be avoided, such as McKenzie (1978) or that illustrate a particular idea or nuance, such as Coney *et al.* (1980).

Plate tectonics (McKenzie & Parker 1967) is an efficient, dominantly convective method for the earth to lose its heat by the generation, cooling, subsidence and subduction of a boundary conduction layer segmented into a mosaic of plates (Oxburgh 1974). The continental crust represents about 4 billion years of differentiation and modification of lighter and radiogenic elements forming a weak, buoyant, largely non-subductable cap. Definition of the lithosphere as a rigid outer layer depends upon strain rate and, therefore, time. Instantaneously, at high strain rates, during the transmission of shear waves, the whole earth with the exception of the outer core behaves rigidly. For times approaching its age, the viscous figure of the earth implies a vanishingly thin lithosphere. Definitions of the lithosphere that are tectonically more useful involve geological strain rates of $10^{-13}\ s^{-1}$ to $10^{-16}\ s^{-1}$ for deformation times up to a few hundred million years. A particularly useful definition of the base of the short-term lithosphere is the sudden decrease in velocity of shear waves, which is probably an

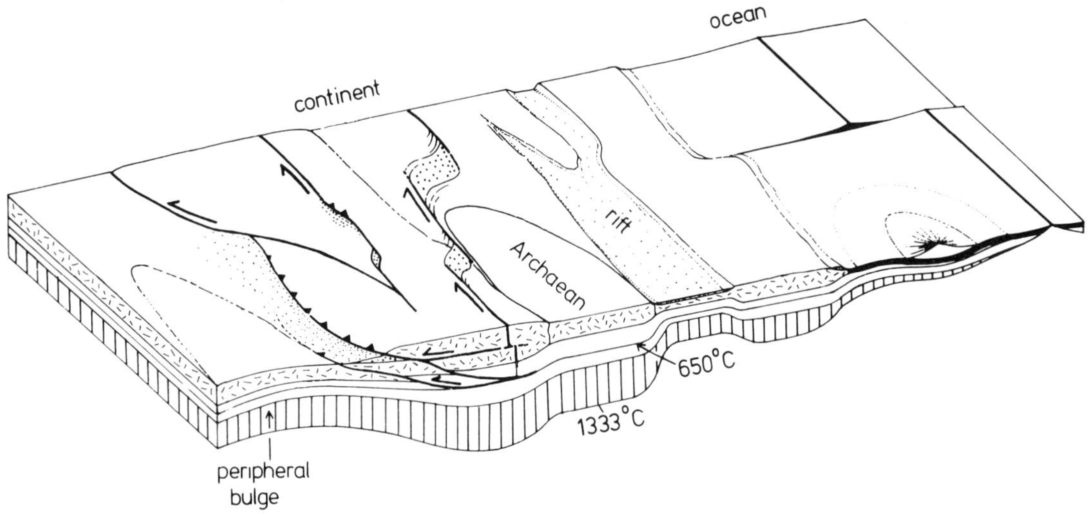

FIG. 1. Schematic representation of the continental and oceanic lithosphere and its deformation by extension, shortening and flexure. Vertical lined ornament = lherzolite.

isotherm of about 1333°C that defines the peridotite solidus. A lithosphere thus defined is a boundary conduction layer of short-term (< a few years) strength which may be subdivided into longer term lower viscoelastic and upper elastic sublayers, the former characterized by power law creep, and the latter by diffusional elastic creep, jointing, fissuring and faulting. The upper elastic portion, referred to hereafter as the elastic lid, may be an effective stress concentrater or fault-promoter, as creep relieves stress in the viscoelastic lower lithosphere and thus reduces the thickness of the layer in which force is converted into stress in both extension and compression. The elastic lid, or thinner, longer-term lithosphere is the portion in which earthquakes occur, i.e. the portion in which elastic stresses and strains can be stored for earthquake repeat times of about 10^2 a. This separation of the thermal boundary conduction layer lithosphere into viscoelastic and elastic layers is of considerable importance in flexure, extension, and compression, because penetrative rock fabrics may develop in the viscoelastic portion while less penetrative shear zones, faults, joints and fissure characterize various levels of the elastic lid. Furthermore, it seems clear from both geological field studies and theoretical considerations that the strain evolution of the elastic lid is bewilderingly complex. Even though it may be dominated by plane strains during large-scale short-term deformations, the elastic lid responds irregularly to longer-term, more uniform progressive strains of the viscoelastic lower lithosphere. It is not appropriate to simply define the lower boundary of the elastic lid as a particular isotherm because rheology is also conditioned by strain rate and confining pressure. For high strain rates, faults within the continents may rip through to a 12 s Moho, e.g. the Wind River Thrust (Brewer & Oliver 1980). These concepts are schematically expressed in Fig. 1, in which the lithosphere is depicted as a boundary conduction layer varying from zero thickness at ridge axes, then thickening and subsiding, following a square root of time law, to about 100 km under the oldest oceans with a lithosphere as thick as perhaps 300 km under the oldest Archaean nuclei, but averaging about 125 km under the Proterozoic/Palaeozoic continents. In areas in which the lithosphere is stretched or shortened by plane strain at strain rates larger than about $10^{-14}\ s^{-1}$ it is thinned and thickened, respectively. Where the lithosphere is stretched and thinned (Fig. 2A), isotherms are packed together, the geothermal gradient and heatflow are increased and strain rates decline during stretching if the same cross-sectioned area of the lithosphere remains involved in the deformation. After an infinite amount of time, thermal recovery of the lithosphere restores it to its original thickness. Thus, stretching has the effect of bringing rocks to higher levels, reducing confining pressure on these rocks, and allowing them to cool with time. Therefore, rocks starting at a particular level in the crust and mantle increase their strength with time. Where sediments fill the furrow resulting from stretching, they, and part of the basement, warm under increasing confining pressure and therefore weaken. Where the lithosphere is shortened and thickened (Fig. 2B) at strain rates larger than about $10^{-14}\ s^{-1}$, isotherms are pushed apart, the geothermal gradient and heatflow are reduced and strain rates increase during shortening if the same cross-sectional area of the lithosphere remains involved in the deformation. Similarly, thermal recovery eventually restores the lithosphere to its original thickness. Shortening has the effect, ignoring denudation, of pushing rocks to greater depth and thus increasing

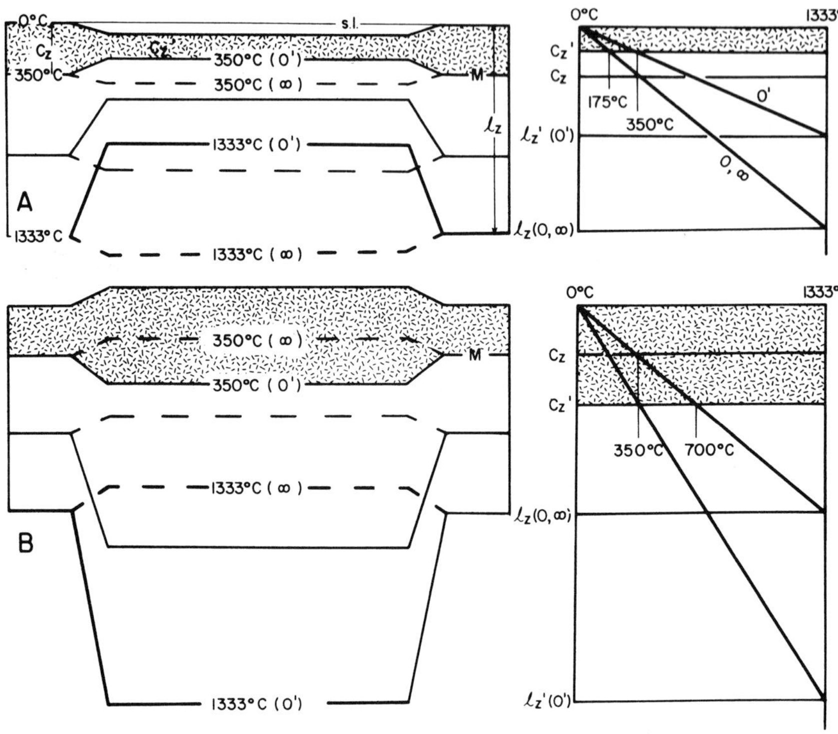

FIG. 2. Geometric and thermal consequences of the plane strain extension (A) and shortening (B) of the lithosphere.

confining pressure and causing them to warm up with time. Therefore, rocks starting at a particular level in the crust or mantle weaken with time. Shortening may be accomplished by various combinations of finite homogeneous strain and thrust stacking, the former dominating the viscoelastic, the latter generally characterizing the elastic lithosphere. During and after shortening, denudation acts to reduce crustal thickness to remove heat and to strip the more radiogenic upper crust.

Plate tectonics, as contrasted with its harbingers prior to 1967, depends for its kinematic success (e.g. Atwater 1970; Dewey *et al.* 1973; Isacks *et al.* 1968; Le Pichon 1968; McKenzie & Morgan 1969; McKenzie & Parker 1967; Morgan 1968; Wilson 1965) on the long-term torsional coherence of plates, i.e. upon the notion that they do not really distort to any significant extent as they evolve and that most of the relative motion between plates is taken up by slip at their edges. The evidence for this consists of widespread, thin, flat-lying, undistorted, sedimentary sequences in ocean basins and on the cratons and the fact that congruent rifted continental margins and magnetic anomalies may be fitted together with such precision to allow palaeotectonic reconstructions.

The small thickness/width aspect ratio of the lithosphere with a stress-diffusing lower portion and a stress concentrating upper portion gives it a high degree of torsional coherence or rigidity but a low flexural rigidity (Walcott 1970). The lithosphere is easily depressed by a point or line load, the width of the zone over which the load is supported and the amplitude and wavelength of the depression and peripheral bulge being dependent upon the elastic thickness of the lithosphere (Beaumont 1978; Watts *et al.* 1980) (Fig. 3). For an infinitesimally thin lithosphere, support and compensation are according to an Airy model, while for a whole earth lithosphere, support is total and no depression would occur beyond that allowed by elastic compressibility. Beaumont (1978) argued that flexure of the elastic lithosphere is short-term and that, with time, stress relaxation by diffusional creep causes a viscoelastic decay from flexural load support towards a local Airy load support (Fig. 3). As we shall see below, the evidence from the loading of rifted basins does not support the viscoelastic model.

Although the evidence for long term large-scale torsional coherence of the lithosphere is clear, significant departures from this ideal occur in plate boundary zones and, to a small but geologically important extent, within plates. The oceanic and continental portions of plates, although similar in short-term flexure, behave very differently in a torsional sense. Intraplate earthquakes, deformation zones such as rift valleys and mountain belts, and volcanic hotspots are much commoner in the continents than in the oceans (Fig. 4). This strength contrast between at least the upper part of the continental and oceanic lithosphere

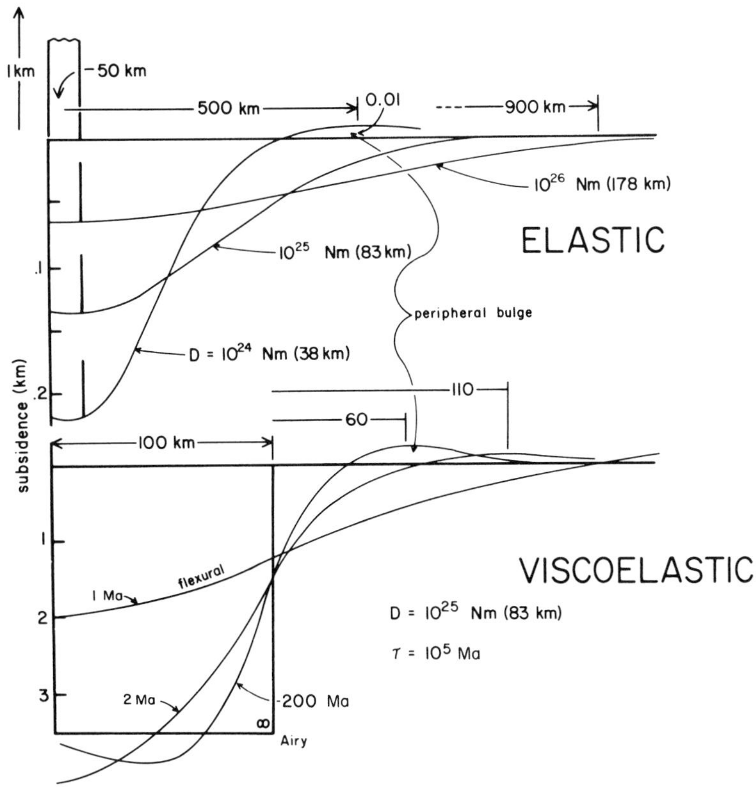

FIG. 3. Flexure of a loaded elastic lithosphere of various effective elastic thicknesses of flexural rigidities (D) and the decay from flexural to Airy isostatic compensation for an 83 km viscoelastic lithosphere with dissipation decay constant of 10^5 Ma (from Beaumont 1978).

FIG. 4. Global map from 70°N to 70°S showing plate boundaries and deformation zones. Black dots = major intraplate earthquakes; open circles = hotspots not obviously at major plate boundaries, believed to *result* from intraplate cracking and extension; solid circles = longer-lived hotspots believed to have sub-lithospheric origin; encircled dots = hotspots at oceanic triple junctions; vertical lines = wide zones of deformation mainly constituting intracontinental plate boundary zones; lines with adjacent *cc* enclose areas of deviatoric compression; lines with adjacent *tt* enclose areas of deviatoric tension; granite symbol = areas stabilized as cratonic interiors by about 1.7×10^9 a.

is caused by a combination of weakness and buoyancy of quartz-rich, and strength and negative buoyancy of olivine-rich rocks, respectively. In the oceans, plate boundaries are generally narrow and well-defined, whereas in the continents, they may form wide, rather diffuse zones (e.g. the Himalayas and Tibet) that grade into what, in another context, might be called intra-plate deformation; this point will be taken up below in a discussion of Mesozoic/Cenozoic foreland Europe. The only parts of the continents that appear to be largely immune from intra-plate deformation are those Archaean 'nuclei' that were not remobilized after about 1.7×10^9 a (Fig. 4).

A geologically useful view of the lithosphere is not one of large strong 'rigid' plates but one in which it is viewed as a complex delicate thin membrane in which stress diffusion from plate boundaries, from body forces generated by lithosphere thickness variations and perhaps from convective driving forces, results, preferentially, in complex deformation of a thin continental elastic lid rather than in oceanic deformation. Hence, the continents are appallingly complex repositories of a geological history approaching 4×10^9 a, while the stronger easily subductable oceanic lithosphere, with a life span of about 180×10^6 a is the principal agent of global heat loss. As we shall see below, the implications of this view of the lithosphere are profound for its secular evolution by thickening and the resultant secular evolution of the continental crust (Dewey & Windley 1981).

2. Rationalism of plate tectonics

Plate tectonics is not a simple panacea for geology that has suddenly allowed us to explain the structure and evolution of the upper lithosphere. It is an extremely complicated tool whose geological implications are not fully understood. Plate tectonics does not make geology any easier; on the contrary, it makes it a great deal more complicated in that we are constrained to build and test models within a rational plate tectonic framework rather than to erect *ad hoc* constructions such as cryptoblemes and geosynclines. It is this rationalism, predictability and testability in accounting for complex geological relationships, in addition to its prime power to explain bathymetry, magnetics and seismicity, that gives plate tectonics its paradigmatic position in the earth sciences. In this section, the rational predictive power of plate tectonics is illustrated and then tempered with a discussion of some of the difficulties.

a. Plate kinematics and geology; the larger view

The evolution of tectonically-coherent plate mosaics must generate very complicated geological sequences at plate boundaries (Dewey 1975). Even a simple two-plate boundary evolving around a fixed pole can produce complex diachronous facies patterns (Dewey 1976). In Fig. 5, the evolution of a simple three-plate mosaic is illustrated (modified from Dewey 1975, fig. 7, in which a full discussion will be found). Two important principles are illustrated. First, the whole of the *bc* boundary slowly changes its position with respect to the instantaneous relative motion pole *bc*. This involves both steady change of, for example, the rate and direction of subduction at point 2 and the progressive change of the subduction zone to a transform, then a ridge, as it changes its orientation with respect to rotational latitudes near the *bc* pole. Secondly, the *abc* triple junction is unstable (McKenzie & Morgan 1969) and changes its position in the frames of reference of all three plates and their relative motion poles and evolves, from a ridge/ridge/trench junction, through several alternating phases of transform/transform/trench, trench/trench/transform, and trench/trench/ridge. This has a profound effect on the geological history of plate margin localities that are fortunate in not being subducted. Consider point 1, which has a history, from A to D, consisting of 9 discrete phases. At first, 1 is on a *bc* subduction zone with fast head-on convergence but gradually slowing and becoming more oblique. Then a trench/trench/ridge junction sweeps past to the left causing a brief cessation of arc volcanism and mafic igneous activity in the

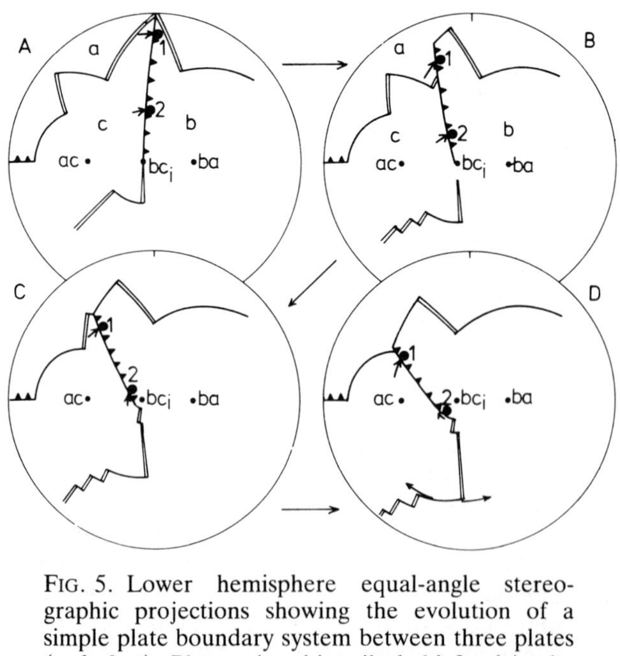

FIG. 5. Lower hemisphere equal-angle stereographic projections showing the evolution of a simple plate boundary system between three plates (*a*, *b* & *c*). Plate *a* is arbitrarily held fixed in the stereographic reference frame and plates *b* and *c* rotate at the same velocity, anticlockwise and clockwise, respectively, around finite motion poles *ba* and *ac*, which are arbitrarily held fixed in the *a* frame of reference.

fore-arc region. 1 is now on an *ab* subduction with slower, oblique, but constant-rate convergence. Then a trench/trench/transform junction sweeps past to the right, returning point 1 to a *bc* subduction zone at which the convergence rate is slower and slowing. Then follow the leftward-sweeping of a trench/trench/ridge junction with a return to *ab* subduction, followed by a rightward passage of a trench/trench/transform junction giving a return of point 1 to *bc* subduction. Therefore, we see a basis for three common features in the geological evolution of many areas; progressive change, abrupt punctuation of events, and polyphase events. It is possible that some polyphase deformations in orogenic belts have this kind of origin. This rationale was used by Atwater (1970) in her classic pioneering work on the evolution of Cenozoic western North America, in which she used oceanic magnetic, lineament and some palaeomagnetic, data to make an evolutionary reconstruction of relative motion across the Cordilleran borderland and to predict with considerable accuracy its evolution. A similar method was used by Smith (1971) and Dewey *et al.* (1973) in their analyses of the gross tectonic evolution of the Alpine System in relation to the relative motion of Europe and Africa deduced from the opening histories of the Central and North Atlantic and the Labrador Sea. This quantitative approach is one of the main ultimate goals of attempting to characterize the geological history of any area in terms of relative plate motion, is only theoretically possible for the Mesozoic/Cenozoic, and is fraught with severe problems.

b. Boundary evolution and geology: the smaller scale

Although the geology of plate boundaries is, in detail, poorly known (Smith 1976), two examples will suffice to illustrate that plate tectonics makes sense of geological relationships that only ten years ago seemed to defy a rational actualistic explanation. Both examples involve oceanic fracture zones and allow pleasingly fine-scale geological interpretations.

No class of plate boundary could be apparently simpler than a ridge to ridge transform with its fracture zone continuations, yet it produces fundamental and complex modifications in the oceanic crust and mantle of the kind that may explain some fine details of ophiolite geology, such as some of the internal complications of the Unst and Ballantrae Complexes. In the Bay of Islands Ophiolite complex of western Newfoundland, Karson & Dewey (1978) showed that a complicated segment, the Coastal Complex, trending approximately at right angles to dykes of the sheeted complex, possesses a polyphase history that can be explained in the context of the evolution of an oceanic fracture zone, summarized sequentially and briefly as follows. A segment of oceanic lithosphere is accreted at a ridge axis near a transform offset along which it then passes to acquire a wide range of shear zone fabrics and in which various levels of the crust and mantle are juxtaposed in shear lenticles. This assemblage then passes the other ridge axis termination where it is reheated and intruded by ultramafic and mafic magmas that cross-cut earlier fabrics. The package is now embedded in a fracture zone along which only dip-slip motion occurs and along which alkalic seamounts unconformably develop.

Common features of turbidite sequences in the internal flysch zones of orogenic belts, and in modern and ancient subduction accretion prisms (Leggett *et al.* 1979) are longitudinal turbidite supply, sudden along-strike changes in thicknesses of turbidite groups and widths of accretionary units, episodic appearance of ophiolite and volcanic complex accretionary terrains with or within mélanges, and facies changes from quartz-rich to mafic-rich sands. Fig. 6 is a schematic attempt to relate and explain these features as a result of the subduction of a fracture zone. Subduction-accretion prisms are common in neutral arcs (Dewey 1980) in which oblique relative motion (UvO) is resolved into components of subduction (FvU) and intra-arc strike slip (OvF) so that a frontal platelet of the arc (F) is detached from the overriding plate (Fig. 6).

Consider the case where FvU is parallel with a fracture zone being subducted, that separates older and younger ocean floor both of which slope from the trench towards the ocean. Oblique subduction of magnetic anomalies or isodepth lines in the trench causes it to possess a longitudinal slope and therefore to funnel longitudinal turbidity currents. The fracture zone scarp, however, blocks turbidite flow until the thickness of trench turbidites can build up sufficiently to overflow it. This causes an along-strike facies contrast between greater and lesser subduction-accretion in shallow filled and deep empty trench segments. The segment to which sediment supply is large develops a thick wedge of accreted turbidites with no ophiolites (Fig. 33D) and the empty segment accretes a mélange of abyssal/hemipelagic sediments with slices of clipped-off seamount (Fig. 33C), causing an offset of the inner trench wall. Seamounts along the fracture zone play the dual sedimentological role of providing basaltic sand dispersal fans and of blocking turbidite flow when they reach the trench.

c. Problem of vertical motions

A variety of mechanisms have been proposed for generating *primary* vertical motions, such as the rise and fall of asthenoliths and antiasthenoliths. In my view, all these mechanisms, apart from those related to major primary hotspots (Fig. 4) and the bending of the lithosphere under load, are fallacious. Vertical motions are primarily a response to extending and shortening of the lithosphere, i.e. a secondary response to lateral motions. Vertical motions resulting

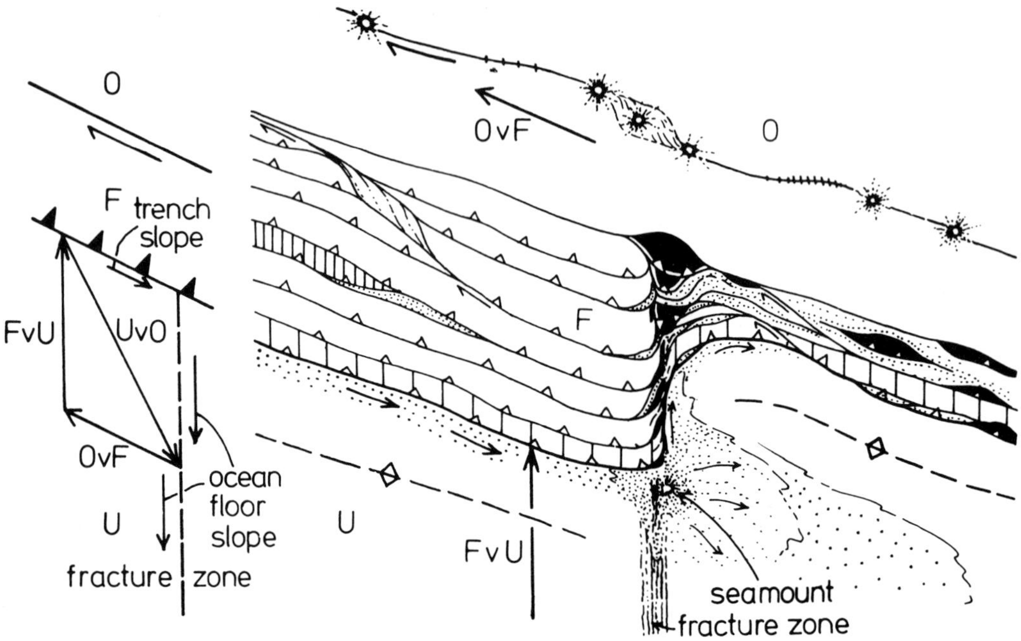

FIG. 6. Schematic illustration of a fracture zone being subducted at a trench with longitudinal supply to a subduction-accretion prism that comprises a fore-arc region to an arc with transform faulting along the volcanic axis. U = underriding plate, O = overriding plate, F = frontal arc plate.

from lithospheric loading can generate differential reliefs, between the floors of basins and the crests of peripheral bulges, of up to 10 km while vertical reliefs generated by lateral motions can attain 22 km.

d. Sea-level changes

It has long been recognized that a distinction can be drawn between local and global transgressions and regressions. World-wide transgressions are recognized in the early Cambrian, the early Silurian, the early Carboniferous, the Rhaetian/Hettangian and the Aptian/Albian. Plate-tectonically controlled mechanisms for large changes in sea-level include the accretion/decretion of continents (Dewey & Windley 1981), shortening and extending continental crust during times of major continental collision (Devonian, late Carboniferous to Triassic) and fragmentation, and changes in ridge volume caused by changes in ridge length and plate-accretion rates (Pitman 1978). Pitman (1978) was the first to show that global sea-level changes may result in both regression and transgression on a continental or basin margin depending on the position of the strand line with respect to the line along which the rate of sea-level fall is equal to the rate of subsidence (Fig. 7).

e. Rates of processes

Relative plate motion occurs typically at rates varying between 0.1 cm a to 20 cm a based upon average motion measured by finite difference solutions measured over 10 to 180×10^6 a. These rates allow us to derive a range of rates of geological processes as follows: uplift and subsidence (non-eroded and non-sediment loaded) by shortening and stretching of the lithosphere up to about 0.5 cm a, thermal subsidence by cooling of the lithosphere up to about 0.3 cm a, sea-level changes up to about 0.001 cm a and strain rates of 10^{-13} to 10^{-16} s^{-1}.

f. Episodicity

The evolution of a mosaic of continent-carrying plates provides a logical explanation for episodic, random, continuous, discontinuous, diachronous and

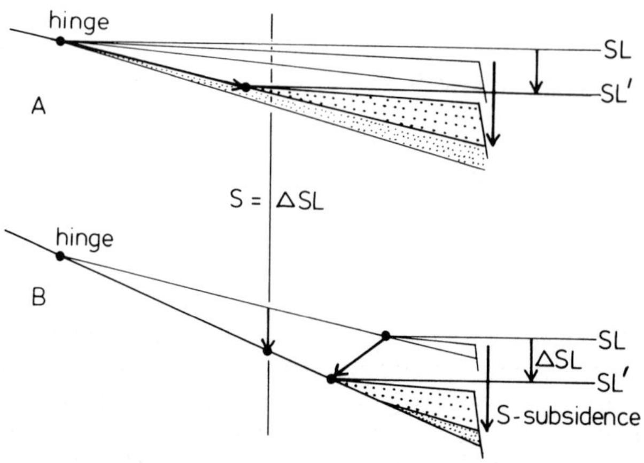

FIG. 7. Relationship between falling sea level, transgression and regression at a basin margin.

polyphase events. Plate tectonics does not give a simple immediate basis for periodicity. Collision between upstanding and buoyant objects varying in size from seamounts to continents (Nur & Ben-Avraham 1978) cause deformations commensurate with their size and shape. A long held view of deformation has been one of global episodicity (Stille 1924) and there is no doubt that deformation phases are not purely random around the globe. The question, however, is to what extent events are globally synchronous; that is, do we lump or split. The late Palaeozoic saw a great deal of orogeny, through Carboniferous to middle Permian times, that was associated with the assembly of Pangea, but Stille's Bretonic, Sudetic, Asturic and Saalic phases are restricted to the Hercynian orogen. Such phases may be widespread and synchronous, within the limits of biostratigraphical resolution, particularly as events that affect the forelands of orogenic belts, as we shall see below. Supposed global synchroneity of major orogenic events, such as the Laramide 'Phase' of late Cretaceous age may be partly due to a lumping tendency. However, it may be interpreted also in plate tectonic terms related to the increased rate of plate motion during the Cretaceous, which would have increased the arrival and collision rate of oceanic plateaus and seamounts at subduction zones, caused overriding plates to over-run trenches more often (Dewey 1980), and increased the likelihood of global plate boundary reorganization. Evidence of increased rates of relative motion comes primarily from magnetic data (Pitman 1978) but corroborating geological corollaries are seen in marine transgression (Pitman 1978), and a higher incidence of ophiolite obduction, and in the generation of granodiorites and blueschists. Periods of rapid continental dispersal and breakup like those beginning in the early Cambrian and early Cretaceous herald the beginning of major phases of marine transgression, widespread carbonates and continental margin orogeny, and alternate episodically, perhaps even periodically, with continental assembly phases such as the late Silurian/Devonian and the late Carboniferous/Permian, during which orogeny results from major continental collisions and when falling sea-level leads to widespread emergence and red-bed deposition.

3. Difficulties of plate tectonics

a. Inherent complexities of boundary evolution

Even if plate boundaries were simple clean lines across which relative slip occurred, plate tectonics, by its very nature, must generate boundary geology of great spatial and temporal complexity as already shown above (Fig. 5); it is capable of accounting for geological histories at least as complicated as those deduced from geological mapping.

b. Inversion and non-uniqueness problems

This is a problem that plagues all sciences, but geology perhaps more than most. We can build theoretical models of plate boundary behaviour to show how complicated things can be and, in great detail, the kinds of things that can happen (Fig. 5) but it is not so easy, perhaps completely impossible, to reconstruct, for example, a unique and convincing plate tectonic history of the Caledonides. Some further reasons for this will be developed below but the principal reason is the problem of recognizing the imprint of particular kinds of plate boundary, and the slip across them, from their geology. For example, an important aim of structural geology is the detailed and continuous integration of strain and incremental strain history across a deformation zone and to then relate this to the history of relative motion between the bounding plates. We appear to be a long distance from this goal, even in areas such as the Alpine System where we have some independent idea of the relative motion of the larger bounding plate (Fig. 42, inset). A pessimistic view is that the behaviour of the long-term elastic lid in deformation zones is so complicated, and capriciously related to old inhomogenities that its deformation does not accurately reflect the gross history of shape change in the viscoelastic lower lithosphere, which is driven in turn by motions of the bounding plates. A more optimistic view is that deformations of the upper lithosphere are roughly plane strains at least for any deformation phase, and therefore that smaller-scale deformation may be integrated in this context. I lean towards the latter view, though which is correct will take a great deal of painstaking fieldwork to determine.

c. Scale problem

The wide diffuse nature of most continental plate boundary zones means that the scale at which plate tectonics operates is often very much larger than the scale at which we are trying to solve geological problems and therefore there is a serious problem in relating what we see in the field to plate tectonics in often a qualitative, let alone quantitative, way. For example, if we were to spend our lives mapping in even a large region of central Asia *within* the wide plate boundary zone of Tibet, Sinkiang and Mongolia (Fig. 8) we would probably learn little or nothing about the relative motion of the Indian and Eurasian plates or perhaps even not know of their existence. We would know that an important regional event occurred at about 40 Ma ago but we would not know that it was the collision of peninsular India with Tibet. The plate boundary tectonic 'climate' is one of convergence but the local tectonic 'weather' varies from convergence, through strike-slip to extension. It is generally not possible to reconstruct tectonic climate

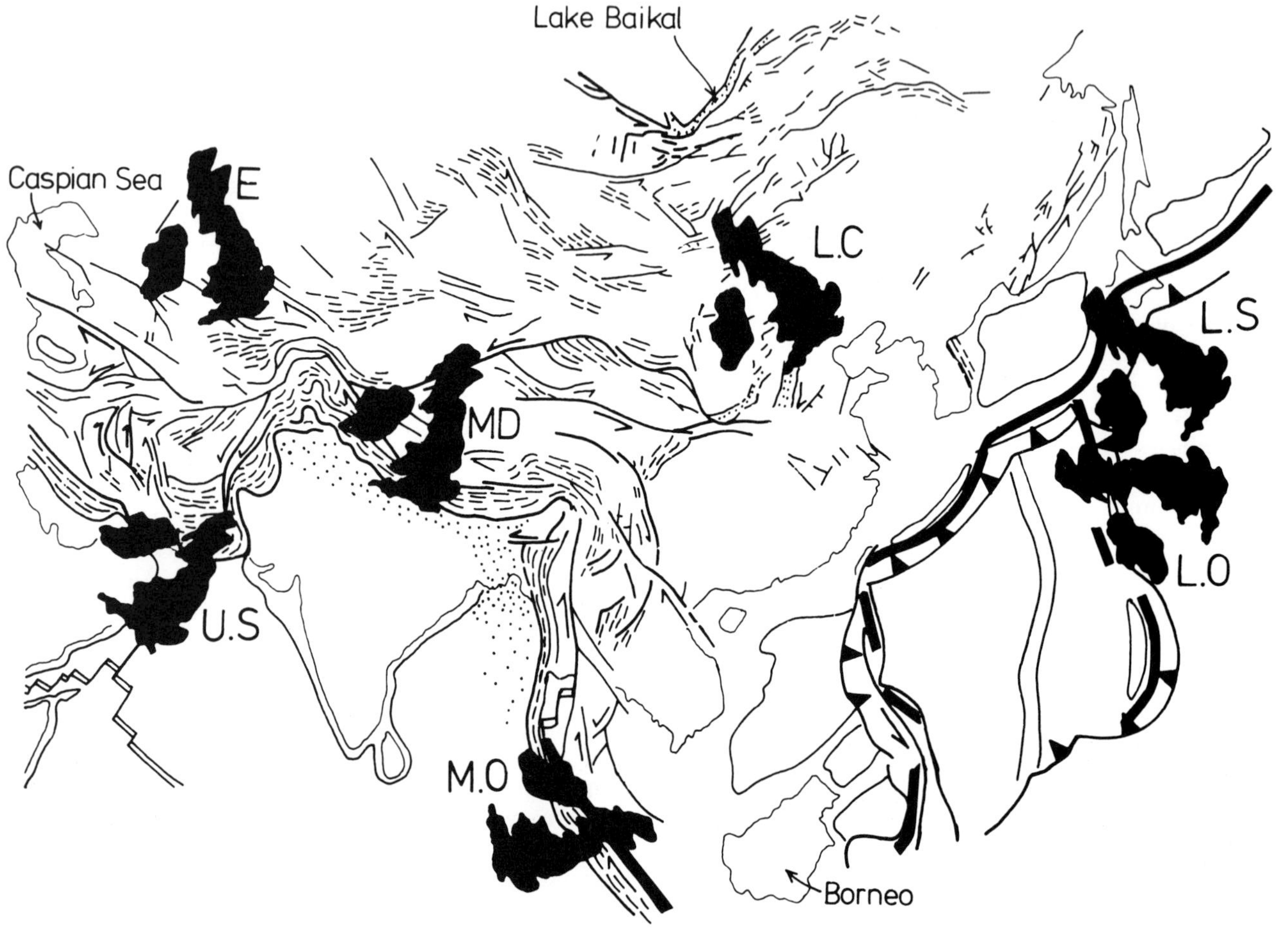

FIG. 8. Instantaneous neotectonic map of Asia with superposed British Isles to show the inherent scale problem in dealing with the plate boundary behaviour of small areas and to illustrate the importance of seeking neotectonic analogies for palaeotectonic synthesis. E = Eocene, LC = Lower Carboniferous, LO = Lower Ordovician, LS = Lower Silurian, MD = Middle Devonian, MO = Middle Ordovician, US = Upper Silurian.

from local geological weather without synthesizing extremely large regions. Therefore, any plate tectonic analysis of the British Isles must put them in a much larger regional and even, occasionally, global context before very much sense can be extracted. This point is emphasized by placing an outline British Isles in various positions and orientations in Asia to emphasize the scale problem and to suggest some tectonic analogies that are perhaps appropriate for various episodes in their history. A plate boundary *within* the British Isles can be last recognized in late Silurian times [Iapetus Suture through Ireland and the Solway Firth (Phillips *et al.* 1976)] and during the left lateral Great Glen phase when a cumulative relative motion of up to perhaps 2000 km (Kent & Opdyke 1979) was taken up across the Highlands probably in mid-Devonian times.

d. Palinspastic problems

The implications of the mobilism inherent in plate tectonics are more profound than the enormous rotations between plates, because large displacements generate palinspastic problems within plate boundary systems and produce special difficulties for the geologist intent upon regional palaeogeographic reconstructions. First, within orogenic zones, shortening of up to 300 and perhaps 400 km can be taken up by basement intracrustal flaking and underthrusting. This has long been recognized from stratigraphic and structural observations that demand basement subduction or swallowing (Ampferer & Hammer 1911; Bally *et al.* 1966; Gee 1975) but the recent results of COCORP deep seismic reflection profiling in the Carolina Appalachians have now characterized the typical geometry as one of shallow flat mega-flakes up to 12 km thick overthrust for distances up to at least 200 km. Secondly, the geological work of Coney *et al.* (1980) and the palaeomagnetic work of, for example, Beck (1980) has shown that the Cordilleran Belt of western North America W of the pre-Antler shelf edge is a collage of displaced, rotated, and mangled terrains, some of the larger of which are shown in Fig.

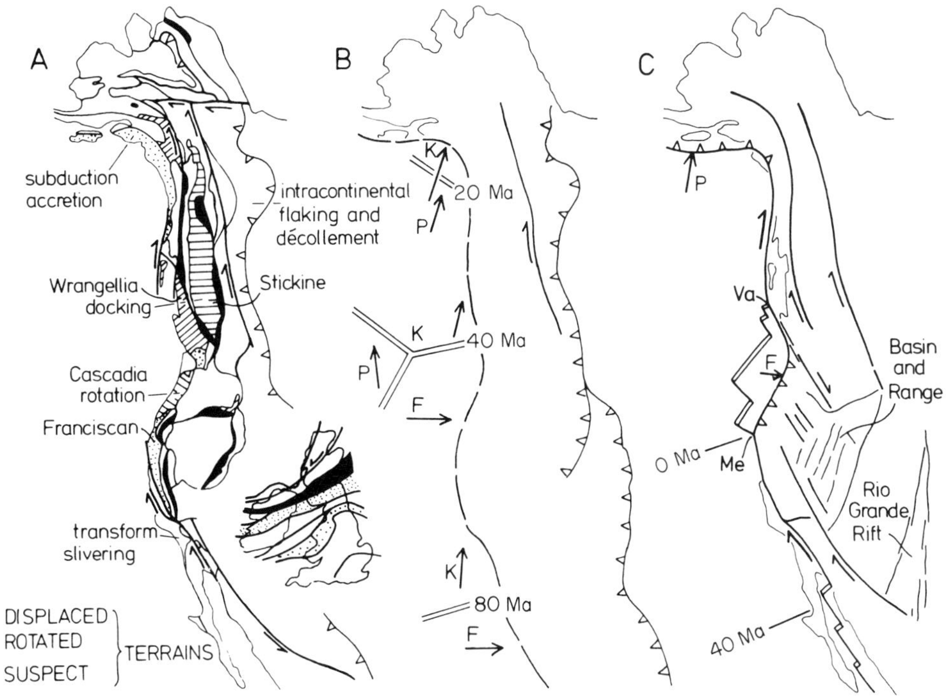

FIG. 9. A. Simplified terrain map of the U.S. and Canadian Cordillera to highlight some displaced, rotated, and accreted terrains. B. Northward motion of the Kula/Farallon/North American triple junction from 80 Ma to 20 Ma. C. Present plate boundary system along the U.S. and Canadian Cordillera. P = Pacific Plate, F = Farallon Plate, K = Kula Plate, Me = Mendocino triple junction, Va = Vancouver triple junction. Small British Isles inset to highlight some Caledonian terrains.

9A. Many of the larger terrains, such as Wrangellia, have an oceanic affinity, supporting the contention of Nur & Ben-Avraham (1978) that they are exotic oceanic plateau and arc complexes. The palaeomagnetic and faunal data suggest motions of thousands of kilometres of several of these terrains with respect to 'stable' North America along lineaments, some of which are sutures modified by transform slivering. Some of the motion, whereby terrains sidled northwards along the western edge of North America, often to end up in the 'collisional graveyard' of Alaska, occurred as a result of Pacific/North America transform motion (Fig. 9C) as the Farallan/North America subduction interaction diminished (Atwater 1970), but much could have been accomplished during Kula/North America oblique transform interaction N of a northward-sweeping Kula/Farallan/North America triple junction (Fig. 9B). Because transform faults are not simple, long-lived, single faults but anastomosing, braided systems of faults that constantly change their position, the displaced terrains are a composite of slices, that, like goods wagons, may be left in a siding to be picked up later; the northward motion may involve many stages of docking, sidling and strike-slip dissection. Palaeomagnetic data (Beck 1980) shows conclusively that many fairly substantial terrains have rotated by up to 70° (Fig. 10B), thereby changing the orientation of intra-block palaeogeographic elements and causing severe structural compatibility problems at block margins (Fig. 10C). It is not inconceivable that some terrains have been turned 'back to front' with successive waves of extension and compression along block margins during rotation (Fig. 10C). Although the displaced terrains are characteristically elongate parallel with the western margin of North America, many may have changed their overall shape by motion along, typically hard to recognize, internal strike-slip faults (Fig. 10A). Lastly, superimposed upon these complications, are the inherent complexities related to the evolution of strike-slip systems (Fig. 10D–I).

Gapais & Le Corre (1980) and Badham (1982) have taken the view, following the model of Arthaud & Matte (1977) which links the southern end of the Urals with the northern end of the late Palaeozoic Appalachians with a Hercynian right-lateral megashear, that the Hercynides are basically a strike-slip orogen. In this context, some 'orogenic' events may result from the short-duration opening, rotation, flaking and obduction of pull-apart basins (Fig. 10D–I), which may show all transitions from a graben floor of slightly stretched continental crust with or without volcanism, through severely-attenuated continental crust with ultramafic diapirs (?Lizard ultramafic/mafic complex)

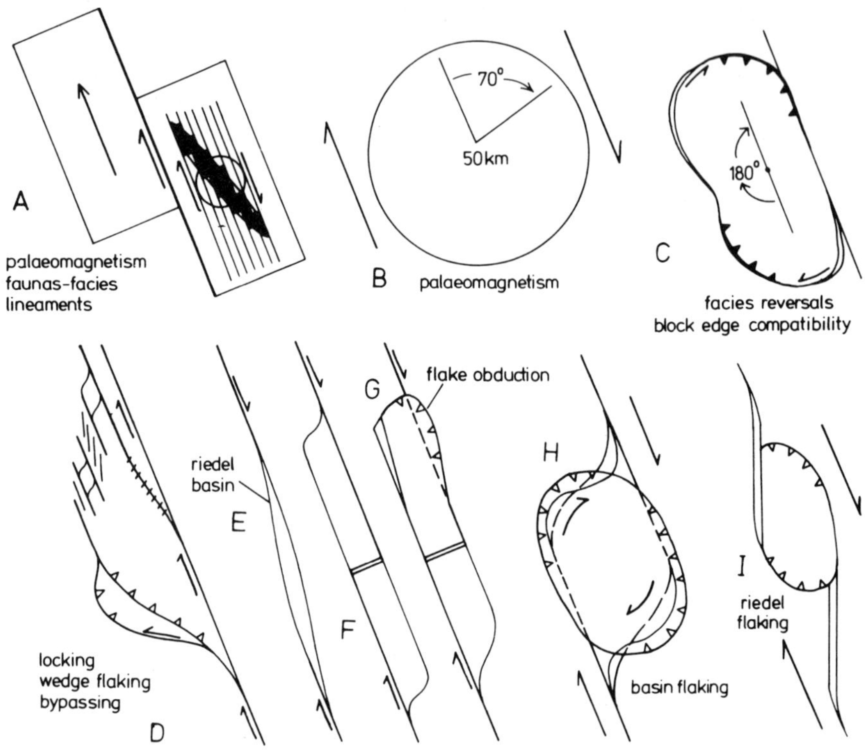

FIG. 10. Schematic illustration of some of the facets and complexities of the transform tectonics of displaced terrains in continental margin megashears.

to an oceanic floor. This wide variety of basin floors causes a wide and complex range of strengths with time relative to unstretched surrounding crust (see following section on basins).

Therefore, when we try to solve problems of Caledonian evolution in a plate tectonic setting, we are faced with what, at first glance, seem insurmountable difficulties. The probable 2000 km left-lateral displacement during the Upper Palaeozoic seemingly demanded by the palaeomagnetic evidence (Kent & Opdyke 1977) indicates that we should look in northern Norway for terrains that were adjacent to Scotland in early Devonian times. Moreover, we cannot be certain of the position of any Caledonian terrain relative to its present neighbours (Fig. 9A) prior to the time at which they are absolutely unequivocally joined by 'linking stratigraphical sequences' such as the Upper Llandoverian sediments of the Welsh Borderlands that link the Shelve to the Pontesford/Longmynd basement by mid-Silurian times. We can have little idea of the relative positions of the Midland Valley, the Lake District, the Mona Complex and the Welsh Basin during mid-Ordovician times, which ignorance prevents our making anything beyond cartoon reconstructions; such a Caledonian cartoon is attempted below. Again, the scale problem (Fig. 9A, inset) demands that we consider the whole Appalachian/Caledonian belt in attempting to solve the displaced terrain problem in the British Caledonides. A subtle philosophical point in tectonic analysis and synthesis of orogenic belts is involved here. The tectono-stratigraphical *zone* approach (Dewey 1969; Williams 1978) has encouraged a view of along-strike cylindrism that tends to mask along-strike variation that is so critical to the more useful stratigraphic terrain approach. Although it is extremely appealing to draw, for example, cross-sections illustrating the Lower Palaeozoic evolution of Scotland, it is only useful insofar as it reminds one of the difficulties of doing so.

e. Preservation problems

There are some tectonic questions to which we probably can never know the answer and therefore, which it may be fruitless to ask, relating to the way in which plate tectonics allows only a partial preservation of critical data. First, oceanic plates with most of their vital information are subducted leading to a quantum jump in the difficulty of plate tectonic analysis prior to about 170×10^6 a. Oceanic scraps of sediment and seamount may be preserved in subduction-accretion prisms but any attempt to palinspastically unravel such prisms, to determine the amount of lost oceanic basement from which they were stripped, would be absurd. Unlike Rocky Mountain-style foreland thrust belts, where such a practice may be viable, subduction-accretion prisms are not susceptible to the drawing of balanced cross-sections because of the penetrative nature of their strain and the fact that much of the layer 1 sediment is subducted. Secondly, continental terrains may be lost beneath intracontinental overthrust zones. Where a suitable lubricating horizon allows cover *décollement*, the 'lost' basement may be almost wholly represented in cover nappes (e.g. Helvetides) but in the absence of such a horizon, hundreds of kilometres of sediment may be 'swallowed' with the basement (e.g. Southern Appalachians) leaving only a mylonite zone as evidence of its earlier existence. Thirdly, a particularly invidious lack of preservation is the possibility of subduction-truncation either by strike-slip removal or by the subduction of slices of the overriding fore-arc, processes that could have been responsible for the removal of late Palaeozoic/early Mesozoic terrains from the Peruvian/Chilean margin of South America (Dewey 1980). Lastly, exotic displaced terrains in one area, such as the North American Cordillera, mean lost terrains elsewhere, perhaps the western margin of South America.

Finally, what is not lost by one or more of the foregoing processes may become, eventually, hopelessly mangled and smeared out to an extent that precludes or makes very difficult the unravelling of original geological relationships. This will, unfortunately, produce a cylindroidal simplicity in an orogen with a long history of ocean closure and complex marginal arcs, such as the Urals or Caledonides. The simplicity and along-strike conformity of Palaeozoic orogens may be more apparent than real having been generated, possibly, by large longitudinal and shear strains that caused very large internal rotations. Consider what might be the eventual geometric fate of the complex oroclinal arc, small ocean basins, and continental fragment tectonics of SE Asia if Australia continues to move northwards and flattens the whole collage against the Asian continent. There is no particular reason to suppose that the British Caledonides, for example, had a very much simpler history.

f. Slip vector problems

If we wish to make rational plate tectonic reconstructions to account for the geological history of particular regions such as the British Isles, we need to be able to define the slip vector, that is the direction and rate of relative motion, of bounding plates for the time interval concerned and then to convincingly relate the geological history to the relative motion. For the Mesozoic and Cenozoic, slip vectors may be determined from finite difference solutions using continental margin and magnetic anomaly fitting and fracture zone data. Slip vectors thus defined may be used as a backdrop for palaeotectonic and palaeostructural analysis and reconstruction. An ultimate aim of such studies might be the deduction of slip vectors from the geology. Attempts of this kind have been made for subduction-accretion prisms (Moore & Connelly 1977) and may eventually be applied successfully to older prisms like the Southern Uplands. However, the complexity of rock response in most plate boundary zones inhibits, if not prevents, the development of such rock-strain/slip-vector relationships. There are, however, several useful approaches to the slip-vector problem, most of which depend upon data synthesis over a very wide region and yield direction but, rarely, rate of slip, and several of these will be used in later sections.

1. The intensity of most plate boundary activity increases to lower rotational latitudes. Near-pole activity is slow, complicated, often diffuse, and somewhat confusing; a good example is the near-pole portion of the Eurasian/North American boundary in Siberia.

2. The distribution of synchronous, but different, kinds of plate boundary activity such as ophiolite generation (ridge), major shear zones (transforms) and calc-alkaline zones (subduction) allows a crude depiction of plate boundary geometry and an approximate guess at the direction of relative motion; this method has been shown in the Alpine System to yield good correlations with the slip-direction determined independently from Atlantic finite-difference solutions (Smith 1971; Dewey *et al.* 1973).

3. The broad geometric arrangement of deformational facies patterns along colliding continental margins from intense to less intense, with and without strike-slip motion, and the timing of terminal collisional events along the suture complex, may provide clues to the relative motion direction during collision.

4. The internal structures of ophiolites, particularly sheeted-dyke complexes, give direct information on the rate and direction of plate-accretion at the ridge at which they were formed (Kidd 1977; Kidd & Cann 1974; Gass & Smewing 1981). The NW orientation of dykes normal to an ancient fracture zone in the Bay of Islands Ophiolite Complex in Newfoundland suggests right-lateral motion on a ridge/transform system and, as argued below, a right-lateral pull-apart origin for the South Mayo Trough. Similarly, the present approximately N–S orientation of dykes in the Troodos Complex is consistent with the anticlockwise rotation of Africa relative to Eurasia during the Cretaceous, although this may be fortuitous, as palaeomagnetic data suggest that Cyprus has been rotated. This raises the main difficulty with using the internal structures of ophiolites to determine spreading rates and directions; most ophiolite complexes form obducted sheets up to about 10 km thick, many of which may have suffered significant, but difficult to determine, rotations about vertical axes during structural emplacement. The problem is made particularly acute because the motions of small plates, blocks and flakes in collisional systems, although ultimately controlled by the relative motions of larger bounding plates, generally do not bear a simple relationship to them. Also, when sub-

ducting oceanic plates become very small the dynamics change to allow rather fast rotations by as much as 40° (Menard & Atwater 1968).

5. Major intracontinental transform lineaments, such as the Great Glen Fault, the Tornquist Line and the Elbe/Dobrogea Line, when viewed with other similar faults in a major regional context, may give direct information on the slip direction and, in favourable circumstances, slip rate between major plates. This approach is used in late Palaeozoic reconstructions below.

6. APW paths, particularly when integrated with lineament and facies data, are a critical tool in evaluating pre-Jurassic relative motion (McElhinny 1973). Integrated analyses of this type have been made by Smith *et al.* (1973).

7. The tectonic style of subduction complexes, whether dominated by compression, strike-slip or extension, may be related directly to the motion of the overriding plate in an asthenosphere reference frame (Dewey 1980) and therefore may form an important element in evaluating relative motion.

8. However, the most promising approach is that developed by A. M. Ziegler and his Chicago school. This is an iterative global analysis and synthesis method in which palaeomagnetic, faunal, facies, palaeoclimatic, igneous, metamorphic and tectonic data are used to develop global palaeogeographic reconstructions from which can be derived finite difference relative motion solutions.

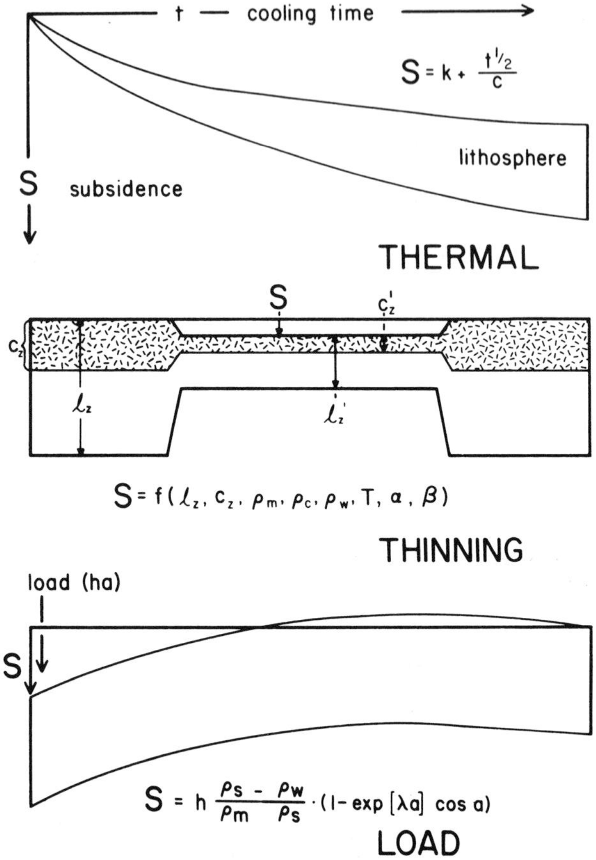

FIG. 11. Three principal mechanisms of subsidence; thermal due to cooling and thickening of the lithosphere, lithospheric attenuation, and load-induced.

4. Basins

Sedimentary basins are a sensitive barometer of tectonic climate and a subtle indicator of deformation mechanisms that affect the lithosphere. The post-Silurian history of the British Isles and surrounding areas is, to a large extent, one of a delicate interplay of basins, troughs, platforms, horsts, ridges, axes, depressions and swells, with or without fault control, without or with magmatism. It is therefore of interest to examine, briefly, the physics of vertical motions of the lithosphere illustrated by some British examples. For simplicity, vertical plane (biaxial) strain is assumed throughout; that is, cross-sectional area is conserved and no volume change occurs.

The three principal 'driving mechanisms' for vertical motions of the lithosphere are thermal, crustal thickness change, and the addition or removal of load. Cooling and, therefore, thickening of the lithosphere, thinning of the crust, and loading, cause subsidence (Fig. 11). Heating and, therefore, thinning of the lithosphere, thickening of the crust, and load removal cause uplift. Subsidence mechanisms such as meteorite impact, salt and carbonate solution and coal removal are ignored here; the first was only important in earlier Precambrian times, and the rest are only of local minor interest. Metamorphic phase changes in the lower crust may also produce subsidence, but are hard to model. Thermal subsidence occurs by cooling and thickening of the lithosphere, after an event that thins the lithosphere by raising the 1333°C isotherm (Fig. 11). Subsidence, by the addition of a load such as a volcano, an ice sheet, a pile of thrust sheets or sediments, depends upon the elastic thickness of the lithosphere. Subsidence by sediment loading alone cannot initiate a sedimentary basin. Cooling or crustal thinning must precede or accompany sediment loading. Thick volcanic piles, however, with their capacity for being laid down above sea-level can alone drive the subsidence of basins. Load subsidence varies from a local Airy-type compensation with no peripheral bulge, where the lithosphere is effectively zero, to load distribution over a wide area with a small, far-distant peripheral bulge (Fig. 11). Examples of the role of the peripheral bulge may occur in the geology of the NW Highlands (Fig. 12). On a palinspastic reconstruction, with the effects of telescoping between the Sole Thrust and the Moine Thrust removed, the Eriboll Quartzite oversteps the Torridonian onto a broad swell or ridge, probably a peripheral bulge caused by Moine/Appin/

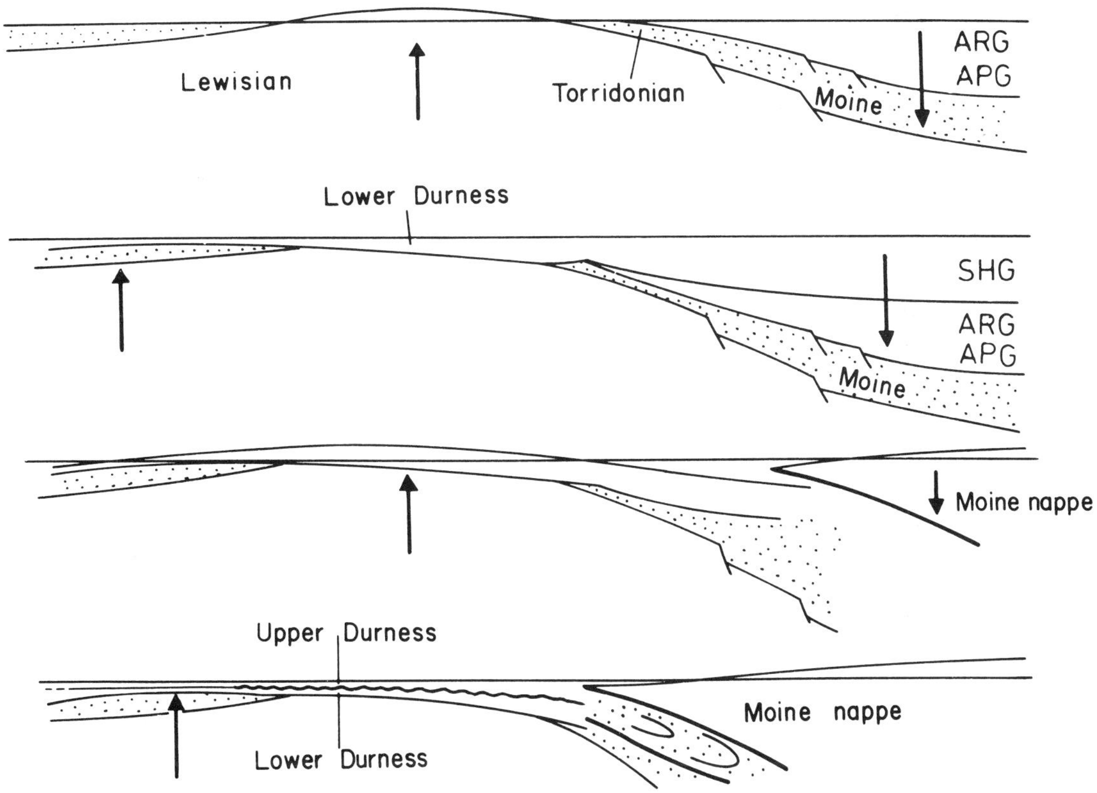

FIG. 12. Schematic interpretation of the Eriboll quartzite overstep of the Torridonian and the intra-Durness disconformity (Palmer *et al.* 1980) as the result of late Precambrian an early Ordovician peripheral bulges related to sedimentary loading and thrust loading respectively in the Highlands. APG = Appin Group, ARG = Argyll Group, SHG = South Highlands Group.

Argyll Group loading in the Highlands. The intra-Durness disconformity, recognized by Palmer *et al.* (1980), may represent a peripheral bulge caused by loading by an advancing Moine Nappe. Similarly, the growth of the Bristol Channel landmass during mid-Devonian times, with the consequent flood of Hangman Grits into N Devon, may have been caused by sediment loading in the Cornubian Trough, although a thermal bulge marginal to the Cornubian rifted basin may have been a possible origin. The rise of St. Georges Land itself may have been enhanced by loading in Carboniferous troughs to the N and S. The late Carboniferous/early Permian Mid North Sea High has about the correct wavelength and amplitude for a peripheral bulge due to loading by Hercynian marginal thrust sheets and Coal Measure/Rotliegendes sediments in the Southern North Sea Basin.

Crustal thinning may be accomplished by one or several of three mechanisms. Subcrustal erosion has been promoted, particularly, by the Russian school, but seems fanciful and unlikely as a primary mode because it needs the lithosphere to be thinned by some asthenospheric blowtorch mechanism so that the 1333°C isotherm is at or within the crust. This would produce crustal melting on a prolific scale and enormous uplift, phenomena that generally do not precede basin subsidence but may have been important during the earlier Precambrian. Uplift and, consequently, supracrustal erosion result from raising the 1333°C isotherm whenever the crust is not simultaneously thinned or is thinned by a smaller amount. The early Cretaceous separation of the Iberian sub-continent from the Brittany margin and the Grand Banks of Newfoundland was preceded by a doming event, judging from the eastward tilting and dispersal of the Wealden facies, and westward Aptian overstep. However, we should be careful in ascribing such an origin to the westerly Wealden provenance and the westward Aptian overstep because rift-shoulder, thermal-bulge mechanisms may have been involved and the early Cretaceous was a time of low sea-level followed by a major mid-Cretaceous rise. Following doming and denudation, isostatic recovery subsidence produces a basin at the base of which the earliest basin sediments rest unconformably upon deeply-eroded basement, pre-basin sediments having been stripped during the doming process. This kind of basin would be more likely to have a 'bull's eye' pattern, would have a flat or slightly depressed Moho, and would not have bounding faults; I know of no such basins.

The only basin driving mechanism that systematically satisfies the geological history of most intracratonic and rifted margin basins, aside from those foreland basins obviously caused by thrust loading, such as the Westphalian coal belt of NW Europe, is the lithospheric attenuation model of McKenzie (1978) whereby stretching (by the factor β) causes lithospheric thinning (by the reciprocal of β). Subsidence or uplift results from stretching depending upon the ratio of crustal to lithospheric thickness (C_z/l_z). Where $C_z/l_z < 0.1375$ uplift results, where $C_z/l_z > 0.1375$ subsidence results (Fig. 13). The subsidence (S_i) that results directly from stretching is given by the expression (from McKenzie 1978) in Fig. 13, which assumes a perfect Airy compensation with no flexural strength for the lithosphere; where β is large and stretching occurs rapidly, this is a reasonable approximation. The two principal variables in determining the amount of initial subsidence are C_z and l_z, which may vary considerably between various plate boundary tectonic environments. Island arcs have large C_z/l_z ratios and, therefore, small β values produce large subsidences. Old cratons have low C_z/l_z ratios such that uplift may result from stretching. The effect of varying lithospheric thickness with crustal thicknesses of 31.2 and 35 km are shown in Fig. 13, together with the effect of varying crustal thickness with a lithospheric thickness of 125 km. 31.2 km is a typical crustal thickness in areas of stable, little modified, continental crust where determined seismically and is a thickness that isostatically allows the surface to lie at mean sea level; in the ensuing analysis, thicknesses of 31.2 km and 125 km (Sclater *et al.* 1980) are adopted for the crust and lithosphere respectively. Assuming geologically instantaneous stretching, stretching and its synchronous subsidence are followed by a subsidence (S_t)

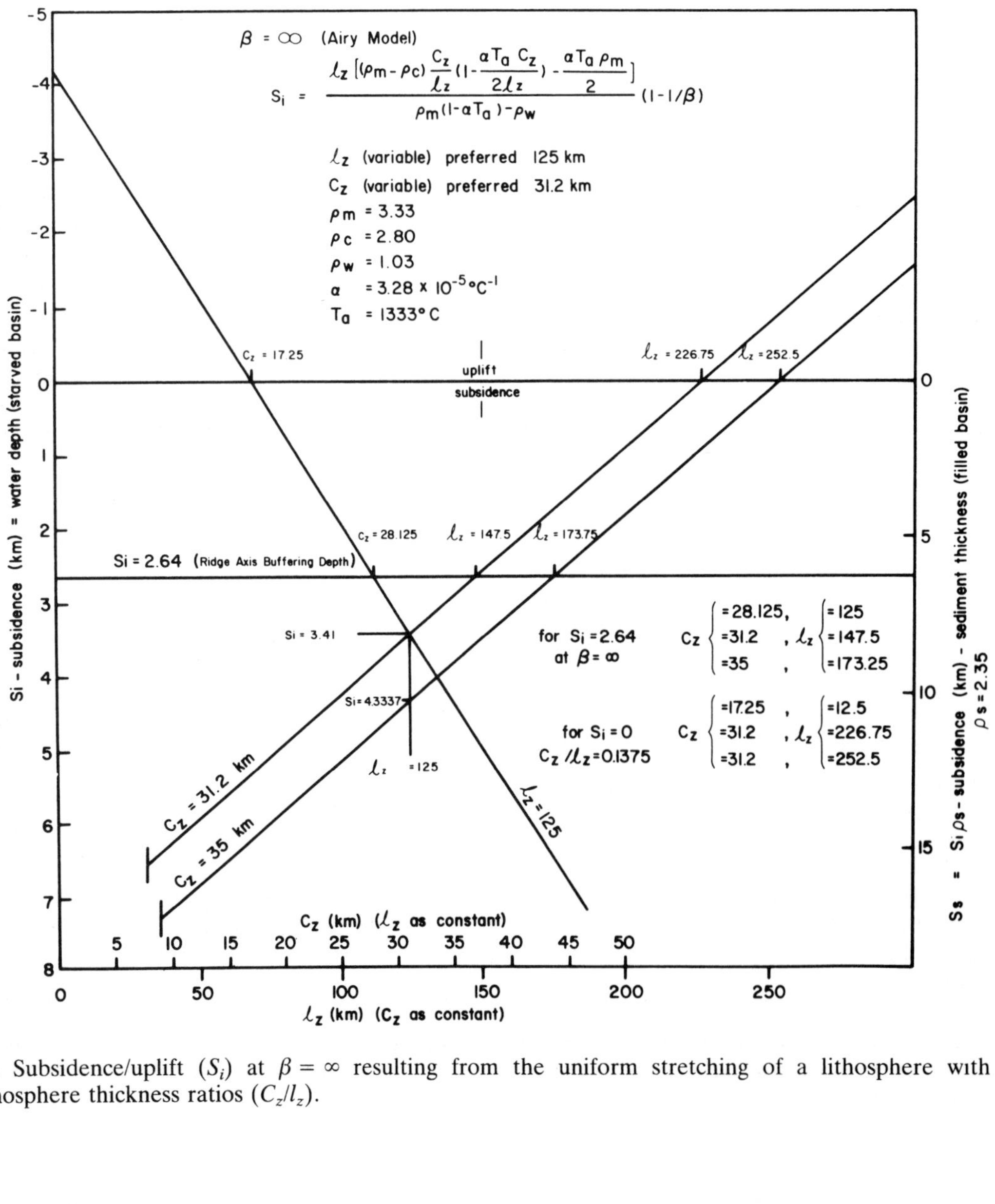

FIG. 13. Subsidence/uplift (S_i) at $\beta = \infty$ resulting from the uniform stretching of a lithosphere with various crust/lithosphere thickness ratios (C_z/l_z).

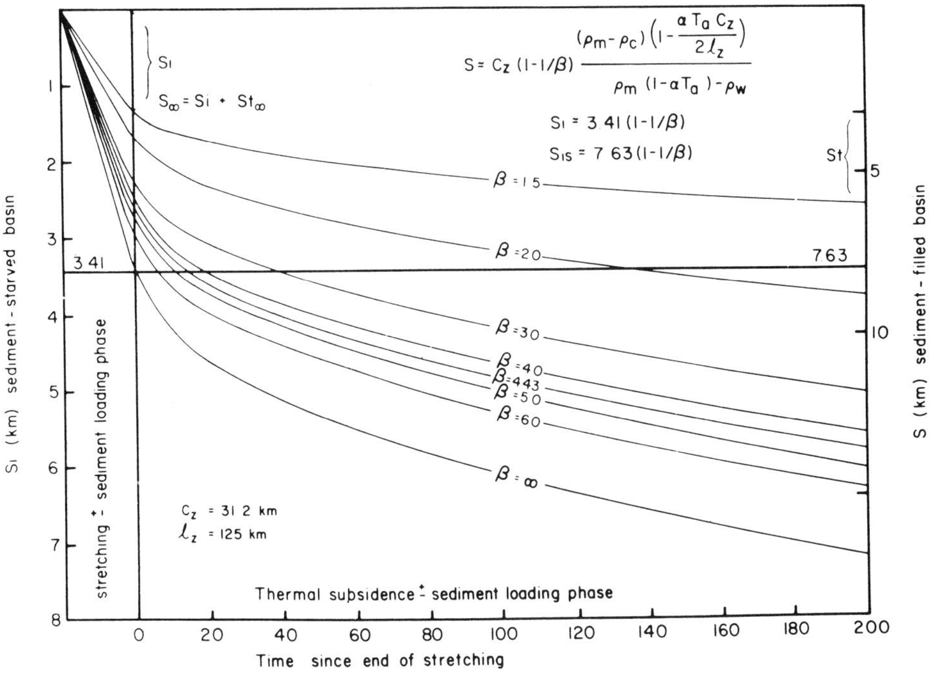

FIG. 14. Subsidence during uniform lithospheric stretching (S_i) and during thermal recovery (S_t) for initial crustal and lithosphere thicknesses of 31.2 and 125 km, respectively, for β values of 1.5 to ∞.

produced by cooling and thickening of the lithosphere (Fig. 14). The driving subsidence S_i and S_t may be increased by sediment loading or may simply produce, in the absence of sediment supply, a deep trough. If stretching occurs over a long period, part of the thermal subsidence, S_t, is transferred back into the stretching phase. The amount and rate of subsidence during the stretching phase are thus determined by β and the strain rate $\dot{\varepsilon}$ (Fig. 15). Using the common (geologically determined) range of stretching times of between 10 and 30 Ma, a typical range of final graben-complex widths of between 30 and 300 km, and the assumption that the same cross-sectional area of crust is involved, minimum strain rates of about $10^{-16}\ s^{-1}$ result. However, the assumption of constant cross-sectional area is incorrect; the width of the stretching zone decreases or remains constant with time and strain rates probably vary with time from about $10^{-13}\ s^{-1}$ to $10^{-15}\ s^{-1}$. Graphic presentations of some of the main features of McKenzie's (1978) model are shown in Figs 16 and 17. A particular feature of the model is that the upper elastic lid is characterized, during stretching, by

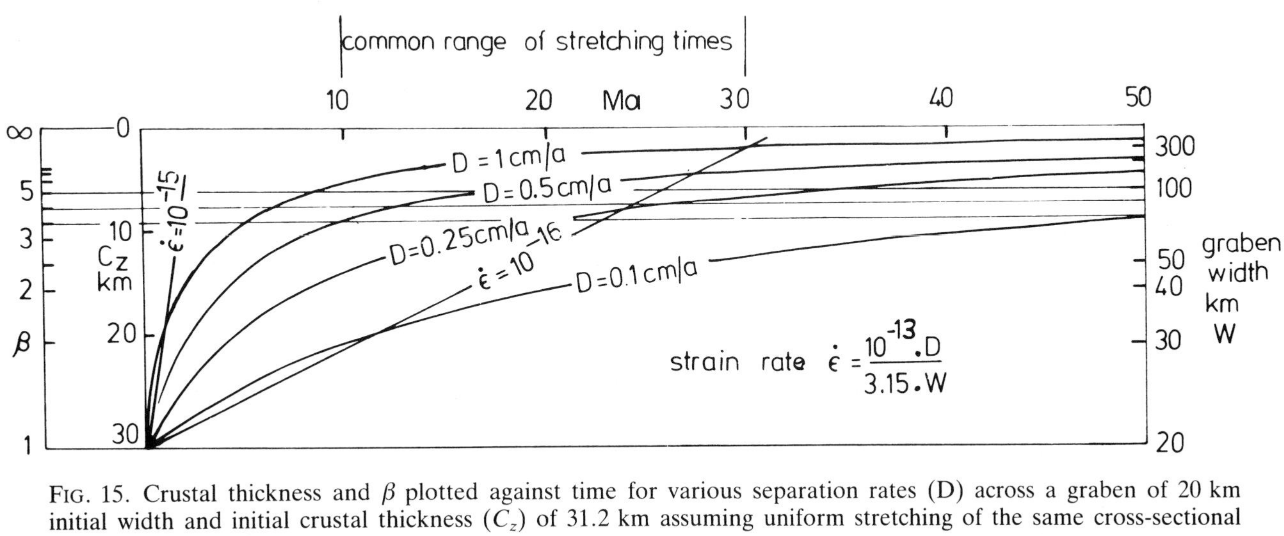

FIG. 15. Crustal thickness and β plotted against time for various separation rates (D) across a graben of 20 km initial width and initial crustal thickness (C_z) of 31.2 km assuming uniform stretching of the same cross-sectional area of crust with time.

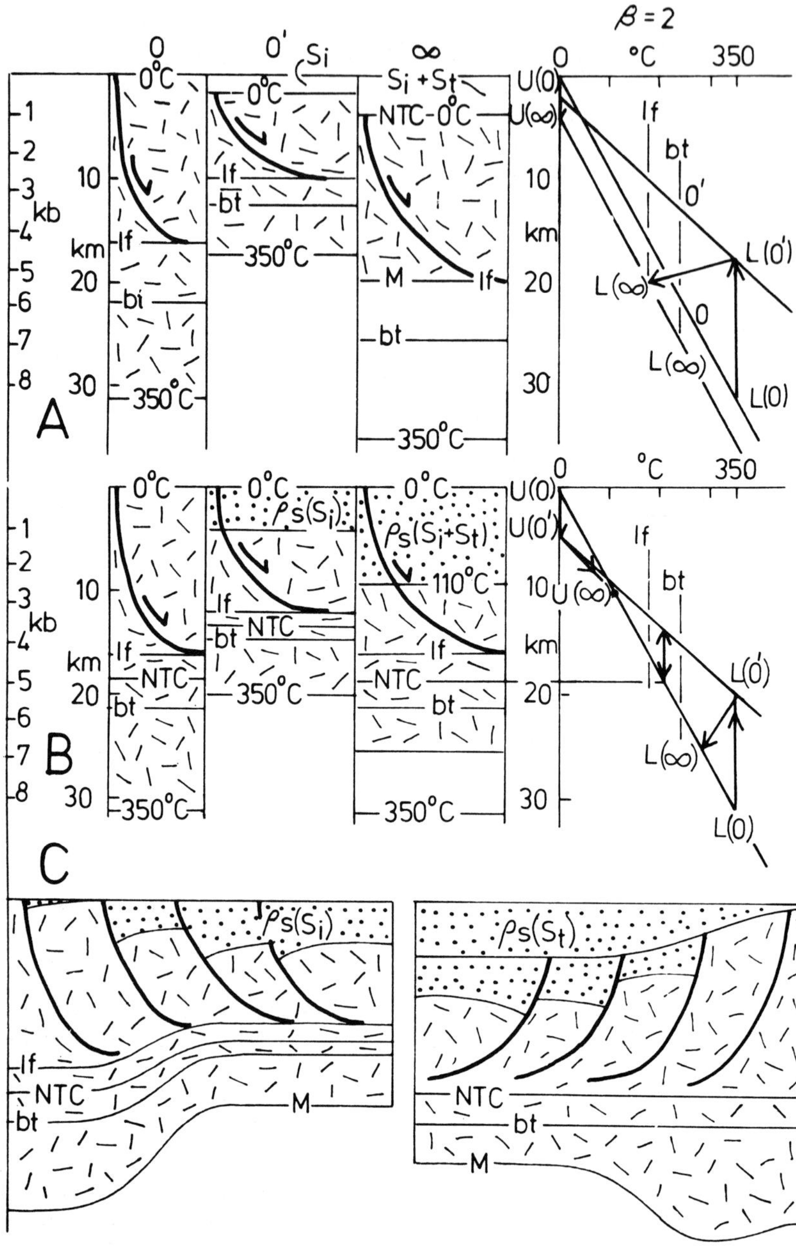

FIG. 16. Crustal structure, temperature and confining pressure assuming an instantaneous lithospheric stretching ($\beta = 2$) model (McKenzie 1978) before stretching (0), at the end of stretching (0′), and at an infinite time (∞) after stretching, for a starved basin (A) and a continuously filled basin (B). An initial Moho temperature of 350° is assumed with, for simplicity, a linear geothermal gradient. NTC = level above which rocks are hotter and below are colder than prior to stretching; *bt* = arbitrary argon blocking temperature for a particular mineral species; *lf* = temperature level in crust at which listric faults flatten and below which diffusional creep occurs; S_i = subsidence during stretching phase; S_t = subsidence during thermal phase; ρ_s = sediment density; L = point on Moho; U = point on original upper surface of crust.

listric and/or rotational normal faults of the kind described by Proffett (1977) with a major décollement (1*f*), beneath which creep mechanisms predominate. As the lithosphere thins, particularly where β and $\dot{\varepsilon}$ are large, the elastic lid is likely to become an effective stress concentrator, enhancing fault, joint, and fissure development.

An alternative or complementary mechanism of pull-apart basin development has been suggested by Royden *et al.* (1980), the main features of which are illustrated in Figs 18 and 19. This mechanism involves separative motion by the emplacement of mafic/ultramafic dyke complexes into a constant thickness lithosphere. Both the McKenzie (M) and Royden *et al.* (R)

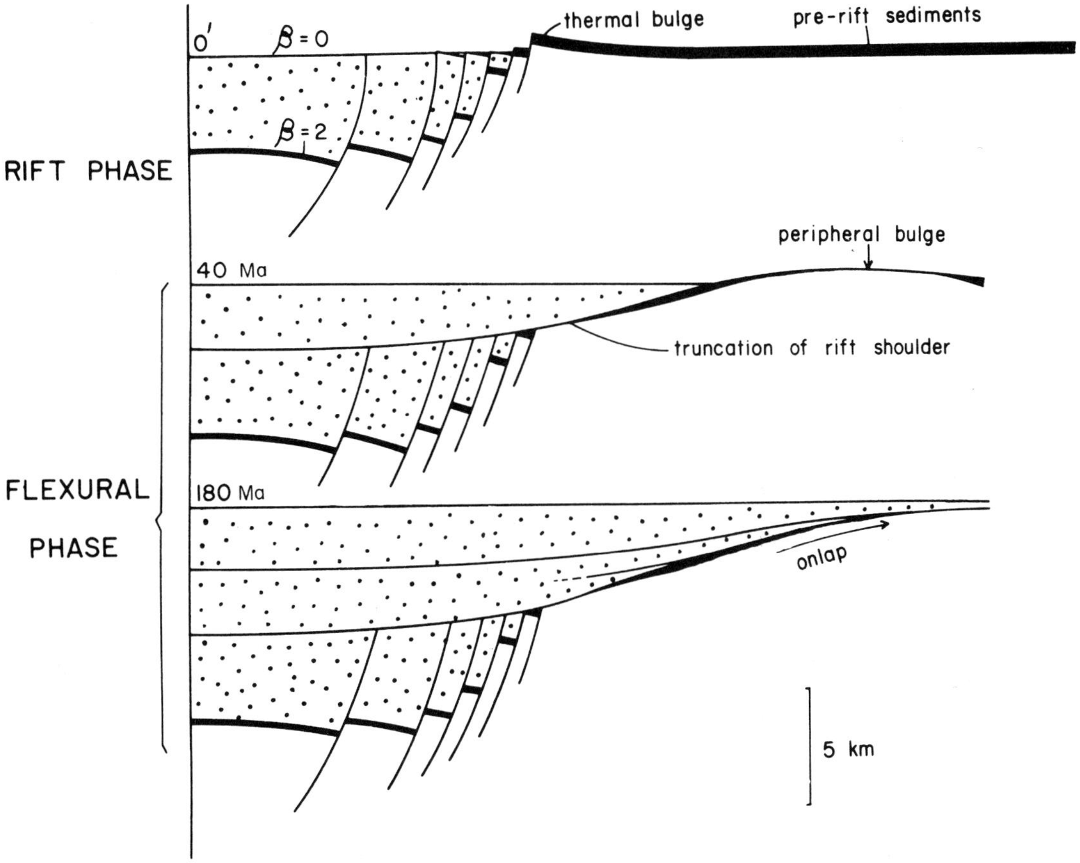

FIG. 17. Evolution of a sedimentary basin during rift and flexural phases assuming a lithospheric stretching model (McKenzie 1978).

mechanisms (compare Figs 16 and 17 with Figs 18 and 19) produce a rapid initial subsidence, extensional faulting and fissuring, and approximately the same initial (S_i) and thermal (S_t) subsidence for particular β values. In both R and M models, fault-controlled, rift subsidence approximating an Airy-model (at least for β values greater than 2), with a peripheral thermal bulge, is followed by sediment-load-enhanced thermal subsidence supported over a larger and larger area with the classic basin margin onlap following a migrating peripheral flexural bulge, producing the common 'Steer's Head' or 'Texas Longhorn' geometry (Figs 17, 19).

There are, however, a number of fundamental differences between the M and R models, most of which can be rapidly determined geologically. First, the Moho rises beneath the basin in the M model, and falls slightly in the R model. In the M model, crustal seismic velocities remain the same but are more closely spaced, whereas in the R model, large amounts of mafic/ultramafic material in a slightly thickened crust cause increased seismic velocities after cooling. Transition from the fault-controlled rift stage to the flexural stage is short in the M model but more protracted in the R model because the lithosphere remains warmer and weaker longer, allowing local compensation to linger on into the post-stretching phase. This may be thought of as a contrast in the speed of 'coupling' between the stretched and unstretched lithosphere in the M and R models so that peripheral bulges will develop earlier and more strongly in the M model, but marginal thermal bulges caused by lateral heat transfer at basin margins will be more prominent for the R model causing early, fault-controlled rift shoulders. The history of basinal geothermal gradients is fundamentally different in the M and R models. In the M model, for an unfilled basin (Fig. 16A), isotherm spacing is narrowed by $1/\beta$ (for geologically instantaneous stretching) and all rocks in the lithosphere became gradually colder with time. For a filled basin (Fig. 16B), rocks above the line of no thermal change (NTC) become warmer with time and rocks below NTC become colder. In the R model, magma at 1333°C comes to the surface, the geothermal gradient is reduced, heat is pumped into the crust and all rocks in the lithosphere become initially hotter than they were originally, then gradually cool to equilibrium temperatures at $t = \infty$. Thus, in the M model, individual rock units in the lithosphere not only cool but, during stretching, enjoy a reducing overburden, both

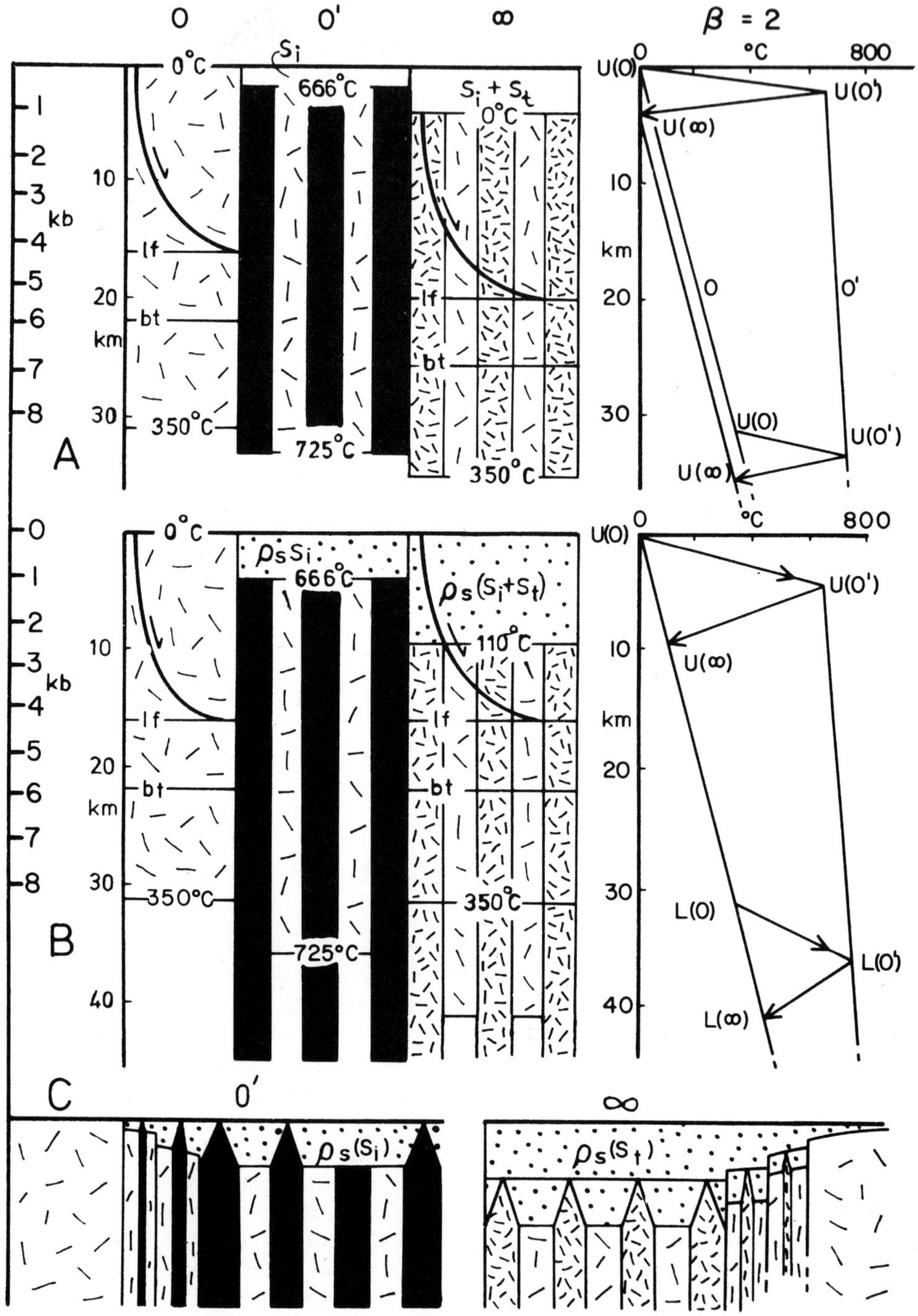

FIG. 18. Crustal structure, temperature and confining pressure assuming an instantaneous lithospheric dyke model (Royden *et al.* 1980) before dyking (0), at the end of dyking (0′), and at an infinite time (∞) after dyking, for a starved basin (A) and a continuously-filled basin (B). Nomenclature and symbols are as in Fig. 16.

of which factors increase rock strength and reduce ductility; declining strain rate may in the early stages of stretching work against these factors. In the R model, individual rock units in the original lithosphere are warmer, slightly deeper and therefore weaker than they were originally, having passed through a very hot, very weak stage during stretching. A further corollary of contrasting temperature histories in the M and R models is a difference in metamorphic history and in the setting of isotopic clocks as temperature for a particular mineral or rock species falls below the blocking temperature. In the M model, falling temperatures and pressures cause rapid retrogression during stretching and isotopic clocks may be set as blocking temperature thermochrons descend from their immediately post-stretching levels. In the R model, rapid

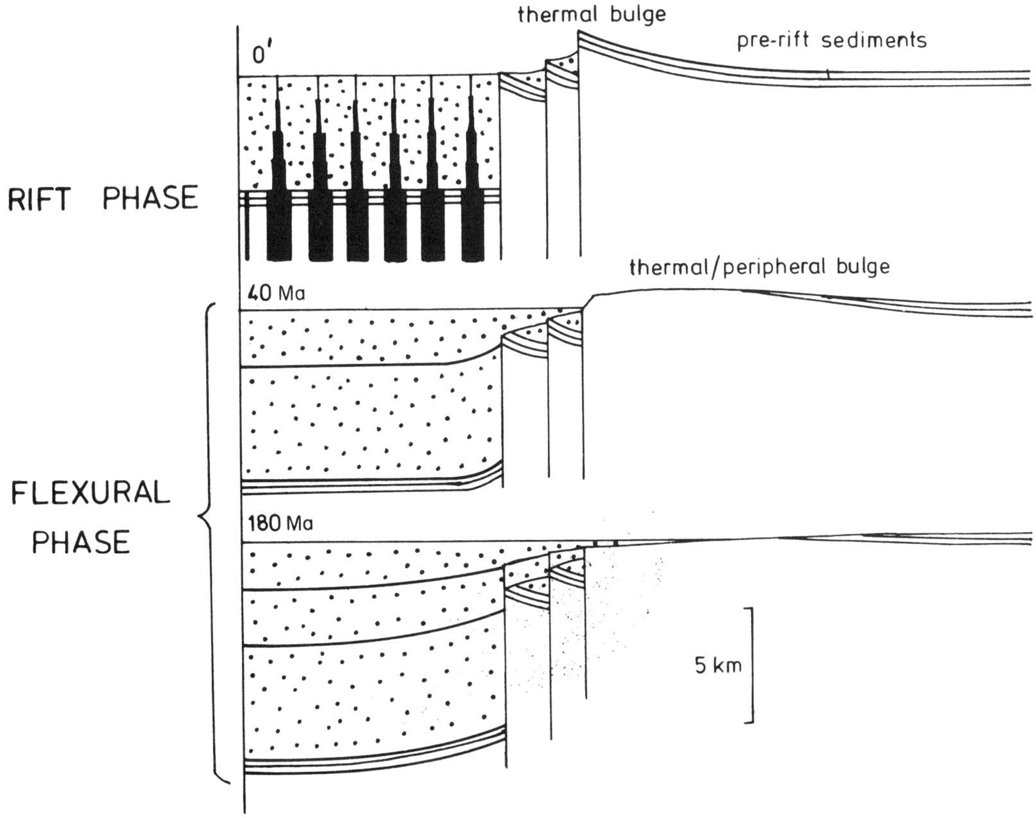

FIG. 19. Evolution of a sedimentary basin during dyking and flexural phases assuming a lithospheric dyking model (Royden *et al.* 1980).

prograde thermal metamorphism during extension is followed by retrograde metamorphism at roughly constant pressure; the same set of blocking temperature thermochrons are carried to much higher levels during extension and, as they descend, cause the progressive resetting of isotopic clocks throughout the crust. In the M model, injection of mafic magmas into the lithosphere cannot occur, with an initial lithosphere thickness of 125 km, until β reaches about 2 (discussed more fully below) and, thus, early basin evolution occurs without volcanism; crustal melting caused by heating from mafic magmas can only take place at an advanced stage of basin evolution. By contrast, the R model *assumes* mafic magmatism and, therefore, mafic volcanism and crustal melting characterize the whole history of stretching and the early parts of the thermal subsidence phase. Lastly, listric, normal growth-faulting and normal fault block rotation characterize the M model, whereas steep or vertical basin margin normal faults, nucleating hypabyssal intrusion and volcanism, persisting into the post-stretching phase, typify the R model.

The M and R models are not exclusive alternatives. Although there are many basins that accord to an M-type history throughout their evolution, β values for these basins appear always to be less than 2. It will be argued below that, for a 125 km initial lithosphere thickness, the M mechanism is exclusive until β reaches a value of about 2, that from $\beta = 2$ to a β value of about 3 there is an increasing chance of an R component and that, from $\beta = 3$, R becomes progressively dominant until at $\beta = 4.43$, R operates exclusively and continental rupture takes place.

If one adopts the view that stretching cannot occur to an extent that initial subsidence, S_i, is greater than 2.64 km (the average height of oceanic ridges), stretching, or β, is buffered at this value. Fig. 20 illustrates such β buffering values for a range of initial crustal thicknesses (C_z) where l_z = 125 km, from which it will be seen that $\beta_{\max} = \infty$ where C_z = 28.125, and $\beta_{\max}$ = 4.43 where C_z = 31.2. At $\beta = 2$, whatever the value of C_z, alkali basalt is produced by a few percent partial melting of mantle lherzolite (Figs 20, 21A) but is unlikely to reach the surface because the small volume of magma will rapidly freeze at depth. As β increases, progressively larger volumes of less alkaline mafic magma are generated as the 1333°C isotherm rises until, at about $\beta = 4$, tholeiite is produced in sufficient volumes that its eruption at the surface is extremely likely. Thus, as β increases from about 2 to about 4, a progression occurs from the intrusion of small high-pressure alkaline mafic plutons

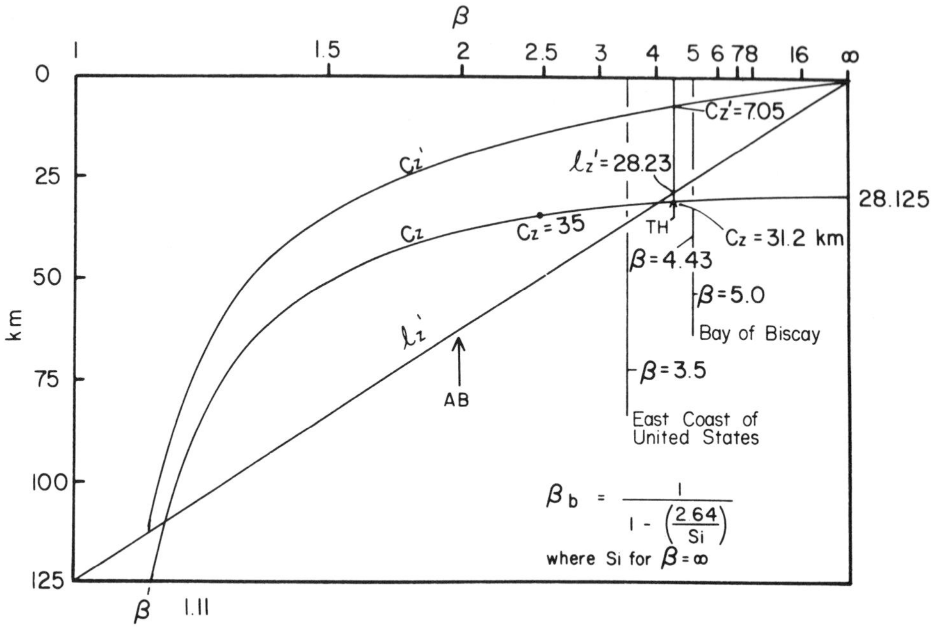

FIG. 20. Initial (C_z) and final (C_z') crustal thickness with a 125 km lithosphere, plotted against β values buffered by stretching to an S_i value of 2.64 km (oceanic ridge height).

deep in the lithosphere, through the intrusion of less alkaline mafic intrusions at deep crustal levels, finally to mafic dyke swarms and flood basalts. This is illustrated schematically with an increasing sediment thickness, as β increases, in Fig. 21B. Where β is less than about 2, a basin section without volcanics develops; at $\beta = 2$, a small amount of alkali basalt could appear in the section at the end of the stretching phase and, at $\beta = 4.43$, we see a section in which the lower half contains no volcanics, followed conformably by a section with alkali basalts and, later, tholeiites. However, if the strain rate is very low, such that thermal thickening of the lithosphere occurs as it is stretched, β could reach very large values without any magmatism developing, or only small amounts of alkali basalt.

Another way in which β is likely to be buffered at or limited to somewhere just in excess of 4, particularly during fast stretching at strain rates greater than 10^{-14} s^{-1}, is that the rise of large volumes of tholeiite in continuous dyke swarms will heat and weaken the crustal rocks and provide magmatic surfaces across which a stress continuum cannot be maintained; thus continental rupture will ensue (Fig. 22). The rationale for β buffering at $S_i = 2.64$ is illustrated in Fig. 21C; if β were to reach infinity very rapidly, a ridge would suddenly develop about 0.78 km higher than the adjacent, subsided, stretched continental crust. As argued above, it is difficult to see how large volumes of tholeiite with a density of about 2.6 could be kept down. We might envisage a situation in which an alkali basalt–tholeiite sequence could be laid down as a wedge on the stretched crust with a lateral transition into the oceanic crust either non-sediment loaded (Fig. 21D) or sediment loaded (Fig. 21E). In the latter case, the basalt wedge would be likely to occur as a sill complex.

Further β-limiting mechanisms relate to changing strain rate and the rising and cooling of increasingly strong olivine-rich mantle rocks beneath the basin. A progressive increase in β is indicated by the oceanward thickening sediment wedge at most continental margins. This suggests that, either the same cross-sectional area of crust is involved in an increasingly wide graben complex with a strain rate and subsidence rate decreasing with time at any point but increasing at any time towards higher β values, or that strain becomes concentrated into a smaller cross-sectional area of the widening graben complex. In the latter case, cessation of stretching, perhaps indicated by the end of fault control, is diachronous, becoming younger towards higher β values, in which case strain rates could remain constant or even increase in a progressively thinning and embrittling lithosphere in which increasing tensional stress eventually causes fissuring, dyking and rupture.

Where extension, rifting and magmatism affects older Precambrian shield areas, a distinctive basin margin style is developed, such as the SE Greenland and western Indian continental margins. In these areas, uplift, extensional faulting, alkaline magmatism and planation were followed by the development of an oceanward thickening wedge of tholeiitic flood basalt occupying a monocline. In the Thulean province of NW Scotland, late Palaeocene/Eocene extension and

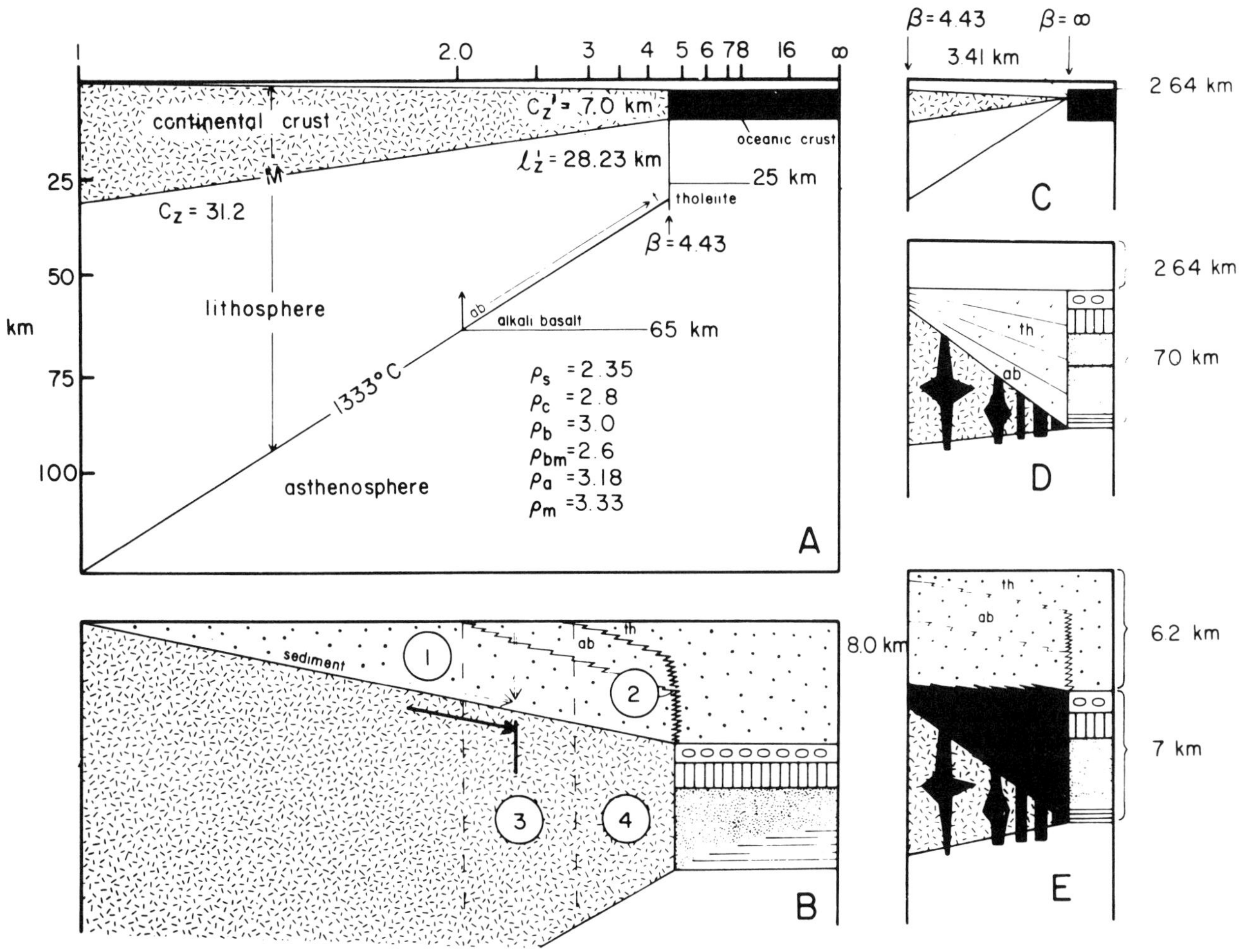

FIG. 21. A. β plotted against crustal thickness (C_z, C_z') lithosphere thickness (l_z, l_z') and initial subsidence (S_i) where $C_z = 31.2$, $l_z = 125$, and where β is buffered at 4.43 by $S_i = 2.64$ km (oceanic ridge crest depth) and where stretching takes place over a sufficiently short period at sufficiently high strain rates that thermal recovery of the lithosphere may be ignored. This may be considered as a time sequence for a locality where $\beta_{max} = 4.43$ or as an idealized section across a half graben or continental margin from unstretched ($\beta = 1$) to fully stretched ($\beta = 4.43$) crust. B. Idealized geological section, based upon A, which also may be considered as a temporal or spatial sequence, assuming that the basin is filled continuously to sea-level, and that β is buffered by $S_i = 2.64$. *ab* = section containing alkali basalt; *th* = section containing tholeiite; 1.2 = first possible appearance of alkali basalt in the section; 3 = thinned continental crust with alkalic intrusions; 4 = extremely thinned continental crust with alkalic, cut by tholeiitic, intrusions. Double-lined arrows in sedimentary section and solid arrow in basement indicate direction in which β increases; vertical solid line in basement indicates uniform stretching. C. Non-sediment-loading configuration of a continental margin where β is not buffered by ridge axis depth and reaches ∞ by purely lithospheric attenuation without igneous activity. D. Same as C, with increased vertical scale and with wedge of oceanward-dipping submarine basalts effecting the transition from continental to oceanic crust. E. Same as D, but continuously sediment-loaded where most of the basalt forms a sill-complex wedge at the base of the sedimentary section.

magmatism was accompanied by minor uplift, with no subsequent basin development. Such sequences may be explained by an M to R mechanism progression during the very rapid stretching of an old thick cratonic lithosphere (Fig. 23). For a 31.2-km initial crustal thickness, initial uplift occurs where initial lithosphere thickness exceeds about 227 km. A 287 km lithosphere is chosen, to produce uplift of a few kilometres in the advanced stages of stretching. Uplift, erosion and denudation without magmatism proceed until $\beta = 4.42$, after which there is an increased likelihood of the eruption of decreasingly alkaline volcanics. Not until β reaches a value of about 10 does the base of the lithosphere rise sufficiently to produce large volumes of tholeiite; 10 therefore would act as a buffering value. The oceanward-dipping monocline (Fig. 23C)

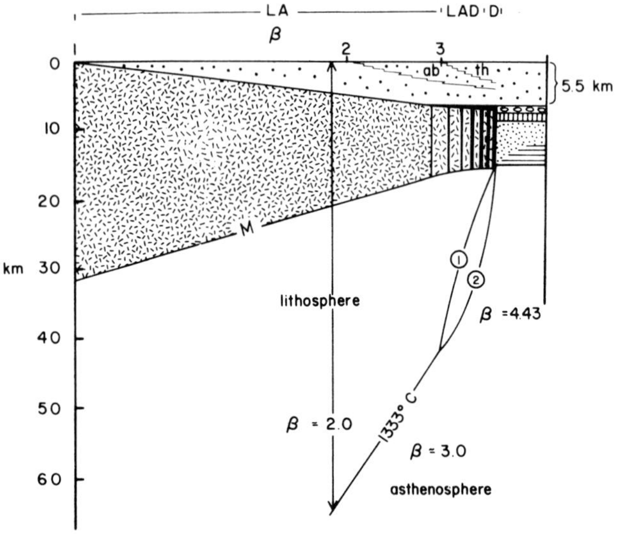

FIG. 22. β plotted against crustal thickness and initial subsidence with same initial parameters as for Fig. 21, where β is buffered at 4.43 by $S_i = 2.64$, but where, at $\beta \simeq 3.0$, a progressive transition to a pure lithospheric dyking mechanism (D) occurs as a result of the liberation of an increasing volume of increasingly tholeiitic magma, with continuous sediment loading to sea-level. LAD = combination of attenuation and dyking.

may result from a transition from uplift to subsidence at the rifted margin caused by advective thinning of the lithosphere that increases the C_z/l_z ratio.

A particularly important question for the tectonic evolution of basins evolved by, predominantly, the M mechanism is the strength of the basin lithosphere relative to adjacent unstretched lithosphere both during stretching and during subsequent reactivation in extension, strike-slip and compression. During instantaneous or very short-term stretching, all rocks in the stretched lithosphere are at about the same temperature or only slightly cooler than originally. Therefore, the thinned lithosphere is a weak stress-concentrator and further stretching is concentrated in the deformed zone. For longer stretching times (lower strain rates) and, as post-stretching thermal re-equilibration occurs, rocks in the stretched lithosphere progressively cool and strengthen until the stretched lithosphere becomes stronger than the unstretched. This effect is particularly important for large β values where olivine-rich mantle rocks rise to shallow levels of lower confining pressure. A further effect in increasing rock strength in the graben basement is the development of linear and planar fabrics during stretching that induce a strength anisotropy. These linear and planar fabrics also have the effect of inducing a thermal conductivity anistropy which enhances the lateral flow of heat towards basin margins and the preferential cooling and thickening of basin margin lithosphere. Thick basinal sequences, particularly shales, have a blanketing effect which further enhances lateral heatflow; these effects are particularly important, even in the absence of fabrics or blanketing sediments, in narrow basins, where a large part of the thermal subsidence phase is transferred back into the stretching phase. This has the further effect of progressively concentrating stretching towards the axes of rift basins to maintain or perhaps increase strain rates. This will be reflected stratigraphically in depocentres progressively migrating from

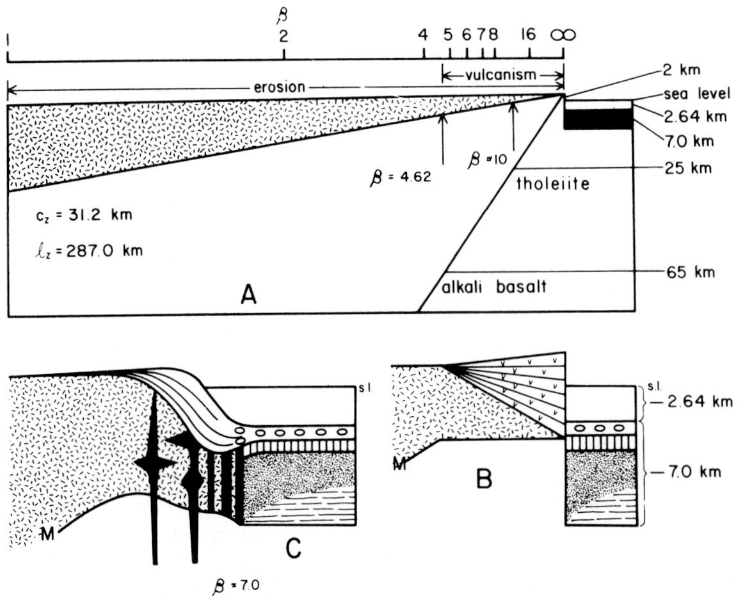

FIG. 23. A. β plotted against crustal thickness for $C_z = 31.2$ km and $l = 287$ km. B. Attenuated continental margin loaded by oceanward dipping basalt wedge where $\beta > 4.42$. C. Continental margin monocline in basalt wedge of East Greenland type.

basin margins and in substantial clastic floods from large thermal rift shoulders. For long stretching times, the strength of the basinal lithosphere may become greater than that of the unstretched lithosphere, in which case, either stretching and basin formation migrate out into hitherto unstretched lithosphere (perhaps illustrated by the northward migrating Devonian trough of Cornubia) or continental separation occurs along listric fault zones of weakness at one of the basin margins. This latter method is a further β-buffering mechanism for margins in which little volcanism precedes separation and where wide subsided platforms on thinned, stretched, continental crust, such as the Exmouth and Vøring plateaus, remain attached to rifted margins, usually producing a contiguous margin asymmetry. Thus, there is a trade-off between decreasing strength by lithospheric thinning and increasing strength caused by thermal re-equilibration. For large β values and high strain-rates, a very much thinner lithosphere results in a weak zone during and for some time after stretching; for small β values and low strain rates the stretched lithosphere rapidly becomes a stronger zone. Another significant effect of slow strain rates is that the C_z/l_z ratio decreases during stretching and hence a gradually slowing subsidence rate could be followed by the cessation of subsidence or even uplift. Similarly, where a basin has been developed by polyphase stretching, as for example the North Sea, each subsequent stretching phase affects a lithosphere with a smaller C_z/l_z ratio than before and is thus reflected in a smaller subsidence or uplift for a particular β value, perhaps causing an underestimation of β. A reduced C_z/l_z ratio would be indicated by uplift, fault truncation and a mild regional unconformity such as, for example, the early Cretaceous sub-Speeton/Cromer Knoll unconformity of the North Sea. Another way in which β might be underestimated is that the thickness of sediment laid down in a basin during the stretching phase is reduced by the stretching; the oldest sediments are thinned by $1/\beta$ with a continuous upward decrease in amount of thinning to no reduction at the top of the S_i section. This is an exceedingly vexing problem because it means, first, that a substantial amount of structural work must be done on microjoints, fissuring, clay fabrics, etc. before restored S_i thicknesses can be estimated and, secondly, that because these small-scale but critical fabrics cannot be seen on seismic sections, even properly migrated seismic sections cannot be used to determine β accurately.

With some of these model parameters, constraints and difficulties, let us now examine several British basins. The general form of the Wessex Basin and its margins conforms well to the 'Steer's Head' form with Jurassic onlap onto the Mendip/London Platform (Fig. 24). It is, though, a complicated basin whose subsidence and tectonic history (Fig. 25) is polyphase, with stretching in the early Triassic, renewed stretching in

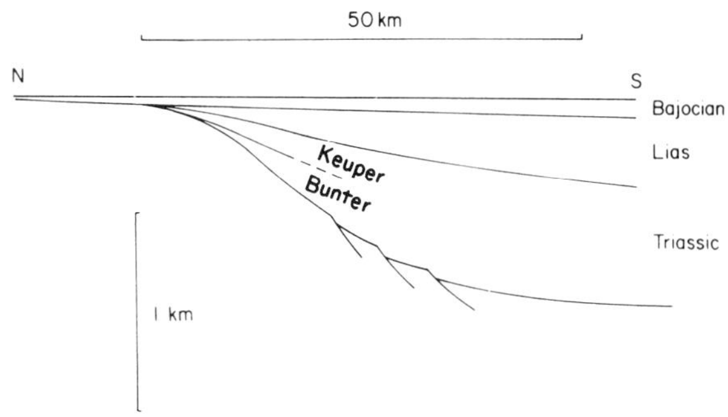

FIG. 24. Restored section at the end of the Bajocian Stage from the Mendips in the N to S Dorset in the S to show Triassic/Jurassic flexural onlap on the northern margin of the Wessex Basin.

the Toarcian (Jenkyns & Senior 1977), an accelerated phase of subsidence in the late Jurassic, a late Neocomian phase of extensional faulting, minor uplift and denudation (Ringstead Phase) and an Eocene phase of accelerated subsidence. The North Sea Basin with its median Viking/Central Graben marginal onlap also follows the 'Steer's Head' pattern. The North Sea grabens show a slightly different polyphase history with a strong mid-Jurassic extensional phase with both alkaline and tholeiitic minor volcanism (Malm *et al.* 1979) and a Palaeocene phase of strong extension, volcanism and turbidite deposition in suddenly deepened troughs (Fig. 25). When the Wessex Basin and Central Graben are backstripped (that is, stripped of the effect of sediment loading, using the integrated density curves of Fig. 26, to reduce the subsidence curves to that portion due only to driving subsidence, S_i and S_t), subsidence curves (Fig. 27) may be directly compared with the β curves of Fig. 14. This comparison, assuming initial crustal and lithospheric thickness of 31.2 km and 125 km respectively, yields cumulative β values of 1.4 for the Wessex Basin and 1.9 for the Central Graben. The lack of volcanism directly associated with stretching in the Wessex Basin is expected with a β value of 1.4 but the mid-Jurassic volcanism of the North Sea is not, unless β has been seriously underestimated for one or all of the reasons cited above.

The Midland Valley underwent rapid subsidence with tholeiitic volcanism in the early Devonian and deformation during the mid-Devonian, followed by declining subsidence in the late Devonian and Carboniferous, with alkalic volcanism in the early Carboniferous (Fig. 28). Progressive Carboniferous onlap of the southern Uplands (Fig. 29) accords well with the notion of a thickening, strengthening lithosphere following an early Devonian stretching phase. Backstripping the northern Midland Valley subsidence curve of Fig. 28 and normalizing the resulting driving-subsidence curve on the β curves of Fig. 14, so that the top of the Lower Devonian section coincides with the

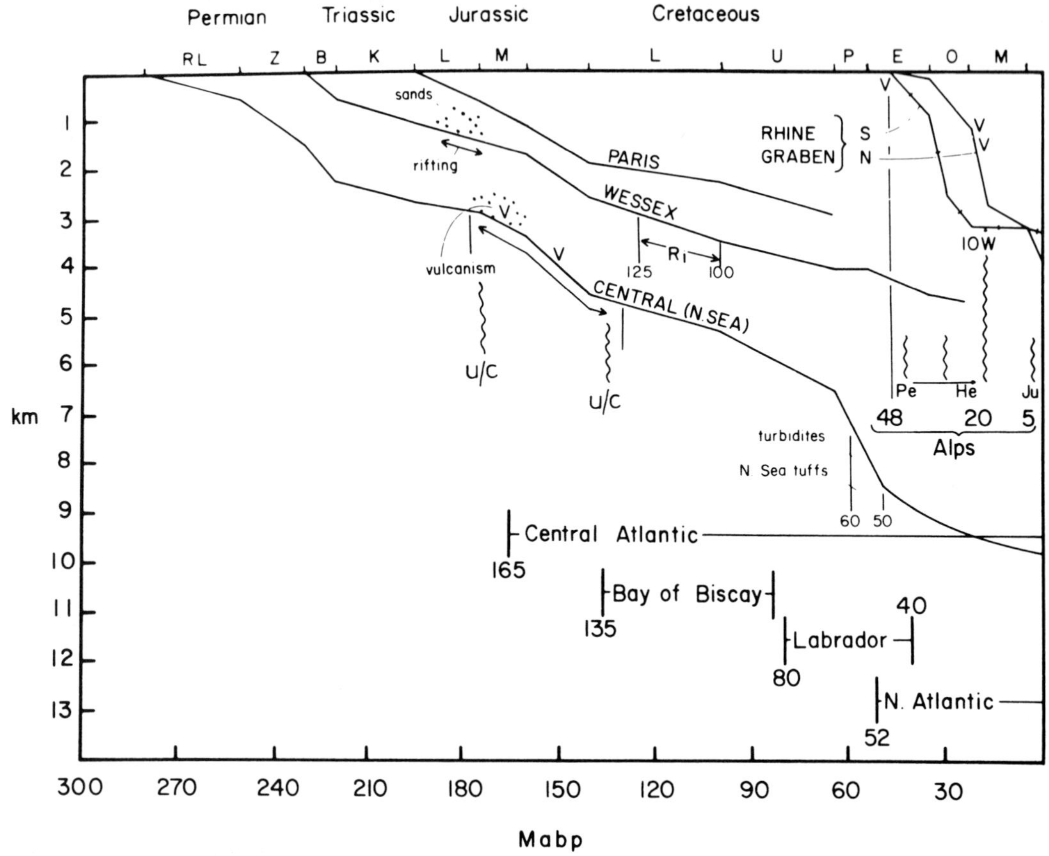

FIG. 25. Sediment thickness against time for the North Sea Central Graben, the Wessex Basin, the Paris Basin and the northern and southern Rhine Graben. Ri = Ringstead deformation phase in south Dorset; IOW = Isle of Wight/Lulworth/Sutton Poyntz deformation phase; Pe = Pennine deformation phase in the Alps; H = Helvetic deformation phase; JU = Jura deformation phase.

end of the S_i phase (Fig. 30), we see that the post-Middle Devonian sequence follows, fairly precisely, the S_t portion of a $\beta = 6$ curve. The Lower Devonian S_i portion, however, is displaced by the mid-Devonian deformation event (concentrated in thinned lithosphere), and also gives an initial subsidence just in excess of $\beta = \infty$. This apparent paradox may result from one or all of three causes. First, β may indeed be equal or approximately equal to infinity and the slight excess is within the limits of error; then, during the mid-Devonian deformation the lithosphere was structurally thickened slightly to transpose the thermal portion of the subsidence curve to a value equivalent to $\beta = 6$. Second, the stretching phase may have been sufficiently long, and in a basin sufficiently small that lateral conduction was significant, to allow part of the thermal subsidence phase to have been transferred back into the stretching phase. Thirdly, it is extremely likely that we cannot assume initial crustal and lithospheric thicknesses of 31.2 and 125 km, respectively, for the earliest Devonian, pre-stretching foundation of the Midland Valley. However, a β factor of 6 and an initial crustal thickness of about 31 km giving a sediment section 13 km thick on a 5-km stretched crust, is compatible with the LISPB profile.

The Carboniferous block and basin tectonic style of central and Northern England is particularly susceptible to analysis using the McKenzie (1978) mechanism (Leeder 1982). Strong Dinantian structural differentiation into limestone platforms, with marginal reefs bordering limestone/shale basins, strong fault control, and an outbreak of alkaline mafic volcanism at the end of the Viséan, was followed by mid and late Carboniferous basin margin onlap. This pattern is consistent with Dinantian lithospheric stretching followed by thermal recovery in the mid and late Carboniferous. Dinantian stretching was probably the result of *en echelon* (e.g. Edale and Gainsborough basins) extensional basin formation in a dextral shear system (Fig. 38). The Craven basin subsidence curve (Fig. 28) and its backstripped equivalent, normalized so that the end of the Dinantian corresponds with the end of the S_i phase, yield a β value of just over 2, but the basin was not continuously filled to sea-level. Projecting a $\beta = 2$ curve back to the end of the S_i phase and substracting the backstripped S_i gives an early Carboniferous water

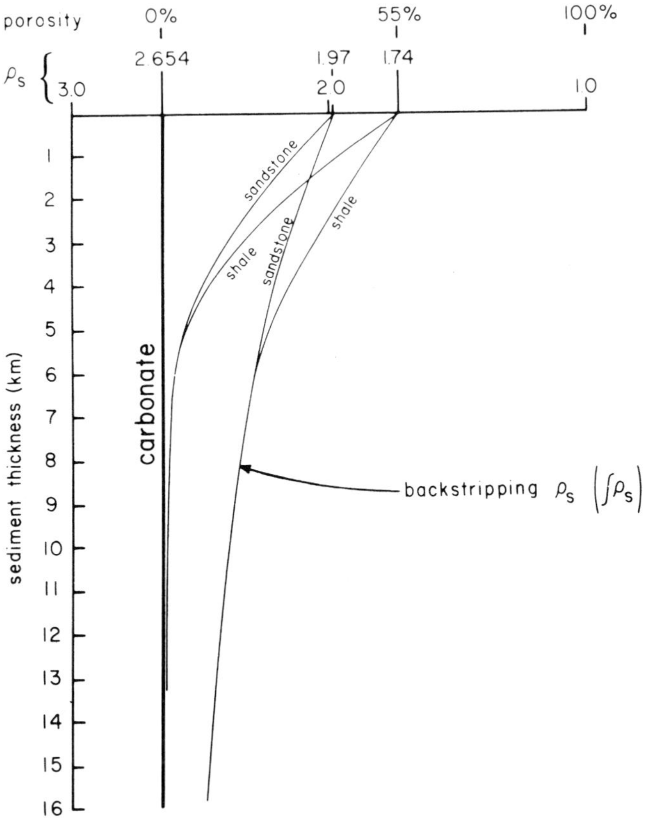

FIG. 26. Sediment porosity and density as a function of depth and integrated porosity and density for backstripping a particular thickness of sediments.

depth of 880 m in the basin during Worston Shale deposition.

Applying the M model to the late Palaeozoic history of Cornubia, Devonian stretching was followed by Carboniferous thermal subsidence. Stretching appears to have been diachronous, having begun during the

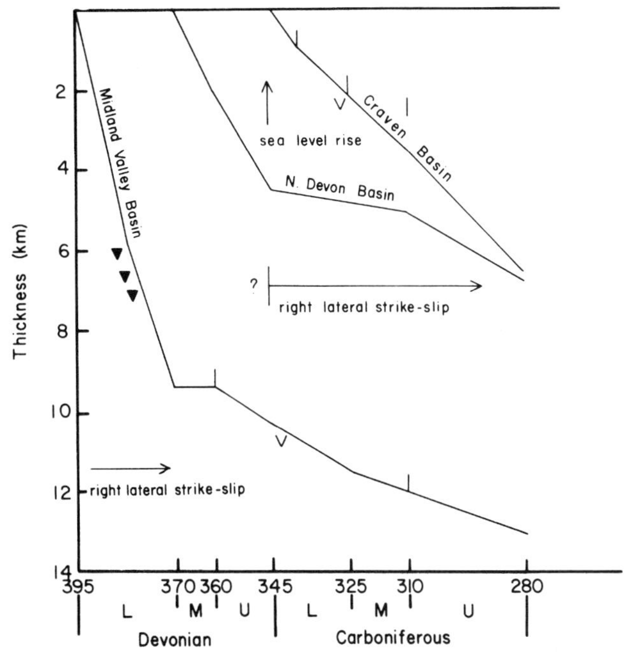

FIG. 28. Sediment thickness against time for the Craven Basin, the Midland Valley and N Devon during the Devonian and Carboniferous. Triangles denote tholeiite, V denotes alkali basalts.

early Devonian in S Cornwall and in mid-Devonian times in N Devon. In N Devon (Fig. 28), a β factor of about 2 is consistent with the form of the subsidence curve (Fig. 30) and with thin beds of alkalic mafic tuff persisting until early Carboniferous times. The difference between the back-projected thermal subsidence curve and the backstripped curve during the mid-Carboniferous suggests a water depth of about 500 m.

The above discussions will suffice to illustrate that the McKenzie (1978) model, modified where β is greater than about 3, by the Royden *et al.* (1980)

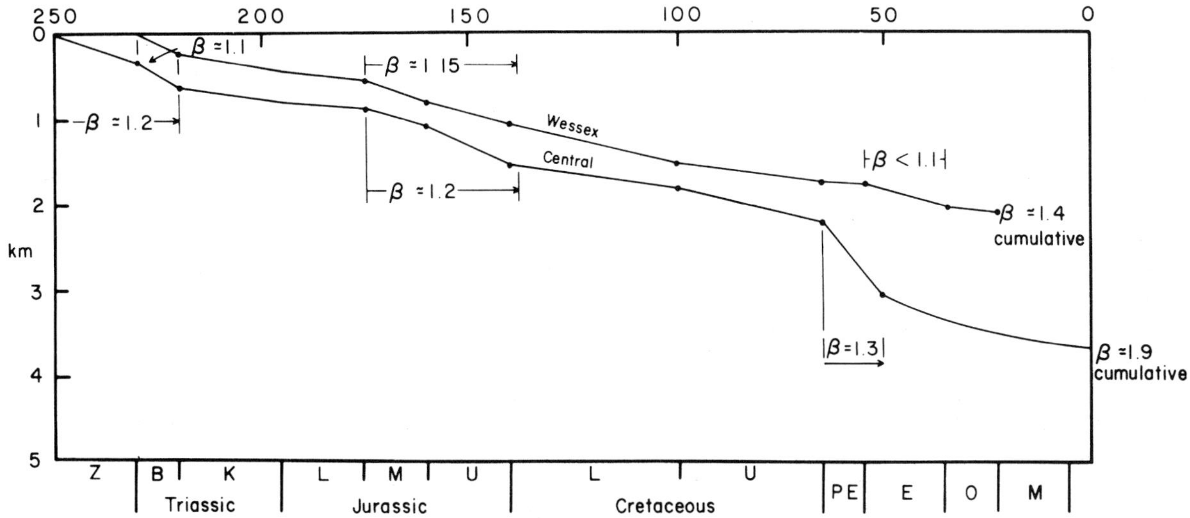

FIG. 27. Backstripped Central Graben and Wessex Basin.

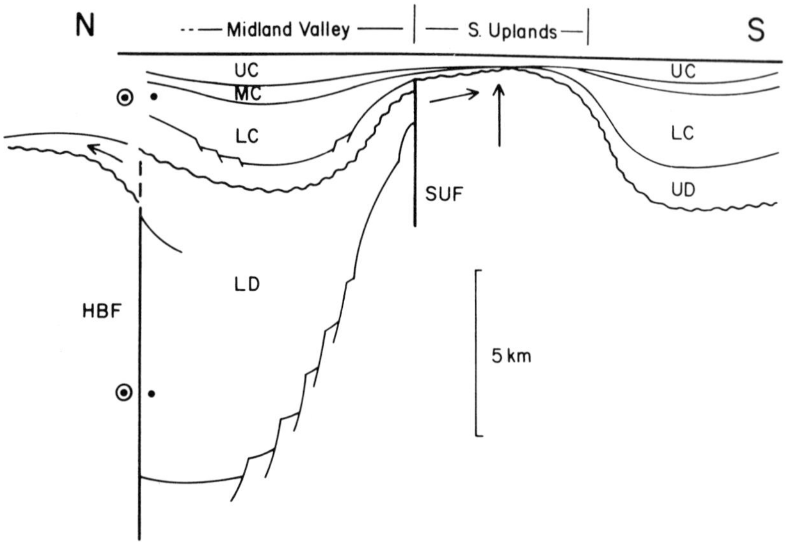

FIG. 29. Schemetic restored section for the end of the Carboniferous across the Midland Valley and Southern Uplands to show thickness variation in Devonian and Carboniferous strata and to emphasize the early/medial Carboniferous onlap across the Southern Uplands in the Sanquhar region.

mechanism, although simple in concept and basic physics, is a powerful tool in understanding rift basins and has fundamental implications for the development of associated rift shoulders, peripheral bulges, intrabasinal unconformities, and more subtle stratigraphical features such as the geometry of onlap, offlap, and disconformities. The principal difficulty in making quantitative statements resides in uncertainties about initial pre-stretching thicknesses of crust and lithosphere. 31.2 and 125 km, respectively, are probably about right for the younger stable cratons stretched

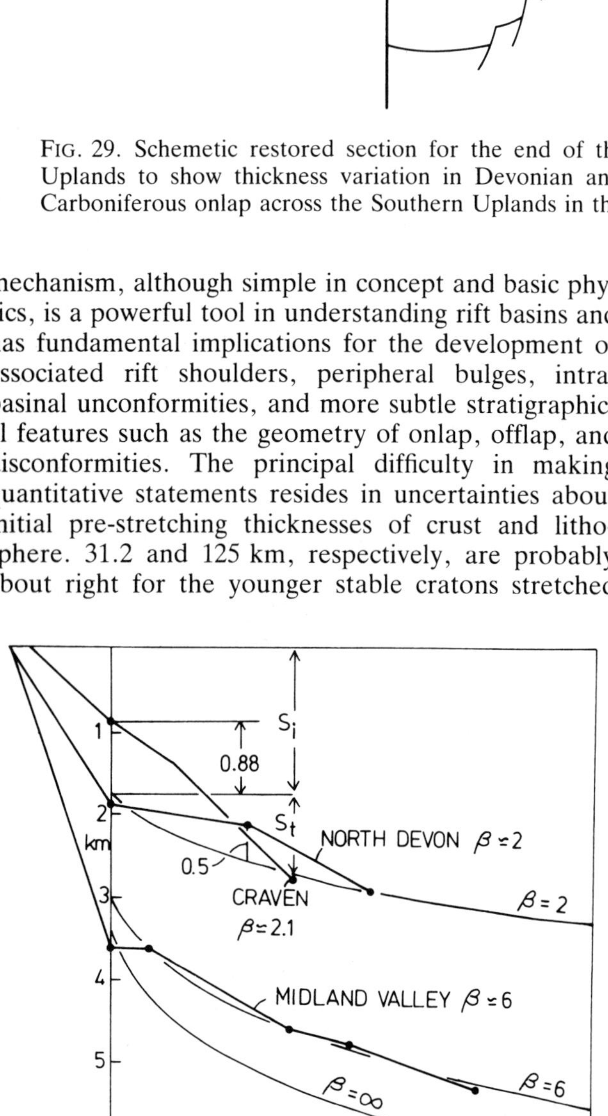

FIG. 30. Backstripped Craven Basin, Midland Valley and North Devon on β/subsidence plot assuming C_z = 31.2 km and l_z = 125 km.

during the Phanerozoic, but many basins develop within arcs where C_z/l_z ratios are much higher and in mountain belts were crustal thicknesses are very much greater. Also, as the average thickness of the lithosphere has increased with time, there has been a corresponding secular decrease in average C_z/l_z ratios. Thus, Precambrian basins have much thicker sections and larger amounts of more tholeiitic volcanism for smaller β values.

5. Basin inversion—mountains

Lithospheric extension by stretching and/or dyking results in rifts, troughs, graben complexes, and basins, within and at the edges of continents, that form long-term lithospheric inhomogeneities. Depending upon β, the relative roles of stretching and dyking, the rate of stretching, and the time elapsed since stretching, these extension zones may be weaker or stronger than the adjacent unstretched lithosphere. In general, the stretching model yields an initially weak zone that rapidly became stronger than adjacent lithosphere, whereas the dyking model produces a zone that remains weak for a very long time. In either case, they commonly comprise regions that, subsequently, nucleate zones of strike-slip and/or shortening deformations. Several Phanerozoic orogenic belts, such as the Alps and the Pyrenees, and many Proterozoic belts, such as the Damarides and the Ubendides, appear to have resulted from largely intracontinental deformations whereby severely attenuated continental lithosphere was shortened and thickened by a variety of mechanisms ranging from thrust restacking to mega-transform shearing. Further, there appears to be a spectrum of orogenic belts, ranging from those that resulted from suturing following a history of oceanic subduction through those with a para-oceanic history of severely attenuated continental crust with diapiric mantle 'windows' and minor semi-organized plate accretion to those with a purely intracontinental graben history (Dewey & Windley 1981).

Three major styles of shortening are common; subduction of basement with cover-stripping (Helvetic style), thrust flaking with basement strongly deformed (Penninic style), and thrust flaking where rigid basement sheets ride on a thin shear carpet (Austro-Alpine style). Listric normal faults commonly rejuvenate listric thrust faults (well displayed in the Alberta/Idaho/Montana Rockies) and the reverse also occurs; Jackson (1980) convincingly demonstrated the present-day thrust rejuvenation of listric normal faults in the Zagros Mountains. Listric normal faults are convenient zones of weakness in stretched crust for thrust nucleation, particularly where the graben floor is stronger than adjacent unstretched lithosphere. The early Miocene Isle of Wight Monocline in the core of the Wessex Basin is a fine example of basin rejuvenation or inversion (Figs 25, 42). The Isle of Wight/Poxwell/Sutton Poyntz Line is a high angle thrust that reverses a high-angle normal fault, which is probably listric (Stoneley 1982). Intracontinental detachment of thrust flakes 5–15 km thick of great lateral extent is known in many orogenic belts (e.g. Scandinavian Caledonides, Gee 1975; Southern Appalachian Piedmont, Brewer & Oliver 1980; Assyrides of eastern Turkey, my observations). Coward & Siddans (1979) argued that much of the Welsh Basin sits upon a listric thrust complex. The Austro-Alpine Nappes are an assemblage of thin sheets, the thickest of which, the Silvretta/Altkristallin consists mainly of basement, with a thin Mesozoic cover, that was little deformed internally during Tertiary emplacement. In the steep belt, S of the Canavese high-strain zone, the Austro-Alpine nappe includes a slice of continental upper mantle (Finero/Ivrea high-pressure lherzolites) above which only 10–15 km of continental crust separates the Moho from Mesozoic supracrustals. This crustal thinning was achieved during the Toarcian by lithosphere stretching, witnessed by rapid subsidence from benthonic to abyssal depths and widespread fissuring and faulting of pre-Toarcian carbonates. The Alps are the classic example of the progressive roughly plane-strain thrust restacking of a thinned continental crust, the earliest slices involving mantle, with a crustal thickening by late Oligocene times to about 70 km, of which some 20 km has now been stripped to form the Molasse of the Swiss Plain, the Rhone delta, the Po Plain and clastics distributed by the Rhine to the North Sea.

Fig. 31 illustrates some of the principal features of the shortening of thinned continental lithosphere. Thrust flaking and restacking involves the upper lithosphere elastic lid whose thickness is dominantly strain-rate and temperature dependent. In unstretched lithosphere with a 30 km crust, thrusts such as the

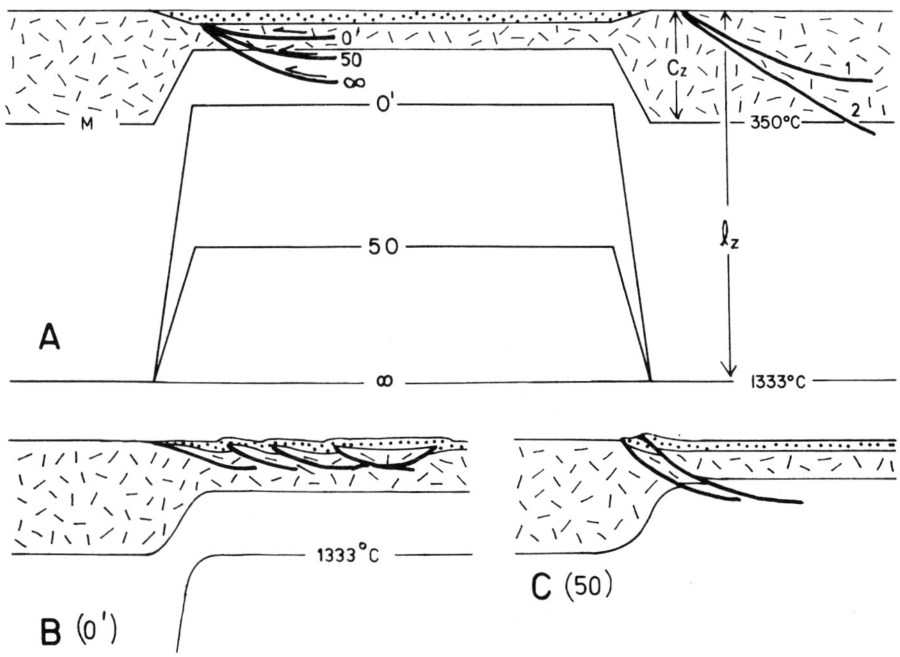

FIG. 31. Shortening of previously-stretched lithosphere by thrust flaking of an elastic lid to a viscoelastic lithosphere. (A) Position of depth to which listric faulting bites immediately after, 50 Ma after, and at an infinite time after, stretching. (B) Listric thrust for moderate strain rate in unstretched craton. (C) Listric fault for high strain rate in unstretched craton.

Wind River Thrust (Brewer & Oliver 1980) may bite to the Moho [Fig. 31A(2)] where strain rates are high. In stretched lithosphere, the depth to which thrust flakes detach is probably mainly dependent upon the time elapsed since the end of stretching, which in turn mainly determines the thickness of the elastic lid. Fig. 31A shows this dependence schematically; immediately after stretching, small thin flakes involve only continental crust but, as time proceeds, the strong elastic lid thickens until, eventually, mantle is included in thicker, larger thrust flakes (Fig. 31, B & C). Below the elastic lid, shortening progresses by other deformation mechanisms such as transient creep, power law creep and superplastic flow. Thrust planes may be nucleated by listric faults at shallow levels and by flat-lying linear/planar fabrics, induced during stretching, at deeper levels. Structural evidence, in the shear carpets upon which thrust flakes move, appears to suggest that the shear zones thicken with time, probably caused by weakening resulting from shear heating, and therefore that shear strain rates diminish as thrusting progresses.

An interesting feature of the early deformation history of most orogenic belts is that shortening, nappe emplacement and, commonly, high pressure metamorphism proceed for a long time before the appearance of those indicators that have been regarded traditionally as indicators of orogeny, particularly the dispersal of clastic wedges. This is not surprising if orogenic belts and inversion zones result predominantly from the compressive and strike-slip reactivation of stretched continental lithosphere. Fig. 32 illustrates the geometric effects of crustal thickening and elevation by the progressive shortening of continental crust of various initial thickness determined by earlier stretching of a 35 km crust by β factors of 1, 2, and 4. The stretched crust is considered to have only a thin sedimentary sequence. 6 km is taken as a buffering mountain elevation above which summit elevations cannot rise because of high erosion rates; this in turn buffers crustal thickness at about 70 km, which allows the denudation of almost 40 km of the upper thickened crust to return, eventually, the crust back to normal cratonic thickness. Shortening of a normal thickness crust ($\beta = 1$) produces uplift and, hence, clastic wedges at the outset of deformation, and a 70 km crust is achieved at about 50% shortening. For a half normal thickness crust ($\beta = 2$), shortening reaches a value of about 50% before uplift above sea-level allows the dispersal of clastics; thus a considerable amount of strain, including slaty cleavage development, takes place prior to emergence of the deforming pile. For a quarter normal thickness crust ($\beta = 4$), shortening must reach about 75% before emergence. This has far-reaching implications for relationships between polyphase deformation, metamorphism, deformation and unconformities that will be explored thoroughly elsewhere; it will suffice here

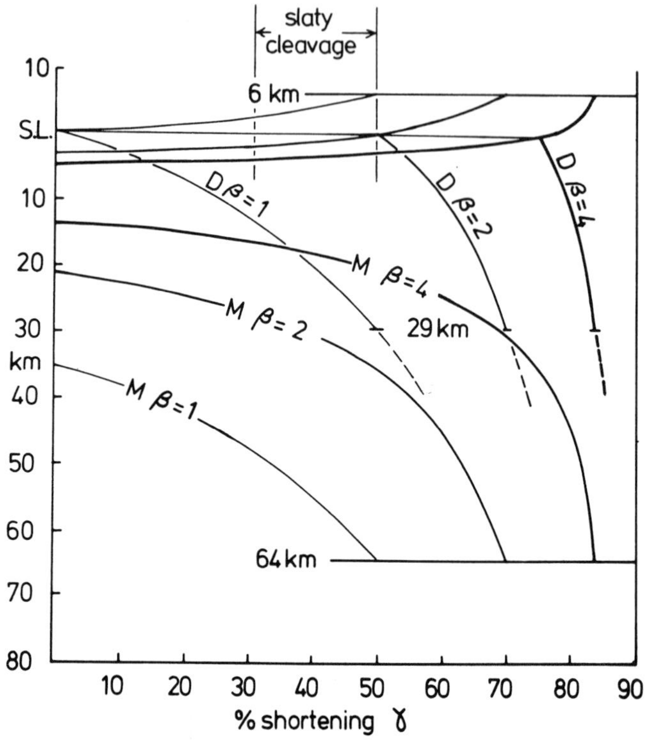

FIG. 32. Plane-strain shortening and thickening of continental crust from three different initial thicknesses (35, 17.5 and 8.75 km, these initial thicknesses having resulted at an earlier stage from stretching of a 35-km crust by β = 1, 2 and 3, respectively). M = Moho, D = level to which denudation can eventually strip the thickened crust. Curves assume a starved initial basin and a buffering mountain elevation or *Gipfelfluh* above which erosion rates are faster than uplift rates. No particular mechanism of shortening, such as homogeneous strain or thrust restacking, is specified.

to remark that clastic wedges and unconformities may appear only towards the end of the long history of polyphase deformation and metamorphism, much of which may take place below sea-level. A good illustration of this concept is the subduction accretion complex (Fig. 33) of which the Southern Uplands are a particularly fine example (Leggett *et al.* 1979). In such a prism, where the trench is fed by turbidites, listric thrust packages are detached from the subducting plate and accreted to the leading edge of the overriding plate, then enjoy a progressive deformation by shortening and thickening; as the deforming wedge reaches a thickness of about 30 km, uplift occurs to form a fore-arc bulge. Thus, again, most deformation occurs below sea-level and is not recorded by clastic wedges and unconformity.

This concept is particularly useful in providing at least a partial explanation for one of the most vexing questions of Caledonian history, i.e. the early history

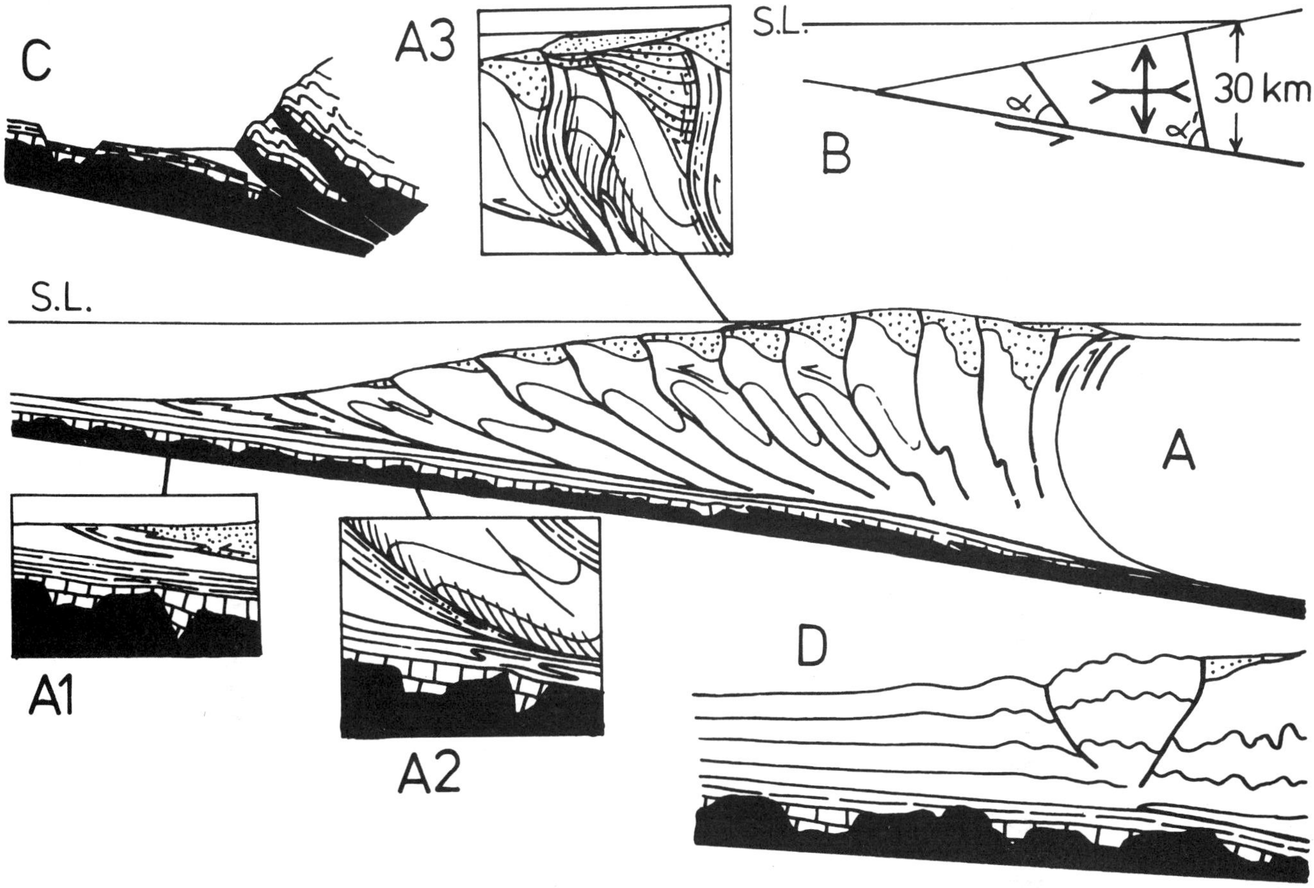

FIG. 33. Schematic representation of some characteristics of subduction-accretion complexes. A. Off-scraping of trench sediments by a thrust/recumbent fold mechanism, followed by progressive shortening, thickening and polyphase deformation of the accretion-prism and the growth of thrust-bounded upper slope basins. Deformation takes place below sea-level until the accretion prism is about 30 km thick. B. Progressive plane-strain thickening of the subduction-accretion prism. C. Subduction-accretion of mélange containing ophiolite slices at an open trench, almost bare of sediments, such that an irregular upper surface of the oceanic crust is not mechanically insulated from the overriding plate. D. Subduction-accretion of a thick turbidite wedge that insulates an irregular upper surface of the oceanic crust from the overriding plate.

of the Grampian orogenic sequence during pre-Caradocian times. It is clear, from geochronological evidence, that deformation, metamorphism and magmatism were occurring perhaps as early as 530 Ma but certainly by Arenigian times, and yet a flood of clastic detritus from denudation of the Highlands does not appear in the Midland Valley and Southern Uplands until Caradocian times with minor harbingers in Llandeilian strata. Fig. 34 illustrates, schematically, a possible plane strain shortening, crustal thickening, emergence and denudation history for the Grampian Highlands capable of accounting for a long history of deformation below sea-level. The model assumes that shortening began 530 Ma ago and ended in early Caradocian times, when a peak crustal thickness of 70 km was reached (or which 30–35 km has now been denuded, Wells 1979) and an initial crustal thickness of about 16 km made up mostly of Moine/Dalradian sediments on a severely attenuated continental (Cowhythe Gneiss) and/or oceanic (Unst/Highland Border ophiolites) crust. This yields a gross shortening of about 80% so that the original width of the deformed zone was about 1000 km and, assuming that the same cross-sectional area of crust was involved throughout strain, an average cross-orogenic displacement rate of about 1 cm/a and average strain rates increasing from 3.17×10^{-17} s^{-1} at the outset, to 1.58×10^{-16} at the end of deformation. It is clear, however, that the constant cross-sectional area assumption throughout deformation may be incorrect. Most orogens exhibit thrust sequences that intermittently prograde to involve more and more foreland cover and, commonly, basement; this progradation progresses at rates of about 0.5 cm/a, which may substantially reduce the rate of increase in strain rate as shortening progresses.

In western Ireland, metamorphic detritus from deformed or deforming Dalradian rocks in Connemara and N Mayo appears much earlier than tectonically

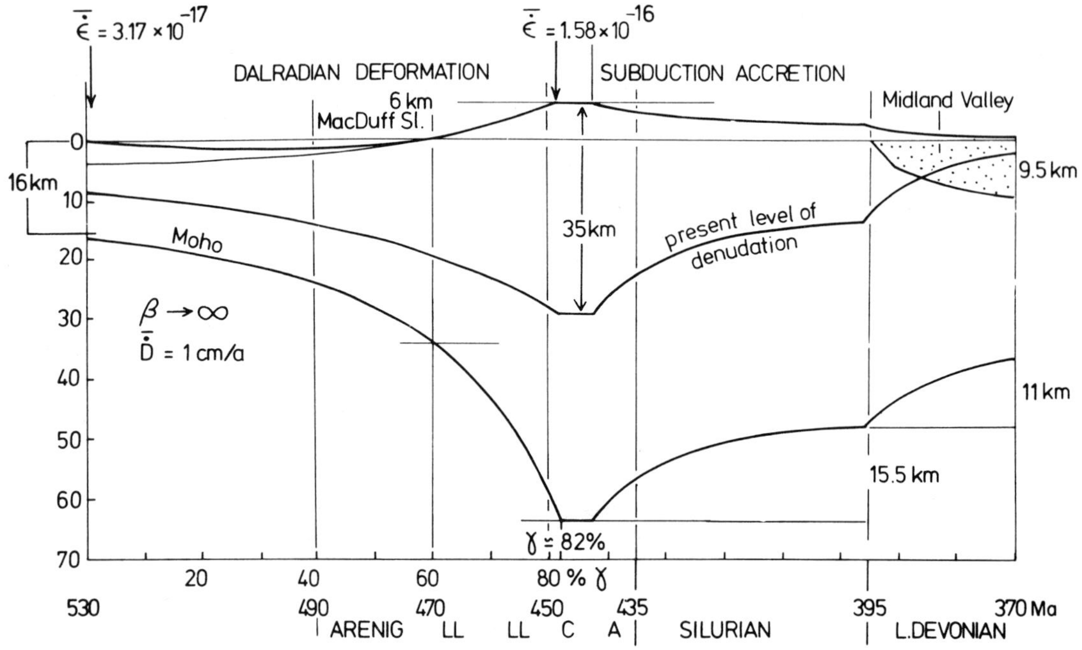

FIG. 34. An interpretation of the Grampian orogeny in Scotland as an effect of continuous or semi-continuous shortening of a thinned continental crust assuming a crustal thickness of 16 km at the onset of deformation. At a displacement rate of 1 cm/a and assuming uniform strain across the mobile zone, strain rate $\dot{\varepsilon}$ increased during deformation from 3.17×10^{-17} s^{-1} in late Cambrian times to 1.58×10^{-16} s^{-1} in mid-Ordovician times.

equivalent clastics in Scotland. In the S Mayo Trough, clastics with a Dalradian provenance occur in mid to late Arenigian strata, indicating emergence of deformed and low grade Dalradian rocks. This may be explained by one or both of two modifications to the Scottish model of Fig. 34; a thicker initial sediment/crustal section and/or a higher strain rate reflecting a faster convergence rate across a narrower zone of deformation. In Fig. 35, a 22 km sediment/crust section reflecting a β value of about 4 is shortened to peak thickness by Llandeilian time with deformation ceasing by the early Caradocian. Emergence occurs in

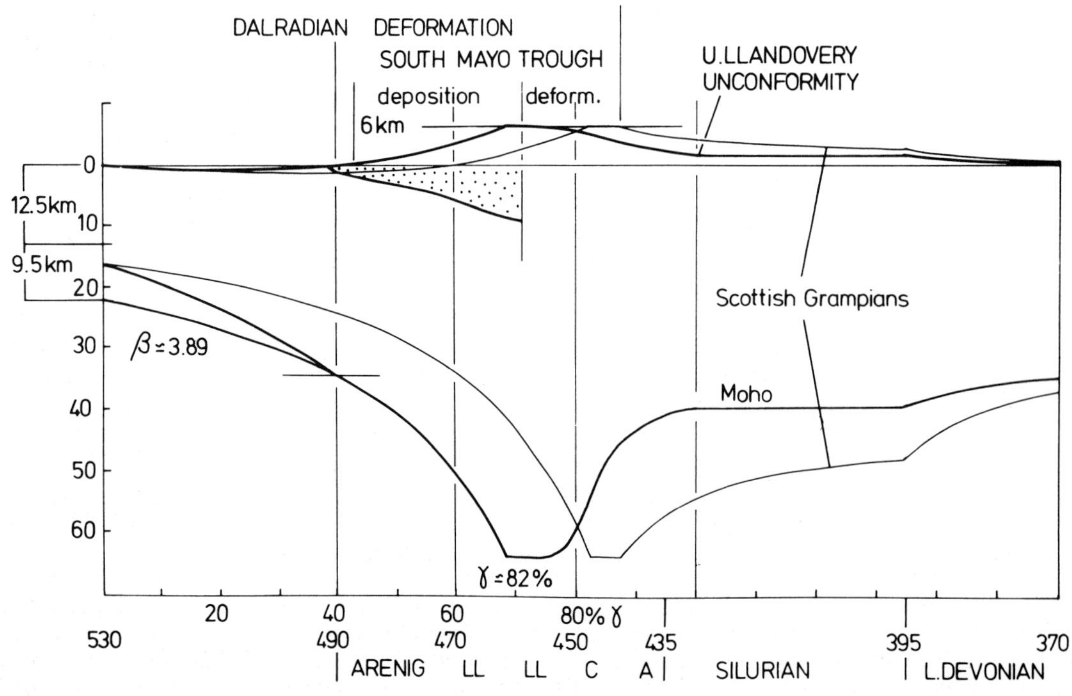

FIG. 35. An interpretation of the Grampian deformation of the Dalradian rocks of western Ireland as an effect of continuous or semicontinuous shortening of a thinned continental crust ($\beta = 3.89$ assuming $C_z = 35$ km).

the Arenigian to provide low grade detritus to the S Mayo Trough but not until Llanvirnian time did thick coarse clastics (Partry Supergroup) flood into the trough during peak mountain height. Ongoing convergence, responsible for the final stage of the Dalradian polyphase strain sequence, also deformed the S Mayo Trough into a broad syncline, which, with the Connemara Dalradian, was sufficiently denuded by late Llandoverian time to allow shallow marine deposits to be laid down unconformably across a stripped Dalradian/Ordovician basement.

6. Palaeotectonics

a. The Caledonides

The broadest features of the Caledonian/Northern Appalachian orogenic belt are consistent with the orogen having evolved, in a plate boundary framework, via a sequence of continential rifting and separation during the late Precambrian, subduction of the Iapetus ocean floor during the early Palaeozoic, and continental collision in the late Silurian/early Devonian, the final Iapetus suture passing from the Solway Firth Line through the Irish paratectonic Caledonides (Phillips *et al.* 1976). N and S of the Solway Line, the Caledonides have quite different histories and cannot be related in any way until the early Devonian collision that produced much of the deformation of the paratectonic Caledonides. In Fig. 36, an attempt is made to illustrate a conceivable, generalized plate boundary evolution for the British Caledonides and Newfoundland Appalachians that is capable of accounting for the broad geometry and timing of events. These cartoons must not be taken too seriously. Although we can recognize elements, terrains, and

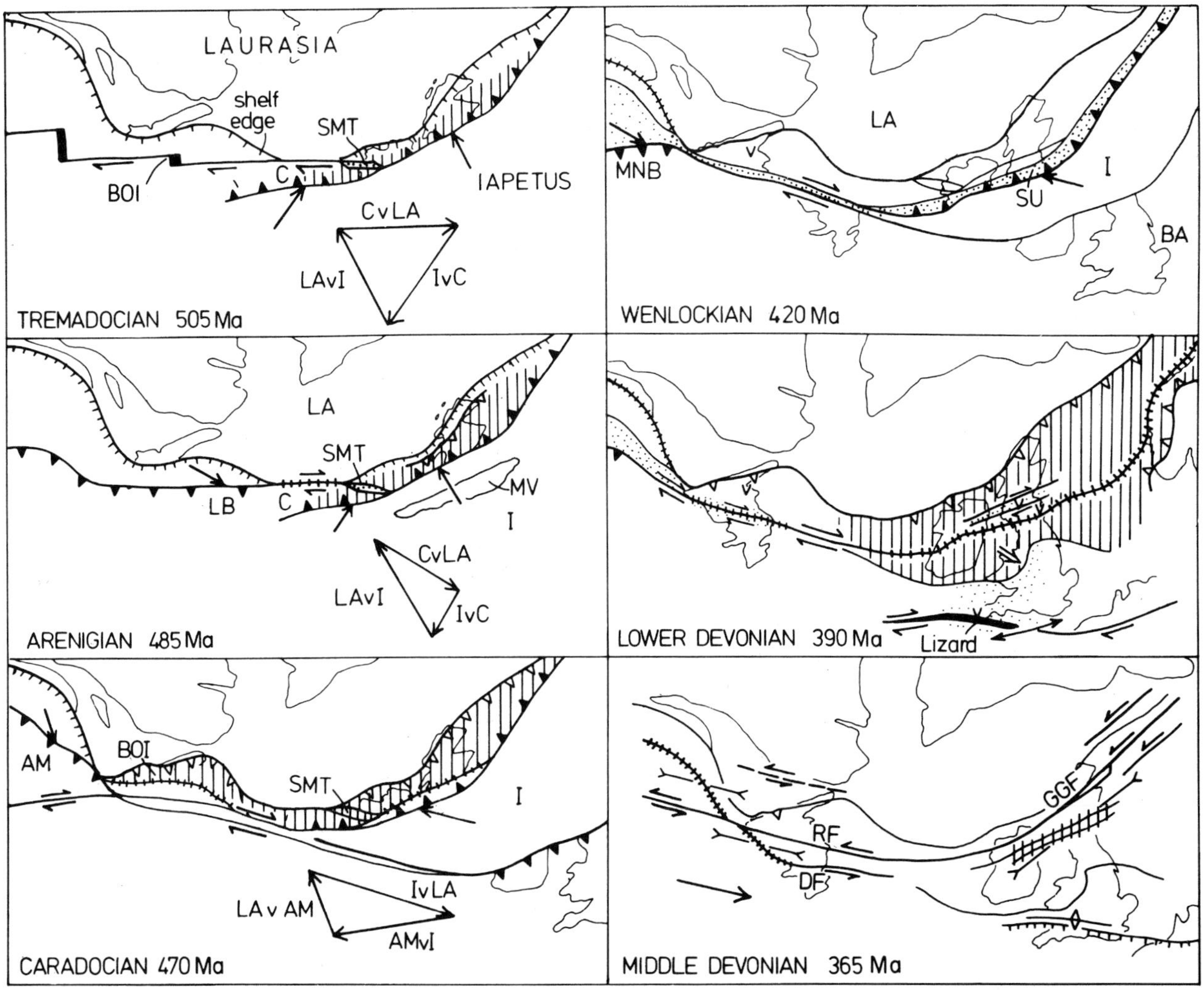

FIG. 36. A speculative and schematic model for the evolution of the British Caledonides and Canadian Maritime Appalachians from Tremadocian to Middle Devonian times. Vertical lined ornament = convergent deformation; fine stipple = subduction-accretion prism; coarse stipple = sands; BA = Baltic Shield; BOI = Bay of Islands Ophiolite Complex; C = Connemara block; DF = Dover Fault; GGF = Great Glen Fault; I = Iapetus; LA = Laurasia; MNB = Maine/New Brunswick Zone; MV = Midland Valley; RF = Reach Fault; SMT = S Mayo Trough; SU = Southern Uplands Fault.

zones between the Laurentian and Baltic/Midland Block forelands that are susceptible to plate boundary interpretations, the original configuration and disposition of these terrains is unknown for the palinspastic reasons outlined previously. We do not know, for example, the original relative positions of the Lake District and the Welsh Basin, the Shelve and the Longmynd, the Midland Valley and the Grampian Highlands, Connemara and N Mayo; a large amount of relative motion, particularly strike-slip, can be arcane and 'concealed' in very narrow, innocent-looking fault zones. For example, the 2000 km sinistral offset in the Northern Appalachians indicated by palaeomagnetic data (Kent & Opdyke 1979) probably occurred during mid-Devonian times (Great Glen phase, Figs 36 & 37) and was distributed on ten or more faults, each with up to about 100 km of displacement. This lasted about 20 Ma, giving an average displacement rate of about 10 cm/a, a rate consistent with present day relative motion rates. The reconstructions of Fig. 36, therefore, are designed as a collage of compatible elements; the instantaneous relative motion vector triangles are no more than a plausible representation of relative motion.

Tremadocian/Arenigian palaeotectonics present a particularly acute problem—the tectonic setting of plate boundaries at which ophiolites were generated. In western Newfoundland, the Bay of Islands Complex was formed during the Tremadocian at a ridge/transform/fracture zone complex with the fracture zone roughly parallel with the rigid continental margin (Karson & Dewey 1978). During Arenigian times, subduction, probably nucleated on the fracture zone, led to continent-arc collision by early Caradocian time with the concurrent deformation and metamorphism of the continental margin sedimentary prism beneath the obducted Bay of Islands Ophiolite Complex. A plausible explanation for the S Mayo Trough is that of a pull-apart basin on a right-lateral intra-arc transform with the generation of the Lough Nafooey Group mafic lavas during Tremadocian/early Arenigian times possibly as the layer 1 cap on an ophiolitic trough floor represented now by the ultramafic and meta-gabbroic rocks of the Clew Bay Fault Zone. The progressive growth of the Dalradian metamorphic complex is revealed, in western Ireland, as argued in the previous section, by the Murrisk Supergroup turbidite and Partry Supergroup 'molasse' fill of the S Mayo Trough. The development of the S Mayo Trough, therefore, is viewed as synchronous with the later stages of Dalra-

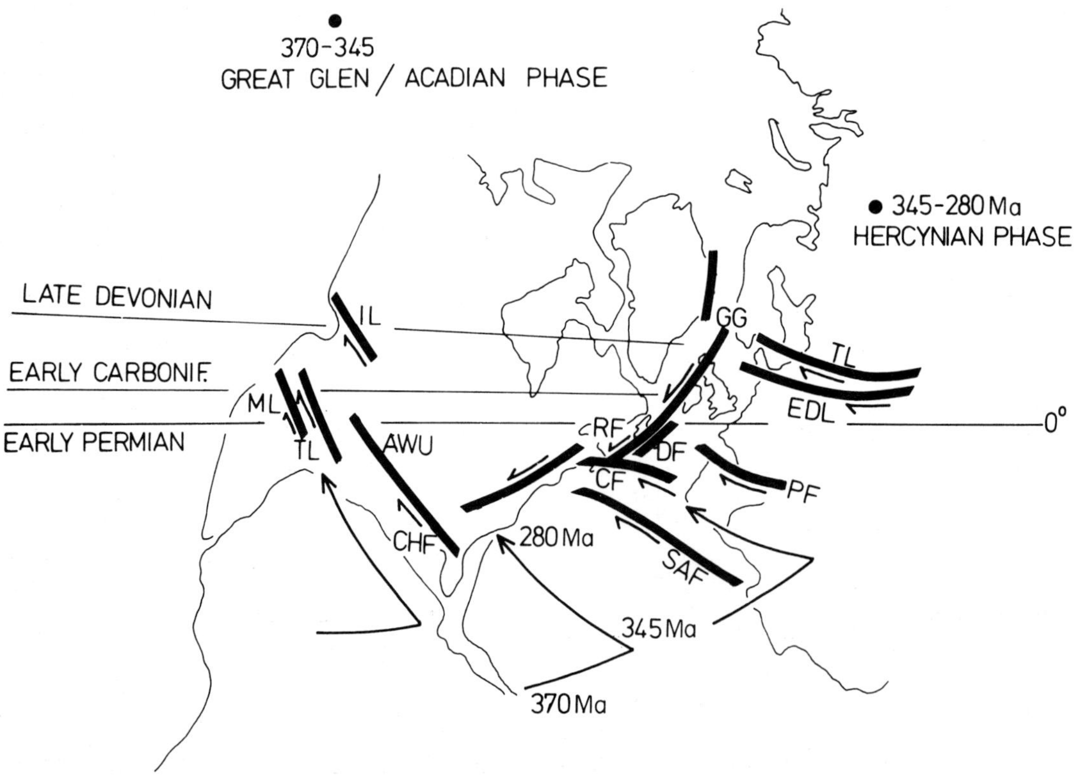

FIG. 37. Early Permian reconstruction of the continents around the Atlantic showing late Devonian to early Permian transform lineaments and relative motion of Gondwana and Laurasia from 370 Ma to 280 Ma. Black circles = finite difference poles of relative motion based upon transform lineaments; AWU = Amarillo–Wichita Uplift; CF = Chedabucto Fault; CHF = Choctaw Fault; DF = Dover Fault; EDL = Elbe/Dobrogea Line; GG = Great Glen and related faults; IL = Idaho Line; ML = Mohave Lineament; PF = Pyrenean Fault trend; RF = Reach and associated faults; SAF = S Atlas Fault; TL = Texas Lineament; TLN = Tornquist Line.

dian deformation and metamorphism and, thus, an unconformity between the Connemara Schists and the S Mayo Trough sequence does not exist. However, the progressive shortening model of Dalradian deformation with Arenigian emergence in Connemara and Llandeilian emergence in Scotland is only one of at least three possible explanations for the Grampian event sequence, all of which may have played a varying role along the Laurasian continental margin during late Cambrian/early Ordovician times. The collision of oceanic plateaus, seamounts and seamount chains (Nur & Ben Avraham 1978) play the dual role of incorporating ophiolites into the subduction complex and inducing a regional deformation, the extent and severity of which depends upon the size of the colliding object. The irregular, slightly diachronous docking of a large oceanic plateau, perhaps represented by all or part of the Midland Valley (Fig. 36), could be invoked as an explanation for the British segment of the Grampian orogeny. A further possibility is a progressive change in tectonic style of a Grampian subduction complex from Tremadocian/Arenigian extensional or neutral, with intra-arc ophiolite-floored basins to Caradocian compressional, caused by a change in motion of the overriding plate in an asthenosphere reference frame and/or the age of subducting lithosphere (Dewey 1980). The uncertainties and difficulties are profound. For example, the tectonic significance of the Unst Ophiolite/Morven–Cabrach zone is not obvious beyond the notion that it is probably of earliest Cambrian age and related temporally to the Tayvallich rifting event. It may have been formed in an early back-arc basin or may be analogous to the Rockall Trough, having separated a small continental block from Laurasia. One can but build plausible models for old orogenic belts like the Grampian zone. The suggestions of Henderson, Robertson and Yardley and their colleagues (this Part) are very helpful in expanding the range of geometric possibilities.

However, by Caradocian time, the Grampian zone was largely stabilized as an Andean-style continental margin orogen, although subsequent strike-slip motions preclude our making definitive statements about the earlier shape and relative position of the zone. From Llandeilian to Ludlovian times, the Southern Uplands accretionary prism grew southwards above a subduction zone dipping beneath the Midland Valley and Highlands (Leggett *et al.* 1979), providing the only portion of the Caledonides that can be interpreted in plate boundary terms (Fig. 36) with confidence. By early Devonian times, the British segment of Iapetus had closed; by mid-Devonian times the northern Appalachian segment had shut (Fig. 36).

b. The Hercynides

The Hercynides and their continuation, the Alleghenides of North America and the Mauritanides of NW Africa, are slightly easier to tackle than the Caledonides, at least on a large scale, because their early Permian stabilization was co-eval with a continental assembly from which Jurassic relative motion began the opening of the Central Atlantic and closure of Tethys. Therefore, a reasonably certain end configuration (Smith *et al.* 1973) may be defined for the Hercynian/Alleghenian tectonic sequence. Fig. 37 illustrates an attempt at defining late Palaeozoic relative motions between Laurasia and Gondwana, from an iterative approach using palaeomagnetic data and the supposition that major intracontinental transform faults, where consistency is displayed, enable us to define latitudes of relative motion and, therefore, an approximate rotation pole. The result, for the period during which the Hercynides evolved from early Carboniferous (345 Ma) to early Permian times (280 Ma), is consistent with the model of Arthaud & Matte (1977) and the view of Badham (1982) that the European Hercynides evolved in a right-lateral megashear that joined the southern end of the Urals to the Alleghenides. Figs 38–41 illustrate a broad-scale plate tectonic interpretation of the Hercynide/Alleghenide zone and its northern foreland for the early Dinantian, Namurian, Westphalian and early Permian. The Hercynides and Alleghenides exhibit a somewhat different tectonic history and orogenic style. The Hercynides have a complex oroclinal geometry, rapid variations in metamorphic grade across shear zones and major thrusts that emplaced thinned flakes of crust and, locally, upper mantle, and a rather complicated capricious history probably having involved the generation and flaking of basins with short-lived ophiolites, pseudo-ophiolites and partially-melted mantle diapirs like the Lizard Complex all in a strike-slip mega-shear tectonic setting. By contrast, the Alleghenides, in particular of the Southern and Central Appalachians S of New York, are more linear and cylindroidal and are dominated by a shortening of at least 200 km achieved largely along the Blue Ridge and Brevard Thrusts (Brewer & Oliver 1980) but also by sub- and intra-cover-stripping thrusts of the Valley and Ridge and Appalachian Plateau. Furthermore, NW of the Southern Appalachians and their continuation in the Ouachita and Marathon mountains, the foreland of Oklahoma and Texas was deformed over an area extending as far as Colorado as a result of a staged and progressive collision between North America and Gondwanaland. The Northern Appalachians show a complex late Palaeozoic history characterized by mainly dextral transforms with major pull-apart basins and minor thrusts (Dwight Bradley, pers. comm.).

During the early Dinantian (Fig. 38), closure of a minor oceanic tract led, by Namurian times, to collision and suturing along a major flat thrust in the Massif Central, the resulting southwest emplacement of the nappes of the Montagne Noir, and collision, with the beginning of Piedmont/Brevard/Blue Ridge shorten-

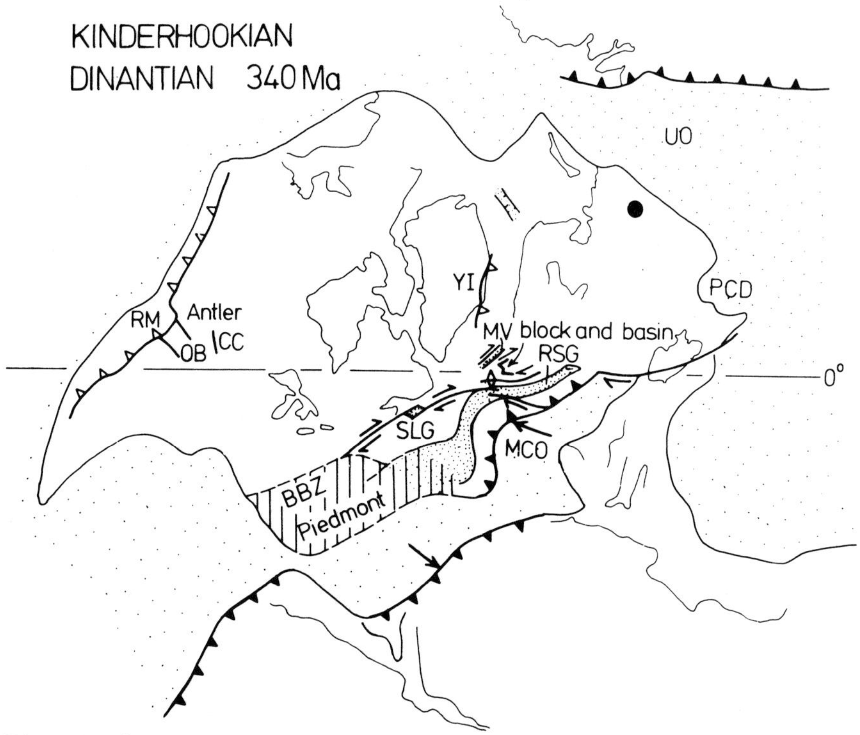

FIG. 38. Early Dinantian (Bretonic Phase) palaeogeography and palaeotectonics of the Appalachian/Hercynian zone. BBZ = Brevard/Blue Ridge Zone; CC = Cedar Creek Uplift; MCO = Massif Central Ocean; MV = Midland Valley; OB = Oquirrh Basin; PCD = PreCaspian Depression; RM = Roberts Mountain Thrust Complex; RSG = Rheinische Schiefergebirge Graben; SLG = St Lawrence Graben; UO = Uralian Ocean; YI = Ymer Island deformation phase; coarse stipple = oceanic crust; vertical lines = deformation; black circle = average Laurasia/Gondwana rotation pole.

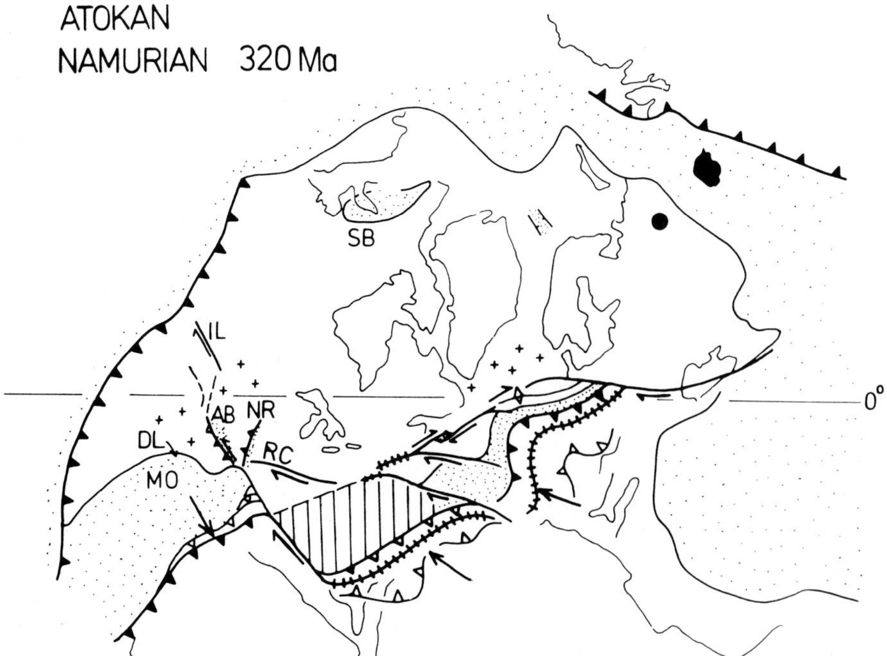

FIG. 39. Namurian (Sudetic Phase) palaeogeography and palaeotectonics of the Appalachian/Hercynian zone. AB = Anadarko Basin; DL = Dimple Limestone; IL = Idaho Lineament; MO = Marathon Ocean; NR = Nemaha Ridge; RC = Rough Creek Fault Zone; SB = Sverdrup Basin; crosses = areas of uplift; other ornaments as in Fig. 37.

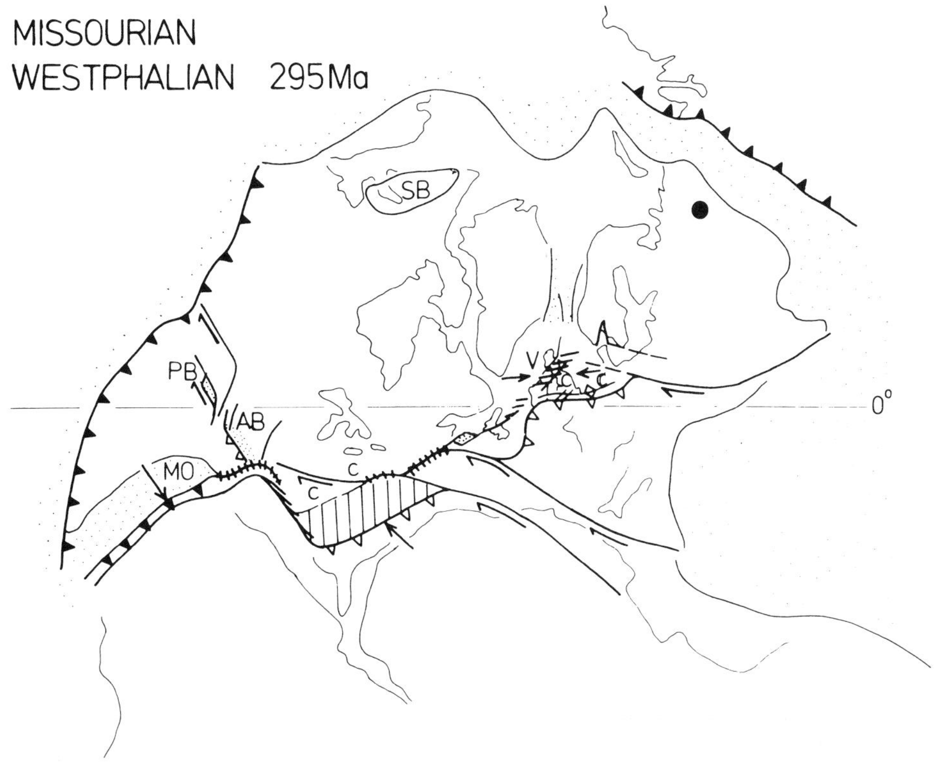

FIG. 40. Westphalian (Asturic Phase) palaeogeography and palaeotectonics of the Appalachian/Hercynian zone. AB = Anadarko Basin; C = coal basins of the Appalachian/Hercynian foreland; MO = Marathon Ocean; PB = Paradox Basin; SB = Sverdrup Basin; V = mafic volcanism and sill intrusion; other ornament as in Fig. 37.

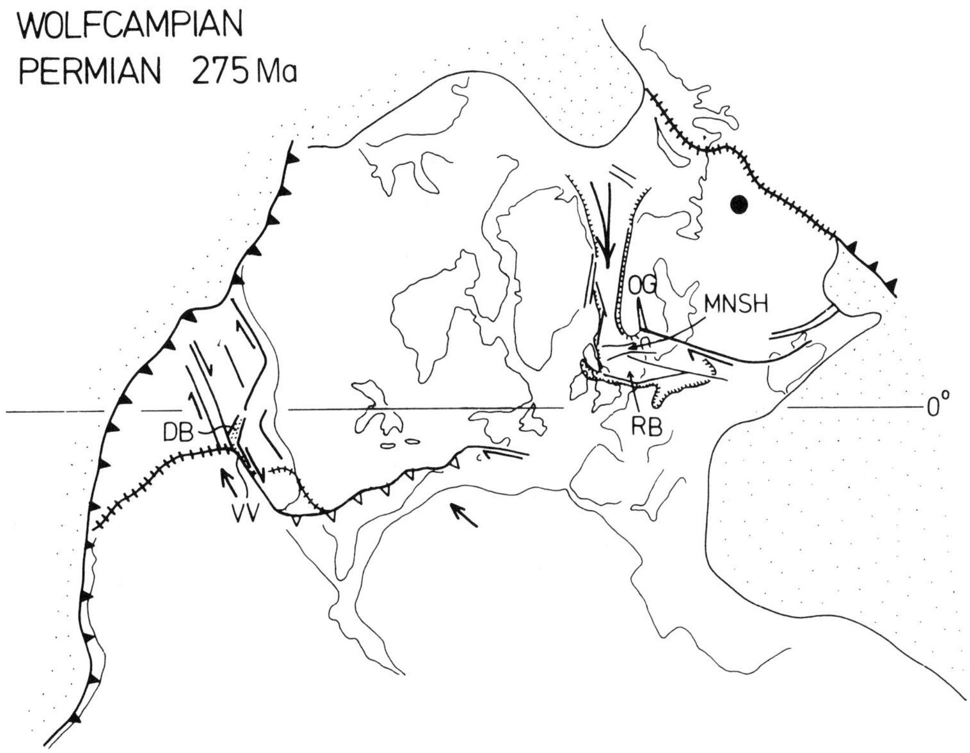

FIG. 41. Early Permian palaeogeography and palaeotectonics of the Appalachian/Hercynian zone. DB = Delaware Basin; MNSH = Mid-North Sea High; OG = Oslo Graben; RB = Rotliegendes Basin; VV = Val Verde Basin; lightly-hachured line = limits of Zechstein marine transgression; broad arrow = boreal provenance of Zechstein transgression; other ornament as in Fig. 37.

ing, in the Appalachians (Fig. 39). During the late Devonian and Dinantian, the Cornubian/Rheinische Schiefergebirge/Harz zone was probably a rift complex with a stretched continental floor; local extreme stretching, during the early and mid-Devonian, allowed the injection of ultramafic diapirs (Lizard) and, possibly, even the development of small oceanic patches such as the Harz Ophiolite. However, the water depth in the trough complex was never very great; N of Wiesbaden, vesicular alkali basalt 'mounds' have caps of shallow water carbonate passing downwards and laterally into radiolarian cherts giving a possible vertical relief of only a few hundred metres. The contraction of this trough, beginning in late Dinantian or early Namurian times (Fig. 39) was accomplished mainly by the listric thrust-stripping of the Upper Palaeozoic sequence from its basement with a northward thrust and Culm turbidite progradation leading, in Westphalian times (Fig. 40), to the development of the exterior coal basins of South Wales, Kent and the Ruhr by flexure induced by thrust loading. The right-lateral transform faulting of the Canadian Maritimes continued northeastwards into the British Isles during Dinantian times, where it was expressed as a distributed zone of shear responsible for the characteristic block and pull-apart basin tectonics of central and northern Britain (Fig. 38).

The mid-Carboniferous warping and uplift of the northern foreland of the Hercynian/Alleghenian belt and the consequent flood of Namurian clastics into the Dinantian pull-apart basins was caused, probably, by collision and suturing in the Massif Central (Fig. 39), which, in turn, probably triggered the beginning of closure of the Cornubian Trough. Similarly, the foreland of the Alleghenides began its complex deformation at this time, again probably as a result of the beginning of Appalachian intracontinental thrust-stacking. Right-lateral strike-slip faulting continued into Westphalian times in the Canadian Maritimes and is expressed in the British Isles by mafic volcanism, major sills and *en echelon* dykes (Francis 1978). During the Westphalian, an accretionary prism on the leading edge of Gondwana collided with North America to form the Ouachitas and to cause intense reactivation of the Oklahoma/Texan/Colorado foreland. In early Permian times (Fig. 41), a similar collision produced the Marathon/Huastecan belt of W Texas and Mexico, causing further reactivation of the North American foreland, and completing Alleghenian/Hercynian suturing. Also, during the early Permian, oceanic subduction was completed by collision to form the Urals but leaving a possible small oceanic remnant, the Pre-Caspian Depression (Fig. 41).

Through Westphalian (Fig. 40) and early Permian (Fig. 41) times, minor rifting occurred in the North Atlantic region between Greenland and Scandinavia and formed the Oslo Graben. These rifts appear to be the method by which the Elbe/Dobrogea and Tornquist right-lateral transforms terminated. The geologically instantaneous transgression that laid down the Marl Slate at the base of the Zechstein sequence, and buried and preserved the dune morphology of the Yellow Sands, may be analogous to the similar rapid transgression that laid down the Aptian evaporities of the South Atlantic to bury a lower Cretaceous non-marine rift sequence. The Aptian salt is restricted to the area between the Niger Delta and the Rio Grande/Walvis Ridges and was probably deposited by a rapid trangression caused by multiple tectonic breaching of the Rio Grande/Walvis ridges to flood sub-sea level grabens. Similarly the Zechstein transgression (at a time of globally falling sea-level) may have resulted from multiple breaching of a barrier in the Spitzbergen region to inundate the Rotliegendes basin complex.

c. Mesozoic/Cenozoic of NW Europe

The complicated system of basins, troughs, swells, lineaments, magmatism, strike-slip faults and inversion zones that characterizes the NW European lithosphere (Fig. 42) owes its origin to the intricate interplay of tectonic events related to several phases of marginal rifting and continental rupture and to foreland compression resulting from the stacking of Alpine nappes. The main causal events, with their timing and relative motion, are summarized in Fig. 42 (inset).

During the Triassic, much of the fundamental rift framework of the British Isles was laid down by stretching to initiate the Fleetwood/Cheshire, Cardigan Bay, Southern Irish, British Channel, Western Approaches, Wessex and Viking Central grabens. Widespread Triassic rifting in Europe preceded and was a harbinger of strong more localized continental stretching beginning in Toarcian time that led, during Bathonian times about 165 Ma, to continental rupture in the Central Atlantic, left-lateral transforming in North Africa/southern Spain and the opening of the Alpine para-oceanic trough (Fig. 42, inset). Triassic stretching of the lithosphere was insufficient or at too slow a strain rate to permit volcanism except in southern Europe, but the mid-Jurassic stretching event was accompanied by tholeiitic volcanism and the development of a local ophiolite floor in the Alpine para-ocean and localized alkalic or tholeiitic volcanism in the North Sea indicating greater, and more rapid, concentrated stretching.

In the early Cretaceous, about 120 Ma, at the end of the Neocomian, the Wessex Basin enjoyed the Ringstead Phase of renewed stretching when normal (mainly down to the S) faulting was concentrated in a broad zone through S Dorset and the southern part of the Hampshire Basin. This phase and the intrusion of the Wolf Rock phonolite are related clearly, temporally and geometrically, to the rotation of the Iberian peninsula away from the Grand Banks and Brittany margins that opened the Bay of Biscay (Fig. 42, inset) and stretched the French Biscay margin by β values up to

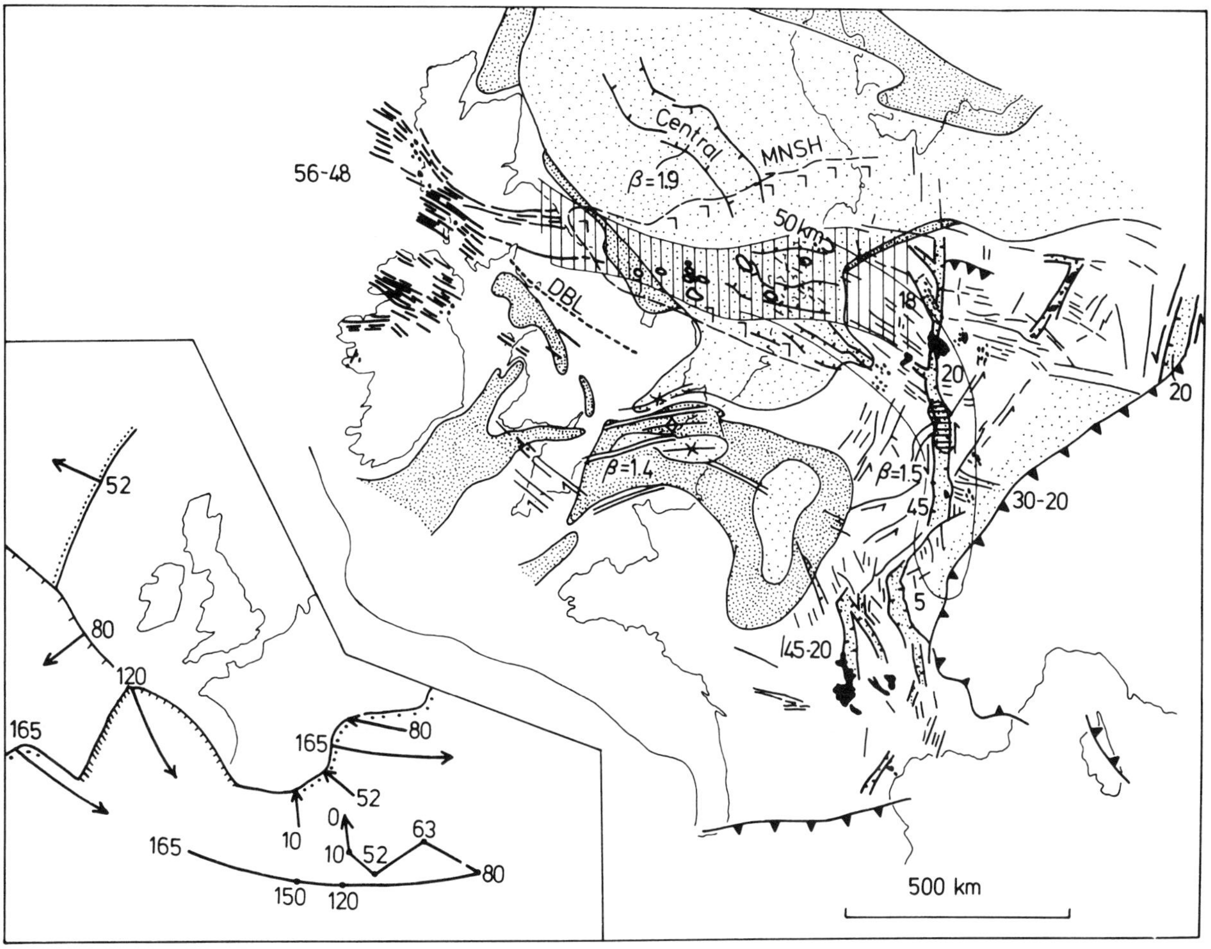

FIG. 42. Outline of the Cenozoic tectonics of NW Europe. DBL = Derby/Bolton Line; MNSH = Mid-North Sea High; fine stipple = Mesozoic basins; coarse stipple = Cenozoic basins; thin lines = structural lineaments, mainly strike-slip faults; thick lines = mafic dykes; double lines = inversion axes, mainly monoclines; black = Cenozoic volcanics; horizontal lines = area of hot springs; vertical lines = area of devolatilized Westphalian coals; small blobs outlined by thick line in southern North Sea = Rotliegendes gas fields; numbers indicate 10^6 a. Inset shows timing, direction and amount of Mesozoic/Cenozoic motion of plates and blocks relative to 'stable' NW Europe.

about 5 (Montadert *et al.* 1979). The North Sea grabens were rejuvenated by a stretching at this time, sufficiently strong to allow the eruption of alkalic tuffs in the southern North Sea.

During the late Cretaceous, at about 80 Ma, in Campanian times, the opening of the Labrador Sea caused a fundamental change in relative motion between Africa and Europe so that the Alpine trough was under compression (Fig. 42, inset). Alpine subduction, and high-pressure metamorphism of the partially subducted Sesia Zone, resulted and the compression was also felt in the NW European foreland. In the southern North Sea, wrench faulting and basin inversion occurred (Ziegler 1975), NNW-striking dolerite dykes were intruded in southern Sweden, and local strong folding affected the Chalk sea floor along the axis of the Wessex Basin (Gale 1980).

During Palaeocene times, relative motion between Africa and Europe was approximately strike-slip in the Alpine region and, consequently, a lull in Alpine shortening occurred. This was accompanied in the North Sea by a renewed phase of stretching and turbidite deposition in the Viking and Central Grabens leading to volcanism in the early Eocene. In the Thulean Province of SE Greenland, Rockall, the Faroes, the northern part of Ireland and NW Scotland, strong rapid stretching across a NW-trending line led to the eruption of tholeiitic plateau lavas from about 66 to 58 Ma, then the intrusion of the great central complexes, then NW-trending tholeiitic dykes from about 56 to 48 Ma. Much of this stretching preceded, along a quite different trend, the NE-trending, 52 Ma opening of the North Atlantic in earliest Eocene times. In the Solepit/southern North Sea area, gas fields in Rotliegendes sandstone reservoirs below a Zechstein seal lie above a zone of devolatilized Westphalian coals, the outgassing of which may have resulted from early Tertiary dyke emplacement in a zone continuing the Cleveland trend from northern England to Holland.

The northward propagation of the Rhine Graben began in Lutetian times about 45 Ma, coinciding with and, therefore, probably related to the thrust stacking of the Austro-Alpine and Penninic nappes, which, in turn was a response to the change in Africa/Europe relative motion at 52 Ma (Fig. 42). Northward rift propagation led to alkaline volcanism about 20 Ma; the lithosphere is still less than 50 km thick beneath the upper and lower Rhine Graben and hot springs are extant between Karlsruhe and Mannheim.

Post-Penninic convergence in the Alps is intermittently recorded in the Oligocene–early Miocene Helvetic and Pliocene Jura phases of cover-stripping and basement subduction. The Helvetic phase of compression is represented in Britain as the Isle of Wight/Sutton Poyntz inversion by the thrust rejuvenation of Ringstead Phase normal faults and as the Sticklepath phase of dextral faulting in Cornubia and minor basin development (e.g. Bovey Tracey marls and lignites). The present state of compressive stress in Britain, resulting from slow Alpine compression by the Adriatic plate, is seen as the recurrence of minor earthquakes, especially centred along lineaments such as the Great Glen Fault and Derby/Bolton Line (fig. 42) (Dunning 1977). The North Sea continues its thermal re-equilibration and subsidence into the present day; the geothermal gradient in the Viking/Central Graben is 3.5°C/100 m and Holland and East Anglia continue their inexorable drowning.

7. Conclusions

The basic key to understanding the tectonic evolution of the continental crust lies in the stretching, shortening and loading of the lithosphere. Inhomogeneities created by stretching and shortening are extremely long-lived, causing frequent tectonic rejuvenation along old lines. Continental crust can be manufactured and modified by plate tectonic processes in many ways, allowing a wide variation in crustal structure (for a thorough review, see Dewey & Windley 1981). Most of the possible variations are seen in the crust of Britain: Laxfordian granulitic shear zones that modified a Scourian crust, an Ordovician/Silurian accretionary prism that grew a continental crust by the off-scarping and thickening of turbidites above a subducting oceanic plate, and Devonian to Palaeocene stretching basins with or without volcanics and sometimes inverted. The stratigraphical record of the British Isles is the record of poorly understood plate boundary effects prior to the late Devonian and the delicate interplay of stretching, loading and compression of the lithosphere in post-Devonian times.

The evolution of the British Isles must be seen against the background of the lithosphere whose thickness secularly thickened from an Archean value almost one-sixth of its present thickness (Dewey & Windley 1981). This smaller lithosphere thickness allowed greater subsidence, a greater incidence of volcanism, and more-tholeiitic volcanism, for particular β values, caused more-local Airy-type, rather than broad flexural compensation, of loading with short-wavelength, sharp peripheral bulges that acted as basin margin source areas. During the Precambrian, the cratons were very much less stable and suffered tectonic reactivation on a very large scale.

We are now entering a phase of total integration in the earth sciences that is not only important in understanding the tectonic history of the crust but provides for the first time a rational predictive basis for the explanation of, and exploration for, hydrocarbon and mineral deposits. We have gone a long way towards rationalism in geology but a difficult path lies ahead. The truth ultimately resides in the rocks and we need a new generation of field work to test hypotheses built from inceptive modelling. The models are critical because the response of the upper lithosphere, or elastic lid, to plate tectonics processes is so complicated that surface geology alone cannot provide definite results from which to reconstruct these processes. The terrain approach to regional tectonic interpretation is proving to be a powerful tool as is deep crustal reflection profiling; perhaps the two most important geophysical measurements that can be made, for tectonic interpretation, are depths to Moho and heatflow. Also the new generation of shallow seismic reflection profiling is yielding new stratigraphic interpretations of basins and their margins and structural interpretations of thrust belts. Much new information, also, will derive from palaeomagnetic studies within deformed zones, such as, for example, the extent to which thrust sheets and flakes have rotated during their emplacement, and limits on the relative latitudinal motion of displaced terrains. Few of these geophysical studies have been undertaken in the British Isles, or anywhere, and are badly needed.

ACKNOWLEDGMENTS. I am deeply indebted to Jack Kirkaldy who encouraged his students to think of stratigraphy in a dynamic tectonic framework, to Gwyn Thomas for introducing me to a fine mapping area in western Ireland, to John Sutton and John Ramsay for their continued encouragement and advice, and to Marshall Kay and Stuart McKerrow for their companionship and help. Lastly, I am particularly grateful to my graduate students at Manchester, Cambridge, and Albany, who taught me much.

8. References

ANDERTON, R., BRIDGES, P. W., LEEDER, M. R. & SELLWOOD, B. W. 1979. *A Dynamic Stratigraphy of the British Isles.* George Allen & Unwin, London, 301 pp.

AMPFERER, O. & HAMMER, W. 1911. Geologische Querschnitt durch die Ostalpen. *Jahrb. geol. Reichsanst.* **61,** 531–710.

ARTHAUD, F. & MATTE, P. 1977. Late Palaeozoic strike-slip faulting in southern Europe and northern Africa: results of a right-lateral shear zone between the Appalachians and the Urals. *Bull. geol. Soc. Am.* **88,** 1305–20.

ATWATER, T. 1970. Implications of plate tectonics for the Cenozoic tectonic evolution of western North America. *Bull. geol. Soc. Am.* **81,** 3513–36.

BADHAM, J. P. N. 1982. Strike-slip orogens—an explanation for the Hercynides. *J. geol. Soc. London,* **139,** 493–504.

BALLY, A. W., GORDY, P. L. & STEWART, G. A. 1966. Structure, seismic data and orogenic evolution of the Southern Canadian Rocky Mountains. *Bull. Can. Pet. Geol.* **14,** 337–81.

BEAUMONT, C. 1978. The evolution of sedimentary basins on a viscoelastic lithosphere: theory and examples. *Geophys. J. R. astron. Soc.* **55,** 471–97.

BECK, M. E. 1980. The paleomagnetic record of plate-margin processes along the western edge of North America. *J. geophys. Res.* **85,** 7115–31.

BONATTI, J. A., SARNTHEIM, M., BOESMA, A., GORINI, M. & HONNOREZ, J. 1977. Neogene crustal emersion and subsidence at the Romanche F. Z., equatorial Atlantic. *Earth planet. Sci. Lett.* **35,** 369–83.

BREWER, J. A. & OLIVER, J. E. 1980. Seismic reflection studies of deep crustal structure. *Ann. Rev. Earth planet. Sci.* **8,** 205–30.

CONEY, P. J., JONES, D. L. & MONGER, J. W. N. 1980. Cordilleran suspect terrains. *Nature, London,* **288,** 329–33.

COWARD, M. P. & SIDDANS, A. W. B. 1979. The tectonic evolution of the Welsh Caledonides. In: HARRIS, A. L., HOLLAND, C. H. & LEAKE, B. E. (eds.). *The Caledonides of the British Isles—reviewed, 187–98.* Spec. Publ. geol. Soc. London, **8,** 768 pp.

DEWEY, J. F. 1969. Evolution of the Appalachian/Caledonian Orogen. *Nature, London,* **222,** 124–9.

—— 1975. Finite plate evolution: some implications for the evolution of rock masses at plate margins. *Am. J. Sci.* **275-A,** 260–84.

—— 1976. Ancient plate margins: some observations. *Tectonophysics,* **33,** 379–85.

—— 1980. Episodicity, sequence and style at convergent plate boundaries. *In:* STRANGEWAY, D. W. (ed.). *The Continental Crust and its Mineral Deposits,* 553–76. Spec. Pap. geol. Assoc. Can. **20,** 804 pp.

——, PITMAN, W. C. III, RYAN, W. B. F. & BONNIN, J. 1973. Plate tectonics and evolution of the Alpine System. *Bull. geol. Soc. Am.* **84,** 3137–80.

—— & WINDLEY, B. F. 1981. Growth and differentiation of the continental crust. *Philos. Trans. R. Soc. London,* **301,** 189–206.

DUNNING, F. W. 1977. Caledonian–Variscan relations in north-west Europe. *Coll. internat. CNRS,* Rennes, No. 243, 165–80.

EVANS, T. R. & COLEMAN, N. C. 1974. North Sca geothermal gradients. *Nature, London,* **247,** 28–30.

FRANCIS, E. W. 1978. Igneous activity in a fractured craton: Carboniferous volcanism in northern Britain. *In:* BOWES, D. R. & LEAKE, B. E. (eds). *Crustal Evolution in Northwestern Britain and Adjacent Regions,* 279–96. Seel House Press, Liverpool, 492 pp.

GALE, A. S. 1980. Penecontemporaneous folding, sedimentation and erosion in Campanian chalk near Portsmouth, England. *Sedimentology,* **27,** 137–51.

GAPAIS, D. & LE CORRE, C. 1980. Is the Hercynian belt of Brittany a major shear zone? *Nature, London,* **288,** 574–6.

GASS, I. G. & SMEWING, J. D. 1981. Ophiolites: obducted oceanic lithosphere. *In:* EMILIANI, C. (ed.). *The Sea* (**7**), 339–62. John Wiley and Sons, New York, 1738 pp.

GEE, D. G. 1975. A tectonic model for the central part of the Scandinavian Caledonides. *Am. J. Sci.* **275A,** 468–515.

HARRIS, A. L., HOLLAND, C. W. & LEAKE, B. E. (eds). 1979). *The Caledonides of the British Isles—reviewed.* Spec. Publ. geol. Soc. London, **8,** 768 pp.

HENDERSON, W. G. & ROBERTSON, A. H. F. 1982. The Highland Border rocks and their relation to marginal basin development in the Scottish Caledonides. *J. geol. Soc. London,* **139,** 433–50.

ISACKS, B., OLIVER, J. & SYKES, L. R. 1968. Seismology and the new global tectonics. *J. geophys. Res.* **73,** 5855–99.

JACKSON, J. A. 1980. Reactivation of basement faults and crustal shortening in orogenic belts. *Nature, London,* **283,** 343–6.

JENKYNS, H. C. & SENIOR, J. R. 1977. A Liassic palaeofault from Dorset. *Geol. Mag.* **114,** 47–52.

KARSON, J. & DEWEY, J. F. 1978. The Coastal Complex, western Newfoundland: an early Ordovician fracture zone. *Bull. geol. Soc. Am.* **89,** 1037–49.

KENT, D. V. & OPDYKE, N. D. 1979. The early Carboniferous palaeomagnetic field of North America and its bearings on the tectonics of the Northern Appalachians. *Earth planet. Sci. Lett.* **44,** 365–72.

KIDD, R. G. W. 1977. A model for the process of formation of the upper oceanic crust. *Geophys. J. R. astron. Soc.* **50,** 149–83.

—— & CANN, J. R. 1974. Chilling statistics indicate an ocean floor spreading origin for the Troodos Complex, Cyprus. *Earth planet. Sci. Lett.* **24,** 151–5.

LEEDER, M. R. 1982. Upper Palaeozoic basins of the British Isles—Caledonide inheritance versus Hercynian plate margin processes. *J. geol. Soc. London,* **139,** 479–81.

LEGGETT, J. K., MCKERROW, W. S. & EALES, M. N. 1979. The Southern Uplands of Scotland: a Lower Palaeozoic accretionary prism. *J. geol. Soc. London,* **136,** 755–70.

LE PICHON, X. 1968. Sea-floor spreading and continental drift. *J. geophys. Res.* **73,** 3661–97.

MALM, O. A., FURNES, W. & BJORLYKKE, K. 1979. Volcaniclastics of Middle Jurassic age in the Statfjord oil-field of the North Sea. *Neues Jahrb. Geol. Palaeontol. Monatshefte,* **10,** 607–18.

MCELHINNY, M. W. 1973. *Palaeomagnetism and Plate Tectonics.* Cambridge Univ. Press, 358 pp.

MCKENZIE, D. P. 1978. Some remarks on the development of sedimentary basins. *Earth planet. Sci. Lett.* **40,** 25–32.

—— & MORGAN, W. J. 1969. Evolution of triple junctions. *Nature, London,* **244,** 125–33.

—— & PARKER, R. L. 1967. The North Pacific: an example of tectonics on a sphere. *Nature, London,* **216,** 1276–80.

MENARD, N. W. & ATWATER, T. 1968. Changes in the direction of sea-floor spreading. *Nature, London,* **219,** 463–67.

MONTADERT, L., ROBERTS, D. G. & DE CHARPAL, O. 1979. Rifting and subsidence of the northern continental margins of the Bay of Biscay: *Initial Rep. Deep Sea drill. Proj.* **48,** 1025–60. U.S. Govt. Printing Office, Washington.

MOORE, J. C. & CONNELLY, W. 1977. Mesozoic tectonics of the Southern Alaskan margin. *Am. Geophys. Union,* Maurice Ewing Series **1,** 71–82.

MORGAN, W. J. 1968. Rises, trenches, great faults and crustal blocks. *J. geophys. Res.* **73,** 1959–82.

NUR, A. & BEN-AVRAHAM, Z. 1978. Speculations on mountain building and the lost Pacific continent. *J. Phys. Earth,* **26,** 21–37.

—— 1974. The plain man's guide to plate tectonics. *Proc. Geol. Assoc. London,* **85,** 299–357.

PALMER, T. J., MCKERROW, W. S. & COWIE, J. W. 1980. Sedimentological evidence for a stratigraphical break in the Durness Group. *Nature, London,* **287,** 720–2.

PHILLIPS, W. E. A., STILLMAN, C. J. & MURPHY, T. 1976. A Caledonian plate tectonic model. *J. geol. Soc. London,* **132,** 579–609.

PITMAN, W. C., III. 1978. Relationship between eustasy and stratigraphic sequences of passive margins. *Bull. geol. Soc. Am.* **89,** 1389–403.

PROFFETT, J. M. 1977. Cenozoic geology of the Yerington district, Nevada and implications for the nature and origin of Basin and Range faulting. *Bull. geol. Soc. Am.* **88,** 247–66.

ROYDEN, L., SCLATER, J. G. & VON HERZEN, R. P. 1980. Continental margin subsidence and heat flow: important parameters in formation of petroleum hydrocarbons. *Bull. Am. Assoc. Petrol. Geol.* **64,** 173–87.

SCLATER, J. G., JAUPART, C. & GALSON, D. 1980. The heat flow through oceanic and continental crust and the heat loss of the earth. *Rev. Geophys. Space Phys.* **18,** 269–311.

SMITH, A. G. 1971. Alpine deformation and the oceanic areas of the Tethys, Mediterranean and Atlantic. *Bull. geol. Soc. Am.* **82,** 2039–70.

—— 1976. Plate tectonics and orogeny: a review. *Tectonophysics,* **33,** 215–85.

——, BRIDEN, J. C. & DREWRY, G. E. 1973. Phanerozoic world maps. *In:* HUGHES, N. F. (ed.). *Organisms and Continents through Time,* 1–42. Spec. Pap. Palaeontol. London, **12.**

STILLE, H. 1924. *Grundfragen der vergleichenden Tektonik.* Borntraeger, Berlin, 443 pp.

STONELEY, R. 1982. The structural development of the Wessex Basin. *J. geol. Soc. London,* **139,** 543–54.

WALCOTT, R. I. 1970. Flexural rigidity, thickness and viscosity of the lithosphere. *J. geophys. Res.* **75,** 3941–54.

WATTS, A. B., BODINE, J. W. & STECKLER, M. S. 1980. Observations of flexure and the state of stress in the oceanic lithosphere. *J. geophys. Res.* **85,** 6369–76.

WELLS, P. R. A. 1979. P-T conditions in the Moines of the Central Highlands, Scotland. *J. geol. Soc. London,* **136,** 663–71.

WILLIAMS, H. 1978. Tectonic lithofacies map of the Appalachian Orogen (1:1,000,000). Dept. Geol., Mem. Univ. Newfoundland, St. Johns. Map. No. 1.

WILSON, J. T. 1965. A new class of faults, and their bearing on continental drift. *Nature, London,* **207,** 343–7.

WOODLAND, A. W. 1975. *Petroleum and the Continental Shelf of North-west Europe,* **1,** *Geology.* Appl. Sci. Publ. Barking, Essex, 501 pp.

YARDLEY, B. W. D., VINE, F. J. & BALDWIN, C. T. 1982. The plate tectonic setting of NW Britain and Ireland in late Cambrian and early Ordovician times, *J. geol. Soc. London,* **139,** 455–73.

ZIEGLER, P. A. 1975. Geologic evolution of the North Sea and its tectonic framework. *Bull. Assoc. Petrol. Geol.* **59,** 1073–97.

Received 12 January 1981; revised typescript received 11 March 1981.

JOHN F. DEWEY, Department of Geological Sciences, Science Laboratories, South Road, Durham SH1 3LE.

Lithospheric stretching, detached normal faulting and footwall uplift

D. Barr

SUMMARY: Lithospheric stretching can successfully account for the overall evolution of many sedimentary basins, and detached normal faulting the detailed geometry of the upper crust. In an instantaneously stretched, isostatically compensated basin, the equations which describe these two processes can be combined to define a 'notional depth to decollement'. Only at this level can the sole to the normal fault system maintain a constant depth below sea-level during extension. The notional depth to decollement depends primarily on the mean density of the basin fill and for a constant-density basin fill (e.g. sea water) coincides with the level of no vertical motion during stretching. In a sediment-filled basin, the notional depth to decollement will increase with the stretching factor β as early-deposited sediments are compacted and the mean density of the sediment column increases. In general, a physical sole fault will not lie at the notional depth to decollement, and must move vertically to maintain isostatic equilibrium: such movement precludes the use of balanced cross-section techniques to determine the physical depth to decollement.

In typical crustal situations, uplift of the sole fault will be more common than subsidence and will in turn cause uplift of any residual, unfaulted basement blocks which rest upon it. Footwall uplift can also be modelled using area-balance constraints, referred to the notional depth to decollement rather than to the physical sole fault. The amount of uplift depends primarily on the initial fault spacing. Three fields can be distinguished: one in which footwalls subside at an increasing rate as β increases, one in which they are uplifted above sea-level then subside below sea-level, and one in which they are uplifted then subside, but always remain above sea-level. Similar relationships exist in an uncompensated basin, where the depth to the physical sole fault replaces the notional depth to decollement.

Curves showing uplift and subsidence versus β have been constructed in dimensionless form (referred to depth to decollement) and for a model basin with an exponential sediment compaction relationship. They agree closely with uplift/subsidence histories inferred from seismic and well data for the North Sea, and with published descriptions of other areas (the Armorican margin, the Aegean Sea).

Models of lithospheric stretching have successfully accounted for the broad geological evolution of several intracontinental basins, marginal seas and passive continental margins (e.g. McKenzie 1978; Steckler & Watts 1978; Sclater *et al.* 1980; Sclater & Christie 1980; Le Pichon & Sibuet 1981; Watts 1982; Hellinger & Sclater 1983). In these interpretations, an initial phase of fault-bounded subsidence S_i reflects the stretching episode. The base of the lithosphere is considered to be thermally controlled and is elevated as isotherm adjustment lags behind stretching. As isotherms recover, the lithosphere cools and thickens as a phase of near-exponential regional (thermal) subsidence S_t ensues. The simplest models (e.g. McKenzie 1978) assume instantaneous stretching. The effect of non-instantaneous stretching is to increase fault-bounded subsidence at the expense of regional subsidence (Jarvis & McKenzie 1980; Cochran 1983). Most estimates of the magnitude of lithospheric stretching have been obtained by modelling the thermal subsidence stage from well data (because there is little faulting, a few wells can be considered representative of a given basin), with the degree of basement thinning being confirmed by interpretation of seismic refraction or gravity data (e.g. Barton & Wood 1984). Upper-crustal extension is usually inhomogeneous and accommodated by detached normal faulting, which may or may not be listric (cf. Chenet *et al.* 1983; Le Pichon *et al.* 1983). Estimates of upper-crustal extension during rifting, as measured on reflection seismic sections (Gibbs 1983), often appear to conflict with those of lithospheric stretching (cf. Christie & Sclater 1980; Smythe *et al.* 1980; Wood & Barton 1983; Ziegler 1983). This paper attempts to clarify the geometrical relationships between subsidence, lithospheric stretching and upper-crustal extensional faulting for the simple case of instantaneous stretching. Further consideration of these relationships permits semi-quantitative modelling of footwall uplift and hanging wall subsidence in extensional basins—a feature of obvious relevance to hydrocarbon exploration.

From Holdsworth, R.E. & Turner, J.P. (compilers) *Extensional Tectonics: Regional-scale Processes.* The Geological Society,
Key Issues in Earth Sciences, **2** (**1**), 55–74.
1476-315X/02/$15.00 © The Geological Society of London 2002
First published in: "Coward, M.P, Dewey, J.F. & Hancock, P.L. (eds) *Continental Extensional Tectonics*, The Geological Society, London, Special Publications **28**, 75–94".

Lithospheric stretching and detached normal faulting

The level of no vertical motion during stretching

Figure 1 is a schematic model of the lithosphere immediately following instantaneous stretching. The upper crust has deformed by faulting above a sub-horizontal decollement and the lower crust and mantle by homogeneous stretching. The average depth to the basin floor corresponds to S_i, the initial subsidence due to stretching. This model is admittedly over-simplified, but can serve as a starting point for more realistic models involving, for example, inhomogeneous stretching.

S_i can be obtained from equation 1 of McKenzie (1978), which simply expresses the requirement that stretched and unstretched lithosphere be in isostatic equilibrium with one another and with the mid-ocean ridge of Parsons & Sclater (1977):

$$S_i = \frac{a\{(\varrho_0 - \varrho_c)t_c/a[1-(\alpha T_1 t_c)/2_a] - \alpha T_i \varrho_0/2\}}{\varrho_0 \times (1-\alpha T_1) - \varrho_w} \times \frac{(1-1/\beta)}{\varrho_0 \times (1-\alpha T_1) - \varrho_w} \qquad (1)$$

Parameters are identified in Table 1. The conventional correction for sediment loading (e.g. Sclater & Christie 1980) makes the assumption that isostatic compensation takes place at the base of the crust. For the purposes of this paper, it is more convenient (and makes little difference to the result) to assume isostatic compensation at the base of the lithosphere, distributed over a horizontal distance comparable in magnitude to the dominant upper-crustal fault spacing. In that case, ϱ_w can be replaced by ϱ_s, the mean density of the sediment + water column after stretching.

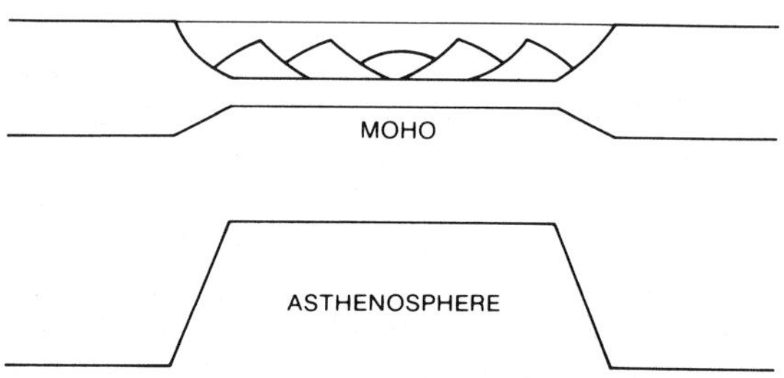

FIG. 1. Schematic model of stretched continental lithosphere, before thermal subsidence. The lower crust and mantle have deformed homogeneously, by ductile flow, while the brittle upper crust has deformed by detached normal faulting.

McKenzie (1978) measured initial subsidence S_i relative to the elevation of unstretched lithosphere and, if the other parameters are kept constant, S_i varies widely with crustal thickness. The thicker the crust the greater the subsidence, and if the crust is thinner than about 18 km, syn-stretching uplift occurs. However, isostatic constraints require such thin crust to lie below sea-level prior to stretching and thick crust to lie above sea level. If on the other hand the pre-stretching crustal elevation is specified, the parameters in equation 1 are no longer independent of one another. For example, a thin crust may be of unusually low density or may be compensated by a high heat flow and thin mantle lithosphere.

If the pre- and post-stretching elevations are measured *relative to a common reference level,* the factor relating S_i to $(1-1/\beta)$ is a constant with units of depth, and corresponds to the *level of no vertical motion within the lithosphere, d_n*. That this level lies within the continental crust is indicated by the general observation that the surface subsides and the Moho rises during stretching. It is convenient to choose present-day sea-level as a reference, in which case d_n corresponds to the predicted depth to a mid-ocean ridge if no oceanic crust were present. This is apparent from Fig. 2. At infinite extension, $(1-1/\beta)=1$, $S_i = d_n$ and a notional column consisting of mantle asthenosphere directly overlain by sea water or sediments of constant density is in isostatic equilibrium with unstretched lithosphere and with all intervening stages (cf. Le

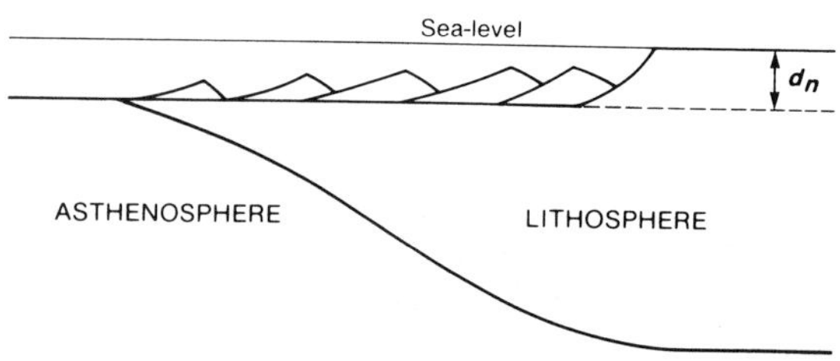

FIG. 2. Physical significance of the level of no vertical motion or 'notional depth to decollement' d_n. It corresponds to the level to which the asthenosphere would rise if loaded to sea-level by water or a sedimentary column of specified density; i.e. in the water-loaded case, to the mantle geoid of Turcotte *et al.* (1977). As the continental lithosphere undergoes instantaneous, homogeneous stretching, its mean density is unchanged and the proportion of its thickness which rises above d_n remains constant. The remaining column to sea-level is filled by water or sediments of the previously specified density. At infinite extension, the sea-water or sediment column would directly overlie the asthenosphere.

TABLE 1. *List of symbols used.*

a	Thickness of lithosphere
t_c	Thickness of continental crust
T_l	Temperature of asthenosphere
α	Thermal expansion coefficient of lithosphere
β	Stretching factor
ϱ_0	Mantle density at 0°C
ϱ_c	Crustal density at 0°C
ϱ_a	Density of asthenosphere
ϱ_l	Mean density of the pre-rifting lithosphere
ϱ_w	Density of sea water
ϱ_s	Mean density of a (sedimentary) basin fill
ϱ_{sg}	Density of sediment grains
f_0	Porosity of a sediment at sea level
f	Porosity of a sediment at depth
k	Post-exponential factor describing sediment compaction
S_i	Initial subsidence due to stretching
S_t	Subsequent (thermal) subsidence
S_∞	Total subsidence at infinite time
d_n	Notional depth to decollement in an isostatically compensated basin (= level of no vertical motion given a basin fill of constant density)
s	Approximate factor relating S_t to $(1-1/\beta)$
d'	Depth to the physical decollement surface in a system of detached normal faults
d_i	Initial (pre-stretching) depth to a physical decollement surface
ϕ_i	Initial (pre-stretching) dip of a 'domino' fault-plane
ϕ	Fault-plane dip after stretching
θ	Dip of the top basement surface after stretching
a'	Initial (pre-stretching) separation between 'domino' faults
c'	Final (post-stretching) separation between 'domino' faults
z_f	Magnitude of footwall uplift
z_h	Magnitude of hanging wall subsidence

TABLE 2. *Estimates of the notional depth to decollement* d_n *and of the thermal subsidence factor* s *for water-loaded and sediment-loaded* *($\varrho_s = 2200$ kgm^{-3}) basins.*

Source for parameters	water-loaded		sediment-loaded	
	d_n	s	d_n	s
McKenzie (1978)	3.07	4.17	6.82	9.24
Sclater *et al.* (1980)	3.24	4.13	7.29	9.31
Le Pichon & Sibuet (1981)	3.61	4.21	7.83	9.12
Cochran (1983)	3.26	4.33	7.16	9.63

Pichon *et al.*'s (1982) discussion of a notional column of mantle asthenosphere directly overlain by sea water, and Turcotte *et al.*'s (1977) water-loaded mantle geoid).

Values of the level of no vertical motion d_n are quoted in Table 2 for various combinations of lithospheric parameters and for $\varrho_s = 1000$ kgm^{-3} (water-filled) and $\varrho_s = 2200$ kgm^{-3} (sediment-filled). The latter density corresponds to that of a quartz sand with 30% water-filled porosity and is a reasonable compromise for a mature basin containing a few kilometres of syn-rifting sediments. For example, Beach (1984) recorded 950 m of syn-rifting sediments with average density 2600 kgm^{-3} in part of the North Sea graben system, and concluded from the geometry of the graben that these had suffered 325 m of post-extensional compaction. Decompacting and assuming water-filled porosity yields a mean density of 2192 kgm^{-3}. For water-filled (sediment-starved) basins, $d_n = 3.1$–3.6 km, while for sediment-filled basins, $d_n = 6.8$–7.8 km. All the values for d_n quoted in Table 2 can be regarded as estimates of the same parameter (the depth to the mantle geoid at a mid-ocean ridge), so for the remainder of this paper, average values of 3.3 and 7.3 km will be used. The differences between the estimates presumably arise from the specific mid-ocean ridge chosen and the correction (if any) made for oceanic crust.

This relationship severely restricts the extent to which variations in crustal parameters can explain inconsistencies between seismically measured β values and basement subsidence (e.g. by inferring an unusually thick crust where subsidence is excessive). As long as the crust was in isostatic equilibrium and close to sea-level before stretching, the amount of subsidence relative to a constant sea-level should be independent of crustal thickness.

Eustatic sea-level changes will have two main effects. If sea-level at the end of stretching is different from that before stretching, this will introduce an error into S_i of the order of the net sea-level change (short-period fluctuations *during* stretching tending to cancel out). If sea-level was constant during stretching but different from that at the present day, d_n will be increased or decreased by an amount equal to the sea-level change. Even changes of a few hundred metres will have only second-order effects in typical sedimentary basins (where S_i = a few kilometres, $d_n = 7.3$ km), so this effect will be ignored for the remainder of the discussion.

Where the mean thickness of syn-rift sediments can be inferred from seismic or other evidence, an approximate indication of the stretching factor can be obtained by assuming that $S_i = 7.3$ $(1-1/\beta)$. This approach is complementary to the more usual practice of modelling the thermal subsidence stage only (cf. Wood 1981; Angelier 1985). However, it is particularly important when dealing with the irregular sediment-thickness distributions found in typical half-grabens to measure *mean* sediment thickness over some area comparable in width to the deeper zone of ductile thinning. For example, much of the discrepancy which Beach (1984) found between β as calculated from the thickness of syn-rifting sediments, and extension ratios measured on reflection seismic sections can probably be attributed to the fact that he used sediment thicknesses from the deepest part of the Witchground graben, rather than the average thickness within the graben as a whole. Beach's β values imply that a 125 km thick lithosphere thins to 90 km across a horizontal distance of less than 10 km, a mechanically unlikely scenario which is contradicted by the distributed lower-crustal thinning recorded in a refraction survey by Christie & Sclater (1980).

Subsequent thermal subsidence will have depressed the level of no (syn-stretching) vertical motion in fossil basins. Following Le Pichon & Sibeut (1981), the total subsidence $S_i + S_t$ at infinite time can be approximated within a few metres or tens of metres by:

$$S_\infty = \frac{t_c(\varrho_0 - \varrho_c)[1-(\alpha T_1 t_c)/2a](1-1/\beta)}{\varrho_0 \times (1-\alpha T_1) - \varrho_s} \cdot \qquad (2)$$

The thermal subsidence $S_t = S_\infty - S_i$ and so:

$$S_t = \frac{(a \times \alpha T_1 \varrho_0) \times (1-1/\beta)}{2[\varrho_0 \times (1-\alpha T_1) - \varrho_s]}$$

$$= \frac{a}{2} \times \left(\frac{\varrho_0 - \varrho_a}{\varrho_a - \varrho_s}\right) \times (1-1/\beta)$$

$$= s \times (1-1/\beta), \qquad (3)$$

and is essentially independent of crustal thickness. Water- and sediment-loaded values of s are quoted in Table 2. Average values of 4.2 and 9.3 km will be used in the remainder of this paper. For a basin in which stretching ceased at about 120 Ma (e.g. the North Sea, the Biscay margin), the thermal subsidence will have reached about 87% of its ultimate value (Le Pichon & Sibuet 1981), and so the subsidence at infinite time can be used with little loss of precision. Present-day depth to the fossil horizon of no vertical motion is given by $(d_n + S_t)$, and is displayed graphically in Fig. 4 for three extreme cases. In the outermost part of a sediment-starved continental margin with $\beta = 4$, this horizon should lie at ≈ 6.5 km below sea-level. Where the initial, rift-stage half-grabens are sediment-filled but subsequent thermal subsidence was largely water-loaded (e.g. the Biscay margin—Le Pichon & Sibuet 1981), this horizon should lie at ≈ 10 km for $\beta = 2$–4. Where sediment loading applied throughout (e.g. the North Sea), this horizon should lie at ≈ 12 km for $\beta = 2$.

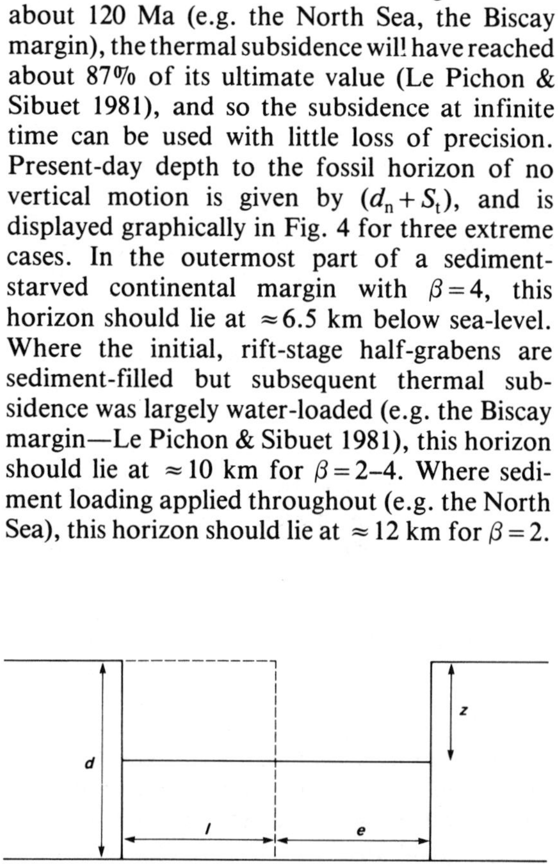

FIG. 3. Schematic balanced cross-section showing the relationship between extension ($\beta = l + e/l$), depth to decollement (d) and mean basin depth (z). See text for further details.

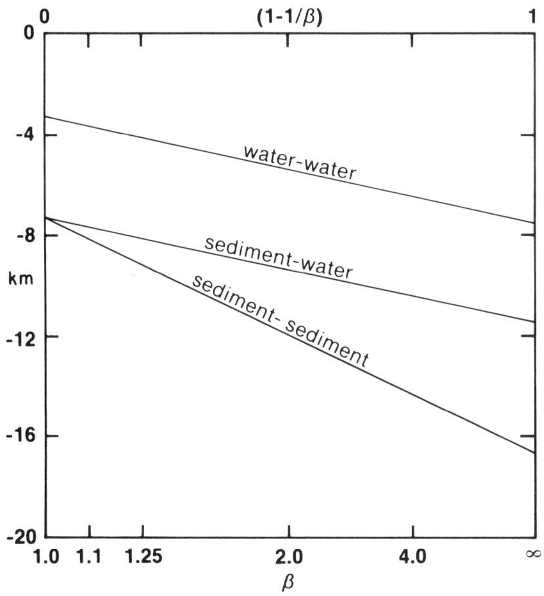

FIG. 4. Notional depth to decollement, after thermal subsidence, for three extreme models: (i) water-loaded throughout; (ii) sediment-loaded ($\varrho_s = 2200$ kgm^{-3}) during stretching, water-loaded during thermal subsidence; (iii) sediment-loaded throughout ($\varrho_s = 2200$ kgm^{-3}).

True and notional depths to decollement

The discussion so far has ignored the detailed geometry of the upper crust. Recent models of extensional basin formation (Wernicke & Burchfiel 1982; Le Pichon *et al.* 1983; Gibbs 1984) have emphasized the role of detached normal faulting above a gently dipping decollement surface, and have applied balanced cross-section techniques originally developed in compressional terraines (Dahlstrom 1969). Following Gibbs (1983), a simple geometric model can be constructed for faulting above a basal decollement at depth d (Fig. 3), whereby a line of original length l is extended to $l+e$, resulting in subsidence z. Assuming preservation of the cross-sectional area within the basement and no vertical movement of the basal decollement:

$$e \times d = (l+e) \times z \tag{4}$$

and the stretching factor β is defined as $(l+e)/l$. Substituting $\beta \times l$ for $l+e$ and $l \times (\beta - 1)$ for e in equation 4, and rearranging:

$$z = d \times (1 - 1/\beta). \tag{5}$$

The analogy with equation 1 is obvious. In an isostatically compensated basin, any attempt to apply the area-balancing technique of Gibbs (1983) will yield a 'depth to decollement' which corresponds to the level of no vertical motion d_n, rather than the depth to a physical sole fault. The depth to the sole fault will only be correctly determined where by coincidence it lies at or near the level of no vertical motion. Conversely, if the amount of basement subsidence is controlled only by the depth to a physical sole fault, the basin will be unable to maintain isostatic equilibrium during extension, its sediment fill will be uncompensated, and the simple stretching model of McKenzie (1978) will be inapplicable (as in the Inner Moray Firth basin, which is associated with a regional negative free-air gravity anomaly and lacks significant thermal subsidence—McQuillin *et al.* 1982; Barr 1985).

Thus the indiscriminate application of area-balancing techniques to syntectonic sediments in extensional basins should be discouraged. Where they have yielded results compatible with reflection seismic evidence, either this must be put down to pure coincidence, or the basin in question must be uncompensated. If in a compensated basin the calculated 'depth to decollement' (= the level of no vertical motion) deviates markedly from the values predicted in Table 2, the basin must have formed by some mechanism other than uniform stretching (e.g. the level of no vertical motion will be shallower if β(mantle) > β(crust), and deeper if β(crust) > β(mantle)). Calculating the level of no vertical motion may therefore help in determining the relative proportions of crust and mantle stretching and hence whether heat flow within the basin was significantly greater during stretching than it is today.

In many respects, the upper part of a compensated basin behaves during stretching as if it were an uncompensated basin with a basal decollement lying at the level of no vertical motion d_n. It is therefore convenient to introduce the alternative term 'notional depth to decollement'. This enables comparisons to be made between basins whose degree of compensation is unclear, but in which a real or notional depth to decollement can nevertheless be determined by area-balancing. As will be shown below, the term 'level of no vertical motion' is inappropriate to most sedimentary basins because the magnitude of d_n increases with β as the sediment density ϱ_s increases through compaction.

Comparison of notional and observed depths to decollement

Information on actual depth to decollement d' can be obtained from seismic reflection data (Fig. 5). Where a basal decollement is observed,

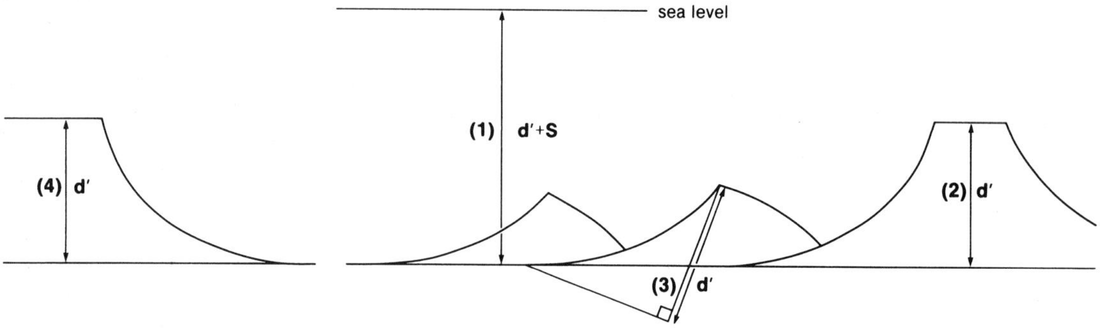

FIG. 5. Techniques for measuring actual depth to decollement: (1) by subtraction of thermal subsidence S_t; (2) from the height of major basement horsts; (3) by reconstructing the initial shape of large basement fault blocks; (4) by direct measurement from an unstretched graben margin.

$d' + S_t$ can be measured directly from sea-level (1) and d' obtained using the curves in Fig. 4, or d' can be reconstructed from large basement blocks whose bounding faults did not intersect above the sole fault (2, 3), or from the overall form of the graben (4). Using equation 5, depth to decollement can also be calculated in uncompensated basins from the decompacted thickness of the syn-rift sediments and a seismically measured β value (5). In several published cross-sections, d' appears to be close to the level of no vertical motion or notional depth to decollement (Table 3). In others, d' is either significantly deeper or shallower. This suggests that there is no mechanical reason for a sole fault to lie at the notional depth to decollement, in either compensated or uncompensated basins, and so in the general case we should expect sole faults in compensated basins to move vertically during stretching.

TABLE 3. *Estimates of the depth in kilometres to a physical decollement, based on seismic data.*

Area	d'	$d' + S_t$	β	method
North Sea (Beach 1984)	2	4.2	1.4	(4), (5)
West Orkney Basin (Brewer & Smythe 1984)	15–18		1.6	(2), (3)
Inner Moray Firth Basin	20–25		1.08	(5)
(Barr 1985)	>10		1.08	(2), (4)
Bay of Biscay	5			(3)
(de Charpal *et al.* 1978)	4	9		(1)
Armorican margin	6	9	2.5	(1)
(Montadert *et al.* 1979)	7		2.5	(3)
Aegean Sea (Jackson & McKenzie 1983)	*c.* 10			(8)
Sevier Desert, USA (McDonald 1976)	3		1.5–2	(2)
Snake Range, USA	6–7		6	(6)
(Miller *et al.* 1983)	1.1		6	(7)

Notes: Method of calculation (cf. Fig. 5 for (1) to (4)): (1) $d' + S_t$; (2) large horsts; (3) reconstructing fault blocks; (4) graben margin; (5) area-balance; (6) pre-rifting stratigraphic thickness in upper plate (a version of (3)); (7) present-day stratigraphic thickness of upper plate; (8) earthquake foci. Methods (1), (4), (5), (7) and (8) measure the present-day depth to decollement. Methods (2), (3) and (6) reconstruct the initial depth to decollement.

Vertical movement of the sole fault during extension

In practice, the physical depth to decollement will often be controlled by externally imposed factors. The 'sole fault' may correspond to a brittle–ductile transition (e.g. Miller *et al.* 1983) and so broadly to an isotherm within the crust, it may be localized in a weak horizon (e.g. evaporites) or at a basement/cover interface, or it may reactivate a pre-existing low-angle fault. In all these cases, it is likely that the sole fault, once initiated, will remain at essentially the same geological horizon during deformation. This horizon will, however, move passively upwards or downwards as isostatic equilibrium is maintained. Figure 6 represents the case where initial depth to decollement d_i is shallower than the notional depth d_n. As before, $z = d_n \times (1 - 1/\beta)$, but that part of the crust lying between d_i and d_n deforms homogeneously so that the decollement subsides to d'. Clearly, $\Delta d = (d_n - d_i) \times (1 - 1/\beta)$ and so:

$$d' = d_i + \Delta d = d_i + d_n - d_i - \frac{d_n}{\beta} + \frac{d_i}{\beta}$$

$$= d_n + \frac{(d_i - d_n)}{\beta}. \qquad (6)$$

The same equation applies if the sole fault is deeper than d_n.

If a sole fault can be observed on seismic records, then from a knowledge of β, the original (pre-stretching) depth to decollement can be calculated from:

$$d_i = d_n + \beta \times (d' - d_n), \qquad (7)$$

d_n being selected according to the nature of the basin fill. For large values of β, d_i is strongly dependent on the chosen value of d_n, which in turn depends on the detailed nature of the sedimentary fill. For this reason it is best to measure the initial depth to decollement directly, using methods (2) and (3) of Fig. 5.

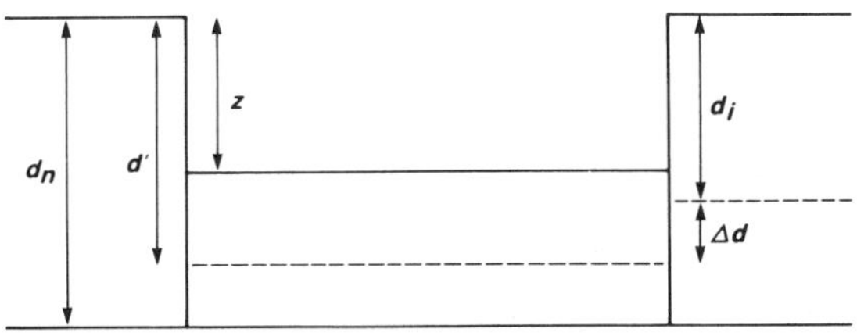

FIG. 6. Construction to show how a physical decollement (broken line) moves relative to the notional decollement d_n during stretching (see text).

In the Bay of Biscay cross-section quoted in Table 3, the volume of the sedimentary fill is insufficient, in relation to the volume of the intervening basement blocks, to accommodate Le Pichon & Sibuet's (1981) β estimate of 2.5. This suggests that, at least in the later stages of extension, the sediment/water interface lay well below sea-level and the system was in part water-loaded. This emphasizes the general point that the density of a sedimentary basin fill will vary with β—e.g. in the early stages of extension, uncompacted sediments will dominate and d_n will have a relatively low value. Even if the decollement surface initiated at the notional depth d_n, the value of d_n will change through time and the decollement surface will either have to switch to this new depth or, more likely, move vertically to maintain isostatic equilibrium. Because such vertical movement displaces the decollement surface and the top basement interface by the same amount, it is geometrically impossible for the physical decollement surface to track the notional depth to decollement, and so the area-balance constraints required for section-balancing cannot be maintained during stretching.

The effect of sediment compaction

An attempt will now be made to semi-rigorously model the instantaneous stretching phase for a sediment-filled, isostatically compensated basin. As the basin extends, early-deposited sediments will be buried and compacted and the mean density of the sediment column (ϱ_s) will increase. At any point during the basin evolution, the notional depth to decollement d_n will be controlled by the density of the basin fill. This in turn depends upon the mean height of the sediment column and the type of sediment involved. For this simple model, the uneven horizontal distribution of sediments implied by an irregular top basement interface will be ignored, so that the basin fill is represented by a sedimentary layer of uniform thickness S_i. An exponential compaction relationship will be assumed, of the form:

$$f = f_0 \times exp(-k \times z), \qquad (8)$$

where f is the porosity at depth z, f_0 porosity at sea-level and k a constant. By a procedure analogous to that followed by Christie & Sclater (1980), it can be shown that:

$$\varrho_s = \varrho_{sg} - \left(\frac{(\varrho_{sg} - \varrho_w) \times f_0 \cdot (1 - \exp(-k \times S_i))}{k \times S_i} \right). \qquad (9)$$

Taking Sclater & Christie's (1980) 'shaly sand' relationship ($\varrho_{sg} = 2680$ kgm^{-3}, $\varrho_w = 1000$ kgm^{-3}, $f = 0.56 \exp(-0.39z)$):

$$\varrho_s = 2680 - \left(\frac{940.8(1 - \exp(-0.39S_i))}{0.39S_i}\right). \quad (10)$$

(Depths are given in kilometres in this case).

For a given value of β, d_n can be obtained iteratively: S_i and hence ϱ_s are calculated for a reasonable value of d_n (e.g. 7 km), and the new value of ϱ_s provides a new estimate of d_n, which yields new values of S_i and ϱ_s, etc. The results of this calculation are displayed in Fig. 7, together with curves relating to footwall uplift which will be described later. d_n increases from 4.95 to 9.53 km, and lies in the range 7–8 km for moderate values of β. ϱ_s increases from 1740 to 2430 kgm^{-3}, and is close to 2200 kgm^{-3} for $\beta = 2$. To a first approximation, the curve representing the mean depth to basement can be used to predict β from a knowledge of the average thickness of syn-rifting sediments, corrected for post-rifting compaction. This is preferable to simply assuming a constant sediment density. For a given basin, a more accurate result could be obtained by constructing a similar set of curves for a local density/depth relationship derived from well data, or for a local extension/subsidence relationship derived from seismic and/or well data.

The behaviour of three physical decollement surfaces of initial depth 2.5, 4.95 and 10 km has also been investigated. Equation 6 was solved numerically by determining d_i after a small increment of $(1 - 1/\beta)$, using this as d_i for the next increment and selecting a new value for d_n from Fig. 7. The resulting curves are plotted on Fig. 7. Note how even the physical decollement which initiated at the notional depth to decollement departs from this depth as the basin evolves, and how the shallowest physical decollement only converges on the notional depth at very large β.

Footwall uplift

Geometric models

Vertical movement of the sole fault provides an additional mechanism for footwall uplift within a stretched basin, beyond those provided by lateral heat flow (Cochran 1983), inhomogeneous stretching (Royden & Keen 1980; Hellinger & Sclater 1983), lithospheric flexure (Watts *et al.* 1982), local isostatic effects (Jackson & McKenzie 1983) or area-balance constraints (Barr *et al.* 1985). The first four mechanisms invoke modifications to the simple stretching model and can generate absolute uplift (above sea-level); the last mechanism only accounts for differential subsidence and applies to basins which do not maintain local isostatic equilibrium.

Consider a residual, unfaulted basement block resting on the sole fault. If isostatic equilibrium is maintained on a horizontal scale greater than that of the basement block, the block will either subside or be uplifted, depending on whether the initial depth to decollement d_i is shallower or deeper than the notional depth d_n. In general, a sole fault controlled by a brittle/ductile transition at 300–350°C (Chen & Molnar 1983; Sibson 1983) will lie at a depth of 10–20 km and so be deeper than d_n, therefore footwall uplift should be the rule. Smaller blocks may experience either net uplift or net subsidence, depending on how their detailed geometry interacts with that of the system of fault planes. The (signed) subsidence, averaged over the basin as a whole, should correspond to that predicted by the lithospheric stretching model (although strictly, a correction should be made for the additional isostatic load of emergent blocks). From Fig. 7, a sole fault initially at 10 km depth would be uplifted by ≈ 800 m for $\beta = 1.25$ and ≈ 1500 m for $\beta = 2$. This mechanism may explain the persistent uplift of major, internally unfaulted basement highs within extensional basins, such as the Halibut Horst within the Moray Firth basin (Johnson & Dingwall 1981). Gravity data indicate that even this 100×25 km horst is not locally compensated.

Uplift of minor blocks bounded by listric normal faults cannot readily be assessed quantitatively. However, footwall uplift in a system of rotated planar ('domino') fault blocks is amenable to calculation (Fig. 8—cf. Wernicke & Burchfiel 1982; Le Pichon *et al.* 1983). Assuming that all the 'dominoes' are the same size, and that some area-conserving, megascopically ductile process operates at the base of each (so that voids do not open up), it is only necessary to consider whether asperities on the top basement interface rise above sea-level. Figure 9 represents the geometry of this interface after stretching by a factor:

$$\beta = \frac{B'A'}{B'C} = \frac{c'}{a'} = \frac{\sin\phi_i}{\sin\phi} = \frac{\sin\phi_i}{\sin(\phi_i - \theta)}.$$

ϕ_i is the initial dip of a fault plane, ϕ its present-day dip and θ the dip of the top basement interface (initially horizontal). AB represents the

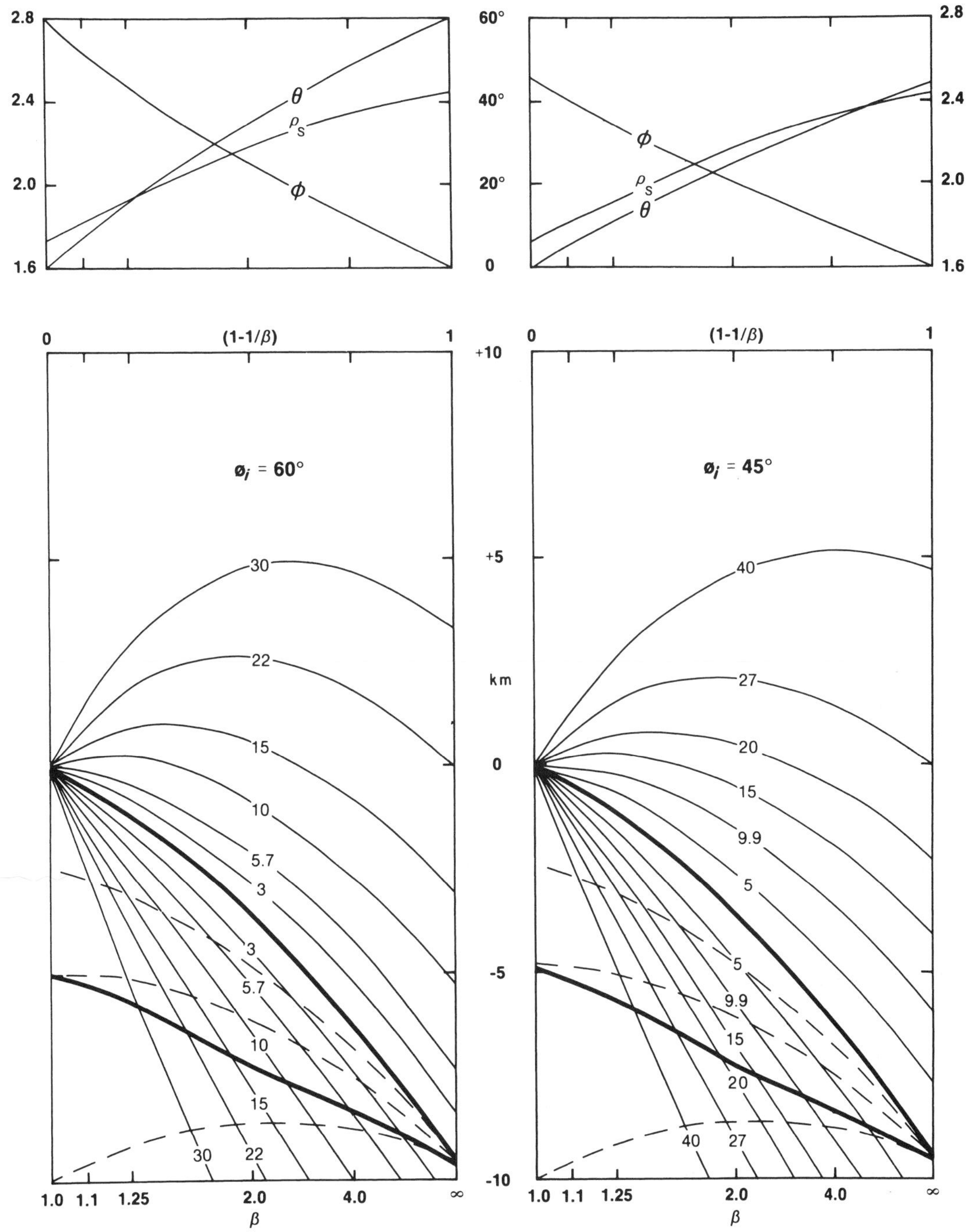

FIG. 7. Model set of curves representing the evolution of an instantaneously stretched, isostatically compensated, sediment-loaded basin with increasing β. Upper-crustal extension is accommodated by rotated planar ('domino') faults. Initial fault-plane dip ϕ_i is assumed to be (a) 60° or (b) 45°. The curves for sediment density, overall subsidence and depths to decollement (notional and physical) are independent of assumptions about the type of upper-crustal extension.

The sediment column is assumed to compact according to Sclater & Christie's (1980) 'shaly sand' relationship:

$$f = 0.56 \exp(-0.39z),$$

where f is porosity at depth z. The sediment grains are assigned a density of 2680 kgm^{-3}, and the pore space filled with water. The mean density of the sediment column is plotted as ϱ_s. The vertical scale is in specific gravity, e.g. 2.0 = 2000 kgm^{-3}. The upper heavy curve shows the mean depth to the basin floor, while the lower heavy curve represents the notional depth to decollement (see text). The broken curves show average basement elevation, such that excess area $AA'D$, is balanced by the deficiency ABC. Clearly, $\Delta h = 0.5h'$. Since

$$\sin\theta = \frac{h'}{a'},$$

$$h' = a' \times \sin\theta = \frac{c'}{\beta} \times \sin\theta.$$

The maximum footwall elevation above the mean basin floor is given by:

$$\Delta h = 0.5h' = 0.5a' \times \sin\theta = \frac{0.5c' \times \sin\theta}{\beta}, \quad (11)$$

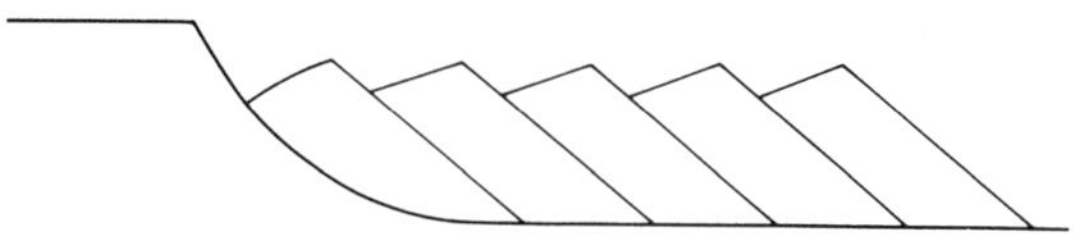

FIG. 8. Schematic representation of a system of rotated planar ('domino' or 'pack of cards') detached normal faults. The fault planes rotate with bedding as extension progresses, and the lower portion of each fault block must undergo ductile deformation. All faults move simultaneously and a listric normal fault is required to separate the system from undeformed basement.

where c' and a' are respectively the present-day and original horizontal separations between major 'domino-bounding' faults.

If isostatic equilibrium is maintained on a horizontal scale greater than the distance between hanging wall and footwall cut-offs, the physical depth to decollement can be ignored since the *shape* of the top basement surface is unaffected by vertical uplift or subsidence. Relative to sea-level, footwall uplift z_f is given by $\Delta h - S_i$; if $\Delta h > S$, footwall blocks will be emergent. The corresponding hanging wall subsidence z_h is given by $\Delta h + S_i$. At large extensions, the space problem at the base of the 'dominoes' becomes more severe. Ultimately, the hanging wall intersects the sole fault and the model is clearly unrealistic. Extension can only continue if the hanging wall blocks slide as individual riders on the sole fault (e.g. the left-most fault-blocks in Fig. 2—cf. Gibbs 1984, fig. 11) or if new, high-angle faults are initiated (cf. Proffett 1977). In the former case, z_f will remain constant at the value which obtained when individual riders became detached from one another.

Since both S_i and θ are functions of β, there is a critical initial fault spacing a'_{crit} above which footwall uplift can give rise to islands:

$$a'_{crit} = \frac{2S_i}{\sin\theta}. \quad (12)$$

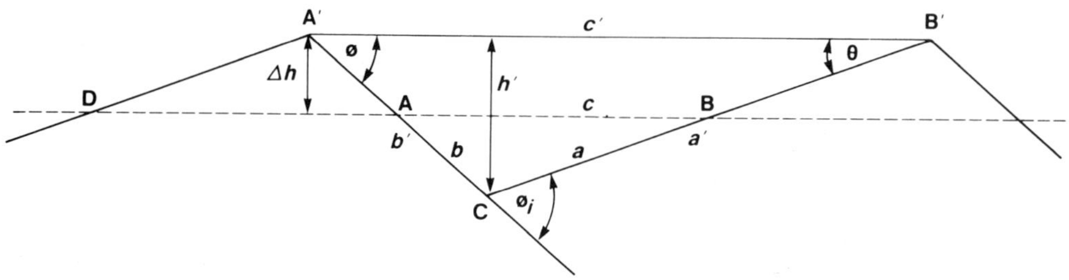

FIG. 9. Geometry of the top basement interface in a system of 'domino' faults (cf. Fig. 8). The broken line represents the mean depth to the basin floor, ϕ_i is the initial fault-plane dip, ϕ the present-day fault-plane dip, θ the dip of bedding, a' the initial fault spacing and c' the present-day fault spacing. For further discussion, see text.

FIG. 7 cont.

the behaviour of three different physical decollement surfaces. Note that they converge on the mean basin floor at large β, rather than on the notional depth to decollement,and how even the physical decollement which initiated at the notional depth, departs from it as extension proceeds. The 'domino-fault' model breaks down where the physical decollement surface appropriate to the basin under discussion intersects the hanging wall cut-off, e.g. at $\beta = 2$ for a sole fault initiated at 10 km depth with an initial 'domino-fault' spacing of 20 km.

The upper panels show how ϕ (fault-plane dip) and θ (bedding-plane dip) vary with β for 'domino-fault' models with $\phi_i = 60$ and 45°. The labelled curves in the lower panels represent footwall uplift/subsidence (upper set) and hanging wall subsidence (lower set) for the indicated initial fault-spacings (labelled in kilometres). See text for details of their derivation. No correction has been made for the additional isostatic load of emergent footwalls.

The value of a'_{crit} for infinitesimal extension can be obtained from a consideration of Fig. 10. For very small values of β, angle $B'DC$ approaches a right angle, h^* approaches h' and $\tan\phi \simeq \frac{h'}{e}$ Thus:

$$\beta = \frac{(a'+e)}{a'} \simeq 1 + \left(\frac{h'}{a' \times \tan\phi}\right),$$

and

$$\Delta h = \frac{h'}{2} \simeq \frac{(\beta - 1) \times a' \times \tan\phi}{2} .$$

Since $S_i = d_n \times \left(1 - \frac{1}{\beta}\right) = \frac{(\beta-1) \times d_n}{\beta}$,

then $\Delta h = S_i$ implies that:

$$a'_{crit} \simeq \frac{2d_n}{\beta \times \tan\phi} . \quad (13)$$

As $\beta \rightarrow 1$, $\phi \rightarrow \phi_i$ and

$$a'_{crit} \rightarrow \frac{2d_n}{\tan \phi_i} . \quad (14)$$

At infinite extension, $\phi \rightarrow 0$ and $\theta \rightarrow \phi_i$, so

$$a'_{crit} = \frac{2d_n \times (1 - 1/\beta)}{\sin\theta} \rightarrow \frac{2d_n}{\sin \phi_i} . \quad (15)$$

In general, a'_{crit} increases with β, so footwall uplift should be most widespread at low extensions, although the magnitude of the uplift on an individual fault will initially increase with β. Uplift is also favoured by a high initial fault-plane dip ϕ_i and a low-density basin fill, e.g. water. The initiation of new, high-angle faults (cf. Proffett 1977) in a sediment-filled basin will reset β and ϕ_i and provoke renewed uplift of both basement and lithified sediments.

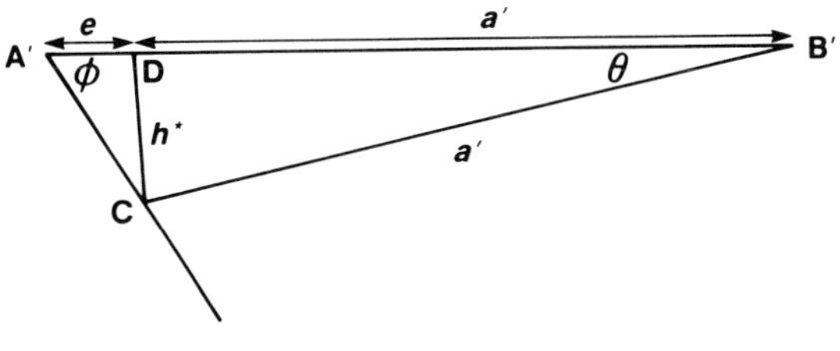

FIG. 10. Construction showing the geometry of a 'domino' fault at infinitesimal extension (labels correspond to those in Fig. 9, but the value of θ has been exaggerated for clarity). See text for further discussion.

In Fig. 11, critical values of a' for infinitestimal and infinite stretching are plotted against ϕ_i. For a particular density of basin fill, they define three fields: one in which footwalls always subside, one in which they are first uplifted then subside below sea-level with increasing extension, and one in which they are uplifted then subside, but always remain above sea-level. The curve for $\beta = 4$ is also shown in Fig. 11, as an alternative to that for $\beta = \infty$, since beyond this point oceanic crust may form and stretching cease (Le Pichon *et al.* 1982). Corresponding values of c'_{crit} define the present-day spacing of islands produced by this mechanism and are in the range 5–25 km for moderate values of β and ϕ_i. These figures are reasonably consistent with the present-day spacing of islands in the Aegean Sea (Jackson *et al.* 1982a, b) and of major fault blocks in the Brent province of the North Sea (Bowen 1975; Hay 1978) and in the Bay of Biscay (de Charpal *et al.* 1978), all areas where footwall uplift and erosion are well documented.

Figure 12 shows representative uplift/subsidence paths for a water-loaded basin. Where $a' < 3.8$ km, footwalls subside more slowly than the basin as a whole (which follows the curve for $a' = 0$). Where 3.8 km $< a' <$ 7.6 km, footwalls are initially uplifted by a few hundred metres then subside below sea-level. Where $a' > 7.6$ km, footwalls are uplifted by between a few

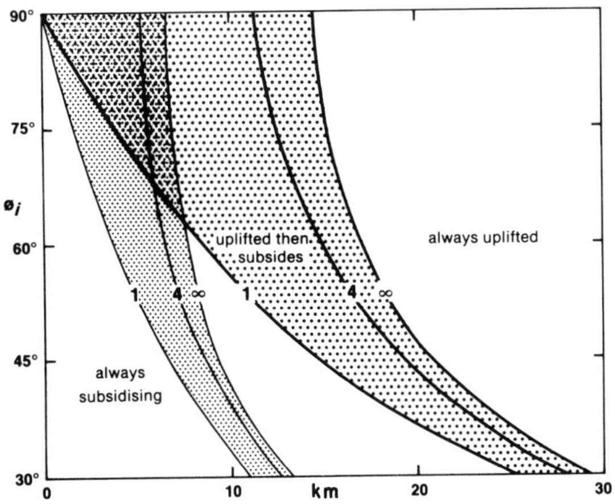

FIG. 11. Graph showing critical values of initial 'domino-fault' spacing for footwall uplift to occur, for initial fault-plane dips ϕ_i in the range 30 to 90°. Two sets of curves are shown, one for water loading (light shading) and one for sediment loading ($\varrho_s = 2200$ kgm^{-3}—dark shading). In each case, the three curves represent critical spacings for $\beta = 1$, $\beta = 4$ and $\beta = \infty$. The shaded and unshaded areas define three fields of contrasting footwall behaviour, which are labelled for the sediment-loaded case. See text for further discussion.

hundred metres and a few kilometres, then subside slightly but always remain above sea-level. Large-scale uplift (sufficient to leave a geological record) requires values of a' not much less than 7.6 km, i.e. the value of a'_{crit} for infinite stretching. Otherwise, footwall uplift is severely limited in both magnitude and duration (cf. the curve for $a' = 5$ km). Sediment-loaded curves will have the same general shape, if both the subsidence scale and a' are expressed in terms of d_n—e.g. for $\varrho_s = 2200$ kgm^{-3}, these will be increased by a factor of 2.206.

The curves in Fig. 12 can be used qualitatively to model the effects of footwall erosion. If erosion is 100% efficient, the footwall cut-offs should sink below sea-level at the apex of each curve rather than at the zero-crossing. However, because the topographic crest of the fault block migrates updip as it is eroded, final subsidence actually takes place where the curve is tangential to one of a family of straight lines radiating from sea-level at $\beta = \infty$. For example, Fig. 13 represents a set of such curves constructed for a basin loaded by sediments with average density 2200 kgm^{-3}—footwalls are actively eroded over that part of each curve where $z_f = 0$, and fault-blocks with $a' > 17$ km are emergent to $\beta = \infty$.

The additional isostatic load imposed by emergent footwall blocks can be modelled as follows. Consider a column of width c' compensated within the asthenophere and filled to sea-level (or some other convenient datum) with sediments. If the top basement surface lies at or below sea-level, the areas and densities are as defined in Fig. 14a. Compare this with a column in which a footwall island of cross-sectional area A_3 rises above sea-level, but all other areas remain unchanged and the base of the lithosphere remains at the same depth (Fig. 14b). This is equivalent to removing basement of area A_3 from below sea-level, and to conserve area replacing it by sediments of area A_3. This column is now out of isostatic equilibrium with that of Fig. 14a. To restore equilibrium the column must subside, displacing a quantity of asthenosphere A_5 (Fig. 14c). An equivalent area of sediments is deposited. Balancing columns (a) and (c):

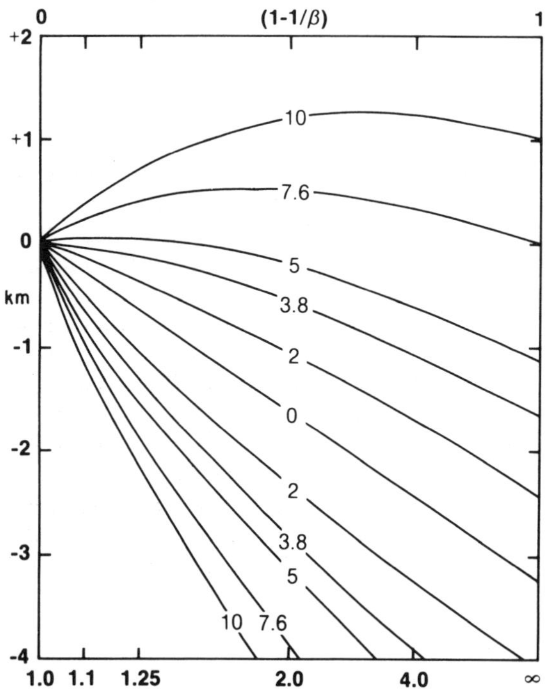

FIG. 12. Set of curves showing footwall uplift (upper set) and hanging wall subsidence (lower set) in an instantaneously stretched, isostatically compensated, water-loaded basin assuming upper-crustal extension to be accommodated by 'domino-faulting' with initial fault-place dip $\phi_i = 60°$. The labels on the curves correspond to the initial fault-plane spacing a' in kilometres. The curve for $a' = 0$ represents the mean basin floor. No correction has been made for the additional isostatic load of emergent footwall blocks.

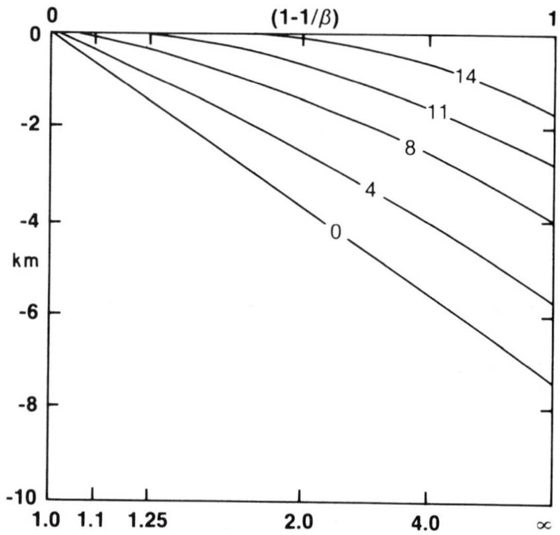

FIG. 13. Set of curves showing footwall uplift in an instantaneously stretched, isostatically compensated basin, loaded by sediments of density 2200 kgm^{-3}, assuming complete erosion of emergent footwalls. Upper-crustal extension is by 'domino-faulting' with initial fault-plane dip $\phi_i = 60°$. The labels on the curves correspond to the initial fault-plane spacing a' in kilometres. The curve for $a' = 0$ represents the mean basin floor. Active erosion takes place where the footwalls are at sea-level.

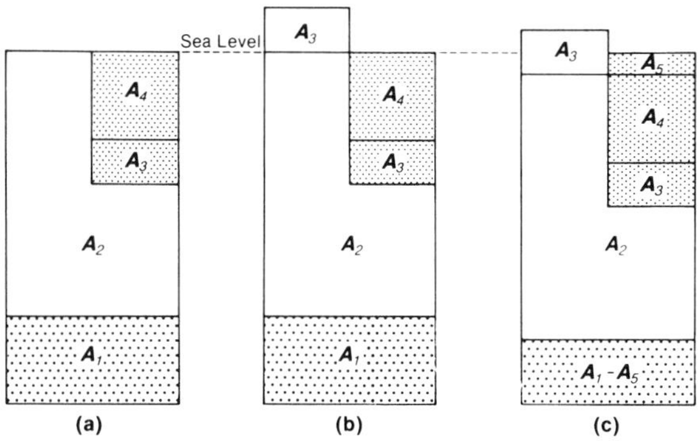

FIG. 14. Construction used to determine the additional isostatic load imposed by an emergent footwall block. Each column is of width c' (the wavelength of the 'domino' faults in Fig. 15), and areas A_1 to A_5 are the same in each case. Pre-stretching lithosphere (basement) of density ϱ_l is unornamented, the asthenosphere of density ϱ_a is represented by coarse stipple and sediments of density ϱ_s by fine stipple. Column (a) is in isostatic balance (distributed over a width c') and reaches to sea-level. Column (b) is identical but for an island of area A_3 rising above sea-level and a correspondingly greater sediment fill. It is out of isostatic equilibrium with column (a). Column (c) is identical to column (b) except that it has been allowed to sink to reach isostatic equilibrium with column (a), displacing asthenosphere of area A_5 and permitting deposition of additional sediments.

$$A_1\varrho_a + A_2\varrho_l + A_3\varrho_l + A_4\varrho_s = A_1\varrho_a - A_5\varrho_\alpha + A_2\varrho_l + A_3\varrho_l + A_3\varrho_s + A_4\varrho_s + A_5\varrho_s,$$

$$\text{i.e. } A_5 = \frac{A_3\varrho_s}{(\varrho_a - \varrho_s)} \; . \qquad (16)$$

Figure 15 shows a situation equivalent to that of Fig. 14b, where an island has risen above sea-level but no allowance has yet been made for its additional isostatic load. From Fig. 15,

$$\frac{AF}{AG} = \frac{BC}{DE}, \text{ i.e. } \frac{z_f}{z_f + S_i} = \frac{BC}{0.5c'},$$

so

$$BC = \frac{c' \times z_f}{2(z_f + S_i)},$$

and the area of triangle ABC (equivalent to A_3 of Fig. 14) is

$$\frac{c' \times z_f^2}{4(z_f + S_i)} \; . \qquad (17)$$

Thus

$$A_5 = \frac{\varrho_s \times c' \times z_f^2}{4(\varrho_a - \varrho_s) \times (z_f + S_i)} \; .$$

Treating A_5 as a uniform sediment blanket of width c', i.e. ignoring the slight interruptions caused by the islands, the additional subsidence caused by the emergent footwalls is given by:

$$z_a = \frac{\varrho_s \times z_f^2}{4(\varrho_a - \varrho_s) \times (z_f + S_i)} \; . \qquad (18)$$

This component of subsidence is typically less than 10% of z_f, and even in the most extreme cases, only reaches $\approx$ 25% of z_f. Given the other simplifications involved, correcting the curves was not considered worthwhile.

In a real basin, the density of the sedimentary fill will evolve through time, increasing as deeper sediments are compacted. This will amplify the subsidence component S_i at large β, and shorten both the amount and the duration of footwall uplift. For example, consider the isostatically compensated basin discussed in the previous section and represented in Fig. 7. Values of footwall uplift z_f and hanging wall subsidence z_h have been calculated for various initial fault spacings

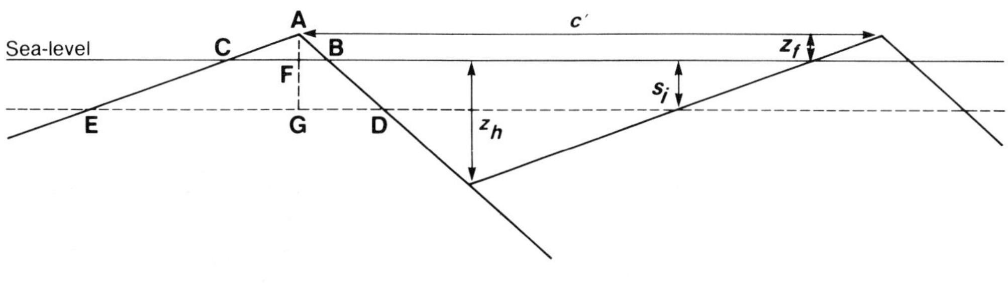

FIG. 15. Geometry of an emergent footwall block ABC. S_i is the initial subsidence due to stretching, c' the present-day fault-spacing, z_f the magnitude of footwall uplift and z_h the magnitude of hanging wall subsidence. See text for further explanation.

a' and for initial fault-plane dips ϕ_i of 45 and 60°. These curves are plotted in Fig. 7, together with those for θ and ϕ, the dips of bedding and the dips of 'domino-bounding' faults after stretching. No correction has been made for the isostatic load of emergent blocks.

Where $\theta_i = 60°$ (Fig. 7a), an initial fault spacing a' of $\approx$ 10 km is sufficient to yield a few hundred metres of footwall uplift, until subsidence ensues at around $\beta = 1.5$. A fault spacing of less than 5.7 km results in continuous subsidence, although the footwall subsides much less rapidly than the hanging wall. Values of a' significantly greater than 10 km result in major footwall uplift (>1 km). In contrast to the accelerating footwall subsidence, hanging wall subsidence is almost linear in $(1 - 1/\beta)$, and even moderate values of a' (e.g. 10 km) permit the hanging wall cut-off to be twice as deep as the average basin floor. This observation emphasizes the importance of using *mean* rather than *maximum* basin depth or sediment thickness in any isostatic or area-balance calculation.

The curves for $\phi_i = 45°$ (Fig. 7b) are similar to those in Fig. 7a, but a' is increased by about 50%. In both cases, geologically reasonable fault spacings permit widespread footwall uplift, which should be particularly evident at moderate values of β.

At high extensions, renewed steep faulting may rejuvenate the system and promote additional uplift. In such cases, the curves in Fig. 7 cannot strictly be re-applied by returning to $\beta = 1$. This is because compaction of the underlying column will create space for new sediments, over and above that produced by extension, and d_n will be depressed to greater depth.

To facilitate comparisons between basins, a final set of curves has been constructed in dimensionless form, referred to the notional depth to decollement (Fig. 16). They emphasize the fact that uplift of footwalls above sea-level requires a fault spacing several times greater than the notional depth to decollement. Figure 16 can also be applied to an uncompensated basin in which subsidence is controlled by the depth to a physical decollement surface—d_n is simply replaced by d', the actual depth to decollement. In that case, the forbidden zone places a strict limitation on the amount of extension which can be accommodated before the hanging wall intersects the sole fault, and d' is independent of the nature of the basin fill.

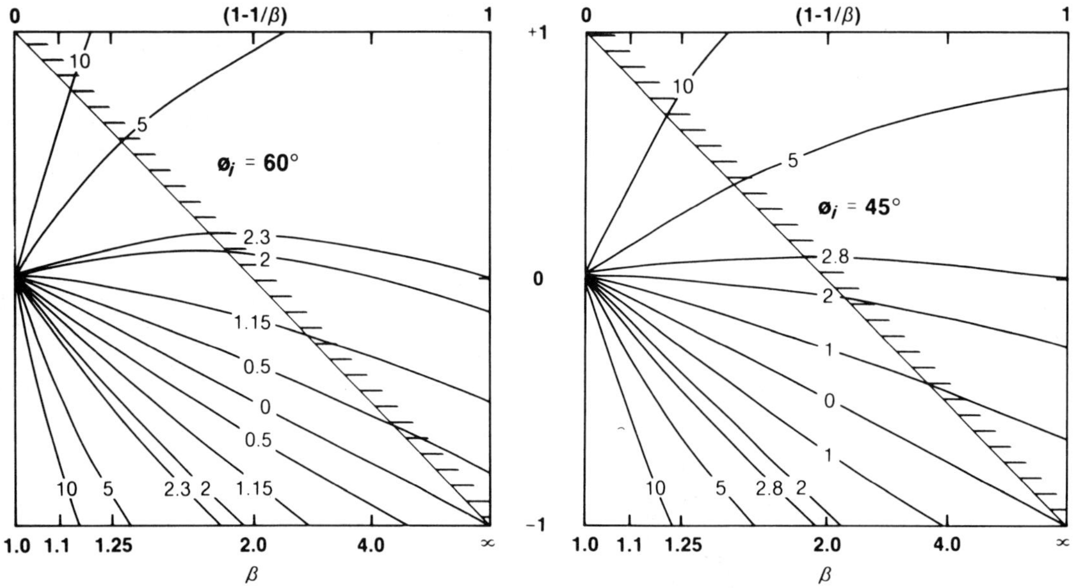

FIG. 16. Dimensionless set of curves showing footwall uplift (upper set) and hanging wall subsidence (lower set) in terms of the notional depth to decollement d_n, for a system of 'domino' faults with initial fault-plane dip $\phi_i = 60°$ (a) and 45° (b). The labels on the curves correspond to the ratio between the initial fault-spacing a' and the notional depth to decollement d_n. The curve for $a' = 0$ represents the mean basin floor. The curves can also be applied to uncompensated basins, where the notional depth to decollement is replaced by the depth to the physical sole fault. In that case, the 'forbidden zone' (to the right of the hatched line) sets an upper limit to the amount of extension which can be accommodated by 'domino' faulting before the hanging wall cut-off intersects the sole fault.

Basins produced by listric normal faulting can also be modelled, to the extent that the geometry of the top basement interface corresponds to that of Fig. 9. In the simplest case (Fig. 17), c'/a' will provide a slight over-estimate of β, but after allowing for this, a similar magnitude of uplift would be expected. Note that in Fig. 17, ϕ represents the dip of the fault plane at relatively shallow levels (down to the hanging wall cut-off), rather than at depth. If minor faulting is present within the hanging wall, this will tend to fill in the basin, at the expense of the height of the next footwall.

Examples from the geological record

North Sea

Inner Moray Firth

The Inner Moray Firth basin forms part of the North Sea graben system (Whiteman *et al.* 1975; McQuillin *et al.* 1982), and was produced by late Jurassic and Early Cretaceous extension. It is not isostatically compensated, and developed by listric normal faulting above a decollement whose depth d' is certainly greater than 10 km (the maximum penetration of available seismic reflection data) and probably lies at 20–25 km below sea-level (Barr 1985). Although footwalls within the basin have undergone relative uplift which is broadly proportional to the magnitude of fault displacement (Jackson & McKenzie 1983; Barr *et al.* 1985), there are no indications of significant emergence during rifting (Linsley *et al.* 1980). Present-day spacing, c', between major half-graben-bounding faults is 20–30 km, and the basin is filled by well-consolidated Jurassic and Cretaceous sediments. Barr (1985) determined a post-Triassic stretching factor β of ≈ 1.1 from reflection seismic data, and major faults have steep dips (45–60°) in their upper few kilometres. Initial fault spacing a' is 18–27 km, i.e. 0.72–1.35 in dimensionless terms (taking the most extreme combinations of c' and d'). Assuming an initial fault-plane dip of 60° (or 45°) the model predicts a final footwall depth of −1000 m to +75 m (−1500 to −800 m) and a maximum basin depth of 3.5 to 3.7 km (2.4 to 3.0 km) at the hanging wall cut-off. The highest values of c' permit a few tens of metres of true uplift (above sea-level) during extension. In the other cases, footwall subsidence is lowest at low values of β. These figures agree reasonably well with observations. Released exploration well 12/23-1, drilled by Total on a footwall block, penetrated ≈ 750 m of syn-rifting sediments, most of which are of lower Cretaceous age and represent the later stages of rifting. A full Jurassic sequence is present, although the Lower to Middle Jurassic is thin and includes some hiatuses indicative of emergence or very shallow marine conditions. Jurassic and lower Cretaceous sediments typically attain 3 km in major half-grabens and locally reach 4 km (Chesher & Bacon 1975; McQuillin *et al.* 1982; Barr 1985). Thus the model successfully explains the accelerating footwall subsidence but broadly linear hanging wall subsidence, and attributes the general absence of footwall uplift to an abnormally deep decollement.

East Shetland basin

The East Shetland basin or Brent Province forms the northwestern part of the Viking graben and comprises a system of down-to-the-E tilted fault blocks which were active during the late Jurassic and perhaps the Early Cretaceous (Hay 1978; Johnson & Dingwall 1981). The sedimentary fill is isostatically compensated by thin crust (Donato & Tully 1981) and the basin was probably produced by lithospheric stretching. About 2.5 km of post-rifting sediments are present, suggesting a β factor of ≈ 1.5, in agreement with Donato & Tully's

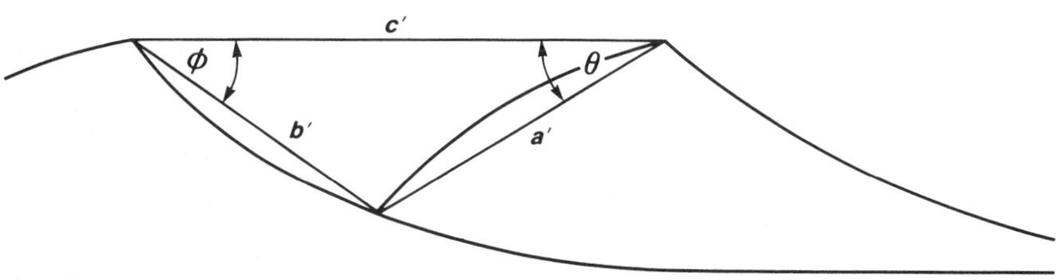

FIG. 17. Schematic representation of a system of listric normal faults. The curved top basement interface is approximated by the straight line segments a', b' and c'. To estimate footwall uplift, the listric normal faults are represented by a set of 'domino' faults with bedding-plane dip θ, fault-plane dip ϕ, and extension ratio $\beta = c'/a'$.

(1981) conclusions. The Brent oil field (Bowen 1975) occupies a footwall block which was emergent during extension. Upper Jurassic and lowermost Cretaceous sediments are absent from the crest of the structure, and the pre-rift Lower to Middle Jurassic section is deeply eroded. The amount of post-Middle-Jurassic erosion (600–1000 m) was estimated by taking Bowen's (1975) cross-section and projecting the top reservoir horizon beyond the unconformity until it intersected the main bounding fault.

In the Brent area, fault spacing c' is ≈ 25 km (Hay 1978) so a' is ≈ 17km. From Fig. 7a ($\phi_i = 60°$), footwall uplift $z_f = 1.5$ km and hanging wall subsidence z_h is in excess of 5 km. The peak of the uplift curve occurs at $\beta = 1.5$, so even if an erosional model like that of Fig. 13 were preferred, footwalls would be emergent throughout the stretching episode. An erosional model would require stretching to continue until the Albian/Aptian, when marine deposition resumed over the crest of the structure, but a model in which islands rose a significant height above sea-level would permit extension to cease somewhat earlier. Fig. 7b ($\phi_i = 45°$) predicts rather less uplift (≈ 500 m), but the curve for $a' = 17$ km remains above sea-level at $\beta = 1.5$.

The Heather oil field (Gray & Barnes 1981) lies 70 km W of Brent, and although the footwall block has suffered erosion, a thinned but continuous sequence of uppermost Jurassic and lower Cretaceous sediments is present. Fault spacing in the Heather area is 15–20 km (Hay 1978), i.e. $a' = 10$–13 km. For $\phi_i = 60°$ (Fig. 7a), a β value of 1.5 is past the crest of the uplift curve, and on an erosional model, footwalls would subside during the later stages of extension. The effect is even more marked for $\phi_i = 45°$ (Fig. 7b).

Although fault spacing in the Brent Province is similar to that in the Inner Moray Firth, isostatic compensation means that subsidence is controlled by a relatively shallow notional depth to decollement rather than a deep physical sole fault, so favouring footwall uplift. There is a suggestion (in comparing Heather and Brent) that variations in the severity of footwall erosion can be attributed to local variations in fault spacing.

Aegean Sea

The area around the Aegean Sea has undergone recent (post-middle Miocene) extension, with a stretching factor β close to 2 (Jackson & McKenzie 1983, and references therein). Islands within the Aegean correspond to the footwall blocks of major active normal faults (Jackson *et al.* 1982a). Jackson & McKenzie (1983) attributed the absolute uplift (above sea-level) of these footwall blocks to a local isostatic mechanism (after Heiskanen & Vening Meinesz 1958). Their geometry can also be described in terms of the uplift model presented in this paper. Jackson & McKenzie (1983) state that the present-day fault spacing c' is 30–40 km, so the initial spacing a' was 15–20 km. They suggest that footwall uplift, z_f, is 500–1000 m and total fault throw 5–10 km, so hanging wall subsidence z_h is 4.5–9 km. Because the Aegean contains deep-water troughs (up to ≈ 1 km), subsidence must be partly water-loaded. The notional depth to decollement d_n should therefore be intermediate between those for sediment-loaded (Fig. 7) and water-loaded (Fig. 12) basins. A reasonable round-number estimate is 6 km, in which case a' corresponds to 2.5–3.5 times d_n, and S_i, the average sediment + water thickness, should be ≈ 3 km. If the fault blocks have a simple 'saw-tooth' geometry, Jackson & McKenzie's (1983) figures imply values for S_i of 2–4 km, and for $\beta = 2$, $d_n = 4$–8 km. From Fig. 16, ($\phi_i = 60°$, $a' = 3 \times d_n$), $z_f = 1.8$ km, $z_h = 4.8$ km, and the total throw on each fault is 6.6 km. For $\phi_i = 45°$, $z_f = 1$ km, $z_h = 4$ km and the total throw is 5 km. These results are reasonably consistent with Jackson & McKenzie's (1983) observations.

The 'forbidden zone' of Fig. 16 does not apply in this compensated basin, since earthquake foci imply that fault planes extend to ≈ 10 km depth (Jackson *et al.* 1982a), and so any physical decollement must be deeper than ≈ 10 km. However, the 'domino' model implies that the fault planes have shallow present-day dips ($\phi = 20$–$30°$), unless an unreasonably high value is chosen for ϕ_i. Evidence from the Corinth area (Jackson *et al.* 1982a) suggests a steeper present-day dip to 10 km depth ($\phi = 45°$), although it is of course possible that β is greater and ϕ lower in the Aegean proper. By analogy with Proffett's (1977) interpretation of the Basin and Range, Jackson & McKenzie (1983) suggested that old fault planes have been rotated to low dips and abandoned, and that new, steep fault planes dominate the topography. A crude assessment of the effects of this process can be made by assuming an initial dip ϕ_i of 60° and a present-day dip ϕ of 45°. From Fig. 7a, $\beta = 1.25$, and so three generations of faulting would be required to accommodate a total extension of $\beta = 2$ (since $1.25^3 = 1.95$). For the most recent set of faults, $c' = 30$–40 km, so $a' =$ 24–32 km, i.e. 4 to 5 times d_n. From Fig. 16, $z_f = 2$–3 km, $z_h = 3.2$–4.2 km, and $S_i = 1.2$ km. This is probably too much uplift, even allowing for reduction by erosion.

Choosing a slightly higher value of d_n (e.g. 7 km, to allow for compaction of sediments deposited during earlier episodes of extension) decreases this uplift figure slightly ($a' = 3.5–4.5 \times d_n$, $z_f = 1.8–2.8$ km, $z_h = 3.2–4.2$ km, $S_i = 1.4$ km).

Major faults in the Corinth area are rather more closely spaced than 30 km (≈ 18 km—Jackson *et al.* 1982a). If a' is set to 15 km for $\beta = 1.25$ and $\phi_1 = 60°$, then $z_f = 900$ m, $z_h = 2.1$ km and $S_i = 1.2$ km; figures in close agreement with local topography. As a final example, consider the Platea area, NE of Corinth. An old erosion surface is uplifted to 300 m at Platea and downfaulted to 900 m below sea-level in the Gulf of Corinth (Jackson *et al.* 1982a, p. 395). The mean depth of 300 m ($= S_i$) suggests that, for this increment of stretching, $\beta = 1.055$. Fault planes dip at 40–45° today, so given the low value of β, the curves for $\phi_i = 45°$ will be used ($\phi = 42°$ for $\beta = 1.055$). For a present-day fault spacing of ≈ 25 km, $a' = 24$ km $= 4 \times d_n$, $z_f = 300$ m and $z_h = 900$ m, in excellent agreement with the observations of Jackson *et al.* (1982a).

Thus the simple model of footwall uplift presented here can be applied in the Aegean area. Note that its success is without prejudice to other, mechanistic models based on an isostatic or elastic response (e.g. Jackson & McKenzie 1983). In its simplest terms, the model is purely geometric and should apply to any isostatically compensated basin with a regular, 'sawtooth-shaped' top basement interface. The breakdown of 'domino-fault' systems at large β, with initiation of new, steeper faults, precludes rigorous application of the model to predict a detailed history of uplift and subsidence versus β.

Armorican margin

Le Pichon & Sibuet (1981) and Le Pichon *et al.* (1983) have shown that the development of the Armorican continental margin is consistent with a model of lithospheric stretching. Extension ratios (β) range from 1.1 to 2.7, in a transect from continent to ocean, and subsidence has been largely water-loaded with a small sedimentary contribution. Le Pichon *et al.* (1983) have suggested that a 'domino-faulting' model can explain the geometry of upper-crustal fault blocks, the initial fault-plane dip ϕ_i being 45°. Le Pichon *et al.*'s (1983) fig. 2, based on Montadert *et al.*'s (1979) depth-converted seismic reflection section, implies a β value of $\approx$ 2.5. On the Armorican margin (unlike the Aegean?), it would appear that simple 'domino-faulting' has been effective even at large β, although as Le Pichon *et al.* (1983) pointed out, it requires substantial ductile strain across a broad brittle–ductile transition, rather than a simple sole fault.

The thermal component of subsidence can be stripped off by applying equation 3, and assuming (following Le Pichon & Sibuet 1981) that S_t has reached 87% of its ultimate value. For water loading:

$$S_t = 3.65\,(1 - 1/\beta) = 2.19 \text{ km.}$$

A ≈ 1 km thick layer of post-rift sediments is present; from Fig. 7, these probably have an average density of ≈ 1900 kgm^{-3}, so the average density of the actual post-rifting column (1 km of sediments + 1.19 km of water) is 1411 kgm^{-3}. By using this new density ($= \varrho_s$) to insert a new value of s into equation 3, and proceeding iteratively, it can be shown that $S_t = 4.34\,(1 - 1/\beta) = 2.6$ km. The mean density of this sediment + water column is 1348 kgm^{-3}. On removing 2.6 km of thermal subsidence, 2.5 km of water + syn-rift sediments remains; from equation 1, $d_n = 4.17$ km, and ϱ_s (the mean density of the sediment + water column during stretching) is 1456 kgm^{-3}. These figures are consistent with the observation that subsidence was largely water-loaded, and with Le Pichon *et al.*'s independent reconstructions (1983, fig. 1).

Present-day fault-spacing c' is 18 km, so $a' = 7.2$ km $= 1.73 \times d_n$. From Fig. 16, this fault-spacing is not quite sufficient to permit footwall uplift. However, if the early stages of extension were water-loaded (Fig. 12), several hundred metres of uplift would be possible, and if β was somewhat lower (e.g. 2), even sediment-loaded uplift would be possible. Given the uncertainties involved, the uplift model is reasonably consistent with Montadert *et al.*'s (1979) cross-section, which shows only minor footwall erosion, and that probably at an early stage in the extension process.

The Armorican data can be used to assess the relationship between the physical depth to decollement d' and the notional depth d_n. Le Pichon *et al.* (1983, fig. 1) provide three reconstructions of the Armorican margin, prior to thermal subsidence, and imply that the initial depth to the 'sole fault' (actually a broad brittle–ductile transition) was 9 km. Their models have $\beta = 1.1$, 1.7 and 2.7, and from equation 6 ($d_n = 4.17$ km), $d' = 8.6$, 7 and 6 km; depths in close agreement with Le Pichon *et al.*'s reconstructions. On Montadert *et al.*'s (1979) cross-section, the physical decollement surface lies at ≈ 9 km below sea-level, and prior to 2.6 km of thermal subsidence, lay at a depth $d' = 6.4$ km. From equation 7, with $\beta = 2.5$ and $d_n = 4.17$ km, its initial depth d_i was 9.75 km. By inverting equation 5, it is simple to demonstrate the inapplicability of simple area-balancing techniques to this area (cf. Gibbs 1983). After stripping off the thermal subsidence, $z = S_i = 2.5$ km, $\beta = 2.5$ and the calculated (notional) depth to decollement is 4.17 km ($= d_n$). This is very different to the physical depth to decollement, which ranges from $\approx$ 6 km to $\approx$ 10 km, depending on the magnitude of β.

Discussion

By comparing the equations for balancing cross-sections above a basal decollement with those for subsidence during lithospheric stretching, a geometric relationship has been identified between detached normal faulting and stretching. For an instantaneously stretched, isostatically compensated basin, area/bed length balancing yields a notional depth to decollement which is related to the 'mantle geoid' of Turcotte *et al.* (1977) and for a constant-density basin fill coincides with the level of no vertical motion within the lithosphere. Prior to thermal subsidence, this notional depth to decollement lies $\approx$ 3.3 km below sea-level for water-loaded basins. Sediment-loaded basins are most readily treated if the subsidence equations of McKenzie (1978) are modified to take explicit account of sediment density; the notional depth to decollement ranges from $\approx$5–10 km, increasing with the stretching factor β. The physical sole to a system of detached normal faults can take any depth, and is not predictable from extension and subsidence data; in general, it will move vertically, as extension proceeds, to maintain isostatic equilibrium. Paradoxically, the physical depth to decollement is more likely to be incorrectly calculated where a reliable sea-level reference is preserved. Where, as in many fossil basins, a datum surface is constructed from major fault blocks which have moved vertically with the sole fault, this datum will not correspond to the contemporary sea-level but may yield a correct depth to decollement.

Vertical movement of the sole fault provides a mechanism for the uplift of major basement horsts within extensional basins, since in general a sole fault localized in a brittle–ductile transition will lie below the notional depth to decollement and will be uplifted as stretching proceeds (provided isostatic compensation operates on a horizontal scale greater than that of the uplifted block). In a system of rotated planar ('domino' or 'pack of cards') detached normal faults, footwall uplift should be widespread for large initial fault spacings (several times the notional depth to decollement). Several sets of curves have been presented, which predict (with increasing β) either accelerating footwall subsidence, uplift followed by subsidence below sea-level, or uplift followed by subsidence to a final elevation above sea-level. These curves can be used to predict and explain footwall uplift/subsidence patterns in a variety of basins. Major fault systems in several basins appear to have had an initial fault spacing of 10–20 km, i.e. about the depth to the brittle–ductile transition. This may have implications for the mechanical behaviour of the brittle upper crust, and if it can be generalized, implies that footwall uplift should be a feature of most extensional basins.

The recognition that footwall uplift and erosion can take place during extension has implications for both neotectonic and palaeotectonic interpretations. It provides support for the suggestion of Jackson *et al.* (1982b) that recent uplift of islands in the Aegean Sea does not imply an episode of compression. Individual footwalls can be uplifted then subside during progressive extension, and footwall blocks from different parts of the same basin can record markedly different subsidence histories. This observation gains increased importance because information about the stratigraphy of most extensional basins is derived mainly from hydrocarbon exploration wells sited on footwall blocks. For example, the preservation in the Inner Moray Firth basin of a continuous sedimentary sequence through the end-Jurassic 'Cimmerian unconformity' is primarily attributable to the fact that the basin is not isostatically compensated, and has a relatively deep sole fault which inhibits footwall uplift. Elsewhere (e.g. the East Shetland basin), isostatic compensation was maintained, the notional depth to decollement remained shallower than 10 km regardless of the depth to the physical sole fault, and footwalls were uplifted above sea-level during extension.

The uplift model described in this paper is essentially geometric, and independent of mechanism—it requires only that the extended top basement interface approximates to the 'sawtooth' form produced by 'domino-faulting'. The local isostatic effects discussed by Jackson & McKenzie (1983) could be superimposed on this model, and listric normal faults can be accommodated if bedding surfaces in their hanging wall blocks are not excessively curved. The uplift model can be applied to detached normal faulting in uncompensated basins, where the notional depth to decollement is replaced by the constant depth to a physical sole fault. It can be extended to more complex basins, such as those produced by non-uniform or non-instantaneous stretching (Royden & Keen 1980; Jarvis & McKenzie 1980; Hellinger & Sclater 1983). For example, it is possible to envisage a situation where the level of no vertical motion lay *above* sea-level. Extreme attenuation of the mantle lithosphere relative to the continental crust (e.g. by heating during stretching) would give rise to overall uplift (cf. Royden & Keen 1980), implying the operation of a *negative* notional depth to decollement. Physical sole faults or decollement surfaces would be progressively uplifted and eroded, to eventually be exposed at the surface, as appears to have happened in the Basin and Range area (e.g. Miller *et al.* 1983). For any stretching model, it should be possible to define an instantaneous notional depth to decollement, based on the relationship between basement subsidence and β, and from this a set of curves like those of Fig. 7 could be generated. The observed history of individual fault blocks would then provide additional constraints on the validity of the basic stretching model.

ACKNOWLEDGMENTS: The ideas developed in this paper have benefited from discussions with colleagues in the British Geological Survey and their presentation from the comments of two anonymous referees. This work is published with the permission of the Director, British Geological Survey (N.E.R.C.).

References

ANGELIER, J. 1985. Extension and rifting: the Zeit region, Gulf of Suez. *J. struct. Geol.* **7**, 605–12.

BARR, D. 1985. 3-D palinspastic restoration of normal faults in the Inner Moray Firth: implications for extensional basin development. *Earth planet. Sci. Lett.* **75**, 191–203.

—— , MCQUILLIN, R & DONATO, J.A. 1985. Footwall uplift in the Inner Moray Firth basin, offshore Scotland. *J. struct. Geol.* 7, 267–8.

BARTON, P. & WOOD, R. 1984. Tectonic evolution of the North Sea basin: crustal stretching and subsidence. *Geophys. J.R. astron. Soc.* **79**, 987–1022.

BEACH, A. 1984. The structural evolution of the Witch Ground Graben. *J. geol. Soc. London,* **141**, 621–8.

BOWEN, J.M. 1975. The Brent oil-field. *In*: WOODLAND, A.W. (ed.) *Petroleum and the continental shelf of North-west Europe*, 353–60. Applied Science Publishers, London.

BREWER, J.A. & SMYTHE, D.K. 1984. MOIST and the continuity of crustal reflector geometry along the Caledonian–Appalachian orogen. *J. geol. Soc. London*, **141**, 105–20.

DE CHARPAL, O., MONTADERT, L., GUENNOC, P. & ROBERTS, D.G. 1978. Rifting, crustal attenuation and subsidence in the Bay of Biscay. *Nature*, Lond., **275**, 706–10.

CHEN, W-P. & MOLNAR, P. 1983. Focal depths of intracontinental and intraplate earthquakes and their implications for the thermal and mechanical properties of the lithosphere. *J. geophys. Res.* **88**, 4183–214.

CHENET, P., MONTADERT, L., GAIRAUD, H. & ROBERTS, D. 1983. Extension ratio measurements on the Galicia, Portugal and northern Biscay continental margins: implications for evolutionary models of passive continental margins. *In*: WATKINS, J.S. & DRAKE, C.L. (eds) *Studies in continental margin geology. Mem. Am. Assoc. Pet. Geol.* **34**, 703–15.

CHESHER, J.A. & BACON, M. 1975. A deep seismic survey of the Moray Firth. *Rep. Inst. geol. Sci. London, 75/11.*

CHRISTIE, P.A.F. & SCLATER, J.G. 1980. An extensional origin for the Buchan and Witchground graben in the North Sea. *Nature, Lond.,* **283**, 729–32.

COCHRAN, J.R. 1983. Effects of finite rifting times on the development of sedimentary basins. *Earth planet. Sci. Lett.* **66**, 289–302.

DAHLSTROM, C.D.A. 1969. Balanced cross-sections. *Can. J. Earth Sci.* **6**, 743–57.

DONATO, J.A. & TULLY, M.C. 1981. A regional interpretation of North Sea gravity data. *In*: ILLING, L.V. & HOBSON, G.D. (eds) *Petroleum Geology of the Continental Shelf of North-west Europe*, pp. 65–75. Institute of Petroleum, London.

GIBBS, A.D. 1983. Balanced cross-section construction from seismic sections in areas of extensional tectonics. *J. struct. Geol.* **5**, 471–82.

—— 1984. Structural evolution of extensional basin margins. *J. geol. Soc. London*, **141**, 609–20.

GRAY, W.D.T. & BARNES, G. 1981. The Heather oil field. *In*: ILLING, L.V. & HOBSON, G.D. (eds) *Petroleum Geology of the Continental Shelf of North-West Europe*, pp. 335–41. Institute of Petroleum, London.

HAY, J.T.C. 1978. Structural development in the Northern North Sea. *J. Pet. Geol.* **1**, 65–77.

HEISKANEN, W.A. & VENING MEINESZ, F.A. 1958. *The Earth and its Gravity Field.* McGraw-Hill, New York.

HELLINGER, S.J. & SCLATER, J.G. 1983. Some comments on two-layer extensional models for the evolution of sedimentary basins. *J. geophys. Res.* **88**, 8251–69.

JACKSON, J.A. & MCKENZIE, D. 1983. The geometrical evolution of normal fault systems. *J. struct. Geol.* **5**, 471–82.

—— , GAGNEPAIN, J., HOUSEMAN, G., KING, G.C.P., PAPADIMITRIOU, P., SOUFLERIS, C. & VIRIEUX. J. 1982a. Seismicity, normal faulting and the geomorphological development of the Gulf of Corinth (Greece): the Corinth earthquakes of February and March 1981. *Earth planet. Sci. Lett.* **57**, 377–97.

—— , KING, G. & VITA-FINZI, C. 1982b. The neotectonics of the Aegean: an alternative view. *Earth planet. Sci. Lett.* **61**, 303–18.

JARVIS, G.T. & MCKENZIE, D.P. 1980. Sedimentary basin formation with finite extension rates. *Earth planet. Sci. Lett.* **48**, 42–52.

JOHNSON, R.J. & DINGWALL, R.G. 1981. The Caledonides: their influence on the stratigraphy of the North-west European continental shelf. *In*: ILLING, L.V. & HOBSON, G.D. (eds) *Petroleum Geology of the Continental Shelf of North-west Europe*, pp. 85–97. Institute of Petroleum, London.

LE PICHON, X. & SIBUET, J-C. 1981. Passive margins: a model of formation. *J. geophys. Res.* **86**, 3708–20.

——, ANGELIER, J. & SIBUET, J-C. 1982. Plate boundaries and extensional tectonics. *Tectonophysics*, **81**, 239–56.

——, —— & —— 1983. Subsidence and stretching. *In*: WATKINS, J.S. & DRAKE, C.L. (eds) Studies in Continental Margin Geology. *Mem. Am. Assoc. Pet. Geol.* **34**, 731–41.

LINSLEY, P.N., POTTER, H.C., MCNAB, J. & RACHER, D. 1980. The Beatrice Field, Inner Moray Firth. *In*: HALBOUTY, M.T. (ed.) *Giant Oil and Gas Fields of the Decade 1968–1978*, pp. 117–29. Am. Assoc. Pet. Geol. Tulsa.

MCDONALD, R.E. 1976. Tertiary tectonics and sedimentary rocks along the transition, Basin and Range province to plateau and thrust province. *In*: HILL, J.G. (ed.) *Symposium on Geology of the Cordilleran Hingeline*, pp. 281–318. Rocky Mtn Assoc. Geol. Denver.

MCKENZIE, D. 1978. Some remarks on the development of sedimentary basins. *Earth planet. Sci. Lett.* **40**, 25–32.

MCQUILLIN, R., DONATO, J.A. & TULSTRUP, J. 1982. Development of basins in the Inner Moray Firth and the North Sea by crustal extension and dextral displacement of the Great Glen Fault. *Earth planet. Sci. Lett.* **60**, 127–39.

MILLER, E.L. GANS, P.B. & GARING, J. 1983. The Snake Range decollement: an exhumed mid-Tertiary ductile-brittle transition. *Tectonics*, **2**, 239–63.

MONTADERT, L., ROBERTS, D.G., DE CHARPAL, O. & GUENNOC, P. 1979. Rifting and subsidence of the northern continental margin of the Bay of Biscay. *In*: MONTADERT, L. & ROBERTS, D.G. (eds) *Init. Rep. D.S.D.P.* **48**, 1025–60.

PARSONS, B. & SCLATER, J.G. 1977. An analysis of the variation of ocean floor bathymetry and heat flow with age. *J. geophys. Res.* **82**, 803–27.

PROFFETT, J.M. 1977. Cenozoic geology of the Yerington district, Nevada, and implications for the nature of Basin and Range faulting. *Bull. geol. Soc. Am.* **88**, 247–66.

ROYDEN, L. & KEEN, C.E. 1980. Rifting process and thermal evolution of the continental margin of eastern Canada determined from subsidence curves. *Earth planet. Sci. Lett.* **51**, 343–61.

SCLATER, J.G. & CHRISTIE, P.A.F. 1980. Continental stretching: an explanation of the post-mid-Cretaceous subsidence of the Central North Sea. *J. geophys. Res.* **85**, 3711–39.

——, ROYDEN, L., HORVATH, F., BURCHFIEL, B.C., SEMKEN, S. & STEGENA, L. 1980. The formation of the intra-Carpathian basins as determined from subsidence data. *Earth planet. Sci. Lett.* **51**, 139–62.

SIBSON, R.H. 1983. Continental fault structure and the shallow earthquake source. *J. geol. Soc. London*, **140**, 741–67.

SMYTHE, D.K., SKUCE, A.G. & DONATO, J.A. 1980. Geological objections to an extensional origin for the Buchan and Withground Graben in the North Sea. *Nature, Lond.* **287**, 467.

STECKLER, M.S. & WATTS, A.B. 1978. Subsidence of the Atlantic-type continental margin off New York. *Earth planet. Sci. Lett.* **41**, 1–13.

TURCOTTE, D.L., HAXBY, W.F. & OCKENDOW, J.R. 1977. Lithospheric instabilities. *In*: TALWANI, M. & PITMAN, W.C. (eds) *Island Arcs, Deep-Sea Trenches and Back-Arc Basins. Am. geophys. Union Maurice Ewing Series*, **1**, 63–9.

WATTS, A.B. 1982. Tectonic subsidence, flexure and global changes in sea level. *Nature, Lond.* **297**, 469–74.

——, KARNER, G.D. & STECKLER, M.S. 1982. Lithospheric flexure and the evolution of sedimentary basins. *Phil. Trans. R. Soc. London*, **A305**, 249–81.

WERNICKE, B. & BURCHFIEL, B.C. 1982. Modes of extensional tectonics. *J. struct. Geol.* **4**, 105–15.

WHITEMAN, A., NAYLOR, D., PEGRUM, R. & REES, G. 1975. North Sea troughs and plate tectonics. *Tectonophysics*, **26**, 39–54.

WOOD, R. 1981. The subsidence history of Conoco well 15/30-1, central North Sea. *Earth planet. Sci. Lett.* **54**, 306–12.

—— & BARTON, P. 1983. Crustal thinning and subsidence in the North Sea. *Nature, Lond.* **302**, 134–6.

ZIEGLER, P. 1983. Crustal thinning and subsidence in the North Sea. *Nature, Lond.* **304**, 562.

DAVID BARR, British Geological Survey, 19 Grange Terrace, Edinburgh EH9 2LF, UK.
Present address: Britoil plc, 150 St Vincent Street, Glasgow G2 5LJ, UK.

Crustal extension in the Basin and Range Province, southwestern United States

W. Hamilton

SUMMARY: Cenozoic extension of areally varying ages and amounts has on average doubled the width of the Basin and Range Province. Extensional structures that formed at all depths down to 20 km, and which range in age from Oligocene to Holocene, are widely exposed and are here interpreted in terms of a model of depth-varying deformation. The middle crust is extended by discontinuous ductile shear as internally undeformed lenses slide apart along gently dipping zones of mylonite. The tops of these lenses are undulating detachment faults, the composite area of which increases with time as deep lenses slide out from underneath shallower ones. Brittle blocks of upper-crust bedrock above the detachments respond first by rotating between range-front faults, the same direction of rotation being maintained across a series of lenses, and then by pulling completely apart, while basinal strata fill the gaps and are dragged directly on detachment faults. Some faults rise gently from the main detachment zones and surface as range-front faults. Most tilted-block ranges are isolated atop detachments.

Detachment faults cut out crust. Beneath them are mid-crustal rocks of any age and type and above them are mostly upper-crustal rocks, including extensive syndeformational basin sediments rotated to steep or moderate dips. As attenuation proceeds and components rise, detachment faults evolve from ductile to brittle, develop splays, and are themselves broken by steep brittle structures related to new, deeper detachments. Parts of detachment faults remain active even after exposure at the surface, but slip on them is then limited to the down-dip direction. It is inferred from seismic reflection profiles and rock-mechanic considerations that the unexposed lower crust is extended by more pervasive ductile flattening.

This paper develops a model of depth-varing extensional styles to account for the common denominators in Cenozoic geological structures exposed by erosion and tectonic denudation throughout the Basin and Range Province. The general model was suggested earlier (Hamilton 1982), and resembles in important aspects that of Davis (1980).

Oligocene to Quaternary extension of the style discussed in this paper has affected the Great Basin part of the province (Fig. 1), and also a region N of the Snake River Plain in E-central Idaho and SW Montana. Most of the similar extension in the Mojave and Sonoran Desert sectors of the province was of Oligocene to middle-Miocene age, correlating with early extension in the N. Middle-crust extensional structures are widely exposed in both N and S sub-provinces, whereas shallow structures are seen primarily in the N one. Topographic relief is high throughout the province. Exposures are excellent in the arid SW part, and good in the semi-arid SE and N parts.

Extension has about doubled the pre-Oligocene width of the province, now 350–700 km wide. Palinspastic reconstruction of Mesozoic tectonic and magmatic belts requires such extension (Hamilton 1978). Relationships between strike-slip and extensional faults in the southern Great Basin indicate late-Cenozoic extension to be at least 65% (Wernicke *et al.* 1982). Pre-Tertiary rocks and Cenozoic basin-fill materials comprise sub-equal amounts above exposed detachment faults, requiring 100% extension in many areas. Extension has, however, been erratic in space, time, and amount.

How has the hundreds of kilometres of extension been transmitted through the continental crust? The field data needed to answer this have been greatly expanded by many geologists, particularly those working with detachment structures within the last decade, but there is still much disagreement regarding extensional mechanisms. Although Ransome, Emmons & Garrey (1910) recognized that Basin and Range faulting required crustal extension (they also mapped a detachment fault), more than 50 years passed before extension was broadly accepted over alternatives of vertical jostling or even compression and it was not until almost 1980 that extension of more than 5 or 10% was widely regarded as likely. Among those with more advanced ideas were Hunt & Mabey (1966), probably first to recognize a complete domiform fault as extensional, Hamilton & Myers (1966), who argued for 100% extension and Armstrong (1972), first to see the regional extent of middle-crust extensional structures.

From Holdsworth, R.E. & Turner, J.P. (compilers) *Extensional Tectonics: Regional-scale Processes.* The Geological Society,
Key Issues in Earth Sciences, **2** (**1**), 75–96.
1476-315X/02/$15.00 © The Geological Society of London 2002
First published in: "Coward, M.P., Dewey, J.F. & Hancock, P.L. (eds) *Continental Extensional Tectonics*, The Geological Society, London, Special Publications **28**, 155–176".

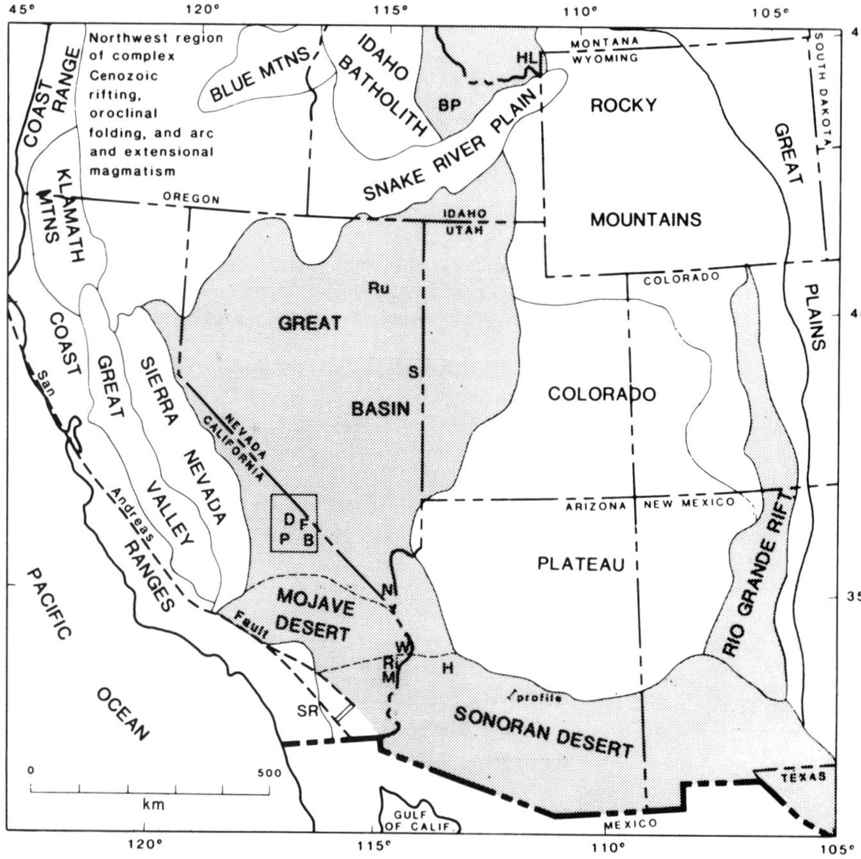

FIG. 1. Index map of part of the western United States. Basin and Range Province of extensional faulting is shaded, as are areas of similar structure in the Rio Grande rift system and in the region N of the Snake River Plain. Extensional faulting also affects much of the 'NW region', and the W and S margins of the Colorado Plateau; most boundaries shown are arbitrary. Death Valley region (Fig. 7) is located by a rectangle; reflection profile (Fig. 24) is located in Arizona. Localities noted in text and figures: B = Black Mts; BP = Borah Peak; D = Death Valley; F = Funeral Mts; H = Harcuvar Mts; HL = Hebgen Lake; M = Big Maria Mts; N = Newberry Mts; P = Panamint Mts; Ri = Riverside Mts; Ru = Ruby Mts; S = Snake Range; SR = Santa Rosa Mts; W = Whipple Mts.

Only a few of the many authors whose reports are integrated here can be cited. I also draw heavily on discussions and field trips with many geologists and on my own detailed and reconnaissance fieldwork. As the structures described here include types unfamiliar to many readers, I illustrate them with a number of photographs.

The model

The model sketched here (Fig. 2) is derived from the systematic variations of observed structures with the inferred depths and temperatures of their formation. Extension is accommodated within the brittle upper crust by rotation about sub-horizontal axes and fragmentation of large blocks that end abruptly downward against undulating regional *detachment faults* that mostly initiated at depths greater than about 10 km. 'Detachment fault' has been used in this context by Davis *et al.* (1980) and many others. 'Denudation fault' (Armstrong 1972) and 'decollement' (Miller *et al.* 1983) are used with approximate synonymity. Upper-crust bedrock blocks are bounded laterally by downward-flattening listric faults, by rotated planar faults and by basin fills deposited in widening gaps between separating blocks. The detachment faults define the tops of great lenses that are separated by gently dipping, anastomosing ductile faults and that retain pre-faulting fabrics in their interiors. As the lenses slide apart the area of their composite top, the detachment faults, is increased and the middle crust is extended by discontinuous ductile flow. Although most faults above the typically mid-crust detachments are probably listric, curving into the detachments, in

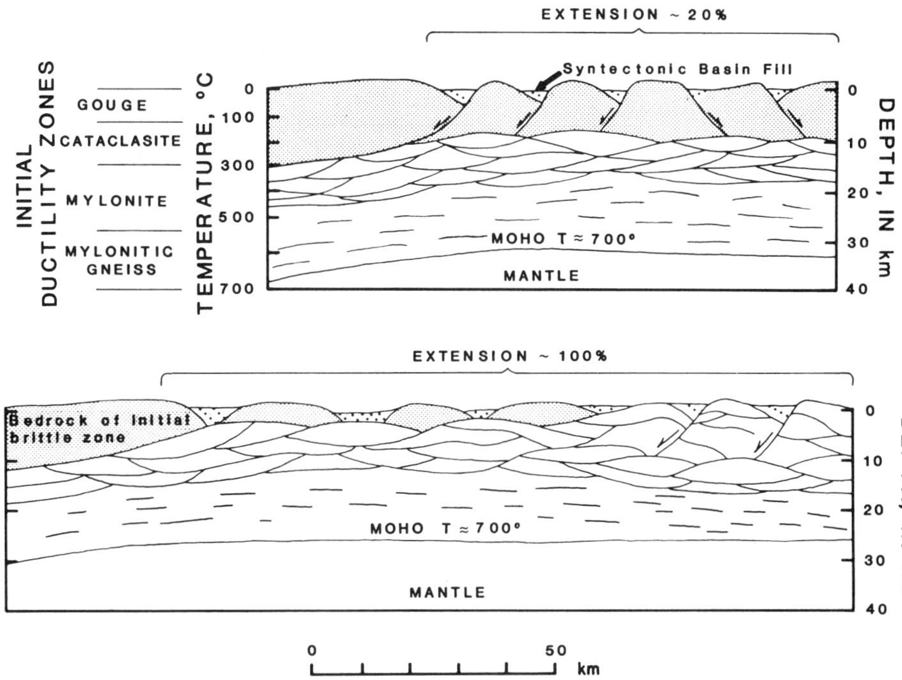

FIG. 2. Cross-sections of extending crust. Brittle upper-crust blocks rotate and separate. Middle-crust lenses slide apart along ductile shear zones; composite upper surface of lenses forms detachment faults that increase in total area with time. Lower crust flattens pervasively. Structural styles are superimposed as components rise to the surface with continuing attenuation. Attenuating crust is partly rebuilt by magmatism, and possibly also by phase change, so crust is thinned by a factor less than extension ratio. Ductility zones after Sibson (1983).

places major gently dipping faults rise for tens of kilometres laterally through upper-crust rocks and curve to steeper dips only at shallow depth.

COCORP reflection profiles (Allmendinger *et al.* 1983) across western Utah show a detachment fault of this upper-crust type, which crops out in the Canyon Range, dipping gently westward for a distance of 100 km or more in the sub-surface, to a depth of at least 13 km. In outcrop and at shallow depth, the fault separates Tertiary strata, rotated to moderate dips, from Palaeozoic strata. The fault most likely steepened into a range-front fault in its upper part, now removed by erosion. The broadly undulating Canyon Range detachment passes westward beneath the Sevier Desert, House Range, Tule Valley, Confusion Range, and Snake Valley, all of which are thus allochthonous above it, and appears to project westward beneath the thick crustal lens whose top is the Snake Range detachment fault. Discontinuous reflections beneath the Canyon Range detachment fault may in part image subdetachment ductile shear zones.

Detachment faults evolve from zones of ductile slip initially 10–20 km deep. Local variations in temperature, strain rate and lithology are reflected in variations in initial depth and configuration of the ductile faults that evolve into detachment faults. As extension progresses and the crust is thinned, most parts of the system rise toward the surface; ductility-facies boundaries rise as extension and possibly also shear heating, outpaces thermal conduction, but in general superimposed structures and fabrics indicate progressions from deformation under deep, hot conditions to deformation under shallow, cold conditions. Ductile faults become semi-ductile, then brittle and splays split away. Still further extension can no longer be accommodated on undulating faults, which then are broken by steep, brittle faults, related to new ductile faults at depth, but limited slip continues on detachment faults even after they are exposed at the surface. Extension in the lower crust is accomplished by more pervasive ductile flattening.

Middle-crust crystalline rocks beneath detachment faults are widely termed core complexes. I do not use the term, for it often incorporates the implication that detachments are necessarily related to local thermal anomalies which is not the case. Some detachments may indeed be related to synextensional magmatism, but most sub-detachment complexes represent nearly random samples of the middle crust, albeit in a region subjected to much synextensional magmatism.

Mathematical and mechanical analysis of depth-varying crustal deformation was presented by Kligfield *et al.* (1984) and Sibson (1983).

Tectonic setting

The middle Tertiary structures of the Basin and Range Province record extension mostly W-southwestward in the S and approximately westward in the N, relative to the continental interior. This extension was perpendicular to the continental margin of the time (Zoback *et al.* 1981) and was synchronous with subduction of Pacific lithosphere beneath most of that margin (Eaton 1984). There was thus great extension of the overriding continental plate in a convergent-plate regime. A common misconception is that overriding lithosphere plates generally undergo shortening and compression; actually, most undergo extension, although not generally as extreme as in the southwestern US. Trenches retreat and roll back into subducting plates. The inclination of a subducting slab marks a transient position, not a trajectory down a fixed slot; subduction occurs at an angle steeper than that inclination and the common regime above a slab is extensional, not compressional. The mid-Tertiary extension occurred after the slowing of the previously rapid westward motion of North America (as viewed in a whole-Earth, zero-sum frame).

Extension of similar style, mostly of mid-Tertiary age, but continuing through the Neogene, affected the Rio Grande rift. Extension there can be visualized as due to the clockwise rotation of the Colorado Plateau about 3° relative to interior North America, about a Euler pole near the Colorado–Wyoming border.

Late Cenozoic extension represents oblique fragmentation of the crust in the NW direction of relative slip between the Pacific and North American Plates, due presumably to partial coupling between those plates (Atwater 1970, Eaton 1984). The structural style of extensional response is similar. There was much inheritance of mid-Tertiary structural grain from Mesozoic structures and from those by late-Cenozoic structures; structures of all of these ages typically have northerly trends in the Great Basin and northwesterly ones in the Mojave and Sonoran Deserts.

Sophisticated climatic analysis of palaeofloras by Axelrod (1985) indicates that surface altitude in the Great Basin, the sector now undergoing broad extension, has increased markedly during the late Cenozoic. This increase has been coincident both with extension and with the rise of much of western North America. Such palaeo-altitude data appear incompatible with speculation by Coney & Harms (1984), that high-standing thick crust spread gravitationally to produce extensional structures.

Magmatism

Mid-Tertiary extension was broadly coextensive in space and time with magmatism both of arc type and of more silicic type intermediate between typical arc and extensional assemblages. Volcanism of both types is now active above subducting lithosphere in the rapidly widening North Island, New Zealand. Thermal softening of the crust by magmatism probably expedited extension. The Colorado Plateau was bypassed by magmatism and left unextended, whereas the Rio Grande rift to the E was the site of extensive magmatism and severe extension. Integration of the high heat flow and widespread late-Cenozoic magmatism with the rates and amounts of extension indicates that these thermal manifestations are largely effects, not causes, of extension (Lachenbruch & Sass 1978) and are enhanced by asthenospheric convection as a by-product of the extension (cf. Steckler 1985). Late-Cenozoic magmatism has been mostly of bimodal basalt-and-rhyolite type that probably reflects the melting of lower-crustal materials by rift-related mantle magmas. Deep-crystallization products of magmas of both arc and rift types must have partly rebuilt attenuating crust.

Structures of the upper crust

The basic structural unit of the Basin and Range Province is commonly but mistakenly perceived to be a tilted bedrock block, which includes a mountain range and the basement of the adjacent basin and which rotates against the next block along a fault that either is flattened with time by rotation, or else flattens downward as a curving, listric surface and that ends as a detachment fault. Such juxtaposed blocks probably formed during an early stage of extension, but most of the province has evolved beyond this stage. The majority of ranges in the province represent a mature stage and are either tilted panels of resistant rocks isolated from one another atop detachment faults or else are raised sub-detachment complexes. Wernicke *et al.* (1985) came to similar conclusions. Examples from the Death Valley region are discussed below.

Major upper-plate panels typically are 70–150 km long and 10–25 km wide in the Great Basin. The width of a mature bedrock panel is markedly less than the distance between range-front faults of adjacent mountain blocks (mostly 20–35 km). Most bedrock blocks are separated from one another, the gaps being filled by syntectonic basin strata in direct contact with gently dipping faults that pass beneath the tilted blocks. Dips in basin fills increase markedly downward, both stratigraphically and along the beds, toward bounding and underlying faults (Allmendinger *et al.* 1983; Effimoff & Pinezich 1981; Smith & Bruhn 1984). Details are complex, for basin fill deposited during early extension can become the lithified bedrock of subsequent deformation. Blocks and basin fills are broken by lesser normal faults. Upper plates can be shattered into small rotated-domino subblocks (e.g. Ransome *et al.*. 1910; Miller *et al.* 1983), or they can be tilted blocks of crust 10-km thick and little deformed internally (Howard *et al.* 1982).

Upper-plate panels are tilted randomly on a provincial scale, but tend to be tilted mostly in single directions over domains of 10,000 km^2 or more (Stewart 1980, 1983b). These domain tilts may record initial coherence of blocks rotating in the same sense. Non-tilted horsts and grabens are uncommon and where present they generally separate domains of blocks of opposite tilt directions.

Young normal faults have dips mostly of 45–60° where they cut bedrock at the surface (Fig. 3). Faults refract upward to steeper dips in unconsolidated materials, in which extensional collapse features are developed (Myers & Hamilton 1964). Erosion of fast-rising scarps to slopes recording their angle of repose (Fig. 4) occurs largely by landsliding that results in the formation of extensive megabreccias along scarps (Fig. 5). In settings of rapid tilting and faulting, lacustrine strata also can be deposited almost directly against high scarps.

Rates and amounts of extension have been non-uniform in space and time. Extreme late-Miocene to Quaternary extension characterizes the Death Valley region. Most deformation more than 30 km across strike to the E of Death Valley pre-dates the late-Miocene, whereas the faulting that has defined the modern ranges across strike to the W of Death Valley is of Pliocene and Quaternary age.

FIG. 3. Normal fault (arrow) between Proterozoic metadolomite (right) and slightly tilted Quaternary alluvium. N end Tucki Mountain, Panamint Mountains, California. The fault cuts only the upper plate, above the detachment fault (not in view). Exposure is 10 m high.

FIG. 4. Late Quaternary normal-fault scarp cutting middle- or upper-Miocene volcanic and tuffaceous rocks. The late-Holocene scarp, 6 m high, along base of mountain, displays structural dip, but the rest of face has been eroded back, mostly by landsliding. Miocene rocks dip steeply E (left) and are truncated at shallow depth by a detachment fault. Slide megabreccia (right foreground, blocky) is separated by broad collapse moat (low ground) from frontal fault. View SE—Artist's Drive fault, Black Mountains, California.

Gently dipping extensional faults formed in upper-crust, low-temperature rocks in a number of places in the Great Basin and commonly truncate rotated upper plates against lower plates. The western Utah example plumbed by COCORP was discussed previously. Wernicke *et al.*

FIG. 5. Small, Holocene slide megabreccia (centre), and upper edge of large late-Pleistocene slide megabreccia (far left), derived from middle- or upper-Miocene volcanic rocks of the Artist's Drive fault. View N—upper and lower plates and the fault are all truncated downward against the Badwater detachment fault in exposures to the S.

(1985) described a domiform example from southern Nevada, similar geometrically to the more common detachment faults atop mid-crust crystalline rocks but with almost unmetamorphosed Palaeozoic strata in the footwall. Some such high-level faults appear to have reactivated faults that initially were thrusts. Howard & John (this volume) show a detachment fault that splays to shallow normal faults that bound crystalline-block ranges in headwall fashion. It is not yet clear whether such headwall blocks are anchored to the deep crust or are adrift above detachment faults with which the shallow faults merge.

Earthquakes

Most small earthquakes in the Great Basin occur shallower than 8 km, whereas large ones typically nucleate at 10–15 km. Shear strength apparently increases below 8 km, then decreases rapidly across the brittle–ductile transition, which represents a temperature of 300–350°C and is no deeper than 15 km (Sibson 1983; Smith & Bruhn 1984). A major unresolved problem is the much greater depth extent suggested for steep faults by studies of seismicity than is permitted by the continuity of detachment faults seen on reflection seismic profiles.

Tilted blocks of bedrock commonly end downward against detachment faults which pass beneath the ranges, and generally only basin fill is in contact with frontal faults. Analyses of coseismic strain, that incorporate the assumption that adjacent tilted blocks of elastic bedrock are in direct contact at sub-surface frontal faults (e.g. Stein & Barrientos 1985) are probably incorrect. The best-documented deformation accompanying a normal-fault earthquake, that at Hebgen Lake, Montana, in 1958, $Ms = 7.5$, consisted largely of absolute subsidence which represented a net decrease of about 1 km^3 of crustal volume (Myers & Hamilton 1964). Coseismic subsidence also greatly excluded rise during the 1973 Borah Peak earthquake, $Ms = 7.3$, in Idaho (Stein & Barrientos 1985). Pre-seismic extensional strain high in the crust apparently is stored primarily as dilatation and released mostly as compaction.

Structures initiated in the middle crust

Detachment faults

Domiform or undulating detachment faults have been recognized in perhaps 50 ranges scattered far beyond the narrow zone depicted in some relatively early reports (e.g. Rehrig & Reynolds 1980). These faults are not thrusts, for they cut out, not duplicate, crust and their upper plates display abundant extensional features. Any middle-crust rocks can lie beneath detachment faults: rocks of various lithologies e.g. mid-crust granites, gneisses and migmatites, or metasedimentary or meta-igneous rocks of any grade from lowest greenschist, even slate, to highest amphibolite, and of ages ranging from Precambrian to Miocene. Non-metamorphosed supracrustal rocks occur primarily in upper plates and commonly are rotated steeply downward against the faults with truncation angles near 50°. Granitic rocks of Phanerozoic, mostly late-Mesozoic, age within upper plates are primarily of upper-crust types.

Upper-plate rocks tend to show the same sense of rotation over large domains. The direction of relative slip along a detachment surface, as defined by extensional structures low in the upper plate, is typically in the same sense as that rotation. In a domain of about 20,000 km^2 in SE California and SW Arizona (e.g. Fig. 6; Davis *et al.* 1980; Howard & John, this volume), upper-plate blocks slipped relatively NE over sub-detachment rocks and have general SW dips. As the unfaulted Colorado Plateau lies NE of this tract and as upper-plate structure requires extreme extension, upper-plate blocks must have moved relatively southwestward away from the plateau by amounts increasing with distance from the plateau. The southwestward offset of lower-plate rocks relative to the plateau must be still greater.

The amount of slip between juxtaposed upper and lower plates is commonly difficult to determine; half the province is covered by Quaternary

FIG. 6. Undulating Oligocene detachment fault, Riverside Mountains, SE California. Upper plate: light-coloured middle-Proterozoic granite (Ᵽg) and dark Oligocene slide megabreccias and redbeds (T; direction of steep dip indicated); broken by spoon-shaped faults. Lower plate: (Pz), Palaeozoic metasediments; (Mz), Mesozoic metasedimentary and metavolcanic rock; and (Ᵽm), retrograded and variably mylonitized middle-Proterozoic gneiss. View N—slip of upper plate relatively NE. Range mapped by author.

basin materials and too little of the other half displays optimal mixes of upper and lower plates, necessary for comparison. The outcrop overlap of 6 km of quite different crystalline assemblages in the Riverside Mountains (Fig. 6) requires an offset substantially greater than that distance; the geology of lower-plate ranges to the SW, from which relative direction the upper plate came, best fits an offset of 20–30 km. Howard *et al.* (1982) inferred 30–40 km of offset of a dyke swarm between upper and lower plates in the same tilt domain, 70 km NE across strike from the Riverside Mountains. The extreme extension within the intervening upper plate is apparently not matched by variations in slip on the detachment, so the area of the detachment surface probably increased with time. Bartley & Wernicke (1984) made a permissive argument for 60 km of slip on the Snake Range detachment, eastern Nevada. Note that in arguing for little slip on the same structure, Miller *et al.* (1983) invoked such improbabilities as regional metamorphism that ended abruptly at the level of the initial fault, and the formation of kyanite at a depth of only 6 km.

Equivalence of deep and shallow structures

Basin and Range detachment faults are regarded by many investigators (e.g. Davis 1980; Miller *et al.* 1983; Rehrig & Reynolds 1980; Reynolds 1985; Snoke *et al.* 1984) as having evolved from deep ductile structures to shallow brittle ones. Explanations diverge widely beyond this limited consensus. Some models infer that shallow deformation above the detachment faults was quite unlike that of presently active parts of the Great Basin and was typified, perhaps, by close-spaced normal faults of small individual displacements and by depositional basins larger than those of the modern system (e.g. Stewart 1983b). I infer, as do many other geologists (e.g. Wernike *et al.* 1985) that on the contrary, tilted Great Basin blocks are forming above detachments like the older ones now widely exposed and that those older structures had basin–range pairs above them.

Dimensions of basement and basin-fill assemblages above detachment faults (Fig. 7) resemble those of basin-range pairs. There is broad temporal and geographical overlap of structures

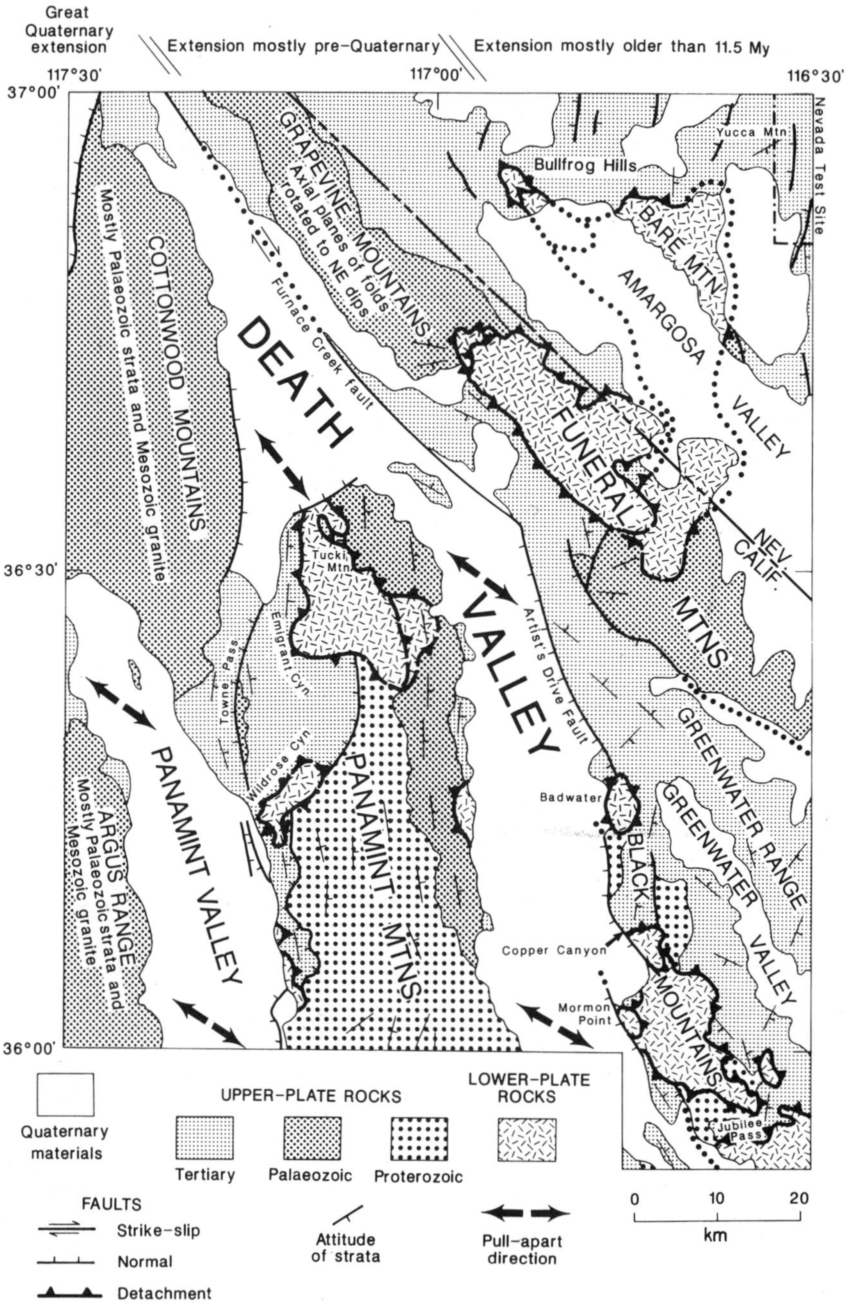

FIG. 7. Tectonic map of Death Valley region, California and Nevada. Most normal faults are omitted. Adapted from published maps, including those cited in text.

exposed as tilted blocks and as detachments. Continuing extension superimposed block structures on detachments at transition times that varied from early-Miocene to Quaternary. There are exposed detachment faults as young as Holocene, and block-bounding faults as old as early-Miocene, nearby within the southern Great Basin. Early-Miocene floras of Nevada indicate that basins and ranges then existing had a topographic relief comparable to modern basin–range pairs (Axelrod 1985). Domains of same-direction tilting of upper-plate rocks above detachment faults are of sizes, typically 10,000–20,000 km^2, similar to those of Great Basin tilted-block domains (Stewart 1980; Fig. 7; Hamilton, unpubl. work). Scarp-facies sediments are widely present in basin strata truncated downard against detachment faults. Abundant, more or less monolithological, megabreccias (Figs 8, 9 & 10) must have slid

FIG. 8 Rotated normal fault. Stripped fault surface (smooth slope, right and foreground) on Cretaceous muscovite granite dips 30°W; light-coloured landslide megabreccias of granite and Proterozoic metasediments dip 20° E into fault. Pliocene or early Pleistocene range-front fault with 3 km of slip, rotated 25° with upper-plate block above detachment fault (not in view). Jail Canyon, Panamint Mountains; mapped by Albee *et al.* (1981).

FIG. 10. Mid-Tertiary detachment fault, SE Whipple Mountains, California. Oligocene volcanic rocks and slide megabreccias dip left (SW) to truncation against horizontal fault at top of Proterozoic (?) plutonic rocks. Mylonitic carapace shows as layering beneath fault.

FIG. 9. Oligocene landslide megabreccia, dip 60° right (SW), 100 m above Riverside Mountains detachment fault.

FIG. 11. View N across Sacramento Pass to subhorizontal mid-Tertiary detachment fault (marked), Snake Range, Nevada. Upper plate consists of rotated blocks of unmetamorphosed Palaeozoic strata that dip to the left. Lower plate (including foreground) is of Cambrian and upper-Proterozoic strata metamorphosed to amphibolite facies (kyanite present) and intruded by muscovite granite. Light-coloured low ground is underlain by upper-plate Oligocene redbeds, which also dip moderately left above detachment fault. Area described by Miller *et al.* (1983).

from high, fast-rising scarps like those of Figs 4 & 5. Truncation angles of Cenozoic strata against faults are typically within the range 40–60°, regardless of the dips of strata and faults (Figs 3, 6, 8, 10, 11, 12, 13, 14); rocks now truncated by gently dipping faults have been rotated down from truncations against normal faults.

Rise with time

Most components in an extending system must rise toward the surface with time. Many investigators have cited evidence for the great rise of detachment faults from middle to upper crustal levels as slip on them progressed (e.g. Davis 1980, 1983; Rehrig & Reynolds 1980; Reynolds 1985; Snoke *et al.* 1984). Final major slip has always been at low temperatures and

FIG. 12. Late-Neogene Tucki Mountain detachment fault. Smooth bedrock slope is stripped fault, dipping gently W, on upper Proterozoic metasedimentary rocks and Cretaceous muscovite granite. Unconsolidated Pliocene alluvium dips (lines) E to truncation against detachment. Emigrant Wash, Panamint Mountains.

shallow depths. The faults are marked by brittle gouge and incoherent breccia, typically a few decimetres (Fig. 15) to a few metres thick, but ranging from a centimetre (Fig. 16) to tens of metres. Many gouge zones grade upward into decreasingly brecciated upper-plate rocks and small-scale extensional structures are widespread low down in upper plates (Fig. 17). Severe brecciation can extend 100 m into the upper plate.

The depth and temperature of slip decrease with time (cf. Fig. 2). Progressively older and more ductile-strain rocks, in the sequence gouge–cataclasite–mylonite–mylonitic gneiss, are displayed downward from many detachment faults. A detachment-fault gouge zone is commonly bounded sharply against a shear-polished pavement (Fig. 15) on a hard, chloritic

FIG. 13. Neogene detachment fault, W side Funeral Mountains, California. Sheared carapace (layering, right) was developed concordantly below detachment in middle-grade Proterozoic metasedimentary rocks. Arrow indicates dip of upper-plate mid-Tertiary redbeds beneath late Quaternary pediment. View N from Keane Wonder Mine.

FIG. 14. Truncation of upper-plate structures against active detachment fault, Black Mountains, California. Interbedded Pliocene clastic strata, slide breccias, and volcanic rocks (lower left) dip obliquely back and right towards the Quaternary normal-fault scarp along high-standing plutonic rocks (left and centre rear). Hanging wall, footwall, and fault are all truncated downward against the active Copper Canyon turtleback (detachment) fault, which forms the smooth face dipping obliquely left through the centre of the view. Light-coloured rocks on the turtleback are a discontinuous carapace of mylonitized dolomite. An active range-front normal fault truncates both upper and lower plates at the top of the alluvial fan.

microbreccia that displays partial reconstitution at low greenschist facies. This cataclasite is typically a few metres thick and it may grade downward into a zone of sheared chloritic, epidotic rocks, tens of metres thick, Beneath these, or directly beneath the microbreccia in other cases, is rock typically recording more ductile shear. Such rock is laminated mylonite where quartzofeldspathic rocks are involved (Fig. 10) and foliated greenschist-facies rock where metasediments are present (Figs 13, 18, 19). These ductile zones typically fade out downward through thicknesses of tens or even hundreds of metres and define carapaces semi-concordant beneath the detachment faults. Mylonitic gneisses are widespread in the most-uplifted lower plates. The mylonitic carapaces and progressively lower-temperature cata-

FIG. 15. Close up of fault of Fig. 10. Hammer is on zone of breccia 0.4 m thick, above which (in shade) is a thin zone of gouge. Upper-plate Oligocene slide breccia dips steeply left. Lower plate displays several levels of shear-polished pavement in coherent cataclastic microbreccia.

FIG. 17. Small normal faults in variably brecciated Tertiary redbeds, 50 m above Riverside Mountains detachment fault.

FIG. 16. Mid-Miocene detachment fault, Newberry Mountains, Nevada. Proterozoic monzogranite both above and below. Upper-plate rock is altered red (in part by Precambrian (?) retrogression from granulite facies) and variably brecciated; lower-plate rock is grey and is stripped along its mylonitic carapace. Gouge zone is 1 cm thick.

FIG. 18. Calcite mylonite derived from marble, in carapace beneath Ruby Mountains detachment fault, Nevada. Quartzite and amphibolite form boudins and rotated blocks. Area studied by Snoke *et al.* (1984).

clasites are developed in some cases on Tertiary mid-crust granites, barely older than the detachment faults that cut them (e.g. Reynolds 1985; Snoke *et al.* 1984). Much debate centres on whether the ductile carapaces represent old zones of any origin affected latterly by extensional slip, or early slip on mid-crust zones that evolved into the shallow detachment faults. My conclusion, shared with many other investigators, is that the mylonites generally belong to the detachment systems, whereas the more regional and higher-grade mylonitic gneisses are in many cases older than extensional deformation.

Crustal level

There is typically great crustal omission across detachment faults, although large upper-plate blocks rotated down against detachment faults show much omission at their toes but little or none at their heels (Howard *et al.* 1982) Rocks exposed beneath detachment faults mostly dis-

FIG. 19. Retrograde carapace in right-central part of Fig. 13. Z folds (arrows to hinges) and small fault (offsets contact ×2 m) indicate shear of top-to-W (right) sense, same as slip of upper plate. b = boudin.

play petrological evidence for crystallization in the middle crust, whereas unmetamorphosed supracrustal rocks of any age are largely confined to upper plates of detachments younger than those rocks. Mesozoic and Tertiary granitic rocks of upper plates are mostly of upper-crustal type in cross-cutting plutons with contact-metamorphosed wall rocks, whereas correlating rocks beneath detachments include abundant mid-crust migmatites and muscovite granite. I have discussed elsewhere depth variations in granitic and migmatitic systems (Hamilton 1981). Kyanite occurs beneath detachment faults in many ranges, whereas andalusite is uncommon. Sedimentary rocks beneath detachment faults may be metamorphosed only to slate, but even this requires a temperature of 300°C and hence generally a mid-crustal depth. Quartzose mylonites, also requiring a minimum temperature of about 300°C, are widespread beneath detachment faults and uncommon in the young rocks above them. These and other indicators of temperature and pressure require a common minimum depth of erosion of 10 km for rocks beneath detachment faults and of 15 km for many of them, even where those rocks are as young as Miocene.

Rapid uplift generally accompanied extension. K–Ar and fission-track ages of rocks beneath detachments commonly approximate the age of tectonic denudation and record rapid uplift and cooling from ambient temperatures of the middle crust. The dates do not generally record mysterious 'thermal events' despite the over-use of that term in the literature. Slip on detachment faults, and hence local rates of extension, must correspondingly have reached at least 10 km 1,000,000 yr^{-1}, or 1 cm yr^{-1}. At any one time, extension was rapid in parts of the province, as in the modern Death Valley region, and much slower or inactive elsewhere. High strain rates may have resulted in the shear heating of ductile shear zones, expediting further strain.

Precambrian basement rocks were widely eroded to mid-crustal depths before the deposition of upper-Proterozoic and younger strata and so occur in both upper and lower plates. Precambrian basement rocks of upper plates, however, tend to be variably altered, brecciated, and oxidized (Fig. 16) whereas those of lower plates generally lack such low-grade alteration and show instead variable carapace retrogression to mineral assemblages of lower greenschist facies and of higher grades at greater depths.

Wernicke (1985) postulated that detachment faults are segments of ramps that cut through the entire crust. The prediction that depths of origin of terrains, juxtaposed by correlating detachments, vary systematically across-strike on a regional scale is not fulfilled, although local sectors of several ranges each, may show such progressions (e.g. Howard & John, this volume). The superb imaging by Klemperer (Klemperer *et al.* 1986) of the Mohorovicic discontinuity across northern Nevada shows the seismic base of the crust to be continuous and gently undulating, unbroken by ramps as postulated by Wernicke (1985). Upper-crust ramps do, however, apparently connect between mid-crust lenses and the surface, as in the Utah COCORP example.

Davis (1983) inferred that gently dipping detachment faults develop by domino rotation of blocks between faults of initially steep dips. The lack of unmetamorphosed basin fills and other upper-crust rocks beneath most detachment faults is evidence against this view.

Late-Cenozoic detachments of the Death Valley region

Most of the detachment-fault literature has described mid-Tertiary complexes and this has led to the erroneous impression that structures of that age are different from younger ones. Accordingly, I emphasize here the detachment faults of late-Neogene age in the Death Valley region (Fig. 7), which is within the most actively extending part of the Great Basin. Here are the highest steep scarps, the greatest deformation of Quaternary materials, the deepest topographic closures of structural depressions and the youngest exposed detachment faults of late-

Miocene to Holocene age. Unresistant Neogene upper-plate materials are widely preserved above the detachments and late-Quaternary valley fill hides much less structure than in less active regions. Mica K–Ar ages of 10–15 Ma for sub-detachment rocks in several ranges presumably date the early stage of vigorous tectonic denudation here as of late-middle-Miocene and early-late-Miocene age.

The domiform Tucki Mountain detachment fault in the Panamint Mountains has flank dips of 10–30° and a broad crest (Figs 20 & 12; Hunt & Mabey 1966). The fault is sub-parallel to the layering in lower-plate rocks, which are upper-Proterozoic metasedimentary rocks and Cretaceous muscovite granites. A retrograde, mylonitic carapace on these materials parallels the domiform fault, which is marked by gouge and incoherent breccia. The upper plate consists of unmetamorphosed Palaeozoic strata in the E, Pliocene (?) slide megabreccias on the crest and Pliocene fluvial strata, slide megabreccias, and basalt to the W, all steeply to moderately E-dipping to truncations downward against the detachment fault and broken by many W-dipping normal faults that end against the detachment.

I found in my own reconnaissance another small detachment dome in the Wildrose Canyon area, where E-dipping Pliocene strata and slide megabreccias are rotated down-to-the-SE against it. Detachment faults, with megabreccias in the upper plate, are exposed in several places to the S along the W base of the Panamint Range, where Albee *et al.* (1981) in part mapped them as 'slide surfaces'. Detachment faults are exposed also at the E foot of the range (the 'Amargosa thrust fault' of Hunt & Mabey 1966), where the lower plate is of mid-Proterozoic augen gneiss intruded by abundant Tertiary dykes. K–Ar biotite ages of 11–14 Ma in the gneiss (Stern *et al.* 1966) presumably reflect rapid Miocene uplift by tectonic denudation. I infer from the alteration of the Tertiary dykes, that they were injected while the wall rocks were still at ambient mid-crust temperatures.

The central and southern Panamint Mountains form a great tipped crustal section, 10–12 km thick and dipping about 20° E, entirely allochthonous above these detachment faults that are exposed to the W, N, and E, as I interpret the reports of Albee *et al.* (1981), Hunt & Mabey (1966), and Labotka *et al.* (1980). The down-section progression of the overlying crustal block begins with upper-Miocene volcanics (Cemen *et al.* 1985) and unmetamorphosed Palaeozoic strata and continues W through increasingly metamorphosed Palaeozoic and Proterozoic rocks, first andalusitic and then sillimanitic (Labotka *et al.* 1980). A shallow, cross-cutting Miocene granite occurs in the middle of the section and a Cretaceous muscovite granite occurs at the same low crustal level as that of sub-detachment rocks. Normal faults dipping gently W through the E-dipping crustal section (Labotka *et al.* 1980), presumably merge with the underlying detachment (cf. Fig. 20). The range-front fault, that bounded this upper-plate has been tilted with the block and with the flanking hanging wall slide breccias (Fig. 5). The palaeo-range-front fault is exposed all along the W slope of the Panamint Mountains (Albee *et al.* 1981), has a slip of about 3 km (Labotka *et al.* 1980) and projects obliquely up the Tucki dome to its crest (Fig. 20). The hanging wall breccias are truncated downward against an exposed detachment fault. The whole assemblage has been raised along young normal faults at the W and N edges of the range, although even the modern faults appear to offset upper-plate

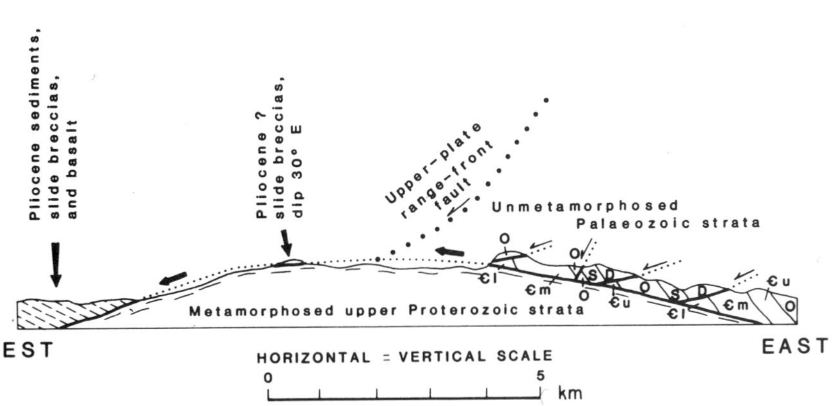

FIG. 20. Section across Tucki Mountain detachment fault, Panamint Mountains, California. Sense of slip of upper plate is the same in both flanks of dome. Palaeozoic strata: €1, €m and €u = lower, middle and upper Cambrian; O = Ordovician; S = Silurian; D = Devonian. Adapted from Hunt & Mabey (1966, their fig. 86).

materials more than lower-plate ones and hence may until recently have ended at the same detachment fault.

What I regard as a domiform detachment fault in the Funeral Mountains was mapped as many faults of diverse types by Hunt & Mabey (1966) McAllister (1971), B.W. Troxel and L.A. Wright (in Labotka 1980), and Giaramita (1984, who recognized its detachment character). Rocks inside the continuous elliptical trace of the detachment, so inferred, are upper-Proterozoic strata and possibly also older-Proterozoic basement rocks, metamorphosed during the Mesozoic to greenschist to uppermost amphibolite facies and intruded by muscovite granite. Kyanite occurs in rocks of all grades, and metamorphism occurred at a uniform depth of about 20 km (Labotka 1980, pers. comm. 1986). Upper-plate rocks are unmetamorphosed uppermost Proterozoic to Miocene strata, the latter including slide megabreccias rotated to steep to moderate E and SE dips and truncated against the detachment (Fig. 13). The Neogene stratigraphy is summarized by Cemen *et al.* (1985), who do not make a detachment interpretation. A greenschist-facies carapace on the lower plate displays isoclinal folding in the same shear sense as the slip direction of upper-plate rocks (Fig. 19) and opposite to vergence in unretrograded deeper lower-plate rocks (Fig. 21). Giaramita (1984) similarly recognized that W-verging folding and shearing (progressing with time from ductile and mylonitic to brittle) affected high lower-plate rocks at the N end of the dome synchronous with Tertiary detachment and denudation.

FIG. 21. Lower amphibolite facies Proterozoic metacarbonates, below carapace near area of Fig. 19. S folds (arrows to hinges) indicate shear of top-up-to-E (left) sense: Cretaceous synmetamorphic shearing had sense opposite to that in Neogene carapace. Exposure is 4m high.

Bare Mountain displays a detachment fault (The Tertiary thrust fault of Cornwall & Kleinhampl 1961; M.D. Carr & S.A. Monsen, unpubl. map; P.P. Orkild pers. comm., 1985) representing either another culmination on an undulating Funeral Mountains fault, or possibly the top of a separate mid-crustal lens. Miocene volcanic and clastic strata as young as about 12 Ma dip moderately to steeply E to their truncation at the gently dipping Bare Mountain fault atop Palaeozoic and upper-Proterozoic strata. These are metamorphosed to greenschist to lower amphibolite facies and are cut by gently dipping Tertiary faults. Biotite K–Ar ages of crystalline rocks beneath the Funeral Mountains and Bare Mountain detachments scatter from 10 to 50 Ma (M.D. Carr, pers. comm. 1985), so variations in synextensional and pre-extensional uplift and erosion are indicated. A composite detachment fault system exposed nearby in the Bullfrog Hills consists of an upper sheet of unmetamorphosed Palaeozoic strata intervening between downward-truncated Miocene rocks above and variably mylonitized gneisses beneath (Florian Maldonado, pers. comm. 1985).

The Black Mountains expose undulating detachment faults, of late-Neogene age, at two structural levels, as I interpret mapping and descriptions by: Drewes (1963—he recognized little extension); Otton (1977—summarized by Wright *et al.* 1974—who recognized great extension); and Wright & Troxel (1984—they inferred mostly imbricate listric normal faults where I identify undulating detachments). My inferences have been shaped also by my own reconnaissance and by discussions with B.W. Troxel. The structurally lower part consists of three small domiform masses, the NW-trending Badwater, Copper Canyon (Fig. 14), and Mormon Point turtlebacks, that are disconnected at present erosion levels and are separated by detachment faults from overlying Cenozoic materials in the W and plutonic rocks in the E. These masses consist of variably mylonitized mid-Proterozoic plutonic rocks capped by discontinuous carapaces of mylonitized and extremely attenuated upper-Proterozoic metasedimentary rocks (Otton 1977). The domiform turtlebacks might represent either giant mullions on a continuous detachment surface, or separate lenses above deeper faults. Drewes (1963) presented a lens-structure interpretation of a different sort. Above the turtleback detachments, to the E is a thick sheet forming a structurally higher lens of diorite and monzogranite of Mesozoic and (?) Tertiary age. Bounding this sheet on the top is the upper detachment fault, marked by breccias and lenses of upper- and lower-plate rocks in a zone as thick as 100 m. Above this zone, or elsewhere truncated sharply against the detachment, are variously mid-Proterozoic plutonic rocks, unmetamorphosed upper-Proterozoic and Cambrian sedimentary rocks and, mostly, mid-Miocene to Quaternary volcanic and sedimentary materials. The Cenozoic sections include extensive intercalated slide megabreccias (most 'chaos' and much 'Palaeozoic sedimentary rock' of Drewes 1963, and most 'megabreccia' and many 'fault slices' of Wright & Troxel 1984), from which I infer that high scarps were present, and hence that extension was very active, while they were being deposited. Around the southern Black Mountains, upper-Miocene rocks generally dip more steeply than do Pliocene rocks, although both are truncated by the same undulating faults, and mid-Quaternary strata are rotated to gentler dips in the same direction (Wright & Troxel 1984), whereas in the N, steep-dipping Pliocene strata are semi-concordant to upper-Miocene strata (Cemen *et al.* 1985; Hunt & Mabey 1966). Rotation synchronous with persistent faulting is indicated.

Each of the three exposed turtleback detachment faults is now active as a low-angle normal fault. A large, active normal fault, limited to upper-plate rocks, ends truncated downward against each turtleback (Fig. 7). Footwall Neogene volcanic rocks, hanging wall Quaternary slide breccias and the late-Quaternary Artist's Drive fault (Figs 4 & 5) are all truncated downward against the gently dipping Badwater turtleback. The Copper Canyon turtleback detachment truncates, from beneath, an upper-plate complex of footwall plutonic rocks, hanging wall tilted Pliocene strata and the intervening large normal fault (Fig. 14). Quaternary strata are rotated to steep truncations against the Mormon Point turtleback fault. In each case, the upper-plate rocks are slipping in the downhill direction on the turtleback detachments; gravitational sliding toward fast-deepening Death Valley is indicated. Faults that originated as zones of ductile slip in the middle crust have risen to the surface and have evolved until they now bound what are in effect giant landslides. Turtleback and upper-plate complexes alike are broken by normal faults that define parts of the present range front (Fig. 14).

Slip on the Bare Mountain detachment was largely completed before about 11.5 Ma, whereas detachments in the Bullfrog Hills continued to be active until perhaps 7 Ma; the Funeral Mountains were probably affected by considerable Pliocene slip and activity has continued throughout the Quaternary in the Black and Panamint Mountains. Late-Quaternary normal faults, with the usual 50° dips, break the Panamint and Black Mountains (Fig. 14) detachments. Major detachment faults must now be active at shallow depth beneath Death Valley, Panamint Valley, and areas W and NW of them.

Above all of these late-Neogene detachment faults, upper-plate strata, Proterozoic to lower-Pliocene, mostly display rotation to general steep to moderate E dips and hence a consistent slip relatively westward across both flanks of each detachment dome, up E flanks and down W flanks. Fault segments of opposite dips belong to continuous domiform or undulating structures, not, as they have often been interpreted, to kinematically unrelated structures sliding down the sides of domes, although the turtleback faults have evolved recently to the condition where such sliding is now occurring. Except for this final phase, which may be exemplified also by late slip on the Tucki Mountain (Fig. 12) and Wildrose Canyon detachments, the domes did not extrude upward, shedding slide masses off their flanks, for regional slip of upper plates requires continuous cover.

Tertiary strata comprise about half of the aggregate width of the upper plates truncated downward against detachment faults exposed in the Death Valley region (Fig. 7), and this requires at least 100% extension of the pre-Tertiary crust. Oligocene and Miocene strata may represent basins of markedly different configuration to that of the Pliocene ones, which in turn were quite different from modern ones (Cemen *et al.* 1985). Basin fills were dragged directly against detachment faults; pre-Tertiary upper-plate rocks did not underlie the basins. B.C. Burchfiel (pers. comm. 1985) regards his mapping as requiring similarly that the Quaternary fill of the actively widening northern Panamint Valley be dragging directly on a shallow detachment fault.

Throughout the region of Fig. 7, high, young fault scarps of the type commonly thought of as typifying Basin and Range structure are developed mostly in the upper-plate assemblages of detachment faults. The lower plates outcrop primarily as domiform masses and are broken into tilted fault blocks to only a minor extent. Relationships between modern and ancient basins and ranges and domiform detachment faults can be seen in Fig. 7. Most of the modern relief correlates with a position relative to the domes rather than with block faulting younger than the exposed detachment faults. The domiform sub-detachment masses and the resistant parts of the tilted upper-plate rocks above them form the modern ranges. The palaeobasins and the palaeoranges represented by, respectively, Tertiary and pre-Tertiary upper-plate materials are distributed randomly with regard to trough or crest positions across the domes. Perhaps initial basins and ranges were positioned systematically above lows and highs in underlying detachment faults (cf. Davis 1980) but were dragged across the expanding composite surface of sub-detachment lenses to their present positions. Where resistant rocks stand high atop a dome, normal faulting toward the adjacent lowlands is facilitated.

Active right-slip faults trend NW and serve as transforms between the modern obliquely separating, N-trending basin-range blocks. I infer that the sub-detachment crustal lenses, now active at depth, are sliding apart in a NW–SE direction beneath the modern blocks, that the northerly trends of those blocks are inherited from older extensional systems and that the strike-slip faults are restricted to upper plates allochthonous above detachments.

The detachment faults disappear beneath upper-plate materials at the E and W sides of the region discussed here. Rock assemblages of upper-plate type are broken by small to very large late-Cenozoic normal faults for about 75 km in both directions. As the age range of normal faulting differs across the region, the age range of major subjacent detachment faulting probably varies correspondingly.

Stewart (1983a) inferred that the part of the Panamint Mountains here regarded as the tilted upper-plate crustal block slipped about 45 km relatively westward from above the Black Mountains turtleback complexes. I concur with this conclusion; Cemen *et al.* (1985) do not.

Increasing area of detachment faults

If the analysis of detachment geometry summarized here is basically correct, then the widening of a normal-faulted terrain must be a response to an increase in area of the underlying detachment faults. The ages and intensities of detachment faulting over broad terrains vary complexly. Upper and lower plates do not display either systematic variations in initial crustal depth nor slips that increase undirectionally, both of which would be required were detachments planar faults cutting through the crust.

Sub-detachment structure

I infer that the middle crust is extended by the sliding apart, along ductile shear zones, of lenses that are little deformed internally. The area of the composite top of these lenses thus increases as deeper lenses are pulled out from beneath shallower ones.

Beneath the cataclastic carapaces beneath detachment faults, lower-plate rocks largely retain pre-detachment fabrics: synextensional ductile shear occurred in discrete zones and was not pervasive. Still deeper, additional gently dipping zones of brittle or ductile shear, discordant metamorphically and structurally with the older rocks, have been recognized (Figs 22 & 23). Such zones divide sub-detachment complexes into lenses. Major faults beneath the upper

FIG. 22. Horizontal sub-detachment Tertiary fault, Big Maria Mountains, California. Permian Kaibab Marble (Pk) and Coconino Quartzite (Pc) are truncated downward against sub-horizontal Mesozoic metasedimentary rocks (Mz), all metamorphosed at upper greenschist facies. Overturning is a product of Cretaceous recumbent folding; slip on the fault is a few km.

FIG. 23. Internally isoclinal mylonite of low greenschist facies, in fault zone near and like that of Fig. 22. Layers and lenses are rich alternately in mylonitic marble (m) and mylonitic greenschist (g; note contortions in lowest mass) but the two types are finely mixed. Kaibab Marble (Pk) at top is boudinaged but not mylonitized.

detachment faults in the Death Valley region were noted previously. John (1982) outlined a large sub-detachment lens with her Mohave Wash fault, and Spencer & Turner (1982) defined a smaller one with their 'lowest detachment fault'. Howard *et al.* (1982) found major sub-horizontal faults just above a detachment. I suspect that two very large lenses are in exposed contact in the Harcuvar Mountains of SW Arizona (cf. Rehrig & Reynolds 1980); a large eastern mass of deep-seated plutonic rocks has a domiform mylonitic carapace, which is overlain in the E by an exposed detachment fault and upper-plate assemblages, but in the W, at Cunningham Pass, by another sub-detachment mass, domiform in its topography, composed of granitic rocks typical of the upper part of the middle crust.

I infer that such lens-defining faults are common although relatively few have yet been mapped. I consider that junctions between detachment domiforms are mostly dihedral angles where deeper lenses emerge from beneath shallower ones; but such junctions in general are hidden beneath upper-plate rocks or beneath the alluvial fill of basins. Such lenses are shown in Fig. 24. In SE California, an undulating detachment fault dips S from the Riverside Mountains (Fig. 6) and another dips N from the Big Maria Mountains, the next range to the S; the distinctive lowerplate rock assemblages of the two ranges are very similar, except that markedly higher-grade Cretaceous metamorphism has affected the southern range, and it is possible that the Riverside lower plate moved about 25 km, relatively NE, from an initial position atop the Big Maria lower plate.

Domal form

Detachment faults can be irregular, undulating surfaces, but many outcrop as broad domes (e.g. Davis *et al.* 1980; Davis 1980; Rehrig & Reynolds 1980; Figs 7 & 20). Flank dips vary from a few to 25 degrees, locally more. Domes can be symmetrical or irregular, and typically have exposed dimensions of 10–20 by 20–40 km. Some are elongate in the direction of slip of overlying plates, as defined by the perpendicular to the axis of rotation of Tertiary rocks and by ductile-slip fabrics and, like parallel second-order corrugations of the surfaces, may be 'porpoising mullion structures' (John 1984; cf. Wright *et al.* 1974). Other domes are elongate at high angles to slip directions, or are irregular and I consider that these shapes record the interactions between diversely shaped lenses as they slid apart.

Other explanations that have been proposed for the domes do not appear to me to be broadly viable. Conjecture that the domes are products of the diapiric rise of hot cores that shrugged off their covers (e.g. Drewes 1981) is disproved because major offsets on bounding faults show no preference for down-dip directions, consistent senses of rotation of upper-plate rocks instead being maintained across series of domes (Fig. 7). Explanations invoking local heat sources (e.g. Miller *et al.* 1983) cannot generally be valid because many sub-detachment complexes were at only ambient middle-crust temperatures during extension. Plutonic rocks are present in many domiforms, as they are throughout the middle crust, but are of all ages from Precambrian to Miocene, whereas the domes were outlined in Cenozoic time and many domes expose little magmatic rock of any age. The common denominator of rocks beneath detachment faults is the depth, not the temperature or age, of their formation. Compressive folding could not have produced the domes, for it is incompatible with the continuing formation of extensional structures shown by many domes. Doming by isostatic uplift following differential tectonic denudation (Spencer 1984) would require, in view of the relatively steep flanks and small dimensions of many of the domes, an improbably low viscosity of the lithosphere.

Dykes

Sub-vertical mafic to silicic dykes of mid-Tertiary age, perpendicular to the direction of extension, are present in many ranges, mostly beneath detachment faults, but are generally sparse. Dyke swarms are present only locally (e.g. John 1982; Spencer & Turner 1982). The intrusion of dykes was therefore not a major contributor to regional extension at exposed crustal levels. Many lower-plate dykes exhibit static alteration to greenschist-facies minerals, which is thought to indicate emplacement at mid-crust depths and retention at greenschist temperatures.

Superimposed structures

Many detachment faults are broken by steep normal and strike-slip faults. Presumably the detachment complexes had risen to high crustal levels where further low-angle slip was not possible, and subsequent extension was accommodated

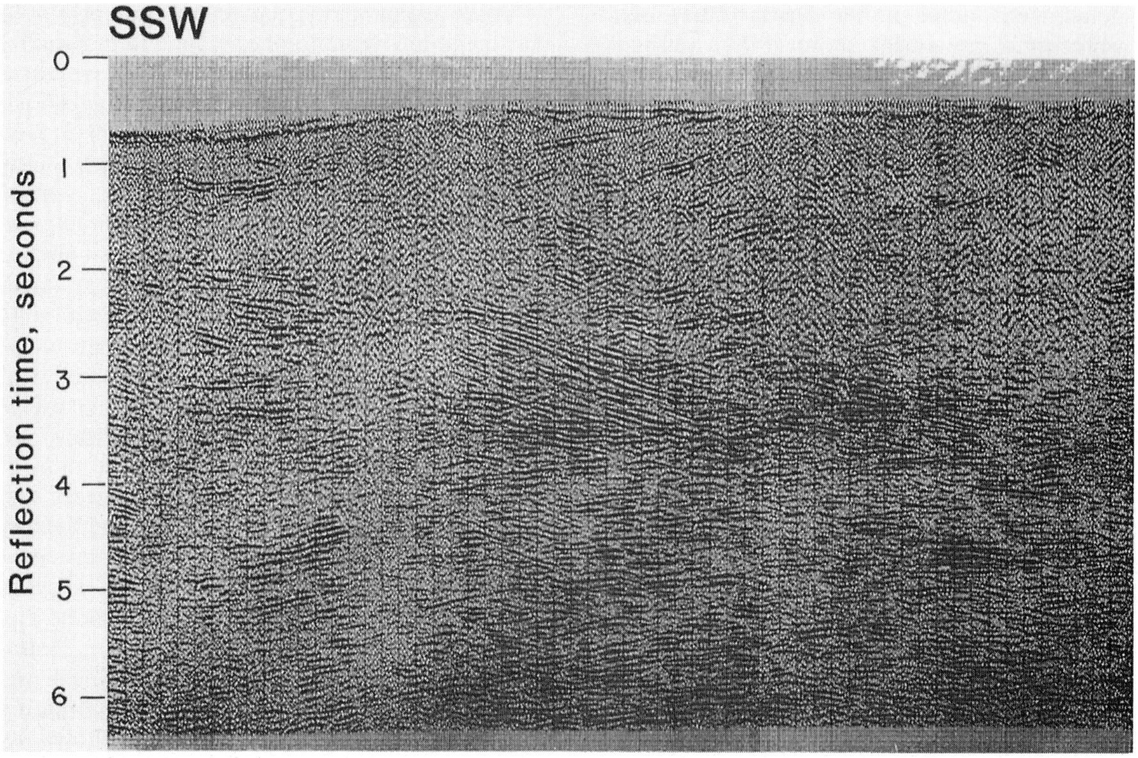

FIG. 24 (Above, and facing page). Deep-crustal seismic reflection profile in S-central Arizona (Fig. 1). The northern two-thirds of the line lies on or near outcrop of lower-plate middle-crust plutonic rocks, structurally beneath a mid-Tertiary detachment fault that outcrops with a northerly dip at the N end of the profile. The southern one-third lies on alluvial basin fill, and underlying bedrock may contain a detachment fault dipping SW. A well a few km W of the S end of the line penetrated 1 km of gravel and arkose, 2 km of upper-plate granitic rocks, a detachment fault at 3 km, and 2.5 km of sub-detachment plutonic rocks, including Tertiary muscovite granite, to bottom (Reif & Robinson 1981). Five biotite and hornblende K–Ar ages between the detachment fault and the bottom of the hole range only from 25 to 31 Ma (Reif & Robinson 1981), indicating rapid uplift from mid-crustal depths at the time of detachment faulting. The depth scale is based on regional refraction velocities. The profile is 40 km long; effective horizontal scale is 1.6 × vertical scale. The N end of line is near 33°05′ N, 111°00′ W; the S-end is near 32°50′ N, 111°10′ W. Profile provided by Pacific West Exploration and the Anschutz Corporation.

Interpretation: Reflectors represent variable mylonitization and shear transposition during Tertiary extension. The top 10 km (present upper crust, but middle crust before extension) is broken into great lenses which have slid apart; few reflections come from inside the lenses. The undulating apparent fabric of the lower 10 km (pre-extension lower crust) may, if real, record pervasive ductile flattening.

in the rotating-block, upper-crustal mode, new detachment faults being developed at depth (Figs 2 & 14).

Structures of the lower crust

The mode of extension deep in the crust cannot yet be inferred from field observations. Lower continental crust is exposed along the SW edge of the province, but deep extensional structures have not been recognized there.

Reflection profiles in Arizona yield information that may prove relevant to the behaviour of the deep crust. These profiles mostly remain proprietary, but a public one is reproduced as Fig. 24. The ground surface along most of this profile is eroded beneath a detachment fault, hence represents a crustal depth of at least 10 km before mid-Tertiary extension. The top 10 km of the profile displays anastomosing zones of apparent reflectors which I consider to be shear zones and to represent, in part, pre-existing lithologies transposed in the shear direction and, in part, velocity anisotropy in variably mylonitized rocks (cf. Fountain *et al.* 1984; Jones & Nur 1984). I conclude that the acoustically transparent rock between reflective zones consists of large lenses that retain pre-shearing massive or heterogeneous fabric. The lower half

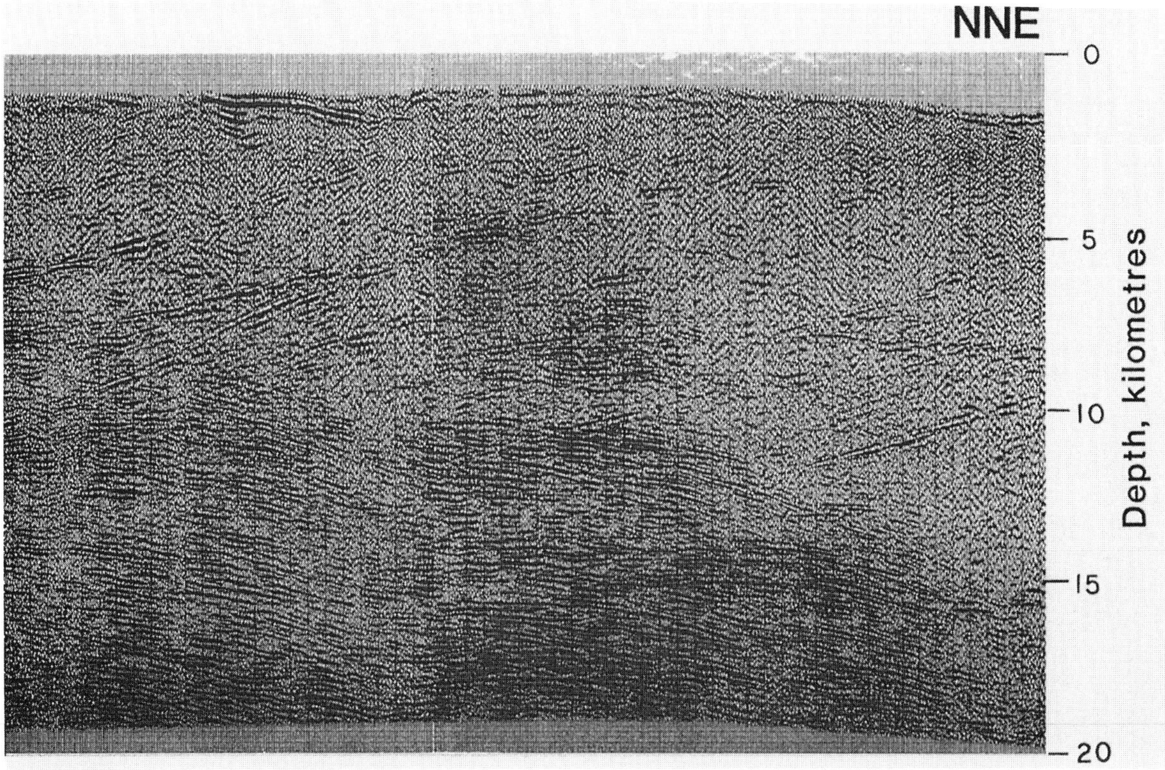

FIG. 24 cont. from facing page.

of this profile, like many of the other Arizona profiles, shows a strong fabric of gently undulating apparent reflectors. If such fabrics truly record varying velocity layering in the deep crust, and are not artefacts due to over-zealous processing or to such acoustic effects as reverberation, then pervasive ductile flattening may be indicated. Such pervasive deformation can be inferred also on theoretical grounds (Sibson 1983).

Other regions

The model advocated here may have a general application to extended continental crust. Domiform detachment faults separating steeply rotated upper-crustal rocks from mid-crustal complexes may be widespread around the world. Most have been interpreted in terms of thrust faulting because of their gentle dips, although this is disproved where thick crustal sections are omitted across them. A few examples are noted here.

The numerous 'Shuswap' domes of mid-crust rocks in NE Washington and interior British Columbia are bounded by detachment faults strikingly like those of the Basin and Range Province, except that those northern faults are of Eocene age. The structures display the same progressions of ductile to brittle structures in carapaces on the lower plates (Lane 1984, and references therein). Lower plates consist of mid-crust crystalline rocks, varying from kyanite greenschist to migmatite and muscovite granite and, locally, granulite, of Proterozoic to Eocene age. Upper plates consist of little-metamorphosed Proterozoic to Eocene strata, often truncated steeply against the gently dipping detachment faults, and upper-crust cross-cutting granitic rocks and their contact-metamorphosed wall rocks. Eocene K–Ar ages in many lower-plate rocks indicate great uplift at that time. Most investigators have interpreted these structures as being due to Mesozoic thrusting or rising of plutons, with minor Tertiary complications. Fox & Beck (1985) are among the few who have explained the structures, largely in terms of Eocene extension, and with this I concur.

The island of Naxos in the Aegean Sea is a detachment complex of Basin and Range type in an extensional setting (Lister *et al.* 1984). Beneath a domiform detachment fault is a lower plate of melange, metamorphosed at blueschist facies in the Palaeogene and then metamorphosed and migmatized under mid-crustal conditions in the Miocene. The upper plate includes unmetamorphosed Palaeogene melange and upper-crustal

Miocene granitic rocks. Both Palaeogene and Miocene indicators thus require great crustal elision across the detachment.

The Betic Cordillera of SE Spain contain high-grade metamorphic rocks separated by gently domiform faults from overlying unmetamorphosed Mesozoic to mid-Tertiary strata, which have steep to moderate dips (Platt 1982; Volk 1967). The faults have generally been assumed to be thrusts despite their elision of thick crustal sections. Platt recognized their extensional nature, although he invoked a rising-dome, gravity-sliding model.

Similarity to thrust deformation

The middle continental crust may deform in the same general lenticular mode under compressive overthrusting as under extension. Late Cretaceous mylonite zones each hundreds of metres thick outline huge, overlapping lenses in a gently dipping stack in the Santa Rosa Mountains of SW California. Reflection profiling shows that the upper-crust basement Palaeogene thrust that bounds the Wind River Mountains of Wyoming flattens downward into a series of lens-defining splays at depths of 17–26 km (Sharry *et al.* 1986) acoustically appropriate for mylonitic zones (Jones & Nur 1984). Such thrust lenses are defined by ductile shear zones in the Proterozoic Grenville crustal-thrust terrain of Ontario, where the pervasiveness of deformation increases with depth of exposure within the lower crust (Davidson 1984).

A great contrast between compressive and extensional regimes is that depth and temperature of mid-crust components increase with time in the former, but decrease in the latter.

ACKNOWLEDGMENTS: I started mapping what are now referred to as detachment faults, in SE California 25 years ago, when I termed them 'gravity thrust faults'; but it was not until I attended the 1977 Geological Society of America Penrose Conference on 'metamorphic core complexes', in Tucson, that I began to appreciate the common characteristics and regional prevalence of these remarkable structures. That fine conference was organized by M.D. Crittenden, Jr, P.J. Coney, and G.H. Davis. Discussions and field trips with, and papers written by, many geologists since then have profoundly influenced my views. My recent debt is greatest to G.A. Davis and K.A. Howard, whose interpretations in many ways overlap my own. This manuscript was much improved as a result of criticism by M.D. Carr, A.M.C. Sengor, and J.H. Stewart.

References

ALBEE, A.L., LABOTKA, T.C., LANPHERE, M.A. & McDOWELL, S.D., 1981. Geologic map of the Telescope Peak quadrangle, California. *U.S. geol. Surv. Map GQ 1532.*

ALLMENDINGER, R.W. *et al.* 1983. Cenozoic and Mesozoic structure of the eastern Basin and Range province, Utah, from COCORP seismic-reflection data. *Geology*, **11**, 532–6.

ARMSTRONG, R.L. 1972. Low-angle (denudation) fault, hinterland of the Sevier orogenic belt, eastern Nevada and western Utah. *Bull. geol. Soc. Am.* **83**, 1729–54.

ATWATER, T. 1970. Implications of plate tectonics for the Cenozoic tectonic evolution of western North America. *Bull. geol. Soc. Am.* **81**, 3513–36.

AXELROD, D.I. 1985. Miocene floras from the Middlegate basin, west-central Nevada. *Univ. Calif. Publ. geol. Sci. 129.*

BARTLEY, J.M. & WERNICKE, B.P. 1984. The Snake Range decollement interpreted as a major extensional shear zone. *Tectonics*, 3, 647–57.

CEMEN, I., WRIGHT, L.A., DRAKE, R.E., & JOHNSON, F.C. 1985. Cenozoic sedimentation and sequence of deformational events at the southeastern end of the strike-slip Furnace Creek fault zone, Death Valley region, California. *Soc. econ. Paleontol. and Mineral., Spec. Publ.* **37**, 127–41.

CONEY, P.J. & HARMS, T.A. 1984. Cordilleran metamorphic core complexes: Cenozoic extensional relics of Mesozoic compression. *Geology*, **12**, 550–4.

CORNWALL, H.R. & KLEINHAMPL, F.J. 1961. Geology of Bare Mountain quadrangle, Nevada. *U.S. geol. Surv. Map GQ 157.*

DAVIDSON, A. 1984. Identification of ductile shear zones in the southwestern Grenville Province of the Canadian Shield. *In*: KRONER, A. & GREILING, R. (eds) *Precambrian tectonics illustrated*, pp. 263–79. Schweizerbart., Stuttgart.

DAVIS, G.A. ANDERSON, J.L., FROST, E.G. & SHACKELFORD, T.J. 1980. Mylonitization and detachment faulting in the Whipple-Buckskin-Rawhide Mountains terrane, southeastern California and western Arizona. *Mem. geol. Soc. Am.* **153**, 79–129.

DAVIS, G.H. 1980. Structural characteristics of metamorphic core complexes, southern Arizona. *Mem. geol. Soc. Am.* **153**, 35–77.

—— 1983. Shear-zone model for the origin of metamorphic core complexes. *Geology*, **11**, 342–7.

DREWES, H.D. 1963. Geology of the Funeral Peak quadrangle, California, on the east flank of Death Valley. *U.S. geol. Surv. Prof. Pap. 413.*

—— 1981. Tectonics of southeastern Arizona. *U.S. geol. Surv. Prof. Pap. 1144.*

EATON, G.P. 1984. The Miocene Great Basin of western North America as an extending back-arc region. *Tectonophysics*, **102**, 275–95.

EFFIMOFF, I. & PINEZICH, A.R. 1981. Tertiary structural development of selected valleys based on seismic data: Basin and Range province, Nevada *Phil. Trans. R. Soc. London,* **A-300**, 435–42.

FOUNTAIN, D.M., HURICH, C.A. & SMITHSON, S.B. 1984. Seismic reflectivity of mylonite zones in the crust. *Geology*, **12**, 195–8.

FOX, K.F. & BECK, M.E., JR, 1985. Paleomagnetic results for Eocene volcanic rocks from northeastern Washington and the Tertiary tectonics of the Pacific Northwest. *Tectonics*, **4**, 323–341.

GIARAMITA, M.J. 1984. Structural evolution and metamorphic petrology of the northern Funeral Mountains, Death Valley, California. Unpubl. MSc. Thesis, Univ. California (Davis).

HAMILTON, W. 1978. Mesozoic tectonics of the western United States. *Soc. econ. Paleontol. and Mineral, Pac. sect. Paleogeog. Symp.* **2**, 33–70.

—— 1981. Crustal evolution by arc magmatism. *Phil. Trans. R. Soc. London*, **A-301**, 279–91.

—— 1982. Structual evolution of the Big Maria Mountains, Northeastern Riverside County, southeastern California. *In*: FROST, E.G. & MARTIN, D.L. (eds) *Mesozoic-Cenozoic ectonic evolution of the Colorado River region, California, Arizona, and Nevada (Anderson-Hamilton volume*), pp. 1–27. Cordilleran Publishers, San Diego.

—— & MYERS, W.B. 1966. Cenozoic tectonics of the western United States. *Rev. Geophys.* **4**, 509–49.

HOWARD, K.A. & JOHN, B.E. This volume. Crustal extension along a rooted system of imbricate low-angle faults: Colorado River extensional corridor, California and Arizona, pp. 299–311.

——, GOODGE, J.W. & JOHN, B.E. 1982. Detached crystalline rocks of the Mohave, Buck, and Bill Williams Mountains, western Arizona. *In*: FROST, G.C. & MARTIN, D.L. (eds) *Mesozoic–Cenozoic tectonic evolution of the Colorado River region, California, Arizona, and Nevada (Anderson–Hamilton volume*), pp. 377–90. Cordilleran Publishers, San Diego.

HUNT, C.B. & MABEY, D.R. 1966. Stratigraphy and structure, Death Valley, California. *U.S. geol. Surv. Prof. Pap. 494-A*.

JOHN, B.E. 1982. Geologic framework of the Chemehuevi Mountains, California. *In*: FROST, E.G. & MARTIN, D.L. (eds) *Mesozoic–Cenozoic tectonic evolution of the Colorado River region, California, Arizona and Nevada (Anderson–Hamilton volume*), pp. 317–25. Cordilleran Publishers, San Diego.

—— 1984. Primary corrugations in Tertiary low-angle normal faults, SE California: porpoising mullion structures? *Abstr. with Programs geol. Soc. Am.* **16**, 291.

JONES, T. & NUR, A. 1984. The nature of seismic reflections from deep crustal fault zones. *J. geophys. Res.* **89**, 3153–71.

KLEMPERER, S.L., HAUGE, T.A., HAUSER, E.C., OLIVER, J.E., & POTTER, C.J. 1986. The Moho in the northern Basin and Range Province, Nevada, along the COCORP 40°N seismic reflection transect. *Bull. geol. Soc. Am.* **97**, 603–18.

KLIGFIELD, R., CRESPI, J., NARUK, S. & DAVIS, G.H. 1984. Displacement and strain patterns of extensional orogens. *Tectonics*, **3**, 577–609.

LABOTKA, T.C. 1980. Petrology of a medium-pressure regional metamorphic terrane, Funeral Mountains, California. *Am. Min.* **65**, 670–89.

—— , ALBEE, A.L., LANPHERE, M.A., & MCDOWELL, S.D. 1980. Stratigraphy, structure, and metamorphism in the central Panamint Mountains (Telescope Peak quadrangle), Death Valley area, California. *Bull. geol. Soc. Am.* **I-91**, 125–9, **II-91**, 843–933.

LACHENBRUCH, A.H. & SASS, J.H. 1978. Models of an extending lithosphere and heat flow in the Basin and Range province. *Mem. geol. Soc. Am.* **152**, 209–50.

LANE, L.S. 1984. Brittle deformation in the Columbia River fault zone near Revelstoke, southeastern British Columbia. *Can. J. Earth Sci.* **21**, 584–98.

LISTER, G.S., BANGA, G. & FEENSTRA, A. 1984. Metamorphic core complexes of Cordilleran type in the Cyclades, Aegean Sea, Greece. *Geology*, **12**, 221–5.

MCALLISTER, J.F. 1971. Preliminary geologic map of the Funeral Mountains in the Ryan quadrangle, Death Valley region, Inyo County, California. *U.S. geol. Surv. Open-File Rep. 71–187.*

MILLER, E.L., GANS, P.B. & GARING, J. 1983. The Snake Range decollement, an exhumed mid-Tertiary ductile-brittle transition. *Tectonics*, **2**, 239–63.

MYERS, W.B. & HAMILTON, W. 1964. Deformation accompanying the Hebgen Lake earthquake of August 17, 1959. *U.S. geol. Surv. Prof. Pap. 435-I.*

OTTON, J.K. 1977. Geology of the central Black Mountains, Death Valley, California: The turtle-back terrane. Unpubl. PhD. Thesis, Pennsylvania State Univ.

PLATT, J.P. 1982. Emplacement of a fold-nappe, Betic orogen, southern Spain. *Geology*, **10**, 97–102.

RANSOME, F.L., EMMONS, W.H. & GARREY, G.H. 1910. Geology and ore deposits of the Bullfrog District, Nevada. *Bull. U.S. geol. Surv. 407.*

REHRIG, W.A. & REYNOLDS, S.J. 1980. Geologic and geochronologic reconnaissance of a northwest-trending zone of metamorphic core complexes in southern and western Arizona. *Mem. geol. Soc. Am.* **153**, 131–57.

REIF, D.M. & ROBINSON, J.P. 1981. Geophysical, geochemical, and petrographic data and regional correlation from the Arizona State A-1 Well, Pinal County, Arizona. *Ariz. geol. Soc. Dig.* **13**, 99–109.

REYNOLDS, S.J. 1985. Geology of the South Mountains, central Arizona. *Ariz. Bur. Geol. Min. Res. Bull. 195.*

SHARRY, J., LANGAN, R.T., JOVANOVICH, D.B., JONES, G.M., HILL, N.R. & GUIDISH, T.M.

1986. Enhanced imaging of the COCORP seismic line, Wind River Mountains. *Am. geophys. Union Geodyn. Ser.* **13**, 223–36.

SIBSON, R.H. 1983. Continental fault structure and the shallow earthquake source. *J. geol. Soc. London*, **140**, 741–67.

SMITH, R.B. & BRUHN, R.L. 1984. Intraplate extensional tectonics of the eastern Basin-Range: Inferences on structural style from seismic reflection data, regional tectonics and thermal mechanical models of brittle–ductile transition. *J. geophys. Res.* **89**, 5733–62.

SNOKE, A.W., LUSH, A.P. & HOWARD, K.A. 1984. Polyphase deformational history of northern Ruby Mountains-East Humboldt Range, Nevada. *In*: LINTZ, J., JR (ed.) *Univ. Nev. Dept. geol. Sci. Western geol. Exc.* **4**, 232–303.

SPENCER, J.E. 1984. Role of tectonic denudation in warping and uplift of low-angle normal faults. *Geology*, **12**, 95–8.

—— & TURNER, R.B. 1982. Dike swarms and low-angle faults, Homer Mountain and the northwestern Sacramento Mountains, southeastern California. *In*: FROST, E.G. & MARTIN, D.L. (eds) *Mesozoic-Cenozoic tectonic evolution of the Colorado River region, California, Arizona and Nevada. (Anderson–Hamilton volume)*, pp. 97–107. Cordilleran Publishers San Diego.

STECKLER, M.S. 1985. Uplift and extension at the Gulf of Suez: Indications of induced mantle convection. *Nature*, **317**, 135–9.

STEIN, R.S. & BARRIENTOS, S.E. 1985. Planar high-angle faulting in the Basin and Range: geodetic analysis of the 1983 Borah Peak, Idaho, earthquake. *J. geophys. Res.* **90**, 11, 355–66.

STERN, T.W., NEWELL, M.F. & HUNT, C.B. 1966. Uranium-lead and potassium-argon ages of parts of the Amargosa thrust complex, Death Valley, California. *U.S. geol. Surv. Prof. Pap. 550-B.* pp. 142–7.

STEWART, J.H. 1980. Regional tilt patterns of late Cenozoic basin-range fault blocks, western United States. *Bull. geol. Soc. Am.* **91**, 460–4.

—— 1983a. Extensional tectonics in the Death Valley area, California: Transport of the Panamint Range structural block 80 km northwestward. *Geology*, **11**, 153–7.

—— 1983b. Cenozoic structure and tectonics of the northern Basin and Range Province, California, Nevada and Utah. *Geotherm. Res. Coun. Spec. Rep.* **13**, 25–40.

VOLK, H.R. 1967. Zur Geologie und Stratigraphie des Neogenbeckens von Vera, südost-Spanien, Unpubl. thesis. Univ. Amsterdam.

WERNICKE, B. 1985. Uniform-sense normal shear of the continental lithosphere. *Can J. Earth Sci.* **22**, 108–125.

——, WALKER, J.D. & BEAUFAIT, M.S. 1985. Structural discordance between Neogene detachments and frontal Sevier thrusts, central Mormon Mountains, southern Nevada. *Tectonics*, **4**, 213–46.

——, SPENCER, J.E., BURCHFIEL, B.C. & GUTH, P.L. 1982. Magnitude of crustal extension in the southern Great Basin. *Geology*, **10**, 499–502.

WRIGHT, L.A. & TROXEL, B.W. 1984. Geology of the north 1/2 Confidence Hills quadrangle, Inyo County, California. *Calif. Div. Mines Geol.* Map Sheet 34.

——, OTTON, J.K. & TROXEL, B.W. 1974. Turtleback surfaces of Death Valley viewed as phenomena of extensional tectonics. *Geology*, **2**, 53–4.

ZOBACK, M.L., ANDERSON, R.E. & THOMPSON, G.A. 1981. Cainozoic evolution of stress and style of tectonism of the Basin and Range province of the western United States. *Phil. Trans. R. Soc. London*, **A-300**, 407–34.

WARREN HAMILTON, Branch of Geophysics, US Geological Survey, Denver, Colorado 80225, USA.

The extensional strength of the continental lithosphere: its dependence on geothermal gradient, and crustal composition and thickness

N.J. Kusznir & R.G. Park

SUMMARY: The response of lithosphere to an applied tectonic tensile force and the resulting stress distribution with depth has been investigated using a mathematical model incorporating the elastic, plastic and brittle behaviour of lithospheric material. Lithospheric strength is shown to be primarily controlled by lithospheric rheology and as a consequence is critically dependent on geothermal gradient and lithospheric composition. The rheologies of the upper crust, lower crust and mantle are assumed to be controlled by dislocation creep in quartz, plagioclase and olivine respectively.

The critical level of tensional tectonic force required to generate geologically significant strains has been calculated as a function of surface heat flow, and the predicted lithosphere strength compared with available levels of tensile tectonic force arising from subduction plate boundaries and isostatically compensated plateau uplift loads. The model predicts significant extensional deformation in regions with surface heat flow >65 mWm^{-2} subjected to a tensile tectonic force of 3×10^{12} N m^{-1} and is in good agreement with observed examples of intraplate extension.

Lithosphere strength is critically controlled by the crustal thickness since the quartzo–feldspathic rheology of the crust is weaker than the olivine rheology of the mantle. A decrease in crustal thickness thus increases the strength of the lithosphere. However, lithospheric extension also increases the geothermal gradient serving to weaken the lithosphere. The rate of extension is critical in determining which of these processes predominates. Fast strain rates ($>5\times10^{-15}$ sec^{-1}) produce a weakening of the lithosphere (i.e. strain softening) while slower strain rates lead to strengthening of the lithosphere (strain hardening). Extensional style is consequently controlled by the lithospheric extension rate; fast extension producing, through strain softening, intense localized lithospheric extension with high (potentially infinite) β values, and slow extension, through strain hardening, giving broader regions of lithosphere extension with finite β values of the order of 1.5.

For intermediate geothermal gradients ($q=55$–70 mWm^{-2}) the model predicts a low stress–low strength region at the base of the crust due to the contrast between plagioclase and olivine rheology at the Moho. Other low-strength regions are predicted within the crust at major compositional (and rheological) boundaries. These low-strength zones are expected to control the location of detachment horizons by which crustai extension occurs particularly at slower strain rates. High geothermal gradients favour the shallower detachment horizons at the expense of the deeper horizons. The reverse is true for low geothermal gradients.

The strength of the continental lithosphere controls both the initiation and subsequent evolution of extensional deformation. The response of a piece of lithosphere to an applied tectonic force is dependent on the vertical distribution of both brittle and ductile strength, which in turn is controlled by the varying rheology of lithosphere material with depth. Whereas brittle strength is controlled primarily by lithostatic pressure and increases with depth, ductile strength is controlled by temperature and decreases with depth because of the geothermal gradient. In tectonically stable lithosphere subjected to an applied force, an upper region of brittle deformation and a lower region of ductile deformation will be separated by a competent elastic region (Fig. 1). With an increase in the magnitude of the applied force, geothermal gradient or with time, the regions of brittle and ductile deformation will extend downwards and upwards respectively, reducing the elastic core to zero and thereby enabling geologically significant strain to take place. At this point, *whole lithosphere failure* (WLF) may be said to occur.

The variation of both brittle and ductile strength with depth controls the vertical distribution of stress resulting from an applied force and defines the critical value of that force which must be applied to the lithosphere in order to produce geologically significant strain rates. The mathematical model used to calculate the stress distribution is more fully described in

From Holdsworth, R.E. & Turner, J.P. (compiler) *Extensional Tectonics: Regional-scale Processes*. The Geological Society, Key Issues in Earth Sciences, **2 (1)**, 97–114.
1476-315X/02/$15.00 © The Geological Society of London 2002
First published in: "Coward, M.P., Dewey, J.F. & Hancock, P.L. (eds) *Continental Extensional Tectonics*, The Geological Society, London, Special Publications **28**, 35–52".

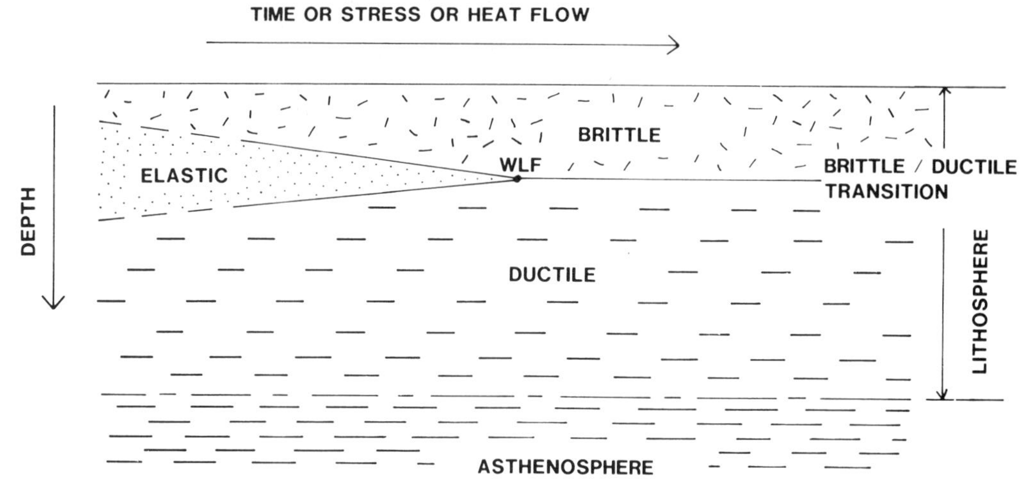

FIG. 1. A diagrammatic representation of the regions of brittle, elastic and ductile deformation within lithosphere subjected to a horizontal tectonic force. With increase in time, the level of the applied stress, or geothermal gradient, the elastic core of the lithosphere may be reduced to zero giving Whole Lithosphere Failure (WLF). Only after WLF can geologically significant strains occur.

Kusznir & Park (1984). The model examines the elastic, ductile and brittle response of the lithosphere to a lateral applied stress, assuming conservation of the total horizontal force arising from that stress, and that the lithosphere undergoes a uniform horizontal strain with depth. A Maxwell visco-elastic model is used for each infinitesimal component of the lithosphere. Brittle failure is predicted by Griffith theory (Griffith 1924) as modified by McClintock & Walsh (1962). Ductile deformation is assumed to be controlled by non-Newtonian power-law creep: dislocation creep in quartz for the upper crust and plagioclase for the lower crust; and a combination of dislocation and Dorn law creep in olivine for the mantle. The main stress-strain equations and values of constants used are given in the Appendix.

In Fig. 2, the strain rates of some minerals and rocks relevant to ductile lithosphere deformation are plotted against temperature. The curves, based on experimental data, assume a dislocation-creep mechanism and correspond to a stress difference of 50 MPa. Temperature is seen to have a marked effect on strain rate. The increase from 600 to 800°C for dry olivine for example, increases the strain rate from 10^{-20} to 10^{-15} sec^{-1}. At the same temperatures, quartz is very much weaker than olivine. At 600°C, the strain rates of dry quartz and dry olivine are roughly 10^{-11} and 10^{-20} sec^{-1} respectively. Typical strain rates for significant tectonic deformation are in the range 10^{-16} to 10^{-14} sec^{-1} (10^{-15} sec^{-1} produces 30% strain in 10 My). For quartz subjected to a differential stress of 50 MPa, a temperature of about 300°C would produce a strain rate of 10^{-15} sec^{-1}, whereas for dry olivine, a temperature of about 800°C would be required.

In the upper crust, deformation will be controlled by the behaviour of quartz (e.g. see White 1976). The rheology of the lower crust, however, is more likely to be controlled by the ductile deformation of plagioclase, which deforms more readily than pyroxene or olivine—quartz is unlikely to be present in sufficient quantity to control the deformation there. A model with a two-layer crustal rheology consisting of 50% wet quartz above 40% plagioclase is therefore used as the basis for the strength calculations. An alternative model with a three-layer rheology consisting of 50% wet quartz overlying 50% dry quartz overlying 40% plagioclase is also investigated.

The critical temperatures required to generate significant strain rates for these different rheologies are reached at different depths depending on the geotherm. For the purpose of the model, the geotherms have been calculated using a steady-state temperature model in which the geothermal gradient is characterized by the surface heat flow, q. The values of conductivity and radiogenic heat productivity, which vary with depth, are as given by Pollack & Chapman (1977).

Figure 3 shows how the model may be used to calculate stress–depth profiles at various times from 10^3 to 10^6 y after application of a tensile force of 10^{12} N m^{-1} to continental lithosphere with a surface heat flow of 60 mWm^{-2} (corresponding to the continental average). As time progresses ductile creed in the lower lithosphere

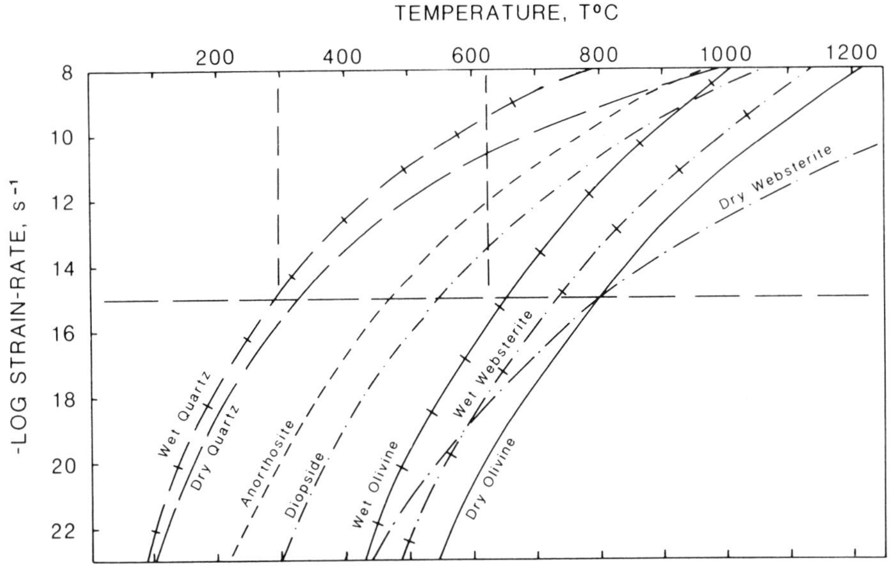

FIG. 2. Strain-rate plotted logarithmically against temperature for a number of rocks and minerals important in lithospheric plastic deformation. The curves assume a dislocation creep mechanism and correspond to a stress difference $(\sigma_1 - \sigma_3)$ of 50 MPa. The curves are derived from experimental data as follows: Quartz (Koch *et al.* 1980), Dry Olivine (Goetze 1978), Wet Olivine (Bodine, Steckler & Watts 1981), Websterite and Diopside (Avé Lallement 1978) and Anorthosite (plagioclase) (Shelton & Tullis 1981).

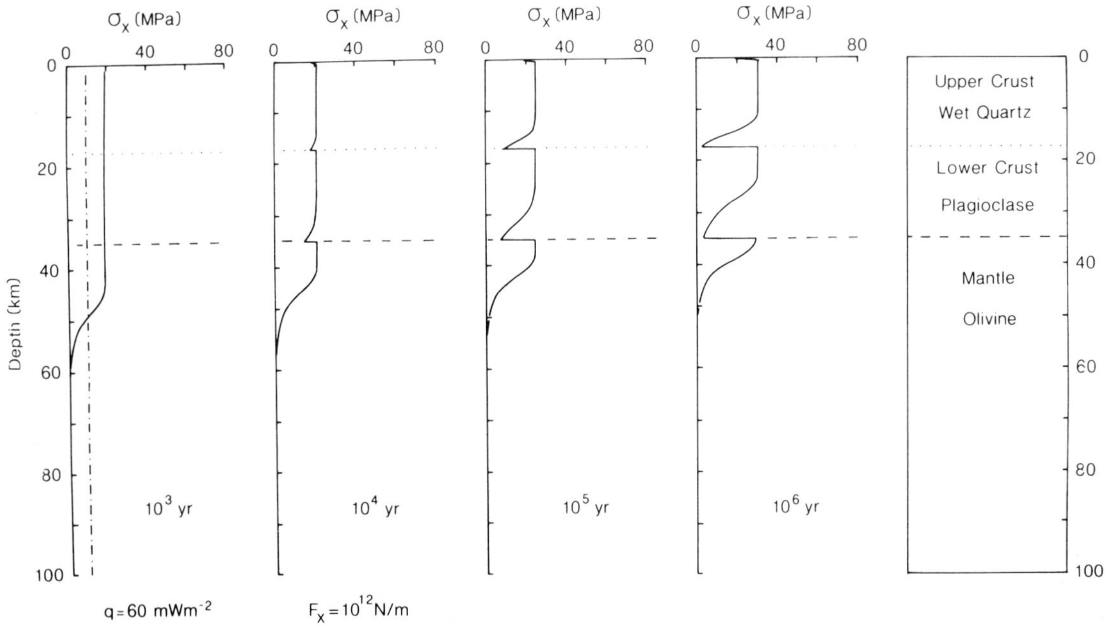

FIG. 3. Stress plotted against depth at various times after the application of a tensile tectonic force of 10^{12} N m^{-1} to continental lithosphere with a surface heat flow of 60 mWm^{-2}. Note the development of low stress–low strength regions above compositional (and therefore rheological) boundaries.

results in the dissipation of stress there and its transfer to the cooler non-ductile upper lithosphere, where it becomes amplified to levels sufficient to generate brittle failure. By 1 Ma after application of the initial force, stress levels in the upper lithosphere have increased by a factor of 4. Two major stress–depth discontinuities are apparent at the rheology changes between the upper and lower crust, and at the Moho.

Since lithosphere rheology is strongly temperature dependent, the geothermal gradient is critical in determining the variation of ductile strength with depth, and also the position and relative importance of the main strength discontinuities on which subsequent major tectonic displacements may take place. The rheological behaviour of the lithosphere is also dependent on the relative proportions of the weaker quartz–plagioclase rheologies of the crust and the olivine rheology of the mantle lithosphere. The relationship between *strength*, *crustal composition* and *thickness*, and *geothermal gradient* is therefore the key to the tectonic behaviour of the continental lithosphere, and the purpose of this paper is to explore and seek to quantify this.

The strength of the lithosphere

The control of strength by the geothermal gradient

The control of the stress–depth relationship by the geothermal gradient is illustrated in Fig. 4 where stress arising from an applied tensile force of 10^{12} N m^{-1} is shown as a function of depth for lithosphere models with a range of geothermal gradients (characterized by surface heat flow). For the coolest lithosphere models with $q = 45$ mWm^{-2}, corresponding to a shield-type geotherm, the applied force is carried to a depth of the order of 75 km leading to relatively small stress levels in the middle and upper lithosphere. As the geothermal gradient (and surface heat flow) increases ductile deformation within the upper mantle and crust extends upwards concentrating stresses within the upper crust and mantle. The large levels of stress associated with the larger heat-flow values are sufficient to produce brittle failure. The model with $q = 80$ mWm^{-2} shows the applied force carried almost completely in the top 10 km of the lithosphere with brittle failure almost completely penetrating the stress-carrying layer. A small elastic core still remains, however. For $q = 90$ mWm^{-2} the elastic core has been annihilated and WLF has occurred. The models with intermediate heat flows in the range 60–80 mWm^{-2} show pronounced low-strength regions in the lower and middle crust respectively, arising from the rheology discontinuities associated with the transition from plagioclase to olivine and wet quartz to plagioclase mineralogy.

The effect of increasing the level of the applied tensile tectonic force is shown in Fig. 5. Stress is shown as a function of depth for a lithosphere model with surface heat flow $q = 70$ mWm^{-2} subjected to a tensile force of 1.0×10^{12}, 1.5×10^{12} and 2.0×10^{12} N m^{-1}. The effect of increasing the value of the applied tensile force is to increase the level of the stress in the upper lithosphere resulting in deeper penetration of the region of brittle failure. For the model subjected to an applied force of 2.0×10^{12} N m^{-1}, the elastic core of the upper lithosphere has been eliminated and WLF has occurred. Only after WLF can extensive brittle and ductile deformation take place and lead to geologically significant strains.

The mathematical model can also be used to calculate the horizontal strains associated with the deformation arising from an applied tectonic force. In Fig. 6(a) horizontal strain is shown as a function of time for lithosphere models, with a range of surface heat flows, subjected to an applied tensile force of 2×10^{12} N m^{-1}. Strains are greater at a given time for models with larger geothermal gradients. Only models with $q > 70$ mWm^{-2} show WLF by 10 My and have geologically significant strain rates. In Fig. 6(b) horizontal strain is shown as a function of time for lithosphere models with a surface heat flow of 70 mWm^{-2} and a range of tensile applied forces. The models subjected to the larger applied forces show larger strains. A force of 2×10^{12} N m^{-1} or greater is required to give WLF by 10 My.

In order to generate WLF within a certain time, a critical level of applied force and a critical geothermal gradient must be exceeded. In Fig. 7 the critical tensile force required to generate WLF within 1 My is plotted as a function of heat flow. This critical force, which can also be regarded as the strength of the lithosphere, shows a very strong dependence on the geothermal gradient and decreases with increase in heat flow. Estimates of the tensile force arising from subduction suction and plateau uplift are also shown together with an estimate of the likely resultant level of tensile tectonic force. It can be seen that only lithosphere with a heat flow >65–70 mWm^{-2} is likely to undergo significant extensional deformation.

These model predictions may be compared with the heat-flow data and tectonic status of a number of representative regions (Fig. 8) divided into, (a) stable continental shields (Precambrian cratons), (b) regions of Palaeozoic orogenic crust, and (c) thermally active regions undergoing current or recent rifting and volcanicity. None of group (a) with heat-flow values in the range 34–49 mWm^{-2} show signs of significant tectonic deformation. In group (b) with heat-flow values in the range 57–75 mWm^{-2} local extensional failure occurs in Central Europe (Rhine–Ruhr rift system—see group (c)) and in

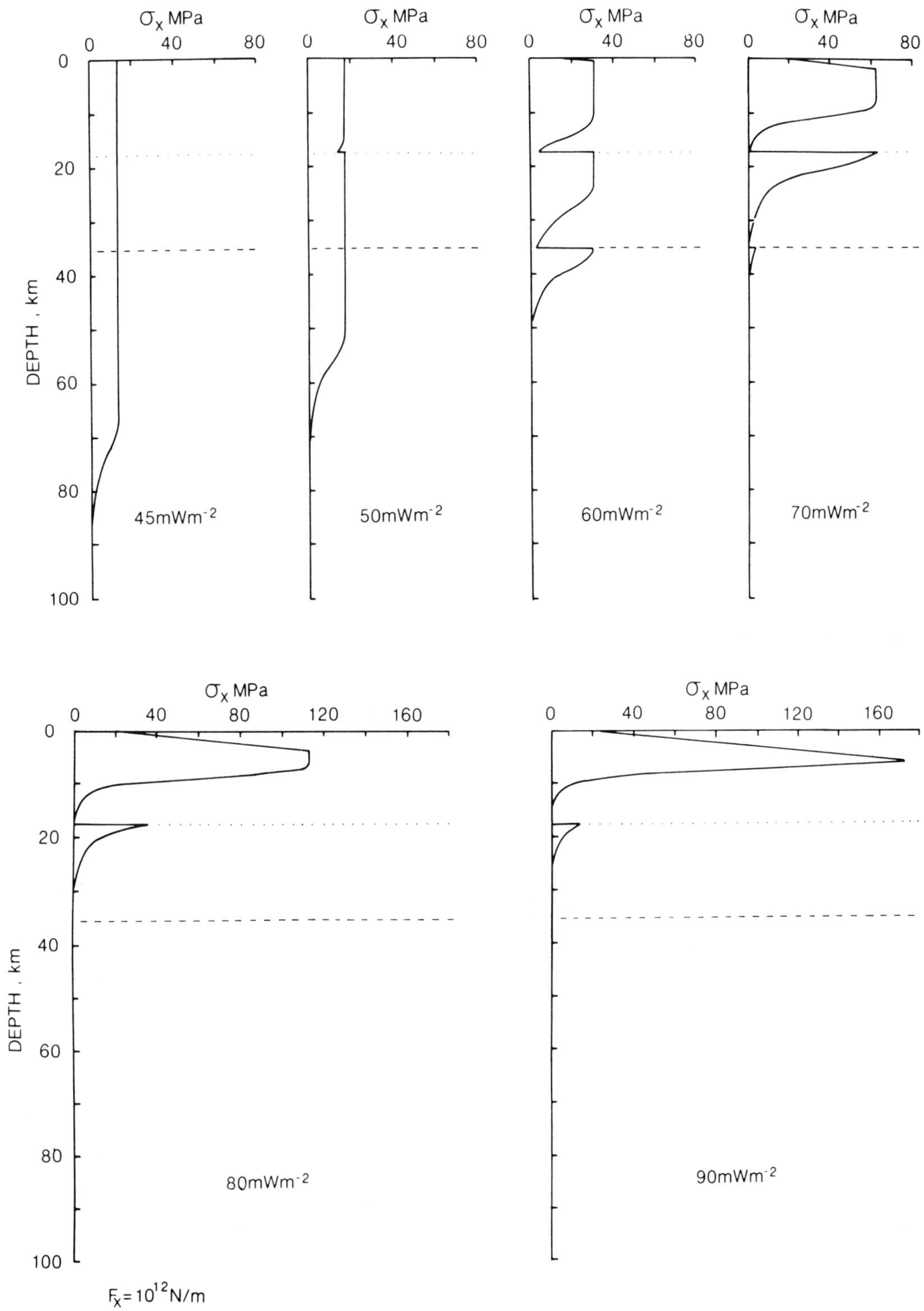

FIG. 4. Stress plotted against depth for a range of geothermal gradients, corresponding to surface heat flows of 45 to 90 mWm^{-2}, at 1 My after the application of a tensile tectonic force of 10^{12} N m^{-1}.

North China (Shansi graben system) both with heat flows of over 70 mWm^{-2}. In group (c) high heat flows (92–107 mWm^{-2}) are associated with the active rifts but lower heat flows occur in the flanking areas.

The model can be strictly applied only to the initial stage of rifting, starting from 'normal' lithosphere. Once crustal thinning has taken place, both the geometry and the rheology of the lithosphere are changed by the emplacement of

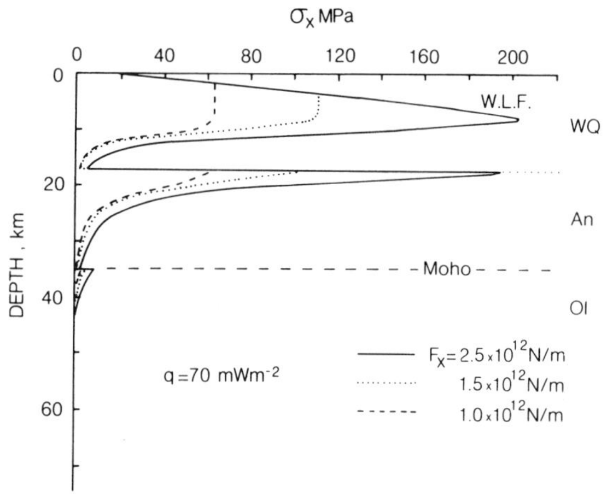

FIG. 5. Stress plotted against depth at 1My after the application of tensile forces of 10^{12}, 1.5×10^{12} and 2×10^{12} N m^{-1} to lithosphere with surface heat flow 70 mWm^{-2}. Increasing the magnitude of the applied force increases stress levels in the upper lithosphere causing brittle deformation to penetrate deeper until WLF occurs as in the case of the 2×10^{12} N m^{-1} model.

warmer asthenospheric material (cf. McKenzie 1978) and by the development of volcanicity. The flank values may thus probably be taken as minimum estimates and the rift values as maximum estimates of the heat flow at the time of rift initiation.

The data of Fig. 8 taken as a whole suggest a critical heat-flow band between 60 and 70 mWm^{-2} above which significant tectonic deformation may take place, which is consistent with the predictions of the model (see Fig. 7).

The control of lithosphere strength by crustal composition

The strength of the lithosphere is strongly controlled by the rheology of the lower crust, which is dependent not only on the geothermal gradient but also on composition and metamorphic grade. The continental crust is an extremely heterogeneous body, and in its lower part may vary from ultrabasic to acidic in composition and from granulite to amphibolite in facies. In Fig. 9(a) stress is plotted against depth for three models with different crustal compositions and rheologies. The effect of introducing a dry-quartz rheology into the lower crust is to considerably weaken the lower crust leading to larger stress levels in the upper crust compared with the model with a plagioclase-rheology lower crust. The stress–depth distribution arising from a three-layer wet-quartz–dry-quartz–plagioclase crust is also shown. In Fig. 9(b) the strength of the lithosphere as a function of heat flow is compared for models with a plagioclase and dry-quartz lower crust. At 70 mWm^{-2} the model with the plagioclase lower crust is approximately twice as strong as that with the dry-quartz rheology. The strengths of all olivine and all wet-quartz models are also shown. For high geothermal gradients ($q>90$ mWm^{-2}) the rheology of the wet-quartz upper crust completely controls lithosphere strength. For low geothermal gradients ($q<50$mWm^{-2}) strength is controlled by the much stronger olivine rheology of the mantle.

The control of lithosphere strength by crustal thickness

The quartzo–feldspathic rheologies of the crust are significantly weaker than the olivine rheology of the mantle. As a consequence, the strength of the lithosphere will be dependent on the relative proportions of quartzo–feldspathic and olivine material that it contains and therefore on crustal thickness. In Fig. 10(a) stress is shown as a function of depth for lithosphere models with crustal thicknesses of 20, 35 and 60 km. The thinner crust models (20 and 35 km) show stress being carried to greater depths than the thick-crust model (60 km) where the weaker quartzo–feldspathic rheology extends deeper. The 60-km crust model shows more than twice the maximum level of stress in the upper crust than the 20-km crust model. In Fig. 10(b) lithosphere strength as a function of heat flow is shown for models with a range of crustal thicknesses. The lithosphere is appreciably stronger for models with thinner crust. For a heat flow of 70 mWm^{-2}, a 20-km thick crust model is three to four times stronger than the 35-km thick model. At higher geothermal gradients, however, (for all but the zero crustal-thickness model) the lithosphere strength is independent of crustal thickness since the brittle–ductile transition has moved out of the olivine mantle and plagioclase lower crust into the quartz rheology upper crust. At low geothermal gradients too the strength of the lithosphere is independent of crustal thickness and is controlled only by the olivine rheology of the mantle. In this region of low geothermal gradients, however, the lithosphere is so strong compared with available levels of tectonic force that it is never likely to fail.

The evolution of lithosphere strength during extension

During extension the stretching of the lithosphere results in thinning of the crust and an

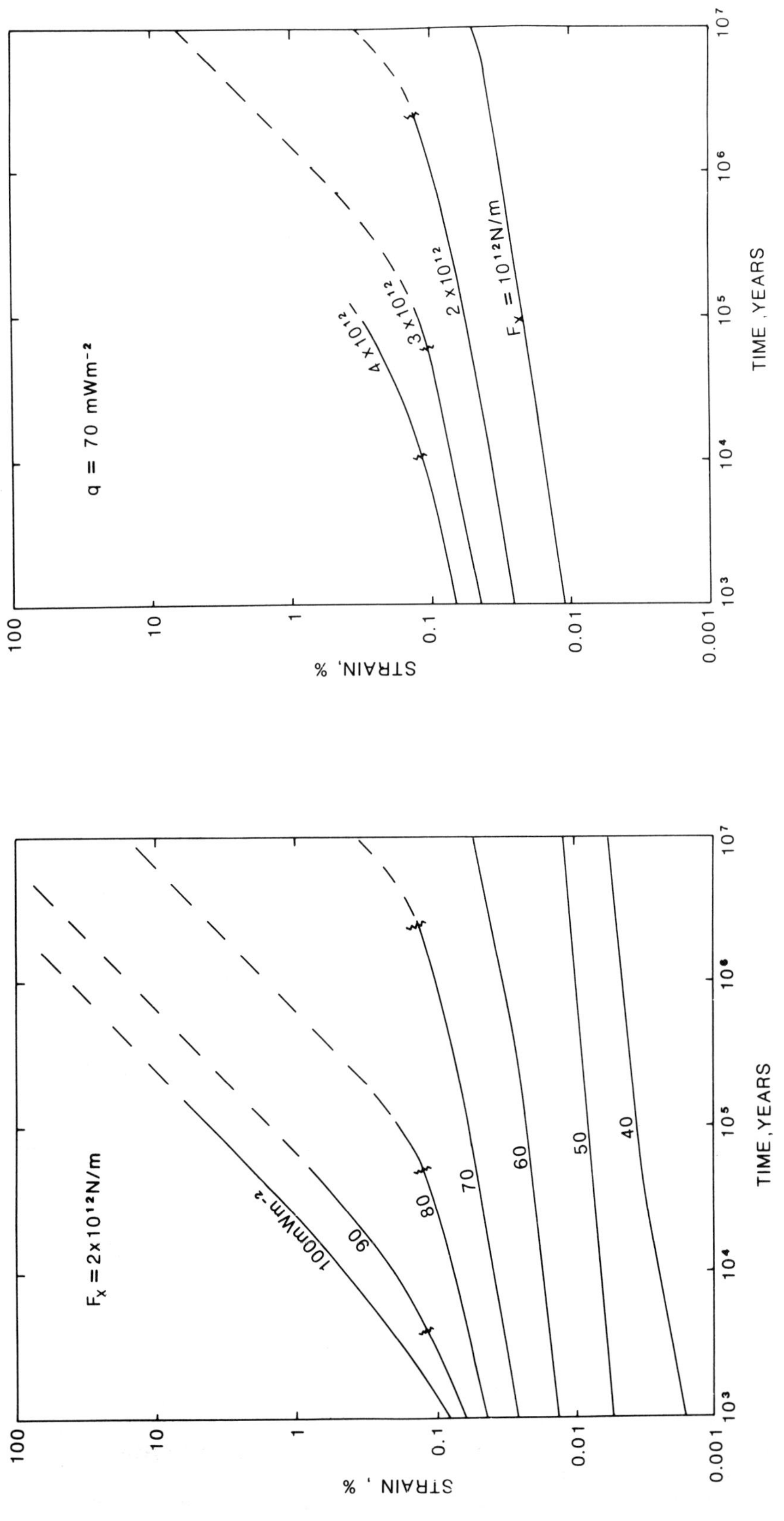

FIG. 6. Horizontal strain (%) plotted against time for: (a) lithosphere with a range of geothermal gradients subjected to a tensile force of 2×10^{12} N m^{-1}. (b) Lithosphere with a geothermal gradient corresponding to $q = 70$ mWm^{-2} subjected to a range of tensile force levels.

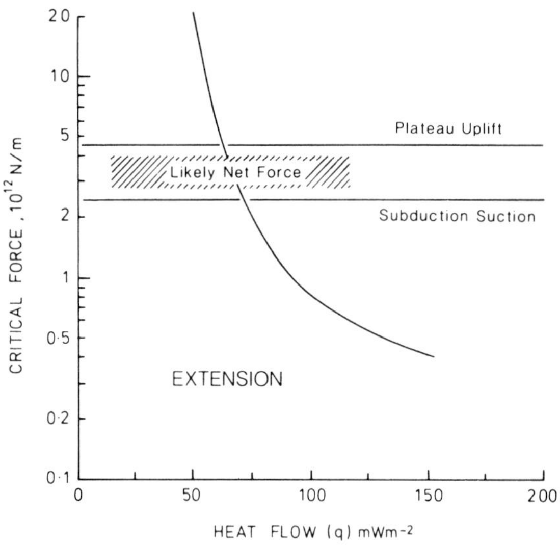

FIG. 7. Lithosphere extensional strength, defined as the critical tensile force required to generate WLF within 1 My, plotted against surface heat flow and compared with available levels of tensile tectonic force. Only lithosphere with a heat flow greater than 65–70 mWm^{-2} is likely to undergo significant extensional deformation.

increase in the geothermal gradient. Both crustal thickness and geothermal gradient have been seen to effect a strong control on lithosphere strength. In Fig. 11(a) the extensional strength of the lithosphere (defined as the applied tensile tectonic force required to generate WLF in 1 My) is shown contoured in crustal thickness–heat-flow space. Increasing the geothermal gradient during extension serves to weaken the lithosphere, while the simultaneous thinning of the crust will act to strengthen it. The lithosphere may show either a net strengthening (i.e. strain 'hardening') or a net weakening (i.e. strain 'softening') during extension depending on which effect dominates.

During extension at a finite strain rate, the increase in geothermal gradient due to lithosphere stretching will be partly offset by heat loss as the geothermal gradient endeavours to re-equilibrate. The behaviour of the geothermal gradient during extension will therefore be controlled by the extensional strain rate; fast strain rates producing a larger increase in geothermal gradient than slow strain rates for the same amount of stretching. In Fig. 11(b) typical tra-

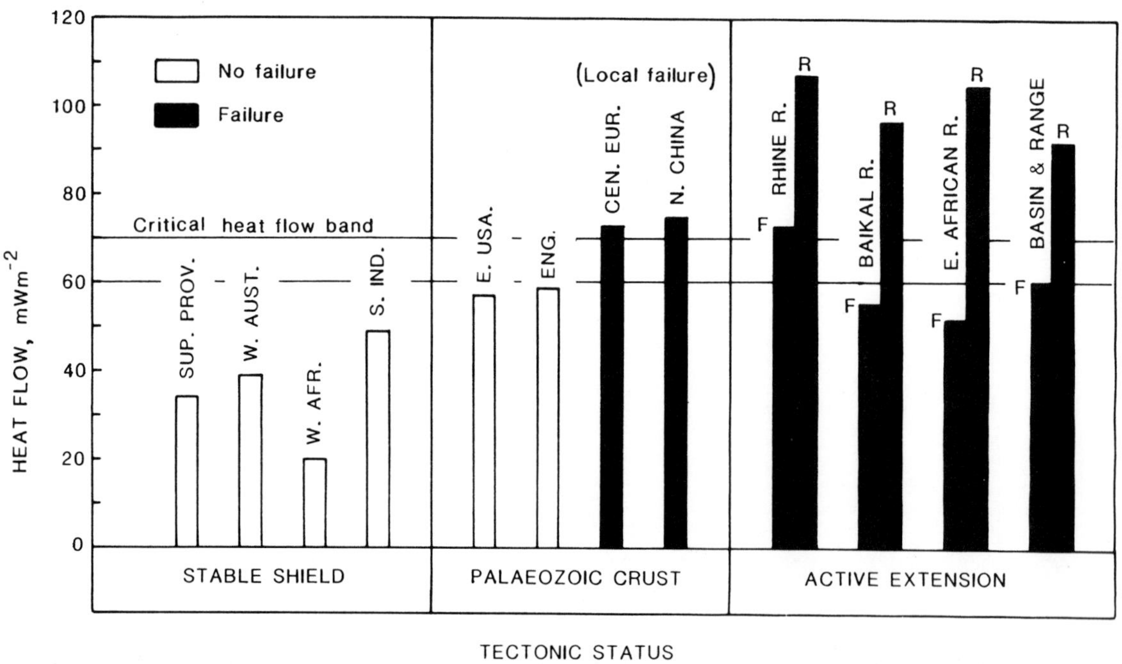

FIG. 8. Comparison of heat-flow values and tectonic status for a number of representative regions grouped into: stable shield (Precambrian cratons), Palaeozoic orogenic crust, and areas of active extensional rifting. Heat-flow data, in mWm^{-2}, are: Superior Province[1] 34±8; West Australia[1] 39±8; West Africa (Niger)[1] 20±8; South India[1] 49±8; E USA[1] 57±17; England and Wales[1] 59±23; Central Europe (Bohemian massif)[1] 73±18; Northern China[2] 75±15; Rhine graben[3] 107±35 (rift) and 73±20 (flanks – Rhenish massif); Baikal rift[3] 97±22 (SE flank 55±10); East African rift[3] 105±51 (flanks 52±17); Basin and Range Province[1] 92±33 (E flank – Colorado plateau[4] 60). Data from Viterello & Pollack (1980)[1], Pollack & Chapman (1977)[2] and Morgan (1982[3]; 1984[4]). For regions undergoing active extension both Rift (R) and Flank (F) heat flows are given.

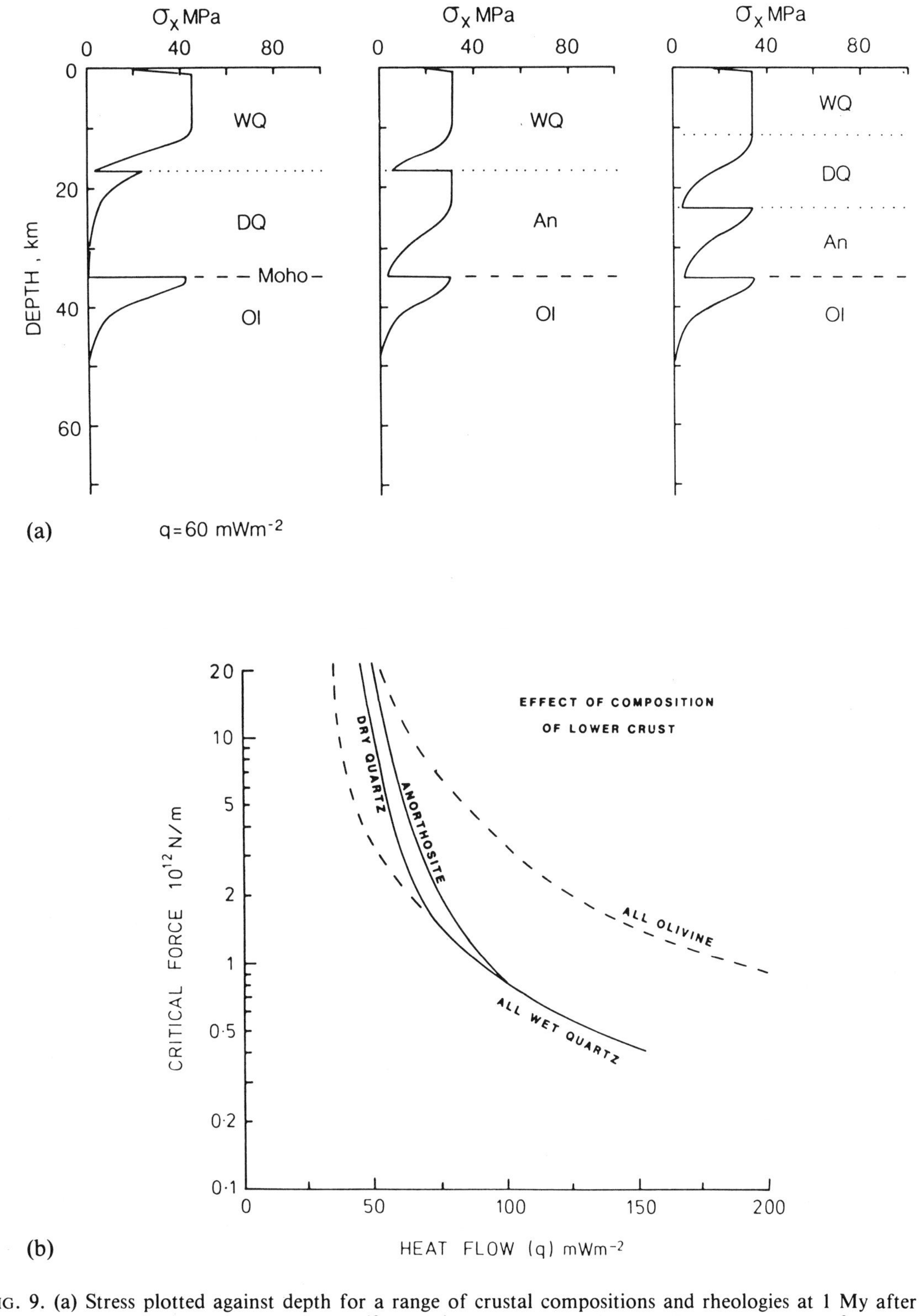

FIG. 9. (a) Stress plotted against depth for a range of crustal compositions and rheologies at 1 My after the application of a tensile tectonic force of 10^{12} N m^{-1} to lithosphere with heat flow $q=60$ mWm^{-2}. WQ = Wet Quartz; DQ = Dry Quartz; An = Anorthosite (Plagioclase); Ol = Olivine. (b) Lithosphere strength plotted against heat flow compared for models with anorthosite (plagioclase) and dry-quartz lower-crustal rheology. Strength curves for lithosphere models with all olivine and all wet-quartz rheology are also shown.

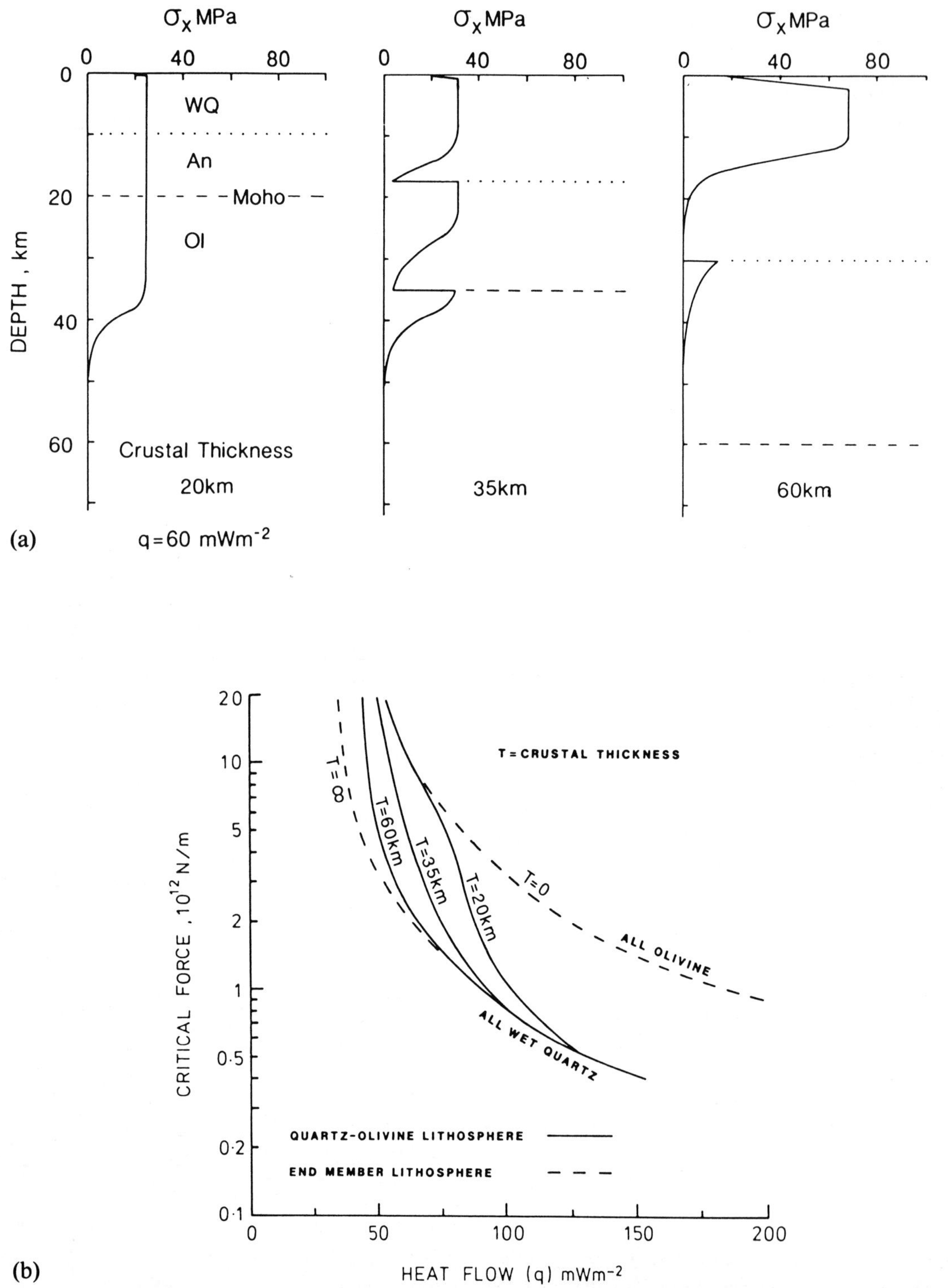

FIG. 10. (a) Stress plotted against depth compared for models with crustal thicknesses of 20, 35 and 60 km at 1 My after the application of a tensile tectonic force of 10^{12} N m^{-1} to lithosphere with heat flow $q = 60$ mWm^{-2}. (b) Lithosphere strength plotted against heat flow compared for models with different crustal thicknesses. Decreasing crustal thickness increases lithosphere strength.

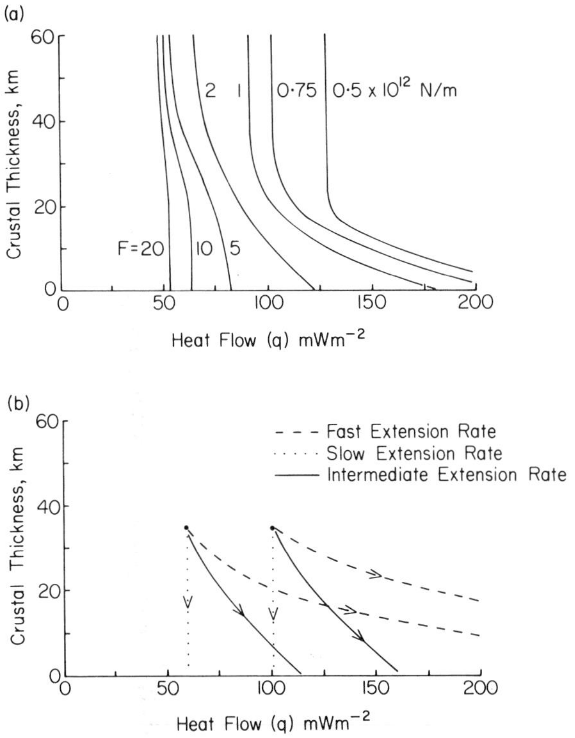

FIG. 11. (a) Contours of lithosphere extensional strength (the critical force required to generate WLF in 1 My) plotted in crustal thickness–heat-flow space. (b) Schematic trajectories in crustal thickness–heat-flow space for fast, slow and intermediate lithosphere extension rates.

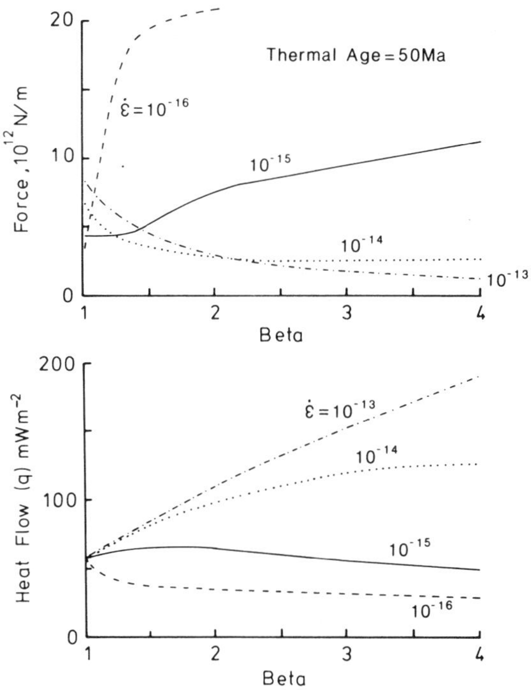

FIG. 12. Lithosphere strength (defined as the force required to maintain the respective strain rate) and surface heat flow plotted against β, the lithosphere extension factor, for a range of strain rates ($\dot{\epsilon} = 10^{-13}$ to 10^{-16} sec^{-1}). The initial model temperature structure corresponds to that of lithosphere with a thermal age of 50 My.

jectories in crustal thickness–heat flow space are shown for very fast and very slow extensional strain rates. Whether or not lithosphere strengthens ('hardens') or weakens ('softens') during extension will depend on whether the actual extension trajectories in crustal thickness–heat flow space ascend or descend the contours of lithosphere strength. As a consequence, the evolution of lithosphere strength during extension will depend critically on the extensional strain rate. For very fast strain rates a net weakening of the lithosphere will occur, while for slow strain rates there will be a net strengthening.

The quantitative control by strain rate on the evolution of lithosphere strength during extension can be calculated using a model which incorporates both the transient behaviour of the geotherm during extension and the thinning of the quartzo-feldspathic crust. In Fig. 12 lithosphere strength and heat flow during extension are plotted against Beta (β), the lithosphere stretching factor (McKenzie 1978), for a range of strain rates. The initial temperature structure of continental lithosphere has been calculated using a cooling-plate thermal model and has a 'thermal age' of 50 Ma. (The thermal structure is determined using a lithosphere thickness of 125 km and an initial uniform temperature at zero thermal age of 1300°C. Crust and mantle thermal conductivities and radiogenic productivities are those given by Sclater *et al.* (1981).) For slow strain rates, $\dot{\epsilon} = 10^{-16}$ sec^{-1}, the lithosphere hardens with extension, while for fast strain rates, $\dot{\epsilon} = 10^{-13}$ and 10^{-14} sec^{-1}, the lithosphere softens. The strain-rate curve for 10^{-15} sec^{-1} shows an almost constant lithosphere strength up to $\beta = 1.5$ followed by hardening. The behaviour of heat flow during extension is also shown to be controlled by strain rate, faster strain rates producing larger heat-flow values. The decrease in heat flow with β, shown for the strain rate of 10^{-16} sec^{-1} is caused by the thinning of the upper-crustal radiogenic heat producing layer during extension.

The effect of the initial geothermal gradient on the behaviour of lithosphere strength during extension is shown in Fig. 13 where lithosphere strength and heat flow are plotted against β, the lithosphere stretching factor, for a range of initial thermal ages of the model. All curves in Fig. 13 correspond to a strain rate of 10^{-15} sec^{-1}

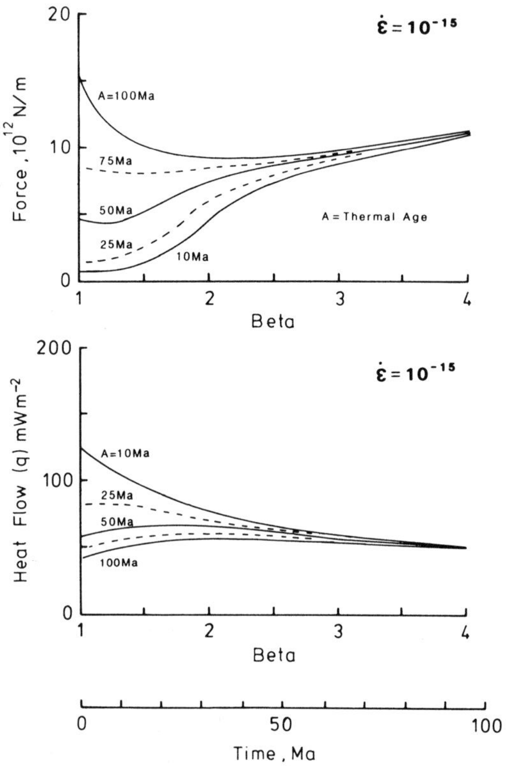

FIG. 13. Lithosphere strength (the force required to maintain the given strain rate) and surface heat flow plotted against β for a strain rate of 10^{-15} sec^{-1} compared for models with different initial temperature structures as characterized by lithosphere thermal age.

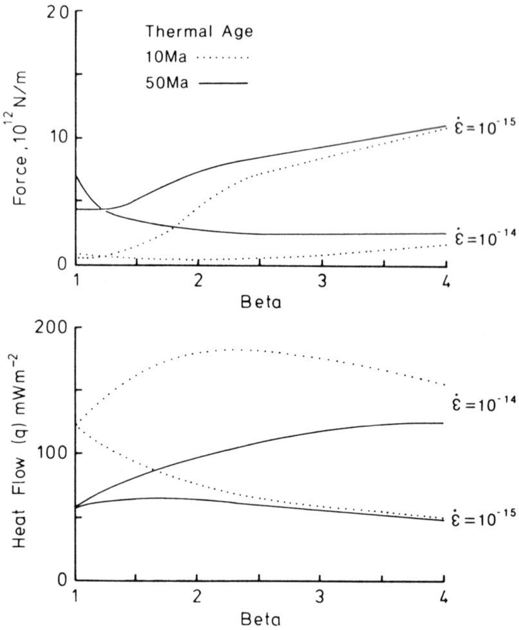

FIG. 14. A summary and comparison of the effect of both strain rate and the initial thermal age (i.e. temperature structure) of the lithosphere on lithosphere strength and heat flow as extension proceeds. β is the lithosphere extension factor. As extension proceeds both lithosphere strength and heat flow depend only on the strain rate not on the initial lithosphere temperature structure.

(and since a constant strain rate is used, time is also shown). The evolution of lithosphere strength during extension is shown to be strongly dependent on the initial thermal state of the lithosphere; thermally young, hot lithosphere tending to harden and thermally old, cool lithosphere to soften as extension proceeds. The cool models with thermal ages of 75 and 100 Ma are unrealistic, however, in that tensile tectonic forces of sufficient magnitude to initiate lithosphere deformation do not exist. The curves for thermal ages of 10, 25 and 50 Ma show an approximately constant lithosphere strength up to $\beta = 1.5$ before hardening occurs. The interrelationship between extensional strength, strain rate and β, and between heat-flow, strain rate and β is summarized in Fig. 14 for the 'fast' (10^{-14} sec^{-1}) and 'slow' (10^{-15} sec^{-1}) strain rates for two models with thermal ages of 10 and 50 Ma respectively. It can be seen that the values of strength and heat flow become increasingly controlled by the strain rate as β increases, and independent of the initial thermal state of the lithosphere.

It is again evident that there is a critical β value of about 1.5 above which slow strain rates lead to rapid strain hardening for both hot and cool thermal models. It is also significant that a fast strain rate can only be initiated with the hotter thermal model since the cooler model requires an unrealistically high force.

The hardening of a piece of lithosphere during extension would be expected to lead to the cessation of extension in that section of lithosphere and its transfer to an adjacent weaker region and would therefore generate finite maximum values of β. The model, as shown in Fig. 13, suggests a maximum β value of ≈ 1.5. This figure is consistent with the β values of a representative set of intracontinental extensional basins with an average of 1.4–1.5 (Table 1). It is clear from the above that the mode of development of lithosphere extension will be dependent on extension rate (see Fig. 15). Fast extension rates are expected to lead to locally intense extension with large β values and large geothermal gradients, whereas slow strain rates are expected to produce smaller β values, distributed over a much wider region, with much lower geothermal gradients (see Fig. 18 also).

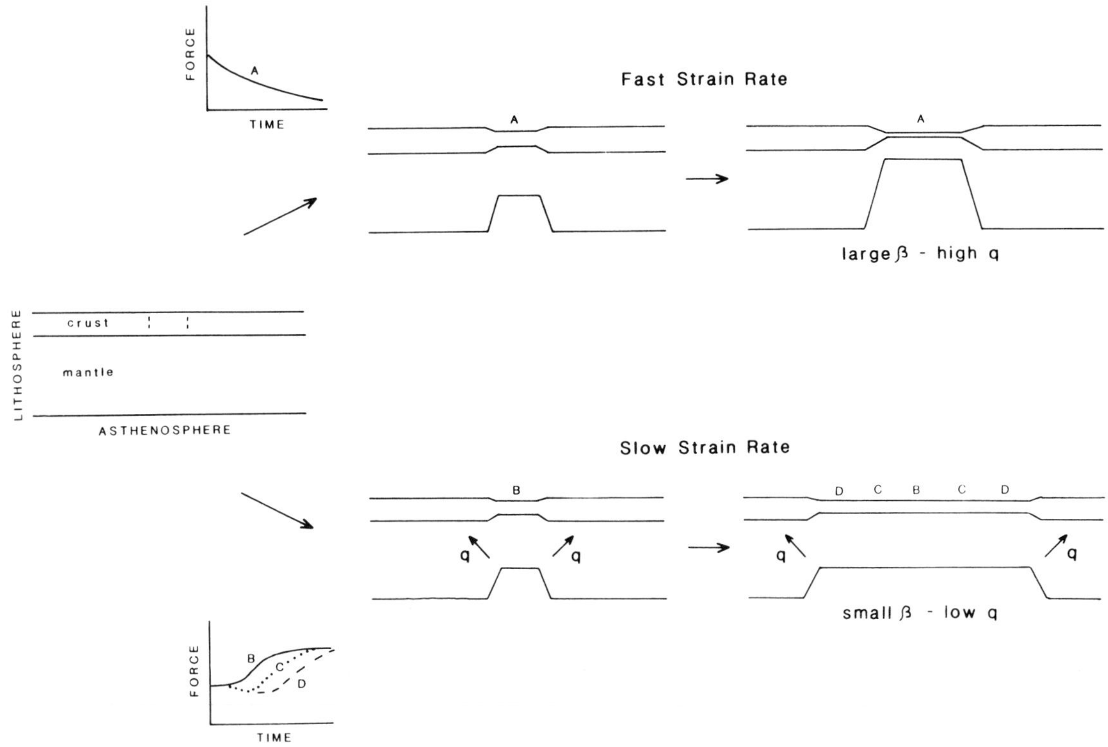

FIG. 15. Schematic diagram showing the effect of extensional strain rate on the style of extension and lithosphere temperature structure. *Fast strain rates*, through lithosphere strain softening, lead to intense localized extension with (locally) large β values and high heat flow. *Slow strain rates*, through lithosphere strain hardening, generates a broad region of extension with small β values and low surface heat flow.

The Rheological control of detachment horizons in extension

For intermediate geothermal gradients ($q = 60–70$ mWm^{-2}) the distribution of stress with depth shows pronounced zones of low stress and low ductile strength in the middle and lower crust (see Figs 4, 9(a) & 16). These zones result from the contrast in rheological strength between adjacent layers of different composition and would be expected to provide sites for the development of detachment horizons during extension. The two-layer crustal model (Fig. 4) shows low-strength zones between the quartz upper crust and plagioclase lower crust, and between the latter and the olivine mantle, respectively. The three-layer crust model (Fig. 16) shows low-strength zones between wet-quartz upper crust and dry-quartz middle crust, between dry-quartz middle crust and plagioclase lower crust, and between plagioclase lower crust and olivine mantle.

An example of such a three-layer rheological crust is found in northern Scotland where the LISPB seismic refraction line (Bamford 1979) is

TABLE 1. *Continental lithosphere extension.*

	β
North Sea basin – Central Graben	1.55–1.9
– Flanks	1.2–1.3
Rhine Graben	1.1–1.3
Pannonian Basin	1.8–2.7
Aegean	1.4
Vienna Basin	1.6–1.8
Paris Basin	1.3
Wessex Basin	1.1–1.25
Worcester Basin	1.2
Bass Basin	1.25–1.5
Gippsland Basin	1.8
East Siberian Basin	<1.5
Michigan Basin	<1.5
Illinois Basin	<1.5
Basin and Range	1.5–2
Representative average	β = 1.4–1.5

Data from G. D. Karner (1985 pers. comm.)

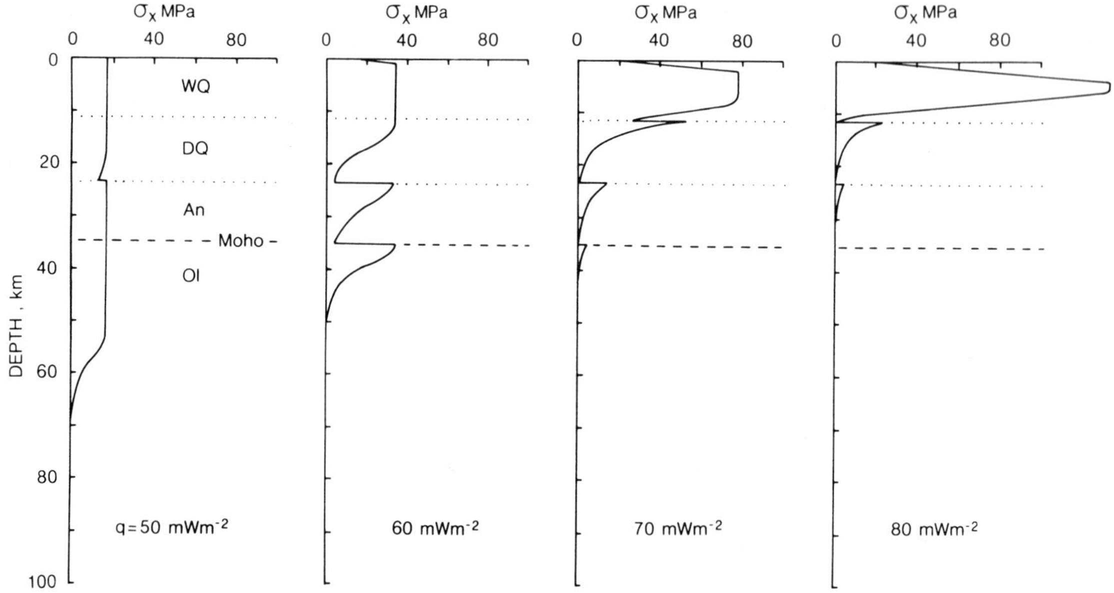

FIG. 16. Stress plotted against depth for a three-layer crustal model with a range of surface heat flows, showing the development of low-stress–low-strength regions in the middle and lower crust. The stress–depth distributions correspond to 1 My after the application of a tensile force of 10^{12} N m^{-1}. The low-strength regions represent the probable sites of detachment horizons. WQ = Wet Quartz; DQ = Dry Quartz; An = Anorthosite (Plagioclase); Ol = Olivine.

interpreted as indicating a Caledonian metamorphic upper crustal layer, above a Lewisian acid gneiss middle crustal layer, overlying a mafic granulite lower crust (Fig. 17). The boundaries between these layers may well represent tectonic detachment levels. The BIRPS MOIST deep reflection line (Brewer & Smythe 1984) shows that the major faults flatten out around 20 km corresponding to the base of the mid-crustal Lewisian acid gneiss layer.

The existence of these low stress, low-strength regions in the lower, middle and upper crust is very sensitive to the geothermal gradient (Fig. 16). For example, increasing the geothermal gradient from 60 to 70 mWm^{-2} introduces a third low-strength zone in the upper crust, while a

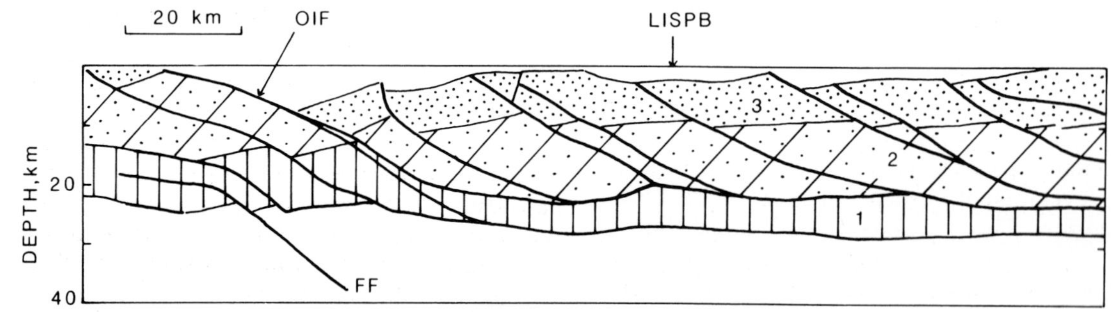

FIG. 17. Model of the MOIST profile across north Scotland after Blundell *et al.* (1984) showing the three main crustal layers recognized by Bamford *et al.* (1979). 1 – *lower crust* (7 km sec^{-1} – probably basic granulite); 2 – *middle crust* (6.4 km sec^{-1} – intermediate granulite-facies Lewisian); 3 – *upper crust* (6.15 km sec^{-1} – amphibolite-facies Caledonian metamorphics – mainly Moine in the E and Lewisian in the W). The uppermost (un-ornamented) layer corresponds to unmetamorphosed sediments from Torridonian to Mesozoic age. FF = Flannan 'fault', OIF = Outer Isles fault. Note that most of the major low-angle extensional faults (or shear zones) sole out along the top of layer 1 at about 20 km depth which therefore represents a key detachment horizon during post-Caledonian extension. A branch of the Outer Isles fault and the fault to the W appear to sole out along the Moho which may also have acted as a detachment horizon in the W (cf. Blundell *et al.* 1984).

further increase from 70 to 80 mWm^{-2} destroys the lowest zone at the base of the crust. Lower geothermal gradients therefore would favour the development of basal and mid-crustal detachment horizons whereas higher geothermal gradients would favour mid- and upper-crustal detachments (see Blundell *et al.* 1985).

Conclusion

The model demonstrates that the extensional strength of intraplate continental lithosphere is critically dependent on geothermal gradient, crustal composition and thickness. Of these, geothermal gradient exerts the strongest control on lithosphere strength; strength decreasing with increase in geothermal gradient. For likely levels of tectonic force, only lithosphere with heat flow >65–70 mWm^{-2} would be expected to undergo significant extension. This is consistent with heat-flow data from currently active and inactive tectonic regions.

Both the bulk strength of the lithosphere and the strength distribution with depth are controlled by crustal composition. Large variations in the predicted strength of the lithosphere are produced by substituting plagioclase rheology for dry quartz in the lower crust, or wet quartz for dry quartz in the upper crust. A three-layer crustal rheology model (wet quartz above dry quartz above plagioclase), corresponding approximately to the present crustal structure of northern Scotland (see Blundell 1985), predicts major strength discontinuities at each compositional boundary. These weak zones represent potential detachment horizons during extension. Lower geothermal gradients (heat flows in the range 60–70 mWm^{-2}) favour the development of basal and mid-crustal detachments, whereas higher geothermal gradients (70–80 mWm^{-2}) favour mid- and upper-crustal detachments. The BIRPS MOIST deep reflection section across northern Scotland indicates that the lower two boundaries appear to have acted as detachment horizons during Mesozoic extension (Fig. 17), which is consistent with the model predictions.

Because the quartzo-feldspathic rheology of the crust is weaker than the olivine rheology of the mantle, crustal thickness exercises a significant control on bulk lithosphere strength. The lithosphere is appreciably stronger for models with thinner crust. For lithosphere with surface heat flow of 70 mWm^{-2}, a 20-km thick crust model is three to four times stronger than a 35-km thick crust. For very high and very low heat flows however, the strength of the lithosphere is independent of crustal thickness being controlled by the quartz rheology of the upper crust and the olivine rheology of the mantle respectively.

The process of lithosphere extension causes both a thinning of the crust, which has a strengthening effect (strain 'hardening'), and an increase in the geothermal gradient, which has a weakening effect (strain 'softening'). The rate of extension is critical in determining which process predominates. Fast strain rates (e.g. $10^{-14} sec^{-1}$) produce a net weakening, while slow strain rates (e.g. $10^{-16} sec^{-1}$) produce a net strengthening. For an intermediate strain rate of $10^{-15} sec^{-1}$, lithosphere with a thermal age of 10–50 Ma shows an almost constant lithosphere strength up to a β value of 1.5, then hardens. This figure corresponds with the average β values of a representative set of intracontinental extensional basins.

It follows that slower extension rates would lead to the cessation of extension at one place when the critical β value was reached, and the spreading of the deformation laterally to an adjoining region, whereas faster strain rates with progressive softening would lead to a concentration of deformation in one section of lithosphere and potentially to infinite β values with crustal separation. Figure 18 is an illustrative model demonstrating the different tectonic styles which might be expected with fast and slow extension rates. Fast strain rates might be expected to lead to narrowing of the zone of active deformation and a progressive increase in β value, whereas slow strain rates should lead to progressive widening of the zone of active deformation as each section reaches the critical β value of around 1.5. The use of major mid- and lower-crustal detachment horizons would be favoured by the slow-extension model.

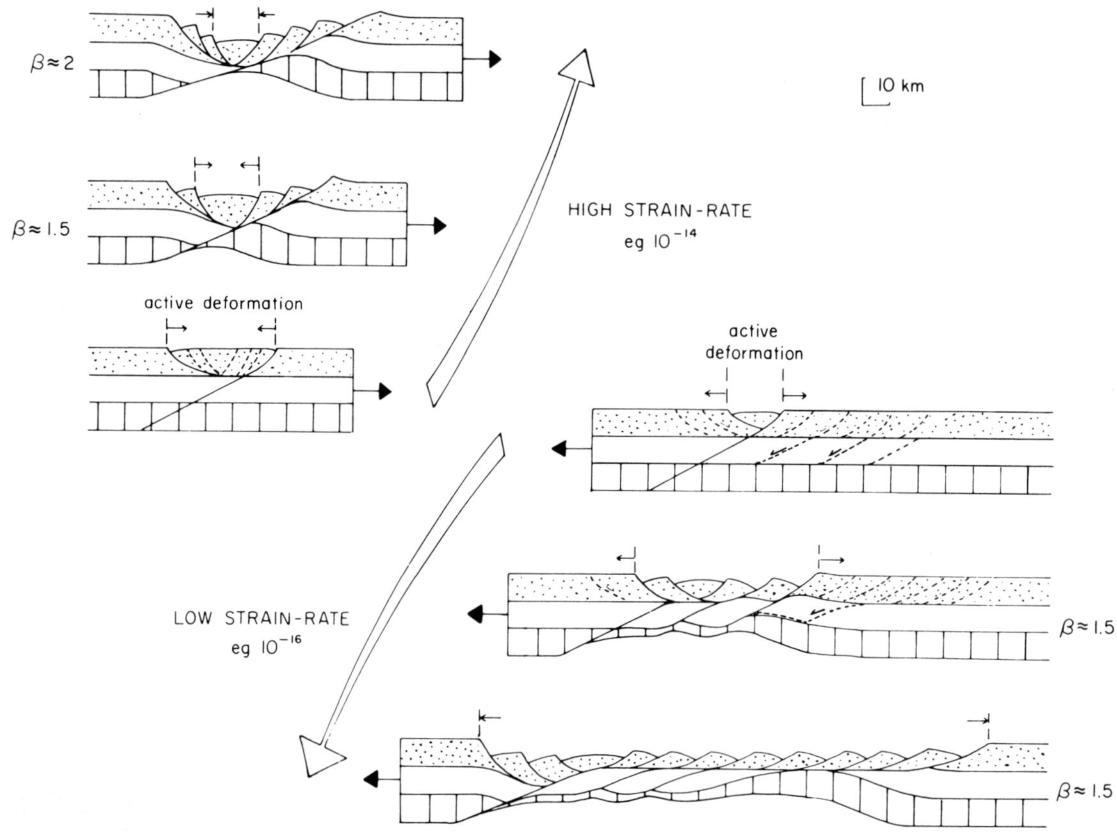

FIG. 18. Illustrative models showing the different styles of extensional deformation expected with fast and slow rates of extension. At *fast strain rates* (e.g. 10^{-14}) strain softening might be expected to localize the deformation near the original site of complete lithosphere failure causing progressive narrowing and intensification of the active deformation and leading potentially to high β values and complete crustal separation. At *slow strain rates* (e.g. 10^{-16}) local strain hardening might be expected to transfer deformation laterally to previously undeformed areas thus progressively widening the zone of active deformation but with a limiting β value of around 1.5. Note in the slow strain-rate model the use of detachment horizons between the crustal layers to transfer the deformation.

References

AVÉ LALLEMENT, H.G. 1978. Experimental deformation of diopside and websterite. *Tectonophysics*, **48**, 1–27.

BAMFORD, D. 1979. Seismic Constraints on the deep geology of the Caledonides of Northern Britain. *In*: HARRIS, A.L., HOLLAND, C.H. & LEAKE, B. E. (eds) *The Caledonides of the British Isles -reviewed. Spec. publ. geol. Soc. London,* **8**, 93–6.

BLUNDELL, D.J., HURICH, C.A. & SMITHSON, S. B. 1985. A model for the Moist seismic reflection profile, N. Scotland. *J. geol. Soc. London*, **142**, 245–58.

BODINE, J.H., STECKLER, M.S. & WATTS, A.B. 1981. Observations of flexure and the rheology of the oceanic lithosphere. *J. geophys. Res.* **86**, 3695–707.

BREWER, J.A. & SMYTHE, D.K. 1984. Moist and the continuity of crustal reflector geometry along the Caledonian-Appalachian orogen. *J. geol. Soc. London*, **141**, 105–20.

GOETZE, C. 1978. The mechanisms of creep in olivine. *Phil. Trans. R. Soc.* **A288**, 99–119.

GRIFFITH, A.A. 1924. Theory of rupture. *Proc. 1st int. Congr. Applied Mechanics*, Delft, **A221**, 163–98.

KOCH, P.S., CHRISTIE, J.M. & GEORGE, R.P. 1980. Flow law of 'wet' quartzite in the α-quartz field. *Eos.* **61**, 376.

KOHLSTEDT, D.L. & GOETZE, C. 1974. Low-stress high-temperature creep in olivine single crystals. *J. geophys. Res.* **79**, 2045–51.

KUSZNIR, N.J. 1982. Lithosphere response to externally and internally derived stresses: a viscoelastic stress guide with amplification. *Geophys. J.R. astron. Soc.* **70**, 399–414.

—— & PARK, R.G. 1984. Intraplate lithosphere deformation and the strength of the lithosphere. *Geophys. J.R. astron. Soc.* **79**, 513–38.

MCCLINTOCK, F.A. & WALSH, J.B. 1962. Friction on Griffith cracks under pressure. *Proc. 4th U.S. Nat. Congr. Applied Mechanics*, pp. 1015–21.

MCKENZIE, D. 1978. Some remarks on the development of sedimentary basins. *Earth planet. Sci. Lett.* **40**, 25–32.

MORGAN, P. 1982. Heat flow in rift zones, In: PALMASON, G. (ed.) *Continental and Oceanic Rifts. Am. Geophys. Union Geodyn. Ser.* 8, 107–22.

MORGAN, P. 1984. The thermal structure and thermal evolution of the Continental Lithosphere. *Phys. Chem. Earth. Preprint.*

POLLACK, H.N. & CHAPMAN, D.S. 1977. On the regional variation of heat flow, geotherms and lithosphere thickness. *Tectonophysics,* **38**, 279–96.

POST, R.L. 1977. High temperature creep of Mt Burnet Dunite. *Tectonophysics*. **42**, 75–110.

SCLATER, J.G., PARSONS, B. & JAUPART, C. 1981. Oceans and Continents: similarities and differences in the mechanism of heat loss. *J. geophys. Res.* **86**, 535–52.

SHELTON, G. & TULLIS, J. 1981. Experimental flow laws for crustal rocks. *Eos.* **62**, 396.

VITORELLO, I. & POLLACK, H.N. 1980. On the variation of continental heat flow with age and the thermal evolution of continents. *J. geophys. Res.* **85**, 983–95.

WHITE, S.H. 1976. The effects of strain on the microstructures, fabrics and deformation mechanisms of quartzites. *Phil. Trans. R. Soc. London,* **A283**, 69–86.

N.J. KUSZNIR, Department of Geological Sciences, University of Liverpool, Liverpool L69 3BX, UK.

R.G. PARK, Department of Geology, University of Keele, Keele, Staffs ST5 5BG, UK.

Appendix

The mathematical model considers the horizontal stresses arising from a tectonic force applied to the lithosphere in the horizontal direction, defined as the x-axis, at time $t=0$. The perpendicular horizontal axis is labelled y and the vertical axis z (measured positive downwards). Fundamental to the model is the conservation of the horizontal tectonic force applied to the lithosphere and this gives the equation:

$$\int_0^L \sigma_x \, dz = \text{constant},$$

where σ_x is the horizontal stress in the lithosphere and L is the lithosphere thickness. Differentiation of this equation with respect to time gives:

$$\int_0^L \dot{\sigma}_x \, dz = 0.$$

The assumption that the various layers of the lithosphere are welded together and that the lithosphere undergoes a uniform horizontal strain with depth gives the equation:

$$\frac{d\epsilon_x}{dz} = 0,$$

or differentiating with respect to time

$$\frac{d\dot{\epsilon}_x}{dz} = 0$$

Each infinitesimal component of the lithosphere is assumed to behave as a Maxwell visco-elastic material capable of brittle deformation. Stress and strain for such a material are linked by the equation:

$$\epsilon_x = \frac{1}{E}(\sigma_x - \sigma_x^0) - \frac{\nu}{E}(\sigma_y - \sigma_y^0) - \frac{\nu}{E}(\sigma_z - \sigma_z^0) + \epsilon_x^v$$

where ϵ_x is the total horizontal strain, σ_x is the total stress in the x-direction, σ_x^0 is the initial stress, ϵ_x^v is the ductile creep in the x-direction, E is Young's modulus and ν is Poisson's ratio. Initial stresses σ_x^0, σ_z^0 and σ_y^0 are used for modelling stress release by brittle fracture. Similar equations exist for strains ϵ_y and ϵ_z.

Plane strain in the y-direction gives the additional equation $\epsilon_y = 0$ while the vertical stress σ_z arising from the applied stress is zero, i.e. $\sigma_z = 0$. Manipulation and integration of the above equations gives the final equations for the temporal and spatial variations of σ_x and σ_y

$$\sigma_x = \int_0^t \left(\frac{1}{L}\int_0^L k\dot{\epsilon}_v dz - k\dot{\epsilon}_v\right) dt' - \frac{1}{L}\int_0^L \sigma_x^0 \times dz + \sigma_x^0,$$

$$\sigma_y = \int_0^t \left[\nu\dot{\sigma}_x - E\,\frac{(2\sigma_y - \sigma_x)}{6\eta}\right] dt' + \sigma_y^0 - \nu\sigma_x^0$$

where $k = E/(1-\nu^2)$.

The derivation of these equations is decribed in greater detail by Kusznir (1982).

Following Kohlstedt & Goetze (1974), Goetze (1978), Post (1977) and Bodine *et al.* (1981) the following relationship between creep rate, stress and temperature for dislocation, and Dorn law creep for the olivine rheology of the mantle is used:

dislocation creep,

$$\dot{\epsilon} = 7 \times 10^{10} \exp\left(\frac{-53030}{T}\right)(\sigma_1 - \sigma_3)^3 \text{ sec}^{-1}$$

for $(\sigma_1 - \sigma_3) < 2$ kb;

Dorn law,

$$\dot{\epsilon}=5.7\times10^{11}\exp\left(\frac{-55556}{T}\left(1-\frac{(\sigma_1-\sigma_3)}{85}\right)^2\right)\sec^{-1}$$

$$\text{for } (\sigma_1-\sigma_3)>2 \text{ kb},$$

where $(\sigma_1-\sigma_3)$ is in kb (1kb = 100 MPa).

The creep rate used for the wet-quartz, dry-quartz and plagioclase rheologies of the crust are as follows and are based on the experimental work of Koch, Christie & George (1980) and Shelton & Tullis (1981):

wet quartz,

$$\dot{\epsilon}=4.36\exp\left(\frac{-19332}{T}\right)(\sigma_1-\sigma_3)^{2.44}\sec^{-1};$$

dry quartz,

$$\dot{\epsilon}=0.126\exp\left(\frac{-18245}{T}\right)(\sigma_1-\sigma_3)^{2.86}\sec^{-1};$$

plagioclase,

$$\dot{\epsilon}=8.2\times10^{2}\exp\left(\frac{-28788}{T}\right)(\sigma_1-\sigma_3)^{3.2}\sec^{-1},$$

where $(\sigma_1-\sigma_3)$ is in kb.

Brittle deformation within the lithosphere has been predicted by the use of Griffith's theory (1924) as modified by McClintock & Walsh (1962). Three failure domains have been used—tensional failure, open-crack compressional, and closed-crack compressional. The physical parameters, within the modified Griffith theory, which control failure are tensile strength, T_0; the frictional coefficient, μ; and the critical stress, σ_{gc}, required to close the Griffith crack. Of these parameters T_0 and μ have the greatest influence on brittle failure.

The following values for these brittle-failure parameters have been used: $\mu=0.5$, $T_0=0.2$ kb (20 MPa) and $\sigma_{gc}=1.0$ kb (100 MPa). The justification for these chosen values and the effect of varying μ and T_0 on lithosphere strength is discussed in Kusznir & Park (1984).

Failure is dependent on stresses σ_x and σ_z. The stress σ_y, being the intermediate stress, is not required in the two-dimensional Griffith–McClintock and Walsh formulation. At failure the stress σ_x is returned to the failure envelope (the stress σ_z is simply lithostatic).

Developments in the structural geology of rifts over the last decade and their impact on hydrocarbon exploration

C. K. MORLEY

Department of Petroleum Geology, University of Brunei, Gadong 3186, Brunei Darussalam

Abstract: Three different aspects of rift geometry and their impact on hydrocarbon exploration are examined: rift architecture, fault angle and basin inversion. Rift architecture is controlled by fault geometry, which itself is heavily influenced by pre-existing fabrics. At the largest scale, rifts are divided into segments tens to hundreds of kilometres long. They can be joined along offset segments, separated by rift jumps and gaps, or terminated in splays. Rift jumps and gaps provide the entry points for large rivers to enter rifts and create along-axis drainage systems, while rift offsets are areas where anomalous structural patterns develop in response to oblique extension. Within rift segments transfer zones between boundary faults mark important changes in rift geometry and are the preferred sites of coarse clastic sedimentation. In the Gulf of Suez, for example, the structures and the syn-rift reservoir units within two major transfer zones are associated with over four billion barrels of reserves out of a total of six billion barrels. The way individual tilted fault blocks terminate include strike ramps, forced folds, rhomb-blocks, transfer faults, and major and minor cross-strike faults. In areas of relatively poor seismic data quality the choice of fault block termination interpreted on geophysical maps can make a considerable difference to structural interpretation. There has been considerable debate as to whether rifts are composed of high-angle (45–75°) faults or a mixture of high- and low-angle faults. Low-angle faults occur in the sedimentary sequence of rifts, where they occur in a variety of structural settings. Such faults can have a significant economic impact in field development. The presence of low-angled basement faults is more problematic. They do apparently occur in rifts, both due to reactivation of older faults and as completely new faults. Rifts can evolve through a variety of low- and high-angle fault structures with time, which impacts trap creation and destruction. Rifts commonly undergo changes in their stress regime which may lead to basin inversion features. This affects trap development and destruction and can lead to a variable subsidence history for basins, as the basin bounding faults change their senses of motion. The recognition of inversion features in many rifts indicates that this is a very important aspect of rift history.

Prior to the late 1970s a relatively small number of workers studied the structural geology of continental rifts; these include Gregory (1921), Quennell (1956), Robson (1971), Baker *et al.* (1972), McConnell (1972), Illies (1974) and Garfunkel & Bartov (1977). Much of this early work focused on the surface geology and geomorphology in the Rhine Graben, East African rift system, Dead Sea Rift and the Gulf of Suez. During the 1980s a considerable amount of new data, especially seismic reflection data, added significantly to the understanding of extension l provinces, although many of the concepts about rift structure that developed during that time existed previously in some form. Some of the main advances and controversies that arose during the 1980s are listed below.

1. Recognition of low-angle faulting in basement rocks of the Basin-and-Range Province (Wright & Troxel 1973; McDonald 1976; Rehrig & Reynolds 1980; Davis & Hardy 1981; Wernicke 1981, 1985).
2. Definition of rift architecture by seismic reflection profiles (e.g. Rosendahl *et al.* 1986; Cheadle *et al.* 1987; Ebinger *et al.* 1987).
3. Recognition of complex evolution of faulting in rifts beyond the simple models of domino faulting.
4. Geometry of basin inversions.
5. Definition of the influence of syn-rift structure on sedimentation.
6. Detailing of fault geometries in rifts.

This paper concentrates on three main themes: rift architecture, fault angle and basin inversion and their impact on hydrocarbon exploration. This is a potentially large topic but it does help limit which important advances are examined, for example, topics dealing with rift mechanics and volcanism are largely ignored. Salt tectonics are also very important in some rifts, particu-

From Holdsworth, R.E. & Turner, J.P. (compilers) *Extensional Tectonics: Regional-scale Processes*. The Geological Society,
Key Issues in Earth Sciences, **2 (1)**, 115–146.
1476-315X/02/$15.00 © The Geological Society of London 2002
First published in: "Lambiase, J.J. (ed.) Hydrocarbon Habitat in Rift Basins, The Geological Society, London, Special Publications **80**, 1–32".

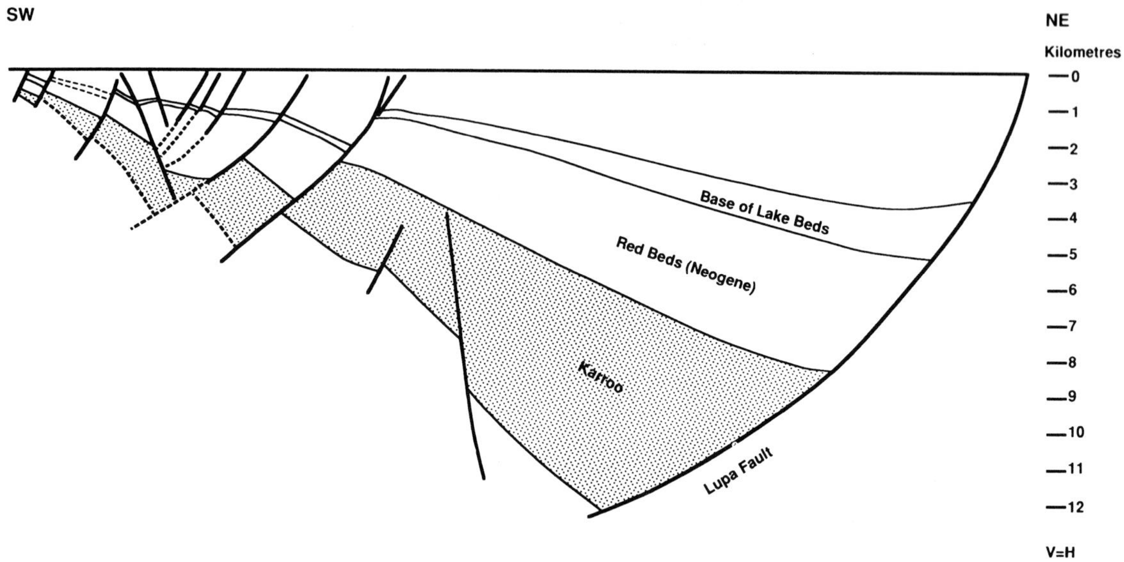

Fig. 1. Cross-section accoss the SE Rukwa rift (see Fig. 4 for location), based on seismic line TVZ-2 (Morley *et al.* 1992*a*). Lake beds are of Upper Miocene–Pleistocene age and Karroo is dominantly Permo-Triassic continental deposits. The following points about half-graben geometry are illustrated by the figure: (1) the major boundary fault (Lupa fault) controls the overall sediment wedge geometry; (2) reactivation of the boundary fault (Karroo and Tertiary times); (3) minor fault, particularly antithetic faults, are best developed near the flexural margin.

larly in the North Sea. However, this is not a fundamental structural style to rifts alone and in the interest of space is ignored in this paper.

Rift architecure

Before the 1980s the basic elements of large-scale rift structure had been established. Half-graben and full-graben geometries were known both from modelling and from natural examples (e.g. Cloos 1936; Dunbar 1949), and changes in structural style at discrete zones (hinge zones, accommodation zones, transfer zones) had also been noted (Moustafa 1976). Variations in rift segment geometry (offsets, jumps, gaps, splays; see Nelson *et al.* 1992 for a summary) had been examined in some detail (e.g. Baker *et al.* 1972; Illies 1974). However, a detailed knowledge of the sub-surface configuration was lacking. These geometries are discussed in this section.

The investigations of the 1980s, in particular those of Project PROBE in East Africa, showed, through high quality seismic reflection data, the details of the changes in geometry (e.g. Rosendahl *et al.* 1986; Ebinger *et al.* 1987; Dunkleman *et al.* 1988). Some boundary faults were large, displaying up to 10 km of extension and up to 7 km of syn-rift fill in the hanging wall (Figs 1–3, and Fig. 4 for location). Such large faults exert a first-order control on extension, subsidence, basin geometry and sedimentation patterns in rifts (Rosendahl *et al.* 1986; Morley 1989; Schlische & Olsen 1990). The two main basin geometries controlled by boundary faults are half-grabens and full-grabens. Some of the basic relationships between minor fault geometry and major fault geometry that are commonly found in these basins are listed in Table 1.

Examination of recent earthquakes is important for understanding how the individual displacement increments are built up on faults (Fig. 5). For large earthquakes (magnitude 6.5 or more on the Richter scale) fault strike lengths of tens of kilometres may be activated, the active surface fault trace commonly forms a series of en-echelon segments (e.g. Stein & Barrientos 1985; King *et al.* 1988). The instantaneous or coseismic deformation associated with such earthquakes produces absolute (not relative) uplift of the footwall and downdropping of the hanging wall in the order of tens of centimetres. The magnitude of the hanging wall displacement is 6–10-times larger than the footwall uplift (Stein & Barrientos 1985). The coseismic (elastic) deformation introduces an isostatic imbalance since lower density sediments have replaced a similar volume of basement rock in the hanging wall (Fig. 5). Erosion of the footwall uplift also contributes to the isostatic imbalance. The isostatic (post-seismic) response takes the form of a gentle vertical uplift across the fault zone and occurs on the time scale of tens to hundreds of thousands of years. This uplift, which decreases in magnitude away from the fault

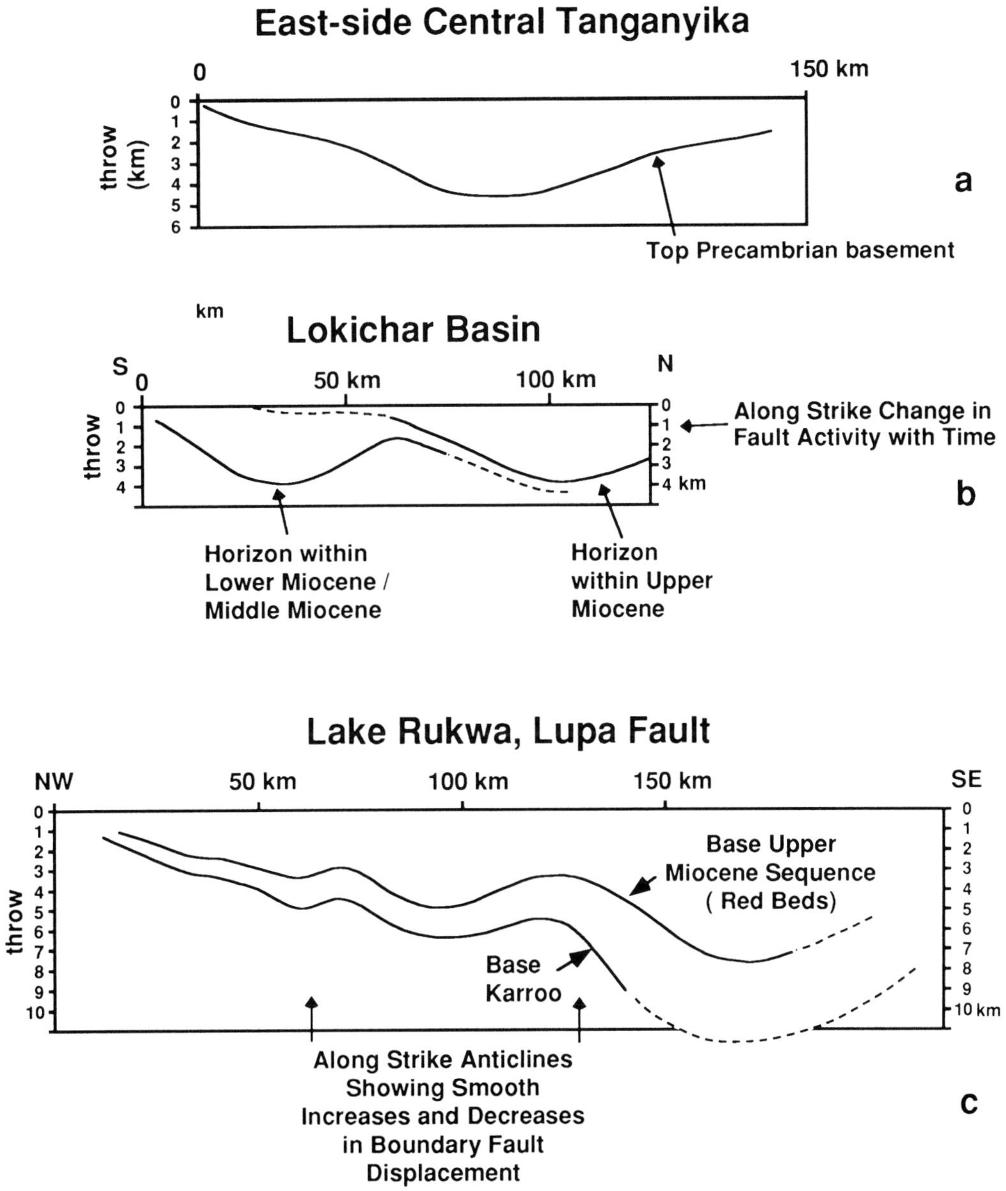

Fig. 2. Examples of major fault geometries from rifts illustrating the displacement patterns of the faults by plotting changes in throw, derived from seismic data, along the strike of the fault (see Fig. 4 for location). The data illustrate a simple case (a) and complex cases where faults are either reactivated (c) or change timing of activity along-strike (b). (a) Lake Tanganyika, east-side boundary fault, illustrating a simple decrease in displacement from a maximum at the centre. (b) Lokichar Basin, Kenya Rift, demonstrating abrupt younging of the main fault activity to the north (this may reflect linkage of younger and older faults). (c) Lake Rukwa, Lupa Fault, displaying skewed displacement maxima, similar displacement trends for both the Karroo and Tertiary age rifting and along-strike anticlines as displacement varies along the fault. Displacement variations may reflect linkage of three separate large faults (each one corresponding to a displacement maxima).

zone, enhances the footwall uplift, reduces hanging wall subsidence and gently warps the fault plane. The result of combined coseismic and post-seismic deformation is to produce footwall uplift that can be very significant (perhaps up to 50% of the hanging wall subsidence; Stein & Barrientos 1985). Since large normal faults that penetrate the upper crust are the result of multiple earthquakes plus aseismic creep, it follows that in order to properly model normal fault behaviour the isostatic and elastic behaviour of the incremental displacements must be taken into account.

Application of flexural-isostatic models to

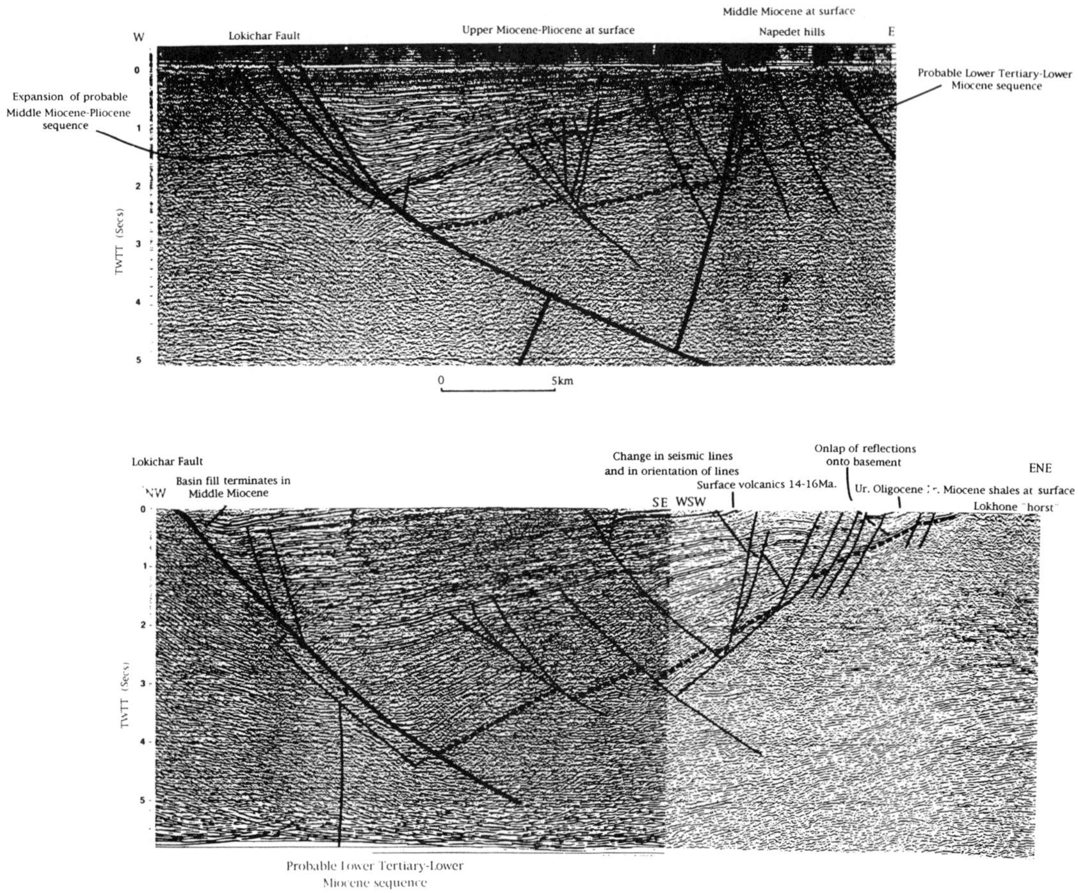

Fig. 3. Seismic lines from the Northern Kenya Rift (after Morley *et al.* 1992*b*) illustrating changes in fault activity with time and along-strike. Top line shows east thickening of probable Lower Tertiary–Lower Miocene section, which is the time equivalent of most of the half-graben wedge in the section to the south (bottom line). The same fault trace at the surface (Lokichar Fault) apparently joins both basins yet the timing of activity youngs significantly to the north (see Fig 2b). The northern line shows a significant reorganization of the basin geometry with time and from two oppositely facing half-graben a composite full-graben is produced.

rifts (e.g. Vening Meinesz 1950; Weissel *et al.* 1987) have recently shown that some large permanent uplifts in rifts may also be attributed to boundary fault geometry. Modification of the McKenzie model (McKenzie 1978) to include flexural-isostatic behaviour, e.g. the flexural cantilever model (Kusznir & Egan 1989), indicates that these uplifts can be very significant, in the order of several kilometres (e.g. Fig. 5f). The flexural cantilever model approximately models the finite displacements on faults as products of elastic dislocations and isostatic responses, such as those described by Stein & Barrientos (1985). Flexural-isostatic models can produce good agreement between the upper crust rifted geometry on seismic lines and those on the computer model, and it is possible to model the effects of the boundary faults on rift geometry and the stepwise evolution of rift stratigraphy (Fig. 5; see Kusznir *et al.* this volume). The model is also conceptually useful because it links uplifted, eroded topography and the basin geometry to the major boundary faults, thus integrating structure and sedimentation patterns (see Transfer zones). In particular, it provides a mechanism for the creation of both synchronous and diachronous unconformities commonly found at the crests of tilted fault blocks that may be difficult to understand in terms of eustatic sea-level change.

Generally, rift stratigraphy in cross-sections through half-grabens evolve as wedge-shaped packages that expand into the boundary fault. With time the wedges prograde and onlap the flexural margin as the basin develops (Fig. 3). In map view the basin area increases and propagates outwards as the boundary fault grows. Full-grabens may display concomitant activity of the boundary faults or activity may switch between the two faults with time giving rise to

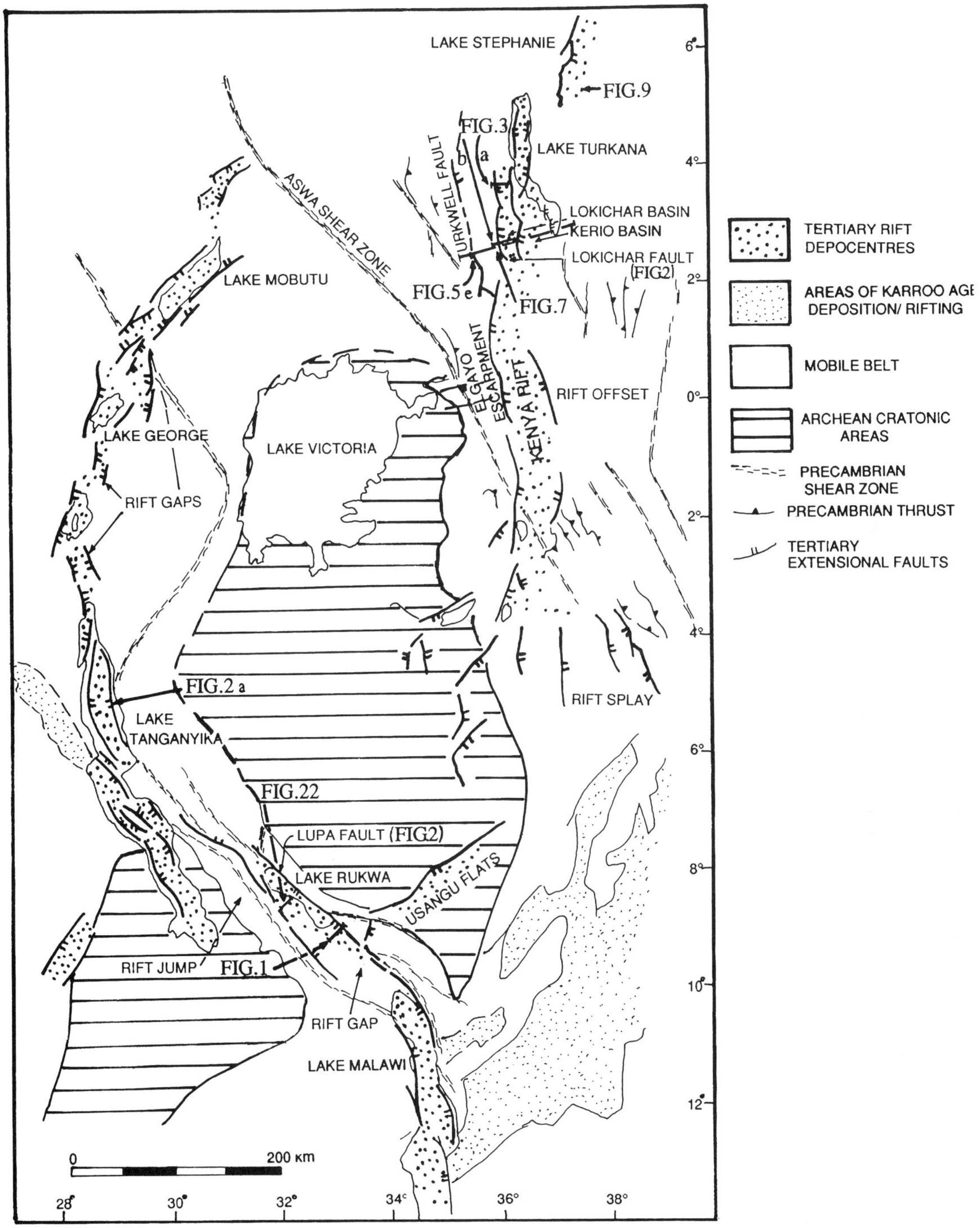

Fig. 4. Regional map of the East African Rift System, illustrating the controls of basement structure on the location of rift structures (compiled from Daly *et al.* 1989; Morley *et al.* 1992*a*; Smith & Mosley 1994.)

wedging of reflectors in alternate directions, vertically in a basin (Fig. 3).

Before going further in the description of rift structures it is necessary to discuss the effects of pre-existing fabrics on fault geometry, because such effects exert a major modifying influence at all scales.

Pre-existing fabrics

Extensional faulting occurs in a relatively low mean stress environment. Hence, mechanical anistropies tend to be reactivated. Youash (1969) has shown that the tensile strength of rocks loaded at 0–60° to pre-existing disconti-

Table 1. *Differences between fault distribution in half- and full-grabens based on data from Lake Tanganyika (Morley 1988).*

	HALF-GRABEN	FULL-GRABEN
	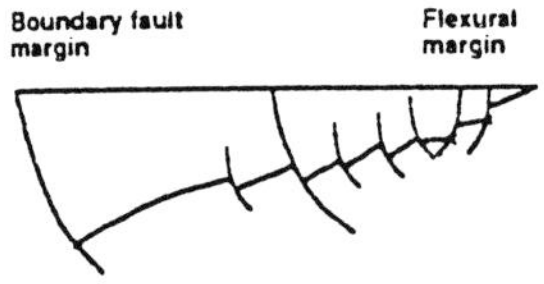	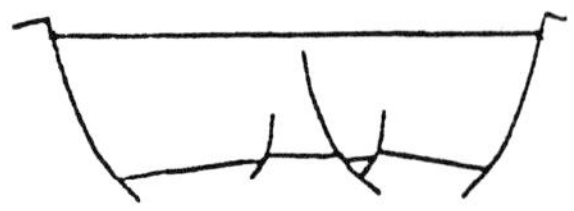
Timing between major and minor faults	Minor faults are commonly abandoned prior to cessation of activity on boundary fault	Minor faults are commonly abandoned prior to cessation of activity on boundary fault
Mixture of antithetic and synthetic minor faults	Synthetic minor faults tend to be more numerous and take up the highest percentage of extension	More even mixture of minor faults dipping in both directions
Partitioning of extension between major and minor faults	Relatively high percentage of extension (20–50%) distributed on minor faults	Relatively low percentage (<20%) of extension distributed on minor faults
Minor fault spacing	Relatively closely spaced	Relatively widely spaced
Minor fault distribution	Minor faulting more intense towards flexural margin (FM) Antithetic faults more common towards FM	Minor faults more frequent in central part of rift, less frequent approaching boundary faults

nuities is only 25–75% of that of similar samples loaded at 90°. Consequently, fabrics, even up to quite high angles (60°) to the extension direction, may be activated. Pre-existing discontinuities may also inhibit fracture propagation if the frictional shear strength of the discontinuity is sufficiently low relative to the tensile strength of the surrounding material (Teufel 1979). Fractures may also be diverted by pre-existing fabrics. In low mean stress environments the angle between the propagation direction and cross-tends is important. Fractures preferentially turn into discontinuities at low angles (30–60°) whilst propagating across them at high angles (Blanton 1982). This experimental evidence suggests that fault orientations up to *c.* 60° to the regional extension direction might frequently occur in rifts. In the Gulf of Suez the fracture and foliation pattern in basement is well exposed (Robson 1971; Fig. 6). Although lineations at a high angle to the main rift trend are very common they are rarely followed by the extensional faults (Fig. 6). One spike in the general low frequency of faults at a high angle to the rift trend is marked 'cross' (Fig. 6), and is a trend used by transfer faults. Trends at a lower angle to the main rift trend (310–340°) are more commonly used, particularly the northwest oblique trend (280–310°) and north oblique trends 350–30° (Fig. 6). If the 325° direction is taken as the mean rift orientation then the main rift faults encompass trends that deviate from this orientation by 45° in an anticlockwise direction and 65° in a clockwise direction. This corresponds well with the experimental limits on pre-existing fabric influence discussed above. Overall, the influence of pre-existing fabric does not produce one dominant direction and only a

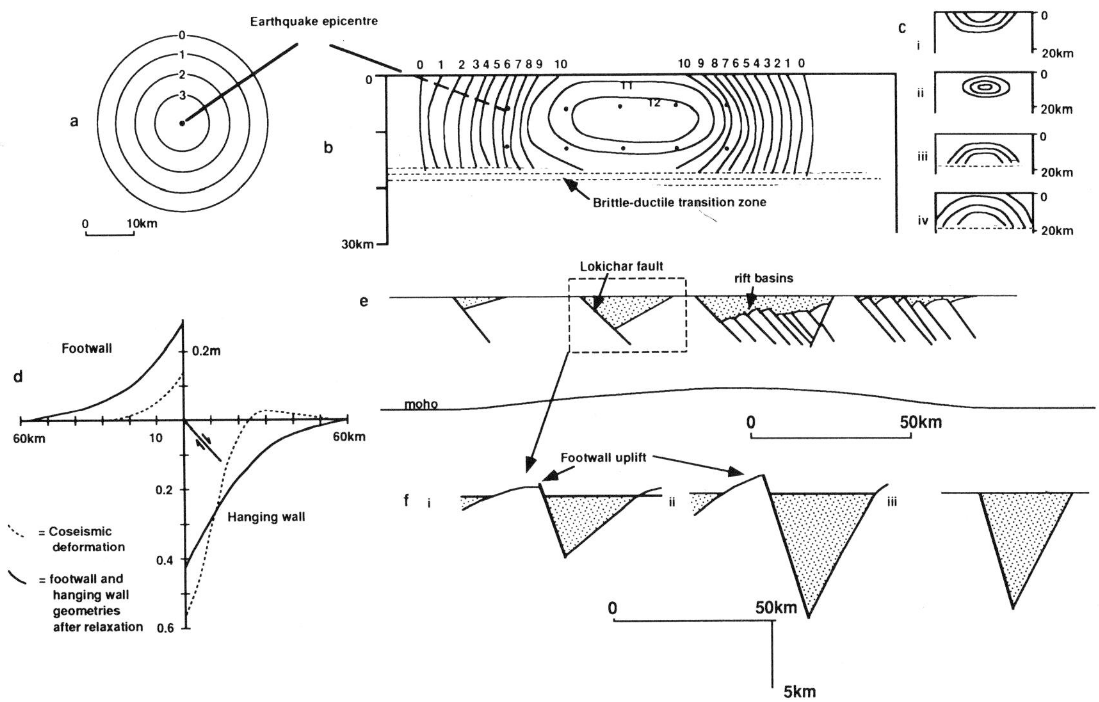

Fig. 5. Illustration of the relationships between individual earthquake displacements on faults and cumulative displacements on faults. (a) Individual earthquake represented as concentric displacement contours lying on the plane of the fault. It is assumed that displacement decreases in all directions away from the epicentre. (b) Displacement gradient on a large fault plane resulting from summing the effects of ten earthquakes similar to (a), one occurring at each of the black dots which represent an epicentre. The displacement contours are halted at the brittle–ductile transition zone and at the Earth's surface. Although very simplified from reality, the resulting displacement field is similar to those mapped from actual faults. The increasing displacement towards the centre of a fault can be seen in Fig. 2a. (c) Within the highly simplified displacement fields like those in (b) it is possible to vary the resulting displacement profile in the following ways: (i)–(iii) represent displacement fields of small faults with smaller associated earthquakes than in (a), e.g. diameter < 10 km. In (i) the epicentres are high in the crust and the fault displacement gradient shows faults with displacements decreasing downwards; (ii) epicentres are located at deeper levels in the upper crust and the fault decreases in displacement in all directions; (iii) epicentres are located near the brittle–ductile transition zone, displacement gradients decrease upwards and laterally. (iv) For large faults with widths of displacement zones larger than those shown in (a) (e.g. diameters > 40km) the displacement gradients will only decrease upwards and laterally, not downwards. (d) Coseismic and relaxed deformation for a 45° dipping normal fault, note that after the effects of relaxation the footwall becomes higher and the hanging wall shallower (after King *et al.* 1988). (e) Cross-section across the Northern Kenya Rift, the displacement on the Lokichar Fault has been modelled in (f) using the flexural cantilever model (Hendry *et al.* in press). In (f) footwall uplift is generated in isostatic response to movement on the fault and the infilling of the hanging wall by lower density sediments. The finite displacements modelled in (fi–fiii) approximate similar lithospheric responses to those for individual earthquakes shown in (d). (fi) Paleogene displacement; (fii) Miocene displacement; (fiii) basin geometry after erosion (note 4-times the vertical exaggeration).

slight asymmetry is present in the rose diagram in Fig. 6. However, locally, the preferential use of the north oblique trends is obvious in the satellite image (Fig. 6).

Basement fabrics can control the location of the rift (e.g. McConnell 1972; Sykes 1978), the geometry of faults within the rift, the location of transfer zones and the location and geometry of rift segments (Robson 1971; Illies 1974; Daly *et al.* 1989; Smith & Mosley 1994). The avoidance of the Tanzanian (Archaen) craton by the East African Rift System is an obvious large-scale control where variations in crustal thickness, deformation style and mineral composition can play their part (McConnell 1972; Fig. 4).

Coming down in scale it becomes harder to predict the nature of the pre-existing fabric control because of multiple competing factors. Commonly, foliations and ductile shear zones exhibit complex folds which preclude large faults following individual structures for long distances. Brittle structures occur relatively late in the evolution of basement rocks hence they are less likely to be folded. Consequently reactiva-

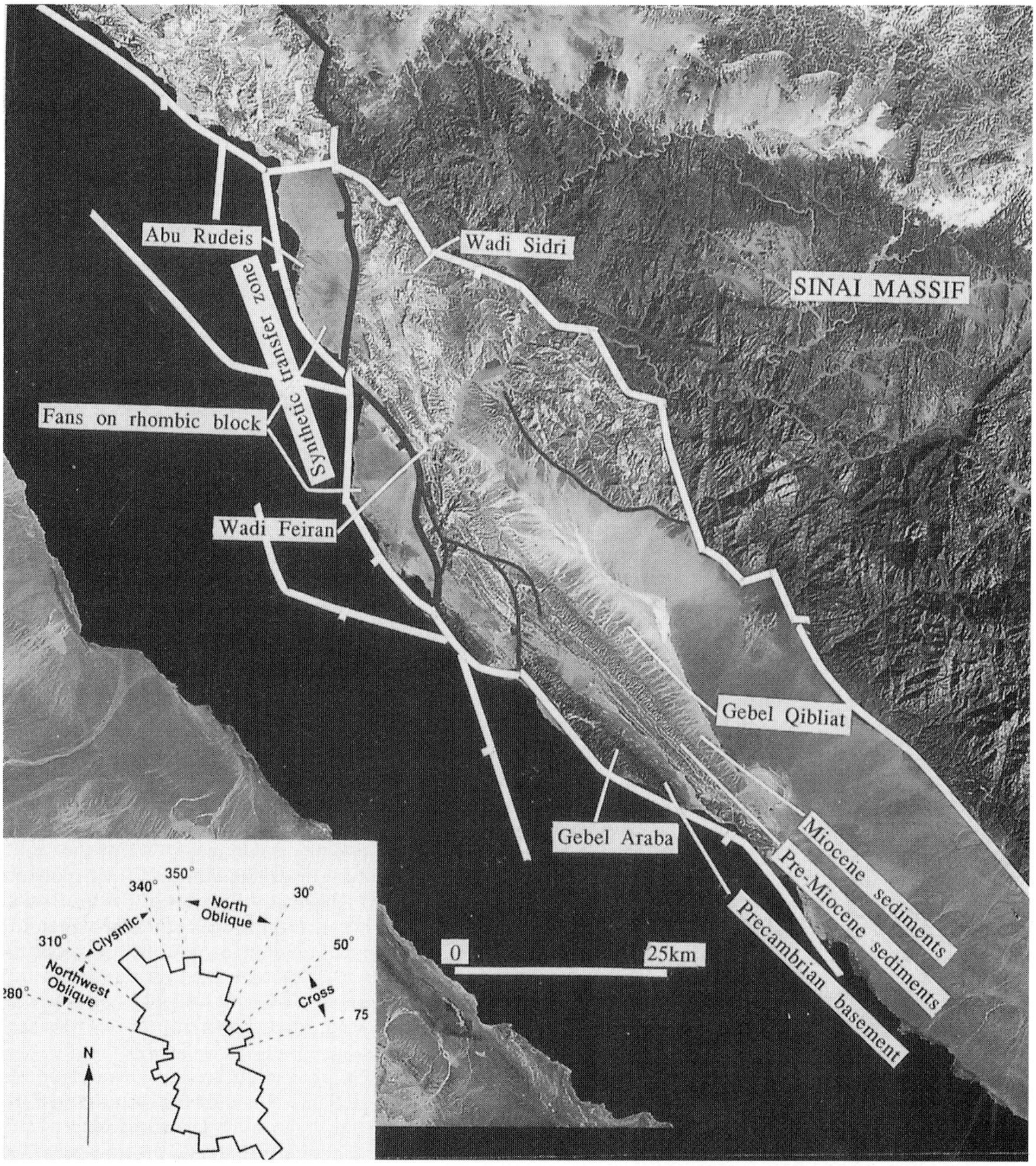

Fig. 6. Landsat image of the Abu Zenima area, Gulf of Suez, illustrating zigzag fault geometries that follow linear brittle fabrics in crystalline basement (Sinai Massif). Input of coarse clastics via fans into the gulf can also be seen associated with a synthetic transfer zone. Fault orientation rose diagram from Patton in Lelek *et al.* (1992).

tion over a large surface area is possible. The Lokichar fault for example (Fig. 7), cross-cuts at an acute angle foliations in Precambrian basement. The fault is curved in map view, but is low-angle and planar in cross-section, and dips some 30–40° lower than the surface foliations. Often, because a fault follows foliations in map view it is mistakenly assumed that it also follows them in cross-section. The Lupa Fault (Fig. 4) follows foliations along strike, but displays a listric shape in dip view (Fig. 1) that is most unlikley to represent the foliation geometry. One possible explanation is illustrated in Fig. 8 where initially high-angle fractures follow the foliation planes in both dip and strike directions. These fractures become linked by lower-angle faults which smooth out the initially jagged fault plane cross-sectional geometry with increasing displacement. This results in the final fault trace following the original strike trace of the fractures but cuts across the fabric in cross-section.

The zigzag fault pattern has been recognized

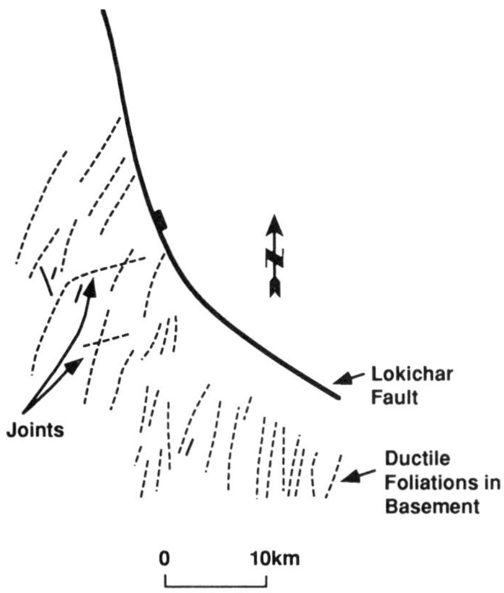

Fig. 7. Map, simplified from Landsat images, illustrating how the Lokichar Fault cross-cuts pre-existing foliations in crystalline Precambrian basement, northern Kenya (see Fig. 4 for location).

as a fundamental pattern of extensional faults (e.g. Ortel 1965; Freund & Merzer 1976; Freund 1982; Krantz 1989). Such fault patterns may be indicative of non-plane strain (3D strain; Reches 1978, 1983). The pattern occurs in finite element models, anisotropic clay physical models (Cloos 1936; Ortel 1965), fractured pavements and at all scales in rifts. Even the rifted coastline of Africa has a strong zigzag shape. This natural pattern can also result from small changes in the stress-state where the intermediate principal stress component parallel to the rift trend approaches the magnitude of the maximum principal stress (Freund & Merzer 1976). This pattern is often enhanced and modified by pre-existing fabrics (Ramberg & Smithson 1975, Figs 6, 9 and 10). This may result in emphasis of certain trends, or deviations, in the angle between the zigzag elements from the idealized anisotropic angle of 30°. Zigzag faults are generally steeply dipping (60–70° in the Gulf of Suez; Colletta *et al.* 1986) and are dominated by dip-slip displacements.

The influence of pre-existing fabrics means that many faults do not display the classic and idealized convex map pattern. Commonly faults form interlocking crooked X shapes or even convex shapes (Fig. 10). This variety can be attributed to smaller faults linking randomly. As the faults grow and link the smaller zigzags and displacement variations may be smoothed out and certain fault branches may remain active whilst others become deactivated. Consequently rules about fault geometry, particularly listric faults, should be treated with caution. Even simple inferences, such as curved surface traces are characteristic of listric faults, can be wrong (see Figs 1 and 4, where the listric Lupa fault in profile is approximately linear in map view).

Rift segments

Continental rifts do not usually occur as single, straight structural entities due to the very strong influence of pre-existing fabrics. Instead they occur initially as an associated system of isolated pockets of extension called segments (along-strike length tens to hundreds of kilometres). The axis of each segment displays a relatively constant strike direction and a constant or smoothly changing amount of extension on successive dip lines along the segment. A segment can be composed of one or more half- and full-grabens. They are separated by a marked structural discontinuity or change in geometry, and may also display differing

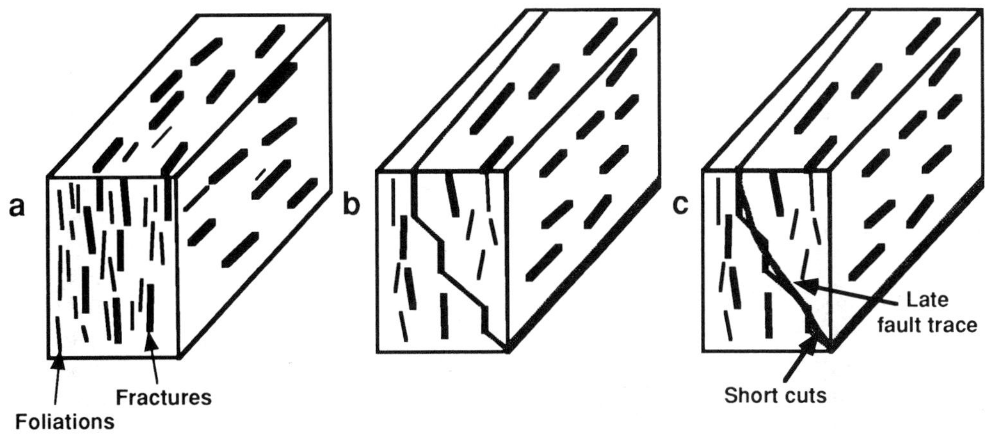

Fig. 8. Possible explanation for faults that follow foliations along strike, but not down dip. (a) Initial extensional fractures follow foliations. (b) Fractures are linked by faults that cut across the foliations. (c) As displacement on the fractures increases the fault becomes smoothed.

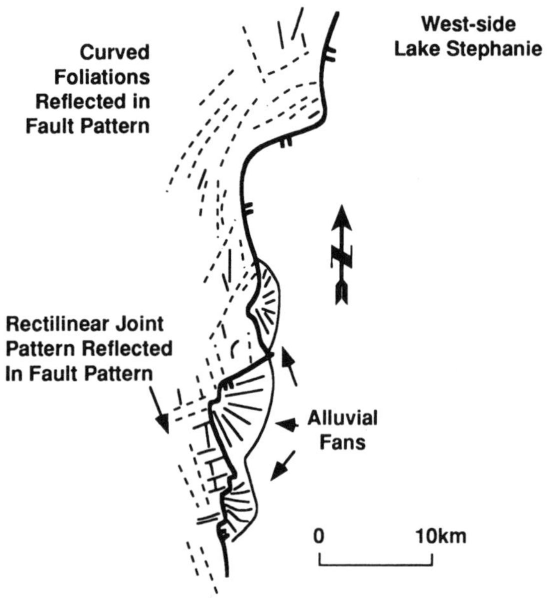

Fig. 9. Map of extensional faults in the Lake Stephanie area, South Ethiopian Rift, illustrating zigzag faults strongly influenced by pre-existing crystalline basement fabric.

structural histories (Nelson *et al.* 1992). Segments grow and interact as the rift system matures, hence the basal sediments in rift basins tend to be young away from numerous minor initial basins and onlap the pre-rift surface. At some point in the extension history, many of the segments overcome the initial influence of pre-existing fabrics and merge to form a continuous rift system, though the initial segment pattern can usually be discerned. The recognition of rift segments is important to regional exploration studies because they can represent major changes and discontinuities in both structural style and history, and in sedimentation patterns. Four main types of rift segment termination and interaction are considered here: rift splays, gaps, jumps and offsets (Fig. 11).

Rift gaps are areas of no extension or reduced extension that occur in line between propagating rift segments (Fig. 11). Hence, rift sedimentation is reduced or absent in these zones. Rift jumps are similar to rift gaps except that the rift segments lie in an en-echelon pattern (Fig. 11). Both gaps and jumps are very obvious in young continental rift systems (Fig. 4) where basement or pre-rift sediments are exposed between many segments. Unfortunately, for the same reasons, such areas are of little exploration value in young rifts. In more mature rift systems, capped by extensive thermal sag basin sequences (e.g. North Sea, Sirte Basin) the high areas between rift segments are often very productive. They provide areas of structural traps for pre-rift sediments, stratigraphic trap potential for syn-rift sediments and commonly areas of antiformal drape structures for post-rift sediments. Gaps are areas where large rivers may penetrate the rift and form along-axis (rift parallel) drainage systems. Such systems tend to be very sand rich (e.g. Rusizi Plain, Lake Tanganyika; Lake Rukwa).

A rift offset is a basinal 'dog-leg' rift segment that lies oblique to the regional extension direction and links en-echelon segments (Fig. 11). Consequently the offset has a structural style different from the segments it joins. The result is an oblique extensional strain pattern that commonly includes anastamosing fault patterns and the dominance of certain fault orientations that are unusual in comparison with the segments it joins. Tilted fault blocks may display less consistent dips and subsidence rates may be lower than in the adjacent segments (Nelson *et al.* 1992). The rift segments that lie perpendicular to the regional extension direction tend to display a relatively symmetrical frequency of fault distributions that lie within *c.* 30° either side of a line normal to the regional extension direction (Fig. 12). In the East African Rift System the rift offset zones are almost certainly present due to basement heterogeneity (Fig. 2) and will, therefore, display certain trends not frequently used in the normal rift segments. The same is probably true for other rifts. Figure 12 shows a comparison between fault patterns in rift offset segments and normal rift segments. It is apparent that the fault pattern in the offset segments is much more asymmetric than the normal rift segments. The rift offsets show differing patterns between each other, for example the central Kenya rift offset zone is dominated by N–S fault trends, like the normal rift segment, but also contains a strong NW–SE trend. The Rukwa Rift displays a much reduced number of N–S faults and is dominated by NW–SE trending faults. The number of faults that occupy a certain orientation does not tell the whole story, certain orientations may be infrequently used but contain long, important faults and vice versa (Fig. 12).

Rift splays are areas where a discrete rift segment terminates into a diffuse zone of faulting. The partitioning of extension from a few discrete large faults in the segment into numerous smaller faults allows strain to be distributed throughout a larger volume of the crust, thus enabling the rift to die out over a short along-strike distance. Subsidence asso-

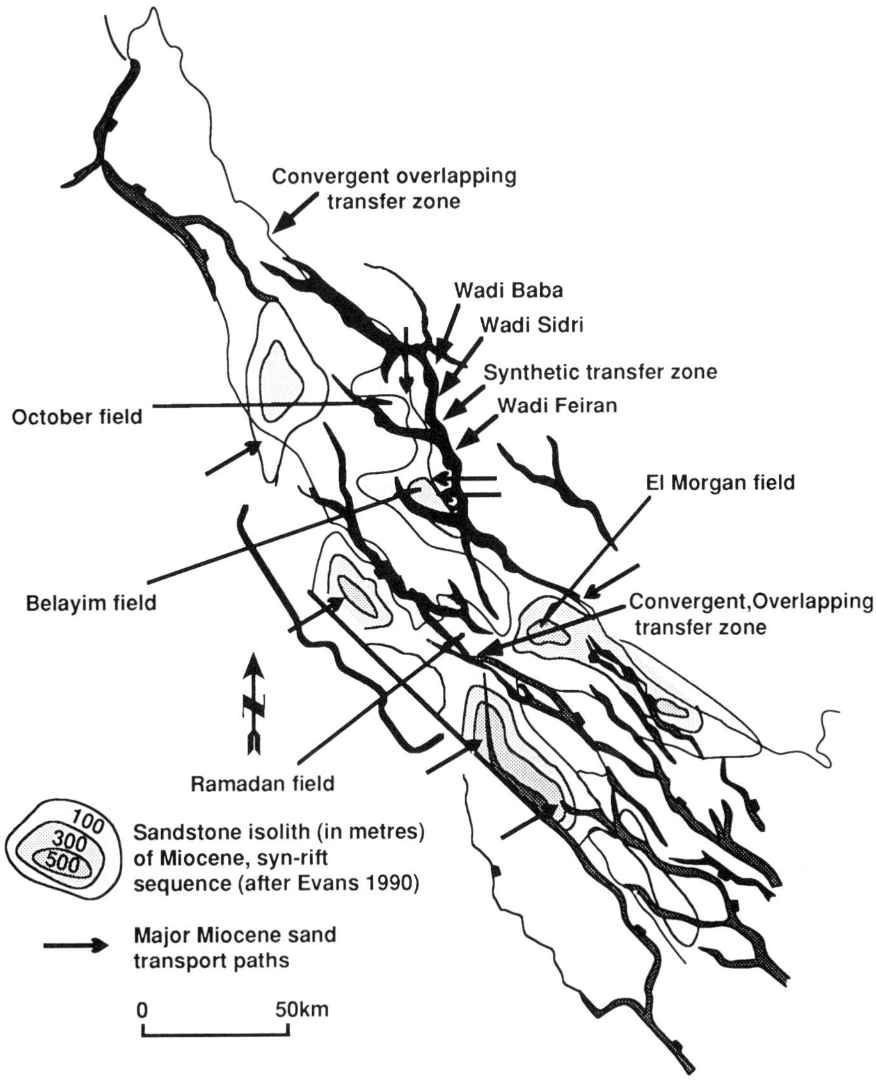

Fig. 10. Map of the Gulf of Suez illustrating the main fault trends and distribution of syn-rift sands. Compiled from Patton *et al.* (1994) and Evans (1990).

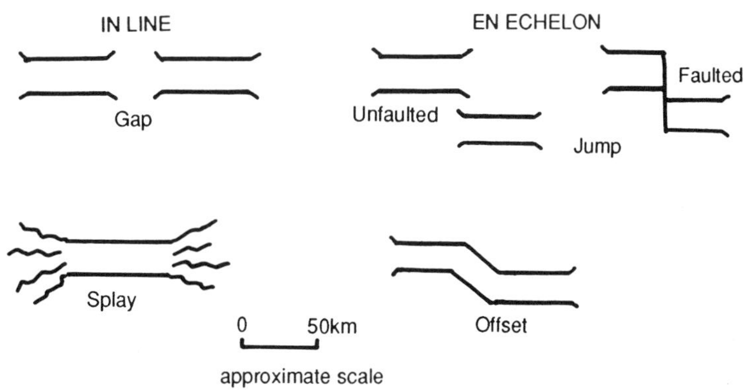

Fig. 11. The main geometries of rift segments interaction (modified from Nelson *et al.* 1992).

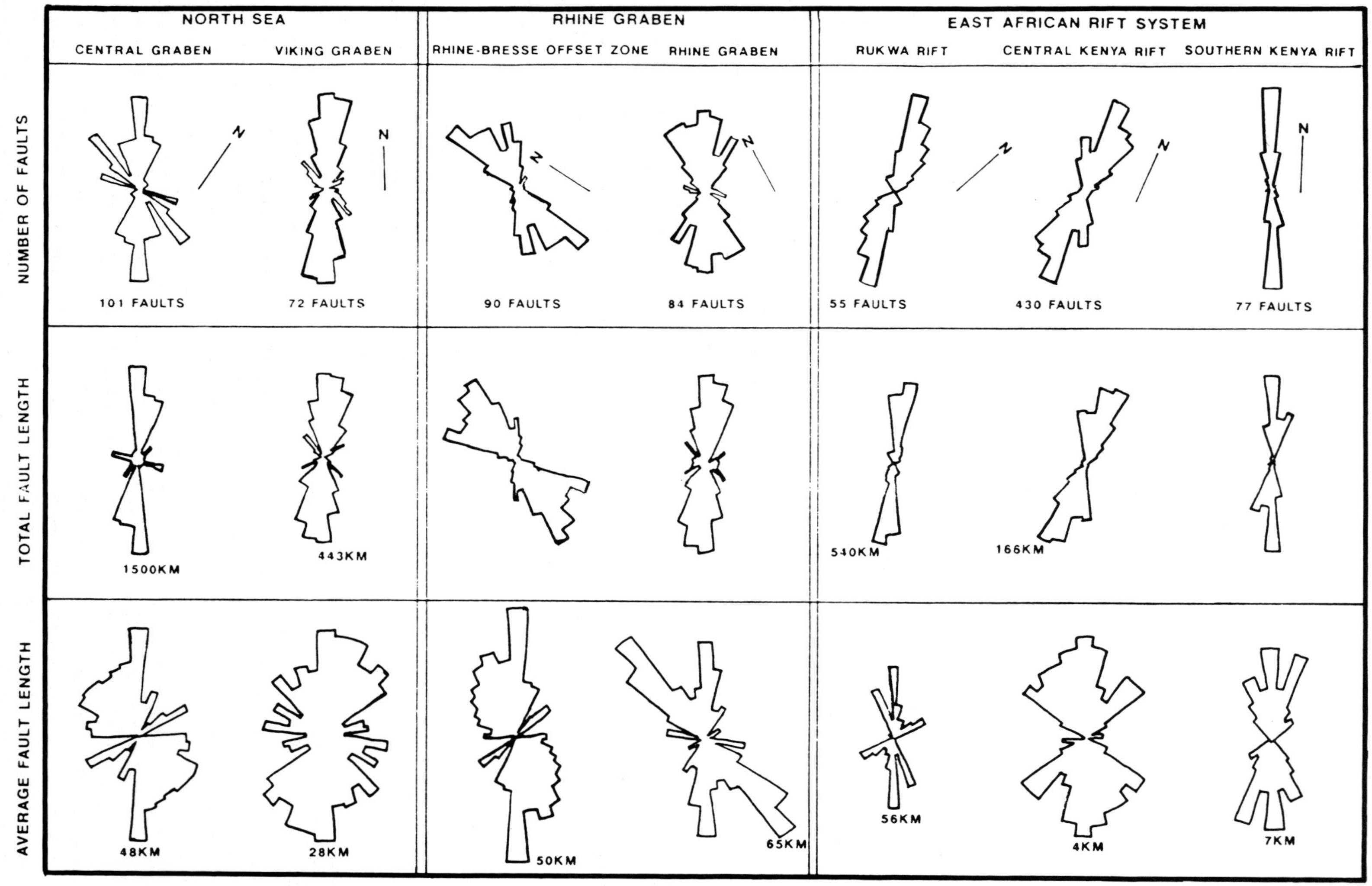

Fig. 12. Diagrams showing the distribution of faults in rift normal and rift offset segments. In order to compare the fault patterns all orientations have been rotated so that the rift segments are oriented vertically on the page. The North Sea has been subject to variable extension directions with time, however, the different patterns between the Viking and the Central Graben shows through.

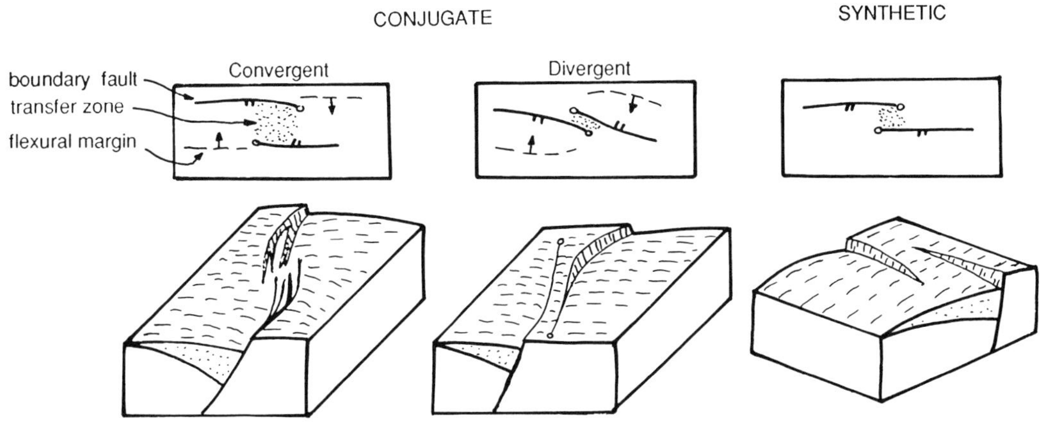

Fig. 13. The three main transfer zone geometries, simplified from Morley *et al.* (1990).

ciated with the small faults tends to be small, and consequently plays are unlikely to be associated with splays in simple continental rifts. However, once a splay is covered by a thermal sag basin the play potential amongst the numerous tilted fault blocks is significantly increased.

Transfer zones

Changes in boundary fault configuration occur in a variety of ways at regions called either accommodation or transfer zones (e.g. Gibbs 1984; Bosworth 1985; Rosendahl 1987; Morley *et al.* 1990). The term 'accommodation zone' (Bosworth 1985) simply describes the zone where a change in geometry occurs between faults, which may be of different ages. The term 'transfer zone' implies that the faults across which the change in geometry occurred were active at the same time and that extension was transferred from one fault, or fault system, to another. In both cases the basic geometric changes are similar: transfer can occur between faults dipping in the same direction, between faults dipping away from each other (back-to-back) or between faults dipping towards each other (Fig. 13). In relatively low-extension rifts (at least up to a β factor of 1.5) displacement transfer commonly occurs in fairly broad zones between overlapping fault systems. Short strike-slip fault segments that permit faults to step backwards or forwards are also common transfer zone features. However, it is unclear at the moment at what point in the extension history of a rift large cross-strike oblique-slip faults dominate the transfer zone geometry. In the East African rift the major boundary faults show gradual changes in displacement with the loss of 1 km of throw occurring over a strike distance of 10–20 km (Fig. 2). The large, gradual changes in fault displacement can produce large along-strike folds with wavelengths of tens of kilometres (Morley *et al.* 1992*a*; Fig. 2c). Only rarely do faults end abruptly at cross-strike faults, suggesting that the transform fault geometry imposed on many rift systems has been greatly over used.

Transfer zones occur at all scales, but because boundary faults can so strongly dominate the rift structure, transfer zones associated with major faults are commonly responsible for wholesale changes in rift geometry. They are important features for understanding the distribution of hydrocarbon reserves since they not only mark changes in large-scale structural style but can also contain structures different from other parts of the rift. For example, convergent overlapping transfer zones may produce an interfingering of faults of opposite dip and break up the rift into numerous small fault blocks (Figs 13 and 14). An anomalously high number of small horst blocks and more numerous, but smaller, tilted blocks may characterize these transfer zones. Such is the case in the central Gulf of Suez where over 2.5 billion barrels of recoverable hydrocarbons have been found both from tilted fault blocks (e.g. Ramadan Field) and from horst blocks (e.g. Morgan Field). The large synthetic transfer zone (Fig. 10) that set up the traps for the Belayim and October Fields in the northern Gulf of Suez (over two billion barrels of recoverable reserves) further illustrates how large reserves can be associated with anomalous structural trends.

The input points for coarse clastic rocks (i.e. potential reservoir rocks) via fluvial systems into rifts and their subsequent redistribution by

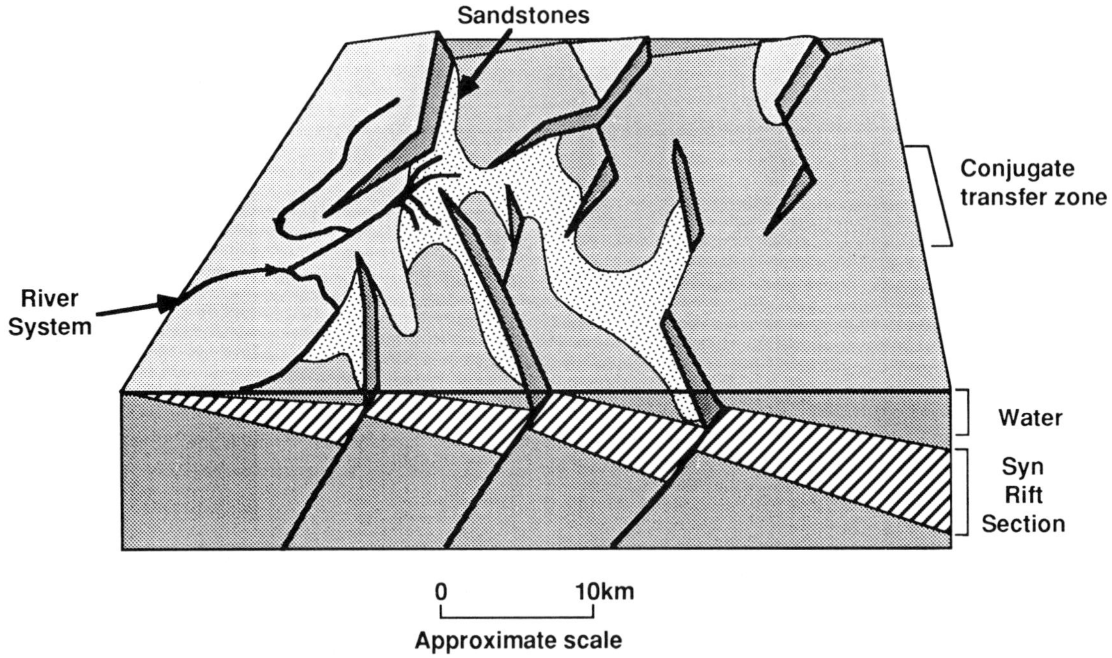

Fig. 14. Block diagram of conjugate overlapping transfer zone illustrating the possible nature of sediment dispersal patterns along topographic lows created by interfingering fault blocks in the transfer zone.

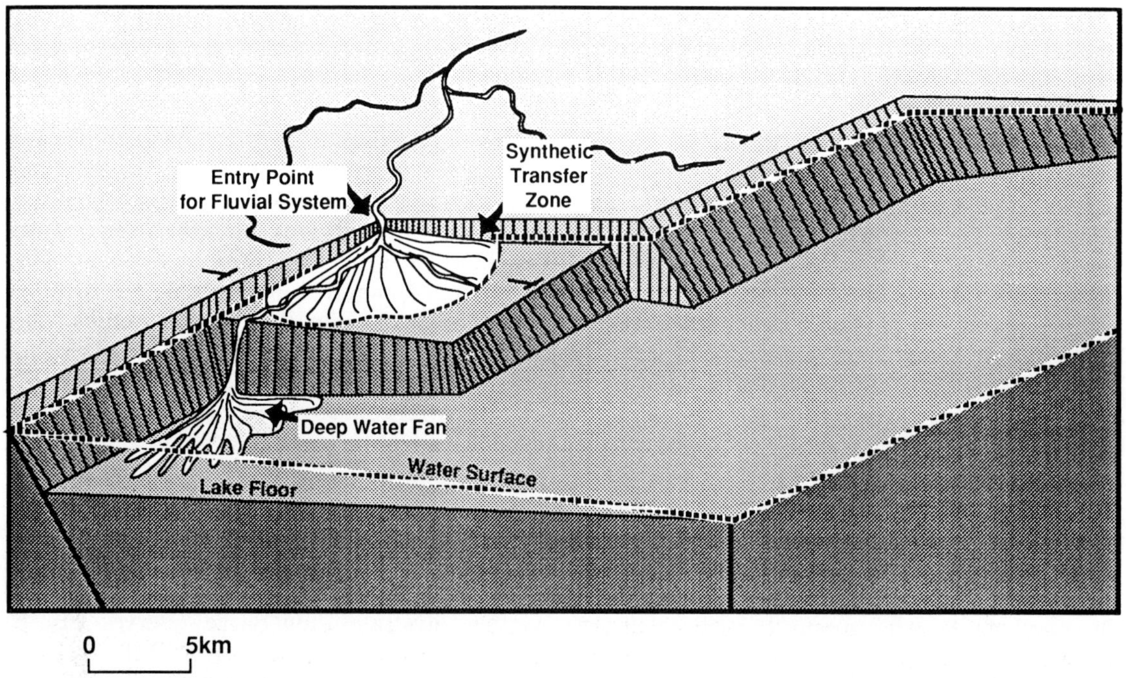

Fig. 15. Block diagram of synthetic transfer zone illustrating the rhombic nature of the fault blocks and the distribution of sediments via intermediate fault blocks.

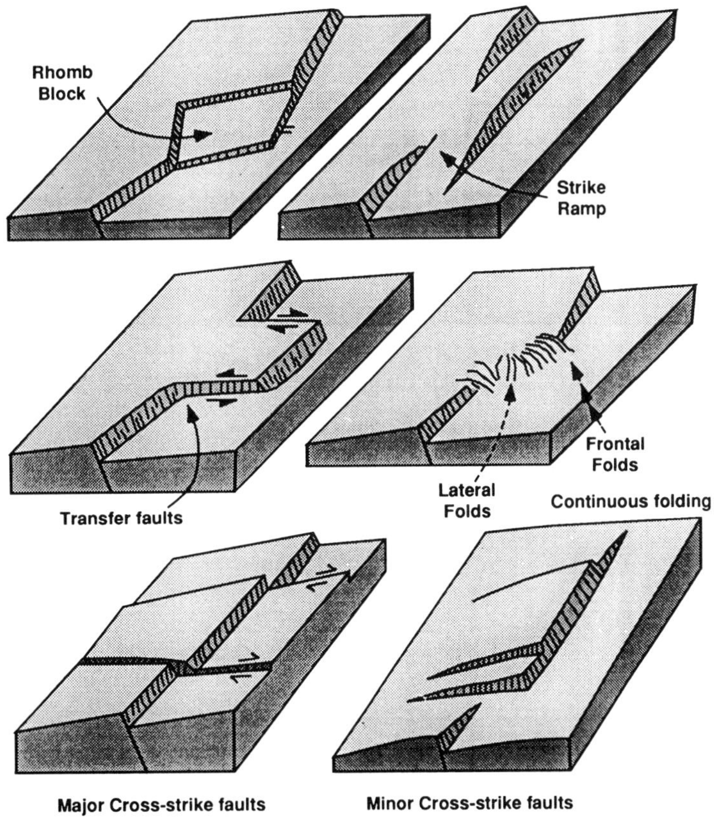

Fig. 16. Block diagrams illustrating the varieties of tilted fault block terminations in rifts.

turbidite flows are greatly influenced by rift structure (Frostick & Reid 1989), in particular by the rift segment (Nelson *et al.* 1992) and large-scale transfer zone geometry (Lambiase 1990; Morley *et al.* 1990). While thermal uplift of the rift shoulders is also possible, the recognition that large extensional faults themselves create an isostatic response that leads to footwall uplift is very significant because it links together structure and drainage pattern in a predictable way. Major faults generate flexural isostatic footwall uplifts that can cause several kilometres of relief (e.g. Kusznir & Egan 1989). The footwall relief is greatest where the displacement on the fault is greatest, as the fault dies out so the footwall topography becomes lower. Since transfer zones and rift segments coincide with changes in structural style (i.e. the terminations of major faults) they are also associated with major topographic breaks, which provide entry points for large fluvial systems (Figs 10, 14 and 15). Commonly, these systems have drained off the back of the footwall uplifts (Cohen 1990). These entry points may persist for a long time and can be associated with both synthetic and conjugate transfer zones (Figs 14 and 15). In the Gulf of Suez, for example, during the periods of Nukhul, Lower Rudeis and Upper Rudeis deposition (Miocene syn-rift deposits) large sandbodies that form significant reservoir units are only known in the vicinity of the major transfer zones (Evans 1990; Hagras & Slocki 1982; Fig. 10). These sandbodies pass from fluvio-deltaic to shallow marine to deep marine environments over relatively short distances due to the strong fault control on bathymetry (Fig. 15). Outside the influence of the transfer zones the deposits are dominated by shales, marls and thin sands. The transfer zones were remarkably persistent sites of coarse clastic sedimentation, and present day fluvial systems and alluvial fans correspond very closely with the location of Miocene marine sand depocentres (cf. Figs. 6 and 10).

Block terminations

As seismic data quality degrades, or lines become widely spaced, the linkage of interpreted fault cuts becomes increasingly speculative, especially for either deep plays or areas where post-rift salt covers the rift sequence (e.g. Gulf of Suez, offshore West Africa). Although the idealized fault pattern in rifts is reasonably well established (e.g. Cloos 1936; Illies 1974; Freund & Merzer 1976) departures from such patterns can be explained by pre-existing basement fabric or a complex history with phases of compressive or strike-slip deformation. Consequently, apparently anomalous aspects of a rift system may be emphasized and significantly change the mapped sub-surface fault pattern. In the Gulf of Suez the poor quality offshore seismic data has meant that the working maps from industry have exhibited great variation in structural geometry, including multiple strike-slip faults that traverse the entire rift to a broken-glass fault pattern where barely more emphasis was given to the rift parallel–sub-parallel faults as to the cross-rift trends. An appreciation of how fault blocks terminate can help resolve such problems. There are six main ways in which blocks terminate or change geometry along-strike, they are as follows (Fig. 16):

1. Zigzag faults, rhomb-blocks. The linking of zigzag faults of similar sizes produces areas where the faults come together and areas where they move away from each other, which tends to produce isolated fault bounded rhombic-shaped blocks (rhomb-blocks; Colletta *et al.* 1986; Fig. 16). Both fault sets have dominantly dip–slip motions. Frontal and oblique closure on the structure is by faulting (e.g. Ninian field: Spencer & Larsen 1990).
2. Strike ramps. These are common in the low-extension portions of rifts where overlapping faults slowly gain and lose displacement passing along-strike (e.g. Griffiths 1980; Fig. 16). They can form simple along-strike dip closures that do not have the risk of seal integrity that lateral fault closures may have. However, strike ramps may also exhibit a mixture of flexure and faulting.
3. Transfer faults. These oblique-slip faults form the lateral or oblique boundaries to frontal faults. They are kinematically part of a continuous fault trace that transfers displacement from one rift parallel fault segment to another and are restricted to the one fault block that they bound (e.g. synthetic transfer zone, Gulf of Suez; Patton *et al.* 1994; Figs. 6 and 16). There are transitional geometries between (1) above and transfer faults.
4. Continuous folding. Narrow zones of forced folding at lateral, oblique and frontal block edges may occur as sedimentary units are draped over a sub-surface fault tip (Fig. 16). Locally the beds can display very high dips of 80–90°. These folds are most likely to form over steeply dipping faults. They are well developed in parts of the onshore gulf of Suez (Patton 1984; Jarrige *et al.* 1990).
5. Major cross-strike faults. These faults form at a high angle to rift parallel structures and cut across multiple fault blocks (Fig. 16). They may be part of a very large oblique-slip transfer zone system between major faults (e.g. Tankard & Welsink 1987) and they may represent a reactivation of pre-existing trends independent of the normal faults they cross-cut. Commonly, large cross-faults are interpreted in areas of poor data quality but, as more data are gathered, the fault zone may be less continuous than previously thought and may be composed of numerous zigzag and transfer faults of different dip directions that are aligned along a pre-existing discontinuity, e.g. Tern Field, North Sea; Pahuys-Sigler *et al.* (1991).
6. Minor cross-faults. These faults form the lateral terminations of fault blocks and probably most commonly originate as reactivation of pre-existing trends. They differ from (3, 5) and (6) above by not being part of a linked fault system. The frontal fault may terminate into the cross-fault, but the cross-fault does not link to another fault (e.g. Brent and Statfjord Fields; Spencer & Larsen 1990; Fig. 16).

The type of block termination that is commonly present varies considerably between individual rifts, depending upon the amount of extension and the type of pre-existing fabrics. The North Sea displays many simple strike ramp geometries but also appears to be very heavily influenced by pre-existing trends that promote cross-fault terminations to block edges [types (3) and (5) above]. By contrast, the Kenya Rift is dominated by strike ramps (low extension geometry). The Gulf of Suez commonly displays types (1)–(4).

Fault angle

Introduction

Traditionally, faults in extensional systems were thought to have been initiated at high angles (e.g. Anderson 1951). Present day low-angle faults were thought to represent highly rotated, originally high-angle faults (e.g. Proffett 1977). The recognition of large, regional, low-angled

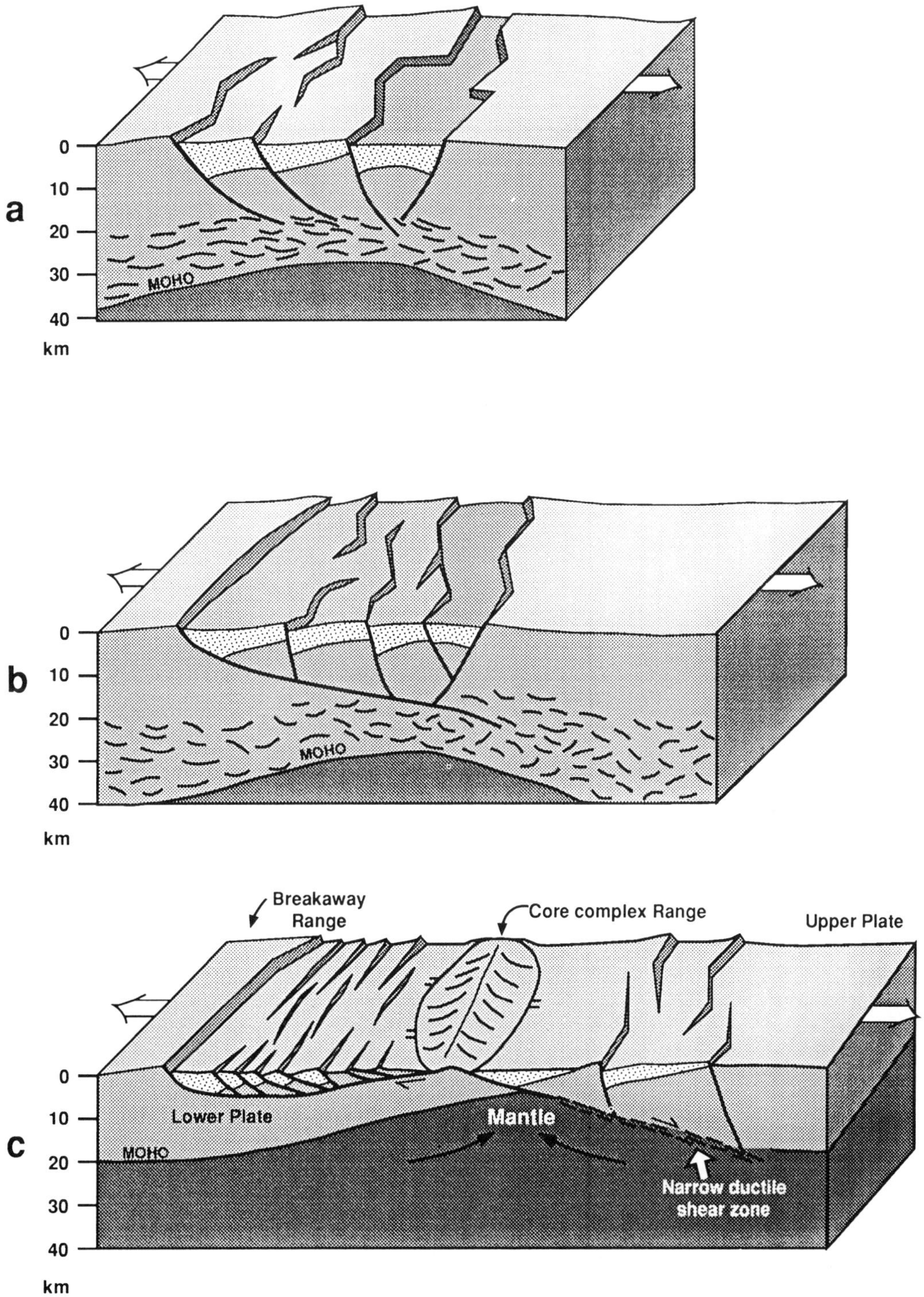

Fig. 17. Diagrams illustrating the main models of crustal extension. (a) Pure shear model (McKenzie 1978). (b) Simple shear–pure shear model. (c) Simple shear model (Wernicke 1981, 1985). Vertical scale = horizontal scale.

extensional fault systems in the Basin-and-Range Province of the USA in the late 1970s, however, suggested that some faults might actually be initiated at low angles (e.g. Rehrig & Reynolds 1980; Davis & Hardy 1981; Allmendinger *et al.* 1983). This province apparently formed in hot, abnormally thick continental crust superimposed on an older thrust belt. Thus, the tectonic setting and great areal extent of the Basin-and-Range makes it an unlikely analogue for structural styles in continental rifts. However, the integration of field observations

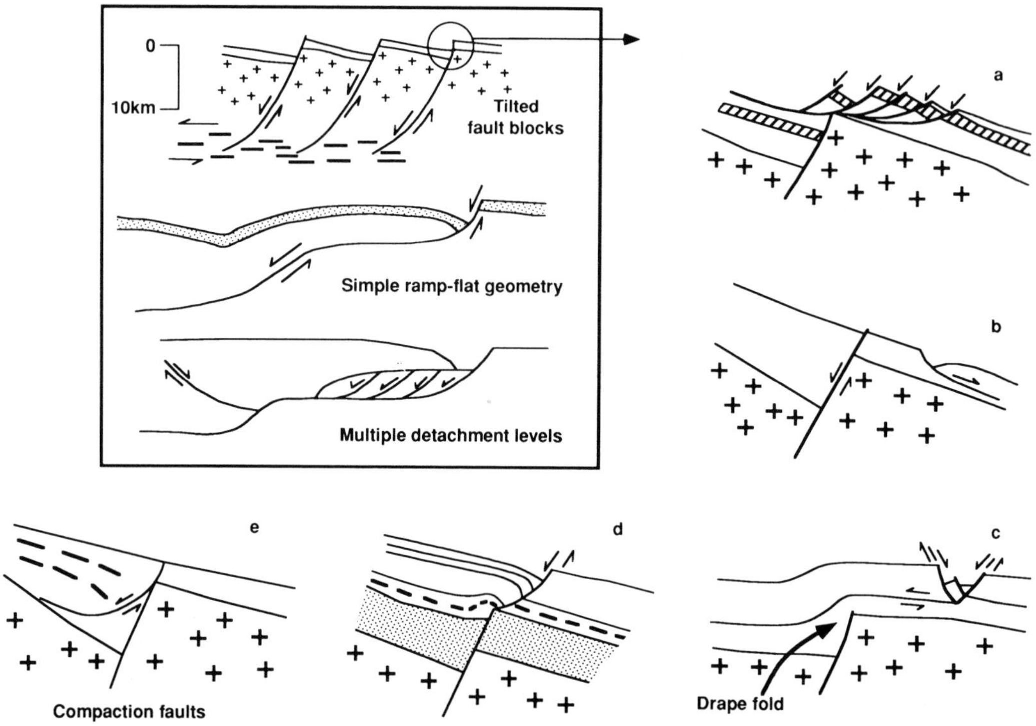

Fig. 18. Different types of major and minor fault geometry in rifts. Fault mechanics indicate that large, basement involved faults in rifts should typically be high angled, however, large, low-angled fault models (some formed along pre-existing structures) have been proposed (inset). These low-angle fault models may display fault bend folds and duplexes analogous to their thrust belt counterparts. Large tilted fault blocks in rifts can have their hanging walls and footwalls modified, in a variety of ways, by detachment faults within the sedimentary section (a–e) (modified from Patton 1984; Withjack *et al.* 1990).

into the low-angle detachment model or simple shear model for extensional faults proposed by Wernicke (1981, 1985) proved a popular model. It helped support similar ideas about the existence of low-angled faults derived from seismic data across rifted continental margins (e.g. Le Pichon & Sibuet 1981). The Wernicke model suggested that high and low-angled extensional fault systems could join into one, or a series of low-angled (less than 20°) faults (brittle faults passing into narrow ductile shear zones) that traversed much or all of the crust. The main zone of thinning (faulting) at the surface would thus be offset tens of kilometres from the zone of lower crustal thinning (Fig. 17c). This contrasted with the McKenzie (pure shear) model of rifting (McKenzie 1978) where ductile lower crustal thinning occurs below the surface rift (Fig. 17a). Several authors have also suggested hybrids of the two models where low-angle faults may pass into a broad zone of bulk pure shear in the lower–middle crust (e.g. Rehrig & Reynolds 1980; Hamilton 1982; combined simple shear–pure shear model, Fig. 17b). The McKenzie and the combined simple shear–pure shear models are probably the most appropriate ones for continental rifts (Buck *et al.* 1988).

The low-angled detachment model has been applied to rifts, passive margins and (extensional collapse) thrust belts throughout the world (e.g. Beach 1986; Castro 1987; Tankard & Welsink 1987; Selverstone 1988; Williamson *et al.* 1990). The recognition of low-angled extensional detachments in basement and basement-cover relationships was a large contributor to the revival of (academic) interest in rifts. The impact on hydrocarbon exploration is less easy to define and the model may not be regarded as very important to finding hydrocarbons by many explorationists. However, the model is important in several ways, including: (1) regional studies of basins; (2) models of tectonic evolution in rifts. These two aspects together with studies on low-angle faults in sedimentary rocks are discussed below.

Application of low-angle fault models

Low-angle detachment faults and linked low- and high-angle fault systems within sedimentary

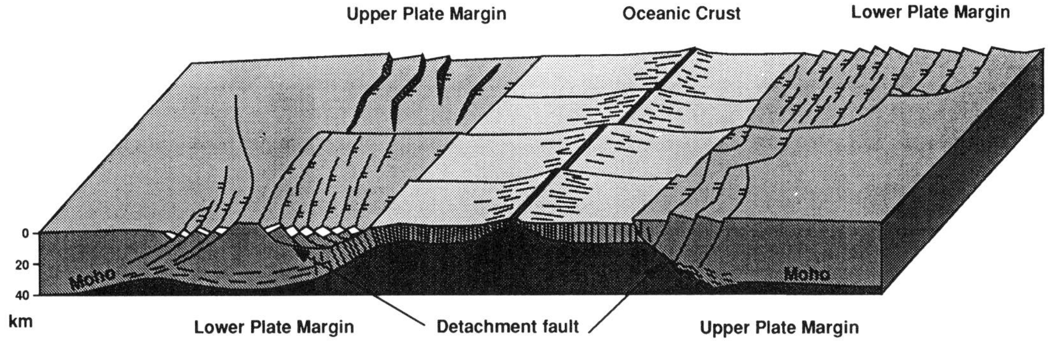

Fig. 19. Application of the simple shear model to basin distribution on rifted continental margins proposed by Lister *et al.* (1986). Lower plate margin corresponds to the footwall of the main detachment faults. The upper plate margin is primarily composed of the hanging wall to the main detachment fault. If the polarity of the detachment fault flips along-strike then the type of margin can also change along-strike.

sequences had been recognized in deltaic areas for a long time by the oil industry (e.g. Woodbury *et al.* 1973; Bruce 1973; Harding & Lowell 1979). Consequently, the impact on exploration of interpretations proposing the existence of low-angled, basement was relatively minor. However, it is a useful concept that ramp and flat geometries on extensional faults could generate broad anticlines both during extensional and inversion phases (Gibbs 1984, 1989; Fig. 18, inset). However, these are hypothesis and have not been proven. It was also recognized that rifts might be composed of multiple detachment levels which considerably increases the potential for structural complexity beyond traditional rift structural models (e.g. Gibbs 1984, 1989; Tankard & Welsink 1987; Fig. 18 inset).

Tilted fault blocks may commonly be modified by minor detachment faults that lie within the sedimentary section of the tilted fault block. These minor detachments exploit bedding surfaces or rock units that possess a high ductility contrast or low shear strength (e.g. evaporites, overpressured shales). They can exhibit a wide variety of geometries (Fig. 18a–e) and have a significant economic impact, since the unexpected occurrence of such faults may affect reserve calculations, field development, well positioning and side-track decisions. Steeply dipping faults with a zig-zag shape at depth can pass upwards into a more shallow dipping, smooth trajectory or listric fault segments within the sedimentary section, particularly if a thick shale or evaporite section is present (Jarrige 1992; Fig. 18d). Well and seismic data also show that the tops of tilted fault blocks are commonly affected by low-angle faults, completely separate from the main fault, that facilitate a basinwards slumping of the fault block crest (Gabrielsen 1986; Fig. 18a). Apparently these faults can occur both early and late in the development of tilted blocks. If sliding occurs when the footwall block has developed significant topographic relief, chaotic deposits are likely to develop in the hanging wall adjacent to the boundary fault.

Detachment faults of opposite dip to the basement fault can also form in the footwalls of major tilted blocks (Fig. 18b) in response to the gravitational instability of the sediments as the block tilts (Patton 1984). Physical experiments in multi-layered sequences suggest that layer-parallel slip associated with forced folding over a deeper normal fault may initiate the formation of small graben structures over the footwall block (Withjack *et al.* 1990; Fig. 18c). This type of structure may be most common when very ductile rocks (e.g. halite) are present. Also, the thickness contrast between syn-rift sediments across a large normal fault may cause compaction related faults to form on the down-thrown side (Fig. 18e). The upper segment of the compaction fault may use a segment of the main boundary fault or activate along eroded fault scarp surfaces (Patton 1984).

In appropriate areas regional, basement involved, low-angled detachment models may be conceptually very important. For example, the low-angled fault model of Lister *et al.* (1986) divides rifted continental margins into upper and lower plate margins based on low-angled detachment fault geometry (Fig. 19). It is an important model for explorationists to either use or dismiss from regional studies of such margins. The model defines two types of rifted margin; one composed of dominantly footwall rocks (rich in sedimentary basins) and the other composed of dominantly the hanging wall sequence (deficient in sedimentary basins), termed lower and upper plate margins, respectively. The model implies that considerable asymmetry in basin distribution exists between the two rifted margins, hence petroliferous trends may not extend from one rifted margin across to the opposite one. Since the detachments may change polarity (every few hundreds of kilometres) passing down the continental margin, the style of rifting will change from upper plate to lower plate and hence cause regional changes in prospectivity. In the south Atlantic the location of major hydrocarbon producing areas passing from north to south switches from the Reconcavo Basin in Brazil, to the Congo–Angolian margin, back to the Campos Basin on the Brazilian margin. This switching makes the changing detachment polarity model an attractive explanation for the distribution of hydrocarbons.

The Southern Atlantic margins have been the subject of several studies that present a simple shear model for the regional margin configuration (e.g. Ussami *et al.* 1986; Castro 1987; Etheridge *et al.* 1989). However, Davison (1988) commenting on Castro (1987) argued that low-angled crustal detachments were not applicable for several reasons, including incompatibility between the detachment models and the crustal structure derived from gravity data. Chang *et al.* (1992) have criticized the Etheridge *et al.* (1989) model for being over simplistic and not matching the existing data. Thus, the detachment model is by no means universally accepted. Nor is the detachment model the only way to generate asymmetric rift margins. It is also possible to generate similar regional changes by asymmetrically rifting a pure-shear model, it is simply a question of where the oceanic crust is initiated with respect to the rift basins.

Rift evolution

In the cold, mildly extended regions beneath portions of the East African Rift System earthquake focal depths are surprisingly deep (25–30 km; Shudofsky *et al.* 1987). These data suggest that in this area extensional faults are steep and have penetrated deeply within the crust at the early stages of rifting (Shudofsky *et al.* 1987; Morley 1989). The main characteristics of the evolution of high-angle, domino-style fault blocks is that they rotate to lower angles with increasing extension. This may cause the faults to deactivate and new, higher-angle faults that cross-cut the older fault set to be initiated (e.g. Morton & Black 1975; Proffett 1977).

With the recognition that low-angle detachments may also exist, and that in many rifts the brittle–ductile transition lies between 20 and 10 km, it has been noted that changing crustal conditions (particularly temperature) during extension will cause changes in rift geometry (e.g. Kusznir & Park 1987). This may also lead to an evolution from high-angle (60–70°) to lower-angle (30–45°) fault geometries with time or vice versa (e.g. Lister & Davis 1989; Morley 1989; Bosworth 1992; Walker *et al.* 1992). This change is independent of block rotation and represents a major change in the rift geometry, it is commonly accompanied by a change in subsidence rate. Higher-angle faults display a larger throw : heave ratio than lower-angle faults, hence, given the same strain rates, higher-angle faults will be regions of relatively higher subsidence (Morley 1989). This is particularly important in continental rifts where deep lacustrine (source rock prone) conditions may be associated with high-angle boundary faults. Whilst lower-angle boundary faults may be associated with more coarse clastic prone sequences (Morley 1989).

Another tendency visible on seismic data in the East African Rift System is for the initial system of alternating half-grabens (spacing < 100 km) to evolve into a more regional mono-detachment orientation with time (several hundred kilometres long). In the Northern Kenya Rift deep reflections on some seismic lines indicate the presence of eastwards- and westwards-thickening basins (Fig. 3) with a spacing < 100 km. These eastward-thickening basins have been overprinted by dominantly westward-thickening basins as a result of the linkage of the east dipping major faults (Elgayo–Turkwell Faults) to form a regional fault trend whose extent is several hundred kilometres (Fig. 4).

As extension increases some fault blocks may become deactivated and sealed by syn-rift sediments (Fig. 3, Table 1). Active faults tend to either rotate to lower angles or be initiated at relatively low angles. Movement of these faults creates large, highly-rotated fault blocks which may reorganize earlier fault block orientations. By increasing the dip of the blocks, the spill point of trapped hydrocarbons will change and the area under closure will decrease. This is seen in the passage from relatively low to high extension rifting passing NW to SE along the Gulf of Suez (Colletta *et al.* 1986; Patton *et al.* 1994). Footwall uplift associated with the extension may also result in significant erosion of the crests of the tilted fault blocks. Thus, rotation leaking of oil along faults, and erosion will tend to destroy or reduce traps in highly

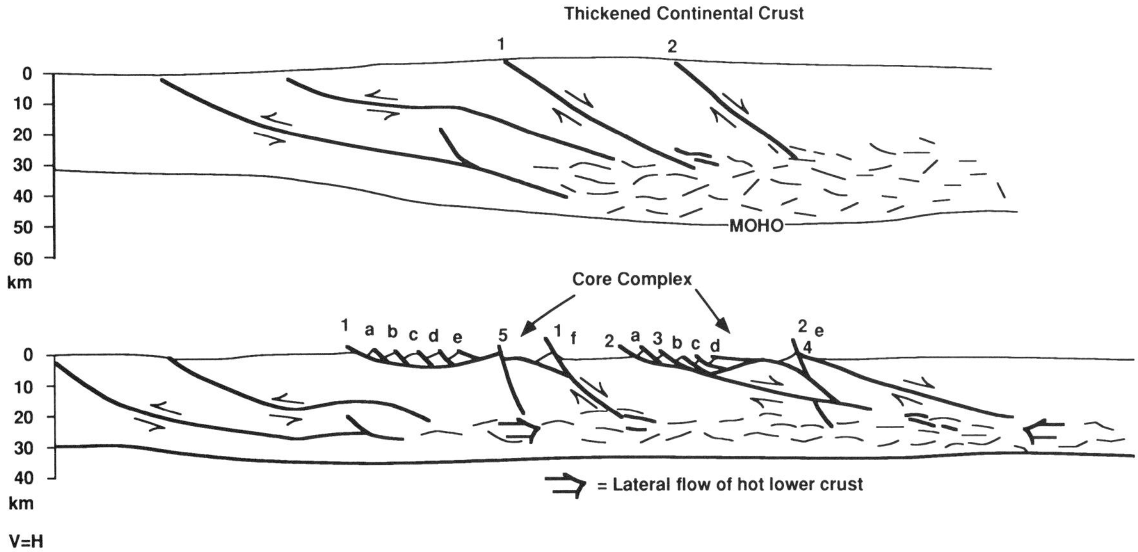

Fig. 20. Diagrams illustrating the evolution of structures associated with extensional collapse of thick, hot crust (collapsed thrust belt, core complex setting). The numbers indicate the relative ages of the faults. Early, large-displacement, high-angled faults (1 and 2) that have become rotated to low angles are shown as being cut by later low-angled faults (3 and 4) and by a late high-angled fault (5). For faults 1 and 2 the a–e lettering denotes the younging direction of deactivated fault slices – a is the oldest and e the youngest.

extended rifts. However, provided a sag basin develops over the eroded titled fault blocks, a sealing facies is deposited and petroleum migration occurred during the sag basin phase, eroded tilted fault blocks can trap large hydrocarbon reserves, as is the case in the North Sea (see summary in Spencer & Larsen 1990).

Controversy of low-angle faults

The low-angle fault model in basement rocks has been the subject of considerable controversy. It has been criticized for the following reasons: (1) low angle faulting is not favoured in theoretical (Andersonian) models of extending brittle layers (Buck 1988); (2) low-angle normal faults are seldom observed seismically (Jackson & McKenzie 1983; Jackson 1987); (3) specifically for the Basin-and-Range, the cooling rates of lower plate rocks in the core complexes require at least 2.5 cm a^{-1} displacement on a fault dipping at 10°, while present day estimates of the extension rates for the entire Basin-and-Range Province are less than 1 cm a^{-1} (Buck 1988); (4) in several cases in the Basin-and-Range the change in metamorphic grade along the footwall of a present day low-angled fault suggests that the fault was actually initiated at a much higher angle (45°) (e.g. Wernicke & Axen 1988).

Deep seismic reflection data indicate that the Moho is flat-lying under the Basin-and-Range (McCarthy & Thompson 1988), suggesting lateral flow of lower crustal material into the region where thinning associated with the upper crustal extension would be expected (Gans 1987; Buck 1988). Such a fluid and weak lower crust would not be able to support large topographic gradients. Integrating this seismic evidence with arguments against low-angled faults, Buck (1988) proposed that the upper portions of high-angle faults became rotated by footwall uplift into low-angled orientations. By successive rotation and deactivation of the normal fault a whole series of low-angle fault blocks could be created, with the lower crust flowing into the region beneath the high-angle fault (Fig. 20). However, the evolution of the Basin-and-Range appears to be more complex than one single fault model permits. There is evidence for both late low-angle faults cross-cutting earlier high-angle faults and also late high-angle faults (Lister & Davis 1989; Fig. 20).

Several workers have argued that the assumption of Andersonian fault mechanics (one of the principal stress axes is vertical) is not necessarily correct (Lister & Davis 1989; Yin 1989; Parsons & Thompson 1993). In order to generate low-angled faults it is necessary to find a mechanism which will cause rotation of the vertical principal stress direction. Such non-Andersonian conditions may be caused by a basal shear stress, such as regional ductile flow in the lower crust or between the crust and mantle (Lister & Davis 1989; Yin 1989), or by igneous intrusions (Parsons & Thompson 1993).

Table 2. *Different types of uplift/inversion features in rifts, with approximate dimensions of the features.*
Style examples; 1, Vejbaek & Anderson 1987; 2, Glennie & Boegner 1981; 3, Lake & Karner 1987; 4, Riis & Fjeldskaar 1992; 5, Skogseid *et al.* 1992; 6, White & McKenzie 1989; 7, Kusznir & Egan 1989.

	Rotational of extensional field	Compression /wrenching	Isostatic rebound	Flexural–isostatic footwall uplift	Magmatic underplating	Thermal effects	
						Mantle plume	Rifted margin
Uplift	10–10^3m	< 2–3 km	1–2 km	< 5 km	Difficult to separate from other effects. < 1 km	1–2 km	1–2 km
Wavelength	Up to a few kilometres	< 10–10^2 km	10^2–10^3 km	< 10–10^2 km	10–10^3 km	> ~10^3 km	10^2–10^3 km
Area	≤10–20 km^2	10–10^3 km^2	10^5–10^6 km^2	10–10^3 km^2	10^4–10^6 km^2	10^5–10^6 km^2	10^5–10^6 km^2
Volume of eroded material	≤20 km^3	10^3–10^4 km^3	10^5–10^6 km^3	10^3–10^4 km^3	10^5–10^6 km^3	10^5–10^6 km^3	10^5–10^6 km^3
Map view	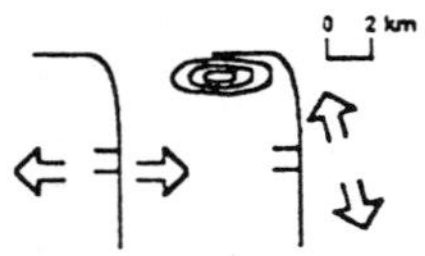	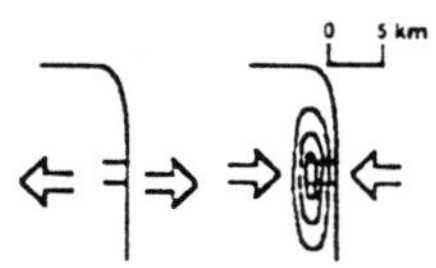	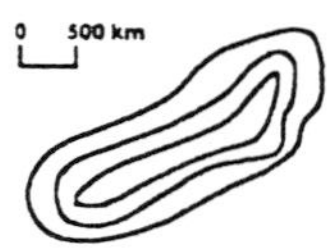	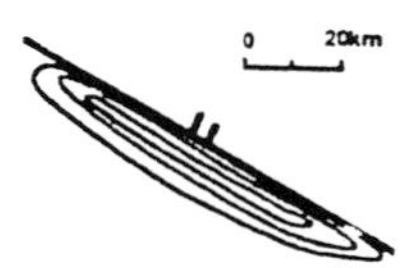	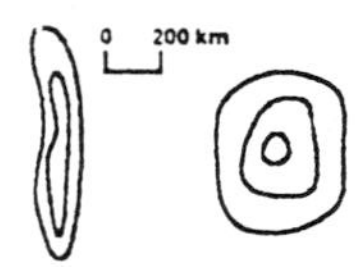	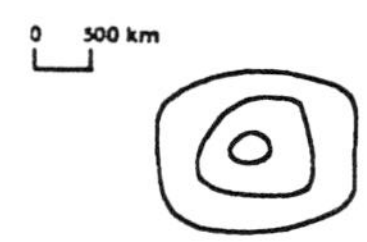	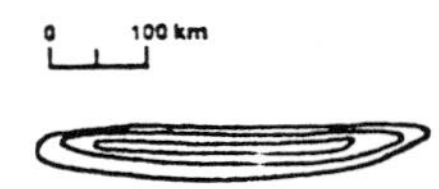
Style Examples	Local folds inverted faults. Jurassic–Early Cretaceous period, North Sea. 1.	Individual folds along faults to whole basins inverted, e.g. Lower Tertiary; North Sea; Sole Pit Basin, Weald 2,3.	Regional uplift large wavelength, low amplitude. Glaciation, e.g. Barents Sea. 4.	Uplift along the footwalls of faults, limited by size of fault and density of hanging wall fill. 7.	Local-regional uplifts, variable geometry, e.g. Voring Basin, Kenya Rift. 5.	Regional circular–elliptical doming, e.g. Afar Dome. 6.	Linear uplift following rifted plate margin, e.g. Red Sea, Voring Basin. 5,6.

Outside of the Basin-and-Range low-angle faults with large displacements of tens of kilometres have been found in the field in collapsed thrust belts (e.g. Tauern window, eastern Alps; Selverstone 1988), but not yet demonstrably in rifts. Deep crustal seismic data across the North Sea (e.g. Cheadle *et al.* 1987) showed a mixture of high- and low-angle extensional faulting. However, the low-angled faulting is apparently associated with pre-existing Palaeozoic structures (see Review in Klemperer & Hurich 1990). The East African Rift Project PROBE and industrial seismic data displays a mixture of high-angled planar and listric faulting and low-angled boundary faults. The low-angled boundary faults (30–40°) cross-cut isoclinal folds in the Precambrian basement and cannot easily be explained as rotated higher angle faults or reactivated thrusts (Morley 1989; Fig. 7). The lower-angled extensional faults are apparently associated with regions of higher heat flow than the higher-angle faults. The current state of knowledge suggests that faults, at least with initial dips as low as 30°, do seem to exist in the upper and middle crust with and without the influence of pre-existing trends. However, initially very low-angled faults do not appear to be characteristic of continental rifts.

Basin inversion

The notion that basins can be inverted has existed for a considerable time (e.g. Lamplugh 1920; Stille 1924). More recently it was recognized in Indonesia that relatively minor (in terms of shortening) but significant (in terms of structures visible on seismic data) reverse faults, strike-slip faults and associated folds developed along reactivated normal faults (Sunda folds: e.g. Mertosono & Nayoan 1974; Mertosono 1975; White & Wing 1978; Eubank & Makki 1981). Basin inversion remained a topic of only minor structural interest until the 1980s. The recognition of inversion features in the North Sea (e.g. Glennie & Boegner 1981; Bally 1984; see papers in Ziegler, 1985; Cooper & Williams 1989) led to more widespread interest. The discussion of Cooper *et al.* (1989) illustrates that the definition of basin inversion is highly debatable. They resolved the definition to (p. 346) 'a basin controlled by a fault system that has been subsequently transpressed–compressed producing uplift and partial extrusion of the basin fill'

The stress regime in many rifts has apparently changed with time, as documented by fault kinematic observations (e.g. Kenya Rift; Strecker *et al.* 1990: western branch of the East African Rift; Ring *et al.* 1993; Delvaux *et al.* 1993), basin inversion features (e.g. Cartwright 1989; Chapman 1989; Charpentier *et al.* 1992) and by changing isopach patterns (e.g. Witch Ground Graben; Boldy & Brealey 1990). Consequently, the original set of faults developed in response to one stress regime may develop non-ideal orientations for extension or any type of reactivation as the rift evolves. Their sense of displacement may change or the faults may be abandoned.

Inversion may be triggered by modification of regional intraplate stresses by deviatoric stresses due to upwelling of asthenospheric convection cells and diapiric intrusions into the lithosphere (Bott 1992). Inversion may also be caused by changes in regional stresses as plate boundaries evolve (Ziegler 1982), or by areal variations in the drag or shear traction forces at the base of the lithosphere (e.g. Doglioni 1990; Bott 1992). More superficial inversion may be initiated by gravitational instability caused by tilting and the subsequent gravity sliding of a portion of the basin fill sequence (commonly on a shale or salt detachment level), or by diapiric salt and shale movements.

Basin inversion is driven by a change in the orientation of the original extensional stress field or a change to compressive or transpressive stress regime (e.g. Cooper *et al.* 1989). There are many other possible causes of basin uplift that are distinctly different from basin inversion, including isostatic rebound after glaciation (e.g. Barents Sea; Riis & Fjeldskaar 1992), mantle thermal anomalies (rift shoulder elevation, mantle plume–thermal doming; McKenzie 1978), magmatic underplating and density changes in the crust due to metamorphism (e.g. Vejbaek 1992). By themselves each mechanism is likely to give rise to a different uplift geometry and magnitude (Table 2). However, in combination it is often difficut to separate out the contribution of the individual effects.

The recent information on stress changes and basin inversion mentioned above suggests that rifts commonly do not evolve under long-lived stable stress conditions. This has implications for trap destruction and creation, and may also help to explain the common divergence of interpretations from the same seismic data set. Most commonly in structural mapping of rifts interpretations tend to become polarized towards wrenching and pure rifting, with the rhombic pattern produced by zigzag faults being a tantalizing target for pull-apart basin interpretations. For example, the different interpretations of Lake Tanganyika (Fig. 21) are based upon applying two different tectonic models to a coarse seismic data set. One model

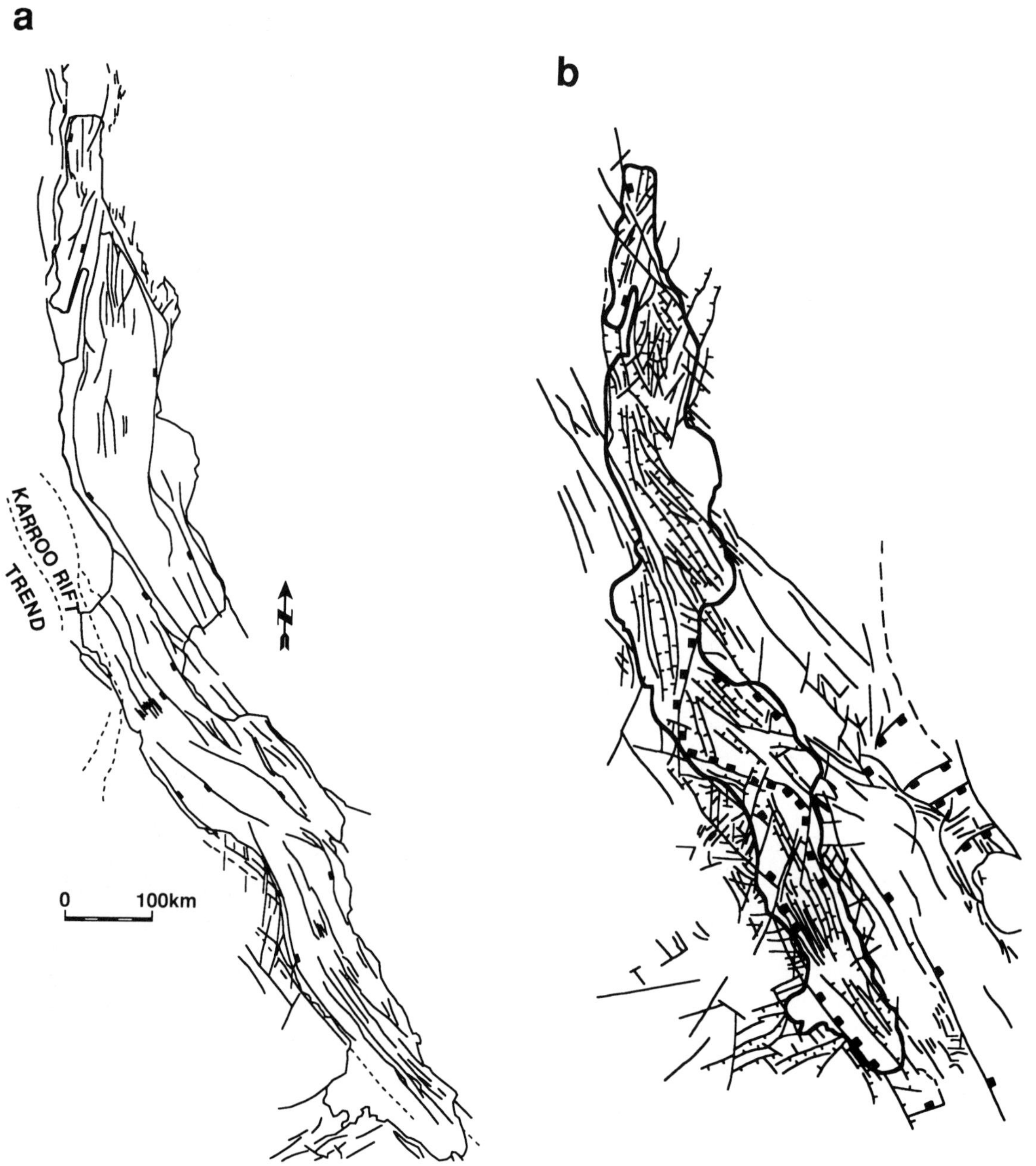

Fig. 21. Comparison of fault pattern interpreted from the same seismic data set over Lake Tanganyika; (a) assuming E–W extension (Morley 1988); (b) assuming NW–SE extension (Rosendahl *et al.* 1986).

assumes NW–SE extension which suggests a strong wrenching component (e.g. Rosendahl *et al.* 1986; Rosendahl 1987), while the other assumes E–W pure to oblique extension (e.g. Morley 1988; Ebinger 1989; Morley *et al.* 1992*a*). The recognition of minor anticlines on seismic data (Fig. 22) and late sub-horizontal slickensides on faults suggested a strong wrench fault component to the Tanganyika–Rukwa–Malawi area (Wheeler & Carson 1989; Kilembe & Rosendahl 1992). However, this evidence did not fit easily with the amount of extension displayed on the major and minor rift bounding faults and the absence of easily demonstrable important wrench features (e.g. Morley *et al.* 1992*a*). The latest fault kinematic results reconcile the contrasting observations by suggesting that a late (2 Ma–Present) reorientation of the regional extension direction from E–W to NW–SE caused reactivation of some originally oblique extensional faults with a strong strike-slip component (e.g. Ring *et al.* 1993). New

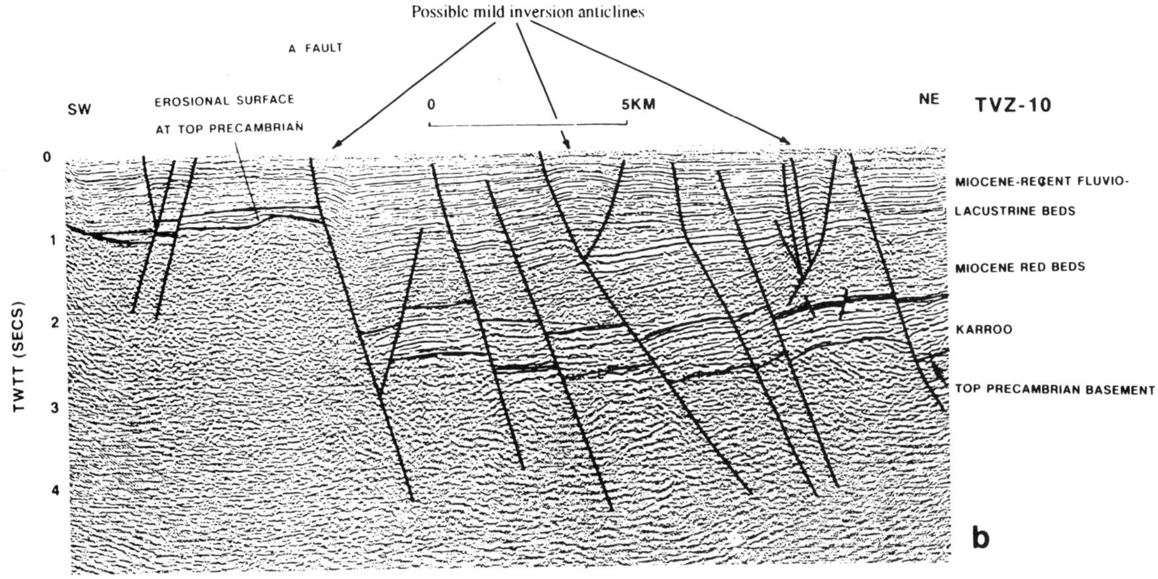

Fig. 22. Detail of seismic section TVZ-7, Lake Rukwa (after Morley *et al.* 1992) illustrating minor, late inversion anticlines.

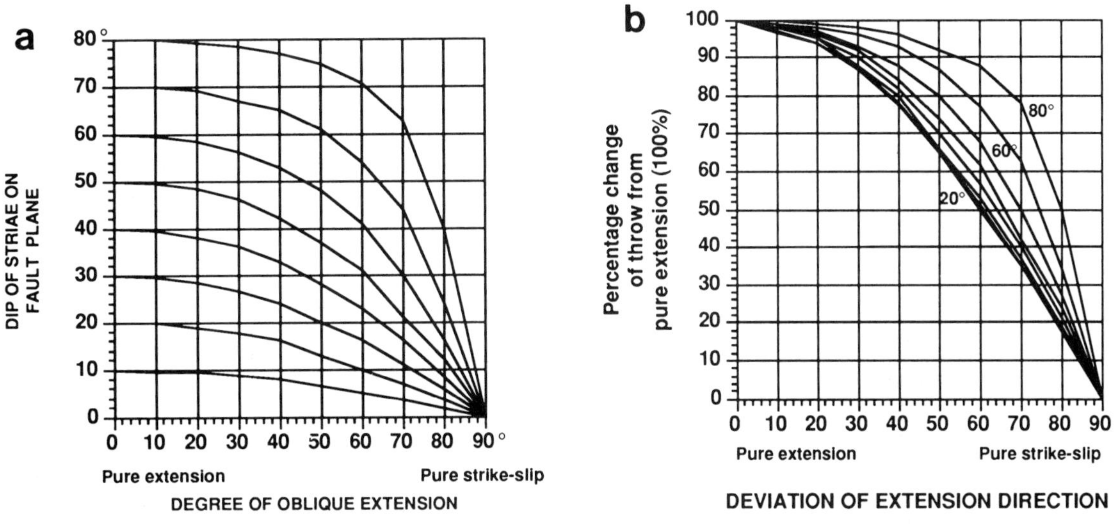

Fig. 23. Graphs illustrating the effects of oblique extension on fault displacement. (a) Change in dip of slickensides on faults of varying dips as a function oblique extension; (b) change in throw (tectonic subsidence) in relation to the degree of oblique extension. Throw is shown as a percentage of the pure dip–slip throw value. The percentage value corresponds to the dip of the fault, the lines correspond to faults dipping at 80, 70, 60, 50, 40, 30 and 20°.

extensional faults were also created at this time, oriented sub-perpendicular to the extension direction, e.g. Usangu Flats (Fig. 4). Where possible, it is important to have fault stress and strain data from outcrop studies early on in the exploration history to help minimize potential errors and controversies over the structural interpretation from seismic data.

It is apparent that rifts commonly evolve from pure-extension to oblique-extension with time, however, since rifts are composed of segments and faults of multiple orientations, such changes will display varying effects in different parts of rifts. The natural tendency for faults to zigzag or the strong influence of basement structure commonly causes some rift offset segments to form at an acute angle to the regional extension direction rather than perpendicular to it (Fig. 4). These oblique rift segments may exhibit much more significant changes in structural style than

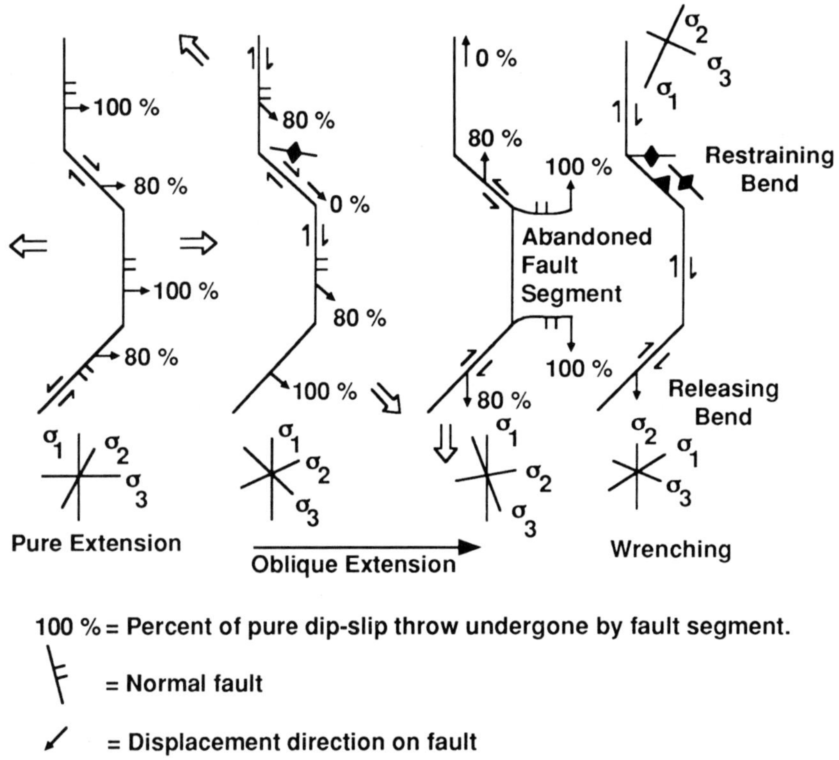

Fig. 24. Examples of how faults change their sense of motion and subsidence rates as a result of changing the extension direction or changing to pure strike–slip (inversion).

rift parallel segments when the stress field rotates during rifting. Figure 23 shows how, for a fault plane, the degree of oblique-displacement on the fault changes the amount of subsidence that its hanging wall undergoes per unit of heave. The effects of oblique opening vary depending upon the dip of the fault but the trends are similar. Generally, up to 40–50° divergence from extension the percentage decrease in subsidence (vertical throw) from the pure extension case is small (*c.* 20%). For higher values of obliquity the amount of subsidence drops rapidly. An example of the effects of these changes is given in Fig. 24; three faults with intersections of 45° are subjected to a change of 45° in the regional extension direction. The change in the amount of subsidence on each fault as a result of this is shown and demonstrates that the segment that went from oblique wrenching to almost pure strike-slip deformation experienced a much greater change in subsidence than the other two orientations. This fault segment is also the only segment likley to develop basin inversion structures. True strike-slip deformation produces a different inversion pattern than highly oblique extension. The opportunities for the creation of extensional fault segments are reduced and inversion may be widespread along the fault trends, not just restricted to minor restraining bends.

Discussion and conclusions

At the beginning of the 1980s the surface geology of many rifts was well known but commonly the sub-surface geology was not. The North Sea, Sirte Basin and Gulf of Suez were highly explored rifts, however, for reasons of seismic data quality (Gulf of Suez), thickness of the post rift sag basin (Sirte Basin), poor or non-existent outcrop (North Sea, Sirte Basin), complex structural history (North Sea) or the complex influence of pre-existing trends (North Sea), plus industrial confidentiality restrictions, no unifying models for rift geometry had been derived from these rifts. The first steps by structural geologists to define rift geometry better in the early 1980s were a mixture of thrust belt derived geometries (extensional duplexes, normal fault ramps and flats, transfer faults), deltaic gravitational tectonics (in particular listric, spoon-shaped faults, low-angled detachments and roll-over anticlines) and Basin-and-Range fault geometries (domino tilted fault blocks, low-angled faults in basement), coupled with known rift geometries (zigzag faults, high-angle faults bounding tilted blocks, half- and full-graben geometries). This fusion of different structural styles into rift provinces proved to be highly controversial, particularly in the North Sea. The definition of rift basin geometries in the East African Rift by Project PROBE (e.g. Rosendahl *et al.* 1986; Ebinger *et al.* 1987) provided the first widely accessible data on the regional stuctural geometry of rifts, and led to the detailing of half-graben geometries, how half-grabens change geometry (accommodation zones; Bosworth 1985) and the influence of boundary fault geometry on stratigraphic packages. Fieldwork in the Gulf of Suez by French teams (e.g. Colletta *et al.* 1986; Jarrige *et al.* 1990) and by GUPCO–Amoco teams (see review in Patton *et al.* 1994); has also provided very important information on the detailed geometries of tilted fault blocks and the relationships of structure and sedimentation

Three different aspects of rift geometry and their impact on hydrocarbon exploration have been examined in this paper: rift architecture, fault angle and basin inversion. Rift architecture is controlled by fault geometry, which is itself heavily influenced by pre-existing fabrics. At the largest scale rifts are divided into segments tens to hundreds of kilometres long. These segments are separated by rift jumps and gaps, they may be joined along offset segments or they may terminate at rift splays. Recognition of rift segment geometry is important for understanding structural trends in basin analysis, differing structural histories and regional sedimentation patterns. For example, rift jumps and gaps provide the entry points for large rivers to enter rifts and create along-axis drainage systems, while rift offsets are areas where anomalous structural patterns develop in response to oblique extension. Transfer zones between boundary faults within rift segments mark important changes in rift geometry. The transfer zones themselves are commonly broad zones of anomalous structural trends and are the preferred sites of coarse clastic sedimentation. In the Gulf of Suez, for example, the structures and the syn-rift reservoir units within two major transfer zones are associated with over four billion barrels of reserves out of a total of six billion barrels for the entire rift. Individual faults bound tilted blocks, the way these blocks end laterally is highly varied, and important in understanding the geometry of individual traps. Often the terminations are influenced by pre-existing fabrics. The common block termination geometries recognized are as follows: strike ramps, forced folds, rhomb-blocks, transfer faults, major and minor cross-strike faults (Fig. 16). In areas of relatively poor seismic data quality the choice of fault block termination interpreted on geophysical maps can make a considerable difference to the structural interpretation.

There has been considerable debate as to whether rifts are composed of high-angle (45–60°) faults or a mixture of high- and low-angle faults. There is little doubt that low-angle faults occur in the sedimentary sequence of rifts, where they occur in a variety of structural settings (Fig. 18). Such faults can have a significant economic impact in field development. The presence of low-angled basement faults is more problematic. They apparently do occur in rifts, both due to reactivation of older faults and as completely new faults. Apparently, rifts can evolve through a variety of low- and high-angle fault structures with time, which impacts trap creation and destruction. Understanding fault geometry greatly impacts the seismic mapping of prospects and projection of regional trends.

Two of the main structural models for rifts that existed at the beginning of the decade (pure shear and simple shear models) have changed. The simple shear model has been largely abandoned, although low-angle faults are a viable interpretation for deformation in the brittle upper and middle crust. The pure shear or combined pure shear–simple shear model remains the most frequently applicable model for continental rifts.

Basin modelling has evolved to the point where rift basins can be modelled sequentially instead of instantaneously. Subsidence on individual faults can be modelled, and the forward modelled section is a balanced cross-section (e.g. Kusznir & Egan 1989). Not only do such models help predict basin subsidence history, extension amount, heat flow and flexural-isostatic uplift, but they also help in understanding the distribution of sediments in rifts by providing a model for the magnitude and distribution of uplifts associated with faulting during rifting. Previously the main uplift mechanisms affecting rifts were considered to be the more regional effects of thermal plumes and magmatic underplating.

Rifts commonly undergo changes in their stress regime which may lead to basin inversion features. This affects trap development and destruction and can lead to a variable subsidence history for basins, as the basin bounding faults change their senses of motion. The recognition of inversion features in many rifts indicates that this is a very important aspect of rift history.

Structural geology plays an important role in

understanding the hydrocarbon geology of rifts in many ways. Not every aspect has been discussed above, for example the role of fractures in reservoir development programmes, faults as seals and fluid conduits, balanced cross-sections and many aspects of structural control on sedimentation. The increasing breadth of structural studies in rifts and the opening up of the former Soviet Union and onshore China suggests that the next decade may be as interesting and innovative as the last.

Amoco provided me with the opportunity to study rifts; much credit for the paper should go to the stimulating discussions I had with former co-workers. In particular, I want to acknowledge Ron Nelson and Tom Patton who were responsible for making me aware of many aspects of rifting, in particular the geometry of tilted fault blocks and transfer zones in the Gulf of Suez. I have also recently benefited from discussions with colleagues at Elf. I thank in particular Jean-Jacques Jarrige for helpful comments made about early versions of the manuscript and the comments made by two anonymous referees.

References

ALLMENDINGER, R. W., SHARP, J. W., VON TISH, D., SERPA, L., BROWN, L., KAUFMAN, S., OLIVER, J. & SMITH, R. B. 1983. Cenozoic and Mesozoic structure of the eastern Basin and Range Province Utah, from COCORP seismic reflection data. *Geology*, **11**, 532–534.

ANDERSON, E. M. 1951. *The Dynamics of Faulting*. Oliver and Boyd. Edinburgh.

BAKER, B. M., MOHR, P. A. & WILLIAMS, L. A. J. 1972. Geology of the eastern rift system of Africa. *Geological Society of America*, Special Paper, **136**, 67 pp.

BALLY, A. W. 1984. Tectonogenese et sismique reflection. *Bulletin de la Societe Geologique de France*, **26**, 279–286.

BEACH, A. 1986. A deep seismic reflection profile across the northern North Sea. *Nature*, **323**, 53–55.

BLANTON, T. L. 1982. An experimental study of interaction between hydraulically-induced and pre-existing fractures. SPE/DOE preprint 10847, *Proceeding of the Unconventional Gas Recovery Symposium*, Pittsburgh, 559–571.

BOLDY, S. A. R. & BREALEY, S. 1990. Timing, nature and sedimentary results of Jurassic tectonism in the Outer Moray Firth. *In:* HARDMAN, R. F. P. & BROOKS, J. (eds) *Tectonic Events Responsible for Britain's Oil and Gas Reserves*. Geological Society, London, Special Publication, **55**, 259–279.

BOSWORTH, W. 1985. Geometry of propagating continental rifts. *Nature*, **316**, 625–627.

——— 1992. Mesozoic and early Tertiary rift tectonics in East Africa. *Tectonophysics*, **209**, 115–137.

BOTT, M. P. H. 1992. Modelling the loading stresses associated with continental rift systems. *In:* ZIEGLER, P. A. (ed). *Geodynamics of Rifting, Vol. III, Thematic Discussions. Tectonophysics*, **170**, 99–115.

BOYER, S. & ELLIOTT, D. 1982. Thrust Systems. *American Association of Petroleum Geologists Bulletin*, **66**, 1196–1230.

BRUCE, C. H. 1973. Pressured shales and related sediment deformation: mechanisms for development of regional contemporaneous faults. *American Association of Petroleum Geologists Bulletin*, **57**, 878–886.

BUCK, R. W. 1988. Flexural rotation of normal faults. *Tectonics*, **7**, 959–973.

———, MARTINEZ, F., STECKLER, M. S. & COCHRAN, J. R. 1988. Thermal consequences of lithospheric extension: Pure and simple. *Tectonics*, **7**, 213–234.

CARTWRIGHT. J. A. 1989. The kinematics of inversion in the Danish Central Graben. *In:* COOPER, M. A. & WILLIAMS, G. D. (eds) *Inversion Tectonics*. Geological Society, London, Special Publication, **44**, 153–175.

CASTRO, A. M. C. 1987. The Northeastern Brazil and Gabon basins: a double rifting system associated with multiple crustal detachment surfaces. *Tectonics*, **6**, 727–738.

CHANG, H. K., KOWSMANN, R. O., FIGUEIREDO, A. M. F. & BENDER, A. A. 1992. Tectonics and stratigraphy of the East Brazil Rift system: an overview. *Tectonophysics*, **213**, 97–138.

CHAPMAN, T. J. 1989. The Permian to Cretaceous structural evolution of the Western Approaches Basin (Melville sub-basin), UK. *In:* COOPER, M. A. & WILLIAMS, G. D. (eds) *Inversion Tectonics*. Geological Society, London, Special Publication, **44**, 177–200.

CHARPENTIER, P., CAILLET, G. & LAURENT, J.-C. 1992. An advanced structural interpretation and fault analysis with Sismage in Snorre and 34/7 Blocks (offshore Norway). *Bulletin des Centres de Recherches Exploration–Production Elf Aquitaine*, **16**, 253–274.

CHEADLE, M. J., MCGEARY, S., WARNER, R. M. & MATTHEWS, D. H. 1987. Extensional structures on the western UK continental shelf: a review of evidence from deep seismic profiling. *In:* COWARD, M. P., DEWEY, J. F. & HANCOCK, P. L. (eds) *Continental Extensional Tectonics*. Geological Society, London, Special Publication, **28**, 445–465.

CLOOS, H. 1936. *Einfuhrung in die Geologie*. Gebruder Borntraeger Verlag, 503 pp.

COHEN, A. S. 1990. Tectono-stratigraphic model for sedimentation in Lake Tanganyika, Africa. *In:* KATZ, B. J. (ed.) *Lacustrine Basin Exploration, Case Studies and Modern Analogs*. American Association of Petroleum Geologists Memoir, **50**, 137–150.

COLLETTA, B., MORETTI, I., CHENET, P. Y., MULLER, C. & GERARD, P. 1986. *The structure of the Gebel Zeit area: A field example of tilted fault block crest in the Gulf of Suez: Field guide*. Institute Francais du Petrol, **34**, 547 pp.

COOPER, M. A. & WILLIAMS, G. D. (eds) 1989.

Inversion Tectonics. Geological Society, London, Special Publication, **44**, 375 pp.

——, ——, DE GRACIANSKY, P. C., MURPHY, R. W. NEEDHAM, T., DE PAOR, D., STONELEY, R., TODD, S. P., TURNER, J. P. & ZIEGLER, P. A. 1989, Inversion tectonics – a discussion. *In:* COOPER, M. A. & WILLIAMS, G. D. (eds) *Inversion Tectonics*. Geological Society, London, Special Publication, **44**, 335–347.

DALY, M. C., CHOROWICZ, J. & FAIRHEAD, D. 1989. Rift basin evolution in Africa: the influence of reactivated steep basement shear zones. *In:* COOPER, M. A. & WILLIAMS, G. D. (eds) *Inversion Tectonics*. Geological Society, London, Special Publication, **44**, 309–334.

DAVIS, D. H. & HARDY, J. J. 1981. The Eagle Pass Detachment; Southeastern Arizona; Product of Mid-Miocene listric normal faulting in the Southern Basin and Range. *Geological Society of America Bulletin*, **93**, 749–762.

DAVISON, I. 1988. Comment on 'The Northeastern Brazil and Gabon Basins: a double rifting system associated with multiple crustal detachment surfaces'. *Tectonics*, **7**, 1385–1391.

DESHENG, L. 1980. Geological structure and hydrocarbon occurrence of the Bohai Gulf oil and gas basin (China). *In:* MASON, J. F. (ed.) *Petroleum Geology in China*. Penwell Books, Tulsa, 180–192.

DELVAUX, D., LEVI, K., KAJARA, R. & SAROTA, J. 1993. Cenozoic Paleostress and kinematic evolution of the Rukwa–North Malawi rift valley (East African Rift System). *Bulletin des Centres de Recherches Exploration–Production Elf-Aquitaine*, **16**, 383–406.

DOGLIONI. C. 1990. The global tectonic pattern. *Journal of Geodynamics*, **12**, 21–38.

DUNBAR, C. O. 1949. *Historical Geology*. John Wiley, 573 pp.

DUNKLEMAN, T. J., KARSON, J. A. & ROSENDAHL, B. R. 1988. Structural style of the Turkana rift, Kenya. *Geology*, **16**, 258–261.

EBINGER, C. J. 1989. Tectonic development of the western branch of the East African rift system. *Geological Society of America Bulletin*, **101**, 885–903.

——, ROSENDAHL, B. R. & REYNOLDS, D. J. 1987. Tectonic model of the Malawi rift, Africa. *In:* BEN-AVRAHAM. Z. (ed.) *Sedimentary Basins within the Dead Sea and Other Rift Zones. Tectonophysics*, **141**, 215–235.

ETHERIDGE, M. A., SYMONDS, P. A. & LISTER, G. S. 1989. Application of the detachment model to reconstruction of continental conjugate passive margins. *In:* TANKARD, A. J. & BALKWILL, H. R. (eds) *Tectonics and the Stratigraphy of the North Atlantic Margins*. American Association of Petroleum Geologists Memoir, **46**, 23–40.

EUBANK, R. T. & MAKKI, A. C. 1981. Structural geology of the central Sumatra back-arc basin. *Proceedings of the 10th Annual Convention of the Indonesian Petroleum Association*, 153–196.

EVANS, A. L. 1990. Miocene sandstone provenance relations in the Gulf of Suez: insights into synrift unroofing and uplift history. *American Association of Petroleum Geologists Bulletin*, **74**, 1386–1400.

FREUND, R. 1982. Role of shear in rifting. *In:* PALMASON, G. (ed.) *Continental and Oceanic Rifts*, Geodynamic Series, American Geophysical Union and the Geological Society of America, **8**, 33–39.

—— & MERZER, A. M. 1976. The formation of rift valley and their zigzag fault patterns. *Geological Magazine*, **113**, 561–568.

FROSTICK, L. & REID, I. 1989. Is structure the main control of river drainage and sedimentation in rifts? *Journal of African Earth Sciences*, **81**, 165–182.

GABRIELSEN, R. H. 1986. Structural elements in graben systems and their influence on hydrocarbon traps and types. *In:* SPENCER, M. A. *et al.* (eds) *Habitat of Hydrocarbons on the Norwegian Continental Shelf*. Graham and Trotman, 55–60.

GANS, P. 1987. An open system, two-layer crustal stretching model for the eastern Great Basin. *Tectonics*, **6**, 1–12.

GARFUNKEL, A. & BARTOV, Y. 1977. Tectonics of the Suez rift. *Geological Survey of Israel Bulletin*, **71**, 44 pp.

GENIK, G. J. 1992. Regional framework, structural and petroleum aspects of rift basins in Niger, Chad and the Central African Republic (C.A.R.). *Tectonophysics*, **213**, 169–185.

GIBBS, A. D. 1983*a*. Balanced cross-section construction from seismic sections in areas of extensional tectonics. *Journal of Structural Geology*, **5**, 153–160.

—— 1983*b*. Secondary detachment above basement faults in North Sea; Clyde Field Growth Fault. *American Association of Petroleum Geologists Bulletin*; **67**, 469–470.

—— 1984. Structural evolution of extensional basin margins. *Journal of the Geological Society, London*, **141**, 609–620.

—— 1989. Structural styles in basin evolution. *In:* TANKARD, A. J. & BALKWILL, H. R. (eds) *Extensional Tectonics and Stratigraphy of the North Atlantic Margins*. American Association of Petroleum Geologists Memoir, **46**, 81–93.

GLENNIE, K. W. & BOEGNER, P. L. F. 1981. Sole pit inversion tectonics. *In:* ILLING, L. V. & HOBSON, G. D. (eds) *Petroleum Geology of the Continental Shelf of Northwest Europe*. Heyden, 110–120.

GREGORY, J. W. 1921. The rift valleys and geology of East Africa. Seely Service, 479 pp.

GRIFFITHS, P. S. 1980. Box-fault systems and ramps: a typical association of structures from the eastern shoulder of the Kenya rift. *Geological Magazine*, **117**, 579–586.

HAGRAS, M. & SLOCKI, S. 1982. Sand distribution of Miocene clastics in the Gulf of Suez. *In:* HANTAR, G. (ed.) *Egyptian General Petroleum Company 6th Exploration Seminar Preprint*.

HAMILTON, W. 1982. Structural evolution of the Big Maria Mountains, northeastern Riverside County, southeastern California. *In:* FROST, E. G. & MARTIN, D. L. *Mesozoic–Cenozoic Tectonic Evolution of the Colorado River Region,*

California, Arizona and Nevada. Cordilleran Publishers, 1–28.

HARDING, T. P. & LOWELL, J. D. 1979. Structural styles, their plate tectonic habitats, and hydrocarbon traps in petroleum provinces. *American Association of Petroleum Geologists Bulletin*, **63**, 2957–2962.

HENDRY, D. B., KUSZNIR, N. J., MORLEY, C. K. & EBINGER, C. J. 1994. *Tectonophysics*, **236**, in press.

ILLIES, H. 1974. Intra-Plattentektonik in Mitteleuropa und der Rheingraben. *Oberrhein. Geol. Abh.*, **23**, 1-24.

JACKSON, J. 1987. Active normal faulting and crustal extension. *In:* COWARD, M. P., DEWEY, J. F. & HANCOCK, P. L. (eds) *Continental Extensional Tectonics*. Geological Society, London, Special Publication, **28**, 3–17.

——— & MCKENZIE, D. 1987. The geometric evolution of normal faults. *Journal of Structural Geology*, **5**, 471–482.

JARRIGE, J.-J. 1992. Variation in extensional fault geometry related to heterogeneity within basement and sedimentary sequences. *Tectonophysics*, **215**, 161–166.

——— D'ESTEVOU, P. O., BOUROLLET, P. F., MONTENAT, C., PRAT, P., RICHERT, J.-P. & THIRIET, J.-P. 1990. The Multistage tectonic evolution of the Gulf of Suez and Northern Red Sea continental rift from field observations. *Tectonics*, **9**, 441–465.

KILEMBE, E. A. & ROSENDAHL, B. R. 1992. Structure and stratigraphy of the Rukwa Rift. *Tectonophysics*, **209**, 143–158.

KING, G. C. P., STEIN, R. S. & RUNDLE, J. B. 1988. The growth of geological structures by repeated earthquakes: 1. Conceptual framework. *Journal of Geophysical Research*, **93**, 13 307–13 318.

KLEMPERER, S. L. & HURICH, C. A. 1990. Lithospheric structure of the North Sea from Deep Seismic Refraction Profiling. *In:* BLUNDELL, D. J. & GIBBS, A. D. (eds) *Tectonic Evolution of the North Sea Rifts*. Clarendon Press, 37–63.

KRANTZ, R. W. 1989. Orthorhombic fault patterns: the odd axis model and slip vector orientations. *Tectonics*, **8**, 483–495.

KRISP Working Group 1991. The Kenya Rift: Pure shear extension above a mantle plume. *Nature*, **345**, 223–227.

KUSZNIR, N. J. & PARK, R. G. 1987. The extensional strength of the continental lithosphere: its dependance on geothermal gradient and crustal composition and thickness. *In:* COWARD, M. P., DEWEY, J. F. & HANCOCK, P. L. (eds) *Continental Extensional Tectonics*. Geological Society, London, Special Publication, **28**, 35–52.

——— & EGAN, S. S. 1989. Simple-shear and pure-shear models of extensional sedimentary basin formation: application to the Jeanne d'Arc Basin, Grand Banks of Newfoundland. *In:* TANKARD, A. J. & BALKWILL, H. R. (eds) *Extensional Tectonics and Stratigraphy of the North Atlantic Margins*. American Association of Petroleum Geologists. Memoir, **46**, 305–322.

LAKE, S. D. & KARNER, G. D. 1987. The structure and evolution of the Wessex Basin, southern England: an example of inversion tectonics. *Tectonophysics*, **137**, 347–378.

LAMBIASE, J. L. 1990. A model for tectonic control of lacustrine stratigraphic sequences in continental rift basins. *In:* KATZ, B. J. (ed.) *Lacustrine Basin Exploration, Case Studies and Modern Analogs*. American Association of Petroleum Geologists Memoir, **50**, 265–276.

LAMPLUGH, G. W. 1920. Structure of the Weald and analogous tracts, *Quarterly Journal Geological Society*, **75**, LXXIII–XCV (Anniversary Address to the President).

LELEK, J. J., SHEPHERD, D. B., STONE, D. M. & ABDINE, S. A. 1992. October Field, the latest giant under development in Egypt's Gulf of Suez. *In:* HALBOUTY, M. T. (ed.) *Giant Oil & Gas Fields of the Decade 1978–1988*. American Association of Petroleum Geologists Memoir, **54**, 231–249.

LE PICHON, X. & SIBUET, J. C. 1981. Passive margins: A model of formation. *Journal of Geophysical Research*, **86**, 3708–3720.

LISTER, G. S., ETHERIDGE, M. A. & SYMONDS, P. A. 1986. Detachment faulting and the evolution of passive continental margins. *Geology*, **14**, 246–250.

——— & DAVIS, G. A. 1989. The origin of metamorphic core complexes and detachment faults formed during Tertiary continental extension in the northern Colorado River region. *USA Journal of Structural Geology*, **7**, 65–93.

MCCARTHY, J. & THOMPSON, G. A. 1988. Seismic imaging of extended crust with emphasis on the western United States. *Geological Society of America Bulletin*, **100**, 1361–1374.

MCCONNEL, R. B. 1972. Geological development of the rift system of eastern Africa. *Geological Society of America Bulletin*, **83**, 2549–2572.

MCDONALD, R. W. 1976. Tertiary tectonics and sedimentary rocks along the transition, Basin and Range province to plateau and thrust belt province. *In:* HILL, V. G. (ed.) *Symposium on Geology of Cordilleran Hingeline*. Rocky Mountain Association of Geologists, Denver, 281–318.

MCKENZIE, D. 1978. Some remarks on the development of sedimentary basins. *Earth & Planetary Science Letters*, **60**, 25–32.

MERTOSONO, S. 1975. Geology of Pungut and Tandum oil fields, Central Sumatra. *Proceedings of the 4th Annual Convention of the Indonesian Petroleum Association*, 165–179.

——— & NAYOAN, G. A. S. 1974. The Tertiary basinal area of Central Sumatra. *Proceedings of the 3rd Annual Convention of the Indonesian Petroleum Association*, 63–76.

MOHR, P. A., GIRNIUS, A. & ROLFF, J. 1978. Present day strain rates at the northern end of the Ethiopian rift valley. *Tectonophysics*, **44**, 141–160.

MORGAN, P. 1983. Constraints on rift thermal processes from heat flow and uplift. *Tectonophysics*, **94**, 277–298.

MORLEY, C. K. 1988. Variable extension in lake Tanganyika. *Tectonics*, **7**, 785–801.

——— 1989. Extension, detachments and sedimentation in continental rifts (with particular reference

to East Africa). *Tectonics*, **8**, 1175–1192.
——— CUNNINGHAM, S. M., HARPER, R. M. & WESCOTT, W. A. 1992*a*. Geology and geophysics of the Rukwa Rift, East Africa. *Tectonics*, **11**, 69–81.
——— NELSON, R. A., PATTON, T. L. & MUNN, S. G. 1990. Transfer zones in the East African Rift System and their relevance to hydrocarbon exploration in rifts. *American Association of Petroleum Geologists Bulletin*, **74**, 1234–1253.
———, WESCOTT, W. A., STONE, D. M., HARPER, R. M., WIGGER, S. T. & KARANJA, F. M. 1992*b*. Tectonic evolution of the northern Kenyan Rift. *Journal of the Geological Society, London*, **149**, 333–348.
MORTON, W. H. & BLACK, R. 1975. Crustal attenuation in Afar. *In:* PILGER, A. & ROSLER, A. (eds) *Afar depression of Ethiopia, Inter-Union Commission on Geodynamics Scientific Report No. 14*. Schweizerbart'sche Verlagsbunchanglung, 55–65.
MOUSTAFA, A. M. 1976. Block faulting in the Gulf of Suez. *5th Egyptian General Petroleum Organization Exploration Seminar*, 19 pp.
NELSON, R. A., PATTON, T. L. & MORLEY, C. K. 1992. Rift–Segment interaction and its relation to hydrocarbon exploration in continental rift systems. *American Association of Petroleum Geologists Bulletin*, **76**, 1153–1169.
ORTEL, G. 1965. Mechanism of faulting in clay experiments. *Tectonophysics*, **2**, 343–393.
PANHUYS-SIGLER, VAN, M., BAUMANN, A. & HOLLAN, T. C. 1991. The Tern Field, Block 210/25*a*, UK. North Sea. *In:* ABBOTTS, I. L. (ed.) *United Kingdom Oil and Gas Fields, 25 Years Commemorative Volume*. Geological Society Memoir, **14**, 191–198.
PARSONS, T. & THOMPSON, G. A. 1993. Does magmatism influence low-angle normal faulting? *Geology*, **21**, 247–250.
PATTON, T. L. 1984. *Normal-fault and fold development in sedimentary rocks above a preexisting basement normal fault*. PhD thesis, Texas A & M University, College Station TX, 164 pp.
———, MOUSTAFA, A. R., NELSON, R. A. & ABDINE, A. S. 1994. *Tectonic Evolution and Structural Setting of the Suez Rift*. American Association of Petroleum Geologists, Special Publication, in press.
PROFFETT JR, J. M. 1977. Cenozoic geology of the Yerington district, Nevada, and implications for the nature and origin of Basin and Range faulting. *Geological Society of America Bulletin*, **88**, 247–266.
QUENNELL, A. M. 1956. Tectonics of the Dead Sea Rift. *Proceedings of the 20th International Geological Congress*. Mexico City, 385–405.
RAMBERG, I. R. & SMITHSON, S. B. 1975. Gravity interpretation of the southern Oslo Graben and adjacent Precambian rocks, Norway. *Tectonophysics*, **11**, 165–216.
RECHES, Z. 1978. Analysis of faulting in three-dimensional strain field. *Tectonophysics*, **47**, 109–129.
——— 1983. Faulting of rocks in three-dimensional strain field. II: A strain theory of faulting. *Tectonophysics*, **95**, 133–156.
REHRIG, W. A. & REYNOLDS, S. J. 1980. Geologic and geochronologic reconnaissance of a northwest-trending zone of metamorphic core complexes in southern Arizona. *Memoir of the Geological Society of America*, **153**, 131–158.
RIIS, F. & FJELDSKAAR, W. 1992. On the magnitude of the Late Tertiary and Quaternary erosion and its significance for the uplift of Scandinavia and the Barents Sea. *In:* LARSEN, R. M., BREKKE, B. T., LARSEN, B. T. & TALLERAAS, E. (eds) *Structural and Tectonic Modeling and its Application to Petroleum Geology*. Norwegian Petroleum Society, Special Publication, **1**, 163–185
RING, U., BETZLER, C. & DELAVAUX, D. 1993. Normal vs. strike-slip faulting during rift development in East Africa: The Malawi Rift. *Geology*, **20**, 1015–1018.
ROBSON, D. A. 1971. Structure of the Gulf of Suez (Clysmic Rift), with special reference to the eastern side. *Geological Society, London, Quarterly Journal*, **115**, 247–276.
ROSENDAHL, B. R. 1987. Architecture of continental rifts with special reference to east Africa. *Annual Review of Earth & Planetary Science*, **15**, 445–503.
——— & LIVINGSTONE, D. A. 1983. Rift lakes of East Africa, new seismic data and implications for future research. *Episodes*, **1**, 14–19.
———, REYNOLDS, D. J., LORBER, P. M., BURGESS, C. F., MCGILL, J., SCOTT, D., LAMBIASE, J. J. & DERKSEN, S. J. 1986. Structural expressions of rifting: lessons from Lake Tanganyika, Africa. *In:* FROSTICK, L. E., RENAUT, R. W., REID, I. & TIERCELIN, J.-J (eds) *Sedimentation in the East African Rifts*. Geological Society, London, Special Publication, **25**, 29–43.
SCHLISCHE, R. W. & OLSEN, P. E. 1990. Quantitative filling model for continental extensional basins with applications to the early Mesozoic rifts of eastern North America. *Journal of Geology*, **98**, 135–155.
SCHULL, T. J. 1988. Rift basins of interior Sudan; Petroleum exploration and discovery. *American Association of Petroleum Geologists Bulletin*, **72**, 1128–1141.
SELVERSTONE, J. 1988. Evidence for east–west crustal extension in the Eastern Alps: implications for the unroofing history of the Tauern Window. *Tectonics*, **7**, 87–105.
SHUDOFSKY, G. N., CLOETINGH, S., STEIN, S. & WORTEL, R. 1987. Unusually deep earthquakes in East Africa: constraints on the thermomechanical structure of a continental rift system. *Geophysical Research Letters*, **14**, 741–744.
SKOGSEID, J., PEDERSEN, T. & LARSEN, V. B. 1992. Voring Basin; subsidence and tectonic evolution. *In:* LARSEN, R. M., BREKKE, B. T., LARSEN, B. T. & TALLERAAS, E. (eds) *Structural and Tectonic Modeling and its Application to Petroleum Geology*. NPF, Special Publication, **1**, 55–82.
SMITH, M. & MOSLEY, P. 1994. Crustal heterogeneity and basement influence on the development of the Kenya Rift, East Africa. *Tectonics*, **12**, 591–606.

SPENCER, A. M. & LARSEN, V. B. 1990 Fault traps in the Northern North Sea. *In:* HARDMAN, R. F. P. & BROOKS, J. (eds) *Tectonic Events Responsible for Britain's Oil and Gas Reserves.* Geological Society, London, Special Publication, **55**, 281–298.

STEIN, R. S. & BARRIENTOS, S. E. 1985. Planar high-angled faulting in the Basin and Range: geodetic analysis of the 1983 Borah Peak Idaho, earthquake. *Journal of Geophysical Research*, **90**, 11 355–11 366.

STILLE, H. 1924. *Grundfragen der Vergleichenden Tektonik.* Borntrager, 443 pp.

STRECKER, M. R., BLISNIUK, P. M. & EISBACHER, G. H. 1990. Rotation of extension direction in the Central Kenya Rift. *Geology*, **18**, 299–302.

SYKES, L. R. 1978. Intraplate seismicity, reactivation of preexisting zones of weakness, alkaline magmatism, and other tectonism postdating continental fragmentation. *Review of Geophysics & Space Physics*, **16**, 621–687.

TANKARD, A. J. & WELSINK, H. J. 1987. Extensional tectonics and stratigraphy of Hibernia oil field, Grand banks, Newfoundland. *American Association of Petroleum Geologist Bulletin*, **17**, 1210–1232.

TEUFEL, L. W. 1979. An experimental study of hydraulic fracture propagation in layered rock, PhD thesis, Texas A & M University, College Station, 99 pp.

USSAMI, N., KARNER, G. D. & BOTT, M. H. P. 1986. Crustal detachment during South Atlantic rifting and formation of Tucano–Gabon rift system. *Nature*, **332**, 629–632.

VEJBAEK, O. V. 1992. Geodynamic modeling of the Danish Central Trough. *In:* LARSEN, R. M., BREKKE, B. T., LARSEN, B. T. & TALLERAAS, E. (eds) *Structural and Tectonic Modeling and its Application to Petroleum Geology.* Norwegian Petroleum Society, Special Publication **1**, 1–17.

——— & ANDERSEN, C. 1987. Cretaceous–Early Tertiary inversion tectonism in the Danish Central Trough. *Tectonophysics*, **137**, 221–238.

VENING MEINESZ, F. A. 1950. Les grabens Africans resultant de compression ou de tension dans la croute terrestre? *Institut du Royal Colonial Belge Bulletin*, **21**, 539–552.

WALKER, C. T., FRANCIS, R. D., DENNIS, J. G. & LUMSDEN, W. W. 1992. Cenozoic attenuation detachment faulting: A possible control on oil and gas accumulation in East–Central Nevada. *American Association of Petroleum Geologists Bulletin*, **76**, 1665–1686.

WANLI, Y. 1985. Daquin oil field, People's Republic of China; a giant field with oil of non-marine origin. *American Association of Petroleum Geologists Bulletin*, **69**, 1101–1111.

WERNICKE, B. 1981. Low-angle normal faults in the Basin and Range province: nappe tectonics in an extending orogen. *Nature*, **291**, 645–648.

——— 1985. Uniform-sense normal simple shear of the continental lithosphere. *Canadian Journal of Earth Science*, **22**, 108–125.

——— & AXEN, G. J. 1988. On the role of isostacy in the evolution of normal fault systems. *Geology*, **16**, 848–851.

WEISSEL, J. K., KARNER, G. D. & COWIE, P. A., 1987. On the causes of rift flank uplift; forward modeling of topography and gravity anomalies over oceanic rifts. *EOS*, **63**, 411.

WHEELER, W. H. & CARSON, J. A. 1989. Structure and kinematics of the Livingstone Mountains border fault zone, Nyasa (Malawi) Rift, southwestern Tanzania. *Journal of African Earth Sciences*, **8**, 393–413.

WHITE, J. M. & WING, R. S. 1978. Structural development of the South China Sea with particular reference to Indonesia. *Proceedings of the 7th Annual Convention of the Indonesian Petroleum Association*, 159–177.

WHITE, R. & MCKENZIE, D. 1989. Magmatism at Rift Zones: the generation of volcanic continental margins and flood basalts. *Journal of Geophysical Research*, **94**, 7685–7729.

WILLIAMSON, P. E., SWIFT, M. G., O'BRIAN, G. W. & FALVEY, D. A. 1990. Two-stage Early Cretaceous rifting of the Otway Basin margin of southeastern Australia: Implications for rifting of the Australian southern margin. *Geology*, **18**, 75–78.

WITHJACK, M. O., OLSEN, J. & PETERSON, E. 1990. Experimental models of extensional forced folds. *American Association of Petroleum Geologists Bulletin*, **74**, 1038–1054.

WOODBURY, H. O., MURRAY JR, I. B., PICKFORD, P. J. & AKERS, W. H. 1973. Pliocene and Pleistocene depocentres, outer continental shelf, Louisiana and Texas. *American Association of Petroleum Geologists Bulletin*, **57**, 2428–2439.

WRIGHT, L. A. & TROXEL, B. W. 1973. Shallow-fault interpretation of Basin and Range structure, southwestern Great Basin. *In:* DE JONG, K. A. & SCHOLTEN, R. (eds) *Gravity and Tectonics.* John Wiley and Sons, 397–407.

YIN, A. 1989. Origin of regional rooted low-angle normal faults: A mechanical model and its tectonic implications. *Tectonics*, **8**, 469–482.

YOUASH, Y. 1969. Tension tests in layered rocks. *Geological Society of America*, **80**, 303–306.

ZIEGLER, P. A. 1982. *Geological Atlas of Western and Central Europe.* Elsevier, 130 pp + enclosures.

—— (ed.) 1985. *Compressional Intra-Plate Deformations in the Alpine Foreland. Tectonophysics*, **137**, 420 pp.

The formation of passive margins: constraints from the crustal structure and segmentation of the deep Galicia margin, Spain

G. BOILLOT,[1] M. O. BESLIER,[1] C. M. KRAWCZYK,[2] D. RAPPIN[3] & T. J. RESTON[2]

[1] *Observatoire Océanologique de Villefranche, Laboratoire de Géodynamique Sous-Marine, B.P. 48, 06230 Villefranche-Sur-Mer, France*

[2] *GEOMAR, Forschungszentrum für Marine Geowissenchaften, Christian-Albrechts Universität, Wischhofstrasse 1-3, 2300 Kiel, Germany*

[3] *Ecole et Observatoire de Physique du Globe de Strasbourg, URA CNRS 323, ULP, 5 rue Descartes, 67084 Strasbourg Cedex; now at Elf Aquitaine Production, Centre Scientifique et Technique, 64018 PAU Cedex, France*

Abstract: The crustal structure of the Mesozoic deep Galicia margin and adjacent ocean–continent boundary (OCB) was investigated by seismic reflection (including pre-stack depth migration and attenuation of seismic waves with time). The seismic data were calibrated using numerous geological samples recovered by drilling and/or by diving with submersible.

The N–S trending margin and OCB are divided in two distinct segments by NE–SW syn-rift transverse faults locally reactivated and inverted by Cenozoic tectonics. The transverse faulting and OCB segmentation result from crustal stretching probably in a NE–SW direction during the rifting stage of the margin in early Cretaceous times. The Cenozoic tectonics are related to Iberia–Eurasia convergence in Palaeogene times (Pyrenean event). In both segments of the deep margin, the seismic crust is made of four horizontal layers: (1) two sedimentary layers corresponding to post- and syn-rift sequences, where velocity ranges from 1.9 to 3.5 $\mathrm{km\,s^{-1}}$, and where the Q factor is low, the two sedimentary layers being separated by a strong reflector marking the break-up unconformity; (2) a faulted layer, where velocity ranges from 4.0 to 5.2 $\mathrm{km\,s^{-1}}$, and where the Q factor is high. This layer corresponds to the margin tilted blocks, where continental basement and lithified pre-rift sediments were sampled; (3) the lower seismic crust, where the velocity (7 $\mathrm{km\,s^{-1}}$ and more) and the Q factor are the highest. This layer, probably made of partly serpentinized peridotite, is roofed by a strong S–S′ seismic reflector, and resting on a scattering, poorly reflective Moho.

A composite model, based both on analogue modelling of lithosphere stretching and on available structural data, accounts for the present structure of the margin and OCB. Stretching and thinning of the lithosphere are accommodated by boudinage of the brittle levels (upper crust and uppermost mantle) and by simple shear in the ductile levels (lower crust and upper lithospheric mantle). Two main conjugate shear zones may account for the observations and seismic data: one (SZ1), located in the lower ductile continental crust, is synthetic to the tilting sense of the margin crustal blocks; another (SZ2), located in the ductile mantle, accounts for the deformation of mantle terranes and their final unroofing and exposure at the continental rift axis (now the OCB). The S–S′ reflector is interpreted as the seismic signature of the tectonic contact between crustal terranes and mantle rocks partly transformed into serpentinite by syn-rift hydrothermal activity. It is probably related to both shear zones SZ1 and SZ2. The seismic Moho is lower within the lithosphere, at the fresh-serpentinized peridotite boundary.

Passive continental margins are the scars of the break-up of continents. Their basement underwent stretching before seafloor spreading started, and contains crucial information about timing and pressure–temperature conditions of lithospheric deformation due to extensional tectonics. Unfortunately, as passive margins are also places where subsidence was important and rapid, in general the crust is covered by a thick sedimentary layer which prevents observation and sampling of the basement.

The West Galicia margin (Fig. 1) is exceptional in that it is a starved margin, covered only by a thin and discontinuous sedimentary layer. These conditions are favourable for imaging by seismic reflection the thinned continental crust and the crustal ocean–continent boundary (OCB), and also to sample the basement by

From Holdsworth, R.E. & Turner, J.P. (compilers) *Extensional Tectonics: Regional-scale Processes*. The Geological Society,
Key Issues in Earth Sciences, **2 (1)**, 147–167.
1476-315X/02/$15.00 © The Geological Society of London 2002
First published in: "Scrutton, R.A., Stoker, M.S., Shimmield, G.B. & Tudhope, A.W. (eds) The Tectonics, Sedimentation and Palaeoceanography of the North Atlantic Region, The Geological Society, London, Special Publications **90**, 71–91".

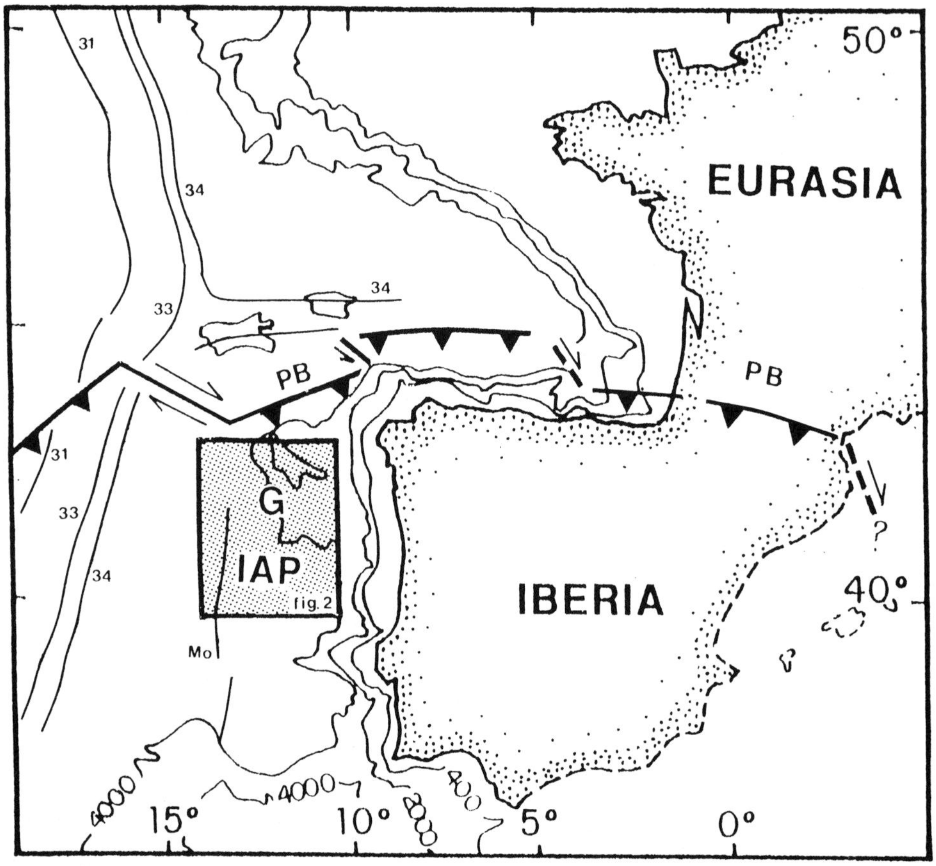

Fig. 1. Location of the studied area. Magnetic anomalies Mo and 31–34 from Srivastava *et al.* (1990). PB, Palaeocene plate boundary after Grimaud *et al.* (1982). G, Galicia margin; IAP, Iberia Abyssal Plain.

drilling or even by diving with the French submersible 'Nautile'. For that reason, the Galicia margin has been intensively studied for 20 years (see recent synthesis in Mauffret & Montadert 1987; Sibuet *et al.* 1987; Boillot *et al.* 1988*b*, 1989*a*).

However, as the OCB and the main extensive structures of the margin are trending N–S, the seismic data were generally recovered along E–W lines. For that reason, transverse structures were poorly imaged, although recognized in some places (Thommeret *et al.* 1988). To fill this gap, in 1990 we recorded (Lusigal cruise) several N–S seismic lines on the eastern, continental side of the OCB (Fig. 2, inset), so discovering that the West Galicia margin is actually made of two distinct segments separated by a major transverse structure. Moreover, a recent study of the Iberia Abyssal Plain (Beslier *et al.* 1993) revealed a segmented structure for the deep margin and OCB in the area located to the south of the Galicia margin (Fig. 2). In this paper we focus on the segmentation and the crustal structure of the deep Galicia margin and adjacent OCB.

Segmentation of the ocean–continent boundary (OCB) offshore Galicia

The West Galicia margin results from lithosphere stretching and rifting during lower Cretaceous times, lasting from 140 to 114 Ma (Boillot *et al.* 1987*b*, 1988*c*; time scale after Kent & Gradstein 1986). Accordingly, the M0 (118 Ma) magnetic anomaly is recognized offshore Portugal, while it is missing offshore Galicia (Fig. 1) where the margin is bounded by the Cretaceous quiet magnetic zone (Srivastava *et al.* 1990).

The segmented ocean–continent boundary

The OCB to the west of Galicia is marked by a basement ridge made of serpentinized peridotite. The ultramafic basement was sampled in several locations by dredging (Boillot *et al.* 1980; Sibuet *et al.* 1987), by drilling (leg ODP 103, drill Site 637; Boillot *et al.* 1987*b*, 1988*c*) or by diving with the Nautile (Boillot *et al.* 1988*a*). It separates two areas with different seismic and structural

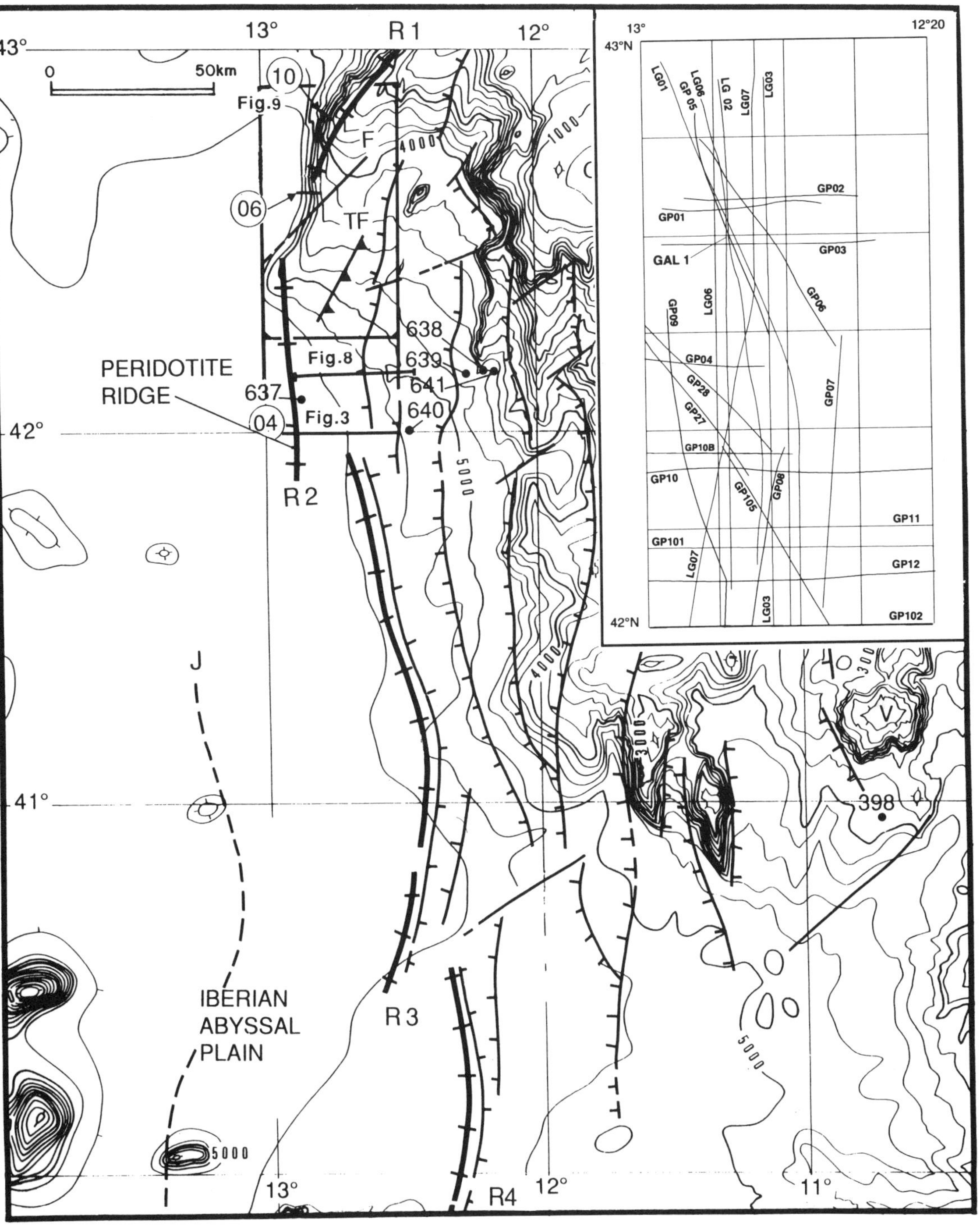

Fig. 2. Structural map of the West Iberia passive margin north of 40°N, after Beslier *et al.* (1993) (location on Fig. 1). J anomaly from Whitmarsh *et al.* (1990). Structural map of the Galicia margin after Thommeret *et al.* (1988) and Murillas *et al.* (1990). R1–R4, segments of the ridge bounding the oceanic and continental domains. Bathymetry after Lallemand *et al.* (1985). V, Vigo seamounts. DSDP Leg 47b and ODP Leg 103 sites, and dive sites (circled numbers) where peridotite was sampled are indicated. Inset: track map of multichannel seismic lines used in the study of the deep Galicia margin.

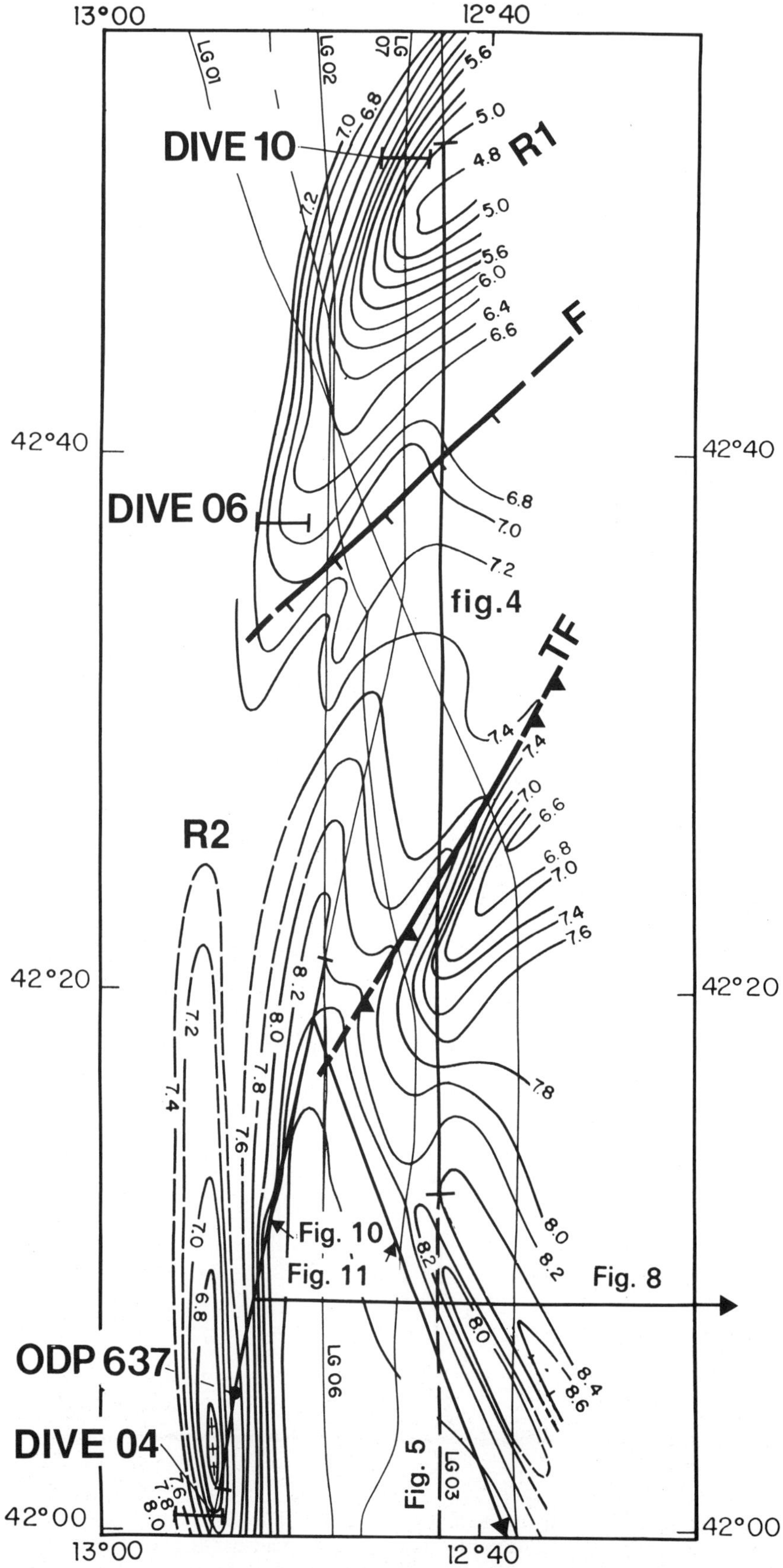

Fig. 3. Depth to the break-up unconformity (BU) (in s twt). Ultramafic basement, faulted enigmatic terrane or syn-rift sediment are resting beneath BU, in different places. Drill and dive sites where samples of serpentinized peridotite were recovered are indicated. F and TF, Mesozoic and Cenozoic transverse faults, respectively.

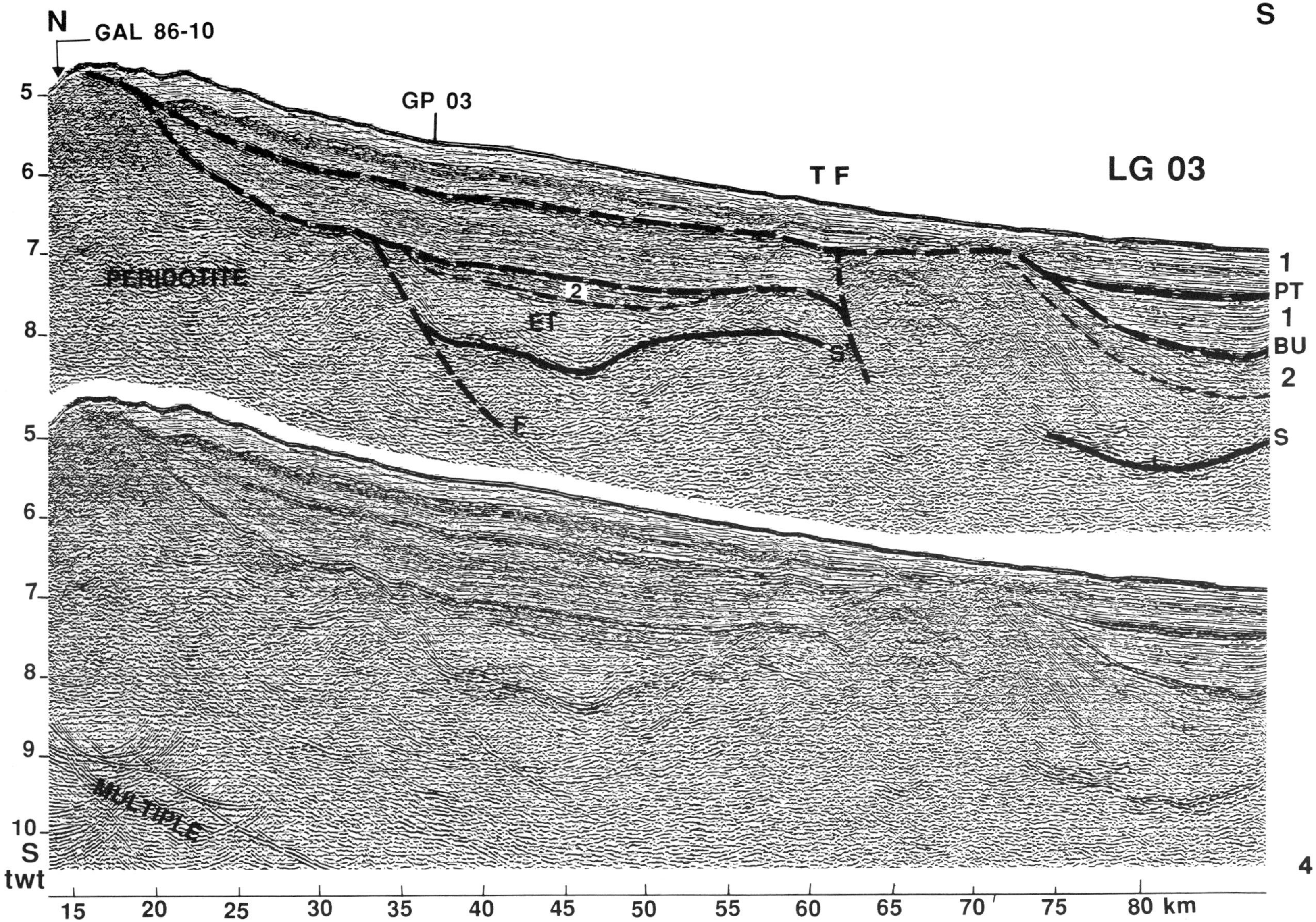

Fig. 4. Image of the crustal structure of the deep Galicia margin along the N–S seismic line LG03. Serpentinized peridotite was sampled at dive site GAL 86.10. ET, enigmatic terrane, including faulted continental basement and lithified sediments. ET is overlain by syn- (2) and post-rift (1) sediments separated by the break-up unconformity BU. F and TF, Mesozoic and Cenozoic transverse faults, respectively. S and S′, S and S′ seismic reflectors, considered in the paper as the seismic signature of the tectonic contact between serpentinized peridotite and crustal terranes, broken by the Cenozoic transverse fault TF. PT, seismic reflector located at the bottom of sediment deposited after the tectonic activity of TF in late Palaeogene time. A better image of TF is shown on Fig. 5. Location on Figs 3, 5 & 9.

characters. On the oceanic side, to the west, the relatively thin sedimentary layer is related to post-rift sediments of the margin (114 Ma to Present). It overlies a diffractive basement interpreted as oceanic crust of the Cretaceous quiet magnetic zone. In fact, Albian oceanic basalt was observed and sampled on the north-western slope of the peridotite ridge (PR) (Malod *et al.* 1993), and infered from refraction data to the west of the Galicia margin PR (Whitmarsh *et al.* 1993). On the eastern continental side the PR is bounded by a sedimentary basin, 10–20 km wide, infilled by syn- and post-rift sediments (Mauffret & Montadert 1987). Hereabouts, syn-rift sediments overlap the flank of the PR, indicating that this part of the ridge is a syn-rift feature.

Figure 3 shows the present morphology of the OCB at the bottom of post-rift sediments. The PR is divided into two segments, R1 and R2, by the transverse faults F and TF, both imaged on Fig. 4. Clearly, TF disturbs the lower part of the post-rift sequence, while it is sealed by the upper part. The oldest deformed sediments can be correlated with Paleogene strata, according to regional seismic stratigraphy calibrated by drilling (Mauffret & Montadert 1987). Thus, TF is a Cenozoic structure, related to Pyrenean tectonics, as are many other faults in that part of the Galicia margin (Boillot *et al.* 1979; Mougenot *et al.* 1984; Malod *et al.* 1993). Its seismic image is confused on Fig. 4 owing to a lot of diffraction events. It is clearer on the record of residual attenuation of seismic waves (Fig. 5A), suggesting a transpressional strike-slip fault or a reverse fault, as expected for structures related to Cenozoic plate convergence. On the contrary, F is sealed by the break-up unconformity. It is clearly a syn-rift normal fault which bounds the northern segment R1 of the peridotite ridge (Fig. 4). However, its setting parallel to TF suggests that both faults are Mesozoic structures, one of them (TF) having been reactivated and inverted by Cenozoic tectonics.

The thinned continental crust of the margin

Further east, the upper crust of the Galicia margin consists of tilted blocks, 16 km across on average and bounded westward by N–S normal faults or sets of normal faults (Montadert *et al.* 1979; Mauffret & Montadert, 1987; Sibuet *et al.* 1987; Thommeret *et al.* 1988). The eastward block tilting involved the formation of half-grabens infilled by syn- and post-rift sediments. The N–S extensional structures are cut and locally shifted by NE–SW transverse faults (Fig. 2; Thommeret *et al.* 1988), which are possibly transfer faults related to the margin segmentation.

Crustal structure: data and methods

Our interpretation of the crustal structure of the segmented margin is based on the following analyses.

Seismic velocities from reflection data

First we derived interval velocities from normal moveout velocities, using the Dix equation (Dix 1955). Because of its limitations this method was applied only to the portions of the seismic lines where the seafloor and the underlying layers are close to horizontal. The second, and more time-consuming, approach adopted was the use of iterative pre-stack depth migration. This has, to date, been applied to two E–W profiles: GP102 and GP03, one from the southern region and one from the northern (Fig. 2, inset). The method is based on depth-focusing analysis (see Reston *et al.* this volume), and provides more meaningful estimates of velocity in the structures.

Seismic refraction

The data recently published by Whitmarsh *et al.* (in press), related to two seismic refraction profiles fired across the OCB of Galicia along 42°10′N, was used.

Amplitude attenuation of recorded waves

This method consists of measuring the amplitude variations of seismic signals with time (see details in Rappin *et al.* in press). The results of a simple modelling of attenuation can be represented in two different images: (1) the distribution of the quality factor Q with time and shot location (Fig. 5A). The Q factor is one of the parameters used to perform the modelling of attenuation v. time. It is related to both absorption and scattering by interfaces and terrane heterogeneities. Its value decreases in particular where heterogeneities have a size close to the wavelength (*c.* 100 m), for example in deformed zones; (2) the residual attenuation of seismic waves with time and shot location (Fig. 5B). This is the difference between the calculated and the measured curves of seismic waves attenuation. As it estimates the most coherent amplitude of the seismic wave it does not depend upon velocity estimation nor upon correlation of signal phases. This method provides an apparent reflectivity of terranes, and gives a more

Table 1. *Values of the Q factor, reflectivity and seismic velocities within layers 1–4 in the northern and southern segments of the deep Galicia margin*

	Seismic velocity (km s^{-1}, twt)		Q factor*		Reflectivity*	
	Northern segment*	Southern segment†	Northern segment	Southern segment	Northern segment	Southern segment
Layer 1: post-rift sediments	2.15–2.66	1.9–3.0	10–11	4.0–4.5	–	–
Break-up unconformity	–	–	–	–	0.4–0.6	0.25–0.35
Layer 2: syn-rift sediments	3.0–4.5	3.5	–	90–200	–	–
Layer 3: pre-rift sediments and basement	4.0–5.7	4.0–5.1	600–650	600–675	–	–
S–S′ seismic reflector	–	–	–	–	0.2–0.25	0.15–0.35
Layer 4: serpentinized peridotite	–	7.0–7.8	800–950	850–950	–	–
Seismic Moho	–	–	–	–	0.1–0.15	0.15–0.2
Layer 5: fresh peridotite	–	8.1	1000	1000	–	–

* Results of this study. Reflectivity values are calculated with respect to a value of 1 at the seafloor.
† After Hoffmann & Reston (1992) for the layers overlying S; Horsfield (1992) and Whitmarsh *et al.* (1993) for layers 4 and 5, respectively.

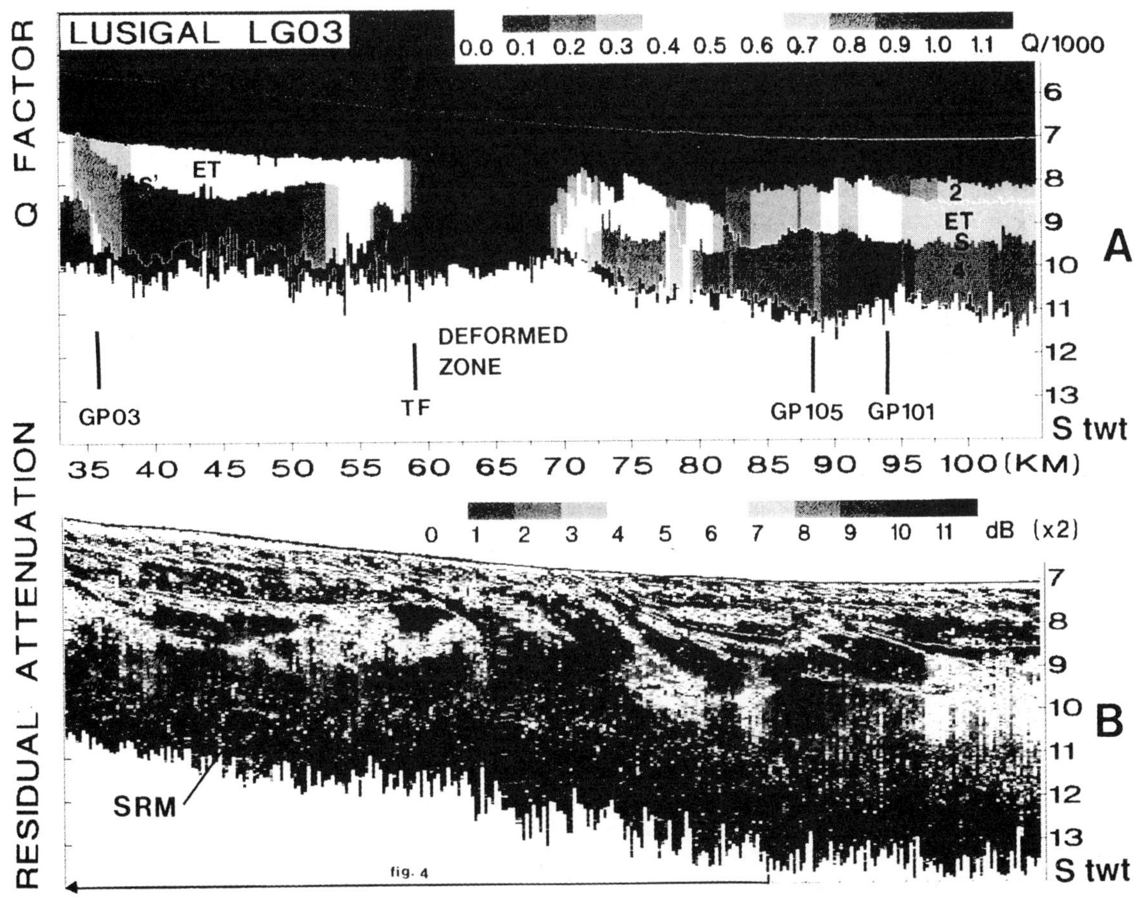

Fig. 5. (**A**) Values of the attenuation factor Q along a section of the seismic line LG03. 1–2, post- and syn-rift sediments; ET, enigmatic terranes, including continental basement and lithified sediments; 4, lower seismic crust, probably made of serpentinized peridotite hereabouts; SRM, Scattering reflective Moho. Location on Fig. 3. (**B**) Values of the residual attenuation of seismic waves along the same seismic line (compare with Figs 4 & 6). Location on Fig. 3.

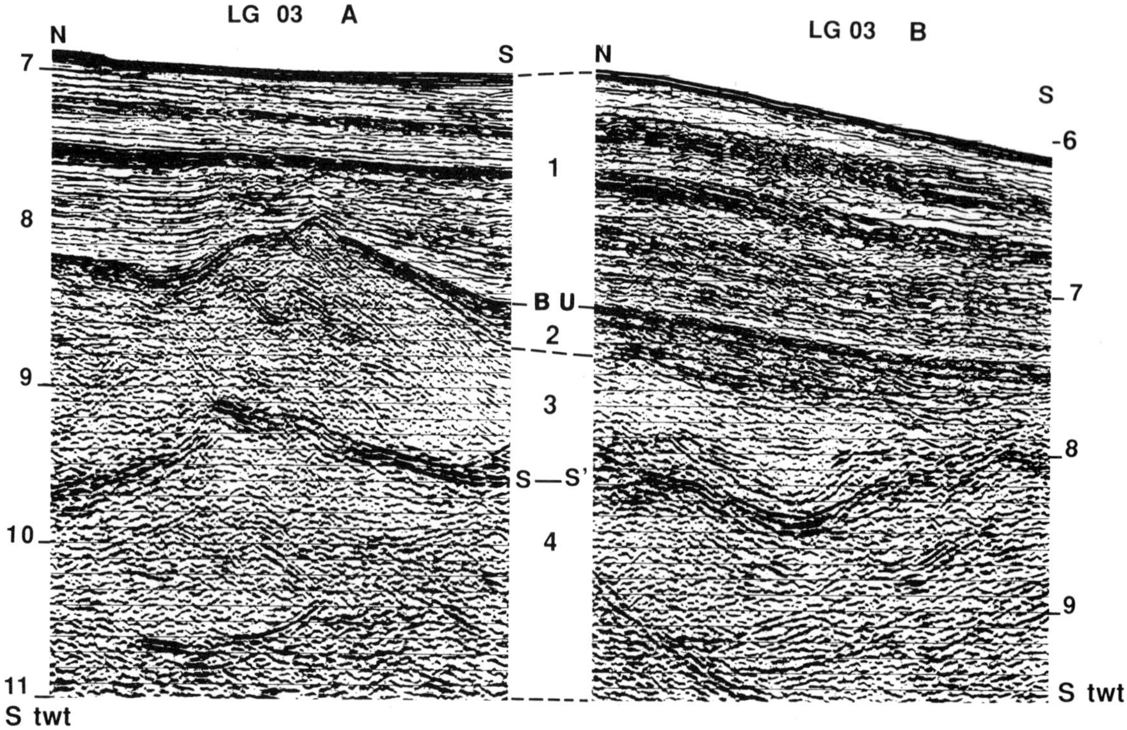

Fig. 6. Comparison of the crustal structure in the southern (**A**) and in the northern (**B**) segments of the deep Galicia margin. 1–2, post- and syn-rift sediments; 3, enigmatic terrane; 4, probable serpentinized peridotite; BU, break-up unconformity; S, S′, S and S′ seismic reflectors interpreted in the paper as the tectonic contact between serpentinized peridotite and continental crust material. The seismic line LG03 is located on Figs 3 & 9.

informative image of reflectors than the classical stack and post-stack migration method (cf. Figs 4 & 5A).

The main result of the combination of these different methods was to show that the crustal structure is identical in the southern and northern segments of the Galicia margin, on each side of the Cenozoic transverse fault TF and associated deformed zone. In both cases the crust consists of four layers, characterized by seismic velocities and Q factors (Table 1). Between these layers are surfaces of large residual attenuation (reflectors), also similar to the north and to the south of TF (Figs 6 & 7).

Upper layers of the continental crust

In this section crustal layers 1–3 are described and their probable geological nature from top to bottom considered.

Layer 1: the post-rift sedimentary sequence.

The high absorption of energy ($Q < 100$) and the low seismic velocities (1.9–2.6 km s^{-1}) measured within layer 1 are in good agreement with the physical properties of non- or poorly-consolidated sediments. Layer 1 was drilled in several places during DSDP Leg 47b (Sibuet *et al.* 1979) and ODP Leg 103 (Boillot *et al.* 1987*b*, 1988*c*). It is made of distal turbidites and pelagic sediments, deposited from 114 Ma to Present, and imaged by reflectors of good continuity but variable amplitude (Mauffret & Montadert, 1988).

Layer 2: the syn-rift sedimentary sequence

The physical properties of the layer 2 ($Q = 100$; seismic velocities ranging from 2.9–3.5 km s^{-1}) are those expected in buried, compacted and progressively lithified sediments. At the drill Site 639 (Leg ODP 103), layer 2 is made of coarse, siliciclastic turbidite and sandstone and by alternating clay and marl, lower Cretaceous in age (135–114 Ma). The seismic facies of layer 2 ranges from chaotic to well layered, with a divergent, fan-like configuration related to the syn-rift tilting of underlying crustal blocks. The divergent structure, however, is poorly or not imaged on NS seismic lines (Figs 4 & 6).

The post-rift or break-up (BU) unconformity

This is marked by a strong reflector and a high positive residual attenuation level located between layers 1 and 2. At drill Site 641 (Leg ODP

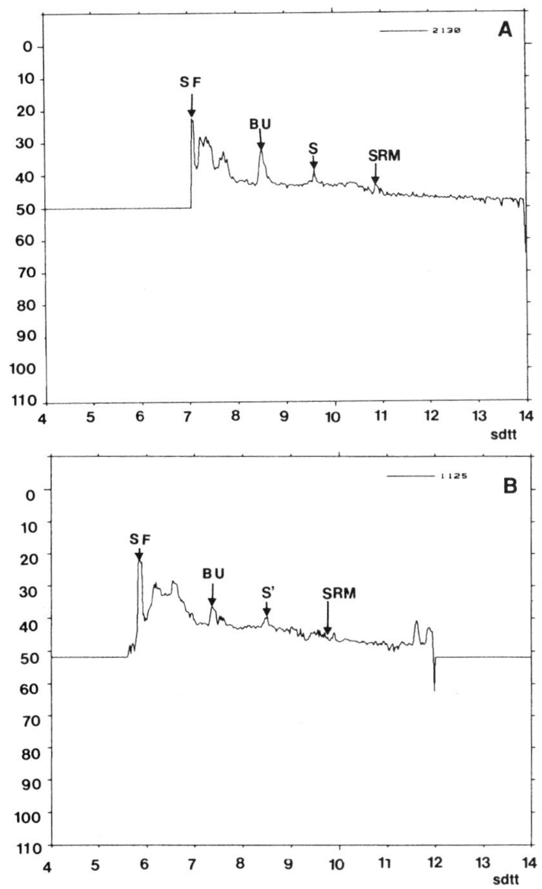

Fig. 7. Attenuation curves of seismic waves through the southern (**A**) and northern (**B**) segments of the Galicia margin. Attenuation rates and reflection patterns are identical on the two curves. SF, seafloor; BU, break-up unconformity; S–S′, S and S′ seismic reflectors; SRM, scattering reflective Moho; vertical scale, attenuation in decibels.

103) it corresponds with a level of coarse, calcareous turbidite deposited at the top of the syn-rift sequence. In many places the BU is also an erosional or non-depositional surface (Mauffret & Montadert 1987).

Layer 3: faulted terrane or enigmatic terrane (ET)

Within layer 3 Q ranges from 600 to 675, and the seismic velocity from 4.0 to 5.7 $\mathrm{km\,s^{-1}}$. The seismic image of ET shows diffractive acoustic basement locally covered by horizontal to gently dipping weak reflectors. Apparent dips in fault blocks range from 40° to few degrees (Hoffmann & Reston 1992). On the N–S seismic line LG 03 (Fig. 4) dipping reflections in layer 3 are also apparent just to the south of TF, and may represent pre-rift sediments tilted by movement along TF.

Layer 3 was sampled in the southern part of the margin, at dive Site 11 (Fig. 8). Here, the westernmost, deepest tilted crustal block is cropping out on the seafloor, allowing the Nautile to recover on a normal fault scarp several samples of granodiorite from the Hercynian upper crust. The continental basement is covered by Mesozoic pre-rift limestones and sandstones (Boillot *et al.* 1988*a*). However, ET may also include other terranes with similar physical properties, for example volcanic rocks, although the margin is devoid of significant magnetic anomaly, or Palaeozoic, poorly metamorphozed sediments as those recovered at another site on the margin by Mamet *et al.* (1991). In our opinion, a correct characterization of the enigmatic terrane remains an important target. It is crucial to verify its geological nature to constrain the interpretation of the underlying S reflector (see the next section).

Enigmatic terranes are thickened within a syn-rift graben bounded by F to the north and TF to the south (Fig. 9). Here, layer 3 was preserved from erosion, or accumulated before the end of the rifting, confirming that TF was a syn-rift structure before its Cenozoic reactivation.

S–S′ seismic reflector, lower seismic crust and reflective Moho

S–S′ seismic reflectors

Crustal layer 3 rests on a strong seismic reflector, S (de Charpal *et al.* 1978; Montadert *et al.* 1979). S was recognized at first in the southern segment of the margin. Here it is either a single, strong reflector, or a sequence of elementary reflectors, horizontal to gently dipping (Hoffmann & Reston 1992). In general, S is located at depths ranging from 0.6 to 1.5 s two-way time (twt) from the top of layer 3 (Mauffret & Montadert 1987). S′ is a similar reflector recently recognized in the northern segment of the margin (Boillot *et al.* 1992). It is characterized by the same seismic signature as S (Fig. 7), and occurs at the same structural level (Fig. 6). More precisely, amplitude analyses show that S and S′ have the same signature in the attenuation curves, i.e. a sharp and significantly high amplitude peak. These observations are in good agreement with the occurrence at depth of an abrupt geological interface between terranes very different in nature, rather than with a progressive geological transition, the expected signature of which being a reflection zone (and not a strong reflector), and a wide (and not a sharp) peak of small amplitude

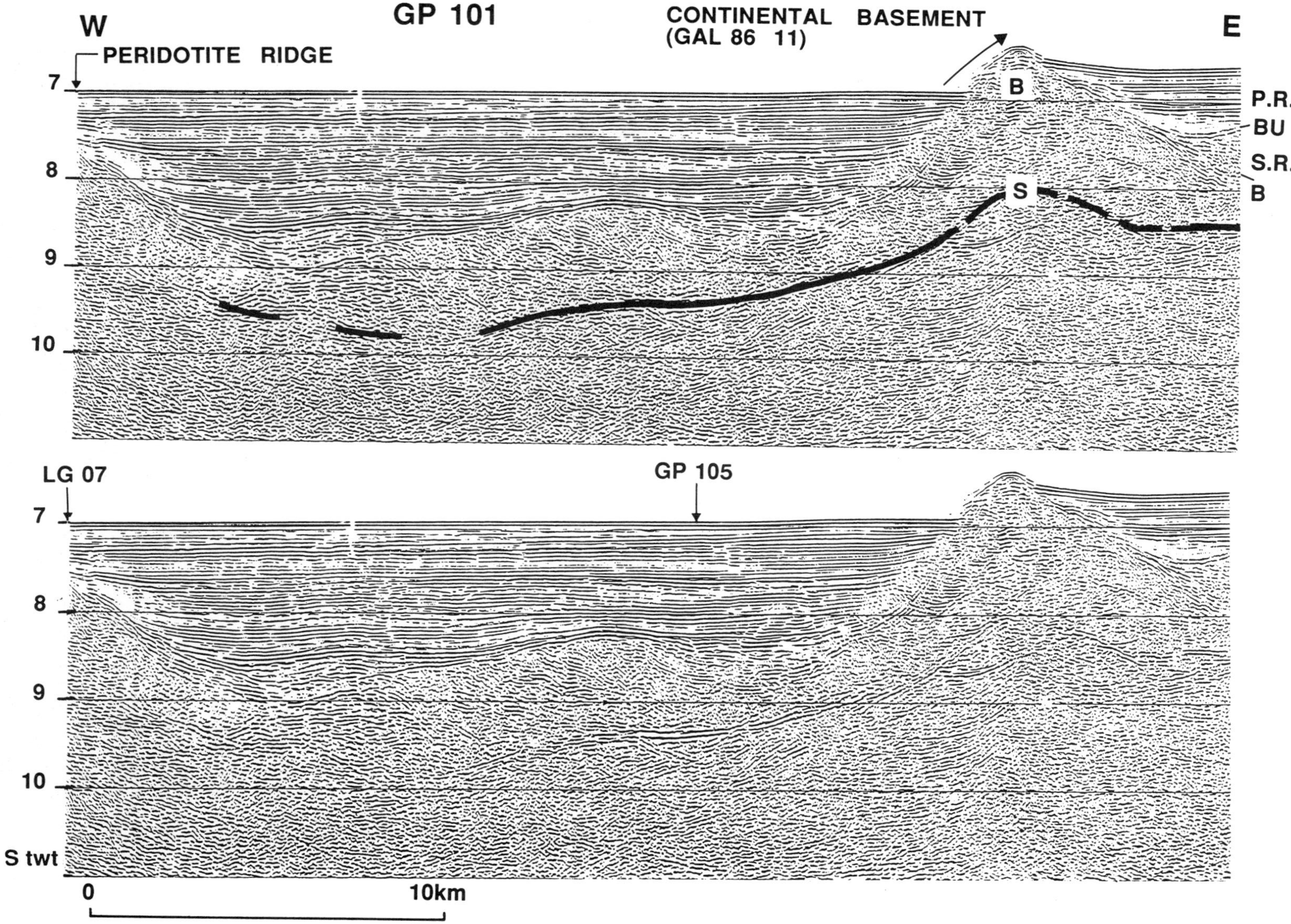

Fig. 8. Section of the migrated GP 101 multichannel seismic line (MCS), from the Institut Français du Pétrole. Granodioritic basement and lithified pre-rift sediments were sampled from faulted layer 3 (enigmatic terrane ET) at dive site 11. S, S seismic reflector. Location on Figs 2 & 3.

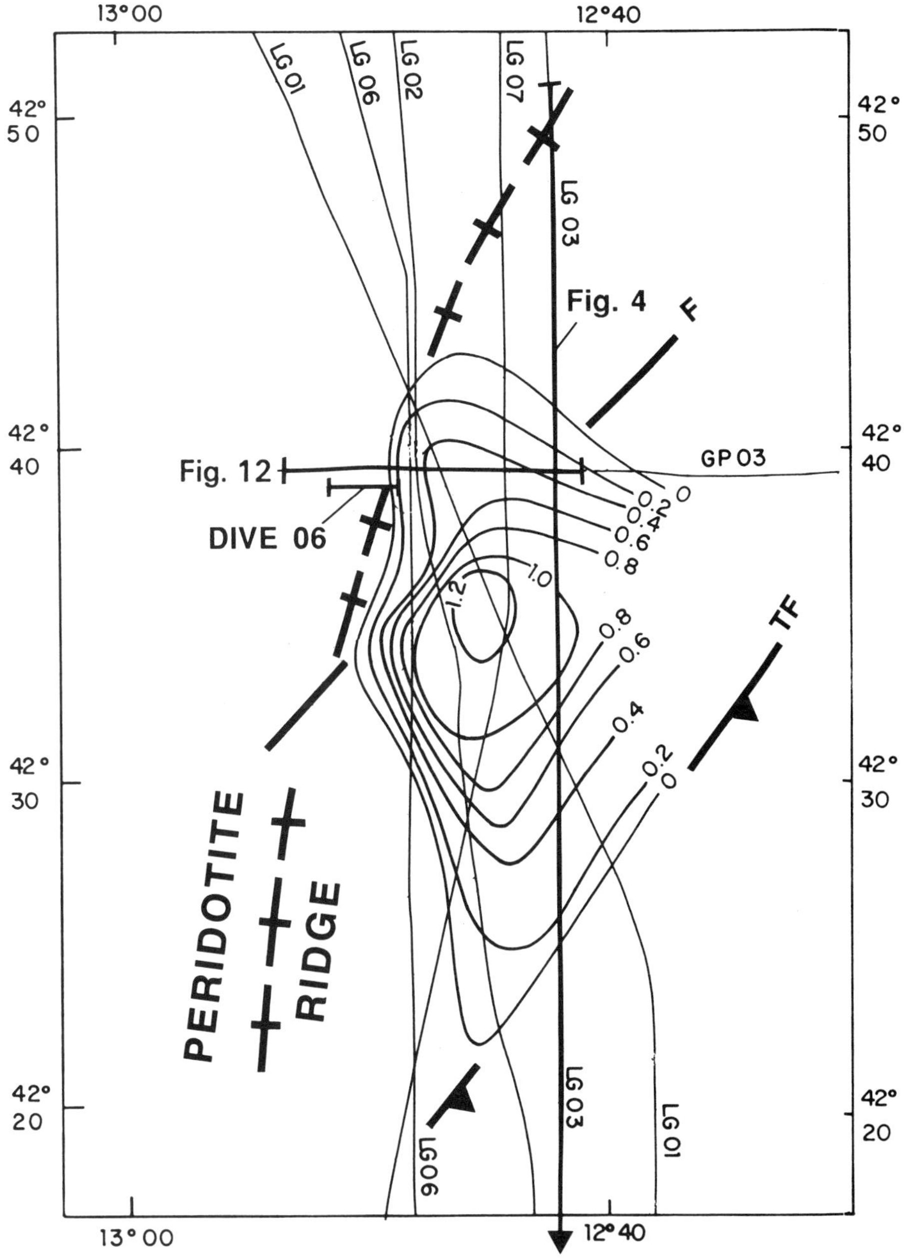

Fig. 9. Thickness of the layer 3 (enigmatic terrane ET) (in s twt). The velocity within layer 3 ranges from 4 to 5.7 km s^{-1}. F and TF, Mesozoic and Cenozoic transverse faults, respectively. Location on Fig. 2.

in the attenuation curves. Moreover, the attenuation variations and parameters within terranes over- and underlying S are remarkably similar to those over- and underlying S′ (Figs 5 & 6).

The significance of S–S′ and the geological nature of related terranes remains controversial (see the discussion in the next section and in Reston *et al.* this volume). However, recent multichannel seismic data allowed the tentative connection of both S and S′ to the top of the serpentinite terranes sampled on the two segments of the peridotite ridge bounding the margin (Boillot *et al.* 1992). For the southern segment, Fig. 10 shows a section of seismic line LG 07 from the top of the PR2, where ultramafic rocks were drilled, to the axis of the sedimentary basin bounding the ridge to the east. The roof of the serpentinite body forms a strong reflector dipping northeast. From south-

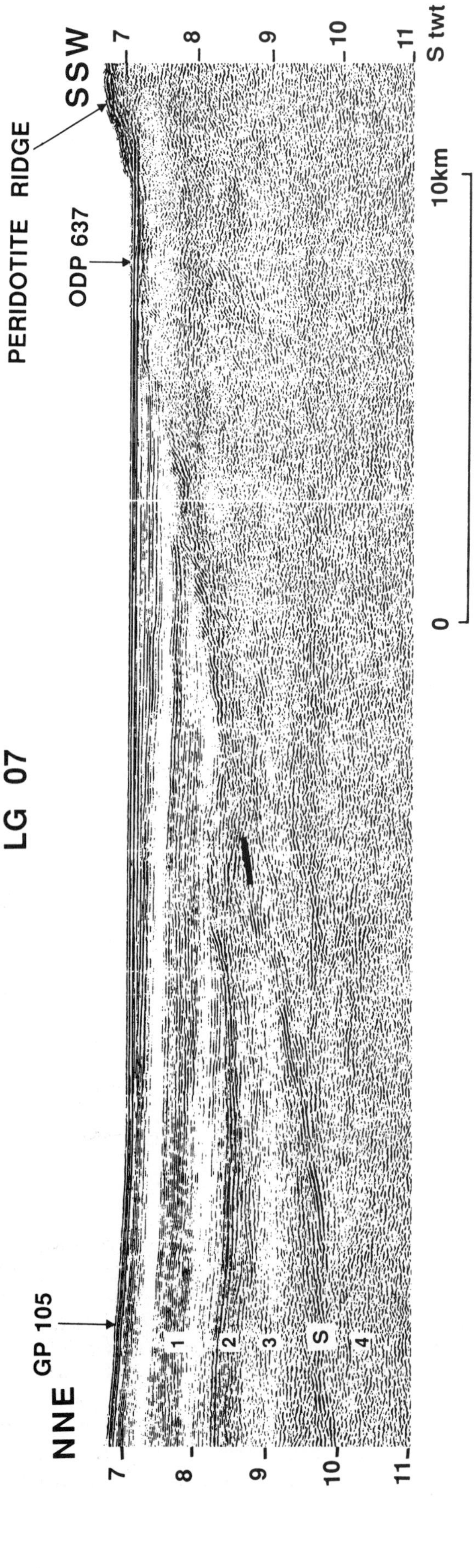

west to northeast it is progressively deepening and covered by post-rift sediments, by syn-rift sediments and by enigmatic terranes at the north-eastern end of the seismic section. Here the reflector is at a depth of 9.8 s twt from the sea surface, i.e. at the same level as S on GP 105 where it disappears from the record, probably owing to Mesozoic deformation along TF (Fig. 11). In our opinion, the connection between S and the top of the peridotite ridge imaged at the same level and at few kilometres distance is highly probable. In the northern segment, the top of the PR1, sampled at dive Sites 6 and 10 (Fig. 2), is clearly connected with the S′ reflector on seismic line GP 03 (Fig. 12).

The lower seismic crust

Beneath S and S′, layer 4 appears to have uniform physical properties. The seismic facies is layered, due to discontinuous horizontal or gently dipping reflectors. The Q factor is very high ($>$ 800). In the core of the PR2, the seismic velocity is 7.2 km s^{-1} (Sibuet, 1992), while it ranges from 7.3 to 7.8 km s^{-1} in the terranes underlying S to the east of PR2 (Horsfield 1992; Whitmarsh *et al.* 1993), suggesting these terranes to be directly connected at depth with the OCB ridge. These data are consistent with a serpentinized peridotite nature for layer 4. It implies that S is the petrological Moho (the boundary between mantle and crust derived terranes).

The seismic Moho

This is necessarily deeper within the lithosphere. In fact, layer 4 is underlain by a thin zone (0.2–0.8 s twt), in which a relatively high attenuation of seismic waves was measured (Q $<$ 100), with systematic, scattered reflectivity. From the north (R1 segment of the margin) to the south (R2), the signature of this zone is constant and typical on amplitude attenuation curves obtained from seismic signal records (Figs 5B & 7).

The weakness and the scattering character of the reflectivity at the bottom of layer 4 is different from other overlying seismic reflectors. The strong attenuation implies the presence of heterogeneities whose size is close to the wavelength of the signal (100–200 m), which induces a maximum of scattering (Herraiz &

Fig. 10. Section of the migrated LG07 MCS line, where the top of the peridotite ridge is imaged and can be connected at depth with the S seismic reflector. 1, 2, 3, 4, post-rift sediments, syn-rift sediments, enigmatic terranes and serpentinized peridotite, respectively. Location on Fig. 3.

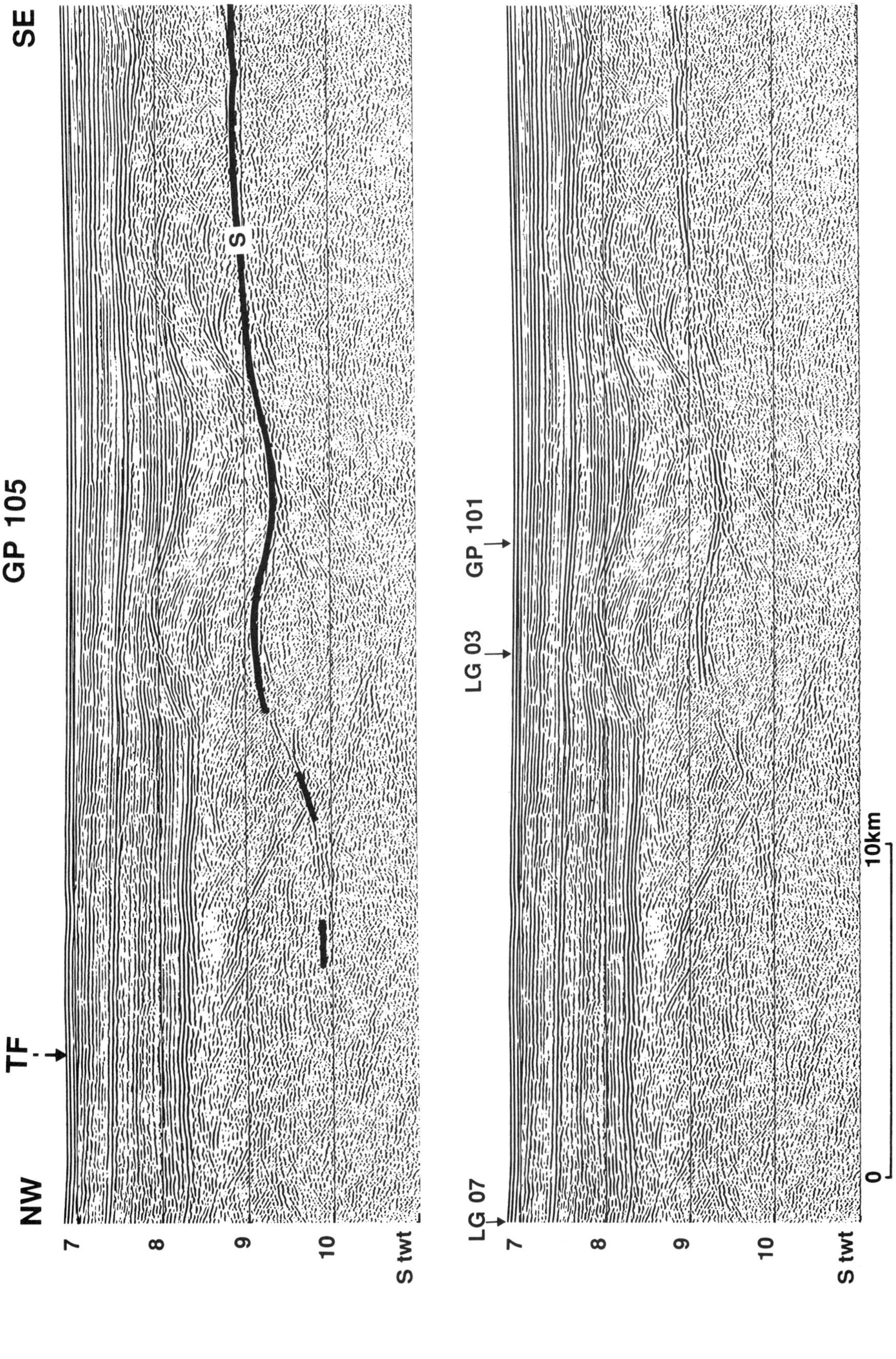

Fig. 11. Section of the GP105 MCS migrated line, from the Institut Français du Pétrole. S, S seismic reflector. Location on Fig. 3.

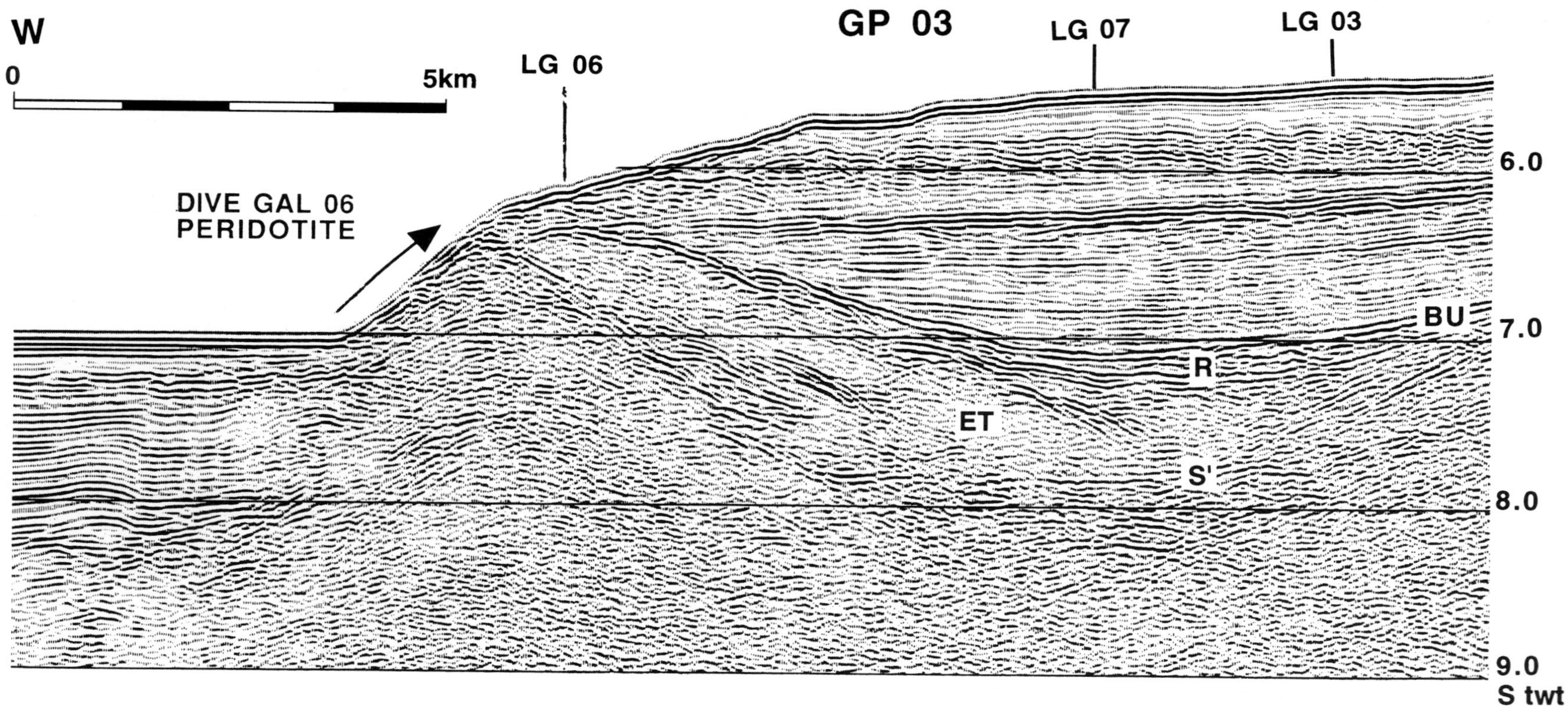

Fig. 12. Section of the GP03 MCS migrated line, from the Institut Français du Pétrole. The S′ seismic reflector is connected with the top of the serpentinite body sampled at dive site 06. BU, break-up unconformity; ET, enigmatic terranes (layer 3); R, syn-rift sediments. See Figs 3 and 9 for location.

Espinosa, 1986). The thickness of the reflective horizon increases considerably beneath the PR, where serpentinized peridotite was unroofed and is now free of crustal cover. To the south of the margin it is located at the level where Recq *et al.* (1991) placed the seismic Moho from their refraction study. All these data are consistent with the idea that the scattering zone of reflectivity is actually the reflective Moho, located at the fresh-serpentinized peridotite boundary.

Discussion

The scattering reflective Moho (SRM)

The fresh serpentinized peridotite boundary is also the palaeohydrothermal front, i.e. the surface where syn-rift hydrothermal serpentinization of mantle rocks stopped after seafloor spreading started and margin lithosphere cooled. The depth of the reflective Moho (7–8 km beneath the seafloor) is a clue for estimating the maximum depth of hydrothermal circulation during rifting within upper, brittle lithosphere covered by sea water. In areas where the continental crust thinned down to 7–8 km the syn-rift hydrothermal circulation probably reached the petrological Moho and the uppermost part of the mantle through the faulted and stretched basement of the margin. Boillot *et al.* (1989*b*) proposed naming this process 'undercrusting': it involves accretion, at the base of highly stretched continental crust, of a layer of serpentinite which belongs to the lower seismic crust by its physical properties, although made of mantle-derived rocks. In the case of the Galicia margin the hypothesis is supported by the occurence of layer 4 along the entire margin with the same physical properties, and its connection with the OCB ridge where serpentinized peridotite was sampled in several places. It is confirmed by the continuity of the SRM which underlies layer 4 in the deep margin and the PR as well. Note that beneath thick continental crust the Moho has a very different signature; around France for example, it is marked by a strong, continuous reflection at the bottom of scattering lower crust (Rappin 1992). In that case, it probably bounds terranes very different in nature, while the high attenuation and scattering reflectivity at the base of the layer 4 of the Galicia margin is in agreement with a transition zone containing both serpentinized and fresh peridotite.

The SRM seems to be not or poorly affected by Cenozoic deformation beneath TF (Fig. 5). However, it shallows regularly northward as does the seafloor. From regional geological studies it is known that the uplift of the northern part of the margin, including Galicia Bank, resulted from Cenozoic tectonics (Boillot *et al.* 1979; Mougenot *et al.* 1984). Accordingly, the reflective Moho is regionally uplifted with the margin, thus accounting for the strong positive gravimetric anomaly located at the northwestern edge of the margin (Lalaut *et al.* 1981).

The S seismic reflector

Since it was discovered beneath the Armorican and Galica margins, S has been considered as a major feature of these passive margins, probably a key for understanding the rifting processes. It was tentatively interpreted as the brittle–ductile transition within the thinned continental crust (de Charpal *et al.* 1978; Montadert *et al.* 1979) or as a major syn-rift detachment fault rooted in the lower ductile crust or in the mantle (Wernicke & Burchfield 1982; Boillot *et al.* 1987; Le Pichon & Barbier 1987; Mauffret & Montadert 1987; Sibuet 1992; Hoffmann & Reston 1992). Another interpretation was recently proposed by Beslier & Brun (1991), who relate S to the development of two conjugate shear zones, located in the ductile crust and in the deeper ductile lithospheric mantle respectively. It is clear that a good understanding of the nature of S would improve our understanding of the processes of crustal extension leading to the creation of the margin.

Sampling the rocks located at the level of, upon and under S is the most efficient way to progress in this discussion. It is of most interest to verify that at first the reflector corresponds to the boundary between crustal and mantle terranes, as proposed in this paper, and to further specify the kinematics of the deformation at the base of the crust and at the top of the mantle. Unfortunately, until now, available petro-structural data come only from mantle rocks recovered in the area where they were unroofed by syn-rift tectonics. In that area both crustal terranes and their contact with mantle material are missing. However, the petrology and fabric of the exposed rocks provide information on the deformation they underwent in the vicinity of the crust–mantle boundary before they were unroofed, at least if they were preserved from superficial erosion since they were exposed on the seafloor. By chance it seems that this was the case at drill Site 637 (Leg ODP 103). Petro-structural studies of cored samples constrained the timing and evolution of the peridotite up to its serpentinization, particularly the pressure–temperature conditions and the kinematics of the ductile deformation (Agrinier *et al.* 1988; Féraud *et al.* 1988; Girardeau *et al.* 1988; Beslier *et al.* 1990). The results show that the evolution of the rocks is compatible with a progressive uplift beneath a continental rift, and moreover that ductile simple shear played a major part in the stretching of the lithosphere before it broke up at 114 Ma. After partial melting under asthenospheric conditions the plagioclase-bearing peridotites experienced intense ductile deformation under lithospheric conditions within a normal shear zone gently dipping toward the continent. From these data it is concluded that a shear zone was actually drilled at ODP Site 637. The kinematics and timing of the deformation are compatible with models of passive margin formation involving simple shear, either along a single normal synrift detachment fault rooted in the mantle (Boillot *et al.* 1987*a*), or in a ductile shear zone within the mantle (Beslier & Brun, 1991; Brun & Beslier in press).

The Galicia margin therefore provides a unique opportunity to study, *in situ*, the shearing of the upper mantle beneath a continental rift. However, the relationship between the drilled shear zone and the S reflector remains questionable (see the next section).

In the previous section, it was stressed that the S′ reflector and the surrounding terranes forming the northern segment of the Galicia margin have the same physical properties as S and layers 1–4 in the southern segment. In both cases, layer 4, roofed by S–S′, is probably made, at least partly, of serpentinized peridotite sampled by drilling and by diving. Moreover, the crustal structure of the OCB and adjacent margin seems to be similar in the Iberia Abyssal Plain (IAP) further south. Here, the OCB is marked by a basement ridge (Fig. 2) with seismic characteristics and tectonic setting comparable to those of the Galicia margin PR. The deepest tilted block of the IAP margin is underlain by a strong seismic reflector S″ similar to S and S′ (Beslier *et al.* 1993), and S″ is located at the level where Whitmarsh *et al.* (1990) placed the boundary between upper and lower seismic crust, with respective seismic velocities of 6.2 and 7 km s^{-1} or more. Such a similarity of the seismic images and seismic velocities with those recorded and measured on the Galicia margin strongly suggests that the terranes resting at the base of the IAP thinned continental crust are also made of serpentinized peridotite (Beslier *et al.* 1993; Whitmarsh *et al.* 1993).

The crustal structure is thus very similar in the different segments of the west Iberia margin and adjacent OCB. The ubiquitous occurrence of a strong reflector (S, S′ or S″) at the same structural level reinforces the interest to investigate it and surrounding terranes.

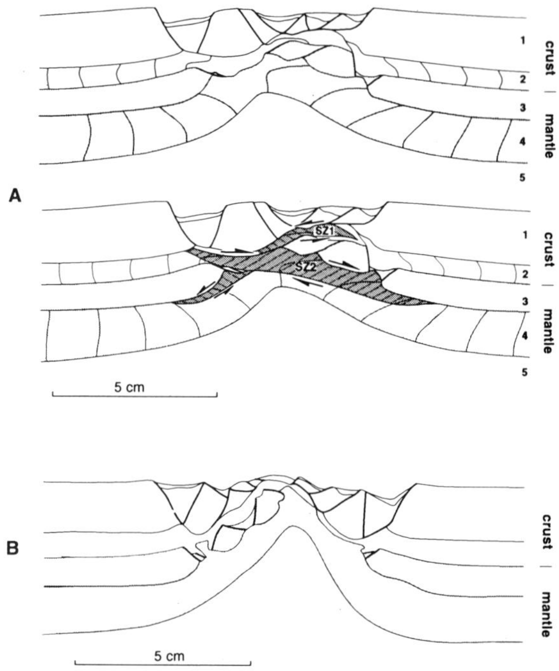

Fig. 13. Analogue modelling of the lithosphere stretching. Brittle continental crust (1) and brittle mantle (3) are modelled by sand, ductile continental crust (2) and ductile lithospheric upper mantle (4) by silicone putties, asthenosphere (5) by golden syrup. Senses of shear are inferred from the deformation of passive markers included in brittle and ductile layers. **(A)** Model 1, moderately stretched (shear zones in dark grey); **(B)** model 2, highly stretched. SZ1 and SZ2, shear zones discussed in the text. From Beslier & Brun (1991).

Conceptual model for the formation of the Galicia passive margin

To build up this conceptual model, we started from two different sets of data.

Firstly, analogue modelling of the lithosphere stretching was used (description of models and their inferences for passive margins formation, including Galicia margin, are detailed in Beslier 1991; Beslier & Brun 1991; Brun & Beslier in press). This takes into account the differential mechanical behaviour of the main lithospheric rheological layers, and allows the study of the stretching mechanisms of the lithosphere and the related crustal thinning processes during a rifting episode (Faugère & Brun, 1984; Vendeville *et al.* 1987; Allemand *et al.* 1989; Allemand & Brun 1991; Beslier & Brun 1991; Brun & Beslier in press). The rheological structure of the lithosphere is simplified in the experiments by a brittle–ductile layered model, which is a good approximation of the structure of a stable continental lithosphere with a normal geothermal gradient (e.g. Ranalli & Murphy 1987; Davy & Cobbold 1991). The experimental analogue materials are dry sand for brittle behaviour and silicone putties for ductile behaviour. The four layers represent the upper brittle crust (sand), the lower ductile crust (silicone putty), the uppermost brittle mantle and the upper lithospheric ductile mantle. This lithospheric structure lays upon golden syrup which simulates the asthenospheric behaviour. The model is submitted to localized horizontal extension (see Beslier 1991; Davy & Cobbold 1991 for details on the method). Figure 13A shows a cross-section of a model at the end of the experiment. Brittle layers (upper crust and uppermost mantle) underwent boudinage, while ductile layers accommodated the boudinage by simple shear along conjugate normal shear zones. With progressive extension, the rupture of the brittle mantle is achieved in one of the necked zones, where the deformation is localized at depth. In the last stage of the rifting two main shear zones accommodate the stretching beneath the rift, the right part of which accounts for the west Galicia margin structure (Fig. 13A): one (SZ1), located in the lower ductile crust, acts with a top-to-the-west sense of shear, and accounts for the continentalward sense of blocks tilting sense in the upper brittle crust; the other one (SZ2), connecting the ductile crust and the ductile mantle in the ruptured zone of the brittle mantle, acts with an opposite sense, i.e. top-to-the-east. Another model (Fig. 13B), identical though extremely thinned, shows that the lower ductile crust can disappear at the rift axis, bringing the mantle directly in contact with the upper crustal blocks, or even with the syn-rift sediments as observed on the Galicia margin.

Secondly, structural data from the Galicia margin have been used. We have already discussed the kinematics of the peridotite ductile deformation established from petro-structural studies of samples recovered on the PR: the mantle rocks have been deformed in a normal ductile shear zone dipping northeast, with a top-to-the-east shear sense. Thus, the results of the analogue models are in good agreement with the available structural data from the margin. Two major shear zones are postulated, SZ1 in the ductile crust, and SZ2 rooted in the ductile lithospheric mantle (Figs 13A & 14A). Currently, only SZ2 has been sampled by drilling and diving on the peridotite ridge, and Boillot *et al.* (1987*a*, *b*, 1988*b*, 1989*b*), developing the initial model of Wernicke (1985), considered it as part of a detachment fault rooted in the mantle. However, the sense of blocks tilting on the margin also implies the existence of SZ1 (Faugère & Brun 1984; Brun & Beslier in press). From this point in the discussion two interpretations are possible, although closely related. Reston *et al.* (this volume) propose S to be related to SZ1, and the tectonic unroofing of PR to SZ2. Some of the current authors (G.B., M.O.B., D.R.) rather believe that the tectonic contact between upper crust material (the tilted blocks of the margin) and mantle material (the serpentinized peridotite) relates to both shear zones (Beslier & Brun 1991). This interpretation is supported by the highly stretched experimental model (Fig. 13B), which accounts for the contact between upper crust and upper mantle terrane. In that case, which corresponds to the final stage of the rifting, the lower crust is extremely thinned and even lost in the vicinity of the rift axis, at the places where the upper brittle mantle is broken (Fig. 13B). As a result, the upper, brittle crust lies either directly over mantle terranes deformed in SZ2 or on interbedded lenses of sheared lower crust initially belonging to SZ1. Therefore, we suspect tectonic melange of various thickness and including sheets of lower crust to be related to the S reflector beneath the deeper crustal tilted blocks (Fig. 14B). Moreover, the serpentinization of mantle rocks beneath the very stretched crust at the rift axis does change the rheological behaviour of mantle terranes. It is thus possible that decollement of tilted blocks occurs at the top of the upper mantle in the final stage of the rifting (Beslier & Brun 1991).

Figure 14C summarizes the crustal structure of the deep Galicia margin from this study. In the deepest part of the margin, thin blocks of upper continental crust are scattered on 'undercrusted' serpentinite, S being the tectonic contact between the blocks and the serpentinized uppermost mantle. However, intercalation between these terranes of sheared lower continental crust is possible locally. Deeper is the fresh-serpentinized peridotite boundary, i.e. the actual seismic Moho. To the east, where the continental crust thickens, 'true' lower continental crust is expected in place of the serpentinite layer. The nature of the lateral transition between the two kinds of lower seismic crust (serpentinized peridotite in the most stretched area; lower continental crust elsewhere) remains a target for further clarification and discussion. Nevertheless, the disappearance of S beneath the tilted blocks of the eastern, upper part of the margin may be a consequence of the transition.

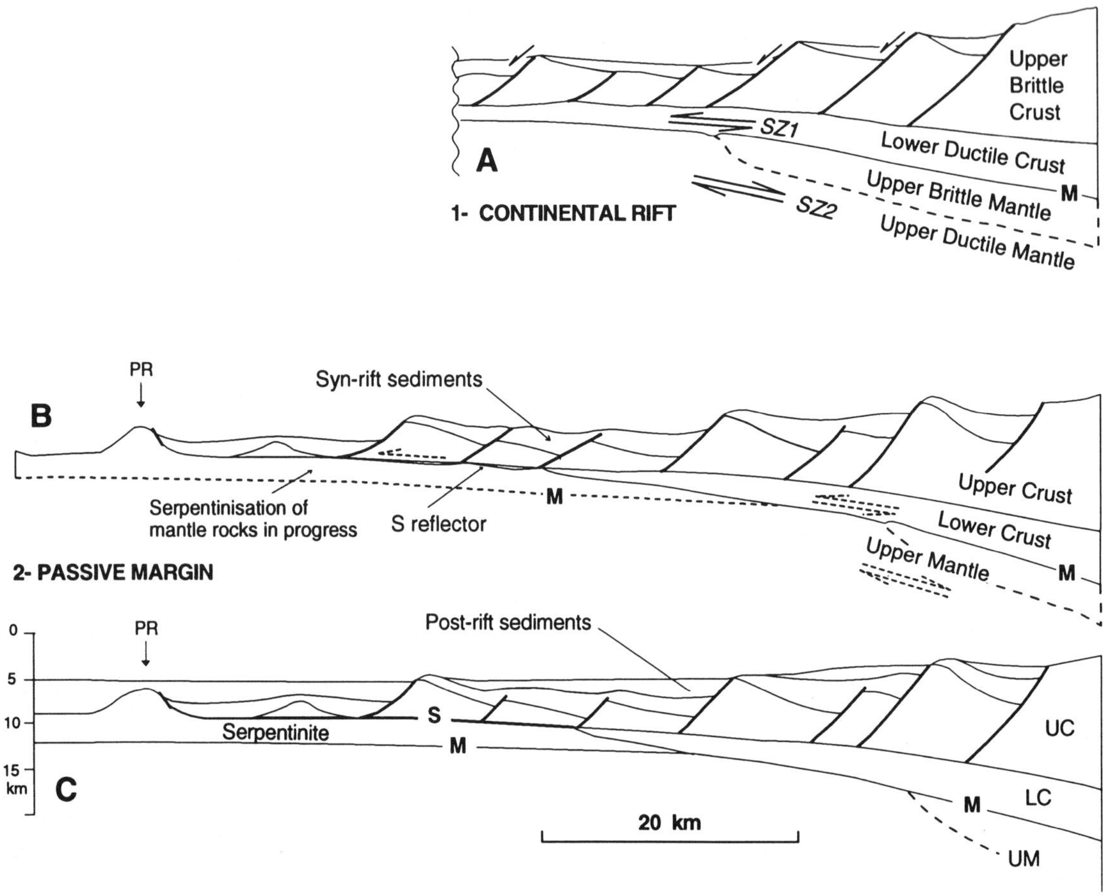

Fig. 14. Conceptual model [(**A**) and (**B**)] and schematic cross-section [(**C**)] of the deep passive Galicia margin. The conceptual model is derived from the analogue modelling depicted in Fig. 13. It describes two representative stages of margin rifting: during continental rifting, and just before seafloor spreading starts (after Beslier & Brun 1991; Brun & Beslier in press). (**C**) is established along 42°N.

Plate kinematic implications of margin segmentation

Beslier *et al.* (1993) suggested that the segmentation of the ridge marking the OCB results from discontinuous, northward propagation of continental break-up and opening of the North Atlantic in early Cretaceous time. Moreover, the SW–NE transverse fault offsetting R1 and R2 (Fig. 3) and R3–R4 (Fig. 2) segments can be interpreted as transfer faults indicating the direction of lithosphere stretching during rifting. In the southern segment of the Galicia margin, SW–NE is also the direction of shearing in mantle rocks recovered on the PR (Girardeau *et al.* 1988; Beslier *et al.* 1990) and the main direction of transverse faults (Thommeret *et al.* 1988). Thus, we suspect that the lithosphere underwent stretching along the present SW–NE direction during rifting, with consequences for the correct location and identification of the conjugate margin in the Newfoundland Basin.

Summary and conclusions

(1) The seismic crust of the deep Galicia Margin is made of four main layers characterized by their seismic facies, seismic velocities, and attenuation of reflected P waves and related reflectivity. Sediments (layers 1 and 2) are classically divided into syn- and post-rift sequences, separated by the post-rift or break-up (BU) unconformity. A reflector of high reflectivity emphasizes the BU. Within the sediments, the attenuation of seismic waves is high, the seismic image is layered, and the velocity ranges from 2.2 to 3.5 km s^{-1}. Faulted layer 3 (enigmatic terrane) rests on the reflector S–S′, another level of high reflectivity. Within ET the velocity ranges from 4 to 5.7 km s^{-1}, and the attenuation of seismic waves is moderate and constant at places where no Cenozoic tectonics occurred.

The layer includes continental basement and lithified pre-rift sediments sampled from the thinned, upper crust of the margin. The lower seismic crust (layer 4) is characterized by low attenuation of seismic waves, high seismic velocities (7.0 km s^{-1} or more), and rests on a scattering, poorly reflective Moho located at the level of the refractive Moho. Correlation with terranes sampled at the OCB shows that the main component of this layer is serpentinized peridotite, resulting from syn-rift hydrothermal alteration of the uppermost mantle. Layer 4 rests on fresh peridotite of the upper mantle at the seismic Moho. (2) The Galicia OCB is marked by a basement ridge made of serpentinized peridotite. It corresponds to the early Cretaceous rift axis, and results from tectonic unroofing of the mantle terranes in the latest stage of margin rifting. Mantle rocks were ductily sheared under lithospheric conditions before their serpentinization and brittle deformation, in agreement with lithosphere stretching models involving simple shear. Two syn-rift shear zones at least are necessary to account for the present margin structure: one (SZ1) located in the lower ductile crust, synthetic to the continentalward tilting of crustal blocks; another (SZ2) rooted in the ductile mantle, a part of which is now exposed beneath sediments at the OCB (PR). (3) The OCB and adjacent margin are divided into two segments by a transverse Mesozoic structure that was partly reactivated and inverted in Cenozoic times. The NE–SW orientation of this transverse structure suggests that the stretching of the lithosphere occurred in early Cretaceous time along the NE–SW direction. (4) The S–S′ reflector is considered to be the seismic signature of the contact beween crustal material of the margin and underlying serpentinized peridotite. The actual nature of this contact constitutes an important target for further research. We propose that it is related to the extreme thinning of the lithosphere in the vicinity of the rift axis (now the deepest part of the margin), a place where the extension tends to localize in the latest stage of the rifting. Here, the top of the mantle deformed in SZ2 is put into contact either with the ductile crust deformed in SZ1 or with the base of the upper crust. Thus, we believe S–S′ to be a major tectonic contact between crustal and mantle material, related to both shear zones SZ1 and SZ2. (5) We conclude that it is of most interest to sample by drilling the terranes located above, at the level of and beneath S–S′. From petro-structural studies of cored samples it is expected that the timing, kinematics and temperature–pressure conditions of the deformation will be investigated; and further, to constrain better the mechanisms of lithosphere stretching and passive margins formation.

We thank Captain J. C. Delmas and the crew of the O. V. Vessel 'Le Suroit'; J. Hervéou and the technical team responsible for the acquisition of multichannel seismic reflection data during the Lusigal cruise (1990); IFREMER and GENAVIR for technical and financial support. The Institut Français du Pétrole, who provided MCS data (GP03 seismic line); CNRS–INSU (IST Program) for financial support of the MCS data processing; Institut de Physique du Globe de Strasbourg (France), and GEOMAR in Kiel (Germany), where MCS data were processed. R. Scrutton helped us to improve the English in the manuscript.

Contribution no. 609 of the Groupe d'Etude de la Marge Continentale et de l'Océan (GEMCO), URA 718 du CNRS et de l'Université P. et M. Curie (Observatoire Océanologique de Villefranche-Sur-Mer).

References

Agrinier, P., Mevel, C. & Girardeau, J. 1988. Hydrothermal alteration of the peridotites cored at the ocean–continent boundary of the Iberian margin: petrologic and stable isotope evidence. *In:* Boillot, G., Winterer, E. L., *et al.* (eds) Proceedings ODP, Science Results, **103**, College Station, TX (Ocean Drilling Program), 225–233.

Allemand, P. & Brun, J. P. 1991. Width of continental rifts and rheological layering of the lithosphere. *Tectonophysics*, **188**, 63–69.

——, ——, Davy, P. & van den Driessche, J. 1989. Symétrie et asymétrie des rifts et mécanismes d'amincissement de la lithosphère. *Bulletin de la Société Géologique de France*, **8**, 445–451.

Beslier, M. O. 1991. Formation des marges passives et remontée du manteau: modélisation expérimentale et exemple de la marge de la Galice. *Mém. Docum. Centre Arm. Et. Struct. Socles, Rennes*, **45**.

—— & Brun, J. P. 1991. Boudinage de la lithosphère et formation des marges passives. *Comptes Rendus de l'Académie des Sciences, Paris*, **313**, 951–958.

——, Ask, M. & Boillot, G. 1993. Ocean–continent boundary in the Iberia Abyssal Plain from multichannel seismic data. *Tectonophysics*, **218**, 383–393.

——, Girardeau, J. & Boillot, G. 1990. Kinematics of peridotite emplacement during North Atlantic continental rifting, Galicia, NW Spain. *Tectonophysics*, **184**, 321–343.

Boillot, G., Beslier, M. O. & Comas, M. 1992. Seismic image of undercrusted serpentinite beneath a rifted margin. *Terra Nova*, **4**, 25–33.

——, Girardeau, J. & Kornprobst, J. 1988*b*. The rifting of the Galicia margin: crustal thinning and emplacement of mantle rocks on the seafloor. *In:* Boillot, G., Winterer, E. L., *et al.* (eds) Proceedings ODP, Science Results, **103**, College Station, TX (Ocean Drilling Program), 741–756.

———, FÉRAUD, G., RECQ, M. & GIRARDEAU, J. 1989*b*. 'Undercrusting' by serpentinite beneath rifted margins: the example of the west Galicia margin (Spain). *Nature*, **341**, 523–525.

———, MOUGENOT, D., GIRARDEAU, J. & WINTERER, E. L. 1989*a*. Rifting processes on the west Galicia margin, Spain. *In:* TANKARD, A. J. & BALKWILL, H. R. (eds) *Extensional Tectonics and Stratigraphy of the North Atlantic Margins*, American Association of Petroleum Geologists Memoir, **46**, 363–377.

———, AUXIÈTRE, J. L., DUNAND, J. P., DUPEUBLE, P. A. & MAUFFRET, A. 1979. The northwestern Iberian Margin: a Cretaceous passive margin deformed during Eocene. *In:* TALWANI, M., HAYET, W. & RYAN, W. B. F. (eds) *Deep Drilling Results in the Atlantic Ocean: Continental Margins and Paleoenvironment.* Maurice Ewings Series 3, Washington DC, American Geophysics Union, 138–153.

———, GRIMAUD, S., MAUFFRET, A., MOUGENOT, D., KORNPROBST, J., MERGOIL-DANIEL, J. & TORRENT, G. 1980. Ocean–continent boundary off the Iberian margin: a serpentinite diapir west of the Galicia Bank. *Earth & Planetary Science Letters*, **48**, 23–34.

———, COMAS, M. C., GIRARDEAU, J. ET AL. 1988*a*. Preliminary results of the Galinaute cruise: dives of the submersible Nautile on the western Galicia margin, Spain. *In:* BOILLOT, G., WINTERER, E. L., ET AL. Proceedings ODP, Science Results, **103**, College Station, TX (Ocean Drilling Program), 37–51.

———, RECQ, M., WINTERER, E. L. ET AL. 1987*a*. Tectonic denudation of the upper mantle along passive margins: a model based on drilling results (ODP leg 103, western Galicia margin, Spain). *Tectonophysics*, **132**, 335–342.

———, WINTERER, E. L., MEYER, A. W. ET AL. 1987*b*. Proceedings, Initial Reports (Pt A), ODP, **103**. College Station, TX (Ocean Drilling Program).

———, ———, ——— ET AL. 1988*c*. Proceedings ODP, Science Results, **103**. College Station, TX (Ocean Drilling Program).

BRUN, J. P. & BESLIER, M. O. (in press). Mantle exhumation at passive margin. *Earth and Planetary Science Letters.*

DE CHARPAL, O., GUENNOC, P., MONTADERT, L. & ROBERTS, D. G. 1978. Rifting, crustal attenuation and subsidence in the Bay of Biscay. *Nature*, **275**, 706–711.

DAVY, P. & COBBOLD, P. 1991. Experiments on shortening of a 4-layer model of the continental lithosphere. *Tectonophysics*, **188**, 1–25.

DIX, C.H. 1955. Seismic velocities from surface measurements. *Geophysics*, **20**, 68–86.

FAUGÈRE, E. & BRUN, J. P. 1984. Modélisation expérimentale de la distension continentale. *Comptes Rendus de l'Académie des Sciences, Paris*, **299**, 365–370.

FÉRAUD, G., GIRARDEAU, J., BESLIER, M. O. & BOILLOT, G. 1988. Datation 39Ar/40Ar de la mise en place des péridotites bordant la marge de la Galice (Espagne). *Comptes Rendus de l'Académie des Sciences, Paris*, **307**, 49–55.

GIRARDEAU, J., EVANS, C. A. & BESLIER, M. O. 1988. Structural analysis of plagioclase-bearing peridotites emplaced at the end of continental rifting: hole 637A, ODP leg 103 on the Galicia margin. *In:* BOILLOT, G., WINTERER, E. L., ET AL. (eds) Proceedings ODP, Science Results, **103**, College Station, TX (Ocean Drilling Program), 209–223.

GRIMAUD, S., BOILLOT, G., COLLETTE, B., MAUFFRET, A., MILES, P. R. & ROBERTS, D. B. 1982. Western extension of the Iberian–European plate boundary during the early Cenozoic (Pyrenean) convergence: a new model. *Marine Geology*, **45**, 63–77.

HERRAIZ, M. & ESPINOSA, A. F. 1986. Scattering and attenuation of high-frequency seismic waves: development of the theory of coda waves. Open File Report, 86-455, US Geological Survey, 1–92.

HOFFMANN, H. J. & RESTON, T. J. 1992. The nature of the S reflector beneath the Galicia Bank rifted margin. Preliminary results from pre-stack depth migration. *Geology*, **20**, 1091–1094.

HORSEFIELD, S. J. 1992. *Crustal structure across the continent–ocean boundary*. PhD Thesis, University of Cambridge, UK.

KENT, D. V. & GRADSTEIN, F. M. 1986. A Jurassic to Recent chronology in the western North Atlantic region. *In:* VOGT, P. R. & TUCHOLKE, B. E. (eds) *Geology of North America*, Geological Society of America, Boulder, CO, vol. M, 45–50.

LALAUT, P., SIBUET, J. C. & WILLIAMS, C. A. 1981. Présentation d'une carte gravimétrique de l'Atlantique du nord-est. *Comptes Rendus de l'Académie des Sciences, Paris*, D, **292**, 597–600.

LALLEMAND, S., MAZÉ, J. P., MONTI, S. & SIBUET, J. C. 1985. Présentation d'une carte bathymétrique de l'Atlantique Nord-Est. *Comptes Rendus de l'Académie Sciences, Paris*, **300**, 145–149.

LE PICHON, X. & BARBIER, F. 1987. Passive margin formation by low-angle faulting within the upper crust: the northern Bay of Biscay margin. *Tectonics*, **6**, 133–150.

MALOD, J. A., MURILLAS, J., KORNPROBST, J. & BOILLOT, G. 1993. Oceanic lithosphere at the edge of a Cenozoic active continental margin (northwest slope of Galicia Bank, Spain). *Tectonophysics*, **221**.

MAMET, B., COMAS, M. C. & BOILLOT, G. 1991. Late Palezoic basin on the west Galicia Atlantic margin. *Geology*, **19**, 738–741.

MAUFFRET, A. & MONTADERT, L. 1987. Rift tectonics on the passive continental margin off Galicia (Spain). *Marine Petroleum Geology*, **40**, 49–70.

——— & ——— 1988. Seismic stratigraphy off Galicia. *In:* BOILLOT, G., WINTERER, E. L. ET AL. (eds) Proceedings ODP, Science Results, **103**. College Station, TX (Ocean Drilling Program), 13–30.

MONTADERT, L., DE CHARPAL, O., ROBERTS, D. G., GUENNOC, P. & SIBUET, J. C. 1979. Northeast Atlantic passive continental margins: rifting and subsidence processes. *In:* TALWANI, M., HAY, W. & RYAN, W. B. F. (eds) *Deep Drilling Results in the Atlantic Ocean: Continental Margins and Paleoenvironments.* American Geophysical Union,

Maurice Ewing Series, **3**, 154–186.

MOUGENOT, D., KIDD, R. B., MAUFFRET, A., REGNAULD, H., ROTHWELL, R. G. & VANNEY, J. R. 1984. Geological interpretation of combined Sea-Beam, Gloria, and seismic data from Porto and Vigo Seamounts, Iberian continental margin. *Marine Geophysics Research*, **6**, 329–363.

MURILLAS, J., MOUGENOT, D., BOILLOT, G., COMAS, M. C., BANDA, E. & MAUFFRET, A. 1990. Structure and evolution of the Galicia interior basin (Atlantic western Iberian continental margin). *Tectonophysics*, **184**, 297–319.

RANALLI, G. & MURPHY, D. C. 1987. Rheological stratification of the lithosphere. *Tectonophysics*, **132**, 281–295.

RAPPIN, D. 1992. Apport des analyses d'amplitude et temps-fréquence à l'exploitation de données de sismique profonde. Thèse de l'Université Louis Pasteur de Strasbourg.

———, MARTHELOT, J. M., DE BAZELAIRE, E. & RAVAT, J. (in press). Analysis of the attenuation of amplitudes on records of the ECORS Pyrenees deep seismic profile. *Geophysical Prospecting*.

RECQ, M., WHITMARSH, R. B. & SIBUET, J. C. 1991. Anatomy of a lherzolitic ridge, Galicia margin. *Terra Abstract*, **3**, 122.

RESTON, T. J., KRAWCZYK, C. M. & HOFFMANN, H. J. 1995. Detachment tectonics during Atlantic rifting: analysis and interpretation of the S reflector, the west Galicia margin. *This volume*.

SIBUET, J. C. 1992. New constraints on the formation of the non-volcanic continental Galicia–Flemish Cap conjugate margins. *Journal of the Geological Society, London*, **149**, 829–840.

———, MAZÉ, J. P., AMORTILA, P. & LE PICHON, X. 1987. Physiography and structure of the western Iberian continental margin off Galicia from Sea-Beam and seismic data. *In:* BOILLOT, G., WINTERER, E. L., MEYER, A. W. *ET AL.* (eds) Proceedings Initial Reports (A). ODP, **103**, 77–97.

———, RYAN, W. B. F. *ET AL.* 1979. *Initial Reports of the Deep Sea Drilling Project*, **47**. US Government Printing Office, Washington, DC.

SRIVASTAVA, S. P., ROEST, W. R., KOVACS, L. C., OAKEY, G., LÉVESQUE, S., VERHOEF, J. & MACNAB, R. 1990. Motion of Iberia since the Late Jurassic: Results from detailed aeromagnetic measurements in the Newfoundland Basin. *Tectonophysics*, **184**, 229–260.

THOMMERET, M., BOILLOT, G. & SIBUET, J. C. 1988. Structural map of the Galicia margin. *In:* BOILLOT, G., WINTERER, E. L. *ET AL.* (eds) Proceedings ODP, Science Results, **103**. College Station, TX (Ocean Drilling Program), 31–36.

VENDEVILLE, B., COBBOLD, P. R., DAVY, P., BRUN, J. P. & CHOUKROUNE, P. 1987. Physical models of extensional tectonics at various scales. *In:* COWARD, M. P., DEWEY, J. F. & HANCOCK, P. L. (eds) Continental extensional tectonics. Geological Society, London, Special Publication, **28**, 95–107.

WERNICKE, B. 1985. Uniform-sense normal simple shear of the continental lithosphere. *Canadian Journal of Earth Sciences*, **22**, 108–125.

——— & BURCHFIELD, B. C. 1982. Modes of extensional tectonics. *Journal of Structural Geology*, **4**, 105–115.

WHITMARSH, R. B., MILES, P. R. & MAUFFRET, A. 1990. The ocean–continent boundary off western continental margin of Iberia – I. Crustal structure at 40°30′N. *Geophysics Journal International*, 509–531.

———, PINHEIRO, L. M., MILES, P. R., RECQ, M. & SIBUET, J. C. 1993. Thin crust at the western Iberia ocean–continent transition and ophiolites. *Tectonics*, **12**, 1230–1239.

Statigraphic consequences of extensional tectonics

Sedimentary models for extensional tilt-block/half-graben basins

M.R. Leeder & R.L. Gawthorpe

SUMMARY: Extensional tectonism produces characteristic half-graben/tilt-block systems whose facies mosaics are influenced by tectonically induced slopes resulting from hanging wall downtilting and footwall uplift. The characteristic asymmetrical subsidence vectors that therefore develop across the graben also exert a fundamental control upon facies distributions. A number of predictive tectono-sedimentary facies models are presented in which these various influences are explored. Alluvial fans and cones react to tilting by becoming segmented, those in the hanging wall showing down-dip hanging wall off-lap and those sourced in the footwall showing progradation from the apex. Lake and coastal waters react instantly to tilting, causing transgression and seiche-induced erosion. Axial through-flowing river channels and delta lobes tend to migrate or avulse towards the axis of maximum subsidence but may be constrained by the toes of footwall-sourced fans. Peat accumulation or soil development are accentuated up the hanging wall dip slope away from the locus of maximum deposition. In coastal areas, fan deltas sourced in the footwall pass offshore into small submarine fans whilst axial fans issue from delta fronts where individual fan lobes may migrate under fault control. In carbonate provinces the footwall scarp may become a bypass margin whilst the hanging wall dip slope may undergo a ramp-to-rimmed shelf evolution with time.

It has become apparent in the last few years that the process of lithospheric extension is characterized by a distinctive development of tilt blocks and half-grabens bounded by major normal faults (Morton & Black 1975; McKenzie 1978; Wernicke & Birchfiel 1982; Jackson *et al.* 1982a, b; Brun & Choukroune 1983; Gibbs 1984). These types of basins develop progressively during extension, and we recognize that during this time they exert a profound and logical influence upon geomorphology and sediment-transfer mechanisms in the vicinity. Examples of active basins of this sort occur in the Aegean back-arc (McKenzie 1978; Jackson *et al.* 1982b; Jackson & McKenzie 1983); in the Basin and Range Province of the western United States (Myers & Hamilton 1964; Wernicke & Birchfiel 1982; Anderson *et al.* 1983); in the lower Rhine basin (Illies & Fuchs 1983); in the Gulf of Suez (Sellwood & Netherwood 1984); and in Afar (Hutchinson & Engels 1970; Morton & Black 1975). Extinct fault-bounded extensional basins occur around the margins of almost all Atlantic-type passive continental margins (e.g. Surlyk 1977, 1978) and in continental areas where extension did not proceed to oceanic separation (failed arms) as in the Central African rift system (Browne & Fairhead 1983). These extinct basins usually lie buried beneath the later deposits of the thermal-contraction phase of extensional subsidence. One of the best known examples is the North Sea basin (Christie & Sclater 1980; Barton & Wood 1984; Ziegler 1982). An increasing number of geologically ancient basins of this type are being recognized, e.g. the early Carboniferous extensional province of the British Isles (Leeder 1982; Dewey 1982; Gawthorpe, in press), and doubtless many others await recognition.

Recently major advances have been made concerning the dynamics and kinematics of normal faulting in areas of active extension (Wernicke & Birchfiel 1982; Jackson *et al.* 1982a, b; Gibbs 1984; Jackson & McKenzie 1983). It is the purpose of this paper to integrate such studies with analysis of sediment transfer and deposition, the resulting three-dimensional tectono–sedimentary-facies models being helpful in the study of ancient basin-fill successions. The models developed herein have been culled from an extensive but scattered geomorphological and sedimentological literature, our unpublished studies in active extensional terrains (Aegean, Basin and Range), our work in the ancient extensional graben-fill successions of the Lower Carboniferous in N Britain (Leeder 1982, 1986; Gawthorpe 1986, in press), and from previous stratigraphic-modelling contributions (Bridge & Leeder 1979).

Tilt-block/half-graben morphology and structure

Originally, Morton & Black (1975) envisaged domino-type fault blocks forming during active

From Holdsworth, R.E. & Turner, J.P. (compilers) *Extensional Tectonics: Regional-scale Processes*. The Geological Society, Key Issues in Earth Sciences, **2 (1)**, 171–184.
1476-315X/02/$15.00 © The Geological Society of London 2002
First published in: "Coward, M.P., Dewey, J.F. & Hancock, P.L. (eds) *Continental Extensional Tectonics*, The Geological Society, London, Special Publications **28**, 139–152".

extension, with the further occurrence of progressive fault rotation causing nucleation of higher angle second- or third-generation normal faults which themselves become progressively rotated. Structural studies in the Basin and Range Province have confirmed this model (Proffett 1977) and led to a number of more detailed tectonic models involving basement thrust reactivation, listric-fault fans and extensional analogues to the ramp and flat models of thrust tectonics (Wernicke & Birchfiel 1982; Brun & Choukroune 1983; Anderson *et al.* 1983; Chamberlin 1983; see also the general papers of Gibbs 1983, 1984). These studies serve to stress the great complexities of extensional faulting, especially on an intrabasinal scale.

Whilst American workers in the Basin and Range Province have stressed the importance of shallow, 'domino-style' faulting, workers in the Aegean extensional province (McKenzie 1978; Jackson *et al.* 1982a, b) have shown the importance of major normal faults of mid-crustal penetration which bound 'bouyant' tilt blocks. Some examples of this type of fault also occur in the Basin and Range Province (Anderson *et al.* 1983). These major faults cause instantaneous unloading along the fault plane during fault motion leading to an instantaneous isostatic upwarp of the footwall block (Heiskanen & Vening Meinesz 1958; Savage & Hastie 1966; Bott 1976). Jackson & McKenzie (1983) calculate this footwall uplift to be around 10% of the hanging wall subsidence.

Whether small- or large-scale faulting develops seems to depend upon the extensional strain rate, the geothermal gradient and the existence of crustal fractures ripe for rejuvenation (Anderson *et al.* 1983; Eaton 1982).

Reduced to their simplest form, tilt-block/half-graben structures can be considered to be bounded by single normal faults which penetrate to mid-crustal levels. As the hanging wall basement detaches from the footwall an asymmetrical basin progressively develops above the hanging wall. The fundamental controls upon geomorphology and sedimentation patterns are as follows (see Figs 1 & 2):

1 Tectonic slopes. These are produced by a combination of footwall uplift and hanging wall subsidence and comprise the steeper footwall scarp slope and the gentler hanging wall dip slope. The footwall area is the main sediment source for the adjacent basin although, due to the asymmetrical nature of the basin, the hanging wall-derived sediment may be spatially more extensive. Recent geomorphological studies (Hanks *et al.* 1984) demonstrate that the gradual decay of the scarp profile with time follows a decay equation of the error-function type. The periodic rejuvenation of the footwall scarp will give rise to important sedimentary consequences in the basin fill (q.v.). In examples where the basin-margin fault has a listric geometry, tilting of the surface during extension is accompanied by rotation and the development of a rollover structure.

2 Asymmetrical subsidence. This is due to the pivot-like motion of the hanging wall after individual extensional episodes. Asymmetrical subsidence following historic earthquakes is best documented for the Hebgen Lake area of Montana, USA. In Fig. 2a we reproduce the

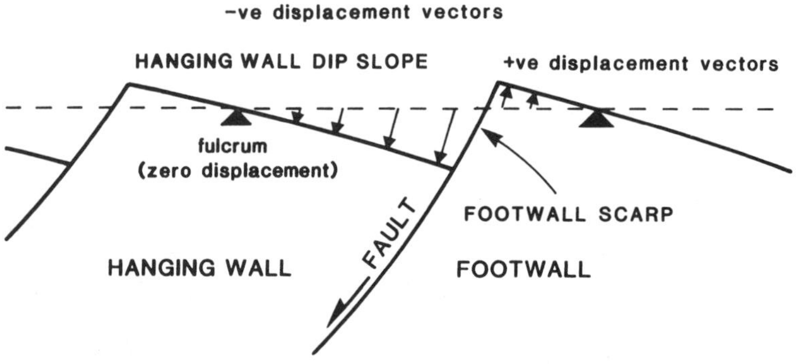

FIG. 1. Nomenclature diagram for tectonic slopes associated with a simple tilt block/half-graben. The main tectonic slopes produced during basin development are the steep, spatially restricted footwall scarp associated with the footwall-to-hanging wall transition, and a broad, gentle slope, the hanging wall dip slope, characteristic of the hanging wall of the basin-forming fault. The fulcrum is the position where displacement of the hanging wall block is zero; either the limit of roll-over in isolate tilt block/half-grabens or the transition from areas of the hanging wall undergoing positive motion due to footwall uplift to areas undergoing negative motion due to hanging wall subsidence. The position of the fulcrum is governed by the relative displacement vectors across the half-graben-bounding fault(s), the presence of antithetic/synthetic intrabasin structures and the subsurface geometry of the faults.

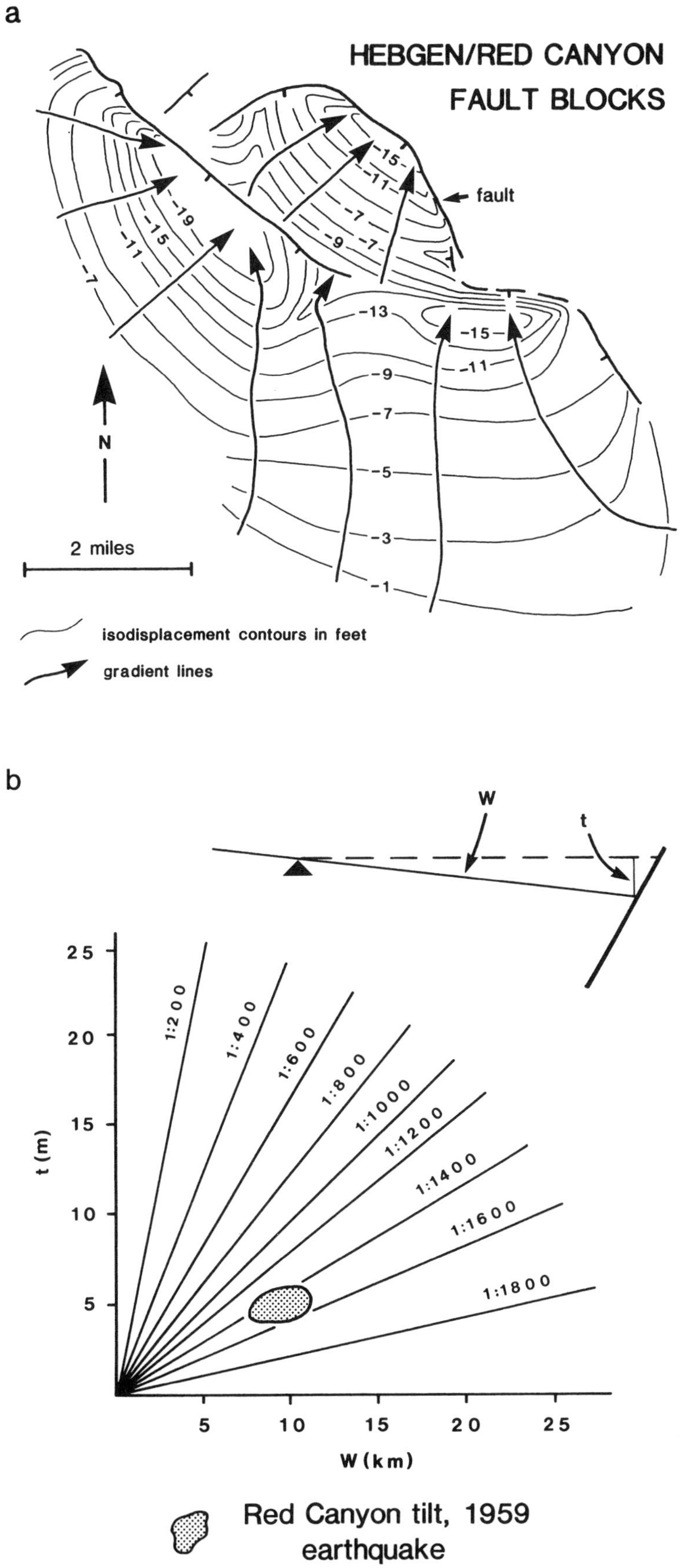

FIG. 2. (a) Example of surface displacement associated with the 1959 earthquake in the Hebgen/Red Canyon tilt block, Montana, USA (after Fraser *et al.* 1964). Contours are of equal ground displacement associated with the earthquake and our arrows represent sketch gradients. (b) Our graph showing the approximate tectonic slopes produced by single motions of vertical magnitude (t) across tilt block of width (W).

isodisplacement contours of Fraser *et al.* (1964) drawn up following the 1959 earthquake, with an indication of the secondary gradients due to tilting. As shown in Fig. 2b, tectonic slopes produced by quite modest fault displacements have a great effect when superimposed on say, typical river gradients. Our fundamental starting point is thus emphasized: all sedimentary processes driven by gravity will be influenced by extensional tilting.

The point where the hanging wall displacement tends to zero may be called the fulcrum. The position of the fulcrum is not fixed but is controlled by the geometry of the basin-margin fault, the presence or absence of antithetic/synthetic intrabasin faults and time. In the simplest case of a tilt block/half-graben bounded by a listric normal fault, progressive extension will lead to the fulcrum moving away from the fault, i.e. up the hanging wall. This effect may be particularly important in the early stages of evolution of tilt blocks/half-grabens, leading to progressive on-lap onto the basement. The seismic section and line drawing across the Fallon basin by Anderson *et al.* (1983, fig. 5) displays such on-lap developed during progressive extension. In a linked series of tilt blocks the fulcrum divides areas undergoing positive motion due to footwall uplift from areas of negative motion arising from hanging wall subsidence. As shown below, we postulate that the location of fulcral lines will have considerable influence upon the sedimentary and geomorphic evolution of a tilt block and should also provide valuable clues concerning crustal dynamics.

In active half-grabens the simple tilt-block concept described above is complicated, to a greater or lesser extent, by the occurrence of both antithetic and synthetic faults which deform both the hanging wall basement and the developing sedimentary fill. Although these various structures have received descriptive attention in seismic studies (e.g. Gibbs 1983, 1984) in ancient basins it is by no means clear how they influenced topography and sedimentation at the depositional surface. Such structures will, however, clearly be of importance in modifying the sub-surface geometry of the sedimentary fill. Major intrabasinal normal faults define important 'mini' horsts and grabens in the Rio Grande rift (Velarde graben of Manley 1979: Ladron horst of Brown *et al.* 1979). Analogous structures are deduced for the extensional Carboniferous basins of northern England (Leeder 1986, in press; Gawthorpe 1986, in press).

At the Earth's surface, extensional normal faults are of finite extent and thus the asymmetrical basins so produced have a distinctive 'scoop' shape, at least in the early stages of extension. This is well seen in the Hebgen Lake area of Montana (Fraser *et al.* 1964, see our Fig. 1) and in the Rio Grande basin system of Colorado/New Mexico where continued extension along a N–S trend has led to the joining up of a series of originally isolated scoops into the present complex linear graben system (Chapin 1979; Bachman & Mehnert 1978).

Another consequence of the finite extent of normal faults at the surface (see Smith & Bruhn 1984) is that overlap zones exist between *en échelon* fault segments. In such zones, where the footwall zone of one fault passes into the hanging wall of an adjacent structure, an oblique monoclinal downbend develops at a high angle to the normal-fault trend. Large drainage systems may exploit the hinterland behind such areas, causing the development of larger-than-average depocentres in the main hanging wall basin (see Fig. 4). We have seen examples in the Greek Aegean, at Karrena Bourla in Locride, and in Montana, USA, along the Madison Front N of 'Quake Lake. Examples have also been described by Crossley (1984) from the East African graben in Malawi. Transform (Bally 1982) or transfer faults (Gibbs 1984) are other structures developed at a high angle to normal faults within half-grabens; indeed, the monoclines described above may be the surface expression of blind transfer faults. These faults separate areas of the basins deforming in different styles or at different rates and, in some instances, the sense of tilting may change polarity across them. Transfer faults also have an important role in sediment transfer, being the sites of major transport of sediment off both the hanging wall and footwall. In the Lower Carboniferous Bowland basin of northern England transfer faults were a major control on the location of coarse-grained carbonate debris-flow deposits that were derived off the hanging wall basin margin (Gawthorpe, in press).

Additional complications also arise because of the occurrence of layers of upwardly mobile partial melts in the shallow crust. These buoyant accumulations cause local areas of minor relative upwarp in an otherwise subsiding regime. Perhaps the most spectacular example occurs beneath the Rio Grande rift at Socorro, New Mexico (Sanford *et al.* 1977; Brown *et al.* 1979). Suspected ancient analogues existed in the Jurassic of the Central North Sea (Leeder 1983) and in the Lower Carboniferous Northumberland graben in the Scottish Borders (Leeder 1976, 1982). Major surface effects will also arise once this magma is erupted, with the production of lava plateaux, large-scale river damming, and local fan formation. Such effects are seen in the Rio Grande rift (Lipman & Mehnert 1979).

A dynamic feature of major importance in extensional basin evolution is the abandonment of one tilt-block/half-graben system because of the abandonment of one crustal-scale normal fault and the development of another (Jackson & McKenzie 1983). If the new fault develops on the hanging wall side of the now-extinct fault then the basin-fill is gradually uplifted in the footwall of the new fault. Examples occur in the Greek Aegean, particularly on the S side of the Gulf of Corinth and in Locride (Jackson *et al.* 1982a; Jackson & McKenzie 1983).

Stratigraphic and sedimentary consequences of extension—general remarks

As noted previously the occurrence of movement on an extensional normal fault causes surface tilting. Thus a tectonic gravity slope arises which will be superimposed upon the existing geomorphic slope. Since many sedimentary processes are gravity controlled it follows *a priori* that erosion, sediment transfer, deposition and soft-sediment deformation (see Leeder (in press) for a full discussion of this latter feature) will then have a tectonic influence, if not control. Once sedimentary systems become established in extensional basins then each slope component will contribute to the basin infill in a distinctive and, to some extent, predictable way. Of major importance are (i) *lateral-transport systems* that deposit sediment in high-gradient fans and low-gradient cones down the footwall and hanging wall slopes respectively, normal to the strike of the main bounding fault, and (ii) low seeking *axial-transport systems* that transfer sediment parallel to the strike of the main bounding faults. The effects of surface tilting on those lateral and axial systems give rise to marked basin-wide variations in lithology, facies and thickness. In the following sections the implications of these various general considerations are applied to particular sedimentary facies associations so that general and predictive basin-fill sequences may be proposed.

Before proceeding to the models it is essential at the outset to stress that there are many other variables (other than tectono–sedimentary, that is) which can affect the details of facies distributions. Thus climate and hinterland geology will also exert profound influences upon facies associations in particular tectonic settings. These two variables will combine to determine the magnitude of clastic-sediment flux into an extensional basin. Clearly, a balance will then be set up between sediment deposition rate and subsidence rate as ultimately determined by the rate of extension. The relative position of sea-level with respect to the basin floor is also of obvious importance. Tilting is expected to have instantaneous and catastrophic effects upon both lake and marine shorelines. Such effects will give rise to a distinctive 'signal' in the basin-fill successions and may cause the production of distinctive sedimentary cycles. Such effects are of major importance to the geologist examining the fills of ancient basins. Here, in the absence of direct evidence for active extensional tectonics, it is essential to integrate the study of observed facies changes with structural information. The following models are proposed as basic templates for such studies.

Tectono-sedimentary facies model A—continental basin with interior drainage (Fig. 3)

This style of basin is a very characteristic one in the Basin and Range Province of the western USA where isolated fault-bounded depressions form local interior-drainage basins with no outlet to adjacent structures. The fundamental slopes are the relatively low-gradient hanging wall dip slope with broad alluvial cones and the relatively high-gradient footwall scarp slope which sources small alluvial fans whose depositional loci occur at the foot of the scarp on the lower hanging wall dip slope. As noted previously, large drainage systems frequently take advantage of transfer-fault zones and areas between *en échelon* fault terminations so that larger-than-average cones may preferentially form in such locations (Fig. 4). Should local climatic conditions allow, permanent or playa lake bodies will form in the basin as close to the locus of maximum subsidence as the footwall-sourced fans will allow. Such triplet cone-lake-fan systems have particularly dynamic interactions at their internal boundaries.

Perhaps the most studied example is Death Valley, California where the meticulous work of Hooke (1972), following on from the studies of Hunt & Mabey (1966), led to the first recognition of the effects of tilt movements on fan evolution and sedimentation. Hooke established beyond reasonable doubt that fan segmentation results from surface tilting and that characteristic off-lap sequences result if the tilting process is carried on periodically. These effects are

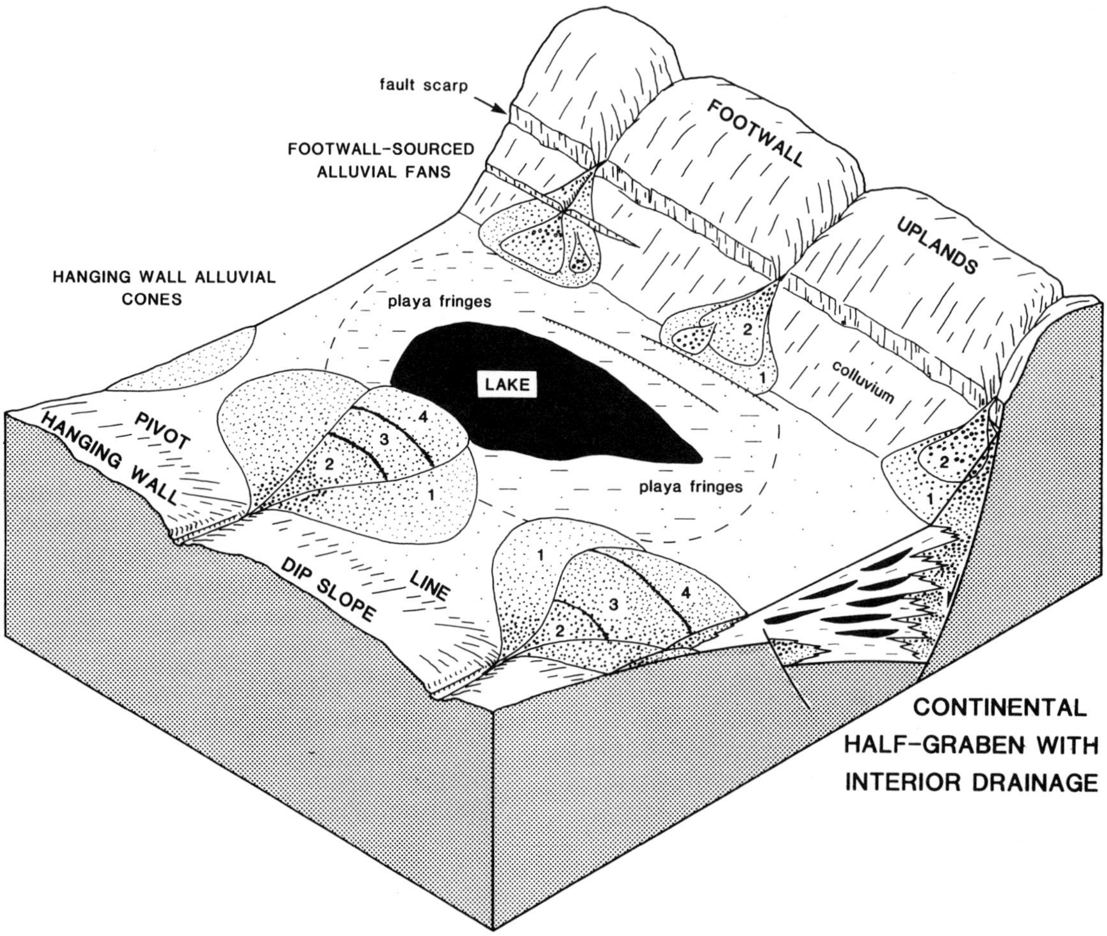

FIG. 3. Isometric diagram showing the main sedimentological features of facies model A: continental basin with interior drainage. Full discussion in text. Note: only the major basin-margin fault is shown; in natural examples the presence of antithetic, synthetic and transfer fault systems strongly modify certain depositional reactions to tilting. In addition, the sub-surface geometry is modified by differential compaction, thinning of the hanging wall associated with development of the roll-over and the presence of antithetic/synthetic fault systems within the sedimentary cover. 1, 2, 3 etc. indicate successive fan lobes.

generally predictable in terms of an extensional tilt-block model and may be amplified as follows.

Periods of fault motion cause the tectonic gradient and length of the footwall scarp to increase instantaneously. Hanging valleys are produced as fan-head channel incision occurs progressively. The gradient of the fan surface in the hanging wall is decreased though, so that renewed sedimentation leads to the construction of a new fan segment close to the fan apex. This is expected to prograde gradually over the old fan segment with time, causing a crude upward-coarsening cycle to develop in response to the progradation of the downslope-thinning clastic wedge. Fault motion will also cause instantaneous lake transgression over the distal portions of the footwall-sourced fans, causing them to shrink substantially in area. Should subsequent gradual fan progradation occur, these transgressive facies will be intercalated at the base of a tectonic fan cycle. Historic observations of lake-tilting, notable after the Hebgen Lake 'quake of 1959 (Myers & Hamilton 1964), show the occurrence of major seiches which may be expected to initiate erosive contacts and debris washover units over a much larger area than that covered by the permanent lake. The alluvial cones of the hanging wall dip slope react to tilting by incising their main fan-apex feeder channels close to the pivot line. The previous equilibrium-cone surface is tilted up by an amount determined by the magnitude of fault motion and a new active cone-lobe surface forms on a lower angle surface, often basinwards of the old cone lobe because of the steepening of the tilted dip-slope surface. Prominent breaks of slope occur at the junction of old and new surfaces, the older surface becoming depositionally defunct and subject to pedogenesis (calcretization in the case of Death Valley). Basinwards, the cones will prograde over the latero–distal portions of older fan surfaces forming off-lap sequences of downslope-thickening and fining wedges (see Hooke 1972, fig. 3). Thin units of lacustrine facies will be incorporated in these successions as progradation occurs over the upslope abandoned portions of the old lake shoreline. Examples of such effects are known from the hanging wall cones of Hebgen Lake (Alexander & Leeder, in press).

It seems clear that fault-induced tilting causes a number of rapid facies changes to occur. It should be borne in mind, however, that fan, cone and lake shrinkage and growth cycles may also be caused by climatic fluctuations, as is frequently seen in the Pleistocene history of the western USA (Hawley *et al.* 1976) and elsewhere (Talbot & Williams 1979). These more gradual sorts of effects must be separated from the purely tectonic effects discussed above. This may only be possible in very well-exposed terrains where good stratigraphic dating is present.

The basin-centre facies in the present model are certainly the most variable since evaporitic deposits may form from playa lakes as well as the non-evaporitic deposits of more permanent, possibly stratified lakes. The primary control here is obviously climatic. A further possibility, where run-off is very low and where regional sand-laden winds blow across the basin, is that small desert ergs may develop, particularly in the axis of maximum subsidence. In such areas the decelerating winds drop portions of their bedload before they accelerate up the hanging wall dip slope or the footwall scarp. Other geometrical possibilities of minor erg formation clearly exist. The Great Sand Dunes of the San Luis basin, Colorado are proposed as an example of this kind of facies association.

Tectono-sedimentary facies model B—continental basin with axial through-drainage (Fig. 4)

Active examples of this style of basin are well developed in the Aegean region, e.g. the River Axios in the Vardar graben, River Speriohos in the Lamia graben, and the Great Meander River in one of the Anatolian grabens. Other examples are the lower Rhine graben (particularly the River Erft in the Erft graben), the Rio Grande rift of Colorado/New Mexico and the Madison River graben of Montana. Certain features are shared in common with model A, namely the occurrence of footwall and hanging wall derived fans and cones respectively. Some basins have, in fact, evolved from interior to through-drainage during their extensional evolution, notably the Rio Grande composite linear graben system which now comprises the once isolated San Luis, Taos, Sante Fe/Espanola and Albuquerque sub-basins (Bachman & Mehnert 1978; Kelley 1977; Manley 1979).

The most fundamental characteristic of this kind of model concerns the reaction of the axial river to episodes of tectonic tilting which influence river migration. The effect to be described was briefly noted by Russell (1954, p. 370) in his Anatolian geomorphological study and independently explored by Bridge & Leeder (1979) in their simulation models of alluvial stratigraphy ('architecture'). The latter authors postulated that periodic abrupt channel movements (avulsions) would cause the 'low-seeking' channel to persistently reoccupy the axis of maximum subsidence in statistical preference to other areas of the half-graben. Thus, with time, a preferential stacking of alluvial sand-bodies would be observed in the axis of the sub-surface basin fill (see discussion in Alexander & Leeder, in press).

A number of other factors will influence this simple model.

1 The actual position of the axial river will be displaced from the axis of maximum subsidence by alluvial fans issuing from the footwall scarp slope. This is well seen in the Anatolian grabens (Russell 1954) and at the margins of the ancestral Rio Grande rift (Hawley *et al.* 1976, p. 250; Bachman & Mehnert 1978, p. 288) where the axial fluvial-channel facies are interdigitated with transverse piedmont facies.

2 Intrabasinal normal faulting may create mini-grabens within the main structure. Such features would tend to 'trap' the axial river and encourage the production of sharply defined fault-bounded sub-surface stacks of axial-channel sand or gravel units. A modern example is represented by the Velarde mini-graben in the Espanola portion of the Rio Grande rift (Reilinger *et al.* 1979, fig. 6; Manley 1979, fig. 2). Ancient examples occur in the axis of the Lower Carboniferous Northumberland graben, as discussed by Leeder (1986, in press).

3 Recent theoretical and field studies (Leeder & Alexander, in press) have shown that axial river channels also gradually respond to tectonic tilting by a process of preferential downslope cut-off and minor avulsion. This leads to the production of an abnormally wide channel belt sand-body,examples of which occur along the South Fork River in the Hebgen area of Montana

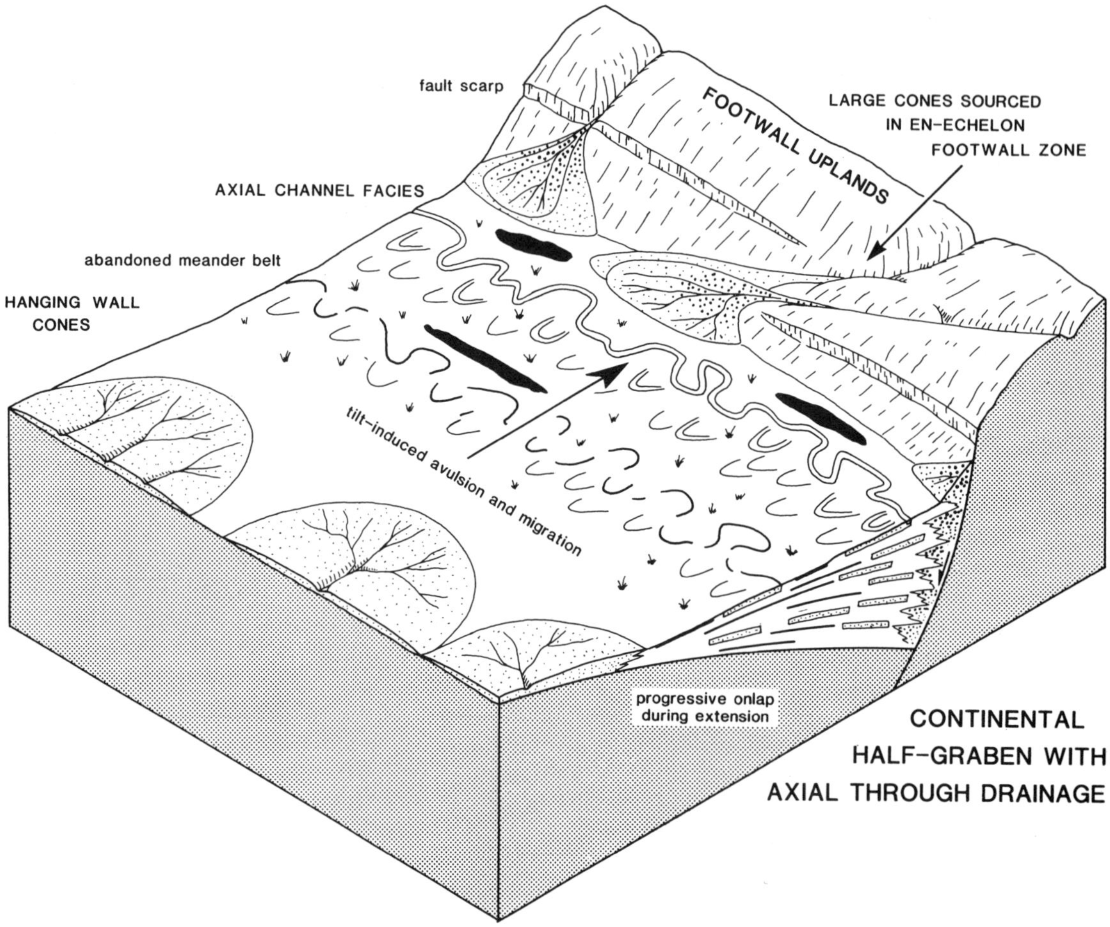

FIG. 4. Isometric diagram for facies model B: continental basin with axial through-drainage. Note the large alluvial fans sourced in the zone between two *en échelon* normal faults and preferential avulsion of the axial river channel(s) towards the faults. Full discussion in text.

(Myers & Hamilton 1964; Leeder & Alexander, in press) and along the Mississippi River in the New Madrid uplift area (Russ 1982; Alexander & Leeder, in press). Very large-scale examples of such behaviour were originally postulated by Mike (1975) in his analysis of channel migration over the Carpathian molasse plain.

4 It is also a possibility that instantaneous, low-seeking avulsion could occur as a direct result of a tilting movement.

A corollary of the tendency for channels to seek out the axis of subsidence by the various means outlined above is that fine-grained out-of-channel flood sediments should be deposited progressively more infrequently as the hanging wall dip slope is ascended. This is because, like the axial-channel system, surface floodwaters will always tend to follow the main basin gradient. In a suitable climatic regime, especially humid tropical, and in areas away from hanging wall cones, this process will cause the formation of a broad belt of hanging wall dip-slope swamplands which will gradually lead to the occurrence of sub-surface peat layers and hence, ultimately, to the formation of coals which will tend to thicken into the basin axis. Minor intrabasinal faulting may cause abrupt thickness changes of the peats. Examples of these sorts of consequences of our model are illustrated by Teichmuller & Teichmuller (1968, figs 5 & 8) and by Brunnacker & Boenigk (1983, fig. 5) from the lower Rhine graben. As the hanging wall is further ascended it may be expected that the lowered water table should encourage the swamplands to peter out and let oxygenated soil horizons develop in their place as the pivot line of the tilted surface is approached. In a more arid climatic regime, the above facies mosaic would be replaced by calcrete-type soil-profile development away from the loci of active hanging wall dip-slope cones. Examples are seen in the Rio Grande rift (Llano de Albuquerque—Bachman & Mehnert 1978, p. 288).

Tectono-sedimentary facies model C—coastal marine gulf basin (Fig. 5)

Active examples of these basins occur at many localities around the margins of the Aegean e.g. the Gulf of Corinth, the Thermaic Gulf, and the Gulf of Euboea. As shown in Fig. 5, the gulf is often actively being infilled by axial drainage systems. Transverse fans and cones are characteristically present in all examples. The gulf morphology is a direct result of half-graben morphology, with the major bounding normal faults determining the position and characteristic cliffed morphology of the linear footwall shoreline. Ancient analogues are discussed by Surlyk & Clemmensen (1983).

Basins with axial drainage will closely follow certain behavioural tendencies proposed for the previous alluvial model. Thus the terminal deltaic lobes to the drainage system will migrate or avulse preferentially towards the axis of maximum tectonic subsidence, thereby leading to the preferential stacking of delta lobes in the sub-surface (Leeder & Strudwick 1986) and a tendency for maximum back-swamp peat development to occur some way up the hanging wall slope. Important additional features will result from instantaneous marine transgression over the delta lobe, and adjacent shorelines following motion along the normal fault. In general, the effects of such transgressions will lessen in importance up the hanging wall slope. In the footwall and on the footwall scarp slope, emergence will occur, as elegantly demonstrated from geomorphological features in the Corinth Gulf by Jackson *et al.* (1982b) after the 1981 earthquakes. The extent of instantaneous transgressions over the low-gradient axial gulf coastal plain may be very large indeed e.g. a 3 m fault throw in an axial coastal plain of gradient 1:500 would flood a maximum of 1.5 km inland.

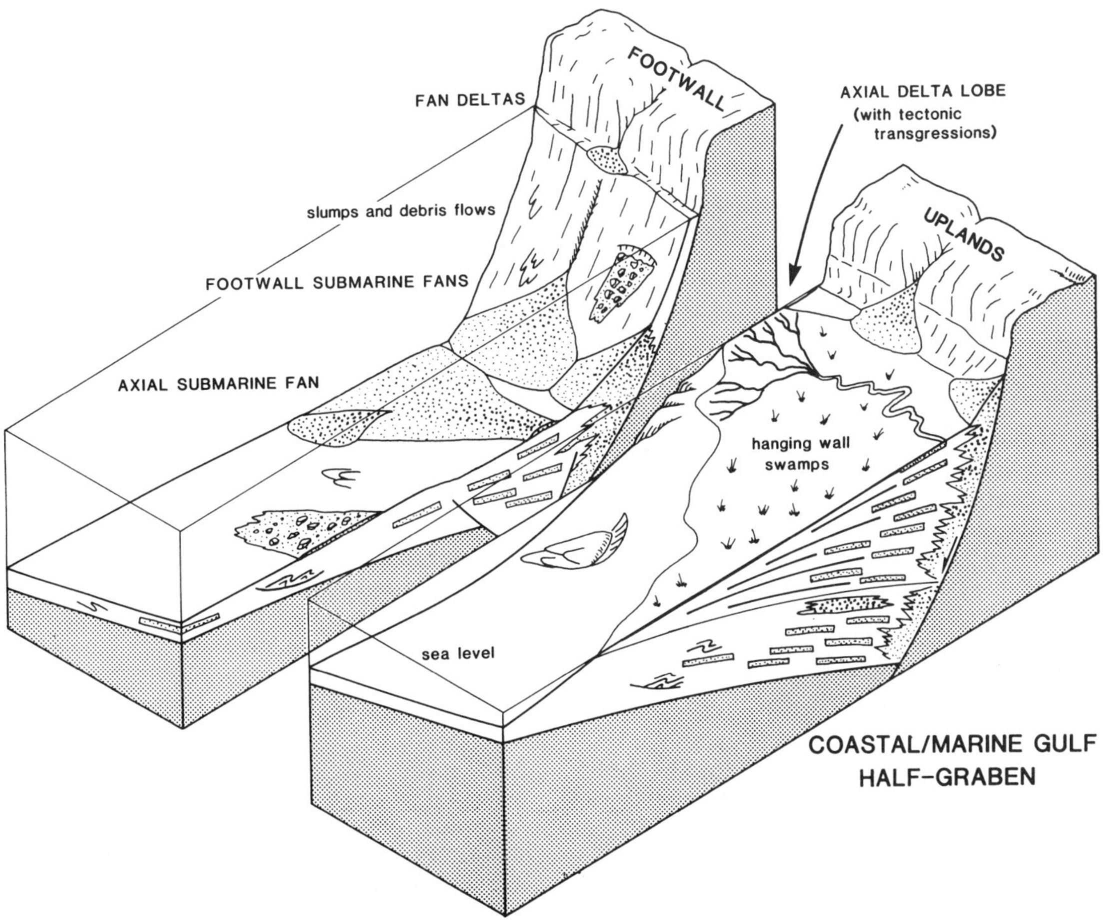

FIG. 5. Isometric diagram showing sedimentological characteristics of facies model C: coastal/marine gulf basin. Location of axial submarine fan, controlled by intrabasin synthetic and antithetic faults, is shown schematically. Full discussion in text.

Renewed clastic progradation after the period of fault motion would then give rise to a very distinctive abrupt transgressive and gradual regressive-facies sequence.

The hanging wall cones and footwall fans sourced in the exposed gulf-perimeter uplands will both show dynamic interactions with the marine-gulf coastal facies after periods of fault-induced subsidence. Footwall-sourced alluvial fans will be subject to the instantaneous effects of marine transgression across their distal aprons. Our observations in the Gulf of Corinth where the fans are largely inactive, being late-Wurmian features, reveal that fan aprons are often cliffed and are subject to marine erosion at present. Eventually, more active fans will prograde once more over such marine transgressive surfaces. Particularly distinctive facies occur in arid climates, where fan aprons may contain sabkha evaporites. Wave-modified shorelines and near-shore carbonate build-ups should provide characteristic markers of fault-generated transgression followed by renewed progradation. Examples of such fans are commonplace around the periphery of the Red Sea and in the Gulf of Eilat (Sellwood & Netherwood 1984).

The footwall alluvial fans may border particularly steep offshore slopes leading into the marine-gulf proper. Such fan systems define the fan-delta complexes of some authors (e.g. Westcott & Etheridge 1982) and these may source important submarine-fan complexes issuing from channels cut into the submarine margins of the faulted gulf margin. Examples were inferred from the Mesozoic tilt blocks of E Greenland by Surlyk (1977, 1978) and are also known in the North Sea (Stowe *et al.* 1982). On the gulf basin floor these transverse fan complexes may interact with axial fans sourced from the axial delta front. These environments are particularly susceptible to earthquake-induced liquefaction and provide a potent and active source for slumps, debris flows and turbidity currents issuing on to the gulf floor via major submarine fans (e.g. the Gulf of Corinth, Perissoratis *et al.* 1984). A major control of fan-lobe location by submarine topography is to be expected since each of these above-mentioned re-sedimentation processes is controlled by the gravity slope. Major episodes of tilting, and internal sub-basins formed by intragraben faults will closely control sand and gravel deposition. Subtle topographic effects on the faulted basin floor will thus cause important lateral changes in turbidite thickness and 'proximality'. Examples are revealed in Sparker surveys across the Gulf of Corinth by Brooks & Ferentinos (1984) and stacked offset fan lobes are seen in seismic reflection charts across Californian borderland fans (Graham & Bachman 1983).

Tectono-sedimentary facies model D—coastal/shelf basins with carbonate facies (Fig. 6)

Once the hinterlands around a half-graben system become more-or-less submerged and the basin becomes part of an actively extending shelf (e.g. modern Aegean Sea) the rate of clastic input may become substantially reduced. Should the graben be located in low-latitude areas of high organic productivity then a purely carbonate facies mosaic will be superimposed upon the structural topographic template discussed previously. Since carbonate production and facies distributions are strongly depth and slope dependent we expect to see marked contrasts in depositional style across a half-graben system. The degree of facies contrast will depend critically upon the relative rates of subsidence versus carbonate production rates. In the following discussion we assume that an adequate differential exists to maintain water depths.

We postulate that the footwall scarp slope will ensure that the footwall-to-hanging wall margin develops into a bypass margin (McIlreath & James 1978) with the scarp fringed by peri-platform talus and the upper slope cut by gullies feeding debris flows and density currents on to the basin floor as allodapic carbonates. More extensive debris-flow units may also form from slumps of the talus fringe, particularly when directly activated by fault motion.

The footwall will be dominated by relatively shallow-water facies such as build-ups and carbonate sand shoals, the exact facies type being dependent upon scarp-slope orientation relative to prevailing winds, and upon the magnitude of the local tidal-current vectors. One important feature that will develop is minor cyclicity related to shallowing during periodic footwall uplift, with the possibility of periodic emergence giving rise to karstic surfaces and calcrete development. The minor cycles themselves should be markedly asymmetrical, with rapid upward shallowing followed by a more gradual deepening trend as subsidence due to thermal decay occurs. The hanging wall dip slope should initially develop into a ramp-type margin (Ahr 1973) deepening (sloping) towards the footwall scarp as a result of the induced tectonic tilt. This will gradually evolve into an up-dip shelf with a rimmed margin, following the evolutionary trend outlined by Read (1982, 1984), and pass down-slope into basinal facies. The rimmed shelf margin may include reef build-ups and sand shoals and may eventually develop into a bypass

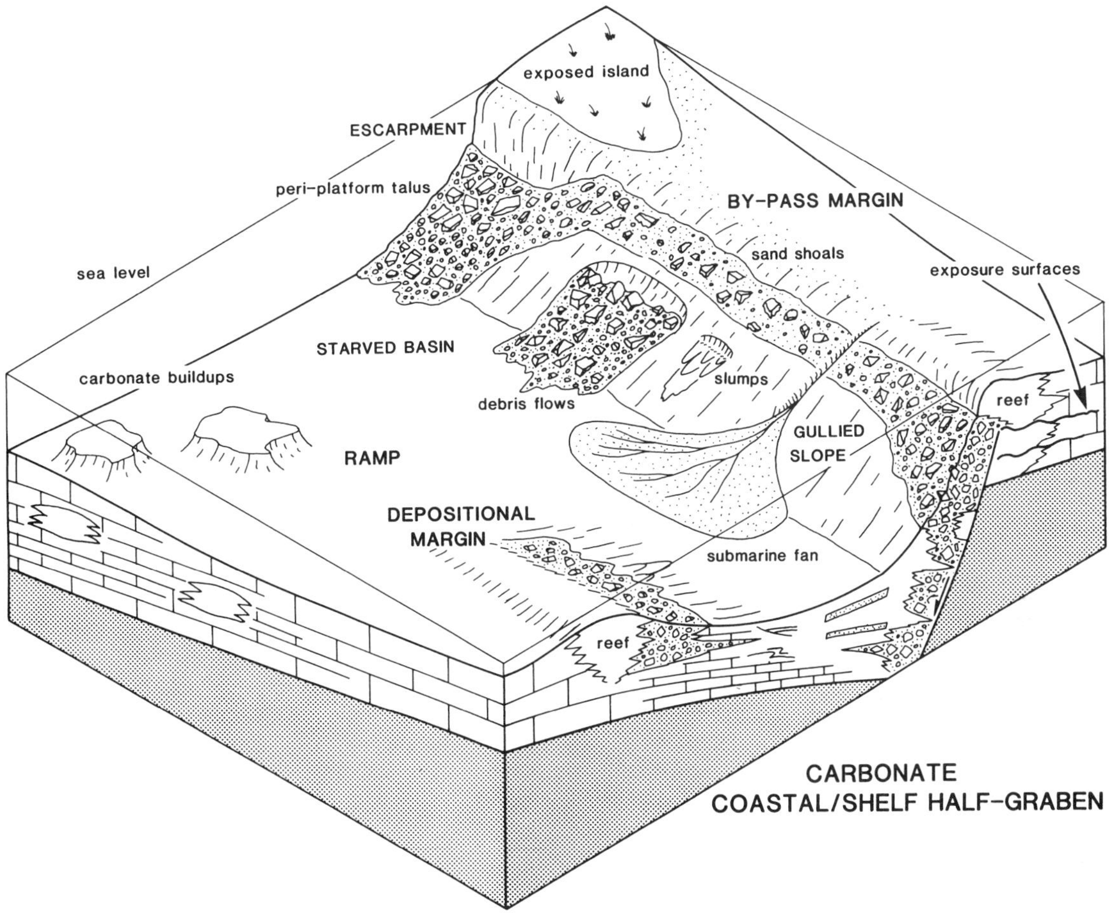

FIG. 6. Isometric diagram showing characteristic sedimentary features of facies model D: coastal/shelf basin with carbonate facies. The variety of carbonate margins associated with tilt block/half-grabens is shown; footwall scarp characterized by a bypass margin of gullied slope or escarpment type while the hanging wall may develop from a carbonate ramp into a rimmed shelf associated with a depositional slope. Full discussion in text.

margin, bordered downslope by talus fans, by now closely resembling the bypass margin on the opposite side of the basin. Thus at an advanced evolutionary stage the carbonate-shelf basin may show a symmetrical geometry, irrespective of the underlying basin geometry. The two margins have markedly different modes of origin, however; the footwall-to-hanging wall margin being predominantly controlled by tectonism and the hanging wall dip-slope margin largely the result of sedimentary processes.

The axial basin floor will in general be starved of sediment and of condensed aspect unless a distant source is available and axial transport of this sediment is possible. Clastic mud plumes may be important in such environments leading to thick mud-dominant sequences in the axial zone of carbonate-shelf basins.

Ancient examples of this type of tilt block/half-graben are present in the Lower Carboniferous (Dinantian) of central/northern England, in particular the Bowland basin and Derbyshire dome (Gawthorpe 1986, in press; Miller & Grayson 1982; Smith *et al.* in press) provide good examples.

Basin evolution and abandonment

During the development of an extensional basin it is clear that an evolutionary sequence of basin fills may develop, starting with continental-interior facies and ending up with shelf/coastal facies. Thus the various characteristic basin-fills noted above may succeed each other stratigraphically. It is becoming increasingly clear, however, following the work of Jackson *et al.* (1982a, b) and Jackson & McKenzie (1983), that entire tilt-block/half-graben systems may become inactive at any stage of their evolution due to a change in position of the major bounding crustal-scale normal-fault system. Should the new fault system propagate on the hanging wall side of the former active basin, then footwall uplift associated with the new fault may cause gradual uplift of the whole basin-fill. Such a process should lead to the occurrence of upward-shallowing trends in the highest sequences of the abandoned basin, followed by continental facies and finally by marked erosional dissection as uplift proceeds above the depositional base-line. Examples of such sequences have been observed by the authors in the footwall-uplifted basin of Alipohori in the Gulf of Corinth and in the Mygdonia area east of Thessaloniki.

Conclusions

We have tried to show how the characteristic kinematic features of extensional half-graben/tilt-block systems give rise to predictable surface effects due to the primary control on depositional processes exerted by imposed tectonic slopes. Such surface depositional controls thus exert a major influence upon resulting facies distributions in the sub-surface. We expect the general models presented above to have applicability in geologically ancient basins, but it is stressed, finally, that basin analysis must involve the tectonic analysis of fault trends and their relation to a mapped lithofacies pattern. Only an integrative study of this kind can truly reconstruct the once-active extensional basin system.

ACKNOWLEDGMENTS: M.R. Leeder wishes to thank Texaco USA and the University of Leeds for financial support. R.L. Gawthorpe wishes to thank BP Minerals International Limited for a postgraduate studentship (1982–1985). We are both indebted to numerous colleagues at Leeds University for frequent tectono-sedimentary discussions, particularly to Harry Clemmey and Jan Alexander. We are grateful to Dan McKenzie and James Jackson who provided the initial encouragement to explore the sedimentary consequences of extensional tectonics. We thank Cath Hunt for help with the figures and typing.

References

AHR, W.M. 1973. The carbonate ramp: an alternative to the shelf model. *Trans. Gulf Coast Ass. geol. Soc.* **23**, 221–5.

ALEXANDER, J. & LEEDER, M.R. In press. Active tectonic control of alluvial architecture. *In*: FLORES, R., ETHRIDGE, F. & HARVEY, M. (eds) *Recent Developments in Fluvial Sedimentology. Spec. Publ. Soc. Econ. Palaeont. Miner.*

ANDERSON, R.E., ZOBACK, M.L. & THOMPSON, G.A. 1983. Implications of selected subsurface data on the structural form and evolution of some basins in the northern Basin & Range province, Nevada & Utah. *Bull. geol. Soc. Am.* **94**, 1055–72.

BACHMAN, G.O. & MEHNERT, H.H. 1978. New K–Ar dates and the late Pliocene to Holocene geomorphic history of the central Rio Grande region, New Mexico. *Bull. geol. Soc. Am.* **89**, 283–92.

BALLY, A.W. 1982. Musings over sedimentary basin evolution. *Phil. Trans. R. Soc. London*, **A305**, 325–37.

BARTON, P. & WOOD, R. 1984. Tectonic evolution of the North Sea basin: crustal stretching and subsidence. *Geophys. J.R. astron. Soc.* **79**, 987–1022.

BOTT, M.H.P. 1976. Formation of sedimentary basins of graben type by extension of the continental crust. *Tectonophysics*, **36**, 77–86.

BRIDGE, J.S. & LEEDER, M.R. 1979. A simulation model of alluvial stratigraphy. *Sedimentology*, **26**, 617–44.

BROOKS, M. & FERENTINOS, G. 1984. Tectonics and sedimentation in the Gulf of Corinth and the Zakynthos and Kefallania channels, Western Greece. *Tectonophysics*, **101**, 25–54.

BROWN, L.D., KRUMHANSL, P.A., CHAPIN, C.E., SANFORD, A.R., COOK, F.A., KAUFMAN, S., OLIVER, J.E. & SCHILT, F.S. 1979. COCORP seismic reflection studies of the Rio Grande rift. *In*: RIEKER, R.E. (ed.) *Rio Grande rift – tectonics & magmatism*, pp. 169–84. Am. geophys. Union.

BROWNE, S.E. & FAIRHEAD, J.D. 1983. Gravity study of the Central African Rift System: a model of continental disruption 1. The Ngaoundere and Abu Gabra Rifts. *Tectonophysics*, **94**, 187–203.

BRUN, J-P, & CHOUKROUNE, P. 1983. Normal faulting, block tilting, and decollement in a stretched crust. *Tectonics*, **2**, 345–56.

BRUNNACKER, K. & BOENIGK, W. 1983. The Rhine Valley between the Neuwied Basin and the Lower Rhenish Embayment. *In*: FUCHS, K., VON GEHLEN, K., HALZER, H., MURAWSKI, H. & SEMMEL, A. (eds) *Plateau Uplift*, pp. 62–72. Springer-Verlag, Berlin.

CHAMBERLIN, R.M. 1983. Cenozoic domino-style crustal extension in the Lemitar Mountains. *In*: CHAPIN, C. (ed.) *Socorro Region II*, pp. 111–18. New Mex. geol. Soc.

CHAPIN, C.E. 1979. Evolution of the Rio Grande rift – a summary. *In*: RIECKER, R.E. (ed.) *Rio Grande rift – tectonics & magmatism*, pp. 1–5. Am. geophys. Union.

CHRISTIE, P.A.F. & SCLATER, J.G. 1980. An extensional origin for the Buchan and Witchground graben in the North Sea. *Nature, London*, **283**, 729–32.

CROSSLEY, R. 1984. Controls of sedimentation in the Malawi Rift Valley, Central Africa. *Sediment. Geol.* **40**, 33–50.

DEWEY, J.F. 1982. Plate tectonics and the evolution of the British Isles. *J. geol. Soc. London*, **139**, 371–412.

EATON, G.P. 1982. The Basin and Range province: origin and tectonic significance. *Ann. Rev. Earth planet. Sci.* **10**, 409–40.

FRASER, G.D., WITKIND, I.J. & NELSON, W.H. 1964. A geological interpretation of the epicentral area - the dual-basin concept. *Prof. Pap. U.S. geol. Surv.* **435**, 99–106.

GAWTHORPE, R.L. 1986. Sedimentation during carbonate ramp-to-slope evolution in a tectonically active area: Bowland Basin (Dinantian), N. England. *Sedimentology*, **33**, 185–206.

—— In press. Tectono-sedimentary evolution of the Bowland Basin, northern England, during the Dinantian. *J. geol. Soc. London.*

GIBBS, A.D. 1983. Balanced cross-section construction from seismic sections in areas of extensional tectonics. *J. struct. Geol.* **5**, 153–60.

—— 1984. Structural evolution of extensional basin margins. *J. geol. Soc. London*, **141**, 609–20.

GRAHAM, S.A. & BACHMAN, S.B. 1983. Structural controls on submarine-fan geometry and internal architecture: Upper La Jolla Fan System, offshore S. California. *Bull. Am. Assoc. Pet. Geol.* **67**, 83–96.

HANKS, T.C., BUCKNAM, R.C., LAJOIE, K.R. & WALLACE, R.E. 1984. Modification of wave-cut and fault-controlled landforms. *J. geophys. Res.* **89**, 5771–90.

HAWLEY, J.W., BACHMAN, G.O. & MANLEY, K. 1976 Quaternary stratigraphy in the Basin and Range and Great Plains provinces, New Mexico and Western Texas. *In*: MAHANEY, W.C. (ed.) *Quarternary stratigraphy of North America*, pp. 235–74. Dowden, Hutchinson & Ross.

HEISKANEN, W.A. & VENING MEINESZ, F.A. 1958. *The Earth and its Gravity Field.* McGraw-Hill, New York.

HOOKE, R. LEB. 1972. Geomorphic evidence for late-Wisconsian and Holocene tectonic deformation, Death Valley, California. *Bull. geol. Soc. Am.* **83**, 2073–98.

HUNT, C.B. & MABEY, D.R. 1966. Stratigraphy and structure, Death Valley, California. *Prop. Pap. U.S. geol. Surv. A494.*

HUTCHINSON, R.W. & ENGELS, G.G. 1970. Tectonic significance of regional geology and evaporite lithofacies in NE Ethiopia. *Phil. Trans. R. Soc. London*, **267**, 313–29.

ILLES, J.H. & FUCHS, K. 1983. Plateau uplift of the Rhenish Massif. *In*: FUCHS, K., VONGEHLEN, K., MALZER, H., MURAWSKI, H. & SEMMEL, A. (eds) *Plateau Uplift*, pp. 62–72. Springer-Verlag, Berlin.

JACKSON, J.A., KING, G. & VITA-FINZI, C. 1982a. The neotectonics of the Aegean: an alternative view. *Earth planet. Sci. Lett.* **61**, 303–18.

—— & MCKENZIE, D.P. 1983. The geometric evolution of normal fault systems. *J. struct. Geol.* **5**, 471–82.

—— , GAGNEPAIN, J., HOUSEMAN, G., KING, G.C.P., PAPDAIMITRIOU, P., SOUFLERIS, C. & VIRIEUX, J. 1982b. Seismicity, normal faulting, and the geomorphological development of the Gulf of Corinth (Greece): the Corinth earthquakes of February and March 1981. *Earth planet. Sci. Lett.* **57**, 377–97.

KELLEY, V.C. 1977. *Geology of the Albuquerque Basin. Mem. 33*, New Mex. Bur. Mines Mineral Res.

LEEDER, M.R. 1976. Palaeogeographic significance of pedogenic carbonates in the topmost Upper Old Red Sandstone of the Scottish Border Basin. *Geol. J.* **11**, 21–8.

—— 1982. Upper Palaeozoic basins of the British Isles - Caledonide inheritance versus Hercynian plate margin processes. *J. geol. Soc. London,* **139**, 479–91.

—— 1983. Lithospheric stretching and North Sea Jurassic clastic sourcelands. *Nature, Lond.* **305**, 510–13.

—— 1986. Tectonic and palaeogeographic models for Lower Carboniferous Europe. *In*: MILLER, J., ADAMS, A.E. & WRIGHT, V.P. (eds) *European Dinantian Environments*, pp. 1–20. Wiley, London.

—— In press. Sediment deformation structures and the palaeotectonic analysis of ancient extensional sedimentary basis. *In*: JONES, N.E. & PRESTON, R.M.F. (eds) *Deformation of sediments and sedimentary rocks. Spec. Publ. geol. Soc. London.*

—— & ALEXANDER, J. In press. The origin and tectonic significance of asymmetrical meander belts. *Sedimentology.*

—— & STRUDWICK, A.E. 1986. Delta-marine interactions: a discussion of sedimentary models for Yoredale-type cyclicity in the Dinantian of N England. *In*: MILLER, J., ADAMS, A.E. & WRIGHT, V.P. (eds) European Dinantian Environments, pp. 115–30. Wiley, London.

LIPMAN, P.W. & MEHNERT, H.H. 1979. The Taos plateau volcanic field, northern Rio Grande rift, New Mexico. *In*: RIECKER, R.E. (ed.) *Rio Grande rift–tectonics & magmatism*, pp. 289–311. American geophysical Union.

MCILREATH, I.A. & JAMES, N.P. 1978. Facies Models 13, Carbonate slopes. *Geoscience Can.*, **5**, 188–99.

MCKENZIE, D.P. 1978. Some remarks on the development of sedimentary basins. *Earth planet. Sci. Lett.* **40**, 25–32.

MANLEY, K. 1979. Stratigraphy and structure of the Espanola basin, Rio Grande rift, New Mexico. *In*: RIECKER, R.E. (ed.) *Rio Grande rift - tectonics & magmatism*, pp. 71–86. American geophysical Union.

MIKE, K. 1975. Utilisation of the analysis of ancient river beds for the detection of Holocene crustal movements. *Tectonophysics*, **29**, 359–68.

MILLER, J. & GRAYSON, R.F. 1982. The regional context of Waulsortian facies in N. England. *In*: BOLTON, K., LANE, H.R. & LEMONE, D.V. (eds) *Symposium on the Environmental Setting and Distribution of the Waulsortian Facies*, pp. 17–33. El Paso geol. Soc. and University of Texas at El Paso.

MORTON, W.H. & BLACK, R. 1975. Crustal attenuation in Afar. *In*: PILGER, A. & ROSSLER, A. (eds) *Afar Depression of Ethopia*, pp. 55–65. Deutsche Forschauna, Stuttgart.

MYERS, W.B. & HAMILTON, W. 1964. Deformation associated with the Hebgen Lake earthquake of August 17, 1959. *Prof. Pap. U.S. geol. Surv.* **435**, 55–98.

PERISSORATIS, C., MITROPOULOS, D. & ANGELOPOULOS, I. 1984. The role of earthquakes in inducing sediment mass movement in the eastern Korinthiakos Gulf. An example from the February 24-March 4, 1981 activity. *Mar. Geol.* **55**, 35–45.

PROFFETT JR, J.M. 1977. Cenozoic geology of the Yerington district, Nevada, and implications for the nature and origin of Basin and Range faulting. *Bull. geol. Soc. Am.* **88**, 247–66.

READ, J.F. 1982. Carbonate platforms of passive (extensional) continental margins: types, characteristics and evolution. *Tectonophysics*, **81**, 195–212.

—— 1984. Carbonate platform facies models. *Bull. Am. Ass. Pet. Geol.* **69**, 1–21.

REILINGER, R.E., BROWN, L.D. & OLIVER, J.E. 1979. Recent vertical crustal movements from levelling observations in the vicinity of the Rio Grande rift. *In*: RIECKER, R.E. (ed.) *Rio Grande rift – tectonics and magmatism*, pp. 223–36. Am. geophys. Union.

—— OLIVER, J., BROWN, L., SANFORD, A. & BALAZS, E. 1980. New measurements of crustal doming over the Socorro magma body, New Mexico. *Geology*, **8**, 291–5.

RUSS, D.P. 1982. Style and significance of surface deformation in the vicinity of New Madrid, Missouri. *Prof. Pap. U.S. geol. Surv.* **H1236**, 95–114.

RUSSELL, R.J. 1954. Alluvial morphology of Anatolian rivers. *Ass. Am. Geog. Ann.* **44**, 363–91.

SANFORD, A.R., MOTT, R.P., SHULESKI, P.J., RINEHART, E.J., CARAVELLA, R., WARD, M. & WALLACE, T.C. 1977. Geophysical evidence for a magma body in the crust in the vicinity of Socorro, New Mexico. *In*: HEACOCK, J. (ed.) *The Earths Crust, Am. geophys. Union Mon.* **20**, 385–403.

SAVAGE, J.C. & HASTIE, L.M. 1966. Surface deformation associated with dip-slip faulting. *J. geophys. Res.* **71**, 4897–904.

SELLWOOD, B.W. & NETHERWOOD, R.E. 1984. Facies evolution in the Gulf of Suez area: Sedimentation history as an indicator of rift initiation and development. *Mod. Geol.* **9**, 43–69.

SMITH, K., SMITH, N.J.P. & HOLLIDAY, D.W. In press. The deep structure of Derbyshire. *Geol. J.*

SMITH, R.B. & BRUHN, R.L. 1984. Intraplate extensional tectonics of the eastern Basin–Range: Inferences on structural style from seismic reflection data, regional tectonics and thermal-mechanical models of brittle-ductile deformation. *J. geophys. Res.* **89**, 5733–62.

STOWE, D.A.V., BISHOP, C.D. & MILLS, S.J. 1982. Sedimentology of the Brae Oil Field, North Sea: fan models & controls. *J. Pet. Geol.* **5**, 129–48.

SURLYK, F. 1977. Stratigraphy, tectonics and palaeogeography of the Jurassic sediments of the areas north of Kong Oscars Fjord, E. Greenland. *Bull. Grønland. Geol. Unders.* **123**, 56pp.

—— 1978. Submarine fan sedimentation along fault scarps on tilted fault blocks (Jurassic–Cretaceous boundary, E. Greenland). *Bull. Grønland. Geol. Unders.* **128**, 108pp.

—— & CLEMMENSEN, L.B. 1983. Rift propagation and eustacy as controlling factors during Jurassic inshore and shelf sedimentation in northern East Greenland. *Sediment. Geol.* **34**, 119–43.

TALBOT, M.R. & WILLIAMS, M.A.S. 1979. Cyclic alluvial fan sedimentation on the flanks of fixed dunes, Janjari, Central Niger. *Catena,* **6**, 43–62.

TEICHMULLER, M. & TEICHMULLER, R. 1968. Cainozoic and Mesozoic coal deposits of Germany. *In*: MURCHISON, D. & WESTOLL, T.S. (eds) *Coal and Coal-bearing Strata*, pp. 347–80. Oliver & Boyd, Edinburgh.

WERNICKE, B. & BURCHFIEL, B.C. 1982. Modes of extensional tectonics. *J. struct. Geol.* **4**, 105–15.

WESTCOTT, W.A. & ETHERIDGE, F.G. 1982. Bathymetry and sediment dispersal dynamics along the Yallahs fan delta front, Jamaica. *Mar. Geol.* **46**, 245–60.

ZIEGLER, P.A. 1982. *Geological Atlas of Western and Central Europe.* Shell Internationale Petroleum Mij. BV, The Hague.

M.R. LEEDER & R.L. GAWTHORPE, Department of Earth Sciences, The University, Leeds LS2 9JT, UK.

Sedimentation and tectonic geomorphology adjacent to major active and inactive normal faults, southern Greece

M. R. LEEDER, M. J. SEGER[1] & C. P. STARK

Department of Earth Sciences, University of Leeds, Leeds LS2 9JT, UK

[1] Present address: Department of Geology, University of Wales College of Cardiff, PO Box 914, Cardiff CF1 3YE, UK

Abstract: The active Psatha–Skinos normal fault is uplifting a contrasting variety of lithologies in its footwall, including Mesozoic limestones, ophiolitic peridotites and poorly-consolidated Plio-Pleistocene sediments of the inactive Megara basin. In the latter basin, linear marginal uplands border the sedimentary fill and are interpreted as degraded and segmented scarp lines which exhibit features such as facetted spurs. *Drainage domains* are defined as groups of drainage basins sharing a common tectono-geological substrate. Contrasting domains in the footwall of the active fault have given rise to spectacular differences in drainage basin growth and form. In the uplifting footwall, raised coastal terrace levels are present at about 10 and 40 m. The latter have yielded corals dated at between 90–126 Ka, and have enabled us to calculate uplift and erosional rates. Extrapolation of these rates indicate that some 350 m of footwall relief has been produced over approximately 1 Ma. Most of the sediment eroded from the uplifting footwall is presumed to have been re-sedimented into the Alkyonides Gulf basin plain during this time. The prominent footwall unconformity so-produced forms a fine modern analogue to features seen in ancient syn-rift tilt blocks such as the Viking Graben. Footwall uplift has also backtilted the former depositional surface of the Megara Basin and caused reversal of flow in its axial drainage domain. A variety of depositional environments are documented in the subsiding hangingwall to the coastal fault, including syn-quake talus cones, alluvial fans, intra-fan marshes and swamps, truncated fans, transgressive barriers and spits.

In the last few years geomorphological and sedimentological criteria have been used to investigate past earthquake frequencies and to assess the probability of future events (e.g. Wallace 1977; Sieh 1984; Hanks *et al.* 1984). The distribution of sedimentary facies in active basins also depends closely upon the nature and degree of tilting and relative uplift produced during surface deformation (e.g. Alexander & Leeder 1987; Leeder & Gawthorpe 1987). There is thus a close relationship between tectonics, geomorphology and sedimentology in areas of active faulting and it is this link that we investigate in this study.

The Aegean Sea and its surrounding areas represents a superb natural laboratory for the study of the surface effects of rapid continental crustal extension. It has recently been estimated (Jackson & McKenzie 1988) that the whole area, from Crete to Bulgaria, is stretching at a maximum mean rate of some 50 mm a^{-1} for the last 5 Ma. During this time the area has been broken up into a series of tilted blocks by major normal faults which have rotated blocks about both vertical and horizontal axes (LePichon & Angelier 1981; McKenzie & Jackson 1983, 1986; Kissell *et al.* 1986; Kissel & Laj 1988). The uplands and basins produced by such rotations yield important information as to the rates of sedimentary and tectonic processes involved.

We present here the results of a study of the faulted margins of the active Alkyonides Gulf and the inactive Megara basin, the latter previously active as a depocentre during the Plio-Pleistocene. The study area (Fig. 1) also includes a portion of the northern margin of the Gerania Mountains, a prominent E–W topographical feature which at present bounds the Alkyonides Gulf along an active fault scarp. The study area was specifically chosen because: (a) the major active normal fault has caused both uplifting and subsiding coastal areas to occur; making it possible to analyse the effects of both uplift and subsidence upon geomorphic and sedimentary systems; (b) a coastal location was considered important since sea level provides a useful datum plane, and because recent advances in our understanding of Pleistocene sea level variations has enabled progress to be made in assessing the relative contributing effects of eustasy, subsidence and uplift (e.g. Keraudren & Sorel 1987; Collier 1990; Collier *et al.* 1990).

The position of some coastal features, including raised beaches, marshes and 'solution' notches, were used by Jackson *et al.* (1982) in their study of the effects of faulting after the 1981 earthquakes which affected the study area. The discovery and radiometric dating of corals in a raised marine terrace has enabled us to estimate the rate of fault footwall uplift during the late Pleistocene.

The effects of tilting on the development of the modern drainage networks in the area was first noted by Mariolakis & Papanikolaou (1982). In the present study, drainage basins have been delimited and analysed with reference to the geological substrate and to the local tectonic environment. We introduce the term *drainage domain* for areas which contain common elements of tectonics and geology.

In addition we have undertaken studies of the margins to the Plio-Pleistocene depocentre of the Megara basin, which we deduce to have been bounded by normal faults during its depositional history. The morphology of the degraded fault line scarps along this basin margin contrasts markedly with the active fault scarps to the north.

From Holdsworth, R.E. & Turner, J.P. (compilers) *Extensional Tectonics: Regional-scale Processes.* The Geological Society, Key Issues in Earth Sciences, **2 (1)**, 185–197.
1476-315X/02/$15.00 © The Geological Society of London 2002
First published in: "*Journal of the Geological Society*, **148**, 331–343".

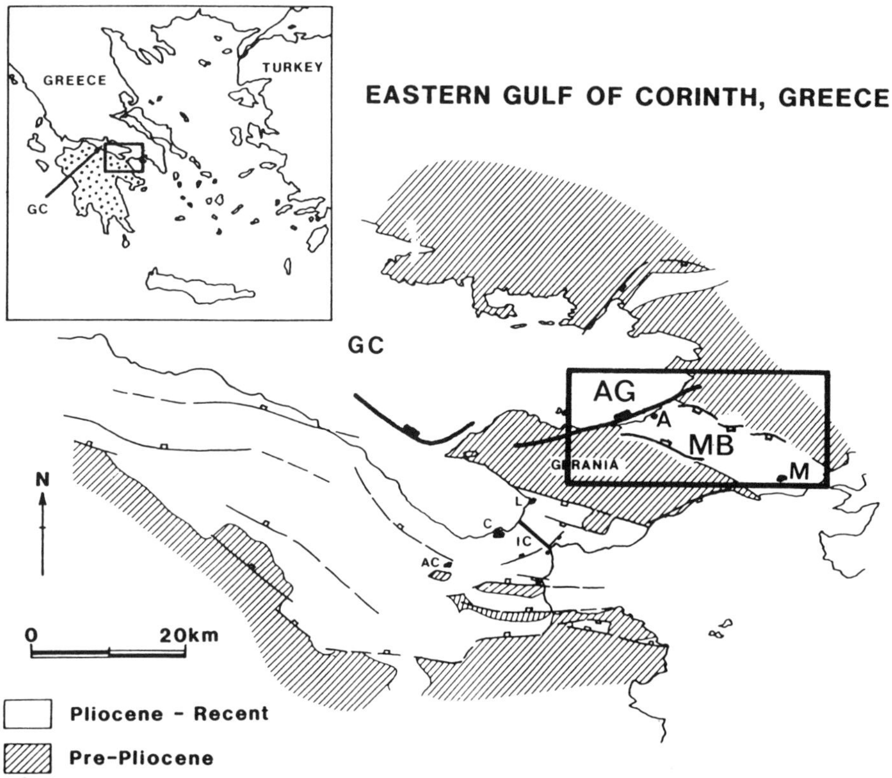

Fig. 1. Map to show the regional context of the study area and its location in relation to the Gulf of Corinth (data from Jackson *et al.* 1982; Collier 1989). The Peloponnisos are stippled on the regional map. Shown on the detailed map are the distribution of Plio-Pleistocene deposits in the abandoned and inverted basins on the southern side of the Gulfs of Corinth (GC) and Alkyonides (AG), including the Megara Basin (MB). Some of the ?inactive normal faults which affect the Corinthian and Megaran basin fills are shown, together with the two normal fault strands (heavy lines) which broke in the great earthquakes of February and March 1981. C, Corinth; AC, Ancient Corinth; L, Loutraki; IC, Isthmus of Corinth; M, Megara; A, Alepochori.

Topography, morphology and solid geology (Figs 2 & 3)

The study area is dominated by two major topographic features.

(1) The NW–SE trending axis of the Megara basin (Theodoropoulos 1968; Mariolakis & Papanikolaou 1982), a now inactive depocentre, lies between the Gerania Mountains to the west and the Pateras Mountains to the east. The altitude of the basin area generally lies below 300 m, whilst the surrounding mountains rise to heights of 1350 m (Gerania) and 1132 m (Pateras).

(2) The Alkyonides Gulf, an E–W marine basin, is bounded to the south by the major, active, Psatha–Skinos coastal fault. It rapidly descends to depths of over 400 m below sea level.

As demonstrated later in this paper, much of the topographic differentiation in the study area (in excess of 1.75 km) is the result of normal faulting. This is initially evident in the strong linear trends to the contours that run NW–SE along the mountain fronts of the Gerania and Pateras massifs. The southern faulted margin to the Alkyonides basin runs approximately E–W to ENE–WSW with steep onshore and submarine gradients and cliffed coastlines where basement Mesozoic limestones and Tethyan ophiolitic peridotites crop out in the fault footwall.

The fill to the Megara basin itself comprises some 1 km+ of largely unconsolidated or poorly consolidated marls, sandstones and conglomerates of Plio-Pleistocene age. These mostly dip at shallow angles (15–25°) to the NE. A narrow coastal plain occurs for some 5 km inland from the village of Alepochori where these deposits have been eroded out of the uplifting footwall to the active Psatha–Skinos fault. Narrow coastal plains are also present in the hanging wall of this fault around the village of Skinos in the far west, and at Psatha Bay in the east of the study area. Depositional environments are restricted to the Skinos–Alepochori–Psatha coastal tract below the 100 m contour, where debris cones, alluvial fans, bays, lagoons separated by offshore bars, beachrock reefs, beach-to shoreface environments and raised beaches are major features. The location of many of these features is closely controlled by faulting.

A prominent drainage divide, in the form of an arcuate ridge, altitude 300–350 m, separates streams which drain NW into the Alkyonides Gulf from those which descend SE to the Saronic Gulf. This was first pointed out by Mariolakis & Papanikolaou (1982). As discussed below, this divide has resulted from the headward migration of drainage networks in response to uplift of the NW margin of the basin, located in the footwall of the active Psatha–Skinos fault.

In terms of basement geology, the Gerania Mountains are dominated by Mesozoic limestones upon which have been thrust large masses of peridotitic ophiolites. The southern parts of the Pateras Mountains are also mainly composed of Mesozoic limestones. It is the contrast in the weathering characteristics of the two basement lithologies which causes local differences in scarp morphology, drainage basin size and alluvial fan development. In general the most striking normal fault scarps occur in the limestones, the preservation of the fault scarps themselves being aided by the process of brecciation and fault scarp cementation described and discussed in detail by Stewart & Hancock (1988) from field areas adjacent to ours.

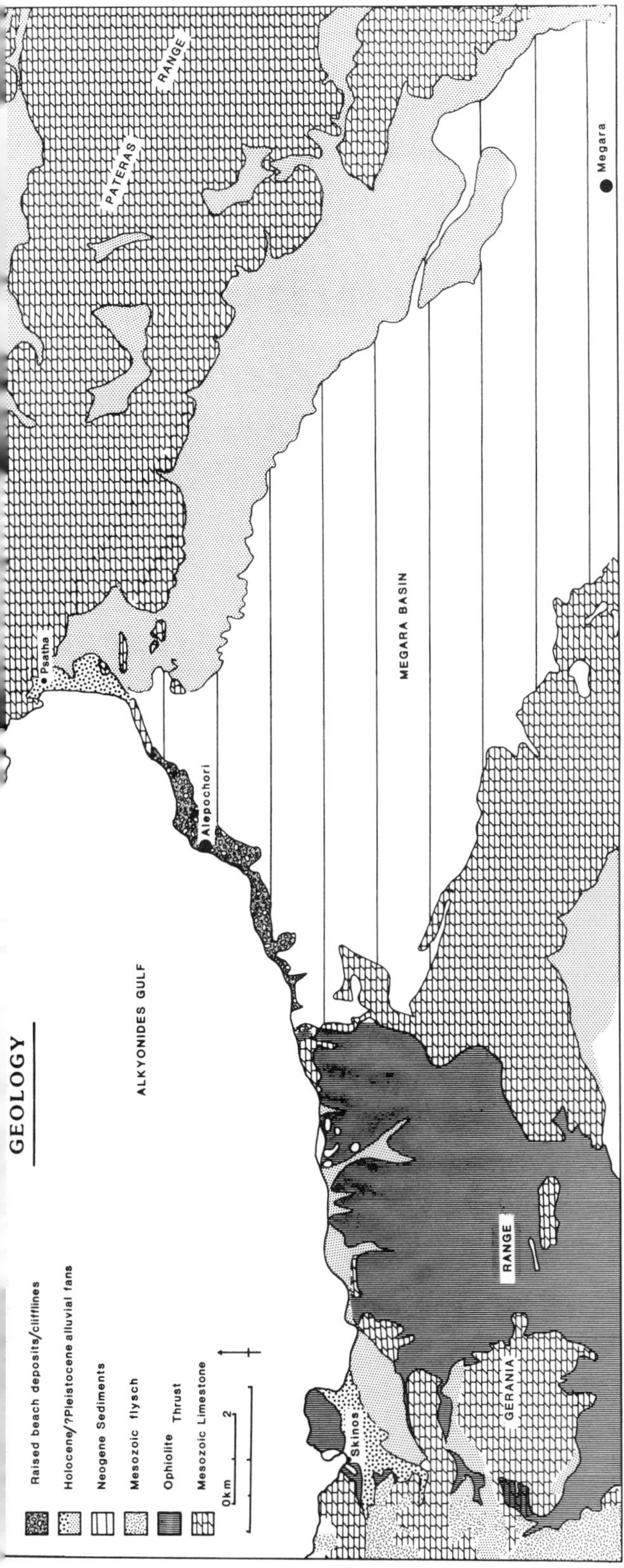

Fig. 2. Geological map of the study area (after Theodoropoulos 1968; IGME 1984). The figure should be studied in conjunction with Figs 3, 4, & 5 which show topography, structure and drainage patterns.

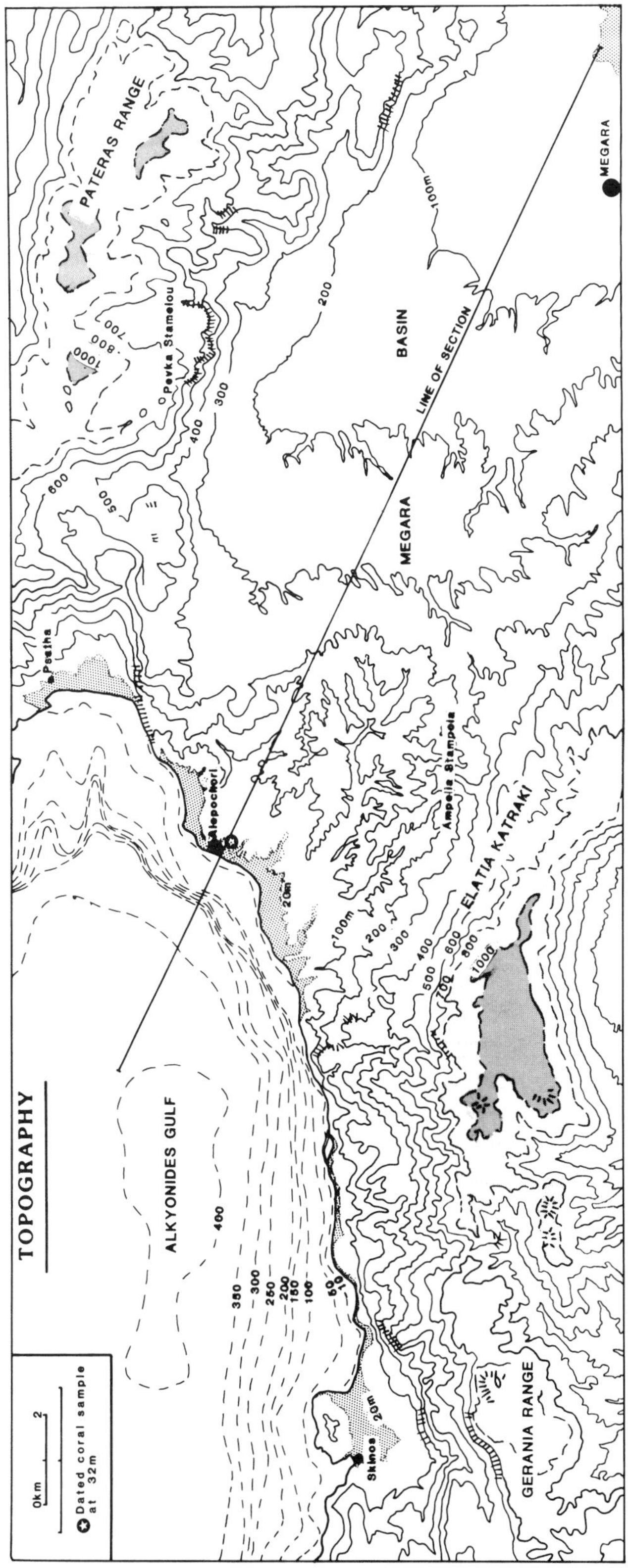

Fig. 3. Topographic map of the study area to show subaerial and generalized submarine contours and the line of the section drawn as Fig. 11 (topography from the Greek OS 1:50 000 map series for Kaparellion and Erithrai).

Geological structures, their morphology and topographic expression (Fig. 4)

The old fault line scarps bounding the Megara Basin

The NW–SE trending Pateras range front extends for some 16 km and forms the most prominent topographic boundary to the Megara basin. The front can be divided into at least three distinct segments, each more-or-less linear but with prominent valley segments causing the front to be locally highly sinuous. The segments are crudely en echelon, stepping to the right. We have found several locations along the length of the front where significant planar outcrops of E–W trending fault surfaces, interpreted as exhumed fault planes, are preserved. In some examples they occurred at the break in slope between a 25–30° dipping limestone bedrock slope and a cemented alluvial fan surface, dipping at 5–15°. Maximum relief of 3 m, towards the south in all cases, can be traced laterally to negligible relief over 100 m or so. One example is more complex, being the freshest of a sequence of fault surfaces spaced at intervals of about 10 m. The footwalls to the scarps expose fine, often multiple generations of tectonic breccias.

Another piece of evidence for the scarp being of fault-line type comes from the occurrence of at least four distinct facetted spurs along the most NW portion of the range front (Fig. 4). The triangular facets bound small drainage basins within the footwall uplands of the fault line. The facets are plano-convex and dip at between 25–30°. The most westerly facet has a spectacular gullied drainage network superimposed on its surface.

The features described above are highly diagnostic of faulted range fronts in neotectonic areas and confirm the deduction that the Pateras range front marks the position of one of the former major basin bounding faults to the Megara Basin. This deduction is supported by the observation, (Roberts & Jackson 1991) that the majority of structural dips in the Neogene fill to the Megara basin are to the NE at angles between 15–25°. This orientation, orthogonal to the strike of the Pateras fault scarp, indicates that the bed dips reflect rotations of originally horizontal strata by the fault during Plio-Pleistocene extension (see the general models developed by Barr 1987 and Jackson *et al.* 1988). The scarp is thus likely to be a few millions of years old, but we are unable to be more precise about this at the present time. The low angles of dip of the range front facets noted above reflect both erosional retreat of the original fault line *and* the 15–25° rotation that has accompanied the extensional faulting along the Plateras line during its active history.

On the eastern side of the Megara basin is another prominent range front that trends WNW–ESE for some 5 km along the base of the prominent linear ridge of Elatia Katraki (Fig. 4). The presence of facetted spurs and the linearity of the feature again strongly supports the proposal that this was also a major basin-bounding fault scarp. The topographic feature dies out to the SE of the deep gorge of Ampelia Stampola wherein the fault juxtaposes Neogene marls against Mesozoic limestones with a large, but indeterminate, throw. In the area to the SE of Ampelia Stampola there is no clear evidence for a faulted basin margin and we follow Theodoropolous (1968) in mapping a non-tectonic, onlapping relationship between the basement and the Neogene from here down to the coast SW of Megara town. At its NW limit the Elatia Katraki fault line dies out into the modern footwall of the Psatha–Skinos fault strand.

The relationship of the Plio-Pleistocene Megara basin to the two major range front scarps of Pateras and Elatia Katraki suggests that the latter was a large antithetic structure relative to the main basin-bounding Pateras Fault. Such a relationship is well seen in the 1981 fault sequence on the south and north sides of the Gulf of Corinth. Here the Pision/Skinos fault is the main structure, to which the Kaparelli Fault is antithetic (see Jackson *et al.* 1988, fig. 20; fig. 1).

The active coastal faults of Psatha/Skinos

The southern shoreline to the Alkyonides Gulf is bounded by a major series of fault scarps, several strands of which were active in the 1981 earthquake sequence (Jackson *et al.* 1982). The fault scarps in the study area are the eastern extension to the major scarps around the villages of Perachora and Pision. Behind the village of Skinos there are two sub-parallel northward-facing scarps.

The southernmost scarp was extensively reactivated in the 1981 earthquake event. Throws of up to 1.5 m were recorded, but values of 0.5–0.6 m were most commonly noted (Jackson *et al.* 1982, fig. 2). The scarp completely loses its topographic expression some 3 km SE of Skinos, coincident with a major change of basement geology from Mesozoic limestones to ophiolitic serpentinites. The 198 ruptures continued about 2 km beyond this point ending in the canyon feeding alluvial fan A (J. Jackson, 1990, pers. comm.).

The northernmost scarp may be traced along a sinuous course to the ENE before it trends offshore just seawards of a prominent sea stack, some 2 km west of the position indicated by Jackson *et al.* (1982 fig. 2). The offshore extension of the fault is clearly indicated by the very high gradients in the submarine contours which trend NE from the point where the fault leaves land. An interesting problem arises concerning the shape of the fault if we make the reasonable assumption, with Jackson *et al.* (1982), that the fault line then swings round to trend eastwards, joining

Fig. 4. Map to show the major structures in the study area. The position and trend of the active coastal fault strands is largely after Jackson *et al.* (1982), the remainder based on our own mapping and observations. Key to numbers: 1, the Skinos fault line is the easternmost portion of a structure which originates some 5 km to the west; the scarp was intermittently reactivated in 1981, with throws of 30–100 cm (Jackson *et al.* 1982, fig. 2); 2, possible submarine extension of 1, drawn along the prominent submarine scarp revealed by soundings (see Fig. 3); 3, the easternmost portion of the fault which extends from Pision, some 6 km west of the study area; as with 1, this fault suffered slip (20–150 cm) in 1981 (Jackson *et al.* 1982); 4, we cannot trace 3 beyond this point, the fault scarp appearing to die out in the peridotite outcrops which begin about here; 5, the spectacular fault scarp on the south side of Psatha Bay; we are unsure of the relationship between this and fault 1/2, there being no particular trend in the submarine soundings to indicate that the two join up offshore; 6, the degraded scarp, with facetted spurs, of Elatia Katraki; we regard this as an important antithetic fault to the main Plateras scarp (7–9) which formed the major faulted boundary to the Megara half-graben during Plio-Pleistocene times; 7–9 right-stepping segments of the Plateras range front scarp with prominent facetted spurs.

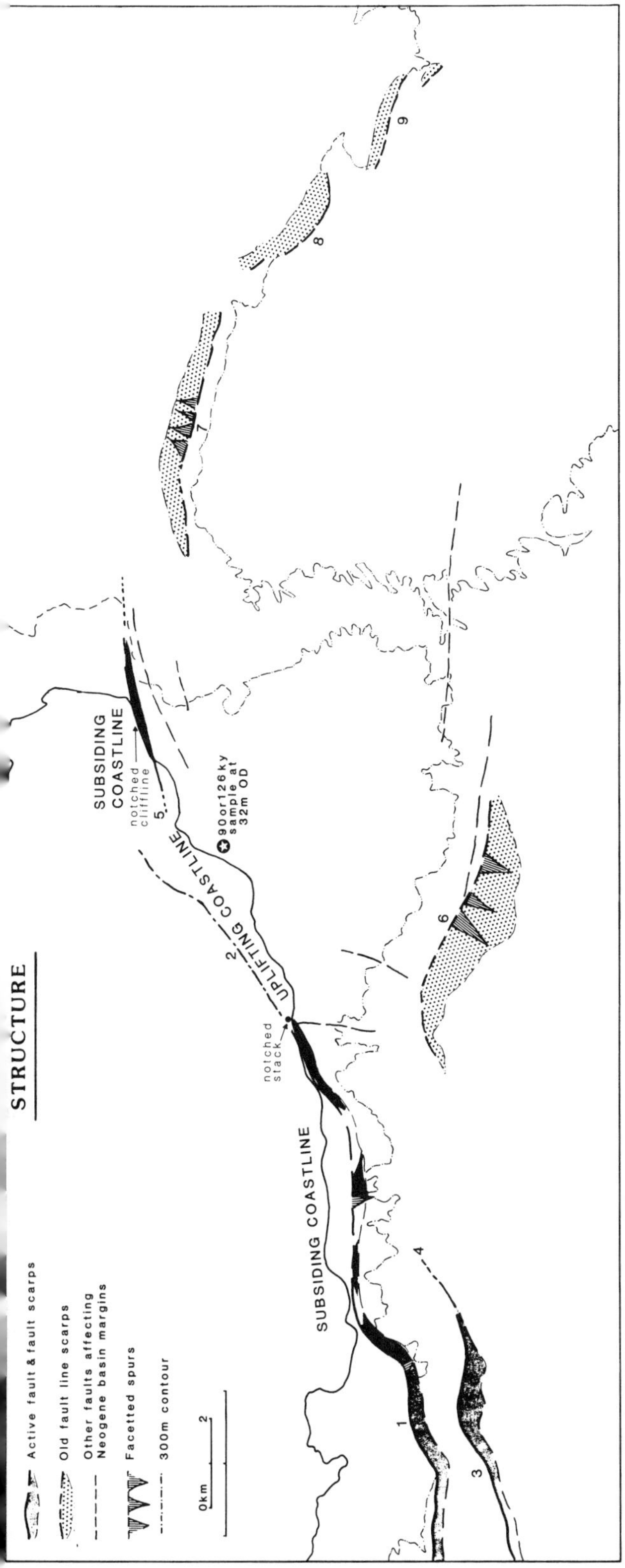

Fig. 4.

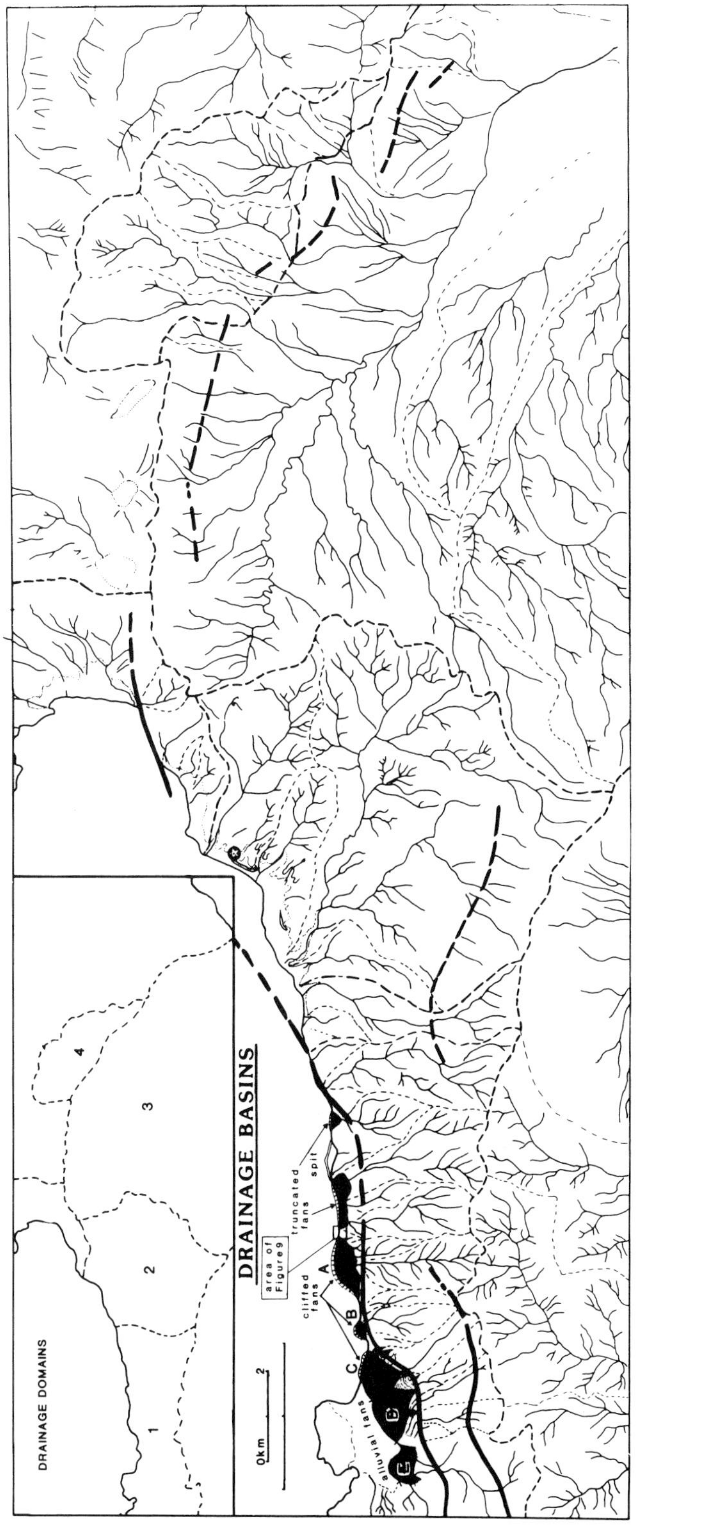

Fig. 5. Map to show the drainage courses, drainage basins (defined by light pecked lines) and faults located in the study area. The main alluvial fans (A–E; shaded black) in the subsiding hanging wall to the Skinos–Psatha fault are also shown. The inset shows the main drainage domains (heavy pecked lines in the main figure) defined in the area. The area of fan development is illustrated in the aerial photograph reproduced in Fig. 8. Fig. 10 shows a detailed sketch map and sedimentary logs from the intra-fan marshes and shoreface environments east of Fan A. At least one of the small talus cones drawn on the east side of Fan C cannot be seen in the 1944 photograph reproduced as Fig. 8. Local people say it formed during the time of the 1981 earthquake.

up with the spectacular 100 m high subaerial fault scarp at the southern end of Psatha Bay. Notwithstanding the relief of this fault footwall, exposed in Mesozoic limestone, there is no trace of the line in the offshore bathymetric contours to the southwest of its onshore outcrop. Either the limestone must die out laterally and be replaced by easily eroded Neogene sediments, or the Psatha fault must die out rapidly. The continuation of the bathymetric escarpment to the NE, past the inferred position of maximum curvature of the fault segment as it goes onshore at Psatha, implies that an important fault junction may occur in the area.

Regarding the relationship between the active coastal faults and the old fault line scarps of the Megara Basin, we note:

(1) the angular disparity in their strike, circa 50°, and
(2) their cross-cutting relationships.

It seems certain that the active faults post-date, truncate and probably de-activated the faults on the SW and NE margins of the Megara Basin. Since the age of the youngest deposits in the basin are probably Pleistocene (Theodoropolous 1968; IGME 1984), the coastal faults may be up to 2 Ma old. Further constraints (to about 1 Ma) are provided by radiometric dates from corals in an uplifted coastal terrace located in the footwall of the coastal fault and from geomorphological deductions (see below).

Drainage basins and domains (Fig. 5)

Since a number of sedimentological features, including alluvial fan size and sediment yield, depend directly upon the characteristics of drainage basins we think it important that analysis of these features be integrated with geological features. We have found it useful to define a number of *drainage domains,* defined as areas of landscape whose drainage basin characteristics such as shape, area and drainage length share a common tectonic and/or lithological substrate.

Drainage Domain 1 occupies an area of some 28.9 km^2 and includes those drainage basins sourced in basement lithologies (peridotite and limestone) in the footwall uplands of the active Pisia and Perachora–Alepochori faults. These young drainage basins, perhaps about 1 Ma old (from the evidence for uplift rates discussed below), are characteristically narrow and elongate with drainage divides at altitudes of between 800 and 1000 m, descending down to sea level in distances of between 1.5 and 5 km. Mean drainage basin area is 1.81 km^2 ($n = 16$; range 0.4–4.71) and mean principal stream length is 1.96 km (range 0.8–5.25). The larger drainage basins in this group are generally found on outcrops of less resistant peridotite basement. In the extreme west of Domain 1 the characteristic double-stepped fault scarps of the Pision and Skinos–Psatha faults occur in a mixed basement area of both limestone and peridotite. The drainage basins here are smaller than those to the east which are sourced in peridotite alone. Headward erosion of drainage divides has evidently proceeded noticeably more slowly in this area.

Drainage Domain 2, occupying an area of some 46.82 km^2, also lies in the footwall of the active Perachora–Alepochori fault. It occurs as a prominent scoop-shaped indentation SE from the coastline. Lithologies here comprise poorly consolidated Plio-Pleistocene marls, sandstones and conglomerates of the infill to the Megara basin. One very large drainage basin with markedly dendritic stream networks dominates the area. Mea drainage basin area is 9.36 km^2 ($n = 4$; range 0.98–32.7) an mean principal stream length is 4.6 km (range 1.0–11 km) On a small scale some individual stream segments are linea and are locally controlled by small and medium scal intrabasinal faults, which cut the basin fill and juxtapos strata of contrasting lithological characteristics. Mor elongate basins occur in deposits at the boundary wit Domain 1, where there is a marked structural an lithological control on the strike-parallel drainage which ha developed between more resistant ridges of conglomerate and breccias and less resistant marls and sand. The drainag divide to the domain is remarkably sharp and lies up to 6 kr inland at a height of between 300–350 m. It is obvious tha rapid headward extension of this divide has occurred (se discussion below) in response to network growth in th easily erodible substrate provided by the Plio-Pleistocen deposits. Calculations concerning the sediment budget an erosion rates from this domain are presented later in thi paper. The southeastern portion of the domain lies alon the inactive fault zone of Elatia Kaatraki, where a numbe of short stream segments occur as remnants to Domain drainage basins or have joined up with the headwar extensions of streams in Domain 2.

Drainage Domain 3 occupies a large area (>120 km^2) i the centre of the Megara basin, the drainage occurring on dipping geomorphic surface that descends gently to the SE This slope and its superimposed drainage directions ar believed to be remnants of an ancestral pattern disrupte and truncated at its headward end by uplift in the footwal of the Psatha–Skinos fault. The three large drainage basin within the domain collect runoff from both the uplands an mountains to the NE and the SW, the trunk stream course being 'axial' with respect to the structural trend of th Megara basin. The largest, most easterly, basin is floored b indurated breccias of unknown, possibly late-Pleistocene age (see below) whilst the other basins are cut int Plio-Pleistocene sands, marls and gravels.

Drainage Domain 4 occupies an area of some 21.8 km in the Mesozoic limestone of the ancient fault line scarp an footwall uplands of the Pateras Mountains, from which th basins supply their discharge to the axial fluvial systems o domain 3. As noted previously these basins may be severa million years old, perhaps being initiated during an earl phase of extension in the Central Greece. The individua basins are markedly elongate and are orientated normal t the line of the fault-controlled gradients of the Pateras rang front. Mean drainage basin area is 3.11 km^2 ($n = 7$; rang 1.1–4.9) and mean principal stream length is 2.9 km (rang 1.0–4.8). Although the basins were clearly initiated an have subsequently evolved in the footwall uplands to th Pateras fault, they differ in some respects from those basin of Domain 1 that occupy the footwall to the activ Perachora/Alepochori fault. Thus the basins are longer wider and contain the beginnings of propagating network of dendritic type in their headlands. The drainage is thu more 'evolved' in this ancient, and now inactive faul compared to the relatively young and active coastal faul bordering Domain 1.

Other areas of drainage

In addition to the major domains of drainage describe above there occur small areas of local drainage related t

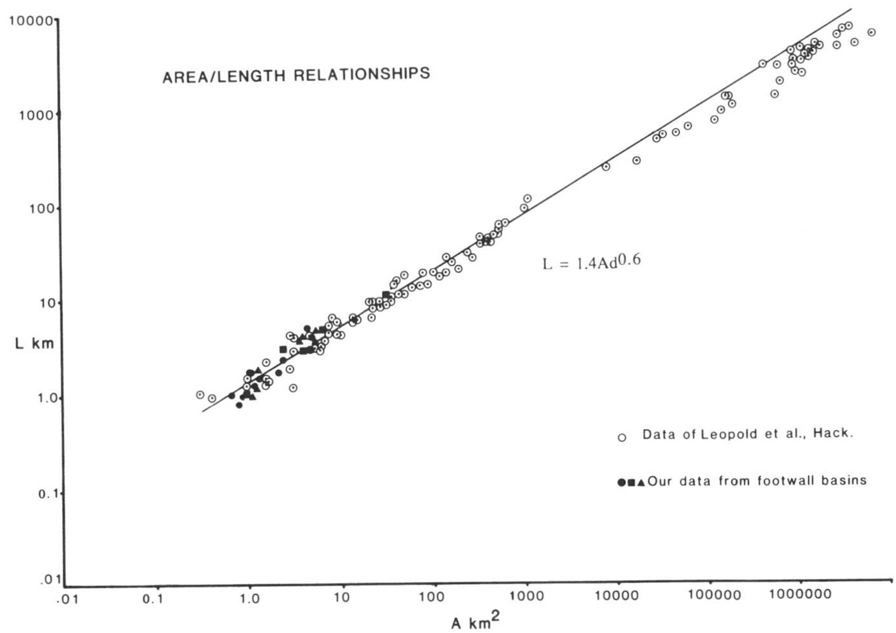

Fig. 6. Graph to show the relationship between drainage basin area (A) and principal stream length (L) established by Hack (1957) and Leopold *et al.* (1964). Also shown are the data from drainage domains 1, 2 and 4 in the present study area.

the course of the active coastal fault. Thus around Psatha bay numerous small drainage basins occur in the hangingwall to the fault, draining the area between the fault and the area of uplands behind the watersheds to Domain 3. Spectacular 'blind' upland depressions of unknown origins occur on trend with the eastern termination of this fault (Figs 4 & 5). Also, minor normal faulting around Skinos headland has created uplands of peridotite that are being drained by small streams oriented parallel to the main trace of the active fault line.

Discussion: drainage basin area/length relationships

From the discussion above it is clear that drainage basin development has proceeded in response to the production of relief and tectonic gradients by normal faulting. The contrasting areas of footwall drainage basins in Domains 1, 2 and 4 reflect both the tectonic environment and the geological substrate. It is known from 'classic' geomorphological investigations of drainage basins (Hack 1957; Leopold *et al.* 1964) that drainage area (A) and drainage basin length, the latter measured as the length of the principal stream course (L), are related by the equation:

$$L = 1.4A^{0.6}$$

As shown in Fig. 6, this relation holds for drainage basin areas over seven orders of magnitude. The relation is of great importance in tectonic geomorphology since the potential length, and hence the area, of drainage down tectonically-produced slopes is strictly controlled by fault spacing, tilting and the nature of the substrate. As shown in Fig. 6, our data from domains 1, 2 and 4 plot well on the trend of the above equation, leading to the hope that future modelling studies of sediment yields from tectonic drainage basins will be aided by the use of this simple scaling law. For example, it is clear that some of the differences in sediment yields and alluvial fan areas sourced in hangingwall and footwall environments (Leeder & Gawthorpe 1987) are explained by the above equation.

The actively uplifting footwall to the Psatha-Skinos Fault (Figs 4 & 5)

Alepochori coastal plain, raised beaches and submarine platform

The crucial feature here is the presence of the easily eroded Plio-Pleistocene sedimentary succession of the Megara basin in the fault footwall. As already discussed the large drainage basins of Domain 2 have evolved in this portion of the footwall since uplift began. The three major drainage outlets transfer sediment into the Alepochori coastal tract where a narrow coastal plain and shoreface environment is present. The construction of this feature has been episodic since the depositional remnants of at least three raised cliffline/beachface deposits and associated alluvial terraces may be mapped inland from the present day coastline (Herforth *et al.* 1971; Kelletat *et al.* 1976; Jackson *et al.* 1982; IGME 1984). Thus the present day coastline in this area has a 'staircase' morphology that is so characteristic of uplifting coastlines (Lajoie 1986).

The highest raised marine deposits (M3) occur between heights of about 25–35 m above sea level. Exposures are generally very poor but in one ravine abundant fragmented scleractinian corals of the genus *Acropora* were discovered (Karen Knapton 1989) in a well-sorted sand below the top of the scarp feature at a height of 32 m above sea level. Abraded and unabraded picks from a large sample of these corals have yielded U-Th dates of 90 ± 4 ka and 126 ± 6 ka respectively (T. Atkinson & P. Rowe pers. comm. 1989). Deposition of the shoreline deposits was evidently broadly coincident with the major pre-Flandrian interglacial (EuTyrrenian) sea level high *5*, as defined from deep sea oxygen isotope stratigraphy. Adopting 126 ka as a maximum age, coincident with isotope stage *5.5.5*, we estimate the approximate minimum rate of uplift in the area as about 0.3 mm a^{-1}.

The second raised beach deposit (M2) is undated at present and lies between 6.5–13 m above sea level. It

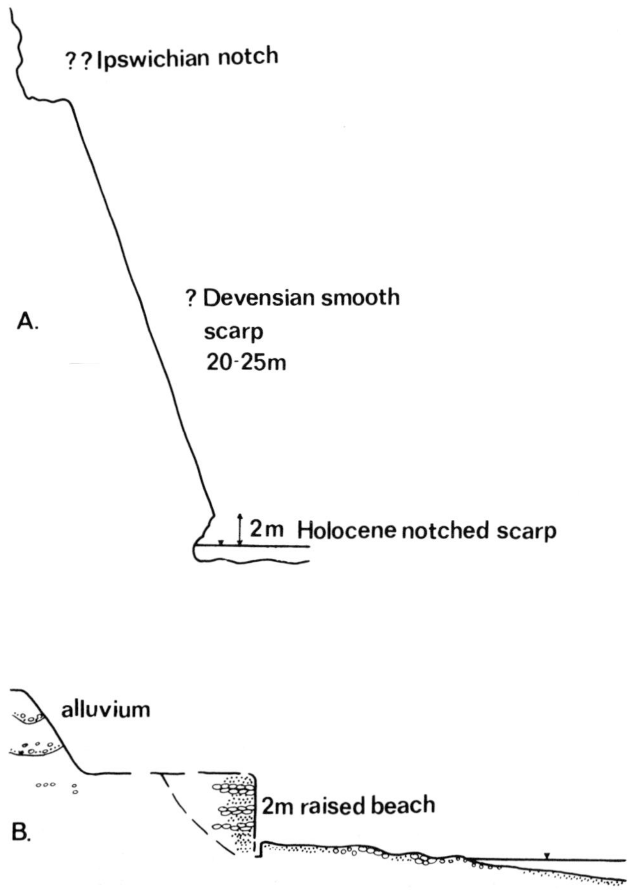

Fig. 7. Uplift features of the post-Flandrian highstand. (**A**) Sketch section through the Psatha scarp and its notched base. (**B**) Sketch section through the 2 m raised beach deposits at Alepochori. Both features yield footwall uplift rates of around 0.3 mm a^{-1}.

defines a prominent scarp and terrace with exposures revealing well-bedded pebbly sandstones and gravels arranged in 5–30 cm units which dip gently and uniformly seawards at angles of between 3–6°. This remnant constructional deposit is onlapped seawards by alluvium.

The lowest raised marine deposits (M1) comprise beachface deposits up to 2 m above sea level. Up to three well-developed levels of imbricated storm berm gravels occur, separated by sandy and shelly horizons of upper beachface/washover origins (Fig. 7B). This sequence evidently represents periods of rapid beachface deposition and progradation followed by uplift, and must have accumulated since the Flandarin sea level highstand some 7 ka ago. In addition to these largely uncemented raised beach deposits, numerous examples of formerly subtidal beachrock occur in the present day beach zone. These comprise well-cemented gravels in 20–30 cm seaward-dipping beds with pitted and bored upper surfaces.

The Holocene beachface deposits offlap an older sequence of poorly sorted, matrix-rich pebbly alluvium which defines terrace A1 (Fig. 7B). It is evident that the first coastal deposits of the most recent highstand of sea level were preceded by an important phase of valley alluviation and coastal plain deposition causing the production of an alluvial terrace in front of the older deposits of M2. This major constructional event was succeeded by renewed valley downcutting, a process whic has continued to the present day as witnessed by th presence of common unpaired terraces cut in up to 15 m c alluvium within the valleys Domain 2. We await the result of ^{14}C dating of wood fragments found in these deposits.

Raised coastal notches

Prominent 'notches' cut into limestone cliffs (see review i Trenhaile 1987) are a prominent feature of man Mediterranean coastlines. They are attributed to biophysica abrasion and biochemical solution by a variety of intertida and shallow subtidal molluscs. Physico-chemical solutio may also play a minor role. Fine examples occur in ou study area on a prominent seastack located just offshore i the footwall of the Psatha–Skinos fault and all along th spectacular footwall scarp at Psatha (Fig. 7A). The chie organisms responsible for notch cutting at the latter localit appear to be the rasping limpet *Patella,* although man other forms contribute, including shallow subtidal echino derms, minute intertidal grazing gastropods and barnacles It is curious to note that in both localities noted above vacated *Lithophaga* borings in the notches are extremel rare, although they can be very common elsewhere in th littoral zones of the Corinth basin.

It is evident that the notches have been uplifted, the li of the highest at Psatha being some 2 m above sea level (Fig 7A). At least 3 subsidiary notches occur at this locality. W interpret the sequence at Psatha as follows. The smootl fault scarp and its thick fault-zone brecciated layer i inferred to have been progressively uplifted as a subaeria feature during the interval between the Ipswichian anc Flandrian highstands of sea level (*c.* 120–7 ka). The coasta notch profile was begun to be cut in the Flandriar highstand. For a total uplift of some 2.0 m in the last 7 ka we obtain a mean uplifting rate of around 0.3 mm a^{-1}, a remarkably similar figure to that obtained from the longe term rate derived from the raised marine deposits a Alepochori. We cannot say at present whether (1) the uplif occurred gradually, (2) several large magnitude earthquake were responsible or (3) a combination of (1) and (2) occurred. It is not known whether the notch at Psatha was uplifted by the 1981 earthquake event.

The subsiding coastal hangingwall to the Psatha–Skinos fault (Figs 5, 7 & 8)

The subsiding coastal hangingwall to the active Skinos–Psatha fault comprises a number of alluvial fans, discontinuous narrow and gently sloping coastal plains, a cliffed bajada, talus cones, coastal lagoons, marshlands and beach/barrier spit and beach shorelines (Fig. 5). These various environments thus comprise a distinctive tectono-sedimentary facies association in the sense of Leeder & Gawthorpe (1987). Sediment is supplied to the coastal tract by the streams that drain the peridotitic and minor limestone hinterlands of Domain 1 discussed above and by reworking of the cliffed toes of the alluvial fans by wave action. As discussed below it is likely that this system has been operative for about the last 1 Ma or so.

Alluvial fans and bajada

Of the 16 drainage basins that comprise Domain 1, the five most westerly examples have alluvial fans at their

Fig. 8. Aerial photograph of the western part of the study area (taken in 1944) to illustrate the form of the prominent alluvial fans issuing from the footwall hinterlands of drainage domain 1 (see also Fig. 5). Note in particular the prominent incised fan channel and its depositional lobe on fan A. Certain talus cones thought to have been caused by motions of the 1981 earthquake sequence are not seen in this photo (see Fig. 5).

downstream terminations. Several of the remainder have alluviated valleys cut abruptly by modern sea cliffs. The five alluvial fans have their apices at the fault scarp, the surface of each one being incised to a greater or lesser degree. Above the intersection of the fault scarp with the surface of fan D, three prominent knick points are seen in the long profile of the channel floor. These are tentatively ascribed to the increased gradients produced by ?periodic footwall uplift.

Those fans that occur at the coastline are cliffed, with vertical faces up to 17 m high (Fig. 9). Levelling surveys along the fan surfaces indicate that substantial coastal retreat has occurred, a maximum of 65 m for Fan A, 37 m for Fan B and 150 m for Fan C. If the cliffed fans have developed due to erosion caused by subsidence, then the cliff heights give a maximum estimate of hangingwall subsidence that has taken place since the fans were constructed, probably since the Flandrian highstand some 7 ka. Alternatively, the cliffing process may be due to coastal retreat caused by a much-reduced sediment supply to the fans in the last few thousand years. Because of the presence of marginal cliffs, the fan channels are deeply incised at their seaward exits, the degree of incision decreasing up-fan. Gravel and wood fragments are delivered to the front of the incised fans in the form of small fan deltas, where major storm events rework and redistribute the material along the beachface. The fan deposits are predominantly poorly sorted and matrix-rich gravels with crude but discernable bedding picked out by 0.5–1 m flow units which show a coarsening-upwards trend from more matrix-rich intervals, rarely channelized, into an apparently elutriated open framework fabric in which the larger, more elongated clasts show imbricate textures. These units are interpreted as the deposits of hyperconcentrated (debris/slurry) flows which were succeeded by a phase of turbulent streamflow.

Fan architecture is revealed by occasional palaeosol horizons. These may be discontinuous, thin (5–10 cm), pale red to brown in colour and are often cut by shallow scours of the flow units above. They are rarely traceable for more than 80 m and are seen to define lobe-like areas within the fan deposits as a whole. Thus on Fan A a thin palaeosol

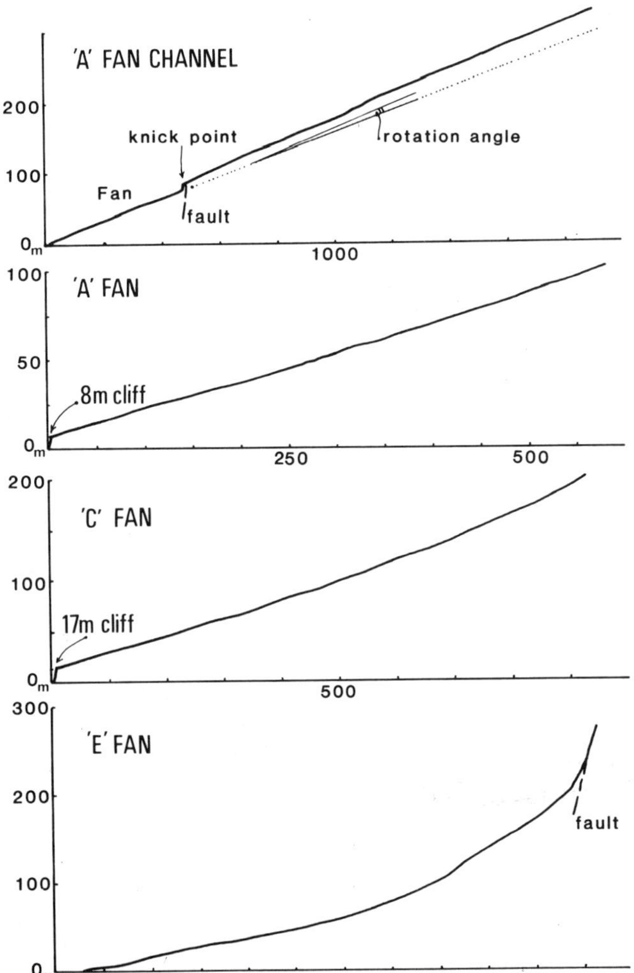

Fig. 9. Longitudinal profiles based on detailed levelling studies taken along some of the alluvial fans and their fan channels in the western part of the study area (see Fig. 5). Note the cliffed profiles of Fans A and C and the asymptote that the toe of Fan E makes with the narrow alluvial plain developed south of Skinos. All elevations above sea level.

horizon is restricted to the western side and clearly represents diversion of only a portion of the whole fan for a period of time sufficient to cause discernable soil formation. By way of contrast a 30 cm brick-red soil of apparent fanwise lateral extent within the deposits of Fan C must be due to lengthy inactivity over the whole fan, possibly due to climatic control.

In contrast to the cliffed margins of drowned fans A–C, the surfaces of alluvial fans D and E merge gradually, as asymptotes, with the alluvium of the Skinos coastal plain (Fig. 9). It is highly significant that the main drainage channels of these latter fans are also incised below the level of the main fan surface, but in marked contrast to the incision of the cliffed fans, the depth of incision decreases to zero downstream. Thus in Fan D, there is up to 30 m of incision at the fan head, decreasing down fan to just 1.5 m some 200 m above the fan outlet where incision has disappeared. These observations lead us to conclude that incision must be due to a non-tectonic mechanism, most probably mid-Holocene climatic change (see discussion of fan incision by Harvey (1989) and for analogous examples from elsewhere in the Mediterranean).

It seems likely that the 'fanless' drainage outlets to the NE of Fan A (Figs 5 & 8) are the proximal remnants of fans that have been completely removed by coastal erosion. It is obvious from Fig. 5 that the degree of erosion increases to the east since the width of the subaerial hangingwall decreases in that direction until the active fault line goes offshore. The linear cliffed margin to the alluvial fills thus defines a *remnant bajada*. The depth of the incised channels along the bajada front decreases inland, with several of the channels having knickpoints in their long profiles.

Talus cones

A number of talus cones occur along the active fault strand where limestones occupy the fault footwall. These features are up to 30 m high and dip at steep angles, 40–50°, resting upon and transgressing over alluvial fan surfaces, lagoonal marshlands and beach ridges. The cones are sourced in the steep footwall uplands to the fault. Sometimes basin-shaped source areas for the talus connect to the cones via long, chute-like, feeder channels. Numerous generations of talus cones with varying degrees of vegetation cover may be discerned along the scarp. They thus form a distinct fringe of well-sorted debris up to 30 m wide. At least two of the youngest cones formed by footwall collapse after the 1981 earthquake event (compare Figs 5 and 8).

Coastal marshes, lagoons and beach/barriers

The drowned alluvial fans discussed above are separated by arcuate areas of coastal saltmarsh and reed beds which extend back to the fault scarp or to talus cones (Figs 8, 10). A large expanse of such marsh occurs landwards of the Psatha Bay beachzone. It appears that the presence of these inter-fan marshlands is a direct result of hangingwall subsidence. The marshlands occur behind gravel beach ridges up to 0.75 m higher than the level of the marsh and it is evident from the subsiding nature of the hangingwall that the marshland sediments are being slowly, and probably episodically, transgressed by backbeach washover deposition.

The marsh deposits east of Skinos (Fig. 10) comprise >0.5 m of organic-rich muds and silty muds rich in leaf litter. The marshes frequently dry out and desiccate during the summer months but at other times are covered by a thin skin of blue-green algal mat. Intercalations of poorly sorted and angular gravels discovered below the marsh surface were evidently derived from adjacent fan lobes at some previous time interval when the fans were more active, since at the present time the marsh deposits are transgressing the fan margins and are themselves being transgressed by beach washover gravels (Fig. 10).

Submarine scarp and basin plain

We have no information at the present time on details of submarine morphology and depositional processes. Bathymetric contours (Fig. 3) reveal the existence of a gently dipping offshore platform which extends down to about 100 m below MSL before a steeper, *c.* 10°, slope leads onto the essentially horizontal floor of the Alkyonides basin plain at a depth of between 350–400 m. The shelving offshore platform narrows to the NE where it eventually intersects

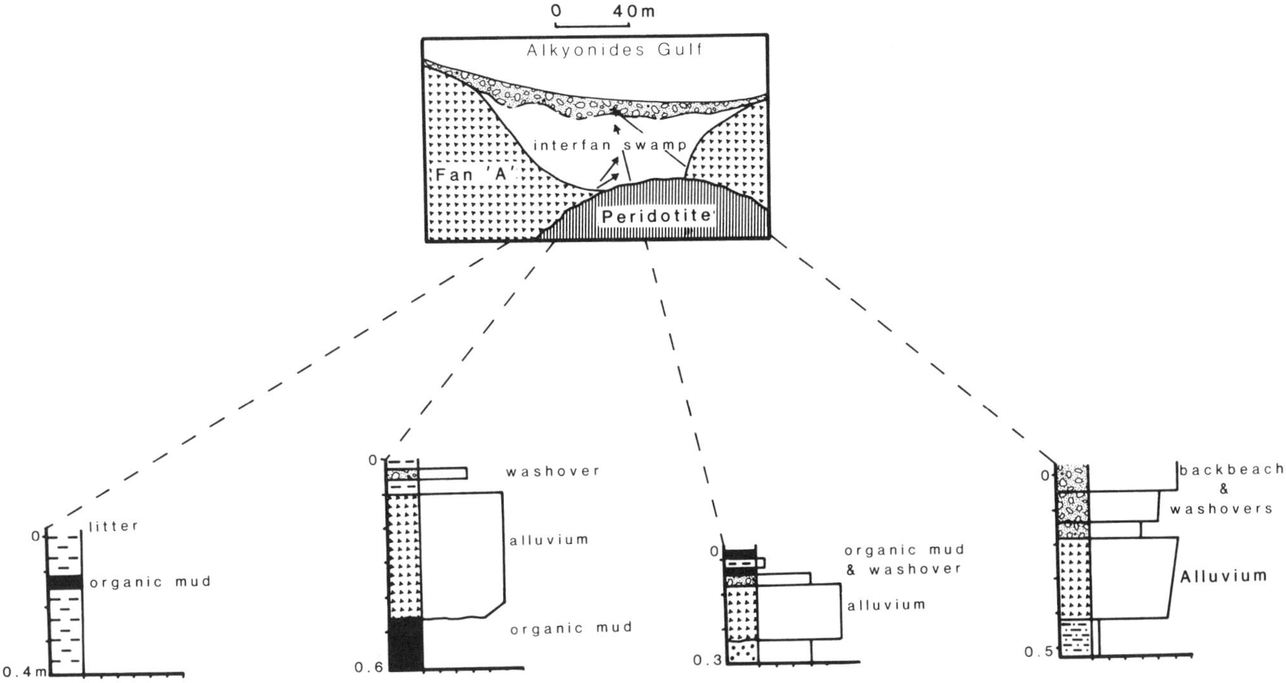

Fig. 10. Detailed sketch map to show the inter-fan marshes developed between Fan A and the truncated fan remnant to the east (For exact location see Fig. 5). The marsh and swamps are separated from the shoreface of the Alkyonides Gulf by a continuous barrier beach. The environments are developed in the subsiding hangingwall to the active coastal fault. The marsh and swamp is being slowly transgressed, with a pre-Recent alluvial wedge recorded in the shallow subsurface.

the steep bathymetric escarpment produced by the active Skinos–Psatha fault. Some part of the platform, perhaps up to 0.5 km wide, has been eroded into the coastal bajada described above as an erosional feature during post-Flandrian subsidence.

The inactive NE faulted margin to the Megara Basin (Figs 3–5)

This relatively old and largely structurally inactive fault-line margin is composed of Mesozoic limestones drained by the basins of Domain 4. These footwall uplands with their truncated spurs and scarp retreat features described previously are bordered by a gently dipping, 1.5–3 km wide, fringe of coalesced alluvial fans which define a continuous bajada front. All modern drainage channels which traverse this bajada are incised to depths ranging from 2–15 m. Sections in these incised channels reveal the ubiquitous presence of angular, unsorted, non-imbricated and cemented gravels arranged in even beds and composed mostly of limestone clasts. Large exposures reveal that the individual units are sheet-like, with gentle (10–20°) dips into the basin to the S and SW. These dips usually appear to have the same azimuth as the surface fan dips seen today, but some large exposures reveal common erosional truncation of the tilted fan breccias by well defined, sometimes convex-up, pediment surfaces. The breccias are interpreted as the products of debris flows which were probably relatively dilute and matrix-poor, being derived from limestone hinterlands in the drainage basins of Domain 4 above. The deposits evidently accumulated rapidly, probably in a more humid climate than seen today in the area. We have no direct evidence of the age of the cemented breccias.

In a prominent gorge some 1 km west of the hillslope named Pevka Stamelou the breccias may be seen in faulted contact with Mesozoic limestone basement. The older cemented breccias are overlain by up to 3 m of largely unconsolidated red-brown, poorly sorted breccias of generally finer grain size. These younger deposits clearly overstep the Pateras fault onto the Mesozoic limestone footwall. We infer that these younger deposits also postdate the cutting of the erosive pediment into the cemented breccias.

Although the alluvial fans along the NE margin are largely inactive, incision of the various fan channels is now taking place. Two examples of fairly recent limestone breccia flow units were mapped in the area south and west of Pevka Stamelou. The elongate debris fans radiate out from the fault-line scarp and contain many depositional lobes of sieve-type and some with levees and overhanging snouts. They appear to have been catastrophically emplaced as variable-density debris flows, probably following a high magnitude flood event which tapped considerable amounts of stored talus and colluvium, as seen at present on many of the drainage basin slopes.

Tectono-geomorphic evolution

The Megara basin was an active depocentre from the Pliocene (possibly Miocene) to the late Middle Pleistocene. The basin form was a half-graben, with the Pateras Fault as the main bounding structure. It is known from several areas of the Aegean (Le Pichon & Angelier 1981; Kissel *et al.*

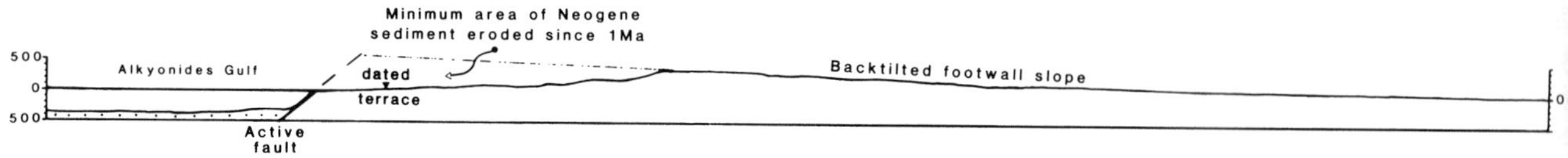

Fig. 11. True-scale section drawn along the line indicated on Figure 2 (see also the panoramic sketch drawn along this line in Fig. 12). The section shows the prominent erosional cirque of Drainage Domain 2 cut into the Plio-Pleistocene deposits of the Megara Basin. These make up the footwall to the active coastal fault in this area. Over the Domain as a whole some 9 km^3 of this sediment have been removed in the last 1 Ma. Note also the rather gentle and steady gradient of the backtilted slope of Dranage Domain 3, interpreted by us as part of a dissected depositional geomorphic surface uplifted in the west by at least 350 m. The position of the raised coastal terrace dated at between 90–126 ka is also shown.

1986; Kissel & Laj 1988) that substantial clockwise rotations of crustal blocks about vertical axes occurred in this time interval. The rates involved, if applicable to the Megara basin, would enable some 25–30° of rotation to have occurred in the time interval concerned, sufficient to have caused reorientation of the Pateras faults from approximately an original E–W orientation to their present 115–125° trend.

The initiation of the Pateras footwall drainage systems of Domain 4 (Fig. 5) must have occurred at the same time as the basin was initiated. The basin fill includes a total volume of some 125 km^3, but much of this was derived axially from drainage systems whose provenance is unknown. A general point is raised here, viz. the impossibility of apportioning sediment volumes eroded from the hangingwall, footwall and axial sources in tilt block basins. Thus it is difficult to determine erosion rates or sediment transfer coefficients for local hinterlands in such basins.

When the Megara Basin was active, the general palaeoslope was down to the NE, in the opposite direction to the present day axial drainage of Domain 3. The depositional geomorphic surface produced is still preserved beneath the deposits of Upper Pleistocene to Holocene age that cap the Domain.

In the Upper Pleistocene, possibly around 1 Ma if our extrapolated rate of uplift for the last 100 ka is reasonable, the coastal faults that presently border the Alkyonides Gulf were initiated. We presume that at this time the Pateras fault became inactive and the Plio-Pleistocene fill to the Megara Basin and the Mesozoic limestones and ophiolites of the Gerania Mountains were uplifted (Figs 11 & 12). Drainage basins in Domain 3 were still active but no further amplification of topography by faulting occurred. Domains 1 and 2 were initiated in the newly-formed footwall uplands to the coastal faults. Drainage network growth proceeded in tandem with the progressive uplift of the footwall, the latter at rates of perhaps 0.3 mm a^{-1}. The basins of Domain 2, excavated in unconsolidated Plio-Pleistocene deposits grew at rates several times those in the limestones and peridotites of Domain 1. At the same time the old geomorphic depositional surface present over the area of Domain 3 was increasingly backtilted towards the SE (Figs 11 & 12). During this time interval a variety of depositional environments evolved in the subsiding hangingwall and uplifting footwall of the coastal fault. The interaction of the uplifting footwall with glacio-eustatic fluctuations led to the production of three coastal terrace levels, the highest of which, at about 40 m, is dated to the last interglacial between 90–125 Ma (Fig. 12). Although we can reasonably estimate the rate of footwall uplift for the last 100 ka as around 0.3 mm a^{-1}, we have no direct means of estimating the complementary subsidence magnitude or rate. An approximate indication may be given by the Jackson & McKenzie (1983) estimate that footwall uplift is about 0.1 times the hangingwall subsidence in normal faults. This would yield about 350 m of post-100 ka subsidence and about 20 m of post-Flandrian subsidence.

Regarding sediment budgets we estimate that the total volume of Plio-Pleistocene sediment eroded from Domain 2 during the last 1 Ma is 8.9 km^3, derived from an area of some 32.5 km^2. This gives a mean sediment yield of some $2.7.10^2$ km^3 km^{-2} a^{-1} or, assuming a realistic bulk sediment density of 2500 kg m^{-3}, 6.85. 10^{-7} tonnes km^{-2} a^{-1}. Mean erosion rates over 1 Ma are estimated as 0.27 mm a^{-1}. The eroded products from the basins of Domain 2 are presumed to have been resedimented, probably as underflows and by other processes, into the Gulf of Alkyonides. The volume of sediment eroded is sufficient to deposit a uniform layer some 180 m thick over the 50 km^2 basin plain of the Alkyonides Gulf below the break of slope at the 350 m isobath.

It is impossible at the present time to estimate the sediment yields from Domain 1, but it is evident from the more resistant nature of the lithologies here, and the small sizes of the drainage basins, that yields are probably much less than those calculated from Domain 2.

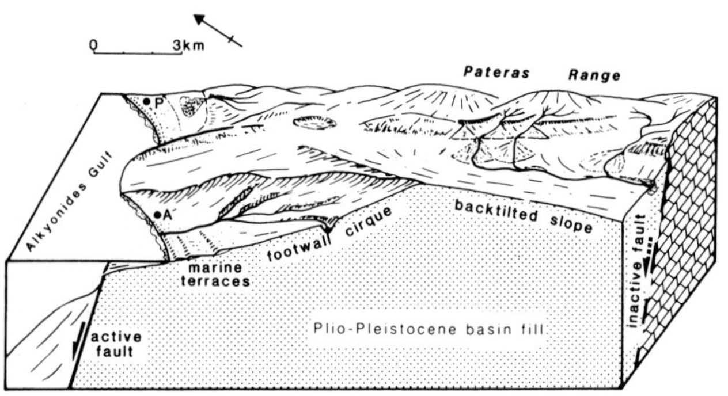

Fig. 12. Sketch to show panoramic view NE across the line of section of Fig. 10, to show: (1) the uplifting footwall to the active coastal fault and the extent of the erosional 'cirque' in the hinterlands of Drainage Domain 2; (2) the raised marine terrace levels in the uplifting footwall; (3) the gently SE-dipping geomorphic surface uplifted and backtilted by post 1 Ma uplift on the coastal fault and the axial drainage channels which flow down this slope towards the Saronic Gulf near Megara town; (4) the ancient fault-line segmented scarps of the Pateras range, with their facetted spurs and largely relict alluvial fans issuing from the Mesozoic Limestone hinterlands of Drainage Domain 4.

Conclusions and economic applications to ancient syn-rift basins

Our integrated approach has shed some light on the rates of tectonic, erosional and depositional processes, and on their interplay, in an area of active normal faulting. We have, for the first time in the Aegean extensional province, obtained information on the mean rate of footwall uplift of a major normal fault over a period of some 100 ka. This has enabled us to contrast the rates of drainage basin growth for different lithological substrates in the fault footwall and to estimate erosion rates. The concept of Drainage Domains may have some useful general applications in continental areas undergoing active tectonic deformation. The spectacular, cirque-like, footwall unconformity cut into the poorly consolidated fill to the abandoned Megara half-graben may provide a useful analogue for similar features seen in seismic sections across ancient syn-rift plays like the Viking Graben in the northern North Sea. Major sand fairways and high reservoir potential may thus exist in adjacent hangingwall sequences. These applications are the subjects of research in progress.

We are grateful to IGME for granting us successive permissions to undertake fieldwork in Greece. Fieldwork funds were provided by NERC, Texaco USA, Mobil North Sea and the University of Leeds. The radiometric age determinations on the coral samples were undertaken by P. Rowe and T. Atkinson at the University of East Anglia; the results and their implications will be discussed fully in a forthcoming paper. P. Harper, K. Knapton, P. Bentham, R.E.Ll. Collier, R. L. Gawthorpe and S. Prossor aided with fieldwork over several years. We are particularly grateful to J. A. Jackson for suggesting to us that the area was an ideal one in which to examine active tectono-sedimentary interactions and we thank him for many illuminating discussions over the years. We thank M. J. Kikby, R. E. Ll. Collier, J. A. Jackson, G. Nicholls and J. R. Underhill for their critical and helpful comments on an earlier version of the manuscript.

References

ALEXANDER, J. A. & LEEDER, M. R. 1987. Active tectonic control of alluvial architecture *In*: ETHERIDGE, F. G. & FLORES, R. M. (eds) *Fluvial sedimentology*. Special Publication, Society of Economic Mineralogists and Petrologists, **39**, 243–252.

BARR, D. 1987. Lithospheric stretching, detached normal faulting and footwall uplift. *In*: COWARD, M. P., DEWEY, J. F. & HANCOCK, P. L. (eds) *Continental Extensional Tectonics*. Geological Society, London, Special Publication, **28**, 75–94.

COLLIER, R. E. Ll. 1990. Eustatic and tectonic controls upon Quaternary coastal sedimentation in the Corinth Basin, Greece. *Journal of the Geological Society, London*, **147**, 301–314.

——, LEEDER, M. R. & MAYNARD, J. R. 1990. Transgressions and regressions: a model for the influence of tectonic subsidence, deposition and eustasy, with application to Quaternary and Carboniferous examples. *Geological Magazine*, **127**, 117–128.

HACK, J. T. 1957. *Studies of longitudinal stream profiles in Virginia and Maryland*. United States Geological Survey Professional Paper 294B.

HANKS, T. C., BUCKHAM, R. C., LAJOIE, K. R. & WALLACE, R. E. 1984. Modification of wave-cut and fault-controlled landforms. *Journal of Geophysical Research*, **89**, 5771–5790.

HARVEY, A. M. 1989. The occurrence and role of arid zone alluvial fans. *In*: THOMAS, D. S. G. (ed.) *Arid Zone Geomorphology*, Belhaven Halsted, 136–158.

HERFORTH, A., SCHRODER, B. & THEODOROPOULOS, D. 1971. Zur jungpleistozanen und Holozanen Kustenmorphologie zwisischen un Sud-Attica. *Bulletin of the Geological Society of Greece*, **8**, 194–198.

IGME 1984. *Kaparellion 1:50 000 Geological Sheet*. Athens.

JACKSON, J. A. & MCKENZIE, D. P. 1983. The geometrical evolution of normal fault systems. *Journal of Structural Geology*, **5**, 471–482.

—— & —— 1988. Rates of active deformation in the Aegean Sea and surrounding areas. *Basin Research*, **1**, 121–128.

——, GAGNEPAIN, J., HOUSEMAN, G., KING, G. C. P., PAPADIMITRIOU, P. SOUFLERIS, C. & VIRIEUX, J. 1982. Seismicity, normal faulting and the geomorphological development of the Gulf of Corinth (Greece): the Corinth earthquakes of February and March 1981. *Earth & Planetary Science Letters*, **57**, 377–397.

——, WHITE, N. J., GARFUNKEL, Z. & ANDERSON, H. 1988. Relations between normal fault geometry, tilting and vertical motions in extensional terrains: an example from the southern Gulf of Suez. *Journal of Structural Geology*, **10**, 155–170.

KELLETAT, D., KOWALCZYK, G., SCHRODER, B. & WINTER, K.-P. 1976. A synoptic view on the neotectonic development of the Peloponnesian coastal regions. *Zeitschrift der Deutschen Geologischen Gesellschaft*, **127**, 447–465.

KERAUDREN, G. & SOREL, D. 1987. The terraces of Corinth (Greece)—a detailed record of eustatic sea-level variations during the last 500 000 years. *Marine Geology*, **77**, 99–107.

KISSEL, C. & LAJ, C. 1988. The Tertiary geodynamical evolution of the Aegean arc: a palaeomagnetic reconstruction. *Tectonophysics*, **146**, 183–210.

——, —— & MAZEAUD, A. 1986. Palaeomagnetic evidence for Neogene rotational deformations in the Aegean domain. *Tectonics*, **5**, 783–795.

LAJOIE, K. R. 1986. Coastal tectonics. *In*: *Active Tectonics*. National Academy Press, 95–124.

LEEDER, M. R. & GAWTHORPE, R. L. 1987. Sedimentary models for extensional tilt block/half graben basins. *In*: COWARD, M. P., DEWEY, J. F. & HANCOCK, P. L. (eds) *Continental Extensional Tectonics*. Geological Society, London, Special Publication, **28**, 139–152.

LE PICHON, X. & ANGELIER, J. 1981. The Aegean Sea. *Philosophical Transactions of the Royal Society of London*, **A300**, 357–372.

LEOPOLD, L. B. WOLMAN, M. G. & MILLER, J. P. 1964. *Fluvial Processes in Geomorphology*. Freeman, San Francisco.

MARIOLAKIS, I. D. & PAPANIKOLAOU, D. J. 1982. The Neogene basins of the Aegean Arc from the palaeogeographic and geodynamic point of view. *Proceedings HEAT Symposium*, Vol. 1, Athens, 383–399.

MCKENZIE, D. P. & JACKSON, J. A. 1983. The relationship between strain rates, crustal thickening, palaeomagnetism, finite strain and fault movements within a deforming zone. *Earth & Planetary Science Letters*, **65**, 182–202.

—— & —— 1986. A block model of distributed deformation by faulting. *Journal of the Geological Society, London*, **143**, 349–353.

ROBERTS, S. & JACKSON, J. A. 1991. Active normal faulting in Central Greece: an overview. *In*: ROBERTS, A. M., YIELDING, G. & FREEMAN, B. (eds) *The Geometry of Normal Faults*. Geological Society, London, Special Publication, **56**, 125–142.

SIEH, K. E. 1984. Lateral offset and revised dates of large earthquakes along the San Andreas Fault system at Pallett Creek, southern California. *Journal of Geophysical Research*, **89**, 7641–7670.

STEWART, I. S. & HANCOCK, P. L. 1988. Normal fault zone evolution and fault scarp degradation in the Aegean region. *Basin Research*, **1**, 139–154.

THEODOROPOULOS, D. 1968. *Stratigraphie und Tectonik des Isthmus von Megara (Griechenland)*. Erlanger Geologische Abhandlungen, 73.

TRENHAILE, A. S. 1987. *The Geomorphology of Rock Coasts*. Clarendon Press, Oxford.

WALLACE, R. E. 1977. Profiles and ages of young fault scarps, north central Nevada. *Geological Society of America Bulletin*, **88**, 1267–1281.

Received 1 February 1990; revised typescript accepted 2 August 1990.

Forward and reverse modelling of rift basin formation

NICK J. KUSZNIR,[1] ALAN M. ROBERTS[2] & CHRIS K. MORLEY[3]

[1]*Department of Earth Sciences, University of Liverpool, Liverpool L69 3BX, UK*
[2]*Badley Earth Sciences, Spilsby, Lincoln, PE23 5NB UK*
[3]*Elf-Aquitaine Production, Rue Laribau, Pau, France*

Abstract: The McKenzie model of continental lithosphere extension describes the first-order lithosphere responses of crustal thinning and geothermal gradient increase following rifting, which lead to syn- and post-rift basin subsidence. At the sub-basin scale, seismic data show the fundamental importance of major basement faults in controlling the geometry and subsidence of rifted sedimentary basins. Reflection and earthquake seismology data show that these faults are generally planar and extend down 10–15 km to mid-crustal level. Below this depth deformation gives way to distributed ductile shear in the lower crust and mantle. Extensional faulting leads not only to hanging wall subsidence but also to footwall uplift. It can be successfully modelled using elastic dislocation, visco-elastic finite element and thin plate flexural-isostasy (flexural cantilever) theories. For extension on multiple faults, interference of footwall uplift and hanging wall collapse gives rise to the familiar block rotations of rift tectonics. Mathematical forward models of rifting, incorporating upper crustal faulting, lower crust/upper mantle plastic stretching, lithosphere thermal perturbation and re-equilibration, sediment loading and flexural isostasy have been developed and applied to both syn- and post-rift stages of basin evolution. The models allow the effects on basin geometry and subsidence of fault spacing, fault polarity and extension magnitude, as well as sediment fill and footwall erosion, to be explored. Reverse post-rift modelling from present day sections may be used to constrain β stretching estimates and to predict palaeobathymetry and topography. Both forward and reverse models have been successfully applied to many rift basins worldwide, including the Viking Graben, North Sea and the East African Rift System, both of which are documented here. The forward model is also capable of simulating the first-order geometry of metamorphic core-complexes and so-called 'detachment faults' documented from the Basin-and-Range Province of the western USA.

The McKenzie model (1978) of continental extension predicts the first-order lithosphere responses to continental extension, notably crustal thinning and geotherm perturbation (Fig. 1a and b). Crustal thinning and its isostatic response give subsidence, while an increase in geothermal gradient and isostasy give uplift. At the time of rifting, the combined effects of crustal thinning and the increase in geothermal gradient generate a net syn-rift subsidence, provided the crustal thickness prior to rifting exceeds 18 km. After rifting the perturbed lithosphere temeperature field cools as the perturbed geotherm re-equilibrates. This cooling occurs in an exponential-like manner, rapidly at first and then increasingly slowly, and has a decay time constant of the order of 60–70 Ma. This post-rift cooling of the lithosphere causes a densification of lithosphere material through thermal contraction, the isostatic response to which generates post-rift subsidence. The McKenzie model predicts rapid syn-rift subsidence, followed by much slower post-rift thermal subsidence (Fig. 1c). This model assumes that the continental lithosphere deforms by distributed uniform plastic deformation, and that the region of crustal thinning and geotherm perturbation are laterally coincident. While the McKenzie model adequately describes the first-order responses of the lithosphere at the regional scale the model has two important limitations, it assumes local (Airy) isostasy, and does not include upper-crustal faults and their control of rift basin geometry.

Seismic reflection data show the importance of major basement faults in controlling extension in the upper part of the continental lithosphere and in influencing rift basin geometry. The geometry of these basement faults is critically important. Within sedimentary sequences both planar and listric faults are seen; however, the debate as to whether extensional basement faults are planar or listric has been the subject of considerable controversy. A detailed review of this discussion is beyond the scope of this paper. Roberts *et al.* (1991) provide a review of the normal fault geometry debate within the context of predominantly European rift basins,

From Holdsworth, R.E. & Turner, J.P. (compiler) *Extensional Tectonics: Regional-scale Processes*. The Geological Society, Key Issues in Earth Sciences, **2 (1)**, 199–222.
1476-315X/02/$15.00 © The Geological Society of London 2002
First published in: "Lambiase, J.J. (ed.) Hydrocarbon Habitat in Rift Basins, The Geological Society, London, Special Publications **80**, 33–56".

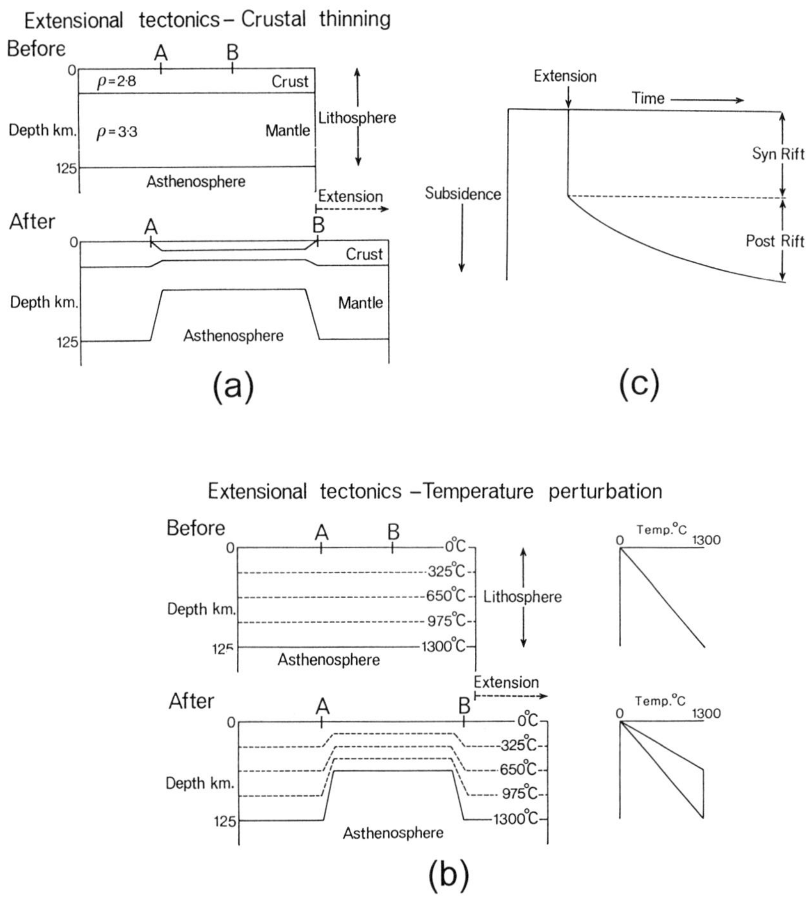

Fig. 1. Extension of the continental lithosphere produces crustal thinning (**a**) and an increase in geothermal gradient (**b**). The isostatic reaction to these produces rapid syn-rift subsidence followed by slower post-rift thermal subsidence, as the perturbed geotherm re-equilibrates (**c**). After McKenzie (1978).

however, their review does not extend to Basin-and-Range-type extensions.

Deep seismic reflection data over rift basins suggest that upper crustal faults extend down through the basement to a depth of *c.* 10–20 km, and that these faults are planar rather than listric (Kusznir *et al.* 1991; Klemperer & Hobbs 1992). Depth conversion and mapping of fault-plane reflections on commercial seismic data also indicate that major basement involved faults are generally planar (e.g. Young *et al.* 1990; Cartwright 1991; Yielding *et al.* 1991). This view is supported by earthquake seismology (Jackson 1987) which shows that in the Aegean, a region experiencing present day continental extension, the angle from the foci of major extensional earthquakes to their co-seismic surface-break makes an angle consistent with their fault plane solutions implying planar rather than listric fault geometry.

Major upper crustal faults seen on deep seismic data do not seem to cause significant offsets in the Moho and it is therefore assumed that the lower crust and mantle deform by distributed plastic deformation, as suggested by the McKenzie model, rather than by faulting. This is consistent with rheological modelling, which suggests that plastic rather than brittle deformation is the dominant mechanism in the lower crust and mantle, for lithosphere with all but the coolest geothermal gradient (Kusznir & Park 1987). Focal depths for the major earthquakes associated with Aegean extension are restricted to the upper 10–15 km of the continental crust, consistent with brittle deformation restricted to the upper crust. However, in the western branch of the East African Rift System earthquakes have been reported with foci in the depth range 20–30 km corresponding to the lower crust (Shudofsky 1985; Shudofsky *et al.* 1987).

Regional crustal sections based on industry and deep seismic reflection data across thermally re-equilibrated rifts, such as the Viking Graben,

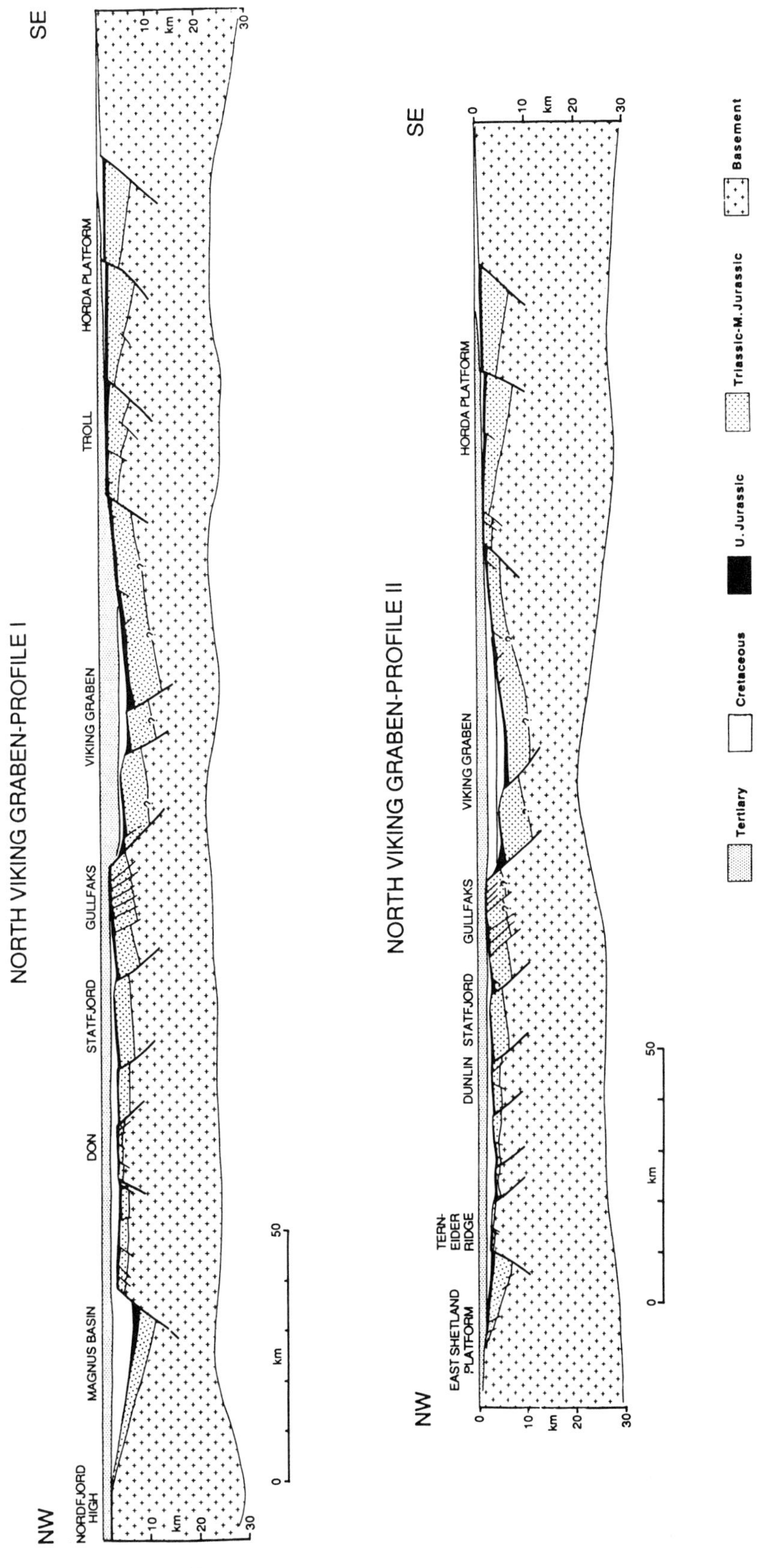

Fig. 2. Regional cross-sections across the Viking Graben, North Sea, based on industrial and deep seismic reflection data, show rift sub-basins, controlled by major basement faults, covered by a regional blanket of post-rift fill. The greatest basin depth coincides with the region of greatest thinning of the pre-rift basement. After Marsden *et al.* (1990).

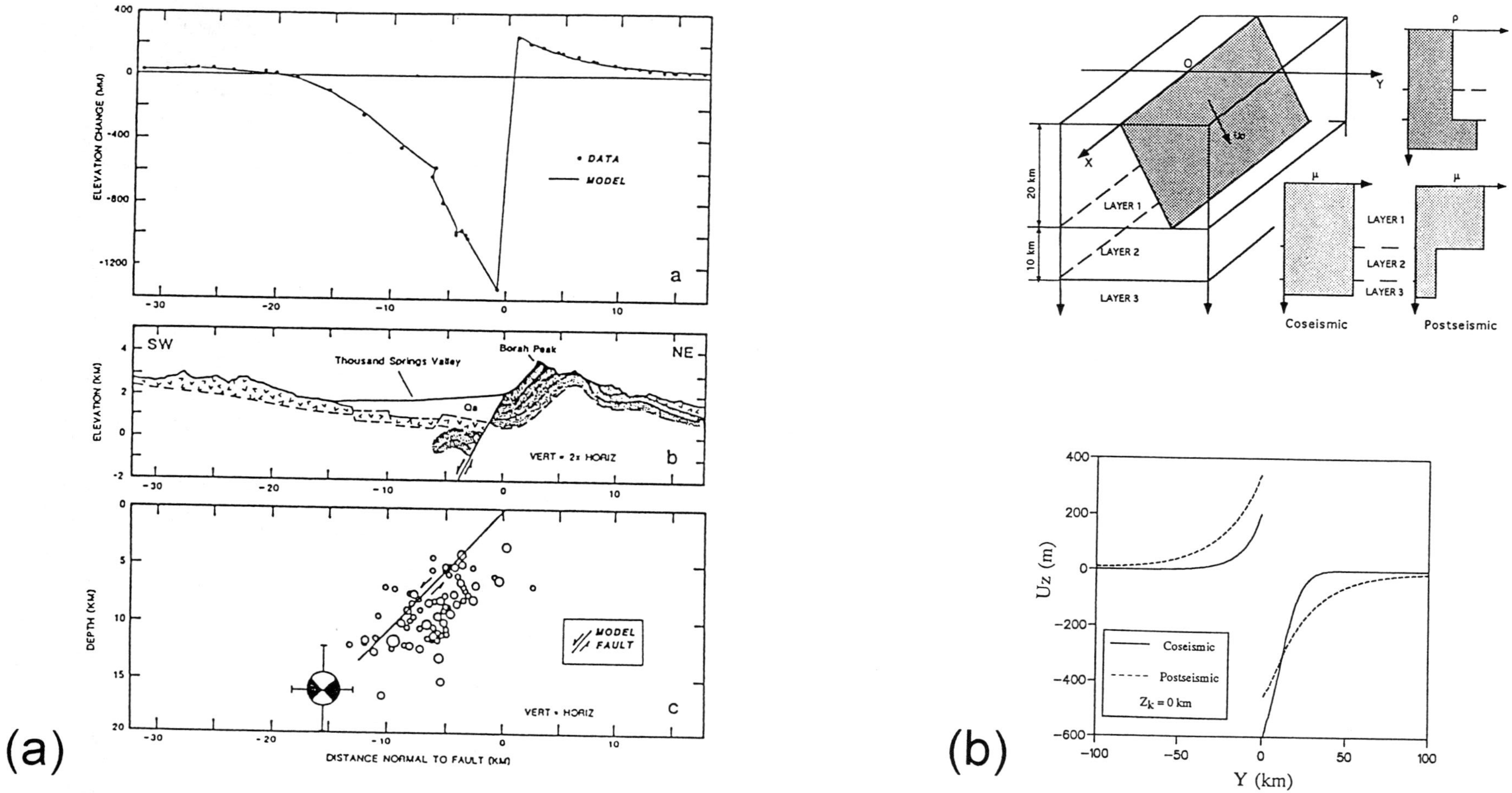

Fig. 3. (**a**) Geodetic observations of coseismic vertical motion associated with major extensional earthquakes involving basement faulting show footwall uplift as well as hanging wall subsidence. From Stein & Barrientos (1985). (**b**) Elastic dislocation theory may be used to predict the footwall uplift and hanging wall subsidence which occurs in response to extensional basement faulting. Coseismic response gives greater hanging wall subsidence than footwall uplift. Subsequent post-seismic relaxation, by plastic creep in the lower crust and mantle, leads to a decrease in hanging wall subsidence and an increase in footwall uplift, and gives similar magnitudes of footwall uplift and hanging wall subsidence. From Ma & Kusznir (1994).

North Sea Basin (Fig. 2), show maximum basin depth where crustal basement is thinnest and Moho shallowest, as predicted by the McKenzie model (Klemperer 1988; Klemperer & White 1989). The syn-rift basin geometry of the Viking Graben is controlled by major basement faults which also generate block rotation tectonics (Jackson & McKenzie 1983). In the case of the Viking Graben, and many other rift basins, syn- and post-rift basin fill sequences are superimposed, suggesting that crustal thinning and lithosphere temperature field perturbations are laterally coincident.

An alternative model for the extension of continental lithosphere has been proposed by Wernicke (1981, 1985), in which the lithosphere deforms on major low-angled detachments which penetrate as continuous structures from the surface to the base of the lithosphere. The Wernicke model would predict a major offset in the Moho, and also a lateral offset between regions of maximum crustal thinning and post-rift thermal subsidence, neither of which is supported by deep seismic reflection or subsidence history data over extensional basins such as the North Sea [Klemperer & White 1989; White 1989 (cf. Beach 1986; Beach *et al.* 1987)]. The evidence for lithosphere-scale detachments is circumstantial (see **Discussion**). No deep seismic reflection section shows an unequivocal example of a dip-slip structure (fault or shear zone) passing continuously from the surface to the base of the lithosphere (or to a depth where the base of the lithosphere might reasonably be assumed to exist). Deep seismic sections do show substantial upper crustal faults with significant top basement extension (heave), however, these sections do not unequivocally show similar large offsets at the level of the Moho. The limitations in resolution of seismic reflection data at the level of the Moho must, however, be acknowledged.

In this paper a quantitative model of continental extension and rift basin formation is described. It is assumed that the upper crust extends by planar faults while the lower crust

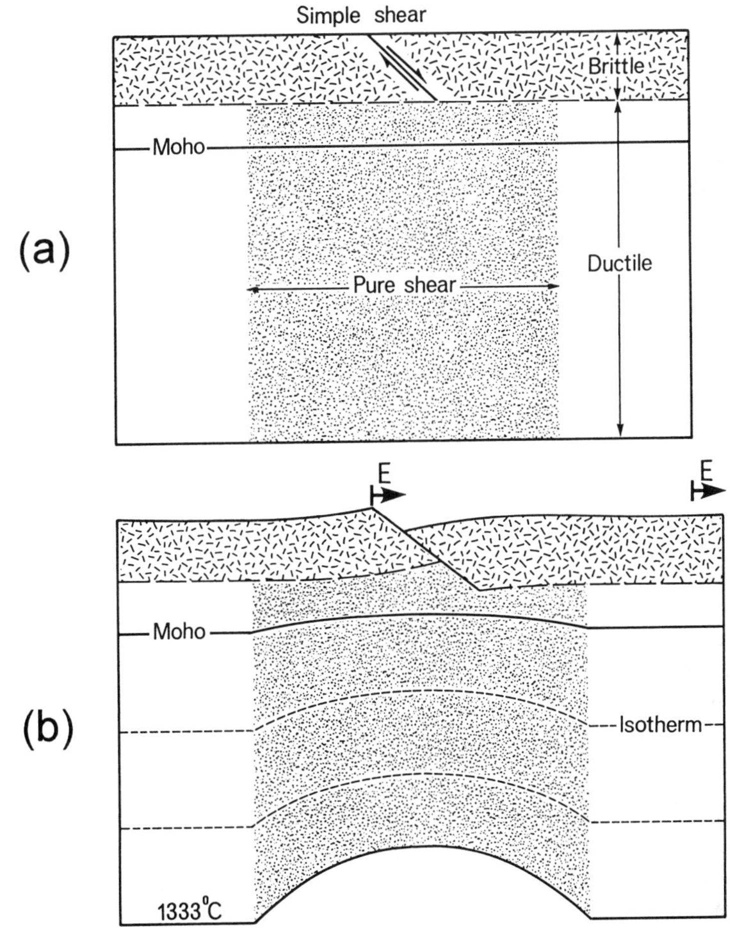

Fig. 4. (**a**) The flexural cantilever model of continental extensional tectonics assumes that the upper crust deforms by faulting on planar faults, while the lower crust and mantle deform by distributed plastic deformation. (**b**) The flexural-isostatic response of the lithosphere to extension produces footwall uplift as well as hanging wall subsidence. From Kusznir *et al.* (1991).

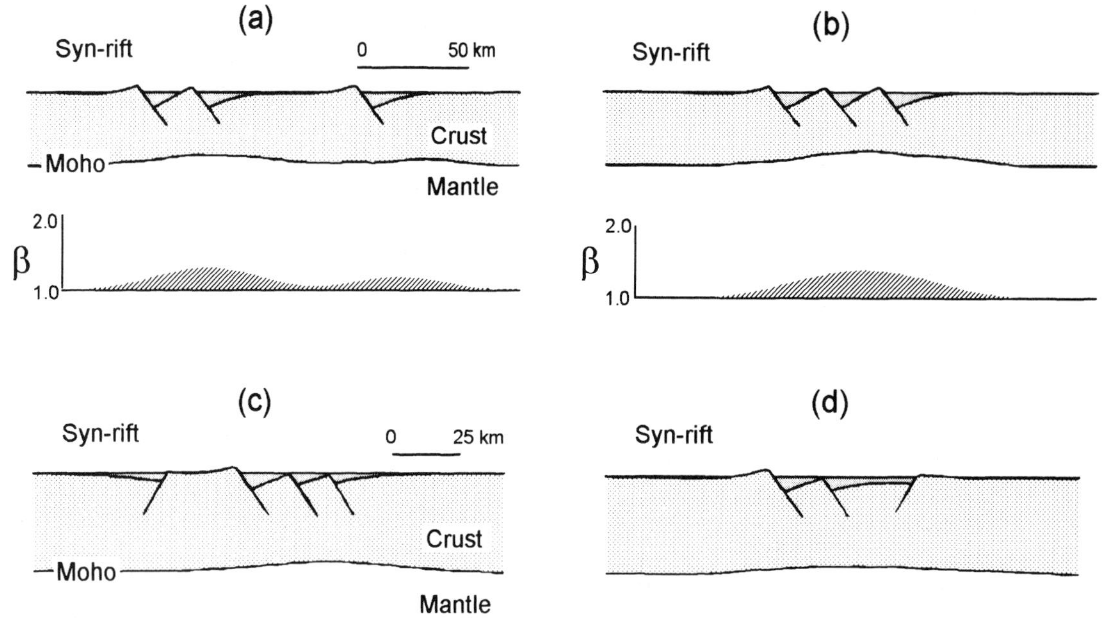

Fig. 5. Crustal structure and basin geometries computed using the flexural cantilever model of continental extension and rift basin formation. The flexural-isostatic lithosphere response to syn- and post-rift thermal loads, sediment fill and erosional denudation are included in the model. (**a**) Syn-rift lithosphere response to extension on a single fault (right) and two adjacent faults (left). The fundamental building block of rift basin formation (right) is a half-graben with footwall uplift underlain by lower crustal thinning and a shallower Moho. Plastic deformation in the lower crust and mantle is described by a 2D β stretching factor profile. (**b**) The remote fault on the right of (a) has been brought closer to those of the left. Superposition of footwall uplift and hanging wall subsidence produce domino-style block rotation tectonics. (**c**) Extension on outward facing faults produces a central horst bounded by half-grabens. (**d**) Extension on inward facing faults produces a full-graben.

and mantle extend by distributed plastic deformation. This model, the flexural cantilever model of continental extension, incorporates the main physical properties of the McKenzie model, but also includes flexural isostasy as well as upper crustal faulting. The flexural cantilever model may be applied to both syn- and post-rift stages of rift basin formation. Example applications of the flexural cantilever model to the formation of the Lake Tanganyika Rift and the Viking Graben of the North Sea Basin are described. In addition, a quantitative method for reverse modelling post-rift basin development – consisting of flexural backstripping, decompaction and reverse thermal subsidence modelling – is described and applied to North Sea case histories.

Syn-rift response of continental lithosphere to extension

Extension on major basement faults not only generates hanging wall subsidence but also gives footwall uplift (Vening-Meinesz 1950; Jackson & McKenzie 1983). Geodetic work following major continental extensional earthquakes allows the coseismic component of this uplift to be measured (e.g. Jankof 1945; Whitten 1957; Richter 1958; Stein & Barrientos 1985). Coseismic footwall uplift and hanging wall subsidence (Fig. 3a) may be quantitatively predicted by elastic dislocation theory (Stein & Barrientos 1985); typically, for extensional faulting coseismic footwall uplift is much less than hanging wall subsidence. For the Borah Peak earthquake, the coseismic ratio of footwall uplift to hanging wall subsidence is of the order of 20%. During the post-seismic interval, stress relaxation within the lower crust and mantle results in a long wavelength uplift which serves to decrease hanging wall subsidence and increase footwall uplift (King *et al.* 1988; Stein *et al.* 1988; Ma & Kusznir 1994). A consequence of this post-seismic relaxation is that in the long term ($>10^5$ a) footwall uplift is predicted to be comparable in amplitude with hanging wall subsidence, in the absence of half-graben isostatic loading by water or sediment (Fig. 3b). The time-scale of this post-seismic relaxation is uncertain. However, while complete relaxation may take in the order of 10^5–10^6 a, significant relaxation may occur on the shorter time-scale of 10^2–10^3 a. In any case, the post-seismic relaxation period is small compared with the time-span of typical rifting events, which typically have durations of 10^6–10^7 a.

An alternative approach to modelling footwall uplift and hanging wall subsidence during continental rifting is to use thin plate flexural-isostasy theory (Walcott 1970). The flexural cantilever model (Kusznir & Egan 1990; Marsden *et al.* 1990; Kusznir *et al.* 1991; Kusznir & Ziegler 1992) of continental extensional tectonics and rift basin formation uses flexural thin plate theory to predict quantitatively the long-term isostatic balance of continental lithosphere following rifting. The flexural cantilever model (Fig. 4) assumes that the upper crust extends by planar faults, and the lower crust and mantle by distributed plastic deformation (pure shear). The flexural response of the upper crust to the lithosphere isostatic loads generated by extension generates footwall uplift, as well as hanging wall subsidence (Fig. 5a): footwall and hanging wall blocks behave as two mutually self-supporting flexural beams.

Within the flexural cantilever model, the plastic deformation in the lower crust and mantle are quantified by a 2D β stretching factor (Fig. 5a): the β stretching factor being defined according to McKenzie (1978). β is the factor by which a unit length of uniformly deforming material is extended and unit thickness is thinned. The syn-rift thermal perturbation of the lithosphere temperature field and its post-rift thermal re-equilibration are computed from the 2D β stretching factor distribution (McKenzie 1978). Loads generated by lithosphere extension associated with crustal thinning, syn- and post-rift thermal effects, sediment fill and erosional denudation are all distributed flexurally. Compaction effects are also incorporated in the model. The assumptions of the flexural cantilever model and its mathematical formulation are described in greater detail in Kusznir *et al.* (1991) and Kusznir & Ziegler (1992).

The fundamental building block of rift basin formation is a half-graben formed in the hanging wall of a major basement fault with associated footwall uplift (Fig. 4). Extension on multiple faults generates superposition of the footwall uplift and hanging wall subsidence of adjacent faults, the resulting interference giving the familiar block rotation tectonics of multi-fault extensional systems (Fig. 5b). In contrast to rigid body block rotation models (e.g. Jackson & McKenzie 1983; Barr 1987*a*,*b*; Mandl 1987; Davison 1989) the flexural cantilever model is able to incorporate faults with both polarities (i.e. directions of dip), position and extension magnitude (Fig. 5c and d). Inward facing faults generate full-grabens (Fig. 5d), while outward facing faults give central horsts (Fig. 5c). The total crustal thinning is a combination of crustal thinning by faulting and lower crustal plastic deformation.

Lake Tanganyika Rift – a syn-rift case history

The Lake Tanganyika Rift, which is *c*. 700 km long and between 30 and 80 km wide, lies in the western branch of the East African Rift System (Fig. 6). The sedimentary fill is poorly dated, but by analogy with drilled sections in the Rukwa Rift, it is probably of Upper Miocene – recent age (Wescott *et al.* 1991). Significant volcanic activity was apparently confined to the northernmost part of the rift, in the Southern Kivu area. The combination of non-volcanic and neotectonic attributes of the Lake Tanganyika Rift, plus a relatively simple and short rift history, provide a good opportunity to investigate and model those crustal and lithospheric processes which occur during continental extension.

Multi-channel seismic reflection data, acquired by project PROBE on a 20–50 km grid, has thrown much light on the upper crustal geometry and structural architecture of the Lake Tanganyika Rift (Rosendahl *et al.* 1986; Rosendahl 1987; Burgess *et al.* 1988). An interpretation of this data is presented in Fig. 6a (Morley 1988). The seismic data reveal a deep rift, with up to 7 km of rift fill, bounded by steep (60–70°) predominantly planar boundary faults (Morley 1988). Maximum observed extension deduced directly from the seismic lines is of the order of 5 km (Morley 1988). The fundamental rift geometry is a single major rift bounding fault controlling the formation of a half-graben (Rosendahl *et al.* 1986). These major faults extend laterally for distances in the order of 50–200 km before transferring throw to another fault. That new fault may display the same polarity (synthetic transfer zone) or change polarity (conjugate transfer zone) (Morley *et al.* 1990) Extensional displacement continuity for the whole rift is accommodated by overlap of these major basin bounding faults and their subsidiaries within transfer zones. Full-graben cross-sections occur within these transfer zones where the polarity of the rift changes along two long overlapping convergent boundary faults (e.g. Profile 210).

On a larger scale, the Lake Tanganyika Rift can be sub-divided into several en-echelon rift segments (Fig. 6b). The two largest segments

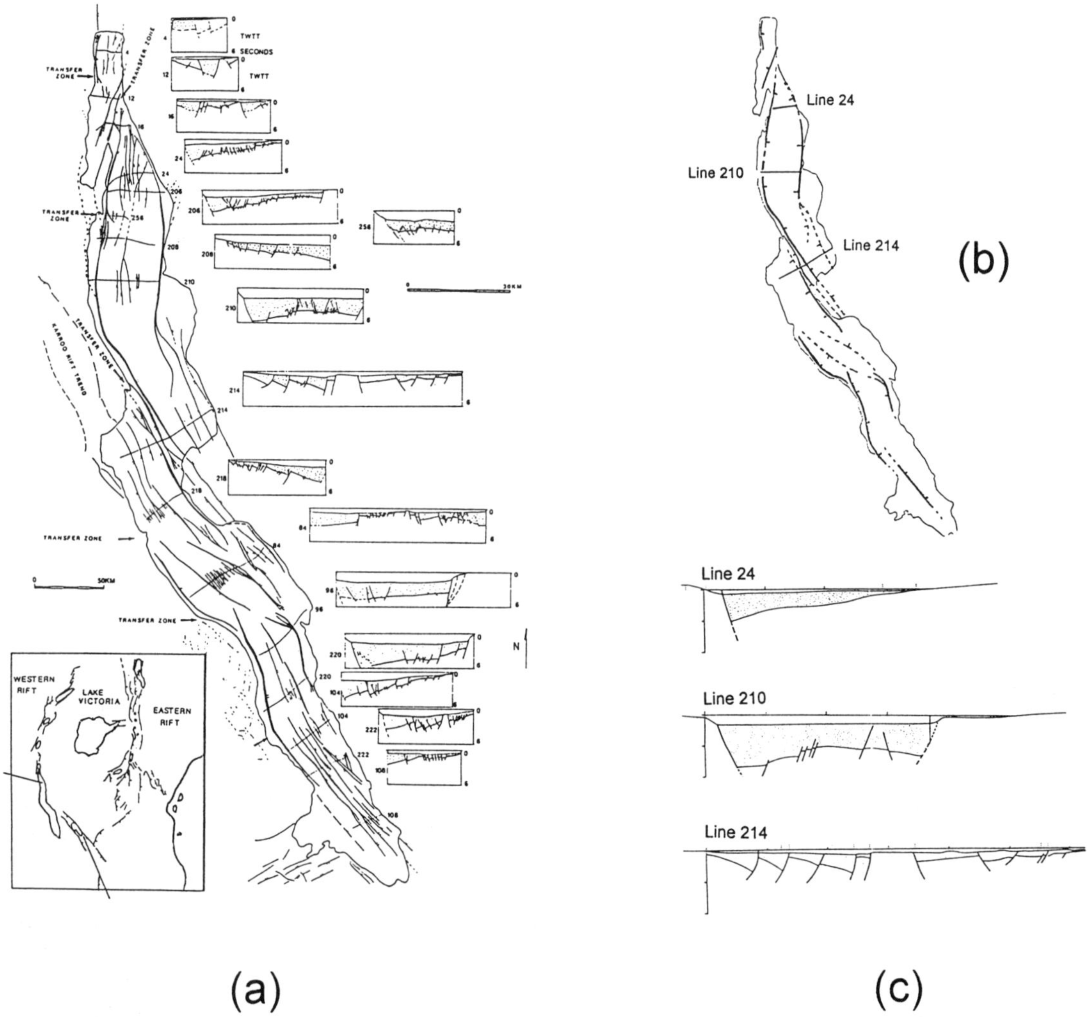

Fig. 6. (**a**) Cross-sections across Lake Tanganyika interpreted by Morley (1986) from seismic reflection data of Rosendahl *et al.* (1986). (**b**) Simplified fault map showing major faults and representative profiles. (**c**) Enlargement of representative profiles across Lake Tanganyika showing half-graben, full-graben and central horst.

meet in the central section of the lake, where overlapping divergent boundary faults create back-to-back graben separated by a central horst (the Kilembi horst – Profile 214). A similar but smaller horst occurs in the northern portion of the lake (the Bay of Burton horst) and the northernmost segment trends onshore into the Rusizi plain. Three representative sections are shown enlarged in Fig. 6c showing half-graben, full-graben and central horst structures. These three sections have been modelled using the flexural cantilever model.

Line 24 – half-graben

Line 24 lies in the northern part of Lake Tanganyika, where there is a single rift bounding fault on the western side of the rift. The rift here takes the form of a half-graben. Fault heaves (extensions) and positions obtained from cross-sections (Fig. 6) interpreted by Morley (1988) from the seismic data of Rosendahl *et al.* (1986) have been input into the flexural cantilever model. Using the present day observed horizontal extension of 3.1 km on the rift-bounding fault, the flexural cantilever model predicts a basin of similar geometry to that observed but insufficiently deep. The model also predicts substantial uplift in the footwall of the main bounding fault (Fig. 7a). Partial erosion of this footwall uplift has reduced the extension observable today on these major rift-bounding faults. In order to model the observed rift basin geometry and depth it is necessary to increase the extension on this rift-bounding fault above that observed today, thus accounting for the

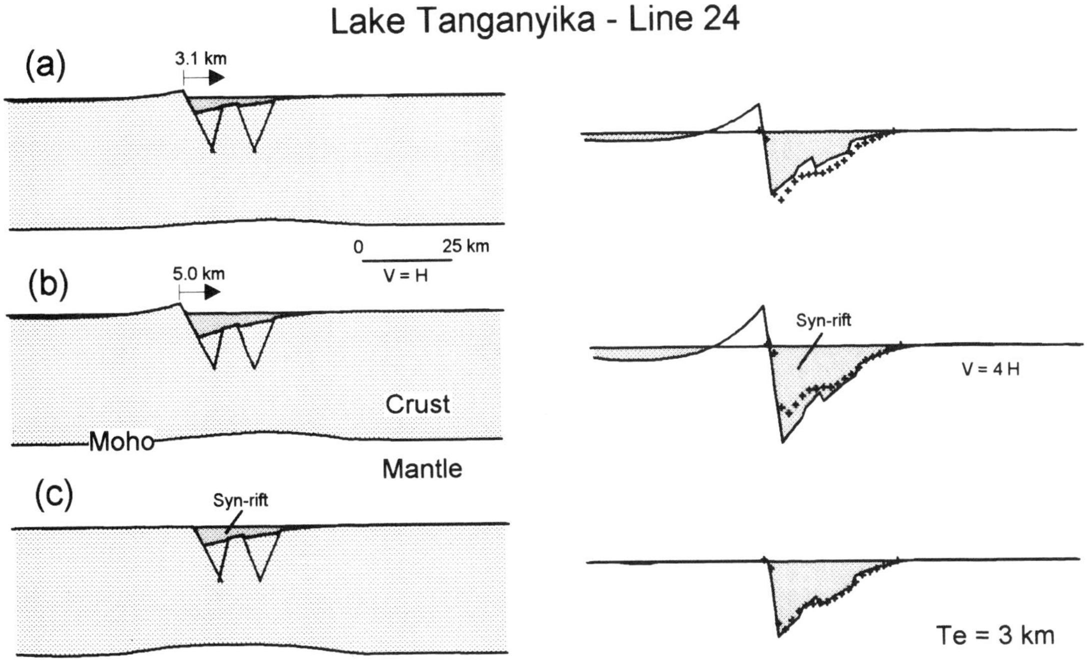

Fig. 7. Crustal structure ($V = H$) and basin geometry ($V = 4H$) predicted by the flexural cantilever model for half-graben profile (Line 24) across Lake Tanganyika: (**a**) 3.1 km extension on major basin bounding fault; (**b**) 5.0 km extension on major basin bounding fault and no footwall erosion; (**c**) 5.0 km extension on major basin bounding fault with footwall erosion. Observed top pre-rift basement is denoted by ***.

erosion. This problem of estimating extension on partially eroded basin margin faults has been discussed in some detail by Roberts & Yielding (1991).

The effect of increasing extension on the rift bounding fault to 5.0 km within the flexural cantilever model is shown in Fig. 7b. The computed rift basin depth is now deeper than that observed. Erosion of footwall uplift removes mass and generates flexural isostatic rebound. This has been computed and is shown in Fig. 7c. After erosion of footwall uplift and isostatic rebound the depth of the rift basin agrees with that observed. Furthermore, the horizontal extension preserved on the rift bounding fault, after erosion, agrees with that observed on the present day section. While the horizontal extension on the rift bounding fault was 5.0 km, that preserved after erosion is 3.1 km. The best-fit model for Line 24 was achieved with an effective elastic thickness (T_e) of 3 km. The total amount of horizontal extension on the section, summed over all faults, is 6 km and gives a maximum β stretching factor of 1.15.

Line 210 – full-graben

Line 210 lies in a region where two rift bounding faults face each other on either side of the rift. Consequently the rift here takes the form of a full-graben. A conspicuous feature of the profile is a central arch, a feature observed in many other rifts for full-graben profiles. The flexural cantilever model of the profile is shown in Fig. 8a. The horizontal extensions on the basin bounding faults are 7.0 and 5.5 km. Locations and extension magnitudes for the internal rift faults are as observed on the seismic section. The model is deeper than the observed basin and shows footwall uplift on either side of the rift. Footwall erosion and flexural isostatic rebound have been modelled and the resulting crustal structure and basin geometry is shown in Fig. 8b. The computed basin depth now agrees well with that observed. The preserved horizontal extensions on the basin bounding faults are in agreement with those observed: erosion has again reduced the amount of preserved extension on the basin bounding faults. The flexural cantilever model is able to replicate the observed central arch within the full-graben, which may be regarded as being generated by the interference of the hanging wall subsidence from each of the two rift bounding faults. The total extension along the profile is 13.5 km, with a maximum β stretching factor of 1.35.

Line 214 – central horst

Line 214 lies in the central region of Lake Tanganyika and shows a central horst with

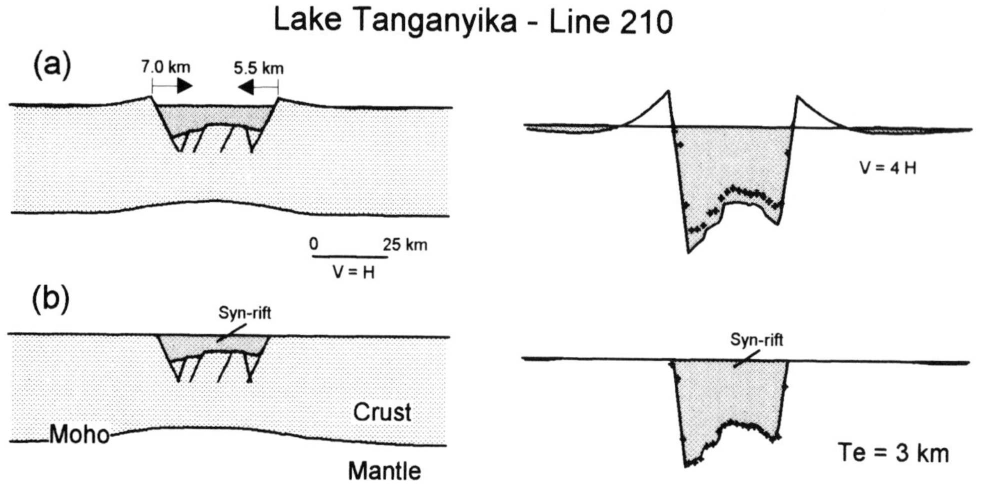

Fig. 8. Crustal structure ($V=H$) and basin geometry ($V=4H$) predicted by the flexural cantilever model for full-graben profile (Line 210) across Lake Tanganyika: (**a**) no footwall erosion; (**b**) with footwall erosion. Observed top pre-rift basement is denoted by ***.

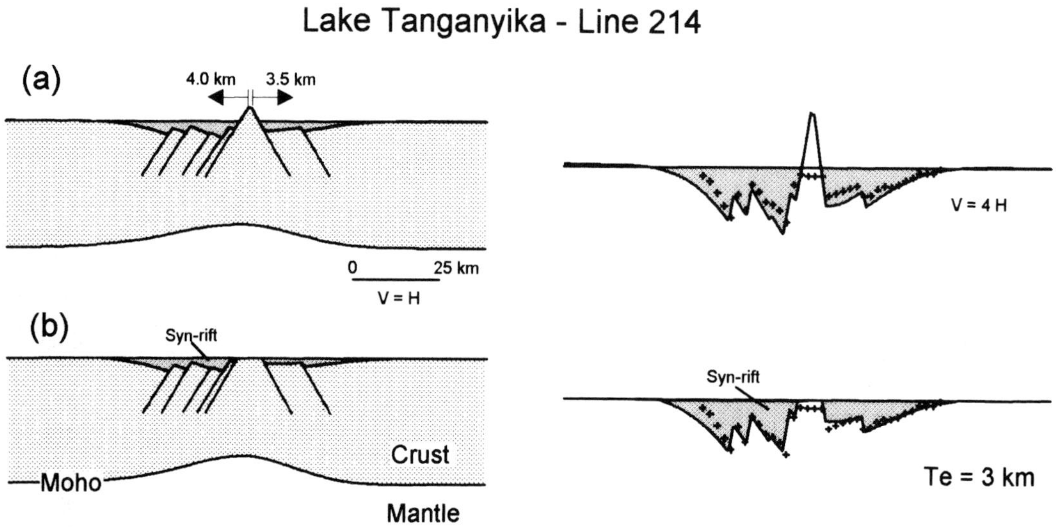

Fig. 9. Crustal structure ($V=H$) and basin geometry ($V=4H$) predicted by the flexural cantilver model for central horst profile (Line 214) across Lake Tanganyika: (**a**) no footwall erosion; (**b**) with footwall erosion. Observed top pre-rift basement is denoted by ***.

outward facing half-grabens on each side. The Tanganyika Rift can be regarded as comprising two large en-echelon sub-rifts, overlapping in the central region of lake (Fig. 6). The central horst of Profile 24 is bounded by two faults of over 100 km extent along strike, which are the major bounding faults of the two sub-rifts. The seismic data shows the central horst to have no sediment cover and pre-rift basement appears to be actively eroding at the lake bed.

The flexural cantilever model of Line 214 prior to erosion is shown in Fig. 9a. The central horst bounding faults are given horizontal extensions of 4.0 and 3.5 km. All other faults are given positions and extensions as observed on the seismic section. The computed rift basin is too deep adjacent to the central horst. The horst itself forms a substantial topographic feature. While the horst is eroded in the lake, along-strike to the SE the horst emerges on to land and forms a distinct, linear, fault-bounded mountain with over 1.5 km of relief above lake level.

The effect of footwall erosion and flexural isostatic rebound has been computed and is shown in Fig. 9b. The horst has been eroded to lake level and the depth of the basin is now in

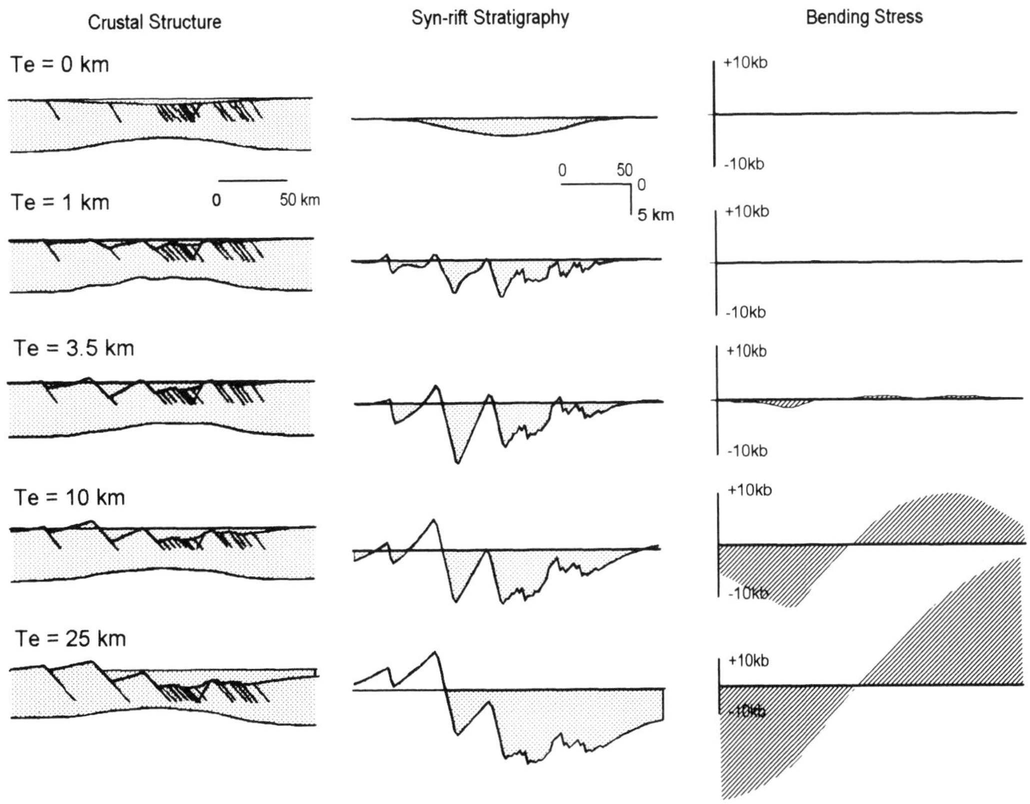

Fig. 10. Model of Lake Turkana from Hendrie *et al.* (1994). Crustal structure and basin geometry predicted by the flexural cantilever model are dependent on the effective elastic thickness, T_e. T_e also controls the maximum horizontal bending stresses.

close agreement with that observed. The preserved horizontal extensions on the horst bounding faults have been reduced to 0.2 and 0.8 km, consistent with those observed on the seismic section. (The discrepancy on the left side of the section, where the model predicts a deeper and wider shoulder to the rift, may be corrected by including a mapped fault to the west of the lake with westward dip and extension of 2 km.)

General comments

The best-fit of modelled profiles to observation was achieved for the Lake Tanganyika Rift using a value of effective elastic thickness (T_e) of 3 km. Effective elastic thickness is a convenient way of parameterizing the flexural strength, i.e. flexural rigidity, of the lithosphere, and may be defined as the thickness of a perfectly elastic plate which has the same flexural strength as the lithosphere. Within the flexural cantilever model fault block rotation is particularly sensitive to T_e. T_e is regarded as an unknown variable within the modelling and the observed basin geometry and stratigraphy is used to determine T_e. Low values of T_e (<5 km) have been obtained from modelling for other rifts, e.g. Viking and Central Graben, North Sea; Aegean; Basin-and-Range; Barrow–Dampier Basins and Exmouth Plateau, N.W. Australia; Tucano–Reconcavo Basin, (Brazil); and Baltimore Canyon (Barton & Wood 1984; Stein *et al.* 1988; Watts 1988; White & McKenzie 1988; Fowler & McKenzie 1989; Kusznir & Egan 1990; Marsden *et al.* 1990; Couzens 1993; Magnavita *et al.* 1994). Large horizontal stresses generate brittle failure in the upper part of the crust and plastic deformation at mid-crustal level (Buck 1988; Kusznir *et al.* 1991; Hendrie *et al.* 1994). This brittle and plastic deformation reduces the flexural bending stresses and bending moment, and therefore the flexural rigidity and effective elastic thickness of the lithosphere. The low values of T_e obtained from modelling of the Tanganyika Rift and other rift examples are controlled by brittle and plastic failure envelopes within the upper crust. A detailed discussion of deformation generated by flexural bending stress and the control of T_e by fault extension is given in Hendrie *et al.* (1994).

The control of crustal structure and basin geometry by T_e is shown in Fig. 10 for a profile

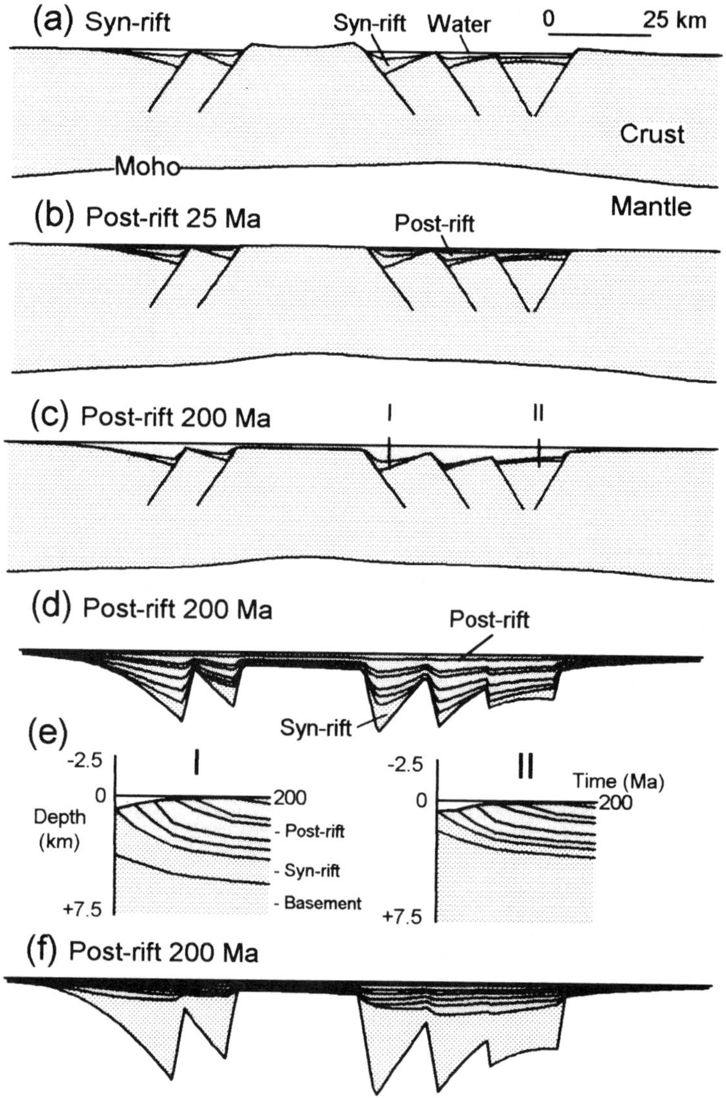

Fig. 11. Syn-rift and post-rift crustal structure and basin stratigraphy computed using the flexural cantilever model. (**a**) Syn-rift stage with partial fill of syn-rift accommodation space and partial erosion of footwall uplift and horst. (**b**) Early post-rift stage following 25 Ma of thermal subsidence, further erosion and sediment fill. (**c**) Mature post-rift stage following 200 Ma of thermal subsidence and sediment fill to sea level. (**d**) Vertically exaggerated section of 200 Ma stage showing basin geometry and stratigraphy. (**e**) Subsidence histories at locations I and II of (c). (**f**) Post-rift stage at 200 Ma but with sediment supply to sea level at all stages.

running across lake Turkana in the Northern Kenya Rift (Hendrie *et al.* 1994). Maximum horizontal bending stresses are also shown. High values of T_e generate too much footwall uplift and rift basins that are too deep, in addition to unrealistically large bending stress. Upper crustal rocks are unlikely to be able to sustain deviatoric stress magnitudes greater than only a few kilobars (Byerlee 1968). For smaller values of T_e the maximum horizontal bending stresses are greatly decreased. The model with $T_e = 0$, corresponding to Airy isostasy, gives no fault block rotation and is equivalent to a 2D McKenzie model. Increasing T_e to 1 km generates fault block rotations but with dip slope inflections which are not observed in the Turkana Rift. The best fit, for the Turkana profile, to observed basin geometry based on seismic reflection data was achieved with a T_e of 3 km.

Post-rift basin development

The flexural cantilever model may also be applied to post-rift basin development. Crustal structure and basin stratigraphy for both syn- and post-rift stages of basin formation are shown in Fig. 11 for a set of faults of varying

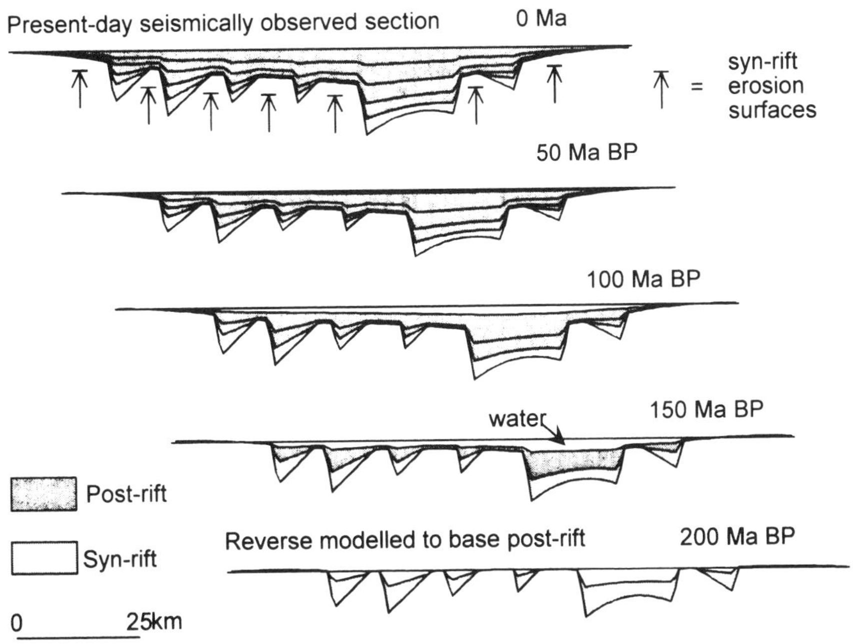

Fig. 12. Schematic series of sequential rift basin cross-sections showing reverse post-rift modelling of post-rift basin development. Reverse post-rift modelling consists of flexural backstripping, decompaction and reverse post-rift thermal subsidence modelling.

polarity and extension. The syn-rift stage of basin formation has only partial sediment fill of available accommodation space (Fig. 11a). Erosion can be expected to occur of sub-aerial topography associated with footwall uplift of basement bounding faults and central horsts (Fig. 11b). Post-rift thermal subsidence and fill of accommodation space leads to a thick post-rift sequence of (Fig. 11c and d) forming a regional blanket over syn-rift sub-basins and onlapping the pre-rift basement of the flanking platforms. Subsidence histories may be computed for any part of the basin (Fig. 11e). Compaction is included in the model and compaction anticlines can be seen to form within the post-rift sequence above deeper rifted fault block crests.

The basin stratigraphy and subsidence is extremely dependent not only on fault spacing and extension but also on sediment supply and erosion. Figure 11f shows the results of a tectonic model identical to that of Fig. 11a–c, but with sediment fill to sea level at all stages. The thickness of syn-rift is greatly increased at the expense of the post-rift.

Reverse modelling of post-rift basin formation

While the flexural cantilever model provides a quantitative forward tectonic model of the syn- and post-rift stages of basin formation, basin subsidence and stratigraphy are extremely dependent on sediment supply. Sediment supply is dependent on many external parameters (e.g. drainage patterns, sediment source type, climate) and as a consequence is extremely difficult to predict quantitatively. Thus, while the syn-rift stage of the development of a basin can be predictively forward modelled with some success, the post-rift development is difficult to model without reference to the actual sediment geometries and lithologies. This information is available in geological cross-sections commonly derived from seismic reflection data. A solution to the problem of sediment supply in modelling the post-rift development of a basin is to use the same physics and mathematics as the forward modelling of the basin, but to start with a present day cross-section and wind time backwards, i.e. reverse model the post-rift history of basin formation.

Reverse modelling post-rift basin development comprises flexural backstripping of the sediment load, sediment decompaction and reverse post-rift thermal subsidence modelling. The post-rift reverse modelling process is illustrated schematically in Fig. 12. Starting with a present day cross-section the post-rift basin thermal subsidence, sediment fill, isostatic response to sediment loading and sediment compaction are rewound backwards in time to the base of post-rift. In more detail, reverse post-rift modelling consists of the following processes:

1. Progressively remove stratigraphic units – youngest first.
2. Decompact (fluff out) lower units.
3. Calculate flexural isostatic uplift due to removal of sediment load.
4. Calculate thermal uplift from estimated β stretching factor.
5. Correct for palaeo-bathymetric water loads.
6. Repeat above from (1) for each unit until the base of the post-rift is reached.

An important test of the post-rift reverse modelling process is that stratigraphic surfaces, once at sea level, should be restored to sea level at the appropriate point in time. Four of the most reliable indicators of palaeo-sea level are wave cut platforms, eroded fault-block crests (Fig. 12), carbonate reef facies and coal-bearing sequences (Bertram & Milton 1989).

Application of the reverse post-rift modelling process is shown in Fig. 13 applied to a section across the Snorre and Visund oil-fields in the

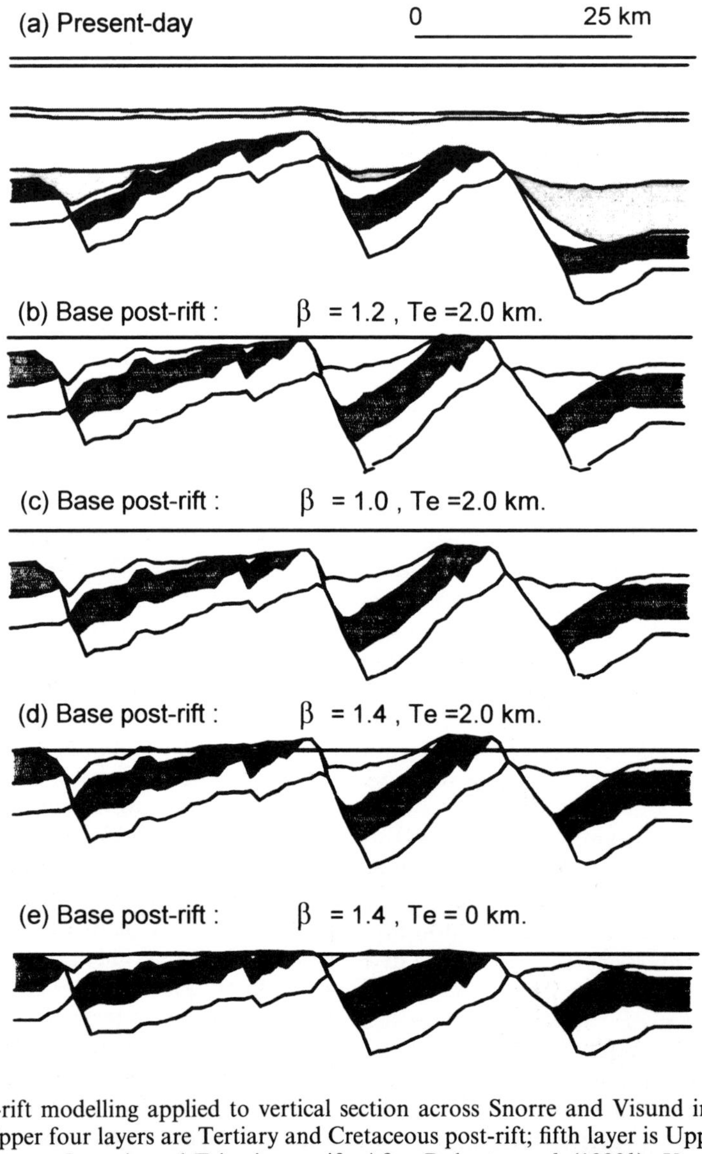

Fig. 13. Reverse post-rift modelling applied to vertical section across Snorre and Visund in the North Viking Graben, North Sea. Upper four layers are Tertiary and Cretaceous post-rift; fifth layer is Upper Jurassic syn-rift; sixth and seventh layers are Jurassic and Triassic pre-rift. After Roberts *et al.* (1993*b*). $V=4H$. (**a**) Present day section. (**b**) Reverse modelled to base post-rift (Base Cretaceous) using $\beta=1.2$ within reverse thermal subsidence modelling. $T_e=2.0$ km. (**c**) As (b) but using $\beta=1.0$ within reverse thermal subsidence modelling, corresponding to no thermal subsidence. (**d**) As (**b**) but using $\beta=1.4$ within reverse thermal subsidence modelling. (**e**) Reverse modelled to base post-rift (Base Cretaceous) using Airy isostasy ($T_e=0$ km) and β 1.4 within reverse thermal subsidence modelling.

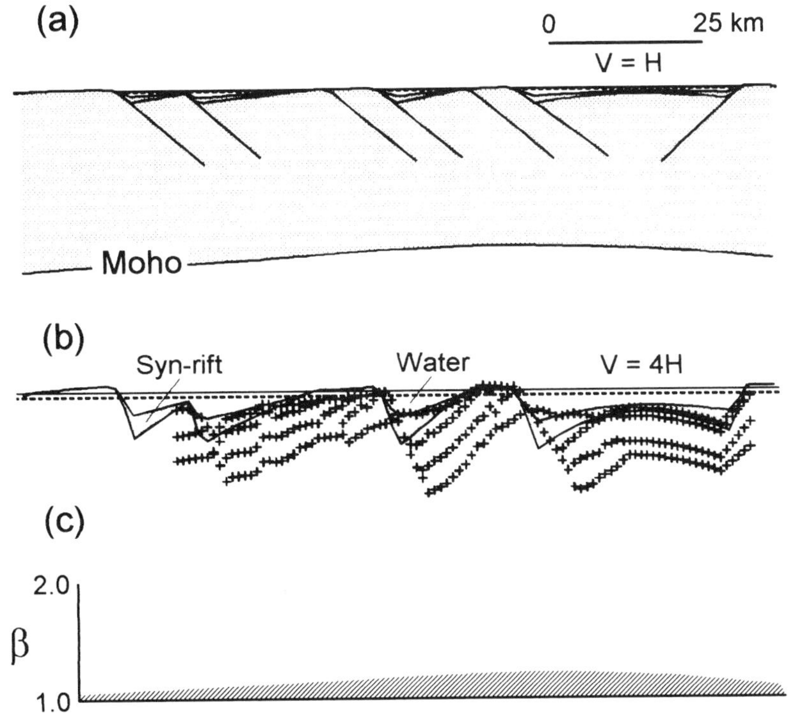

Fig. 14. Syn-rift forward model of Snorre and Visund structures (Viking Graben) using flexural cantilever model. (**a**) Crustal structure at end of Upper Jurassic rifting. (**b**) Basin geometry ($V=4H$) predicted by flexural cantilever model (solid line) after syn-rift partial sediment fill and fault block crest erosion, compared with post-rift reverse model (* * * line) from Fig. 13d. (**c**) β stretching factor profile predicted by flexural cantilever model.

North Viking Graben (after Roberts *et al.* 1993*b*). The present day section (Fig. 13a) shows two rotated fault blocks bounded by major extensional faults. Most recent rifting occurred in the Late Jurassic. Cretaceous and Tertiary sequences represent a post-rift thermal subsidence sequence. The fault block crests show significant erosion of the Lower and Middle Jurassic. In Fig. 13b the section has been reverse modelled to base Cretaceous, near to the base of the post-rift sequence, using a constant β stretching factor of 1.2 to determine the reverse post-rift thermal subsidence. The fault block crests in the model are restored to sea level, as is suggested by the observed geometry and stratigraphy of the fault block crests, which were either at or close to sea level at the end of Jurassic rifting and are both deeply eroded (Hollander 1987; Yielding & Roberts 1992; Yielding *et al.* 1992). The Upper Jurassic, a marine sequence, is submerged in the restored section, with its up-dip pinch-out located at sea level. Within the half-grabens, in the hanging walls of the major faults, substantial bathymetry is predicted to have existed at the end of rifting, with maximum water depths of the order of 1.25 km.

The palaeo-bathymetry and palaeo-emergence predicted by reverse post-rift modelling is dependent on the β stretching factor used to define the reverse post-rift thermal subsidence model. Lower β factors (Fig. 13c) give less uplift and more palaeo-bathymetry, while larger β factors (Fig. 13d) give more uplift, lower palaeo-bathymetry and greater emergence. Reverse modelling of post-rift subsidence provides a powerful technique for estimating β stretching factors from palaeo-bathymetric indicators (such as end rift erosion surfaces). The reverse post-rift models of Fig. 13b–d use T_e of 2.0 km to define the flexural-isostatic response.

Reverse post-rift modelling of profiles using flexural isostasy gives more reliable estimates of β factors than traditional Airy backstripping ($T_e=0$ km) of well data (e.g. Barton & Wood 1984; White & Latin 1993). If Airy isostasy ($T_e=0$ km) is used rather than flexural isostasy, a β stretching factor of 1.4 rather than 1.2 is needed to bring the eroded fault block crests to sea level at the base of post-rift (Fig. 13e). Airy backstripping of a single well on a fault block high greatly over-estimates the β stretching factor (e.g. White & Latin 1993), since subsidence of the fault block highs arising from the

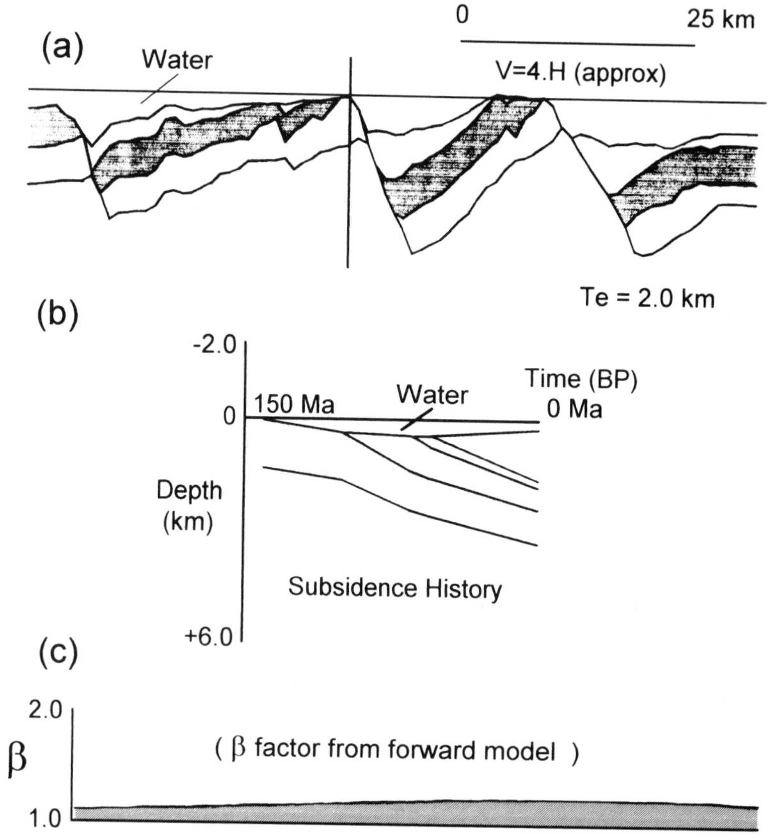

Fig. 15. (**a**) Reverse post-rift model of Snorre and Visund to Base Cretaceous using β stretching factor profile determined by syn-rift flexural cantilever model (Fig. 14c). (**b**) Post-rift subsidence history plot for fault block crest. The reverse model section and subsidence plot show eroded fault block crests returned to sea level at base of post-rift.

sediment infill of the substantial half-graben bathymetry remaining at the end of rifting would be wrongly interpreted as thermal driving subsidence, rather than as a combination of thermal subsidence and flexural-isostatic sediment loading (see also Roberts *et al.* 1993*b*).

Coupled forward and reverse modelling of extensional basin formation

Further constraint on the β stretching factor and effective elastic thickness, T_e, used in reverse post-rift basin modelling may be obtained by forward modelling the syn-rift development of the basin section using the flexural cantilever model (Fig. 14a). The forward modelling introduces further stratigraphic and structural information relating to the syn-rift basin formation stage. Forward modelling gives a laterally-varying β stretching factor which is controlled by the location of the forward model faults and the magnitude of their extension. The amount of syn-rift fault block rotation is particularly sensitive to T_e, which therefore enables T_e to be constrained. Syn-rift modelling results, after syn-rift sediment fill and erosion, may be compared with the reverse modelled section to base post-rift (Fig. 14b). Forward syn-rift modelling of the Snorre/Visund profile gives the laterally varying β factor shown in Fig. 14c and $T_e = 2$ km. The laterally varying β stretching factor has an average value similar to the value of 1.2 shown in the best-fit restoration of Fig. 13b.

The laterally varying β factor profile may then be used to provide an improved post-rift reverse model (Fig. 15). Subsidence history plots (Fig. 15b) may be determined for each point on the cross-section showing palaeo-bathymetry and palaeo-emergence through time.

Forward and reverse rift-basin modelling applied to the Viking Graben, North Sea Basin

Presented in this section are further details of a model originally described by Roberts *et al.* (1993*b*, Fig. 7). Combined forward and reverse

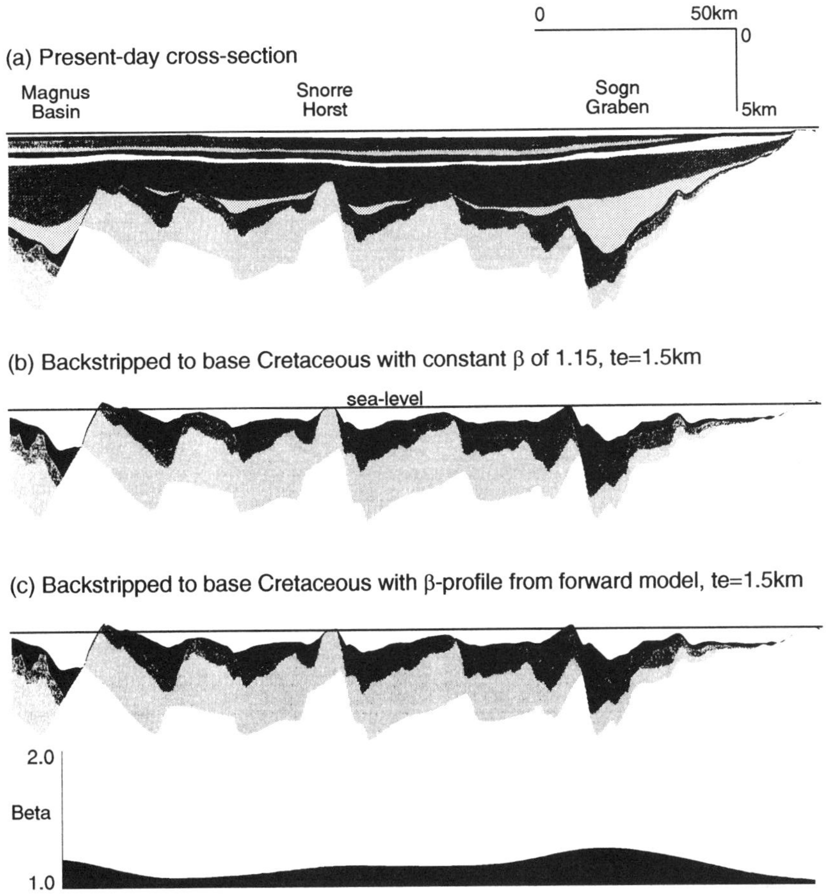

Fig. 16. Northern Viking Graben reverse model $V=4H$ for clarity. (**a**) Present day section. The upper seven layers, below the water layer, are Tertiary and Cretaceous post-rift. The eighth layer (black) is the Upper Jurassic ('syn-rift'). The ninth and tenth layers are Jurassic and Triassic pre-rift. (**b**) Reversed modelled to the base Cretaceous with constant β of 1.15 and $T_e=1.5$ km. (**c**) Backstripped to the base Cretaceous with variable β derived from the forward model (Fig. 17).

modelling has been applied to a 200 km cross-section across the Northern Viking Graben (61° 40′ N) in the North Sea. This example has been chosen for its complexity in both the syn- and post-rift, and brings together in one combined model many of the points raised earlier.

First-pass reverse model

The present day depth section is shown in Fig. 16a. It depicts a complex array of Jurassic tilted fault blocks, of varying polarity, overlain by a thick Cretaceous and Tertiary post-rift sequence. The most prominent features on the section are the Magnus Basin, an asymmetric half-graben, in the west, the eroded Snorre Horst in the centre, and the Sogn Graben in the east.

Figure 16b shows the section backstripped to the base Cretaceous seismic marker (strictly late Ryazanian in age), using a constant value for β of 1.15 and a T_e of 1.5 km. All of the major fault block crests are returned close to sea level in the restoration. In particular, the eroded Snorre Horst is at sea level and the footwall to the Magnus Basin (also known to be eroded; Young 1992) is slightly emergent. Water depth across the restoration is variable and controlled by fault block morphology. Maximum water depths in excess of 2 km are predicted in the Magnus Basin and Sogn Graben.

This first-pass restoration is taken to be geologically acceptable and is used next as a template for forward modelling.

Forward model

The Northern Viking Graben forward model is illustrated in Fig. 17, which was derived using a value for T_e of 1.5 km, the same as used for reverse modelling.

Figure 17a shows a 1 : 1 full-crustal profile of the syn-rift forward model. Note the topography

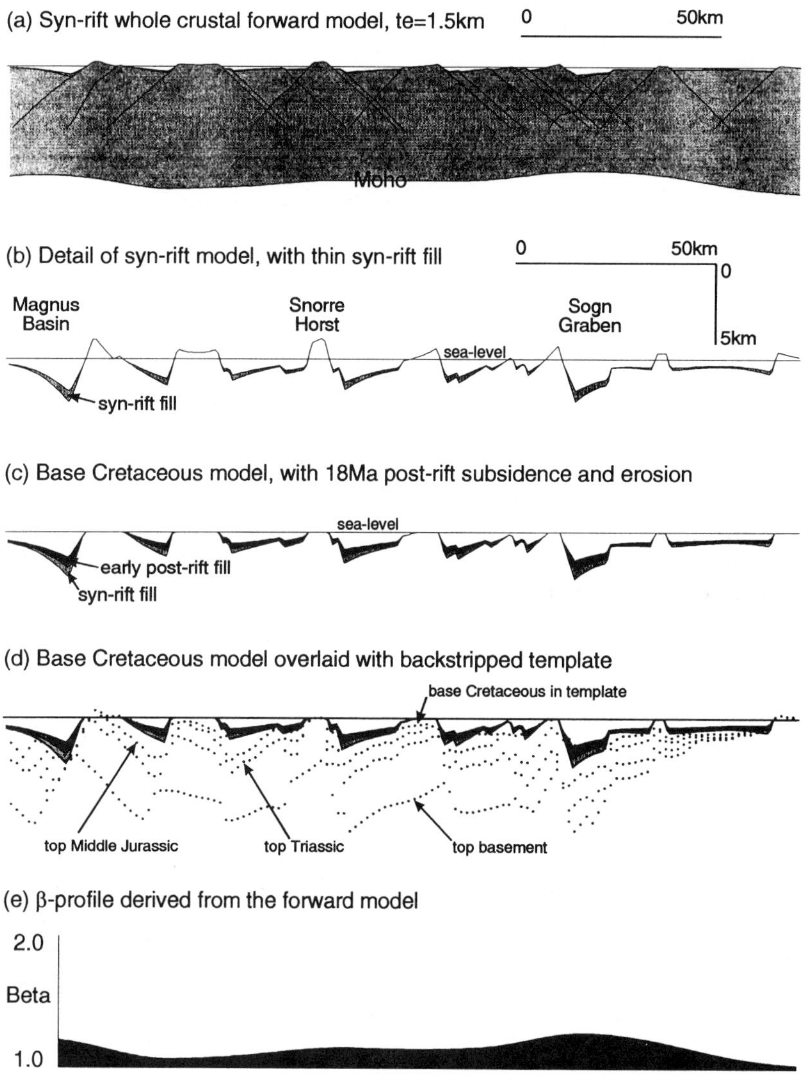

Fig. 17. Northern Viking Graben forward model. (**a**) Syn-rift whole crustal model ($V=H$), $T_e=1.5$ km. (**b**) Detail of syn-rift model. Note the emergent fault-block crests ($V=4H$). (**c**) Detail of base Cretaceous forward model, incorporating erosion, further sedimentation and 18 Ma of thermal relaxation. (**d**) Base Cretaceous model overlain by backstripped template (Fig. 16b). There is a good match between model and template given the complexity of the structure. (**e**) β profile derived from forward model, with maxima below the Magnus Basin and Sogn Graben.

on the Moho and the areas of maximum crustal thinning below the Magnus Basin and Sogn Graben.

Figure 17b shows a vertical exaggeration of the upper part of the syn-rift model. The thin syn-rift fill is shaded. The model predicts emergence above sea level of a number of fault block crests. Palaeogeographically this would appear as an archipelago of small islands. The fault block topography is complex, switching polarity in a number of places. Comparing the syn-rift model with the lake Tanganyika examples, the Magnus Basin is an asymmetric half-graben (cf. Fig. 7), the Sogn Graben approaches a full-graben (cf, Fig. 8) and the Snorre horst is a central horst (cf. Fig. 9).

The rate of Jurassic extension in the northern Viking Graben is thought to have been at its maximum towards the end of the Oxfordian (Roberts *et al.* 1993*a*). Thus, Fig. 17a and b are of a notional Oxfordian model. The flexurally backstripped restoration (Fig. 16b) was taken to the 'base Cretaceous' (late Ryazanian). Therefore, there is a small time gap between the syn-rift model and the backstripped restoration. To eliminate this time gap in the modelling the forward model was next allowed to thermally relax for 18 Ma. At the same time any emergent topography was eroded to sea level and a further interval of condensed deposition was incorporated. The resulting base Cretaceous forward model is shown in Fig. 17c. In stratigraphic terms the early post-rift fill is broadly equivalent to the Kimmeridge Clay source rock interval. The erosion products derived from the uplifted footwall highs have been identifed as locally-developed sandstones within the Kimmeridge Clay (Yielding & Roberts 1992, Fig. 13).

Figure 17d shows the base Cretaceous forward model overlain by the base Cretaceous backstripped template. In terms of matching model and template, top pre-rift in the template is best taken between the base Cretaceous and the top Middle Jurassic. Thus, top pre-rift in the forward model should lie within this interval wherever possible. Differences between the forward and reverse models are probably due to 3D effects. Given the structural complexity of the area, the forward and reverse models provide a good match, not only for the fault block geometry, but also for the base Cretaceous palaeobathymetry across the profile and for the width of the eroded fault block crests. This gives confidence not just in the robustness of the forward and reverse modelling techniques, but also in the initial seismic interpretation and depth conversion from which the section is derived.

The final illustrated aspect of the forward model (Fig. 17e) is the β-profile derived from the model. This is the tight profile of Roberts *et al.* (1993*b*). Predicted values of β reach a peak of 1.23 and 1.29 beneath the Magnus Basin and the Sogn Graben, respectively. This β-profile has been incorporated into a regional map of β by Roberts *et al.* (1993*b*).

Second-pass reverse model

Having achieved a satisfactory match between the forward and reverse models the β-profile from the forward model can be used to constrain further the reverse modelling. The result of backstripping the section to the base Cretaceous with the profile of variable β is shown in Fig. 16c. This restoration is in fact little different from the constant β model because the average β across the profile is close to 1.15. This demonstrates that even in areas of complex upper crustal structure reverse modelling, with the initial assumption of constant β, may not be an unrealistic simplification.

Discussion

The McKenzie model (1978) assumes that the lithosphere extends by plastic distributed deformation which is uniform with depth and provides a description of the regional response of the lithosphere to extension at lateral length scales of 100 km or more. These responses are the isostatic response of the lithosphere to syn-rift crustal thinning and geotherm perturbation followed by post-rift geotherm re-equilibration. A particular success of the McKenzie model is its ability to predict the post-rift basement driving subsidence associated with post-rift lithosphere cooling. However, at the length scale of tens of kilometres and less, rift basin geometry is controlled by basement faulting, a process explicitly not addressed by the McKenzie model. The interaction of hanging wall subsidence and footwall uplift gives rise to 'domino-style' block rotation tectonics. This creates syn-rift accommodation space in the form of half- and full-grabens, together with structural highs in the form of footwall fault block crests, rift flank highs and internal horst blocks. The latter may act as sediment sources through their erosion. The response of the upper lithosphere to faulting may be described quantitatively by elastic dislocation theory or flexural-isostatic thin-plate theory. The flexural cantilever model attempts to combine the regional length-scale responses of the lithosphere to extension described by the McKenzie model with the shorter length-scale processes associated with upper crustal basement extensional faulting. The flexural cantilever model predicts the syn-rift flexural-isostatic response of the lithosphere to crustal thinning by upper crustal faulting and lower crustal distributed plastic deformation, and the increase in geotherm gradient and post-rift geotherm re-equilibration. In addition, it incorporates the flexural-isostatic response to sediment fill and erosional denudation.

Models of continental rifting have been proposed in which regional low-angled detachments cut the whole of the lithosphere, extending as continuous structures from surface to the base of the lithosphere (Wernicke 1981, 1985; Etheridge *et al.* 1990). These models are based largely on field observations within the Basin-and-Range Province of the western USA, where extensional mylonites, associated with metamorphic core complexes within the tectonically exhumed footwalls of low-angled faults, can be traced many tens of kilometres in the direction of slip. These exhumed fault planes are extrapolated downwards on restored sections to imply lithosphere-scale faulting. An excellent review of low-angled detachment faults and metamorphic core complexes in the Basin-and-

Range Province is provided by Lister & Davies (1989). Do such lithosphere scale detachments exist, and does the Basin-and-Range extend by a similar or different lithosphere process to that responsible for the formation of continental rifts such as the Viking Graben and Lake Tanganyika Rift?

The evidence for lithosphere-scale detachments at depth is circumstantial and the existence of such regional low-angled faults within basement has been the subject of considerable debate and controversy. As discussed earlier (see **Introduction**) no deep seismic reflection section shows an unequivocal example of a dip-slip structure (fault or shear zone) passing continuously from the surface to the base of the lithosphere. Additional arguments against the existence of lithosphere-scale low-angled detachments include: low-angled faulting is not favoured in theoretical (Andersonian) models of extending brittle layers (Buck 1988); low-angled normal faults are seldom observed sesimically (Jackson & McKenzie 1983; Jackson 1987); the changes in metamorphic grade along the footwall of exhumed Basin-and-Range fault structures interpreted as lithosphere-scale low-angled faults suggest that the fault was initiated at a much higher angles of the order of 45° (e.g. Wernicke & Axen 1988); and the cooling rates of lower plate rocks in Basin-and-Range core complexes require larger regional extension rates to drive exhumation than is observed (Buck 1988).

Outside the Basin-and-Range Province, basement low-angled detachment faults with large displacements of tens of kilometres, have been found in collapsed mountain belts (e.g. Selverstone 1988), but not definitively in continental rifts. Numerous low-angled detachment models have been proposed for rifts, for example, in the uplifted shoulder of the south eastern Red Sea, low-angled Wernicke-type detachment models have been proposed by Bohannon (1986). However, only low-angled faulting at the base of the sedimentary series has been observed, not within basement rocks. The main low-angled detachment zone proposed by these workers remains an hypothetical zone that underlies high-angled faults at the surface. Furthermore, while asymmetric topography and volcanism have been used as evidence for lithosphere deformation by low-angled detachment in the northern Red Sea, such detachment models cannot account for the observed heat flow (Buck *et al.* 1988).

Can the observations made primarily in the Basin-and-Range Province, which provide the stimulus for lithosphere-scale low-angled detachment models, be reconciled with a lithosphere deformation model in which only the upper crust deforms by faulting, while the remaining lithosphere of the lower crust and mantle deforms by distributed plastic deformation as assumed by the flexural cantilever model? Buck (1988) showed that Basin-and-Range Province type structures, i.e. metamorphic core complexes and large lateral expanses of now horizontal fault plane fabric within exhumed footwall, can be explained without recourse to lithosphere-scale detachments. Buck assumed that high-angled planar extensional faults, similar to those responsible for rift basin formation and to those currently active in the northern Basin-and-Range (Stein & Barrientos 1985; Westaway 1989), cut the upper crust down to depths of no more than 15–20 km, and give way below that depth to distributed plastic deformation in the lower crust and mantle. As extension and crustal thinning proceeds, lower crustal material passes upwards through the plastic to brittle transition so manufacturing a new fault plane, some of which is tectonically exhumed within the footwall of the fault. As a consequence the resulting relict fault plane, exhumed within the footwall, may be much longer in the dip direction than the active fault section, which is restricted to the upper crust. There is, therefore, no need to infer low-angled faults which penetrate the whole of the lithosphere. Buck also showed, using thin-plate flexure theory, that the isostatic response of a tectonically exhumed footwall to large amounts of extension produces rotation about the horizontal axis. Fault plane fabrics, originally at dips of 50–60° when active within the upper crust, are isostatically rotated to the horizontal, as observed within exhumed metamorphic core complexes. Such footwall rotations about the horizontal axis have been confirmed using palaeo-magnetism (Dagley & Piper 1988). Buck inferred a very low T_e of 1–2 km for Basin-and-Range extension, which he suggested was due to pervasive deformation of the upper lithosphere in response to the very large bending stresses associated with footwall and hanging wall flexure.

The flexural cantilever model has been used to explore the formation of low-angled faults and metamorphic core-complexes (Fig. 18). The results of lithosphere extension on a single fault with extension of 10, 25 and 50 km, respectively, are shown in Fig. 18a–c; T_e is 1.5 km, the fault penetrates to a maximum depth of 15 km and has an initial dip of 50°. Plastic deformation within the lower crust and mantle is distributed symmetrically beneath the upper crustal fault

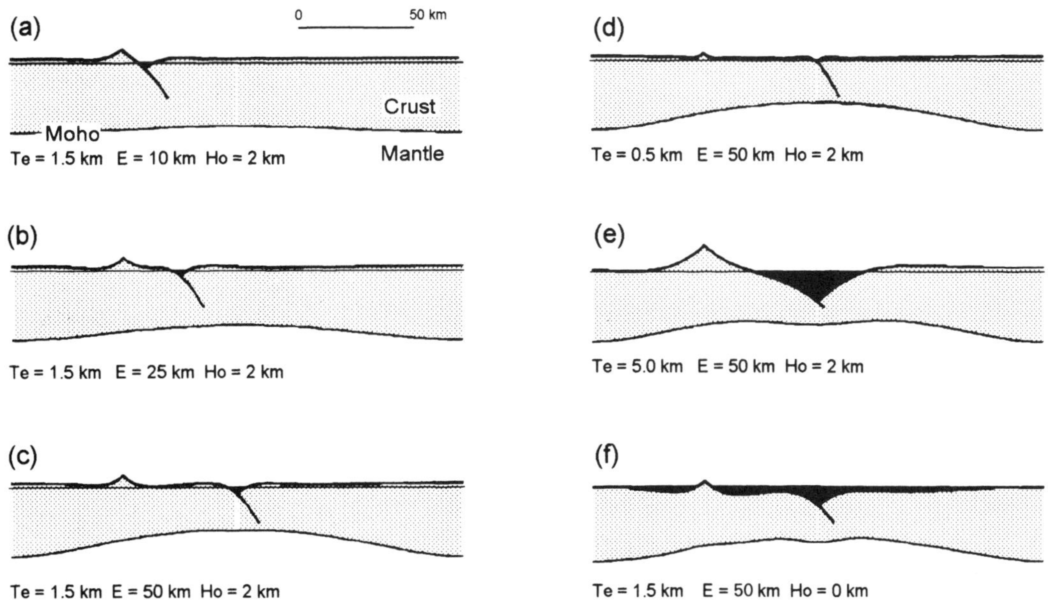

Fig. 18. Application of the flexural cantilever model to Basin-and-Range-style extension. Syn-rift crustal structure generated by extension on single fault. (**a**) Extension = 10 km; T_e = 1.5 km; 200 km width for plastic deformation in lower crust and mantle. Initial pre-rift crustal surface, H_o, 2 km above sea level. Sediment fill to sea level. (**b**) Extension = 25 km; other parameters as in (a). (**c**) Extension = 50 km; other parameters as in (a). (**d**) T_e = 0.5 km; extension = 50 km; other parameters as in (a). (**e**) T_e = 5.0 km; extension = 50 km; other parameters as in (a). (**f**) Initial pre-rift crustal surface at sea-level; extension = 50 km; T_e = 1.5 km; other parameters as in (a).

over a width of 200 km. The initial pre-rift surface of the crust is at 2 km above sea level. Sediment fill is to sea level within the space created by hanging wall subsidence. As extension increases the length of exhumed fault footwall increases. The active buried fault remains near planar and at a steep angle. The rotation of exhumed footwall is pronounced, with the region adjacent to the active fault being rotated through the horizontal so that it dips away from the active fault. The footwall cut-off forms a small dimple which in reality would erode to base level. At 50 km extension a horizontal fault of large areal extent, underlain by a footwall metamorphic core complex, is produced.

The rotation of footwall and hanging wall during extension is stongly controlled by T_e used within the flexural cantilever model. Increasing T_e decreases the amount of isostatic rotation. The effect of using values of T_e = 0.5 and 5 km are shown in Fig. 18d and e for an extension of 50 km. Other parameters are identical to those of Fig. 18c for which T_e = 1.5 km. A T_e of 5 km is clearly too high to produce a low-angled fault or an exposed metamorphic core complex.

The effect of initial elevation of the pre-rift crustal surface exerts a strong control on sediment supply and has a pronounced effect on basin geometry. A model with pre-rift crustal surface at sea level and 50 km extension is shown in Fig. 18f. The resulting basin has been filled to sea level. A substantial basin now exists and the degree of isostatic rotation within the exhumed footwall beneath the basin is diminished. Few faults in continental rifts, such as the East African Rift System or the North Sea Rift, are observed to have extensions/heaves which exceed 10 km. These systems usually exhibit much smaller amounts of extension than observed in the Basin-and-Range (except where sea floor spreading is initiated) and usually involve extension on many faults, typically 10–20 km apart (Fig. 2). The model shown in Fig. 18f is therefore unrealistic.

Conclusions

1. During continental lithosphere extension deformation in the upper continental crust occurs on planar faults extending to a depth of 10–20 km, while the lower crust and mantle deform by distributed plastic deformation. This leads to the formation of rift basins.
2. The McKenzie model of continental lithosphere extension describes the first-order lithosphere consequences of crustal thin-

ning, geotherm perturbation and re-equilibration and isostatic readjustment, leading to rapid syn-rift subsidence followed by slower exponential post-rift subsidence.

3. Rift basin geometry is controlled by major basement faults in the upper continental crust. The fundamental building block of rift basin formation is a half-graben formed in the hanging wall of a major basement fault. Extension on these major basement faults generates not only hanging wall subsidence but also footwall uplift. The interaction of footwall uplift and hanging wall subsidence on multiple faults of similar polarity produces block rotation tectonics, while the interaction of faults of opposite polarity gives full-grabens and fault bounded horsts.
4. Uplift of the footwalls of major extensional basement faults generates erosion, which may be of significant magnitude adjacent to major rift bounding faults.
5. The flexural cantilever model quantifies the flexural-isostatic response of continental lithosphere to crustal thinning by faulting and plastic deformation, and thermal perturbation and re-equilibration. It may be used to predict crustal structure and basin geometry for syn- and post-rift stages of rift basin formation.
6. Rift basin subsidence history is critically controlled by sediment supply, compaction and erosion processes. These serve to amplify basement driving subsidence.
7. Syn-rift forward modelling, using the flexural cantilever model, may be used to predict rift basin geometry, β stretching factors and magnitudes of footwall erosion.
8. Post-rift reverse modelling, starting with present day sections and winding time back to the base of post-rift, may be used to predict β stretching factors, post-rift palaeo-bathymetry, palaeo-emergence and subsidence history.
9. At high values of extension the flexural cantilever model can simulate low-angled faults and footwall metamorphic core-complexes, comparable with those observed in the Basin-and-Range. It is thus a unifying model of rift basin formation.
10. The value of effective elastic thickness (T_e) controlling fault-block geometry and isostatic response to faulting in rift basins appears to be uniformly $\leq$5 km. Although a low value it is significantly not zero. A T_e $\geq$1 km gives a flexural strength for the lithosphere which is sufficient to support fault block topography.

We thank Derek Hendrie, Mark Newall, Gary Marsden, Xin Quan Ma, John Walsh, Juan Watterson and Graham Yielding for discussions, help and encouragement. We also thank John Booth and Bill Bosworth for their constructive reviews of this paper. The modelling of the Viking Graben profiles was carried out in collaboration with Ian Walker, Donald Dorn-Lopez and Tony Doré of Conoco Norway, to whom we are grateful for permission to release the results.

References

BARR, D. 1987*a*. Lithospheric stretching, detached normal faulting and footwall uplift. *In:* COWARD M. P., DEWEY J. F. & HANCOCK, P. L. (eds) *Continental Extensional Tectonics*. Geological Society, London, Special Publication, **28**, 75–94.

—— 1987*b*. Structural/stratigraphic models for extensional basins of half-graben type. *Journal of Structural Geology*, **9**, 491–500

BARTON, P. & WOOD, R. 1984. Tectonic evolution of the North Sea Basin: crustal stretching and subsidence. *Geophysical Journal of the Royal Astronomical Society*, **79**, 987–1022.

BEACH, A. 1986. A deep seismic reflection profile across the northern North Sea. *Nature*, **323**, 53–55.

——, BIRD, T. & GIBBS, A. 1987. Extensional tectonics and crustal structure: deep seismic reflection data from the northern North Sea Viking Graben. *In:* COWARD, M. P., DEWEY, J. F. & HANCOCK, P. L. (eds) *Continental Extensional Tectonics*. Geological Society, London, Special Publication, **28**, 467–76.

BERTRAM, G. T. & MILTON, N. J. 1989. Reconstructing basin evolution from sediment thickness; the importance of palaeo-bathymetric control, with reference to the North Sea. *Basin Research*, **1**, 247–257.

BOHANNEN, P. G. 1986. Tectonic configuration of the western Arabian continental margin, S. Red Sea. *Tectonics*, **5**, 477–499.

BUCK, R. W. 1988. Flexural rotation of normal faults. *Tectonics*, **7**, 959–973.

——, MARTINEZ, F., STECKLER, M. S. & COCHRAN, J. R. 1988. Thermal consequences of lithosphere extension: pure and simple. *Tectonics*, **7**, 213–234.

BURGESS, C. F., ROSENDAHL, B. R., SANDER, S., BURGESS, C. A., LAMBIASE, J., DERSKEN, S. & MEADER, N. 1988. The structure and stratigraphic evolution of Lake Tanganyika. *In:* MANSPEIZER, W. (ed.) *A Case Study of Continental Rifting, Triassic/Jurassic Rifting*, Developments in Geotectonics, **27**, Elsevier, 859–881.

BYERLEE, J. D. 1968. Brittle–ductile transition in rocks. *Journal of Geophysical Research*, **73**, 4741–4750.

CARTWRIGHT, J. A. 1991. The kinematic evolution of the Coffee Soil fault. *In:* ROBERTS, A. M., YIELDING, G. & FREEMAN, B. (eds) *The Geometry of Normal Faults*. Geological Society, London,

Special Publication, **56**, 29–40.

COUZENS, T. 1993. *The rift to drift transition and sequence stratigraphy at passive continental margins*. PhD thesis, University of Liverpool, UK.

DAGLEY, P. & PIPER, J. D. A. 1988. The application of palaeomagnetism to extensional tectonics: a palaeomagnetic study of the Parker District Basin and Range Province Arizona. *In:* KISSEL, C. & LAJ, C. (eds) *Palaeomagnetic Rotations and Continental Deformation. Proceedings of Nato Advanced Research Workshop*. Klower, Netherlands.

DAVISON, I. 1989. Extensional domino fault tectonics: kinematics and geometrical constraints. *Annales Tectonicae*, **3**, 12–24.

ETHERIDGE, M. A., SYMONDS, P. A. & LISTER, G. S. 1990. Application of the detachment model to reconstruction of continental conjugate margins. *In:* TANKARD, A. J. & BALKWILL, H. R. (eds) *Tectonics and Stratigraphy of the North Atlantic Margins*. American Association of Petroleum Geologists Memoir **46**, 23–40.

FOWLER, S. & MCKENZIE, D. P. 1989. Gravity studies of the Rockall and Exmouth Plateaux using SEASAT altimetry. *Basin Research*, **2**, 27–34.

HENDRIE, D., KUSZNIR, N. J., MORLEY, C. & EBINGER, C. 1994. A quantitative model of Cenozoic rift basin development in northern Kenya. *Tectonophysics*, in press.

HOLLANDER, N. B. 1987. Snorre. *In:* SPENCER, A. M. *et al.* (eds) *Geology of the Norwegian Oil and Gas Fields*. Graham & Trotman, 307–318.

JACKSON, J. A. 1987. Active normal faulting and crustal extension. *In:* COWARD, M. P., DEWEY, J. F. & HANCOCK, P. L. (eds) *Continental Extensional Tectonics*. Geological Society, London, Special Publication, **28**, 3–17.

——— & MCKENZIE, D. P. 1983. The geometrical evolution of normal fault systems. *Journal of Structural Geology*, **5**, 471–482.

JANKOF, K. 1945. Changes in ground level produced by the earthquakes of April 14 to 18, 1928 in southern Bulgaria. *In: Tremblements de Terre en Bulgarie*, Nos 29–31. Institut Meterologique Central de Bulgarie, Sofia, 13–136 (in Bulgarian).

KING, G. C. P., STEIN, R. S. & RUNDLE, J. B. 1988. The growth of geophysical structures by repeated earthquakes: 1. Conceptual framework. *Journal of Geophysical Research*, **93**, 13 307–13 318.

KLEMPERER, S. 1988. Crustal thinning and nature of extension in the northern North Sea from deep seismic reflection profiling. *Tectonics*, **7**, 803–822.

——— & HOBBS, R. 1992. The BIRPS Atlas – Deep seismic reflection profiles around the British Isles. Cambridge University Press.

——— & WHITE, N. J. 1989. Coaxial stretching or lithospheric simple shear in the North Sea? Evidence from deep seismic reflection profiling and subsidence. *Memoir of the American Association of Petroleum Geologists*, **46**, 511–521.

KUSZNIR, N. J. & EGAN, S. S. 1990. Simple-shear and pure-shear models of extensional sedimentary basin formation: application to the Jeanne D'Arc basin, Grand Banks of Newfoundland. *Memoir of the American Association of Petroleum Geologists*, **46**, 305–322.

——— & PARK, R. G. 1987. The extensional strength of the continental lithosphere: its dependence on geothermal gradient, crustal composition and thickness. *In:* COWARD, M. P., DEWEY, J. F. & HANCOCK, P. L. (eds) *Continental Extensional Tectonics*. Geological Society, London, Special Publication, **28**, 35–42.

———, MARSDEN, G. & EGAN, S. S. 1991. A flexural cantilever simple-shear/pure-shear model of continental extension. *In:* ROBERTS, A. M., YIELDING, G. & FREEMAN, B. (eds) *The Geometry of Normal Faults*. Geological Society, London, Special Publication, **56**, 41–61.

——— & ZIEGLER, P. A. 1992. The mechanics of continental extension and sedimentary basin format: A simple-shear/pure-shear flexural cantilever model. *Tectonophysics*, **215**, 117–131.

LISTER, G. S. & DAVIES, G. A. 1989. The origin of metamorphic core complexes and detachment faults formed during Tertiary continental extension in the northern Colorado River region, U.S.A. *Journal of Structural Geology*, **11**, 65–94.

MA, X. Q. & KUSZNIR, N. J. 1994. Elastic moduli layering and gravitational effects on 3-D subsurface fault displacement fields. *Geophysical Journal International*, in press.

MAGNAVITA, L., DAVISON, I. & KUSZNIR, N. J. 1994. The erosion and exhumation history of the Reconcavo–Tucano–Jatoba Rift, N.E. Brazil during south Atlantic opening. *Tectonics*, **13**, 367–388.

MANDL, G. 1987. Tectonic deformation by rotating parallel faults – the 'bookshelf' mechanism. *Tectonophysics*, **141**, 277–316.

MARSDEN, G., ROBERTS, A. M., YIELDING, G. & KUSZNIR, N. J. 1990. Application of a flexural cantilever simple-shear/pure-shear model of continental lithosphere extension to the formation of the North Sea Basin. *In:* BLUNDELL, D. J. & GIBBS, A. D. (eds) *Tectonic Evolution of the North Sea Rifts*. Oxford University Press, 236–257.

MCKENZIE, D. P. 1978. Some remarks on the development of sedimentary basins. *Earth & Planetary Science Letters*, **40**, 25–32.

MORLEY, C. 1988. Variable extension in Lake Tanganyika. *Tectonics*, **7**, 785–801.

———, NELSON, R. A., PATTON, T. L. & MUNN, S. G. 1990. Transfer zones in the East African rift system and their relevance to hydrocarbon exploration in rifts. *American Association of Petroleum Geologists Bulletin*, **74**, 1234–1253.

RICHTER, C. F. 1958. *Elementary Seismology*. W. H. Freeman.

ROBERTS, A. M. & YIELDING, G. 1991. Deformation around basin-margin faults in the North Sea/ Norwegian rift. *In:* ROBERTS, A. M., YIELDING, G. & FREEMAN, B. (eds) *The Geometry of Normal Faults*. Geological Society, London, Special Publication, **56**, 61–78.

———, ——— & BADLEY, M. E. 1993*a*. Tectonic and bathymetric controls on stratigraphic sequences within evolving half-graben. *In:* WILLIAMS, G. D.

& DOBB, A. (eds) *Tectonics and Seismic Sequence Stratigraphy*. Geological Society, London, Special Publication, **71**, 87–121.

——, —— & FREEMAN, B. (eds) 1991. *The Geometry of Normal Faults*. Geological Society, London, Special Publication, **56**, 264 pp.

——, ——, KUSZNIR, N. J., WALKER, I. & DORN-LOPEZ, D. 1993*b*. Mesozoic extension in the North Sea: constraints from flexural backstripping, forward modelling and fault populations. *In:* PARKER, J. R. (eds) *Petroleum Geology of Northwest Europe: Proceedings of the 4th Conference*. The Geological Society, London, 1123–1136.

ROSENDAHL, B. R. 1987. Architecture of continental rifts with special reference to East Africa. *Annual Review of Earth & Planetary Sciences*, **15**, 445–503.

——, KILEMBE, E. & KACZMARIK, M. 1992. Comparison of Lake Tanganyika, Malawi, Rukwa, and turkana Rift zones from analysis of reflection seismic data. *Tectonophysics*, **213**, 235–256.

——, REYNOLDS, D. J., LORBER, P. M., BURGESS, C. F., MCGILL, J., SCOTT, D., LAMBIASE, J. & DERKSEN, S. J. 1986. Structural expressions of rifting; lessons from Lake Tanganyika, Africa. *In:* FROSTICK, L. E., RENAUT, R. W., REID, J. & TIERCELIN, J. (eds) *Sedimentation in the East African Rifts*. Geological Society, London, Special Publication, **25**, 29–43.

SELVERSTONE, J. 1988. Evidence for east–west crustal extension in the Eastern Alps: Implications for the unroofing of the Tauern Window. *Tectonics*, **7**, 87–105.

SHUDOFSKY, G. N. 1985. Source mechanisms and focal depths of East African earthquakes using Rayleigh-wave inversion and body wave modelling. *Geophysical. Journal of the Royal Astronomical Society*, **83**, 563–614.

——, CLOETINGH, S., STEIN, S. & WORTEL, R. 1987. Unusually deep earthquakes in East Africa: constraints in the thermo-mechanical structure of a continental rift system. *Geophysical Research Letters*, **14**, 741–744.

STEIN, R. S. & BARRIENTOS, S. E. 1985. Planar high-angle faulting in the Basin and Range; Geodetic analysis of the 1983 Borah Peak, Idaho, earthquake. *Journal of Geophysical Research*, **90**, 11 355–11 366.

——, KING, G. C. P. & RUNDLE, J. B. 1988. The growth of geological structures by repeated earthquakes: 2. Field examples of continental dip-slip faults. *Journal of Geophysical Research*, **93**, 13 319–13 331.

VENING-MEINESZ, F. A. 1950. Les grabens Africains, resultat de compression ou de tension dans le croute terrestre? *Bulletin de l'Institut royal colonial de Belgique*, **21**, 539–552.

WALCOTT, R. I. 1970. Flexural rigidity, thickness and viscosity of the lithosphere. *Journal of Geophysical Research*, **75**, 3941–3954.

WATTS, A. B. 1988. Gravity anomalies, crustal structure and flexure of the lithosphere at the Baltimore Canyon Trough. *Earth & Planetary Science Letters*, **89**, 221–238.

WERNICKE, B. 1981. Low-angle normal faults in the Basin-and-Range province: nappe tectonics in an extending orogen. *Nature*, **291**, 645–648.

—— 1985. Uniform-sense normal sense simple-shear of the continental lithosphere. *Canadian Journal of Earth Sciences*, **22**, 102–125.

—— & AXEN, G. J. 1988. On the role of isostasy in the evolution of normal fault systems. *Geology*, **16**, 848–851.

WESCOTT, W. A., KREBS, W. N., ENGLEHARDT, D. W. & CUNNINGHAM, S. M. 1991. New biostratigraphic age dates from the Lake Rukwa rift basin in Western Tanzania. *American Association of Petroleum Geologists Bulletin*, **75**, 1255–1263.

WESTAWAY, R. W. C. 1989. North-east Basin-and-Range province active tectonics: an alternative view. *Geology*, **17**, 779–783.

WHITE, N. J. 1989. The nature of lithospheric extension in the North Sea. *Geology*, **17**, 111–114.

—— & LATIN, D. M. 1993. Subsidence analyses from the North Sea 'triple junction'. *Journal of the Geological Society, London*, **150**, 473–488.

—— & MCKENZIE, D. P. 1988. Formation of the 'steer's head' geometry of sedimentary basins by differential stretching of the crust and mantle. *Geology*, **16**, 250–253.

WHITTEN, C. A. 1957. Geodetic measurements in the Dixie valley area. *Bulletin of the Seismological Society of America*, **47**, 321–325.

YIELDING, G. 1990. Footwall uplift associated with Late Jurassic normal faulting in the northern North Sea. *Journal of the Geological Society, London*, **147**, 219–222.

—— BADLEY, M. E. & FREEMAN, B., 1991. Seismic reflections from normal faults in the northern North Sea. *In:* ROBERTS, A. M., YIELDING, G. & FREEMAN, B. (eds) *The Geometry of Normal Faults*. Geological Society, London, Special Publication, **56**, 79–89.

——, —— & ROBERTS, A. M. 1992. The structural evolution of the Brent Province. *In:* MORTON, A. C., HASZELDINE, R. S., GILES, M. R. & BROWN, S. (eds) *Geology of the Brent Group*, Geological Society, London, Special Publication, **61**, 27–44.

—— & ROBERTS, A. M. 1992. Footwall uplift during normal faulting – implications for structural geometries in the North Sea. *In:* LARSEN, R. M. (eds) *Structural and Tectonic Modelling and its Application to Petroleum Geology*. Norwegian Petroleum Society, Special Publication, **1**, 289–304.

YOUNG, R. 1992. Restoration of a regional profile across the Magnus Field in the Northern North Sea. *In:* LARSEN, R. M. *et al.* (eds) *Structural and Tectonic Modelling and its Application to Petroleum Geology*. Norwegian Petroleum Society, Special Publication, **1**, 221–229.

YOUNG, R. A., STEWART, S. C., SEAMAN, M. R. & EVANS, B. J. 1990. Fault plane reflection processing and 3D display: the Darling Fault, Western Australia. *Tectonophysics*, **173**, 107–117.

The uplift and erosion of sedimentary basins and the role of overpressure

Some geometrical characteristics of inversion

A. B. Hayward & R. H. Graham

SUMMARY: The inversion of extensional fault systems results in the reversal of slip on the faults and expulsion of the synrift fill. During inversion the beds in the cover sequence shorten before the net extension at the basement level has been cancelled. Shortening of the sedimentary cover generates folding and backthrusting in the still downthrown hanging wall block. Intracratonic inverted basins in different parts of the Alpine Foreland show similar structural geometries with the major extensional faults which controlled basin development reactivated during subsequent compression. We use examples from the Western Approaches and offshore Holland (Broad Fourteens Basin) to illustrate the structural styles developed during inversion. The fundamental control on compressional structural geometry exerted by pre-existing extensional structures is also visible in more complexly deformed orogenic belts, like the Western Alps and the Pyrenees. In these areas inversion also occurs, but more commonly extensional faults which may not have inverted act as an indirect control on the location of ramps, and/or thrust orientation. Seismic data are normally required to establish these effects with certainty. However, as the body of knowledge builds up, it is possible to recognize certain geometrical characteristics which suggest the control of extensional faults in thrust belts. These include footwall shortcuts, out of sequence structures and arcuate thrust–fold traces.

Introduction

The Inversion Tectonics meeting at the Royal Society, London on 3–4 March 1987, provoked lively discussion about the meaning and use of the term inversion. Should it be restricted to 'intracratonic' deformation in the otherwise undeformed foreland regions of orogenic belts, or should it be extended to include the restacking of highly stretched continental margins within the orogenic zones themselves? In this paper we present a series of examples which we believe demonstrate the progressive intensity of inversion ranging from the mild deformational structures of the Western Approaches Basin through to more intense deformation seen in the areas in offshore Holland and the structural styles associated with inversion in the Alpine and Pyrenean areas. In our discussion of inversion within orogenic areas we also review some of the geometrical phenomena which seem to be commonly associated with the presence of pre-existing extensional faults in thrust belts. These may eventually prove to be empirical criteria by which we may come to recognize the presence of extensional faults and half-graben expulsion without seismic, well or complete stratigraphic data.

Theoretical considerations and general principles

In recent years the nature and geometry of extensional fault systems have become increasingly well understood, due both to the much improved resolution of industrial seismic data, and the shooting of a number of deep crustal seismic lines. Most extensional basins are seen to be linkages of asymmetrical half-grabens. The impression of overall full graben symmetry in rift systems is created either by antithetic faults linking in to a controlling fault, or by periodic changes of dip of the controlling faults themselves. This is achieved either by extension parallel transfer faults, (Bally 1981; Gibbs 1983) or by more complex transfer zones (Rosendahl *et al.* 1986).

The 'positive' inversion of a previously formed sedimentary basin has been modelled by Bally (1984) using a simple half-graben geometry which undergoes slip reversal on the main boundary fault. The slip reversal produces volumetric problems in the synrift sediments as they become elevated. The problems are overcome by expelling the synrift sequence both forwards out of the graben towards the

From Holdsworth, R.E. & Turner, J.P. (compiler) *Extensional Tectonics: Regional-scale Processes*. The Geological Society, Key Issues in Earth Sciences, **2 (1)**, 225–247.
1476-315X/02/$15.00 © The Geological Society of London 2002
First published in: "Cooper, M.A. & Williams, G.D. (eds) *Inversion Tectonics*, The Geological Society, London, Special Publications **44**, 17–39".

footwall and backwards on to the old rollover (Figs 1 and 2). In some examples the forward expulsion produces a ramp anticline which is underlain by a flat detachment lying above the footwall (Fig. 1). Backwards expulsion may occur either on a new set of backthrusts (Fig. 2) or on older antithetic extensional faults which become reactivated as reverse faults.

During inversion the synrift sequence is shortened above the prerift surface (Fig. 2). Before the net extension at prerift level has been cancelled shortening of the sedimentary cover may generate backthrusting and folding in the still downthrown hanging wall block. In the case of mild inversion (e.g. the Western Approaches Basin) where there is simple dip reversal of a major extensional fault the contents of the half-graben are extruded as a major fold. In some cases of more severe inversion (e.g. the Broad Fourteens Basin) backthrusts are generated.

In other areas, for example the half-grabens in the Pelvoux Massif of the Western French Alps, there has been rotation of the bounding faults and intense internal deformation of the contents of the half-grabens (Gillcrist *et al. in press*; Gillcrist, Coward *et al. this volume*). Fig. 3 shows possible geometries for half-graben extrusion and deformation. In all of them the basement in either the hanging wall or footwall of the major fault may act as a butress against thrust translation so that a backward propagating deformation sequence may develop (Fig. 3A), an out of sequence thrust may cut the inversion (Fig. 3B) or a shortcut thrust might cut through the extensional fault footwall. Footwall shortcut thrusts are discussed later.

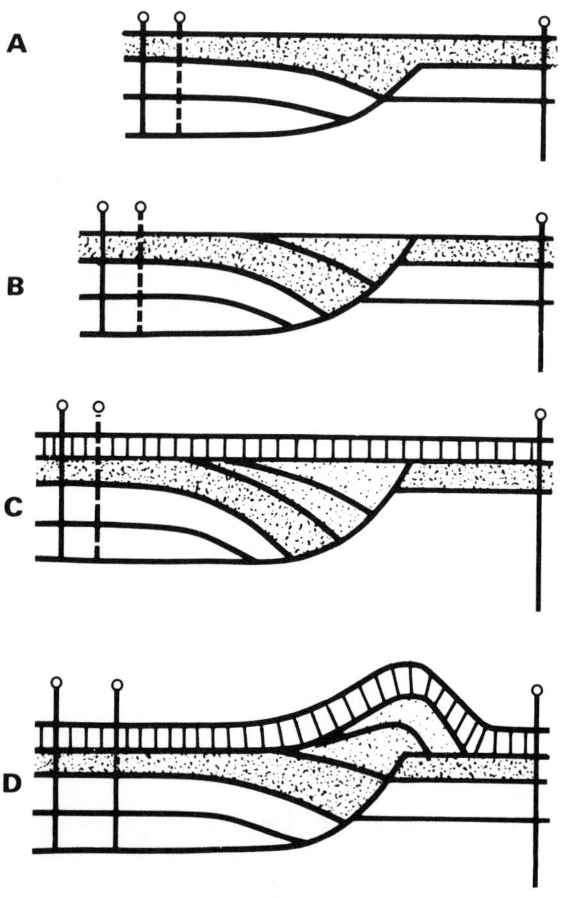

FIG. 1. Graphical construction of a simple listric growth fault subsequently inverted. A, B and C show progressive extension and deposition of a synrift fill (shaded); D shows the subsequent inversion geometry.

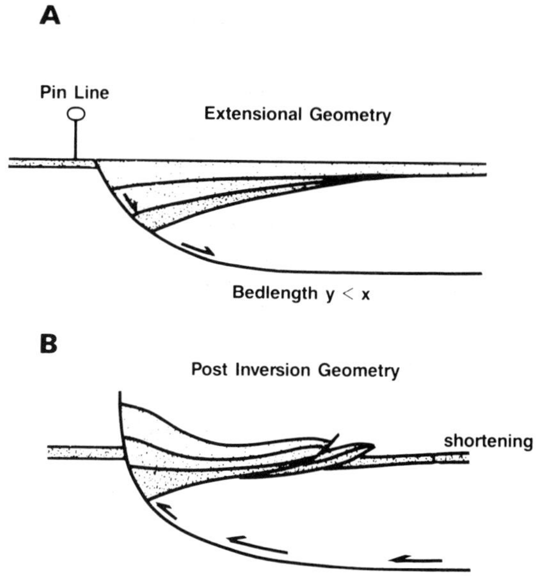

FIG. 2. A, The excessive bed length in the cover sequence to an asymmetric half-graben and B, the accommodation structures that develop during inversion of the half-graben (in part after Bally 1984).

In the following section we use two examples of both mild and moderate inversion in an intracratonic setting to demonstrate some of the geometrical characteristics that we have outlined above.

The Western Approaches Basin: an example of mild inversion

The Western Approaches Basin (Fig. 4) forms a ENE–WSW trending Mesozoic extensional basin of Early Jurassic to Late Cretaceous age. It was inverted during the late Oligocene–Early Miocene.

The basin can be divided into three sub-basins

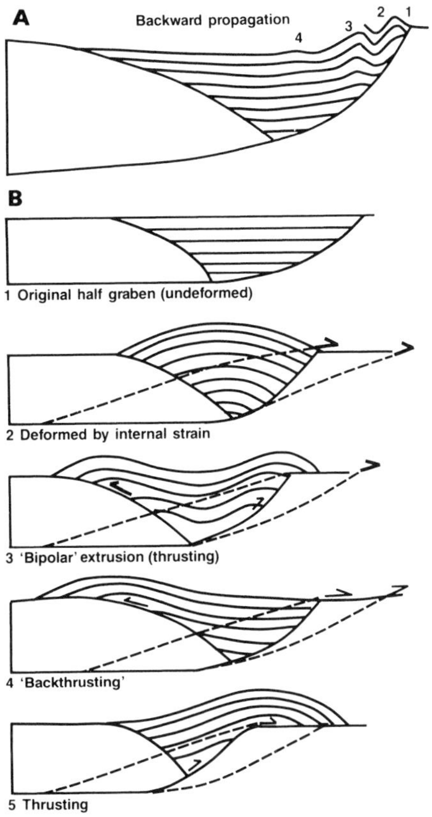

FIG. 3. A, The development of a backward propogating thrust sequence produced by the buttress effect of the footwall fault block during inversion. B, Theoretical models for the extrusion of half-graben synrift fills. Heavy line indicates potential for late out of sequence thrusting.

(Fig. 4) which comprise three inverted half-grabens. In the SW sub-basin a major reverse fault (Fig. 5) along the southern margin of the basin marks the boundary. The NW margin is defined by a number of minor faults (Fig. 5).

The geometry of the sedimentary fill of the half-graben indicates that during the development of the basin different faults were dominant at different times. The Jurassic sequence shows marked thickening to the north into what was clearly an extensional fault of Jurassic age. However, the Lower and Upper Cretaceous section thickens to the S into a major listric fault which controlled the development of the basin throughout the Cretaceous. Restoration of the cross-section in Fig. 5 demonstrates that throughout the Cretaceous the S sub-basin formed an asymmetric half-graben approximately 50 km wide controlled by a major listric extensional fault in the S. The N margin is defined by an extensional fault antithetic to the major basin controlling fault.

During Late Oligocene/Early Miocene inversion both faults were reactivated as reverse faults. The reversal of movement on the major basement faults resulted in uplift, the formation of a series of broad anticlinal culminations in the hanging wall and mild basin-wide inversion. Minor antithetic and synthetic faults were also reactivated.

In contrast to the S sub-basin, the present day geometry of the central sub-basin is controlled by a major reverse fault along the N boundary of the basin (Fig. 6). The Mesozoic stratigraphy

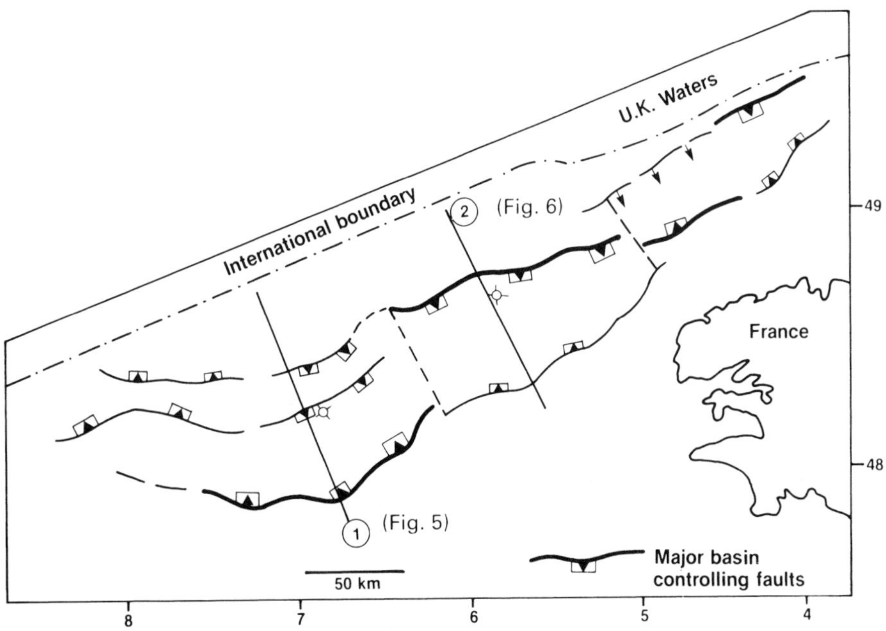

FIG. 4. Structural outline map of the Western Approaches Basin.

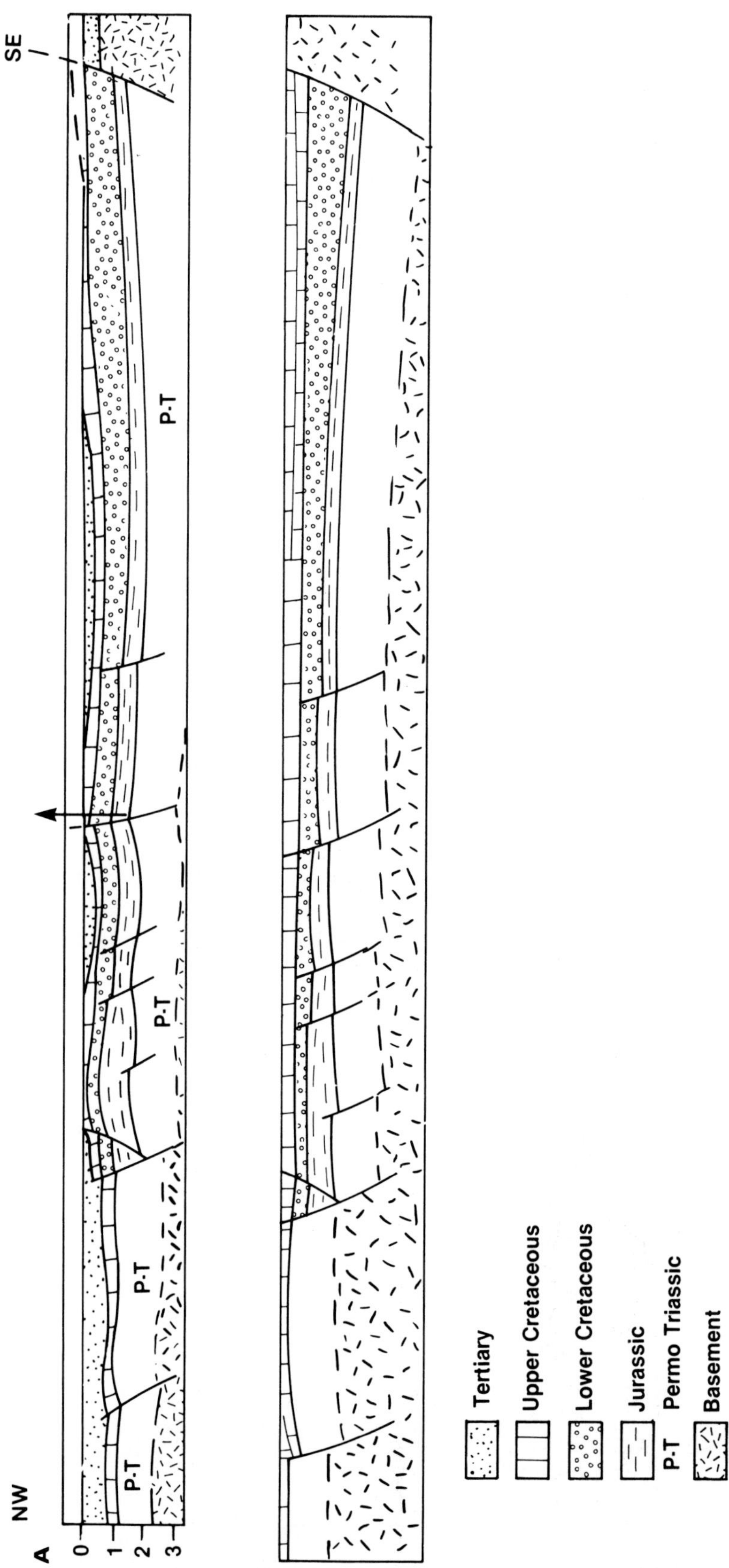

FIG. 5. A, Present day cross-section 1 across the southern sub-basin of the Western Approaches, based on a regional seismic line. Note vertical scale in two way time, $V = H$. B, The same section schematically restored to top Upper Cretaceous.

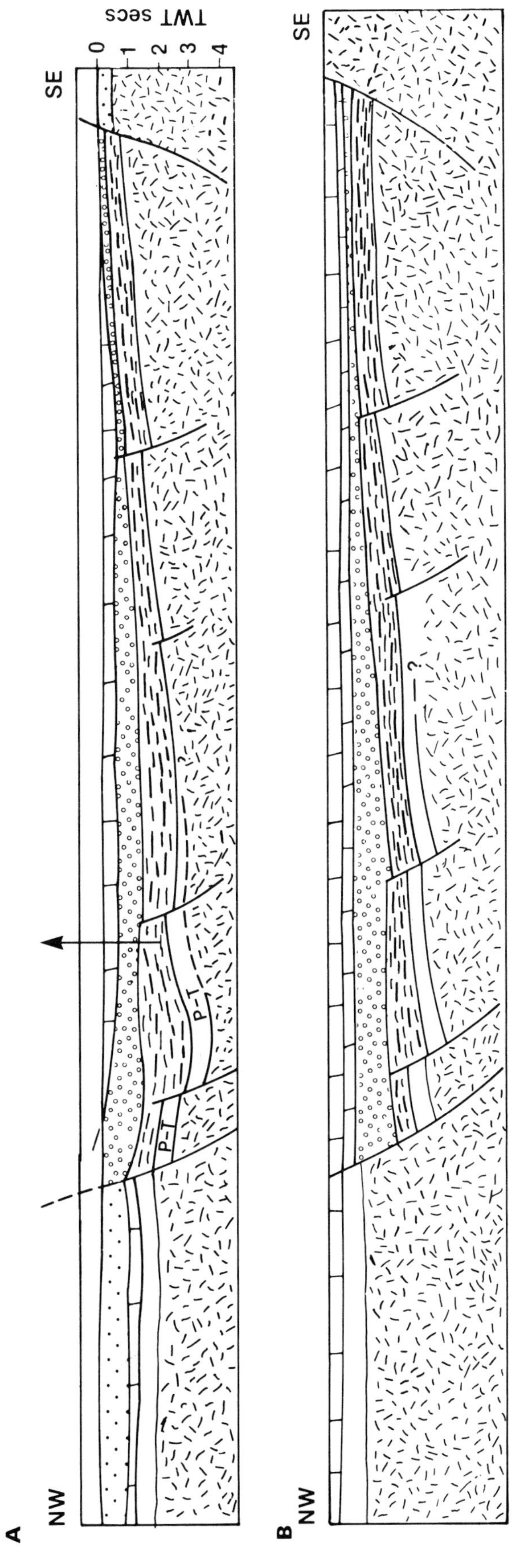

FIG. 6. A, Present day cross-section 2 across the central sub-basin of the Western Approaches, based on a regional seismic line. Note vertical scale in two way time, $V = H$. B, The same section schematically restored to top Upper Cretaceous.

shows a consistent thickening to the north into what was previously a major listric extensional fault. Restoration of the cross-section to the pre-inversion geometry demonstrates that the central sub-basin formed a half-graben approximately 50 km wide controlled by a major listric extensional fault in the N. The S margin of the basin was marked by minor antithetic faults (Fig. 6). During the Late Oligocene/Early Miocene inversion the major extensional fault was reactivated as a reverse fault. Most of the deformation was taken up on the major basement fault producing mild basin-wide inversion.

The N sub-basin also shows a change in asymmetry; the major extensional fault and subsequent reverse faults lie along the S margin of the basin. A change of asymmetry or polarity along strike is typical of many extensional basins (Bally 1981). In this example it clearly influences the subsequent inversion geometry.

The Broad Fourteens Basin: an example of moderate inversion

The Broad Fourteens Basin located offshore of Holland (Fig. 7) forms a NW–SE trending Mesozoic extensional basin of Triassic to Late Cretaceous age. It underwent moderate inversion during the Campanian (Late Cretaceous) (Oele *et al.* 1981).

The present day sub-Tertiary structure (Fig. 8) is controlled by a major bounding fault along the W margin of the basin. Across this fault the Jurassic sediments of the basin are juxtaposed with Cretaceous sediments of the basin margin. In contrast, along the E margin of the basin there is no major basin boundary fault and the Jurassic–Cretaceous sedimentary sequence thins progressively onto the Texel–Ijsselmeer High. Restoration of the cross-section (Fig. 8) demonstrates that the Broad Fourteens Basin formed a large asymmetric half-graben approximately 60 km wide throughout the Jurassic and Cretaceous. It was controlled by a major listric extensional fault along the W margin of the basin. The Jurassic–Cretaceous sedimentary sequence shows a dramatic thickening into the fault and thins onto the hinge area along the E margin of the basin. Minor Jurassic–Cretaceous extensional faults in the hinge area of the basin represent antithetic faults to the major listric fault.

During the Late Cretaceous inversion, the major listric extensional fault and minor antithetic faults were reactivated as thrusts or reverse faults. As in the Western Approaches Basin most of the deformation was taken up on the major basement fault, resulting in the basin-wide inversion of the Broad Fourteens Basin. In contrast to the Western Approaches Basin where there is very little shortening, restored cross-sections across the Broad Fourteens Basin show a total shortening of around ten per cent. The pre-inversion rollover geometry of the basin (Fig. 8) results in the cover sequence of Mesozoic sediments having a greater bed length than the underlying basement surface. The additional shortening required in the cover sequence is accommodated by the backward expulsion of the half-graben fill on a series of reverse faults and listric thrusts. These are detached mostly in the Permian Zechstein evaporite sequence and are developed principally in the hinge area of the graben. A geoseismic section over the rollover (NE) margin of the Broad Fourteens Basin is shown in Fig. 9. Imbricate slices which are detached in Zechstein evaporites end in a triangle zone on the crest of rollover. They were formed during the Late Cretaceous inversion and expulsion of the graben fill towards the rollover shoulder. In areas where there are no Zechstein evaporites to form a suitable detachment, the additional shortening required in the cover sequence is accommodated by the development of a series of small pop-up structures.

Inversion in orogenic belts

Recent years have seen the widespread application of thin skinned structural models, first developed by petroleum geologists working in the E foothills of the Canadian Rockies (e.g. Bally *et al.*, 1966; Dahlstrom 1969), to broad areas of the Alpine fold belt (e.g. Boyer & Elliot 1982; Butler 1983; Butler *et al.* 1986). The application of this type of model has met with varying degrees of success. In the Rocky Mountain foothills the stratigraphy and basement surface is essentially uniform and there are no pre-thrusting extensional faults. Deposition was controlled by the progressive downward flexure of the former passive margin. On the other hand, much of the Alpine area underwent Mesozoic fault-controlled extension (e.g. Trümpy 1980; Lemoine *et al.* 1986). The control exerted by early extensional faults on the compressional geometry during later orogenic deformation is now being recognized from a large number of mountain belts (Jackson 1980; Dewey 1982; Zeigler 1983; Harding 1985; Bally 1984; Etheridge 1986 and Lemoine *et al.* 1986). In the next section we show that in the Western

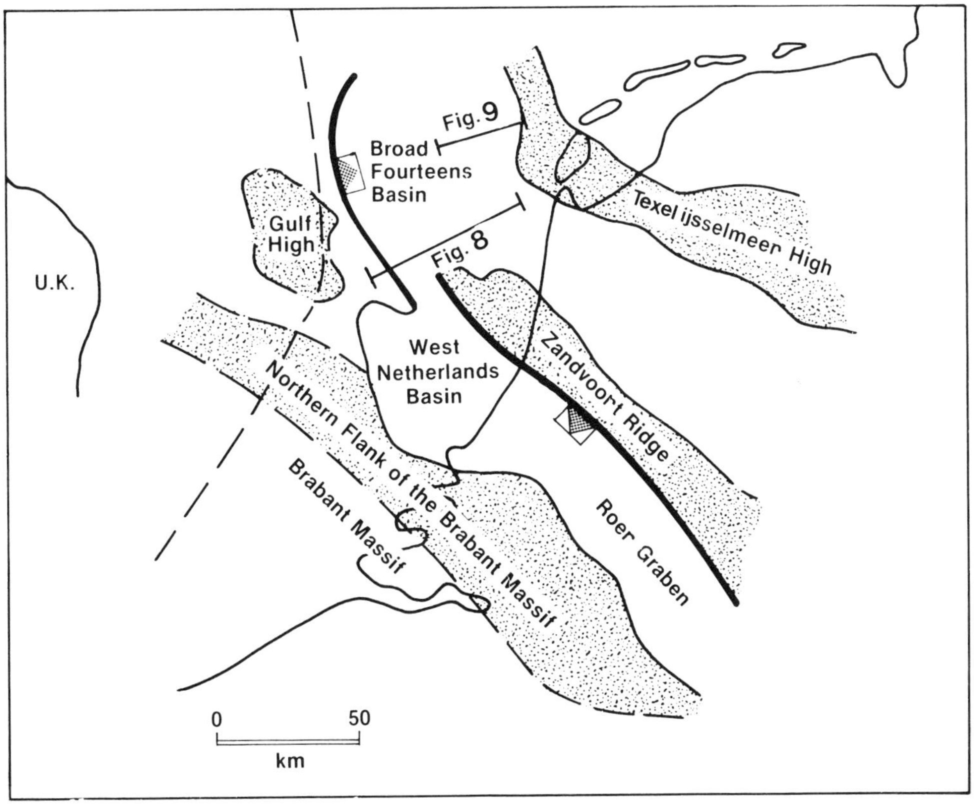

Fig. 7. Structural outline map of offshore Holland.

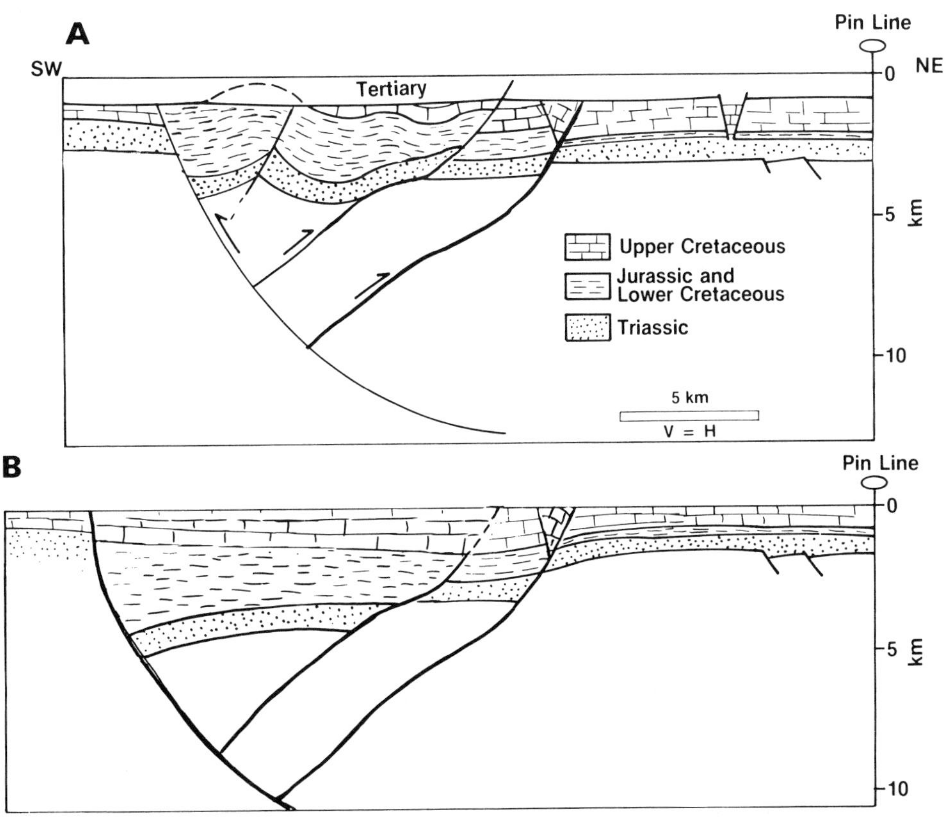

Fig. 8. A, Present day cross-section across the Broad Fourteens basin based on a regional seismic line. See Fig. 7 for location. B, The same section schematically restored to Upper Cretaceous.

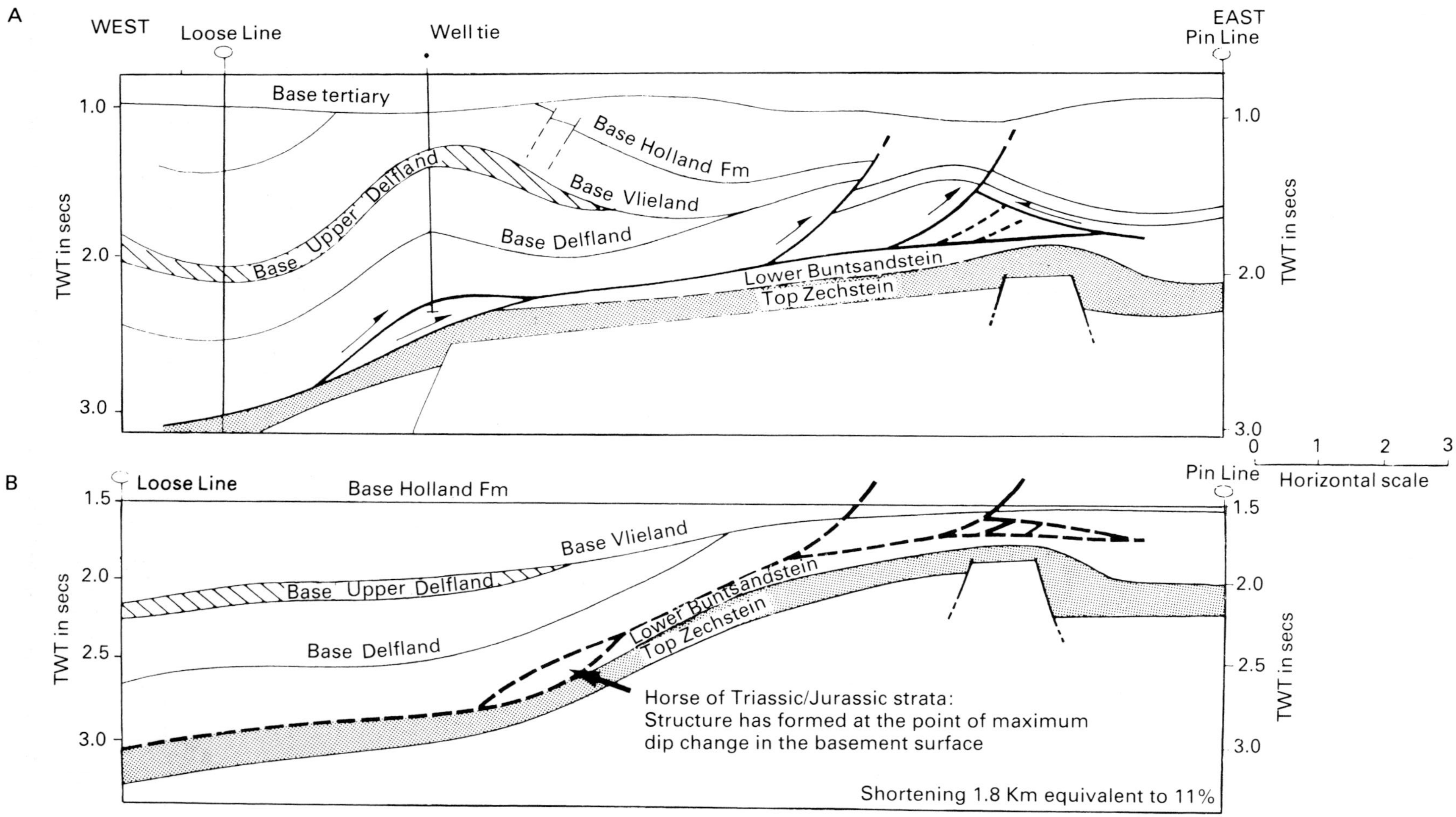

FIG. 9. A, Geoseismic section across the rollover margin of the Broad Fourteens Basin. Vertical scale in two way time, $V = H$. B, section restored in time at Base Chalk. For location see Fig. 7.

Alps and Pyrenees the influence of pre-existing extensional structures in these areas goes beyond the simple inversion seen in the Alpine foreland.

Western Alps

It has long been recognized that the stratigraphic distinctiveness of individual thrust sheets in the Alpine system (the tectono-stratigraphic units of classic Alpine stratigraphy) is related to pre-tectonic palaeogeography. Many authors (Trümpy 1980, Trümpy & Lemoine 1987, Lemoine *et al.* 1986, de Graciansky *et al. this volume* and others) have all demonstrated that this palaeogeography is a reflection of Mesozoic extensional geometry associated with the stretching events which preceeded the development of the Ligurian and Tethyan oceans.

Is this to say that every major thrust sheet is an extruded half-graben and every major thrust an inverted extensional fault? Trumpy and Lemoine (1987) have argued this case for the Helvetic Nappes, though the argument is refuted by Butler, *this volume*.

We briefly outline two Alpine examples of the influence extensional faults on later contractional structures. Both show geometrical relationships that contrast with the inversion described in early sections. The first example is the section across the Belldonne and associated external crystalline massifs (also discussed by Butler, *this volume*; Gillcrist, *et al.* 1988 *this volume* and de Graciansky *et al.*, *this volume*). The second is the Digne thrust in the external Alpine zones of Haute Provence.

Belledonne section

A section from the external zones of the Alps to the Frontal Pennine thrust (leading internal thrust) (Figs 10 and 11) crosses the shelf edge of the Tethyan margin of Europe. The Mesozoic rocks of the Jura are shelf carbonates. The rocks thrust over them in the hanging wall of the leading thrust of the sub-Alpine zone are thicker and more shaley. They are derived from the continental slope. These rocks are deformed into a major syncline W of the elevated crystalline basement which forms the Belledonne and

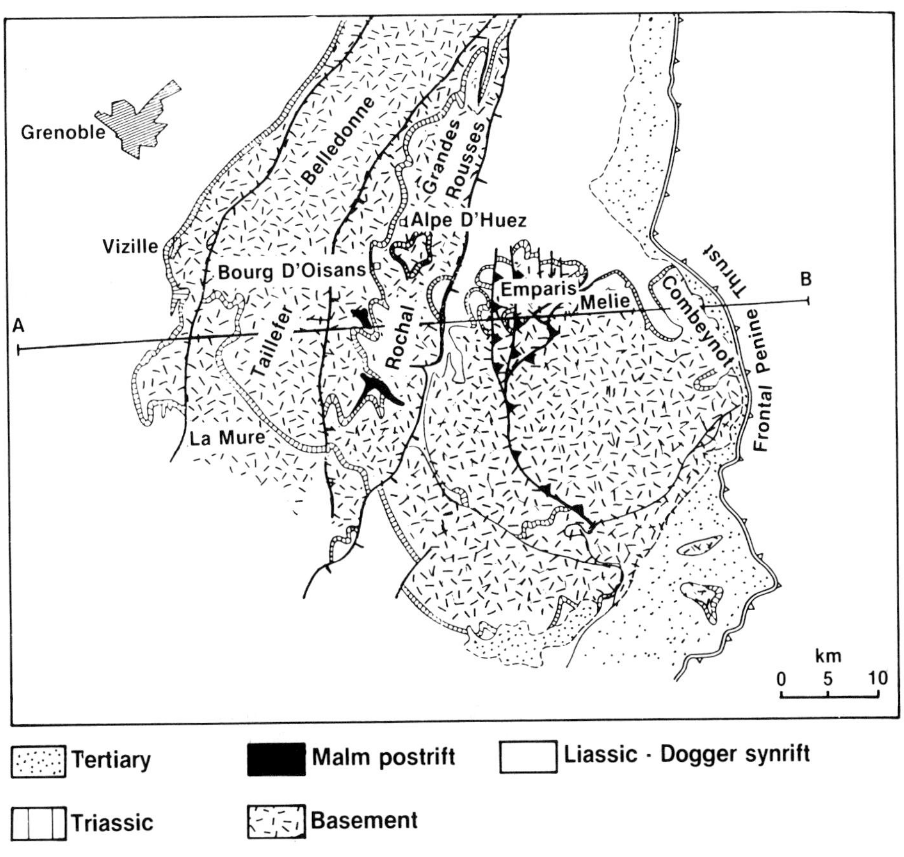

FIG. 10. Simplified structural outline map of the Belledone–Pelvoux area of the Western Alps (after Lemoine *et al.* 1981).

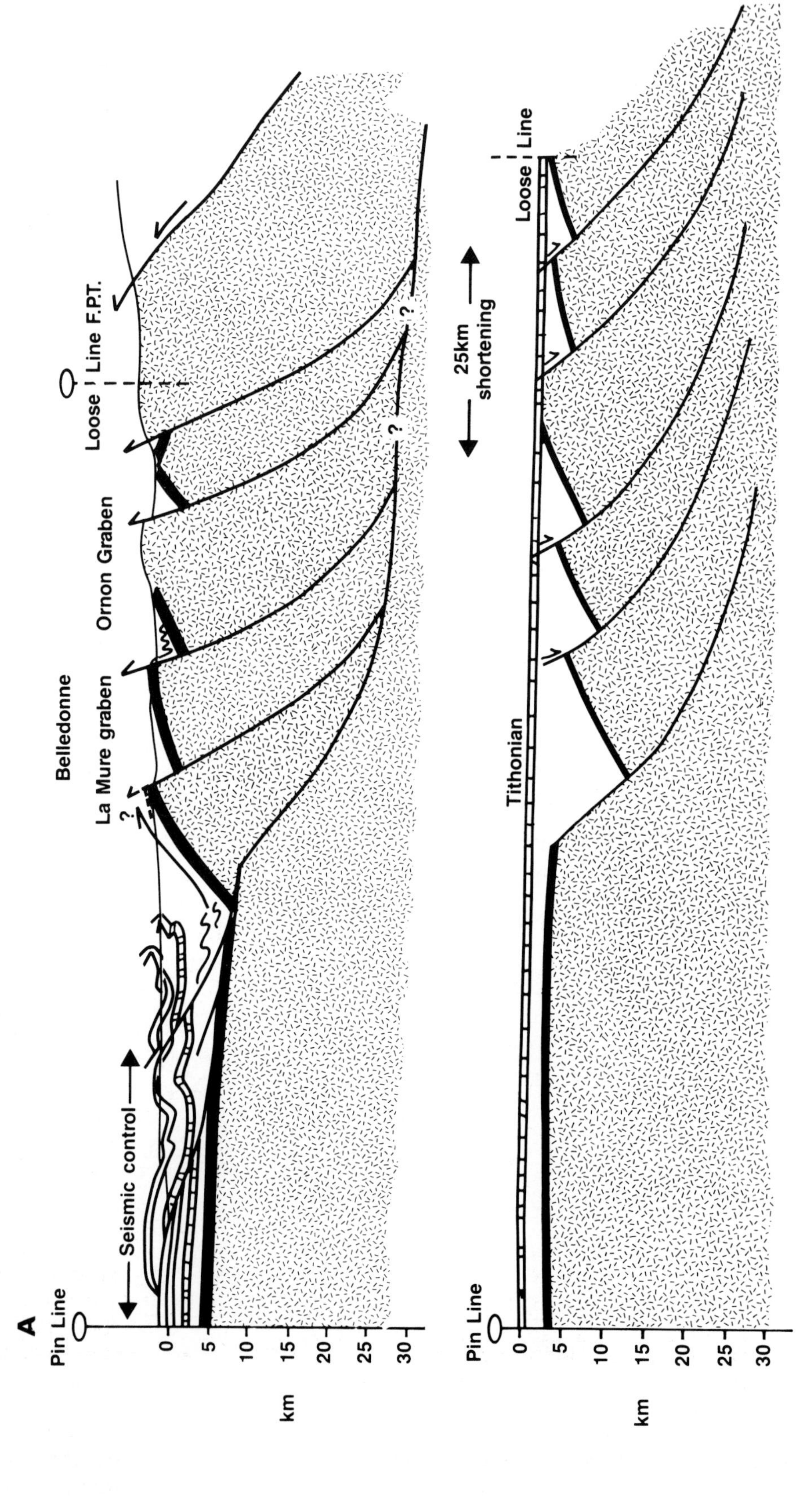
A
Pin Line
Seismic control
La Mure graben
Belledonne
Ornon Graben
Loose Line
F.P.T.
km
0
5
10
15
20
25
30
Pin Line
Tithonian
25km shortening
Loose Line
km
0
5
10
15
20
25
30

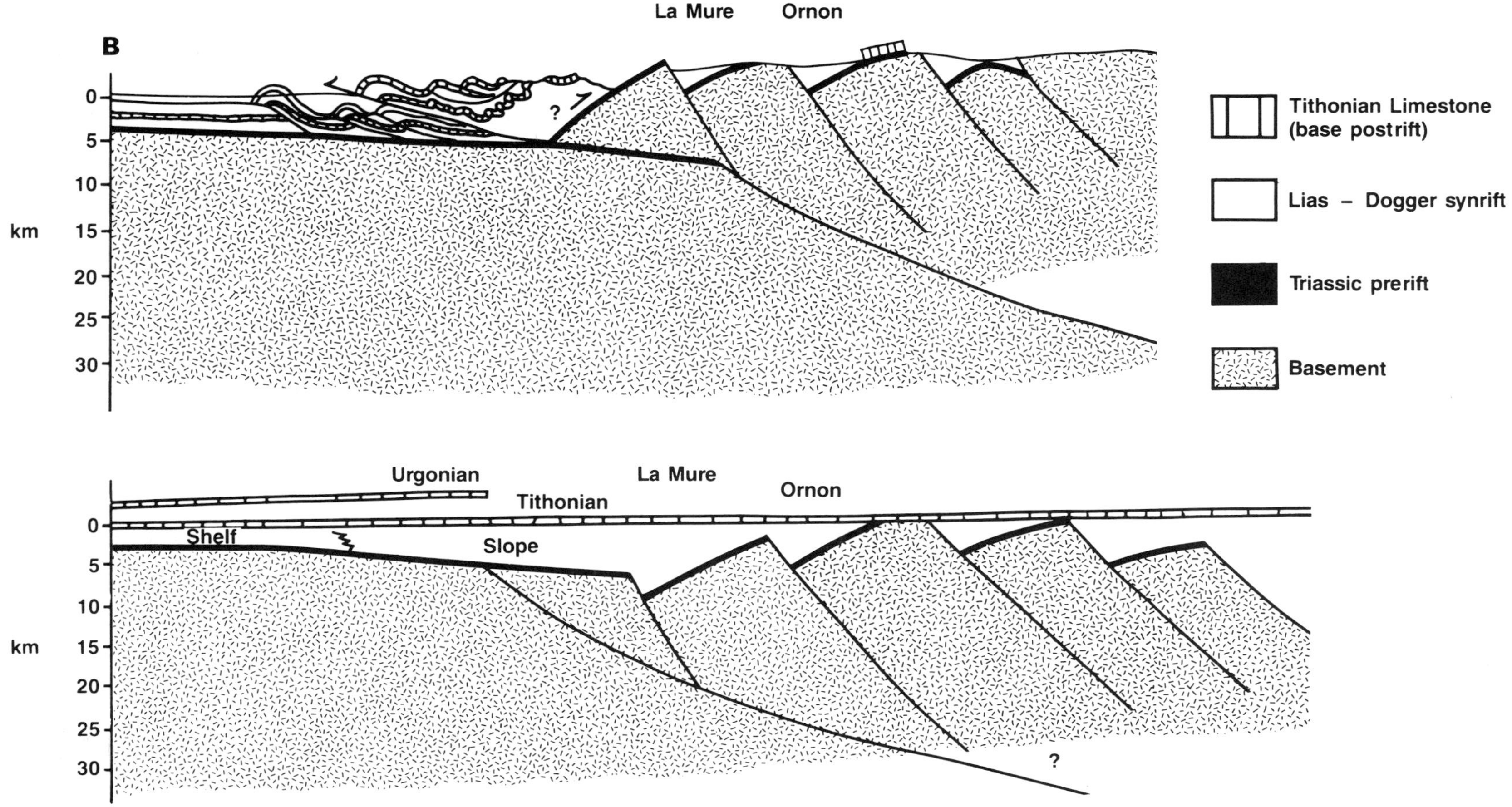

FIG. 11. A, Structural cross-section across the external zones of the Western Alps (in part after Lemoine *et al.* 1981), with restoration (see Fig. 10 for location). Based on the inversion model. B, The same cross-section based on the shortcut model.

associated 'external' massifs. The basement is elevated 12 km above its regional level. Seismic refraction studies (Menard 1976) have revealed a major low velocity zone dipping 30° or so eastwards beneath the Belledonne massif. It seems reasonable to suggest that this is the thrust which brought about the uplift of the massif.

The recent detailed studies of Lemoine, de Graciansky and their colleagues (Lemoine *et al.* 1981; 1985; de Graciansky *et al.*, *this volume*) have convincingly demonstrated the presence of extensional fault blocks and half-grabens within these external crystalline massifs, the detailed structure of which is described by Gillcrist *et al.*, *in press* and Gillcrist *et al.*, *this volume*. Though intensely internally deformed and transformed into nipped-in synclines, the half-grabens are clearly recognizable. The bounding faults of the more W half-graben (La Mure, Bourg d'Oisans) may have been rotated and over steepened, but there is little evidence of actual reversal of movement on these bounding faults, i.e. no demonstrable 'inversion' in the sense in which we have used the term in this paper.

Adjacent to the old extensional faults the synrift sediments exhibit an intensification of internal deformation (folds and cleavage). The implication is that detachment at the base of the cover sequence is transformed into strain in the vicinity of the blockage represented by the old extensional faults. In addition, backthrusts are developed. These represent expulsion of the synrift sequence backwards over the next rotated fault block. Lemoine has used the term 'buffer stop' to describe the buttress effect that produces this deformation.

Only in the E part of this section, not far below the leading internal zone thrust (Frontal Penine Thrust) is there any major 'forward' thrusting of the half-graben sediments (La Grave thrust). Davies (1982) has described small scale inversion in this region, but whether the large scale La Grave structure is genuine inversion of an old extensional fault or a shortcut phenomenon remains unclear.

The Alpine elevation of the deformed half-grabens *en masse* might itself be inversion. Fig. 11 is a section and restoration which allows this possibility. However, an equally viable alternative is the shortcut model (Fig. 11) which assumes that the most westerly original extensional fault was too steep to be reactivated as a thrust first acted as a buttress and was then cut through by the thrust on which the Belledonne massif was elevated.

Digne Thrust System

The Digne Thrust is one of the most important structures in the external Alpine zones of France (Fig. 12). It is 60 or more kilometres long from tip to tip. It transports a thick sequence of basinal Jurassic and Cretaceous black shales (Vocontian facies) southwestwards over thin Mesozoic platform carbonates of the Provencal facies (Figs 13 and 14). South of the S tip, the thrust is replaced by a stack of imbricate thrusts and associated folds whose traces curve through 90° around the so-called 'Arc de Castellane'. The platform facies occurs everywhere S and W of the arc and the thrust system telescopes the facies transition.

The lithological contrast between the footwall and hanging wall is dramatically seen in a window into the footwall of the thrust sheet located in the Gorges de Barles area, 15 km N of the town of Digne (Fig. 12). The Lias and Middle Jurassic is 500 m thick in the footwall compared with 1500 m thick in the hanging wall. Traced southwestwards in the transport direction the hanging wall facies gradually changes and becomes thinner and less shaley until, in the frontal part of the thrust sheet, the facies is not dissimilar to the footwall facies. This gradual change suggests an original structural position on the rollover margin of an extensional basin. Our interpretation of the Digne thrust is that it represents the contents of a basin thrust out over the original extensional fault rollover in a similar style to the Broad Fourteens Basin. In this case the rollover lies below the Valensole plateau and forms the footwall block of the Durance fault, one of the most important palaeofaults of Southeastern France.

The fault or faults which controlled the basin in which the rocks of the Digne thrust sheet were deposited cannot be identified with certainty but are inferred to lie along the margin of the Argentera massif. The structural evolution of the Digne thrust sheet is schematically summarized in Fig. 15. The figure makes the point that although the Digne thrust is an example of an extensional basin 'extruded' as a thrust sheet, there is no simple reactivation of the controlling extensional fault. Indeed, complex internal deformation makes it impossible to recognize the controlling fault with any certainty. It may actually be folded, as indicated in Fig. 15.

The North Pyrenees

The Pyrenean Mountains comprise a central axial zone of Palaeozoic rocks flanked by folded and thrusted Mesozoic and Tertiary strata. They

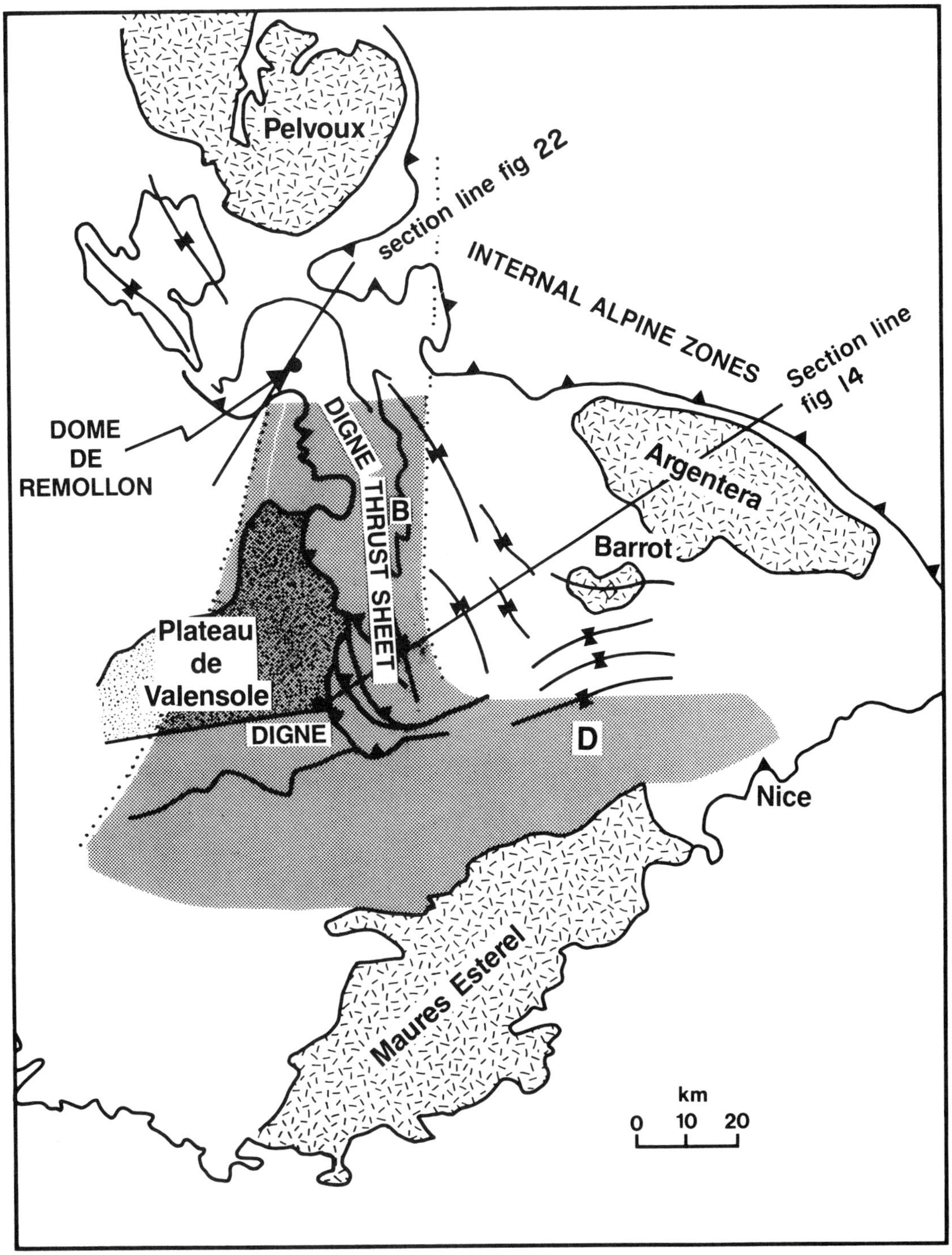

FIG. 12. Map of SE France showing locations of section lines. Stipple is Molasse, hackle ornament is pre-Triassic outcrop. Shaded area is possible extent of Mesozic platform carbonates beneath thrusted basinal rocks with dotted lines as limit (definite fault in west, possible fault in east). The letters A–D refer to contrasted stratigraphies shown in Fig. 13.

have been variously interpreted as both a strike-slip fold belt dominated by steep upright structures (e.g. Choukroune & Matauer 1978) and as a thin skinned fold and thrust belt (Williams & Fischer 1985). Neither interpretation fully takes into account the observed stratigraphic variations or structural style. Recently, improved industrial seismic resolution in the North Pyrenean thrust belt and within the structured foreland zone of the Aquitaine Basin demonstrates convincingly that the Pyrenees formed a major Mesozoic extensional basin controlled by a complex extensional fault system that exerted a major influence during the subsequent Eocene compression.

The major period of crustal extension was related to the opening of the Bay of Biscay in the Lower Cretaceous (Boillot 1984). The northward movement of Iberia throughout the Tertiary resulted in the continental collision

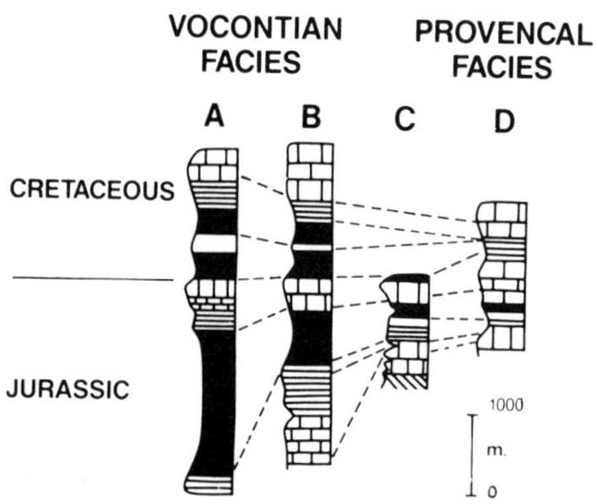

FIG. 13. Contrasting stratigraphy in the basins of SE France (after Debelmas 1974).

and the development of the Pyrenean fold and thrust belt.

We describe below two sections from the central and W area of the North Pyrenees which document both basin inversion and half-graben extrusion. We also discuss the way that earlier extensional faults controlled the location of subsequent thrust ramps.

The first section (Fig. 16) in the extreme W runs from the margin of the Mt Ursuya massif northwards across the North Pyrenean thrust belt and Aquitaine basin. The section is well constrained by seismic and well data. North of the Mt Ursuya massif the Ursuya thrust sheet comprises a Triassic to Upper Cretaceous sequence that has a pronounced wedge shaped geometry. A thick Lower Cretaceous sequence in the north of the thrust sheet passes southwards into a much thinner Lower Cretaceous sequence. Along the N edge of the Ursuya massif the Upper Cretaceous rests directly on basement and is overthrust by a backthrust which carries a complete Mesozoic sequence southwards. This is presumed to represent a half graben that has been completely inverted and extruded both forwards and backwards.

To the north the Came Thrust Sheet drilled

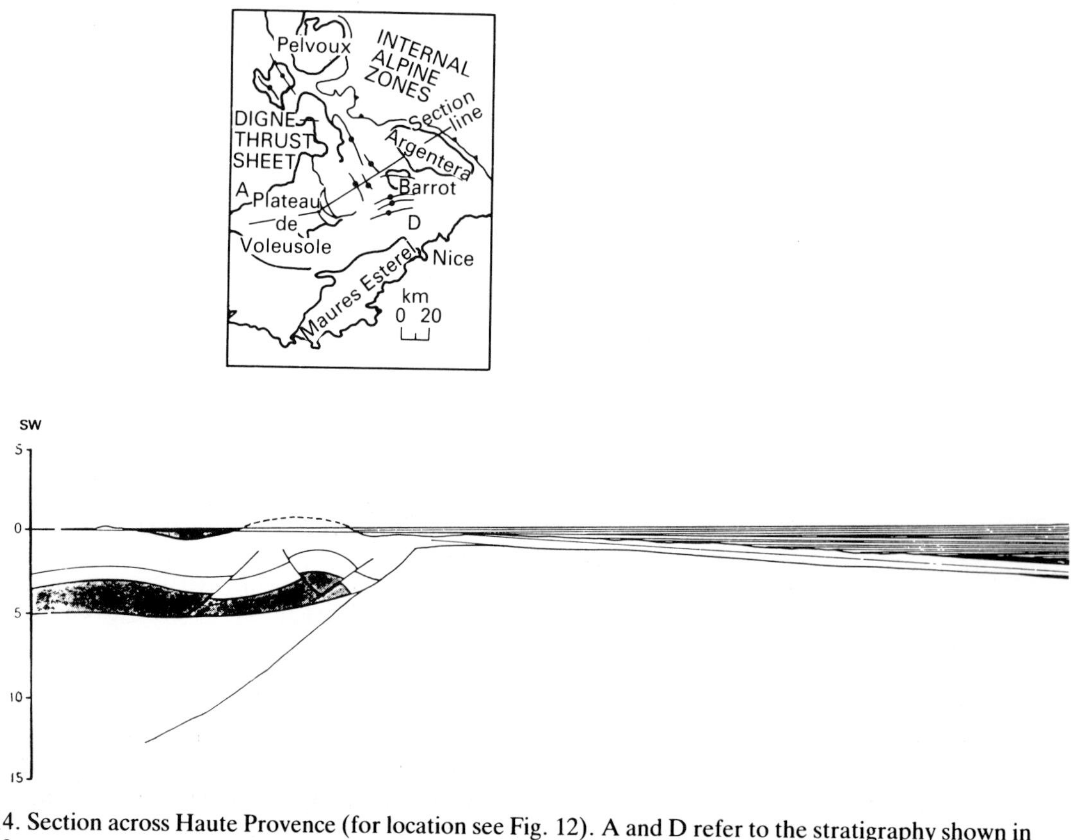

FIG. 14. Section across Haute Provence (for location see Fig. 12). A and D refer to the stratigraphy shown in Fig. 13.

by the Came well (Fig. 16) has a much thicker and more basinal Lower Cretaceous sequence when compared with the Lower Cretaceous carbonate platform sequence penetrated in the St Lons well in the footwall of the thrust sheet. Restoration of this section suggests the presence of a series of extensional half grabens of Lower Cretaceous age. In contrast to those in the Western Alps described earlier, these half grabens have been completely inverted. The second section (Fig. 17) crosses the St Suzanne anticline which is the leading structure in the central W area of the North Pyreenes. The St Suzanne structure is thin skinned and detached within the Upper Triassic evaporite sequence. However, the Upper Cretaceous sequence in the footwall to the structure comprises a thin ~500 m thick carbonate platform sequence that is represented in the hanging wall by a thick basinal shale sequence ~1500 m thick (Fig. 17). Restoration of this section implies the presence of an extensional fault of Upper Cretaceous age that has controlled the location of the subsequent thin skinned thrust ramp.

In the Southern Pyrenees there was no significant Mesozoic extension. The sedimentary sequence comprising a thin Upper Cretaceous carbonate sequence rests unconformably on Triassic evaporites and Palaeozoic basement. The boundary between the North and South Pyrenean zones is marked by a major thrust to the present day, which is likely to be the W continuation of the North Pyrenean fault (J. R. Hossack, personal communication). Restoration of cross-sections implies that this fault was a major extensional fault throughout the Cretaceous. It is tempting to speculate that the reactivation of this fault is responsible for much of the elevation of the axial zone basement massif in this area.

The Southern Pyrenees are a major basement duplex locally with a roof thrust in Triassic evaporites. This represents thrust deformation on the footwall of the old extensional fault. In contrast, in the North Pyrenees much of the deformation can be related to the earlier extensional structures. In some areas reactivation of extensional faults as reverse faults or thrust has resulted in major inversion and graben extrusion. Elsewhere steps in the basement surface control the location of subsequent thin skinned thrust ramps.

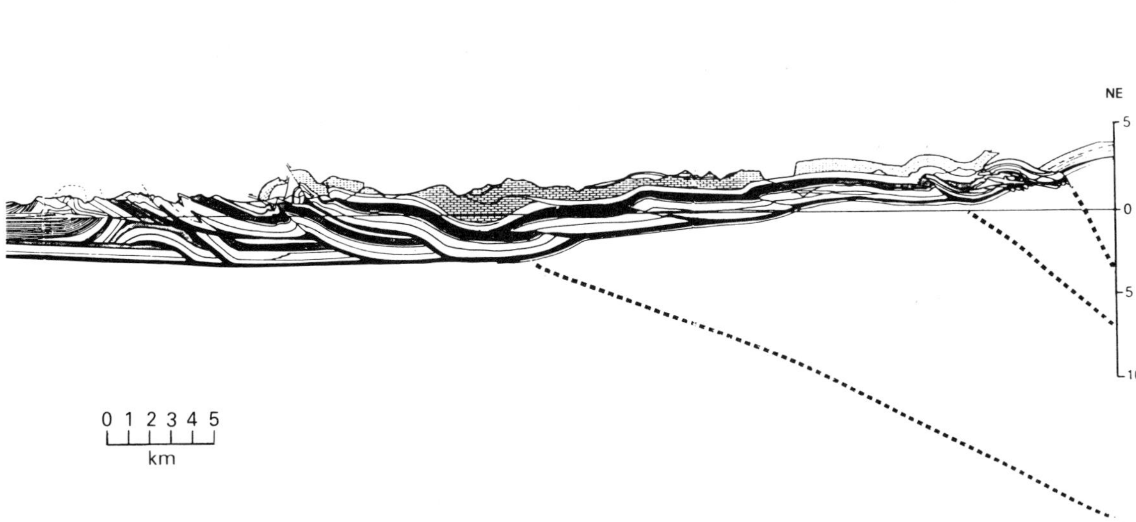

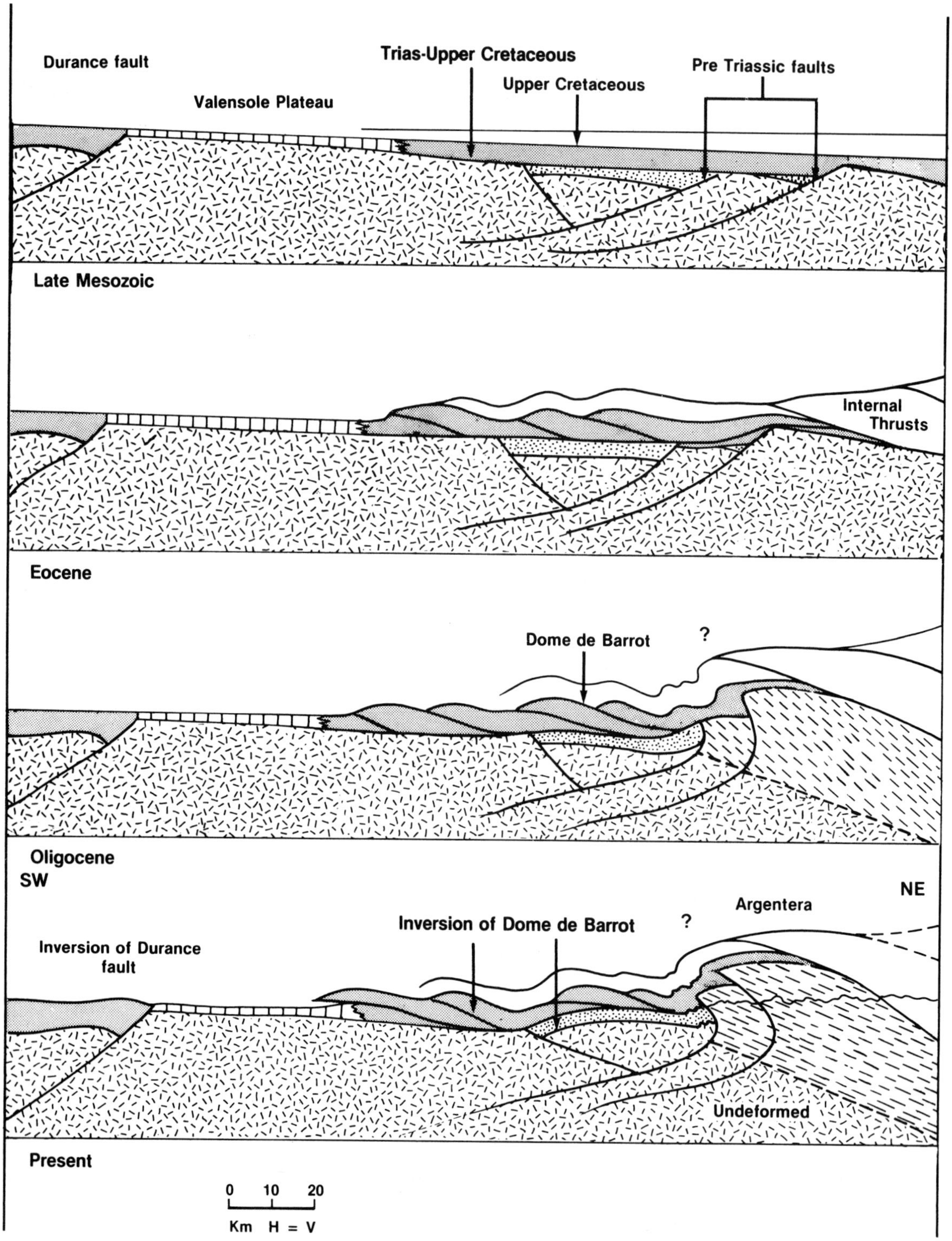

FIG. 15. Possible structural evolution of the Digne Basin.

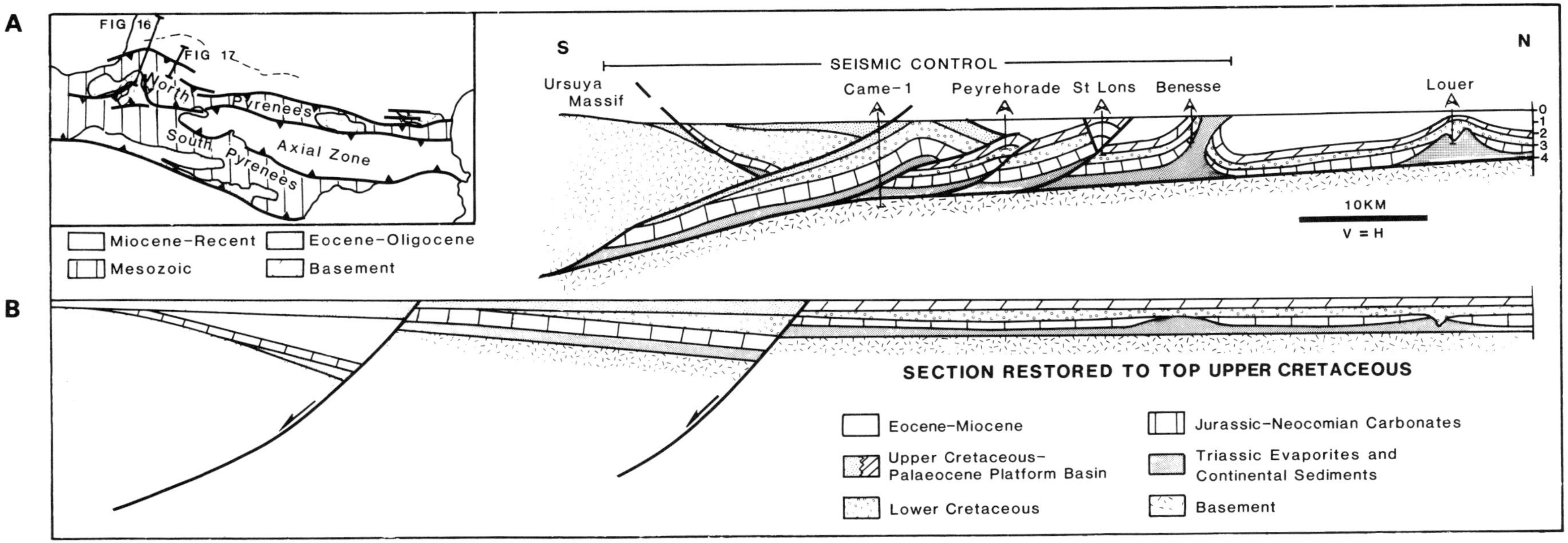

FIG. 16. A, Structural cross-section 1 North Pyrenees. Inset shows location. B, The same section restored to top Upper Cretaceous.

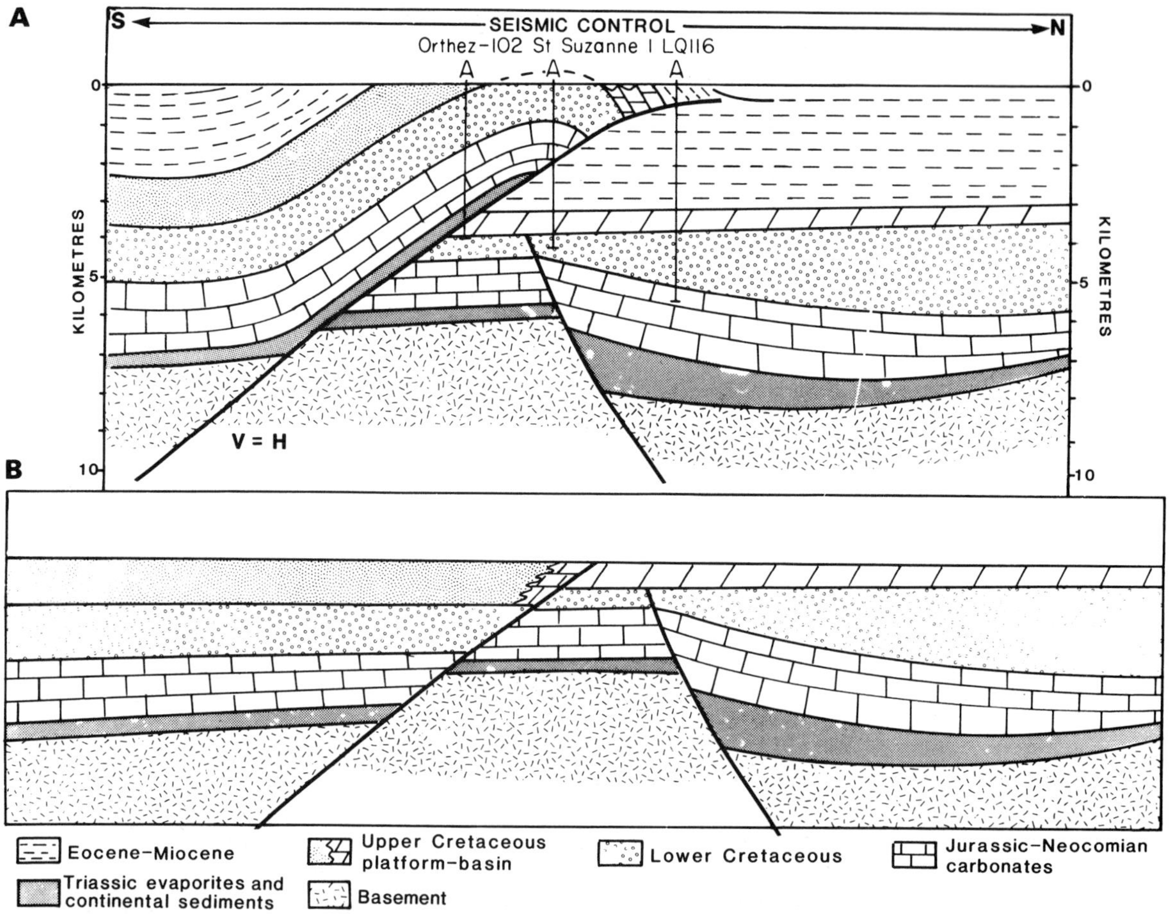

FIG. 17. A, Structural cross-section across the St Suzanne Anticline, North Pyrenees. See Fig. 16 for location. B, The same section restored to top Upper Cretaceous showing control on ramp location by earlier extensional fault. See Fig. 16 for the regional context of this structure.

Indications of the influence of extensional fault systems in orogenic belts

In this section we examine a number of features which are common in the many orogenic belts and which we believe indicate the influence of pre-existing extensional fault systems on the subsequent thrust geometry.

Out of sequence thrusts and arcuate thrust traces

These features seem to be indications of the influence of sub-thrust extensional patterns and sequences. The Digne thrust system described above shows clear evidence of both (Figs 12 and 14). The Castellane arc is almost certainly a response of southwesterly directed thrusts breaking as oblique thrust ramps as they climb sub-thrust obstacles. The trend of the ramps roughly reflects the trend of the sub-thrust structures (Fig. 18) which are most reasonably interpreted as part of the original extensional geometry of this area of the Alpine margin.

The leading part of the Digne thrust shows a complex movement sequence. The thrust is out of sequence in the sense that it cuts both limbs of folds in both its footwall and hanging wall, yet, in turn, it is folded by thrust related folds beneath it (Fig. 14). The geometrical relationship is borne out by the stratigraphic evidence that the last 7 km of displacement on the Digne thrust was across a post-Messinian land surface but earlier movements on the structure were pre-Eocene. The complexity is probably another manifestation of the buttress (buffer stop) effect of the sub-thrust structure. There are numerous other instances of the phenomena; the Sibillini Mountains of Italy (Leveccia 1978) are a particularly good example.

Arcs and out of sequence thrusts should make us suspect the presence of sub-thrust extensional structures (with or without associ-

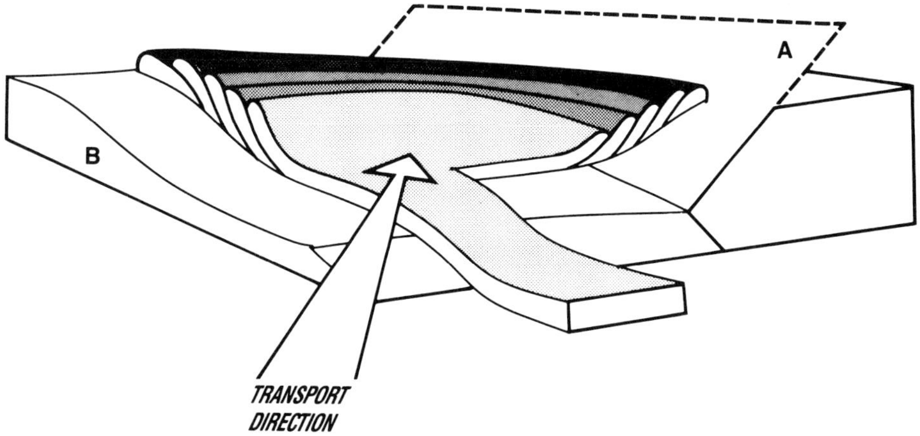

FIG. 18. Exploded block diagram showing oblique ramps and arc formation of thrust related folds associated with oblique inversion. A, a fault, B, a rollover.

ated inversion) with important implications for regional structural interpretation.

Footwall shortcuts

Footwall shortcuts like those suggested in the Belledone section of the Western Alps have been produced in the experiments done by McClay, *this volume* where pre-existing extensional faults are rotated too steeply to be reactivated themselves as reverse faults. In nature footwall shortcut thrusts produce isolated wedges of footwall rocks which may be translated in a thrust hanging wall (Fig. 19). The mechanisms allow isolated lenses of crystalline basement to be emplaced high within an apparently thin skinned sequence. This often presents interpretational difficulties. We know of two relatively large scale possible examples of this phenomenon.

The first example is the Johnstone Thrust/ Benton fault system, (Brooks *et al.* 1983; Hancock *et al.* 1981) in the Hercynian fold belt of Southwest Wales (Figs 20 and 21), where an isolated lens of crystalline basement lies 5 km above regional basement level along the line of a major Caledonian basin controlling fault (Brooks *et al.* 1983). It sits on a shallow thrust (the Johnstone thrust) and is emplaced onto Upper Carboniferous rocks which directly and unconformably overlie Lower Palaeozoic slates. A major extensional fault, the Benton fault, forms the S edge of the crystalline wedge, bounding it from a thick sequence of Devonian rocks (Old Red Sandstone). The fault probably represents the N border of the Old Red Sandstone basin (Hancock *et al.* 1981) shortcut by the thrust during the Hercynian orogeny.

A second example is from the Dome de

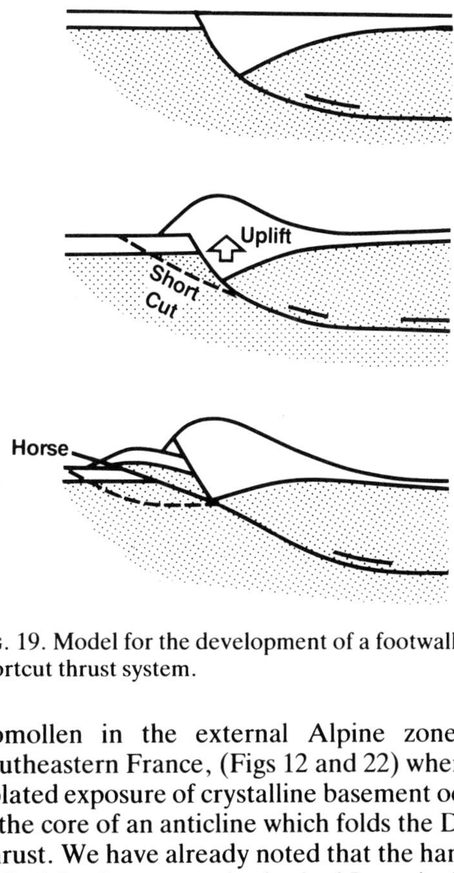

FIG. 19. Model for the development of a footwall shortcut thrust system.

Romollen in the external Alpine zones of Southeastern France, (Figs 12 and 22) where an isolated exposure of crystalline basement occurs in the core of an anticline which folds the Digne Thrust. We have already noted that the hanging wall of the thrust contains basinal Jurassic shales thrusted some 20 km over thinner platform carbonates (Siddans 1978; Gigot *et al.* 1974). The thrust is detached in Triassic evaporites. A 20 km long thin blade cut through basement below this detachment is a most unlikely geometry in the context of the regional geology. A small fragment of footwall from an extensional fault within the original basin is a more reasonable explanation of the basement inlier.

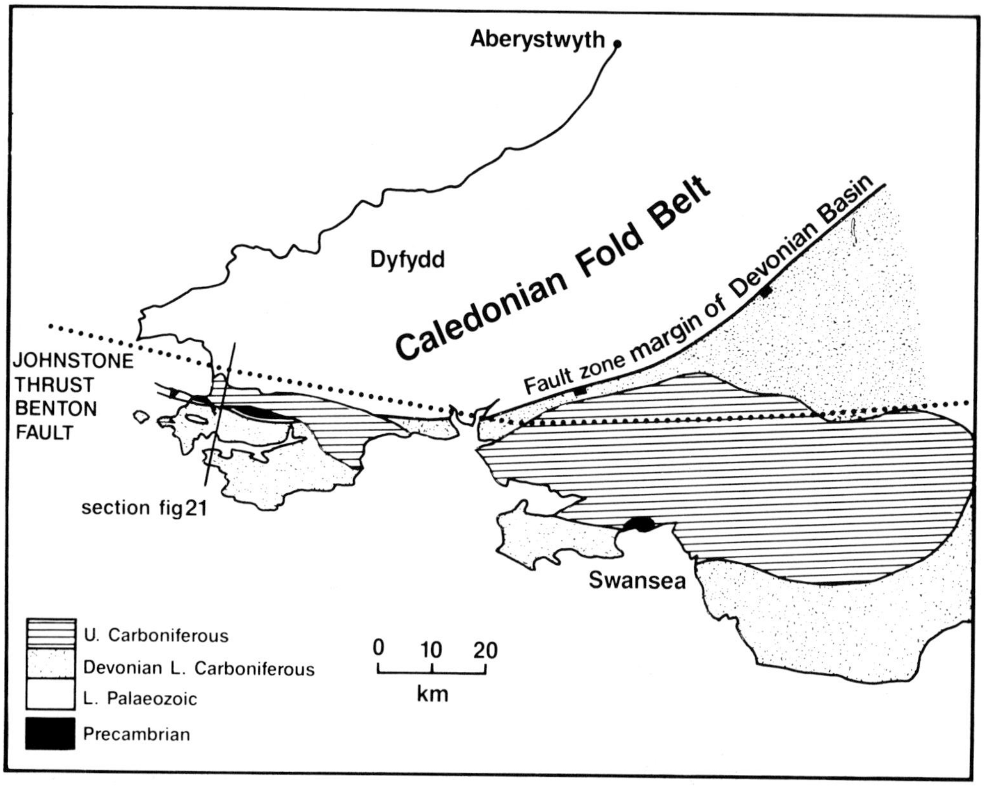

FIG. 20. Map showing the geology of Southwest Wales. Dotted line is limit of Hercynian deformation.

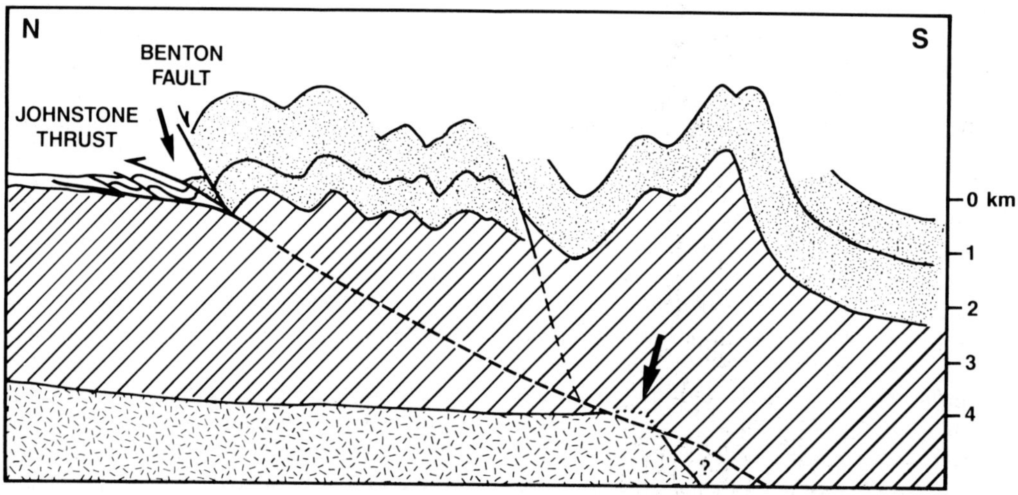

FIG. 21. Sketch section showing the Johnstone/Benton block as a footwall shortcut. Depth to basement is from Mechie 1981. Hackle shading is basement, lined shading is Lower Palaeozoic, dotted is Devonian and Lower Carboniferous. Upper Carboniferous is blank.

Inversion post-dating nappe emplacement

Commonly the floors of orogenic foredeep basins are broken by extensional faults. Some of these faults date from passive margin phases before orogenic collision. Others may have been formed by the foredeep–forebulge deformation wave and show growth sequences of foredeep basin fill. Commonly such structures are inverted late in the deformation history of the mountain belt.

In the foredeep basins of the Apennines and

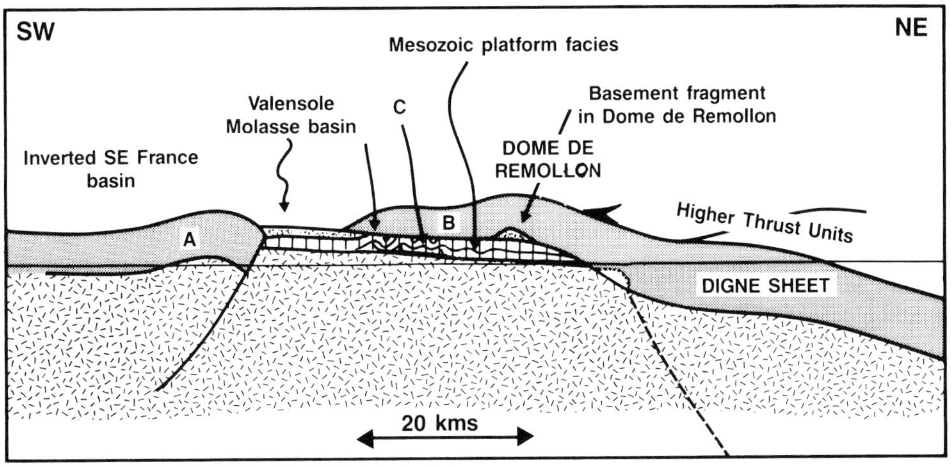

FIG. 22. Conceptual section showing the Dome de Romollon basement outcrop interpreted as a footwall shortcut in the extruded basinal rocks of the Digne thrust sheet. Letters refer to stratigraphy in Fig. 13. For location of section see Fig. 12.

in Oman, inversion of this kind developed after the emplacement of far travelled thrust sheets into the foredeep basins. The thin skinned thrusts have been deformed by 'thick skin' inversion-related anticlines whose scale is greater than earlier thin skinned structures and whose orientation may also be distinct.

Jebel Akdar and Hamrat Duru in Oman and the structure which elevates the main Lagonegro outcrop in the Southern Apennines are possible examples of large scale late inversion structures.

The interplay of thin skinning and inversion — concluding remarks

We have seen that pre-existing extensional faults may trigger ramp generation and complicate thrust sequence and trend. Where inversion occurs later than far travelled thrusting, complex interference of the two types of structure may occur.

The existence of extensional faults also complicates considerably the construction of balanced cross-sections. Although some of the geometric principles still apply, clearly the absolute rigour of the method is lost where there is no table-top basal detachment. Some inversions (e.g. in Southern England, see Simpson *et al.*, *this volume*) 'heal' with the basement once more continuous across the old fault. This creates a potential interpretative pitfall of which we need to beware. It is nevertheless important that we attempt to restore sections. Even though there may be no definitive answer, the technique still provides valuable insights into regional geometry.

The potentially complex inter-relationships between thin skinned thrusting and basin inversion are a prime problem to be tackled by the structural geologist working in orogenic terrains. Inversion is one of a number of possible ways in which pre-existing extensional fault geometry influences later thrusting to produce a more complex geometry than that which we see in areas of pure thin skinned thrusting. If stratigraphic and seismic data are good, interpretative problems are small. However seismic data are commonly equivocal in orogenic regions. Examples like those we have cited may have some empirical interpretative value. Clearly, the application of thin skinned tectonic models without sufficient regard to detailed stratigraphy can be misleading and may result in erroneous structural prediction and shortening estimates.

It is our contention that two broad structural provinces can be recognized in a stretched continental margin which later suffers orogenic contraction. The area of fault-controlled stretching is likely to be the site of basin extrusion, fault block restacking, footwall shortcutting, buttressing and complex structure. It is an area where we should expect bedlength mismatches of cover and basement. Further up the passive margin we may pass into a region which suffered only thermal subsidence with the development of 'steers head' shaped tapering sedimentary wedges. Here, thin skin tectonics on 'table top' basement such as we see in the Rocky Mountain foothills is likely to be dominant. Localized inversions on isolated passive margin faults like those described earlier in this paper spread far into the foreland beyond the limit of any thin skinned deformation. In the external parts of a mountain chain we may see interference of these two sorts of structures, we may also see inversion of extensional faults which formed coevally with foredeep basin development.

ACKNOWLEDGMENTS: We are grateful to BP PLC for permission to publish this paper. Many of the ideas on the offshore Holland area were developed in conjunction with our colleagues in BP Holland. Professors M. Lemoine and P. C. de Graciansky opened our eyes to the geology of the Western Alps. Chris Banks made many useful comments on an earlier version of this manuscript.

References and further reading

BALLY, A. W. 1981. Musings over sedimentary basin evolution *Philosophical Transactions of the Royal Society of London*, **A305**, 325–338.

—— 1984. Tectonogenese et sismique reflection: *Bulletin de la Sociéte Géologique de France*, **26**, 279–286.

——, GORDY, P. L. & STEWARD, G. A. 1966. Structure, seismic data and orogenic evolution of southern Canadian Rockies. *Bulletin of Canadian Petroleum Geology*, **14**, 337–381.

BODENHOUSE, J. W. A. & OTT, W. F. 1981. Habitat of the Rijswijk Oil Province Onshore Netherlands. *In*: ILLING, L. V. & HOBSON, G. D. (eds) *Petroleum Geology of the Continental Shelf of Northwestern Europe*. Institute of Petroleum, London and Heyden & Sons Ltd, pp. 289–300.

BOILLOT, G. 1984. Le Golfe de Gascogne et les Pyrenees. *In*: BOILLOT, G., MONTADERT, L., LEMOINE, M. & BIJU-DUVAL, B. (eds) *Les Marges Continentales Actuelles et Fossiles Autour de la France*. Masson, Paris.

BOYER, S. E. & ELLIOT, D. 1982. Thrust systems. *Bulletin of the American Association of Petroleum Geologists* **66**, 1196–1230.

BROOKS, M. LLEWELYN, D. J. & MECHIE, J. 1983 Geophysical investigations in the Variscides of southwest Britain. *In*: Hancock, P. L. (ed.) *The Variscan Fold Belt in the British Isles*. Adam Hilger, Bristol, 186–197.

BUTLER, R. W. H. 1983. Balanced cross-sections and their implications for the deep structure of the NW Alps. *Journal of Structural Geology*, **5**, 125–137.

—— 1985. The restoration of thrust systems and displacement continuity around the Mont Blanc massif, NW external Alpine thrust belt. *Journal of Structural Geology*, **7**, 569–582.

—— The influence of pre-existing basin structure on thrust system evolution in the western Alps, *This volume*.

——, MATHEWS, S. J. & PARISH, M. 1986. The NW external Alpine thrust belt and its implications for the geometry of the western Alpine orogen. *In*: COWARD, M. P. & RIES, A. C. (eds) & *Collision Tectonics Special Publication of the Geological Society of London*, **19**, 245–960.

CHADWICK, R. A. 1986. Extension tectonics in the Wessex basin, southern England: *Journal of the Geological Society of London*, **143**, 465–486.

CHOUKROUNE, P. & MATTAUER, M. 1978. Tectonique des et Pyrenees: Sur le fonctionnement de la faille transformante nord Pyreneene; comparison avec des modeles actuels. *Bulletin de la Sociéte Geólogique de France*, **7**, 689–700.

DAHLSTROM, C. D. A. 1969. Balanced cross-sections. *Canadian Journal of Earth Sciences*, **6**, 743–57.

DAVIES, V. M. 1982. Interaction of thrust and basement faults in the French external Alps. *Tectonophysics*, **86**, 325–331.

DEBELMAS, J. 1974. *Geologie de France, II, La Chaine Recente*.

DE GRACIANSKY, P., BOURBON, M., CHENET, P. Y. & LEMOINE, M. 1979. Genese et evolution comparees de deux marges continentales passives marge iberique de l'ocean Atlantique et marge Europeane de la Tethys dana les Alpes occidentales *Bulletin de la Société Géologique de France*, **21**, 663–674.

——, DE DARDEAU, G., LEMOINE, M. and TRICART, P. The inverted margin of the French Alps. *This volume*.

DEWEY, J. F. 1982. Plate tectonics and the evolution of the British Isles. *Journal of the Geological Society of London*, **139**, 371–412.

ETHERIDGE, M. A. 1986. On the reactivation of extensional fault systems. *Philosophical Transactions of the Royal Society of London*, **A 317**, 179–194.

GIBBS, A. D. 1983. Balanced cross-section construction from seismic sections in areas of extensional tectonics. *Journal of Structural Geology*, **5**, 153–160.

—— 1984. Structural evolution of extensional basin margins: *Journal of the Geological Society of London*, **141**, 609–619.

GIGOT, P., GRANDJAQUET, C. & HACCARD, D. 1974. Evolution tectono-sedimentaire de la septentrionale du bassin tertaine de Digne depuis l'Eocene. *Bulletin de la Société Géologique de France*, **16**, 128–139.

GILLCRIST, R., COWARD, M. & MUGNIER, J. L. Structural inversion examples from the Alpine foreland and the French Alps. *Geodynamica Acta (Paris)* **1**, 5–34.

——, COWARD, M. P., TRUDGILL, B., PECHER, A. MUNGIER, J. L., Structural inversion in the external French Alps (Abstract). *This volume*.

HANCOCK, P., DUNNE, W. M. & TRINGHAM, M. E. 1981. Variscan structures in South-West Wales. *In*: ZWART, H. J. & DORNIEPEN, V. F. (eds) *The Variscan Orogeny in Europe. Geologie en*

Mijnbouw, **60**, 81–88.

HARDING, T. P. 1985. Seismic characteristics and identification of negative flower structures, positive flower structures, and positive structural inversion: *Bulletin of the American Association of Petroleum Geologists*, **69**, 582–600.

JACKSON, J. A. 1980. Reactivation of basement faults and crustal shortening in orogenic belts. *Nature*, **283**, 343–346.

LEMOINE, M. 1987. Pre oceanic rifting in the Alps. *Tectonophysics*, **133**, 305–320.

——, BAS, T., ARNAUD-VANNEAU, A. *et al.* 1986. The continental margin of the Mesozoic Tethys in the Western Alps. *Marine and Petroleum Geology*, **3**, 179–199.

——, GIDON, M. & BARFETY, J. C. 1981. Les massifs crystalline externes des Alpes occidentaless d'anciens blocs bascules age du Lias Lors du rifting tethysin. *Compte Rendu de l'Academie des Sciences, Paris*. **292**, 917–920.

LEVECCIA, G. 1978. Confronto fra due sistemi a pieghe L'Appennino Umbro-Marchigiano ed il Giura Franco-Svizzero. *Bolletin Société Geologique* **H** **101** 69–76.

MCCLAY, K. R. Analogue models of inversion tectonics, *This volume*.

MECHIE, J. 1981. Seismic studies of deep structure in the British Channel area. Unpublished PhD thesis, University of Wales.

MENARD, 1976. Relations entre structures profondes et structures superficielles dans le Sud-Est de la France. Essai d'utilisation des données geophysiques. Thése 3iéme cycle Grenoble I, 169 pp.

OELE, J. A., HOL, A. C. P. J. & TIEMANS, J. 1981. Some Rotliengend gas fields of the K. and L. blocks, Netherlands offshore (1968–1978) — A case history. *In*: L. V. Illing & G. D. Hobson (eds) *The Petroleum Geology of the Continental Shelf of Northwest Europe*. Institute of Petroleum, London & Heyden & Sons Ltd, 289–300.

ROSENDAHL, B. R., REYNOLDS, D. J., LONBEN, P. M., BURGESS, C. F., MCGILL, J., SCOTT, D., LAMBIASE, J. J. & DENTSEN, S. J. 1986. Structural expressions of rifting: lessons from Lake Tanganyika Africa. *In*: Frostick, L. E. *et al.* (eds) *Sedimentation in African Rifts. Special Publication of the Geological Society of London*, **25**, 19–28.

SIDDANS, A. W. B. 1979. Arcuate fold and thrust patterns in the subalpine chains of south-east France. *Journal of Structural Geology*, **2**, 117–126.

SIMPSON, I. R., GRAVESTOCK, M., HAM, D. LEACH, H. & THOMSON, S. D. 1988, Inversion tectonics of the Wessex Basin, *This volume*.

TRÜMPY, R. 1980. An outline of the geology of Switzerland. *Proceedings of the 26th International Geological Congress Section, G10*, A, 5–105.

WILLIAMS, G. & FISCHER, M. 1985. A balanced section across the Pyrenean orogenic belt. *Tectonics*, **3**, 773–780.

ZIEGLER, P. A. 1983. Inverted basins in the Alpine foreland. *In*: Bally, A. W. (ed.) *Seismic Expression of Structural Styles: A Picture and Work Atlas American Association of Petroleum Geologists Studies in Geology 15*, **3**, 3.3–3–3.3.–12.

A. B. HAYWARD, British Petroleum Development Ltd, Britannic House, Moor Lane, London EC2Y 9BU, UK.

R. H. GRAHAM, British Petroleum Development Ltd, Britannic House, Moor Lane, London EC2Y 9BU, UK.

Geometry and kinematics of inversion tectonics

G. D. Williams, C. M. Powell & M. A. Cooper

SUMMARY: Positive inversion tectonics involves the reversal of extensional fault movement during contractional tectonics. Basin stratigraphy developed before, during and after extensional fault movements may be described as pre-, syn- and postrift sequences. Growth fault activity may be graphically displayed using thickness changes in stratigraphic intervals from fault footwall to hanging wall. Alternatively, it may be recorded using a hanging wall displacement/distance plot. Contractional reactivation of extensional faults puts progressively older synrift markers into net contraction. The point of change from net extension to net contraction is the null point. Its position in the synrift stratigraphy may be used to quantify the inversion ratio, which is defined as the ratio of contractional to extensional movement. Negative inversion is the reactivation in extension of a significant portion of an existing contractional system. Stratigraphic separation diagrams constructed from geological maps may be used to define the null point of individual faults and to quantify their inversion ratio.

Basic nomenclature

Structural inversion occurs when basin-controlling extensional faults reverse their movement during compressional tectonics, and, to varying degrees, basins are turned inside out to become positive features. The result is that individual faults may retain net extension at depth and show net contraction associated with anticline growth in their upper portions (Fig. 1). The converse is also possible. Thus inversion can be considered in terms of both positive (uplift) and negative (subsidence) senses relative to the immediately preceeding history of a fault system (Glennie & Boegner 1981).

In this paper, a fault is considered to be of extensional mode if it involves the displacement of an approximately horizontal marker horizon in the fault hanging wall below its pre-deformational regional elevation, and is of contractional mode if a hanging wall marker is displaced above its regional elevation (Fig. 2). A marker horizon may consist of an originally horizontal or subhorizontal stratigraphic interface, a marker bed on a geological cross-section, or a sequence boundary on a seismic reflection section. This definition is simple; it does not involve discussions of net slip of the fault and closely resembles fault dip separation.

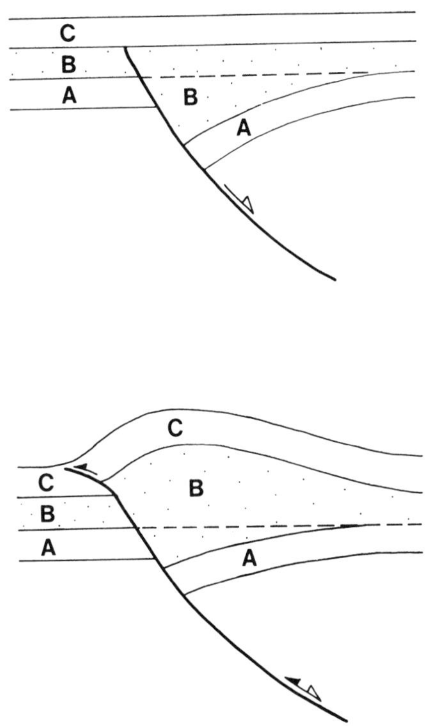

FIG. 1. Schematic diagram of a classical positive inversion structure. A, B and C are stratigraphic sequences. A, prerift; B, synrift; C, postrift sequence.

In our definition we consider apparent dip as portrayed in a cross-section or seismic section, rather than true fault dips. Cross-sections through a fault in any direction will reveal its extensional or contractional mode if our definition is applied. Only a perfectly bed-parallel fault (a flat on flat) and a perfectly strike-slip fault is neither extensional nor contractional and may be considered a neutral mode fault. A fault which appears vertical in the plane of a section cannot be classified in this scheme. Because net slip is not considered in this definition we may apply the contractional/extensional nomenclature to all faults including those

From Holdsworth, R.E. & Turner, J.P. (compilers) *Extensional Tectonics: Regional-scale Processes*. The Geological Society, Key Issues in Earth Sciences, **2 (1)**, 249–261.
1476-315X/02/$15.00 © The Geological Society of London 2002
First published in: "Cooper, M.A. & Williams, G.D. (eds) *Inversion Tectonics*, The Geological Society, London, Special Publications **44**, 3–15".

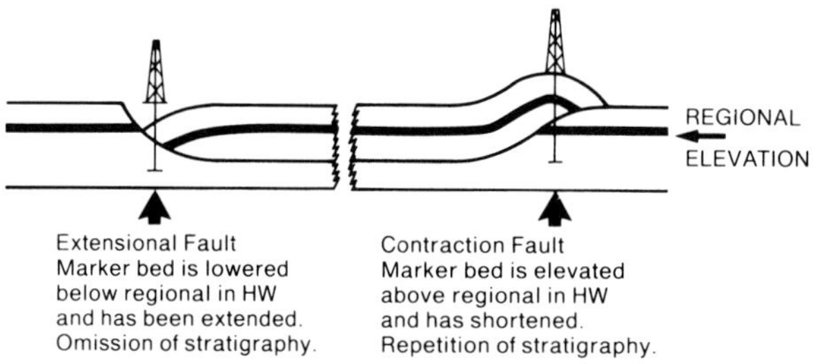

FIG. 2. Definition of A, extensional and B, contractional faults.

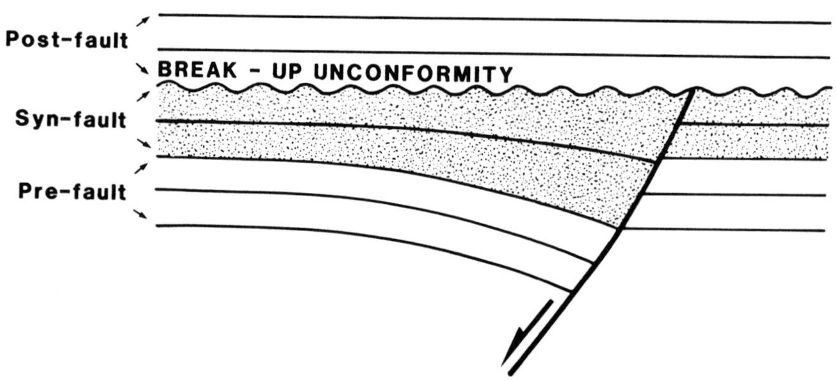

FIG. 3. Schematic diagram of stratigraphy accumulated before, during and after extensional fault movement. The break-up unconformity is shown as a wavy line at the top of the synrift sequence.

generated in a strike slip system at transpressive and transtensional fault bends.

Structural inversion involves fault movements giving rise to a switching of fault mode. A perfect reversal of fault net slip is unlikely and most inversion structures probably result from superimposed oblique slip movements. Our definition may be used to separate individual fault modes and to analyse structural inversion and is preferable to existing, complex, fault nomenclature.

In this paper we shall consider both positive inversion of a stratigraphy accumulated in extensional regimes and the negative inversion of thrust systems.

Extensional faulting

Stratigraphy built up on continental margins or in continental extensional basins, in its simplest form, may exhibit three distinct sequences (Fig. 3).

1 A prerift sequence is deposited prior to any extensional fault movement.

2 A synrift sequence is deposited during extensional faulting. Marked stratigraphic thickness changes from fault footwall to hanging wall are indicative of growth faulting.

3 A postrift sequence is deposited after the cessation of extensional faulting. The postrift sequence may be deposited after a period of non-deposition and/or erosion marked by a break-up unconformity which may remove part of the synrift sequence.

Percentage interval change

A listric extensional fault active during sediment deposition gives rise to a rollover anticline in the hanging wall above the curved portion of the fault (Hamblin 1965; Gibbs 1984). A rotational fault block model dominated by planar faults gives a similar effect. A wedge of synrift sediments thickens towards the fault in either case (Fig. 3). The fault movement history is recorded in stratigraphic thickness changes between footwall and hanging wall. This may be graphically displayed by plotting changes in stratigraphic thickness (footwall to hanging wall) against chronostratigraphy (Fig. 4A) (Beach 1984). Percentage interval change (I) is given by:

$$I = \frac{I_h - I_f}{I_f} \times 100 \qquad (1)$$

where I_f is the thickness of a stratigraphic interval in the footwall of an extension fault and I_h is the thickness of the same interval in the hanging wall. Both I_h and I_f are measured perpendicular to bedding in the plane of the section. They should be measured immediately adjacent to the fault.

Peaks in the resulting curve correspond to periods of maximum fault activity in the synrift sequence. A break-up unconformity will remove part of the synrift sequence. In this case, the fault activity curve will be incomplete and will not return to the ordinate (Fig. 4B). The unconformity will remove a thickness of section which represents a greater time interval in the footwall than in the hanging wall. This is because the footwall represents a condensed stratigraphic sequence.

Displacement/distance

A synrift sedimentary sequence accumulated during growth faulting will show increasing extensional displacements of stratigraphic markers with depth. This may be graphically represented using a modification of the displacement/distance plot of Williams & Chapman (1983). Using the top of the synrift sequence as a reference point, distance along a fault to individual hanging wall markers can be measured. This effectively records hanging wall stratigraphic thicknesses. Hanging wall distance is plotted against displacement (or more correctly, dip separation) of individual markers. Both parameters are measured in the plane of the section and therefore represent *apparent hanging wall distance* and *apparent extensional displacement* (Fig. 5). A curve with a negative gradient results. Changes in gradient relate to variations through time of fault activity. In this respect, the curve is similar to the percentage interval change plot (Figs 4A & B). Because the prerift and postrift sequences were not accumulated during growth faulting, any fault dislocation of these sequences should show no dislocation changes. This yields a curve with an infinite gradient or a slope of 90° on the displacement/distance plot.

In the preceding discussion we have not considered effects such as sediment compaction or displacement gradients along individual faults. These important effects are quantifiable.

Positive inversion of extensional fault systems

Our definition of extensional, contractional and neutral mode faults relies upon the geometry of faults and of marker horizons relative to regional elevation in the plane of a cross-section or seismic reflection section. The contractional reactivation of an extensional mode fault leads to an apparent reversal of movement when considered in the plane of the section. Using a typical pre-, syn- and postrift stratigraphy accumulated before, during and after extensional fault movements, we now consider the geometry of the system during contraction. In this analysis we assume that the stratigraphy is complete and that no break-up unconformity is present.

Contractional fault movement immediately displaces the postrift sequence in the fault hanging wall above its pre-deformational regional elevation. Continued contractional movement on the fault progressively puts synrift marker horizons into contraction (Fig. 6A, B & C). This process occurs sequentially. Markers at the top of the synrift sequence regain their regional elevations and move into contraction before lower synrift markers. If contractional faulting is arrested during this process, the upper part of the synrift sequence may be in net contraction whilst the lower part remains in net extension. The net extensional and net contractional portions are separated by a synrift marker that has regained its regional elevation by contractional movement. This marker now appears unfaulted and this point on the fault is defined as the *null point*. Progressive contractional inversion of an extensional synrift sequence causes the null point to move down the synrift sequence.

Inversion ratio

It is possible to quantify the relative magnitudes of contractional and extensional movement measured in the plane of the section using the position of the null point in a synrift sequence. The inversion ratio (R_i) is defined as the ratio of contractional to extensional displacement. It may be calculated by considering the thickness of the hanging wall synrift sequence measured parallel to the fault and the position of the null point in this sequence (Fig. 7).

$$R_i = \frac{d_c}{d_h} \qquad (2)$$

where d_h is the thickness of synrift sequence

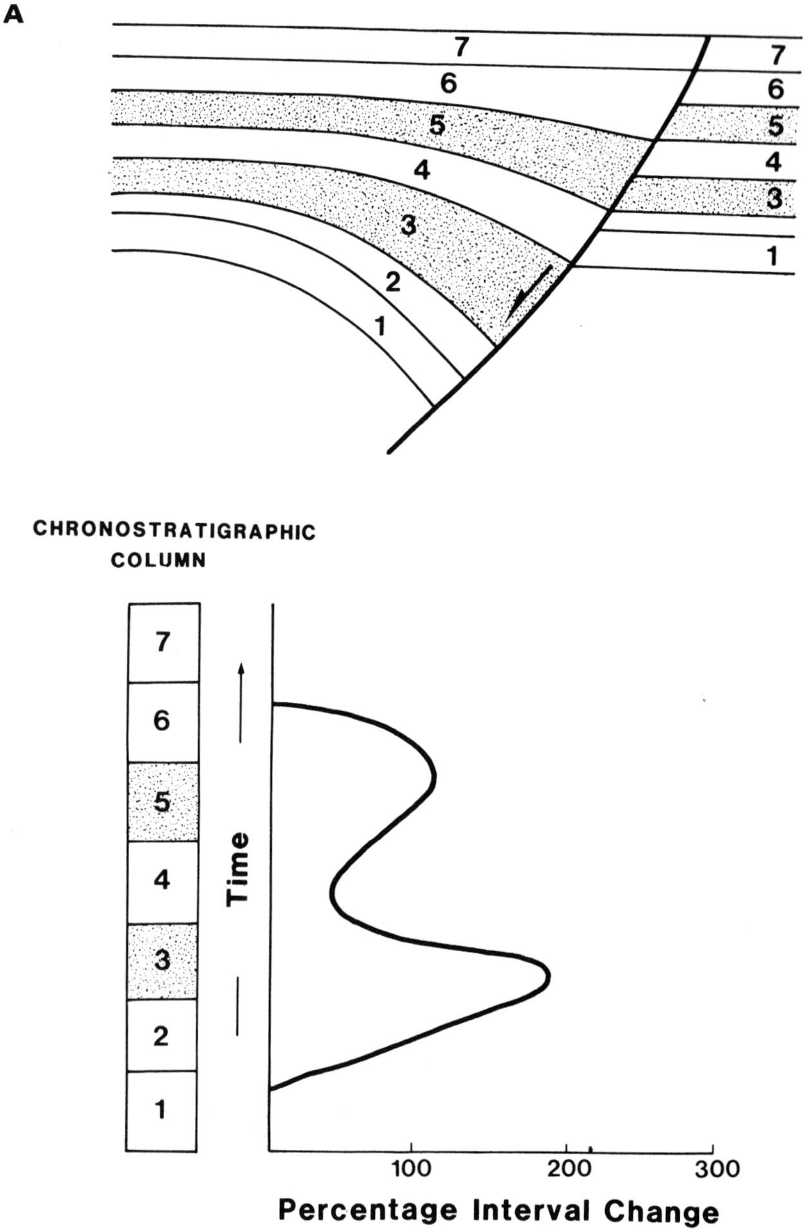

FIG. 4. 'Percentage interval change' plots; A, with no break-up unconformity; B, with break-up unconformity.

parallel to the fault and d_c is the thickness of synrift sequence in contraction (above the null point). We may rewrite this as:

$$R_i = 1 - \frac{d_e}{d_h} \qquad (3)$$

where d_e is the thickness of synrift sequence parallel to the fault below the null point.

From equations (2) and (3), it is seen that if the null point lies at the top of the synrift sequence, $R_i = 0$ ($d_c = 0$ and $d_e = d_h$). That is, no contractional inversion has occurred. If the null point is at the base of the synrift sequence, $R_i = 1$ ($d_c = d_h$ and $d_e = 0$). Total inversion of the synrift sequence has occurred and all markers in the prerift sequence have regained their pre-deformational regional elevations.

Displacement/distance

The displacement/distance plot (Fig. 5) may be further modified to graphically display inversion structures. Using the top of the synrift sequence as a reference datum, distance to

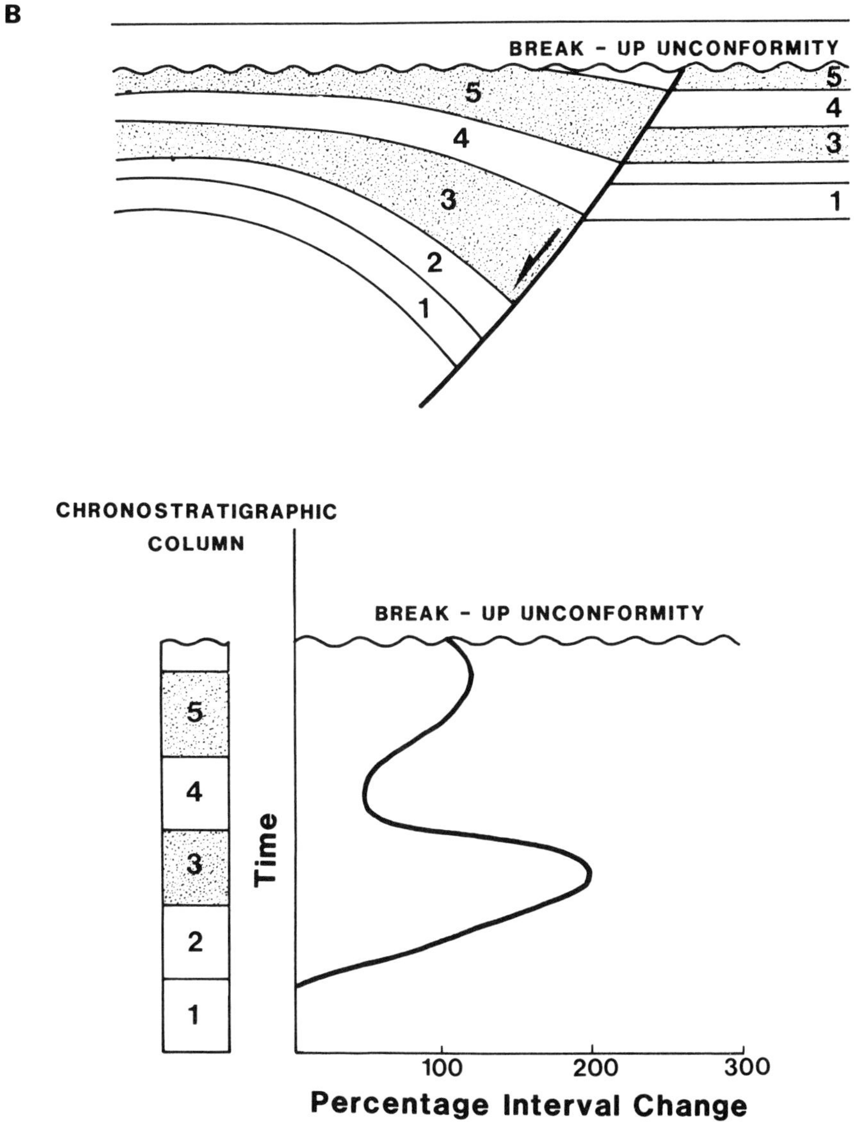

marker horizons is measured parallel to the fault in its hanging wall. Contractional displacement of markers is plotted to the left and extensional displacement is plotted to the right of the ordinate (Fig. 8). A curve with a negative gradient results, and the intercept of the curve with the ordinate defines the null point. The shape of this curve is a result of extensional tectonics, and gradient changes record variations in extensional fault growth. During progressive contractional inversion, the whole curve retains its form, but moves to the left. This causes the null point to move down the ordinate and effectively down section. The parameters d_h, d_c and d_e used in equations (2) and (3) may be read directly from this diagram (Fig. 8). An example of an inverted fault from the Gippsland Basin of SE Australia (Fig. 9) shows inversion of only one fault in an array of extensional faults (Davis 1983). The apparent inversion ratio of this structure is $R_i = 0.7$. However, because this figure is a time section the appropriate velocity function must be applied to convert the inversion ratio to its true value.

Negative inversion of contractional fault systems

The extensional reactivation of originally contractional systems appears to be less common than positive inversion. However, the reactivation may be only partial, and only evident in certain portions of the system. This can produce

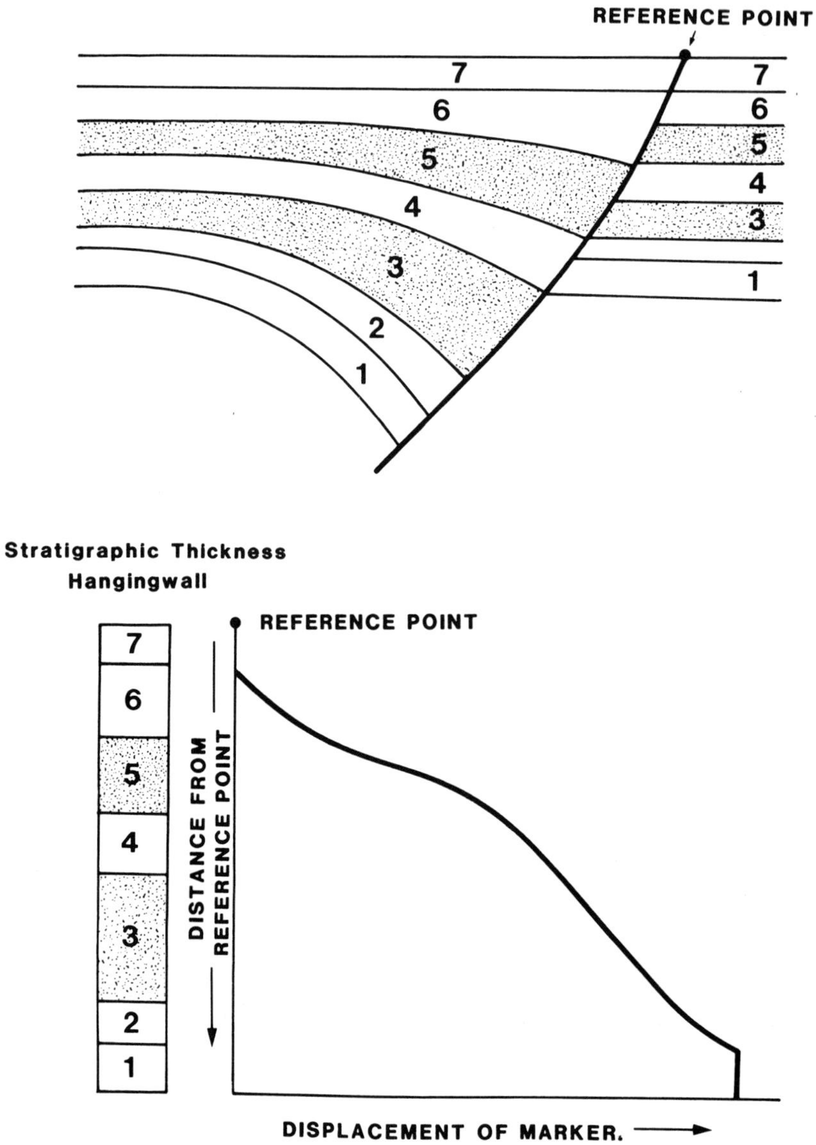

FIG. 5. Displacement/distance plot for the hanging wall of an extensional growth fault.

complex relationships between the portions of faults that are in net extension and those in net contraction. This can be conveniently analysed using the concept of null points and null zones (or neutral faults). These are fault segments which show neither contraction nor extension.

No re-use of the thrust system (Fig. 10)

In this case, an extension fault cuts through a thrust system and may detach at a deeper level. If the extension fault cuts through the thrust system without intersecting repeated stratigraphy across a thrust (which is possible though unlikely) then no null point exists. If the extension fault cross cuts thrust-repeated stratigraphy, then in the early stages of extension the footwall cutoff of the thrust which produces repetition will be the null point. The newly formed extension fault will show contraction below this point. A lower null point which marks the return to extension will be located at the footwall cutoff of the stratigraphically lowest thrusted bed. Both null points will disappear when the hanging wall cutoff of the stratigraphically highest thrusted bed coincides with the upper null point. At this point the extensional fault will be entirely in net extension with no local apparent contraction.

Partial re-use of single thrust (Fig. 11)

In this case an extensional fault has developed with a branch point to the thrust system at the

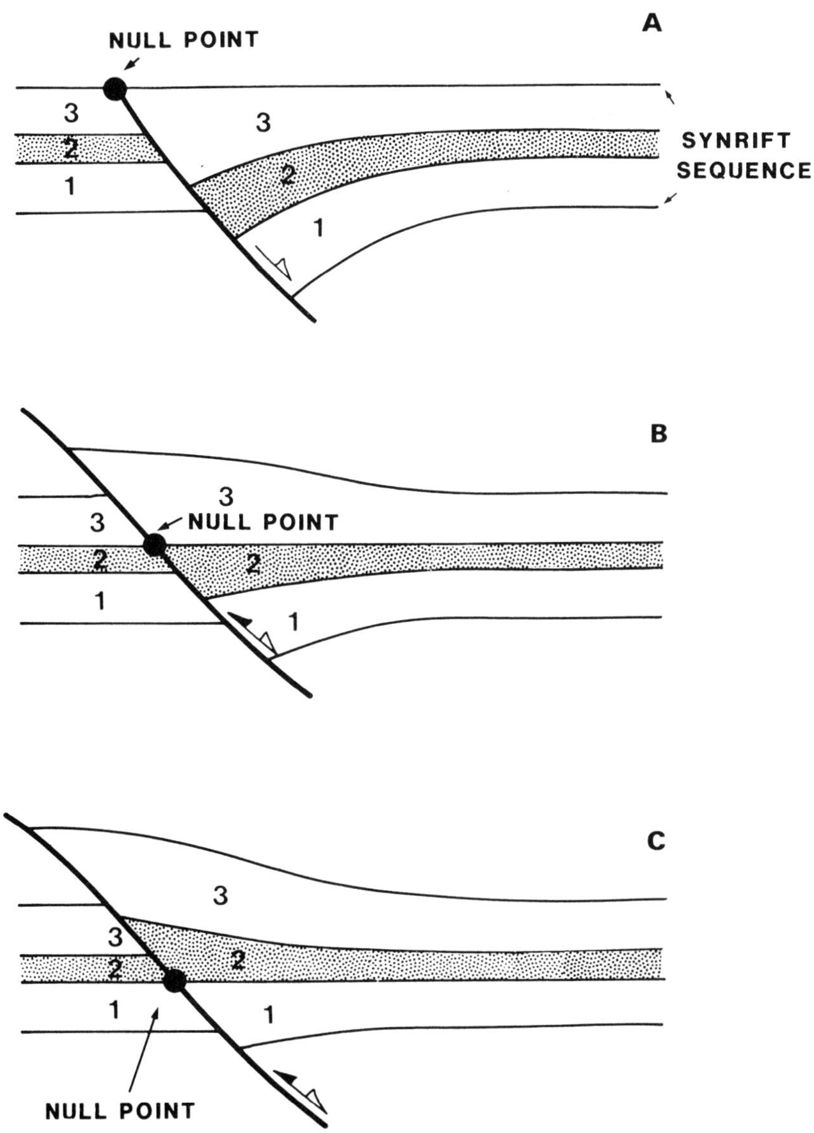

FIG. 6. Sequential diagrams to show the contractional inversion of an extensional fault. The null point moves progressively down the synrift sequence with increased contractional inversion.

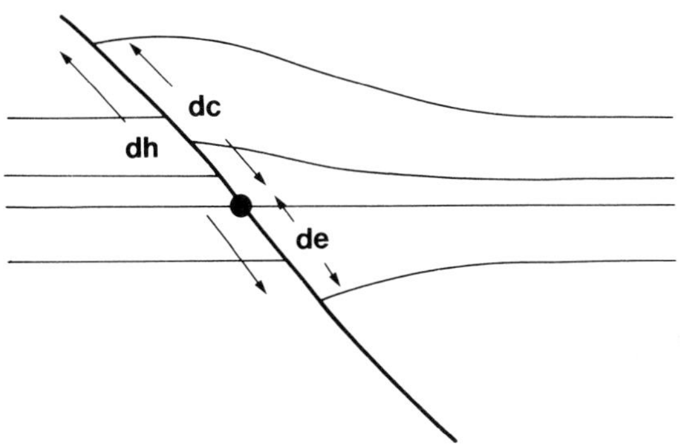

FIG. 7. Parameters used in equations (2) and (3) for the calculation of inversion ratio.

top of the thrust ramp. The null point of the system is this branch point. However, with continued extensional movement, the null point will disappear once the hanging wall cutoff of the stratigraphically highest repeated bed coincides with the branch point. At this position, the null point becomes a null zone which extends down the fault. Any additional extension causes this zone to disappear and the whole thrust ramp becomes extensional.

This does not mean that all the contractional slip has been reversed because the null point is derived from stratigraphic relationships. The extensional fault has shortcut the thrust hanging wall leaving behind a thrusted portion in its

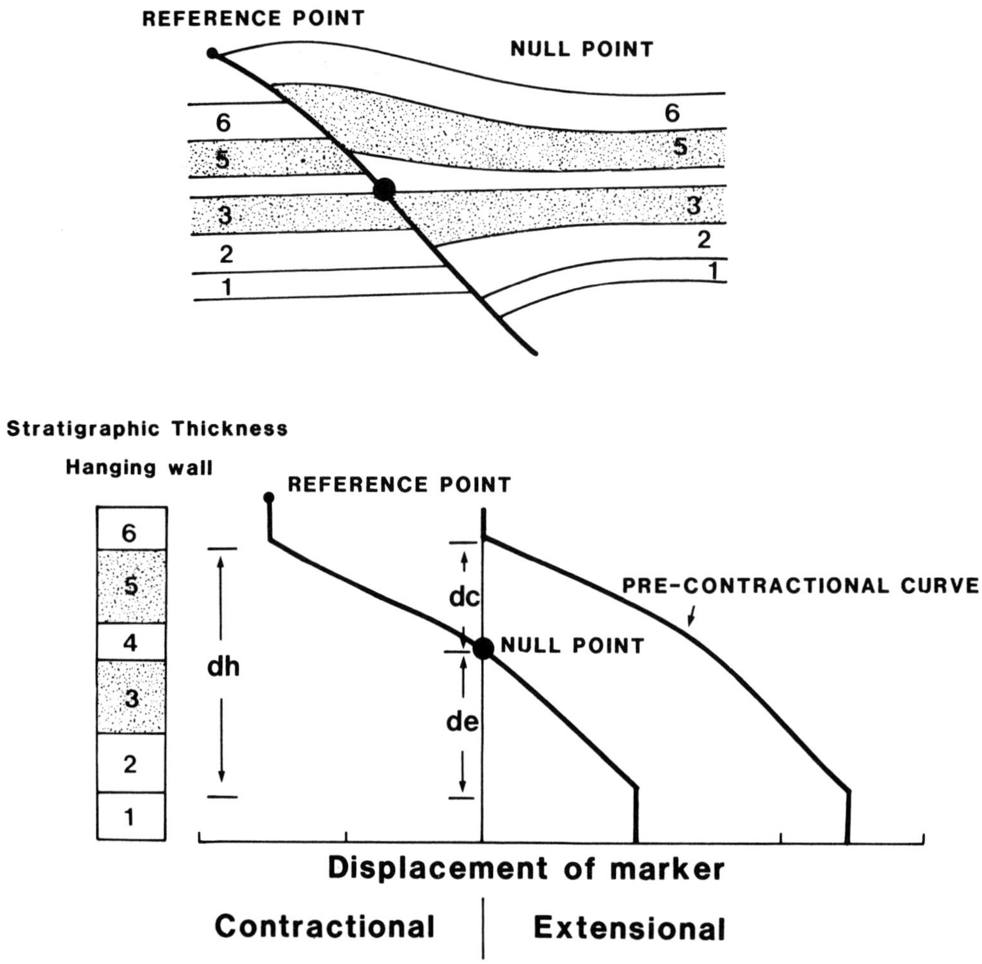

Fig. 8. Displacement/distance plot for a contractionally inverted extensional fault. Curve on right represents the situation prior to inversion. Parameters for equations (2) and (3) are shown.

footwall. If the erosion level was below that of the upper thrust plane then it would be impossible to detect the early thrust phase. The thrust/extension fault plane is a neutral fault showing neither contraction nor extension.

Partial reactivation of a complex thrust system (Fig. 12)

This thrust system comprises a major thrust (thrust 2) with a hanging wall imbricate (thrust 3). Thrust 1 has developed in the footwall of thrust 2 with a branch point at the base of a footwall ramp on thrust 2. Fig. 12 shows two stages in the progressive negative inversion of the thrust system. Thrust 1 is in net extension where it is a ramp, but turns into a neutral fault (a null zone) where it is parallel to bedding. When extensional reactivation has removed all the initial contraction of thrust 1, so that the hanging wall and footwall cutoffs re-match, then a null zone will exist for the whole length of thrust 1. Thrust 2 shows no extensional reactivation and thus has no null point. It remains in net contraction.

After the first increment of extension thrust 3 is still in net contraction until the null point is reached at the base of the extensional synrift package. With greater extension the null point migrates down the ramp until the hanging wall and footwall cutoffs of the repeated section re-match. At this point a null zone is created from the base of the synrift sequence to the next hanging wall ramp cutoff (Fig. 12B). This null zone will remain fixed until the cutoffs of the deeper thrust ramp are re-matched. At this stage, the contractional displacement on thrust 2, which was shortcut by extensional reactivation, has been completely removed by extension. With any additional extension, the whole

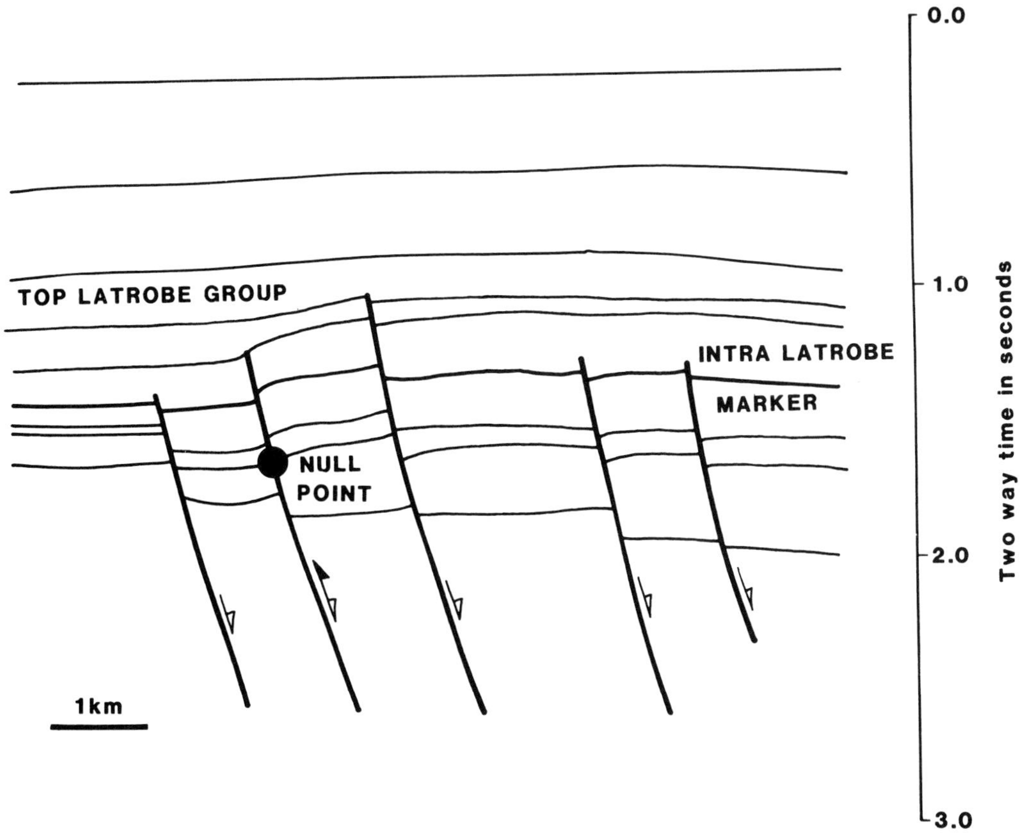

FIG. 9. Line drawing of an interpreted seismic section across the Snapper Field, Gippsland Basin, SE Australia (after Davis 1983).

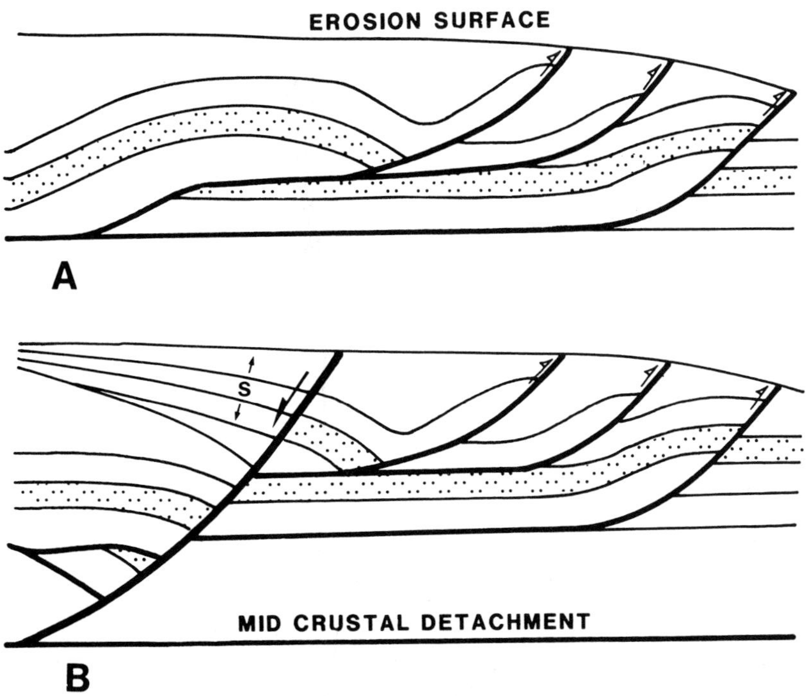

FIG. 10. A, Eroded linked thrust system. B, Extensional fault cross cutting the pre-existing thrust system with no re-use of thrust faults. S represents a synrift extensional sequence.

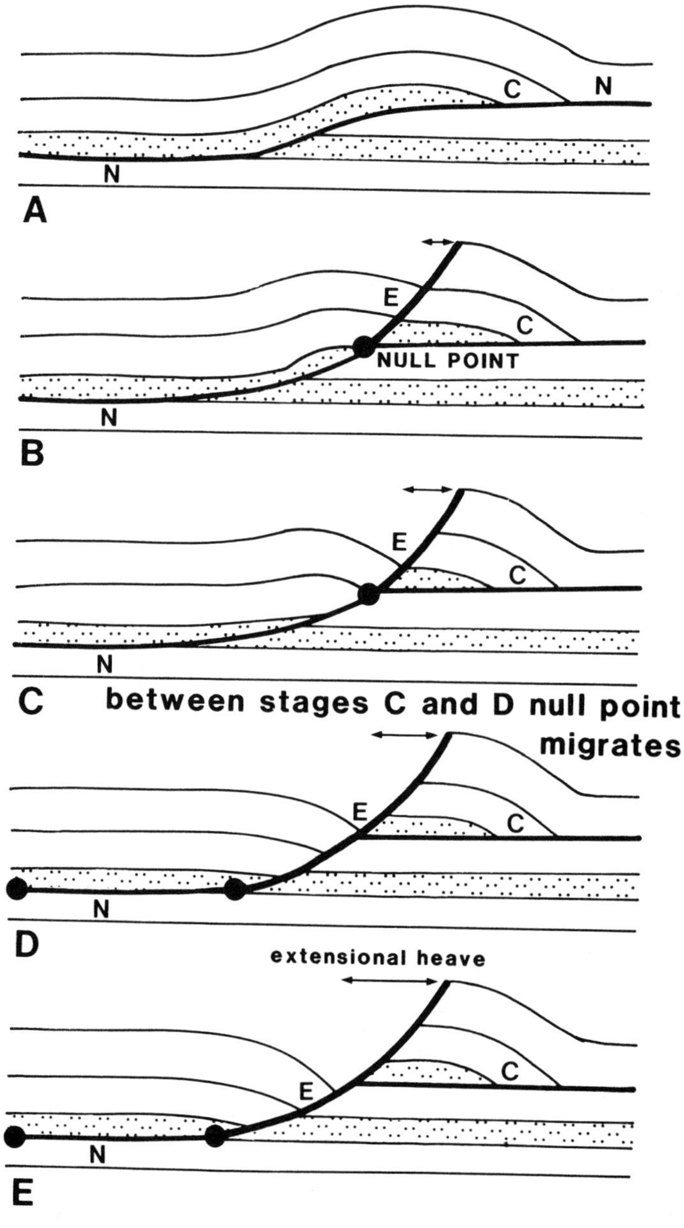

FIG. 11. Sequential diagrams to show the shortcutting and partial reactivation of a single thrust fault by extensional movements. Faults are labelled C, contractional; N, neutral; and E; extensional (represented by thicker lines).

system would be represented by extensional and neutral faults with only the shortcut remnant of thrust 2 showing net contraction.

Recognizing inversion on geological maps

The recognition of inversion tectonics and the quantification of inversion ratio may be performed using geological maps where structures have some degree of plunge. A good stratigraphic control is essential. A region that has undergone positive inversion of a synrift (extensional) stratigraphy will contain faults that apparently change from normal to reverse across the map. In the simplest case, an extensional fault displaces younger material in its hanging wall above older rocks in its footwall. A contractional fault displaces older material on younger rocks. Considerations of net slip are unimportant. A simplified geological map of the Weymouth District (BGS 19) illustrates the

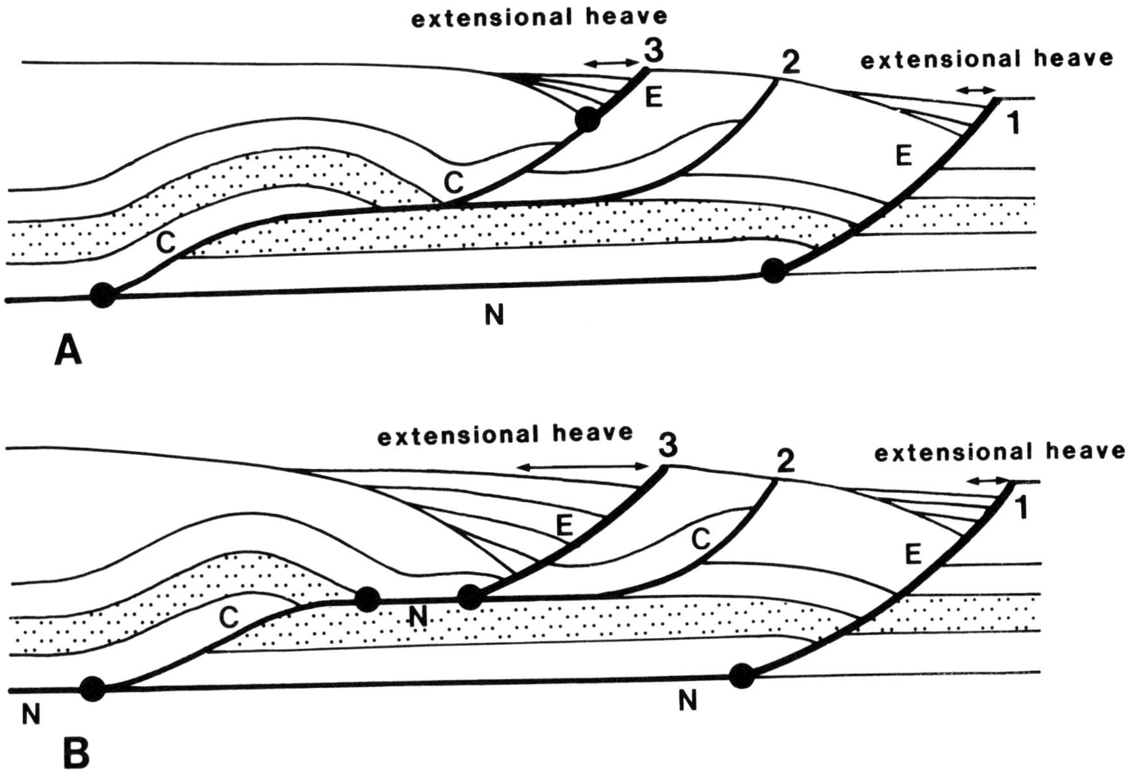

FIG. 12. Thrust system as in Fig. 10, partially reactivated in extension. A and B show two stages in reactivation. Faults are labelled C, contractional; N, neutral, and E, extensional (represented by thicker lines).

concept of changing mode along a single fault (Fig. 13). This effect results from positive inversion tectonics. In this region, the Jurassic and Lower Cretaceous sequences are considered to be synrift. The Upper Cretaceous is a postrift sequence deposited above a significant unconformity (Stoneley 1982). The point of change along the fault from net normal to net reverse displacement is the null point. In this case it occurs at the Upper Cretaceous unconformity.

The geometry of inversion may be analysed and the inversion ratio may be calculated using stratigraphic separation diagrams (Elliott & Johnson 1980) for individual faults. Stratigraphic separation is the displacement of a stratigraphic marker by a fault. Displacement is measured perpendicular to bedding. Stratigraphic separation may be measured in map view along a fault at positions of known stratigraphy. Curves of hanging wall and footwall stratigraphy cut by the fault may be plotted against fault map length. In normal faults, the hanging wall curve plots above that of the footwall. In reverse faults the footwall curve plots above that of the hanging wall. Where footwall and hanging wall stratigraphy differs, both stratigraphic sequences may be incorporated in the stratigraphic separation diagram by plotting individual curves using the appropriate stratigraphy (Fig. 10).

Where faults have undergone positive inversion, the curves cross at the null point. Using the hanging wall synrift stratigraphy, parameters d_h, d_e and d_c may be measured for use in equations (2) and (3) for the calculation of inversion ratio. However, this is not possible in the Weymouth example (Fig. 10) because of erosion and partial omission of Jurassic and Lower Cretaceous stratigraphy in the fault hanging walls. A phase of contractional inversion coupled with erosion took place before the deposition of the Upper Cretaceous/Tertiary section.

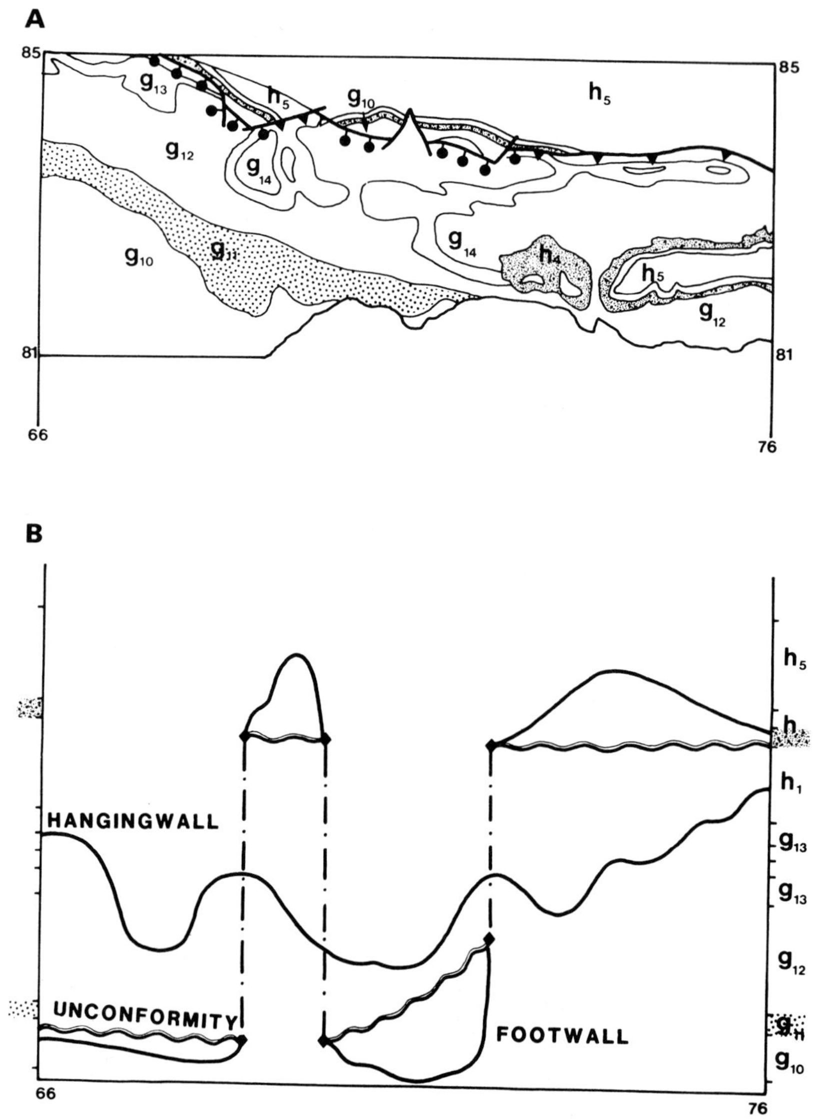

Fig. 13. A, Simplified geological map of the area N of Weymouth (from BGS sheet Weymouth) to show switching in fault mode from normal to reverse as a result of positive inversion. ▼▼, reverse fault; ●●, normal faults. B, Stratigraphic separation diagram constructed from Weymouth sheet. Note that this diagram is plotted on a full stratigraphic frame. Certain parts of the Jurassic/Lower Cretaceous stratigraphy is missing below the Albian (h4) unconformity as a result of early contractional inversion coupled with erosion. The missing stratigraphy is represented by dashed lines in the foot wall curve.

References and further reading

BEACH, A. 1984. Structural evolution of the Witch Ground Graben. *Journal of the Geological Society of London*, **141**, 621–628.

DAVIS, P. N. 1983. Gippsland Basin, SE Australia. *In*: BALLY, A. W. (ed.) *Seismic Expression of Structural Styles. American Association of Petrology Geologists Special Publication*, **3**, 3.3.19–3.3.24.

ELLIOTT, D. & JOHNSON, M. R. W. 1980. Structural evolution in the northern part of the Moine Thrust Belt, Northwest Scotland. *Transactions of the Royal Society of Edinburgh: Earth Sciences*, **71**, 69–96.

GIBBS, A. D. 1984. Structural evolution of extensional basin margins. *Journal of the Geological Society of London*, **141**, 609–619.

GLENNIE, K. W. & BOEGNER, P. L. E. 1981. Sole Pit inversion tectonics. *In*: ILLING, L. V. & HOBSON, G. D. (eds) *Petroleum Geology of the Continental Shelf of NW Europe*. Institute of Petroleum, London, pp. 110–120.

HAMBLIN, W. K. 1965. Origin of "reverse drag" on

the downthrown side of normal faults. *Geological Society of America Bulletin*, **76**, 1145–1164.

STONELEY, R. 1982. The structural development of the Wessex Basin. *Journal of the Geological Society of London*, **139**, 545–554.

WILLIAMS, G. D. and CHAPMAN, T. J. 1983. Strains developed in the hanging walls of thrusts due to their slip/propagation rate: a dislocation mode. *Journal of Structural Geology*, **5**, 563–571.

G. D. WILLIAMS & C. M. POWELL, Department of Geology, University of Keele, Keele, Staffordshire, ST5 5BG, UK.

M. A. COOPER, BP Canada Inc., Oil and Gas Division, 333 Fifth Avenue SW, Calgary, Alberta T2P 3B6, Canada.

Selective fault reactivation during basin inversion: potential for fluid redistribution through fault-valve action

RICHARD H. SIBSON

Department of Geology, University of Otago, PO Box 56, Dunedin, New Zealand

Abstract: Inversion structures, associated with the compressional reactivation of moderate to steeply dipping normal faults inherited from earlier crustal extension, form important structural traps for hydrocarbons. Migration into these traps must occur syn- or post-inversion. Seismic reflection profiles show that inversion is frequently highly selective, with only some of an existing normal fault set being reactivated. Neglecting marked stress-field heterogeneity, frictional mechanics suggests three possible explanations for this selective reactivation: (1) preferential reactivation of shallowest-dipping normal faults in a region that previously underwent the greatest extensional 'dominoing' of fault blocks; (2) the presence of anomalously low friction material along particular faults; and (3) a heterogeneous distribution of fluid overpressures ($P_f >$ hydrostatic) with preferential reactivation occurring in the area of most intense overpressuring. The last possibility is favoured by the likelihood that fluid overpressures develop during inversion as a consequence of the dramatic increase in mean stress that accompanies the transition from an extensional stress regime, with high capacity to store fluids, to a compressional regime with comparatively low storage capacity.

Compressional reactivation of moderately to steeply dipping faults likely requires significant overpressures; in the case of faults with dips in excess of the frictional 'lock-up' angle (typically 50–60°), supralithostatic fluid pressures ($P_f > \sigma_3 = \sigma_v$) are a necessary prefailure condition in rupture nucleation sites. Extreme fault-valve action then becomes possible, with postseismic flushing of fluids upwards along faults from overpressured compartments. Evidence for such activity comes from mesothermal Au-quartz veins hosted in steep reverse faults, where repeated attainment of supralithostatic pressures alternated with discharge episodes along the faults. Episodes of vertical hydrocarbon migration along reverse faults are therefore a likely accompaniment to basin inversion; there are, for instance, historical records of postseismic discharge of aqueous and hydrocarbon fluids from the overpressured basins of the Western Transverse Ranges, California, where steep reverse faults remain active today. Careful assessment of the relative timing of fault reactivation during inversion and hydrocarbon migration is needed to evaluate the hypothesis for ancient inverted basins.

During positive basin inversion, normal fault structures inherited from an earlier phase of crustal extension are reactivated in compression. The importance of such inversion structures as traps for hydrocarbons has become very apparent over the past decade or so (e.g. Bally 1983; Fraser & Gawthorpe 1990). Seismic reflection profiling has revealed the diagnostics of positive basin inversion and a great deal of attention has been paid to the geometrical, stratigraphic and kinematic characteristics of the inversion structures (Ziegler 1987, 1989; Williams *et al.* 1989). Replication of the general geometrical form of inversion structures from analogue modelling (Koopman *et al.* 1987; McClay 1989; Buchanan & McClay 1992; Sassi *et al.* 1993) has yielded additional insights. However, important questions remain related to the mechanics of inversion and the timing of hydrocarbon migration into the inversion structures, which must occur either during or after the phase of tectonic inversion.

In this paper, the frictional mechanics of fault reactivation are explored to investigate the particular stress and fluid pressure conditions under which inherited normal faults may be reactivated in compression. Only the simplest case of coaxial inversion (extensional, followed by contractional pure dip-slip faulting) is considered to keep the analysis tractable. However, in a qualitative sense the results are extendable to more general situations of oblique inversion under triaxial stress. The analysis yields some insights into processes of fluid redistribution likely to accompany inversion. An important aspect is the peculiar ability of high-angle reverse faults of comparatively low displacement to act as major conduits for episodic

From Holdsworth, R.E. & Turner, J.P. (compiler) *Extensional Tectonics: Regional-scale Processes.* The Geological Society, Key Issues in Earth Sciences, **2 (1)**, 263–279.
1476-315X/02/$15.00 © The Geological Society of London 2002
First published in: "Buchanan, J.G. & Buchanan, P.G. (eds) *Basin Inversion*, The Geological Society, London, Special Publications **88**, 3–19".

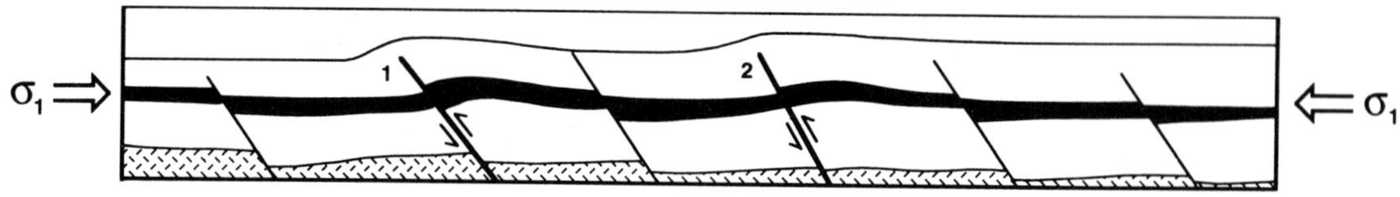

Fig. 1. Cartoon summarizing observations of selective fault reactivation during mild positive basin inversion beneath an unconformity: (1) preferential reverse reactivation of a relatively low-dipping fault; (2) selective reverse reactivation of a fault dipping steeper than its neighbours.

passage of overpressured fluids through fault-valve action, for which evidence exists throughout the geological record (Sibson 1990*a*).

Observations of fault reactivation during inversion

Much of the structural evidence for positive tectonic inversion comes from the detailed seismic reflection profiles available for areas such as the North Sea and related areas in NW Europe, the Taranaki Basin in New Zealand and the Gippsland Basin in southeastern Australia (e.g. Bally 1983; Williams *et al.* 1989). Several important observations can be made from these different studies (Fig. 1).

(i) While it is clear that contraction during the inversion phase is sometimes oblique to the former direction of extension, there are areas such as the Taranaki Basin in New Zealand (King & Thrasher 1993), and the Wessex and Weald Basins of southern England (Stonely 1982; Simpson *et al.* 1989; Butler & Pullan 1990) where extension and contraction appear to have been close to coaxial (but, see also Chadwick 1986).

(ii) In many instances fault reactivation during inversion is highly selective, with only a few structures within an extensive set of pre-existing normal faults, or only individual segments of a normal fault system undergoing compressional reactivation as reverse faults (Badley *et al.* 1989; Hayward & Graham 1989; Williams *et al.* 1989). Nor is it always the apparently more optimally orientated faults (see later) that are selectively reactivated.

(iii) In areas of low to moderate inversion, the reactivated faults have moderate to steep dips (40–60°) and low-angle short-cut thrusts through the footwall are comparatively rare (Hayward & Graham 1989; Simpson *et al.* 1989).

We are therefore faced with a situation where it is apparently easier to reactivate moderate to steeply dipping faults that are not well orientated for reactivation under compression in preference to the formation of new, favourably oriented thrusts. Moreover, such reactivation as occurs is highly selective.

Mechanics of inversion

Topography is unlikely to be extreme in areas undergoing incipient inversion and, while competent strata may act as stress guides to some extent, they are unlikely to deviate significantly from the subhorizontal until inversion is well advanced. The standard 'Andersonian' assumption of horizontal and vertical principal stress trajectories (Anderson 1951) is therefore probably warranted. Under triaxial stress (principal compressive stresses, $\sigma_1 > \sigma_2 > \sigma_3$) with the vertical stress, σ_v (equivalent to the overburden pressure) staying constant, the simplest case of coaxial inversion involves a progressive increase in the horizontal stress from an extensional stress state with $\sigma_v = \sigma_1$ to a compressional state with $\sigma_v = \sigma_3$ (Fig. 2). Normal faults formed in the extensional phase likely initiated as Coulomb shears in planes containing the σ_2 axis with dips of *c.*60° (Anderson 1951), but may since have rotated to lower dips.

Frictional reactivation of faults

In fluid saturated crust, the effective principal compressive stresses are $\sigma_1' = (\sigma_1 - P_f) > \sigma_2' = (\sigma_2 - P_f) > \sigma_3' = (\sigma_3 - P_f)$, and the condition for frictional reactivation of existing cohesionless faults may be represented by the equivalent of Amontons's Law:

$$\tau = \mu_s \sigma_n' = \mu_s(\sigma_n - P_f) \qquad (1)$$

where μ_s is the static coefficient of friction, P_f is the fluid pressure within the rock mass, and τ and σ_n are respectively the resolved shear and normal stress components on the fault. From an extensive series of low-temperature experiments, Byerlee (1978) found the static coefficient of rock friction to be largely independent of rock type and restricted to the range, $0.6 < \mu_s < 0.85$, prominent exceptions being material rich in montmorillonite, allied

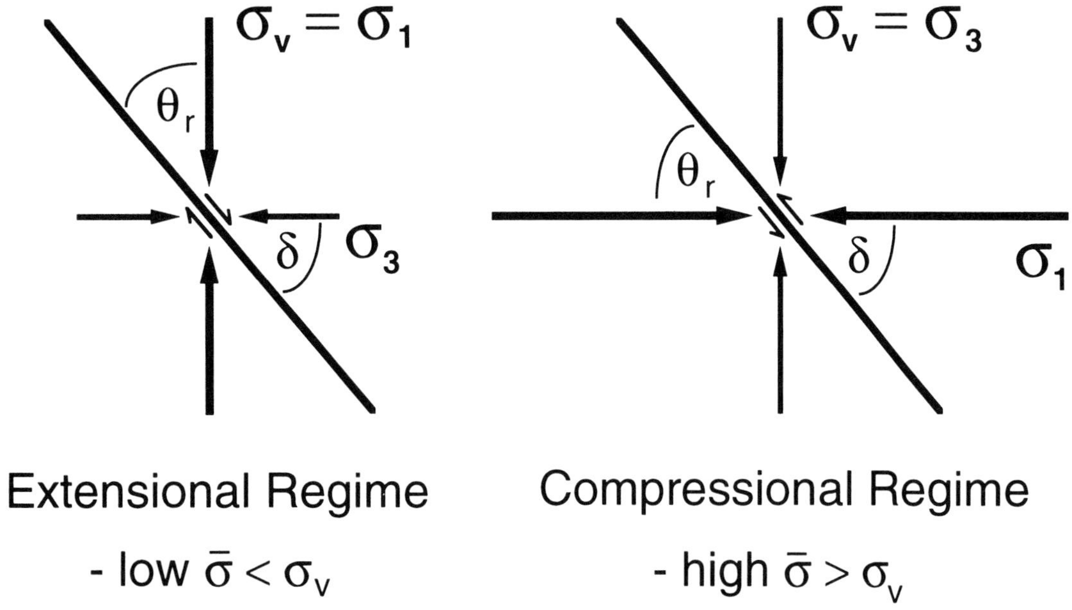

Fig. 2. Coaxial inversion from extensional to compressional stress regimes for a fault with dip, δ where θ_r is the reactivation angle defined with respect to the σ_1 direction.

clay minerals, or evaporites. Elevated temperatures have little effect on rock friction coefficients up to *c*.350°C (Stesky *et al*. 1974). Note that the higher bound ($\mu_s = 0.85$) was derived from comparatively low normal stress experiments ($\sigma_n' < 200$ MPa) most likely to be applicable in the upper few kilometres of the crust.

For the situation where the pole to existing faults lies within the σ_1/σ_3 plane (so that σ_2 has no effect), the reactivation criterion (1) may be recast in terms of the ratio of effective principal stresses as:

$$\frac{\sigma_1'}{\sigma_3'} = \frac{(\sigma_1 - P_f)}{(\sigma_3 - P_f)} = \frac{(1 + \mu_s \cot\theta_r)}{(1 - \mu_s \tan\theta_r)} \qquad (2)$$

where θ_r is the angle of reactivation as defined in Figs 2 & 3 (Sibson 1985, 1990*b*). This expression can be rewritten in terms of the fault dip, δ, by substituting $\theta_r = \delta$ for compressional stress regimes and $\theta_r = (90° - \delta)$ for extensional regimes. The expression is a useful measure of the relative ease of reactivation of differently oriented faults (Fig. 3a). The optimal angle for reactivation, where σ_1'/σ_3' is a positive minimum, occurs when $\theta_r^* = 0.5\tan^{-1}(1/\mu_s)$. For the Byerlee range of coefficients, $25° < \theta_r^* < 30°$, not dissimilar to the orientation of faults during their initiation as Coulomb shears. As θ_r deviates increasingly from θ_r^*, reactivation becomes progressively more difficult, requiring a higher stress ratio. Frictional 'lock-up' with $\sigma_1'/\sigma_3' \to \infty$ occurs as $\theta_r \to 0$ or $2\theta_r^*$, so that either σ_1' has to become very large, or σ_3' has to become very small. The lock-up angle corresponding to $2\theta_r^*$ is given by $\theta_1 = \tan^{-1}(1/\mu_s)$, and for Byerlee friction coefficients should occur in the range, $50° < \theta_1 < 59°$ (Fig. 3b).

The physical basis of frictional lock-up is that at large reactivation angles further boosting of σ_1 adds more to the normal stress 'gluing' the walls of the fault together than to the resolved shear stress promoting instability. As lock-up is approached, the rapid increase in the stress ratio for continued reactivation may lead to the formation of new, favourably orientated faults in accordance with the Coulomb criterion unless $\sigma_3' = (\sigma_3 - P_f)$ is made very low through fluid overpressuring. Reactivation at $\theta_r > 2\theta_r^*$ (the field of severe misorientation) is only possible in special circumstances when the least stress becomes effectively tensile (i.e. $\sigma_3' < 0$, or $P_f > \sigma_3$).

Frictional constraints on inversion

Extensional faulting phase During crustal extension with $\sigma_v = \sigma_1$, normal faults generally initiate with steep dips (typically 55–65°) in accordance with Coulomb failure theory (Anderson 1951). With continued extension, sets of normal faults dipping in the same direction (and their intervening fault-blocks) may rotate progressively to lower dips in a domino-like manner. As domino rotation

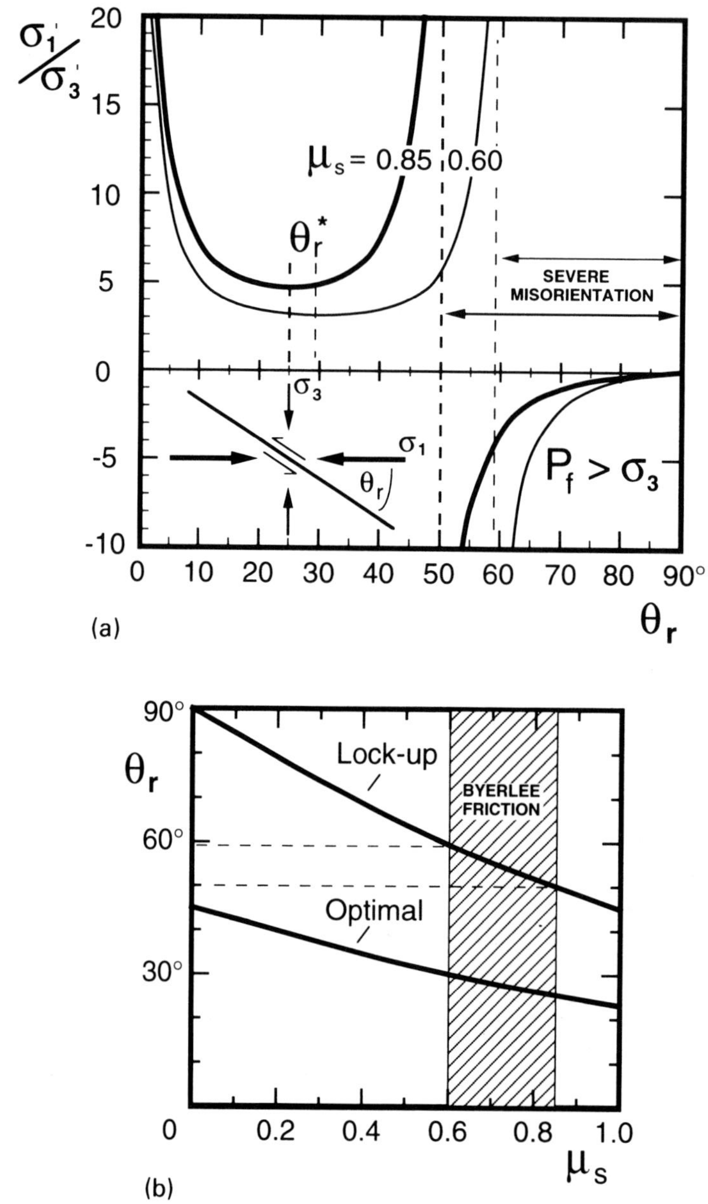

Fig. 3. (**a**) Stress ratio (σ_1'/σ_3') for the frictional reactivation of a cohesionless fault, whose pole lies in the σ_1/σ_3 plane, plotted against the reactivation angle, θ_r, for $\mu_s = 0.6$ (light line) and $\mu_s = 0.85$ (bold line) (after Sibson 1985). (**b**) Values of the optimal angle for reactivation and the frictional lock-up angle for cohesionless faults plotted against the friction coefficient, μ_s.

proceeds, the fault reactivation angle increases so that for Byerlee friction coefficients, lock-up is expected in the dip range 30–40°. The observation that modern, seismically active planar normal faults only remain active down to this dip range (Jackson 1987; Doser & Smith 1989; Roberts & Jackson 1991) supports the application of Byerlee-type friction to natural systems of planar faults.

Compressional faulting during inversion At the onset of a subsequent phase of horizontal compression with $\sigma_v = \sigma_3$, dip angles of faults inherited from the extensional phase (now corresponding to the reactivation angle, Fig. 2) could therefore range from around 30–65°. The ease with which these existing faults are reactivated in compression as reverse faults clearly decreases with increasing dip. In fact, the range of dip values for which fault reactivation can occur within the bounds of frictional lock-up for both extension and compression is quite sensitive to the friction coefficient (Fig. 4). At the high end of the Byerlee range ($\mu_s = 0.85$) only

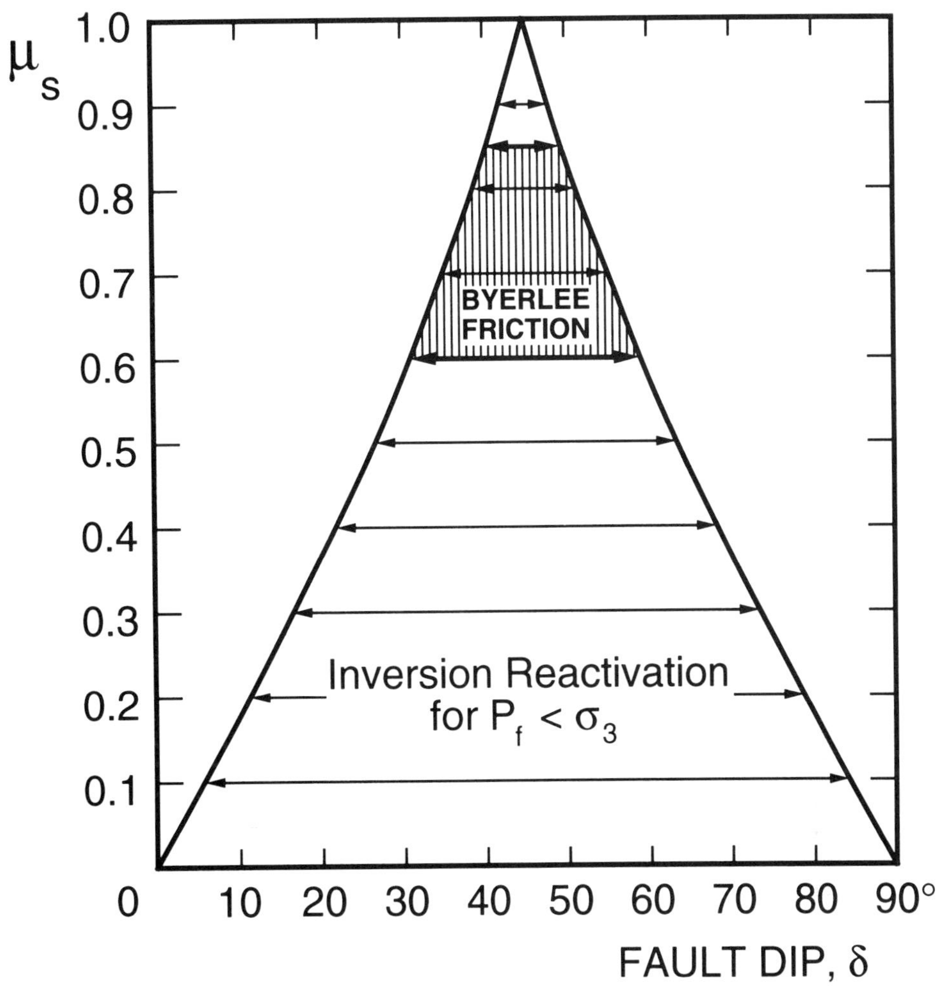

Fig. 4. Permissible range of dips allowing reactivation of cohesionless faults in both extension and compression (provided $\sigma_3' = (\sigma_3 - P_f) > 0$), plotted against static friction coefficient, μ_s. The Byerlee (1978) range of rock friction coefficients is shaded.

faults dipping in the range $40° < \delta < 50°$ can be reactivated in both compression and extension when $P_f < \sigma_3$. At the lower end ($\mu_s = 0.6$), the dip range permitting reactivation in both extension and compression broadens to about $30° < \delta < 60°$ for horizontal σ_3 and σ_1 stress trajectories, respectively.

The magnitude of differential stress required for compressional reactivation varies with overburden pressure and the fluid pressure within the rock mass. At a depth, z, in the crust, the effective overburden pressure is:

$$\sigma_v' = (\sigma_v - P_f) = \rho g z(1 - \lambda_v) \quad (3)$$

where ρ is the average rock density, g is gravitational acceleration, and the pore-fluid factor, $\lambda_v = P_f/\rho g z$. Equation (2) can then be rewritten as:

$$(\sigma_1 - \sigma_3) = \frac{\mu_s(\tan\delta + \cot\delta)}{(1 - \mu_s\tan\delta)}\rho g z(1 - \lambda_v) \quad (4)$$

an expression giving the differential stress required for compressional reactivation of a fault with dip, δ, at a depth, z, for particular values of μ_s, ρ, and λ_v (Sibson 1990b). For ease of scaling, the differential stress required to reactivate a cohesionless fault in compression at 1 km depth at different values of λ_v is plotted against dip angle in Fig. 5, for crust with average density, $\rho = 2350\,kg/m^3$, and with $\mu_s = 0.85$, the most appropriate friction coefficient for the uppermost few kilometres of the crust. For this rock density, the curve with $\lambda_v = 0.4$ corresponds approximately to hydrostatic fluid pressure conditions, higher values representing different degrees of overpressuring. Note how rapidly the required stress level for reactivation rises as the lock-up angle is approached. It is in these fields that new, favourably oriented thrusts are likely to form in preference to continued reactivation of the existing structures. The comparative scarcity of such low-angle thrusts in

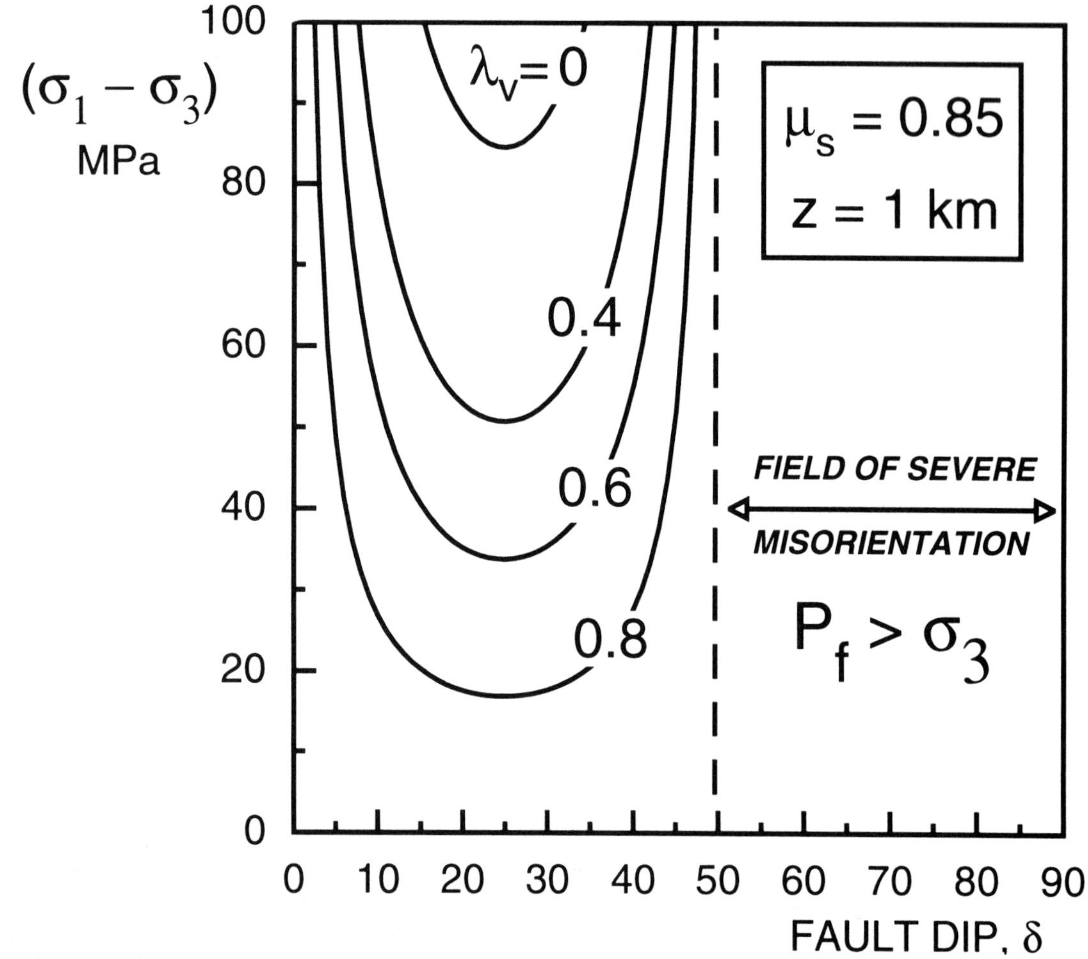

Fig. 5. Differential stress required for compressional reactivation of a cohesionless fault at a depth of 1 km plotted against dip angle, δ, for $\rho = 2350\ kg/m^3$, $\mu_s = 0.85$, and varying λ_v. The field where $P_f > \sigma_3$ is required for reactivation is the field of severe misorientation.

regions of mild inversion where moderate to steep faults have been reactivated in compression has, therefore, the implication that the reactivated faults are extremely weak.

Reactivation of listric faults Because of decreasing dip with depth, originally listric normal faults should experience varying ease of compressional reactivation at different structural levels. Low fault dips at depth allow easy reactivation, but steep dips at higher levels may lead to lock-up and the accommodation of shortening through wallrock strain. This distributed strain at high levels during inversion may contribute to the development of major antiformal buckles in hangingwall strata, perhaps amplifying earlier 'roll-over' structures formed during the extensional phase (Fig. 6). Alternatively, 'short-cut' thrusts may develop in the footwall to accommodate continued shortening.

Analogue modelling Sandbox modelling of inversion tectonics involving the compressional reactivation of earlier extensional faults (McClay 1989; Buchanan & McClay 1992; Sassi *et al.* 1993) lends support to several aspects of the simple frictional mechanics analysis presented above. Lock-up of reverse faults is observed to occur in the dip range 50–60° and is often followed by the formation of new, more favourably oriented low-dipping thrusts as footwall 'short-cuts'.

Mechanisms for selective reactivation

From analogue modelling, Sassi *et al.* (1993) demonstrated that selective reaction could occur within a set of parallel, favourably oriented faults when the faults were closely spaced. However, this form of selective fault reactivation likely arises from stress-field

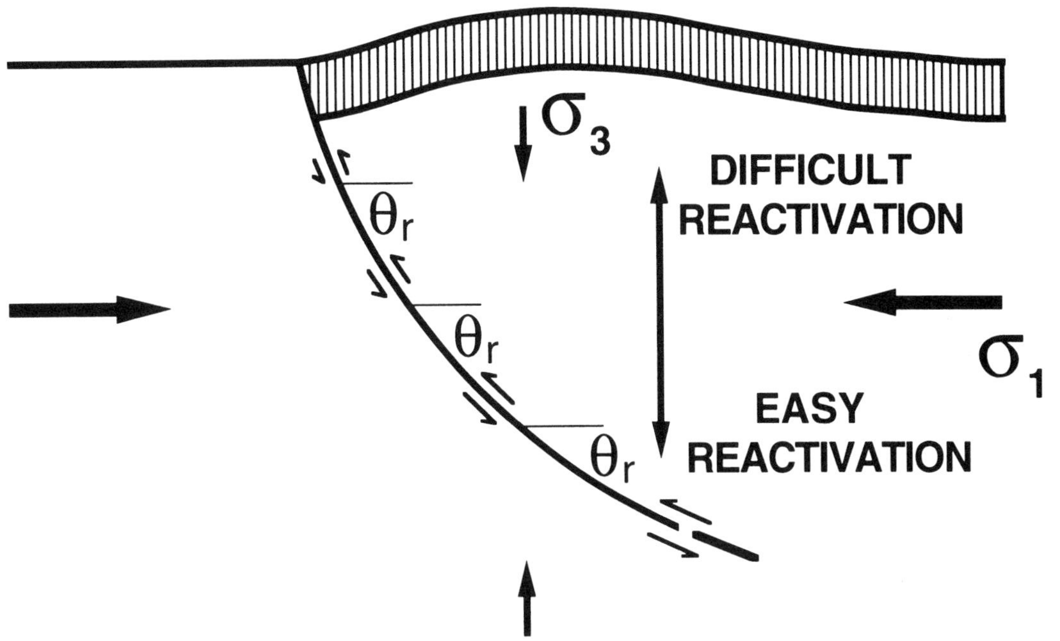

Fig. 6. Cartoon showing how the changing dip of a listric normal fault with depth leads to varying ease of reactivation under compression.

heterogeneity caused by the experimental boundary conditions. On the basis of the 2-D reactivation theory outlined above, and neglecting the possibility of such stress inhomogeneities, three other mechanisms may account for the selective fault reactivation that is observed during inversion.

(i) An obvious expectation is that the lowest dipping faults in a rotated set would be preferentially reactivated in compression at lower stress ratios than steeper, less favourably oriented structures. Inversion should then initiate in the areas of greatest former extension and spread to surrounding areas as the earliest reactivated faults steepen by reverse dominoing during regional contraction. If inversion does not proceed beyond a certain stage, fault reactivation will appear highly selective.

(ii) The presence of anomalously low-friction material on a fault, such as montmorillonite-rich gouge, could lead to its selective reactivation.

(iii) For faults of comparable dip, preferential fault reactivation would occur where fluid pressure levels were locally elevated. This last mechanism takes on special significance given the now widespread recognition of both vertical and lateral compartmentalization of fluid pressures in sedimentary basins (Hunt 1990).

While the first mechanism is undoubtedly important in some areas, there are clear instances where it is not the shallowest dipping fault that has undergone compressional reactivation (e.g. Hayward & Graham 1989; Simpson *et al.* 1989). The second mechanism is difficult to discount but requires rather special pleading because selective reactivation often occurs within a set of faults cutting essentially the same stratified sequence. The third mechanism, invoking locally elevated fluid pressures as a cause of selective reactivation, is of particular interest in view of the fluid overpressuring that is likely to develop during inversion (see below).

Fluid storage in compressional vs. extensional regimes

Both porosity and permeability characteristics of a rock mass are likely to vary with the state of stress, either through the influence of stress-controlled features such as faults, microcracks, hydrofractures and stylolites, or through changes in the level of mean stress, $\bar{\sigma} = (\sigma_1 + \sigma_2 + \sigma_3)/3$, which affects the volumetric elastic strain. Stress-controlled structures may enhance or counteract existing anisotropic permeability depending on the relative attitude of bedding

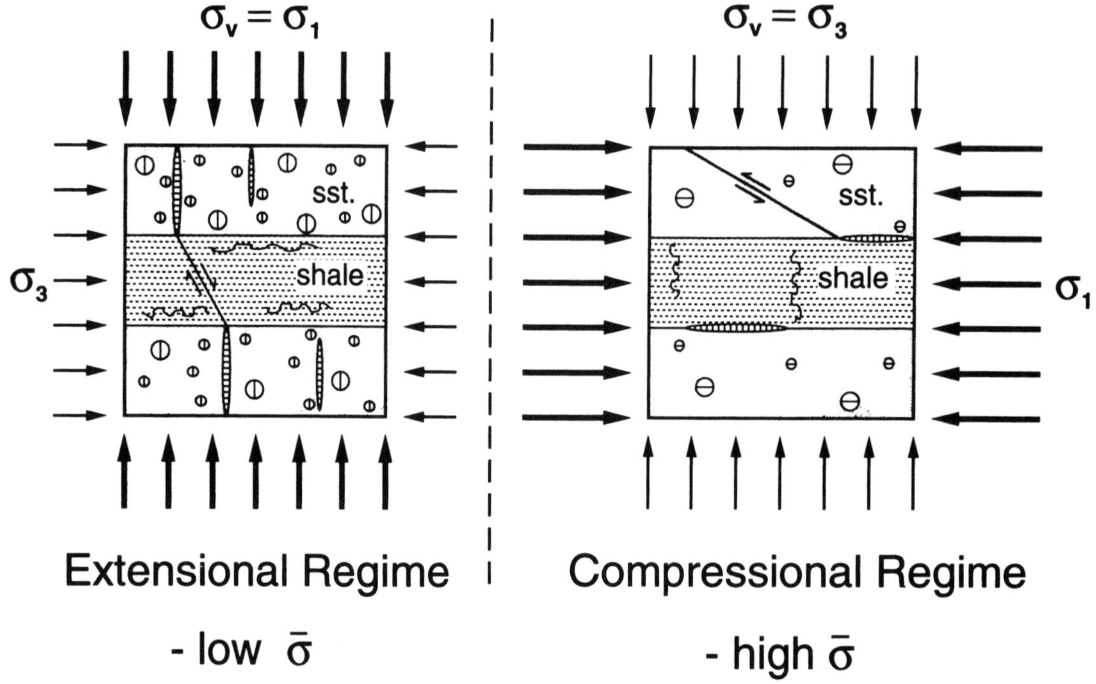

Fig. 7. Schematic illustrating stress-controlled structures affecting the porosity and permeability of the rock mass, and the contrasting fluid storage capacity in extensional and compressional stress regimes (split circles = transgranular microfractures; squiggly lines = stylolites; cross-hatched ellipses = hydraulic extension fractures).

and the principal stresses (e.g. du Rouchet 1980; Sibson 1994).

As illustrated in Fig. 7, vertical permeability is likely to be enhanced in extensional stress regimes ($\sigma_v = \sigma_1$) through the development of subvertical transgranular microcracks and extensional hydraulic fractures (both forming perpendicular to the least compressive stress, σ_3), and by steep normal faults. This may be countered to some extent by the development of flat-lying stylolites forming perpendicular to σ_1 in the finer grained sediments. By contrast, bedding-parallel permeability may be enhanced in compressional settings by microfractures, by hydraulic extension fractures and by low-angle thrusts. The relative development of these various structural features in different stress fields affects the capacity of the rock mass to store fluids.

Conditions for hydraulic extension fracturing

Because of their large aperture and continuity, macroscopic extension fractures formed by hydraulic fracturing are particularly important from the viewpoints of both fluid storage and rock mass permeability. In a rock mass with tensile strength, T, hydrofractures form when:

$$P_f = \sigma_3 + T \tag{5}$$

provided the differential stress, $(\sigma_1 - \sigma_3) < 4T$, so that shear failure is inhibited (Secor 1965). Development of transgranular microcracks through grain impingement is also likely to be enhanced as $\sigma_3' = (\sigma_3 - P_f) \rightarrow 0$ (Palciauskas & Domenico 1980).

Following Secor (1965), the fluid pressure conditions under which hydraulic extension fractures can develop at different depths in extensional (normal faulting) and compressional (reverse faulting) regimes can be represented on a plot of the pore-fluid factor ($\lambda_v = P_f/\rho gz$) against depth (Fig. 8). The curves are plotted for T = 1 MPa and T = 10 MPa, bracketing the common range for long-term rock tensile strength (Etheridge 1983), and represent the λ_v conditions for hydraulic fracturing when differential stress is at the limiting value. In compressional settings, supralithostatic ($\lambda_v > 1$) fluid pressures are required at all depths to induce hydraulic fracturing. In extensional settings, however, hydraulic fracturing can potentially develop under hydrostatic levels of fluid pressure close to the Earth's surface, and at greater

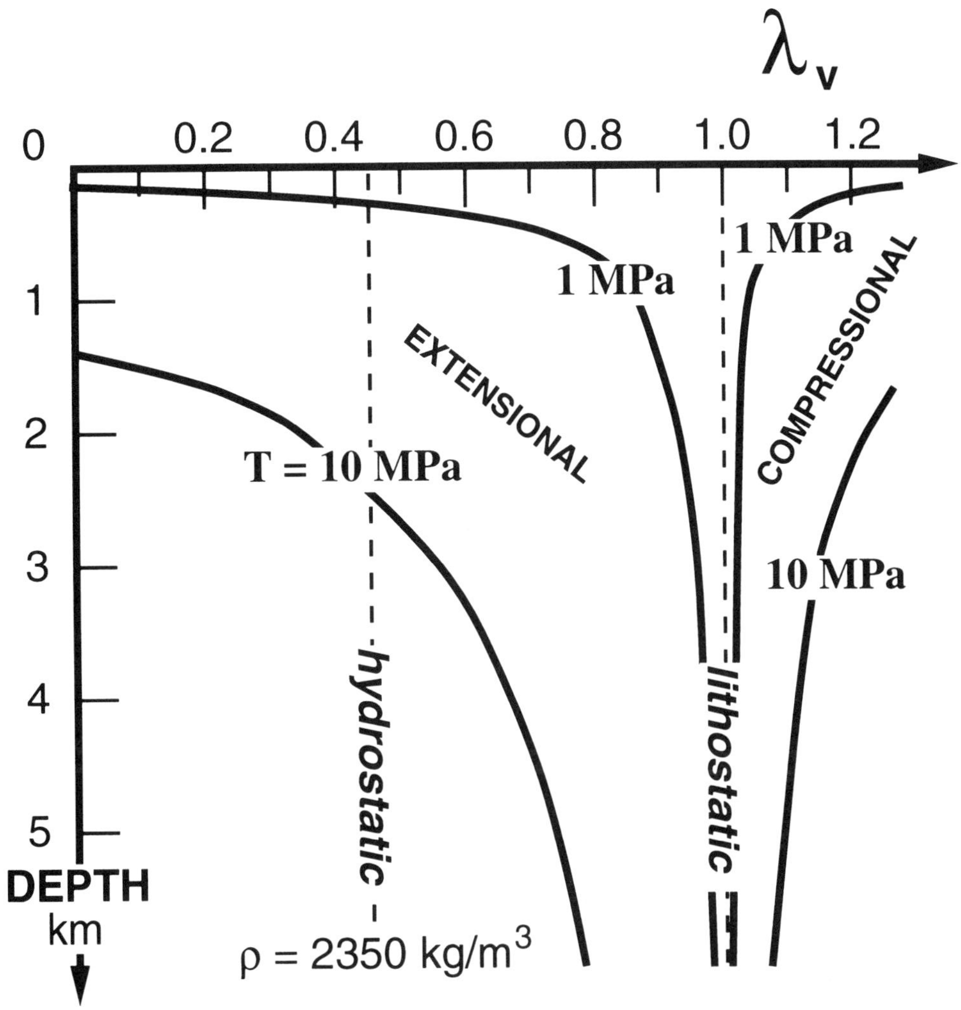

Fig. 8. λ_v plots defining the conditions for hydraulic fracturing at maximum allowable differential stress in extensional and compressional stress regimes for rock tensile strengths, T = 1 MPa and 10 MPa and an average rock density, $\rho = 2350\,kg/m^3$ (after Secor 1965).

depths requires a relatively small degree of overpressuring above hydrostatic. As a consequence, in comparison with a compressional stress regime, the capacity of a rock mass in an extensional tectonic regime to store fluids may be significantly enhanced, especially in the near-surface, through the development of subvertical hydrofractures and transgranular microcracks.

Mean stress effects

Elastic volumetric strain within a rock mass is related to the level of mean stress, $\bar{\sigma}$. In fluid saturated crust, the effective mean stress becomes $\bar{\sigma}' = (\bar{\sigma} - P_f)$ through the principle of effective stress. Increasing the level of compressive mean stress therefore either decreases the volume of the solid framework within the rock mass, causing fluid to be expelled from existing pore space in a manner analogous to squeezing a sponge, or, if fluid cannot escape, boosts the level of fluid pressure. Consider the stress states illustrated in Fig. 2 for coaxial inversion. Clearly, the level of mean stress in an extensional regime is always less than the vertical stress (overburden pressure) while in a compressional regime it must exceed the vertical stress. The question to be addressed, therefore, is the extent to which the mean stress level increases during the transition from an extensional regime with active normal faulting to a compressional regime with the same faults reactivated in reverse mode.

For the 2-D reactivation analysis, where mean stress reduces to $\bar{\sigma} = (\sigma_1 + \sigma_3)/2$, (2) can be rearranged in terms of the effective mean stress, giving:

$$\frac{\bar{\sigma}'}{\sigma_3'} = \frac{2 + \mu_s(\cot\theta_r - \tan\theta_r)}{2(1 - \mu_s\tan\theta_r)} \qquad (6)$$

an expression which gives the ratio of the

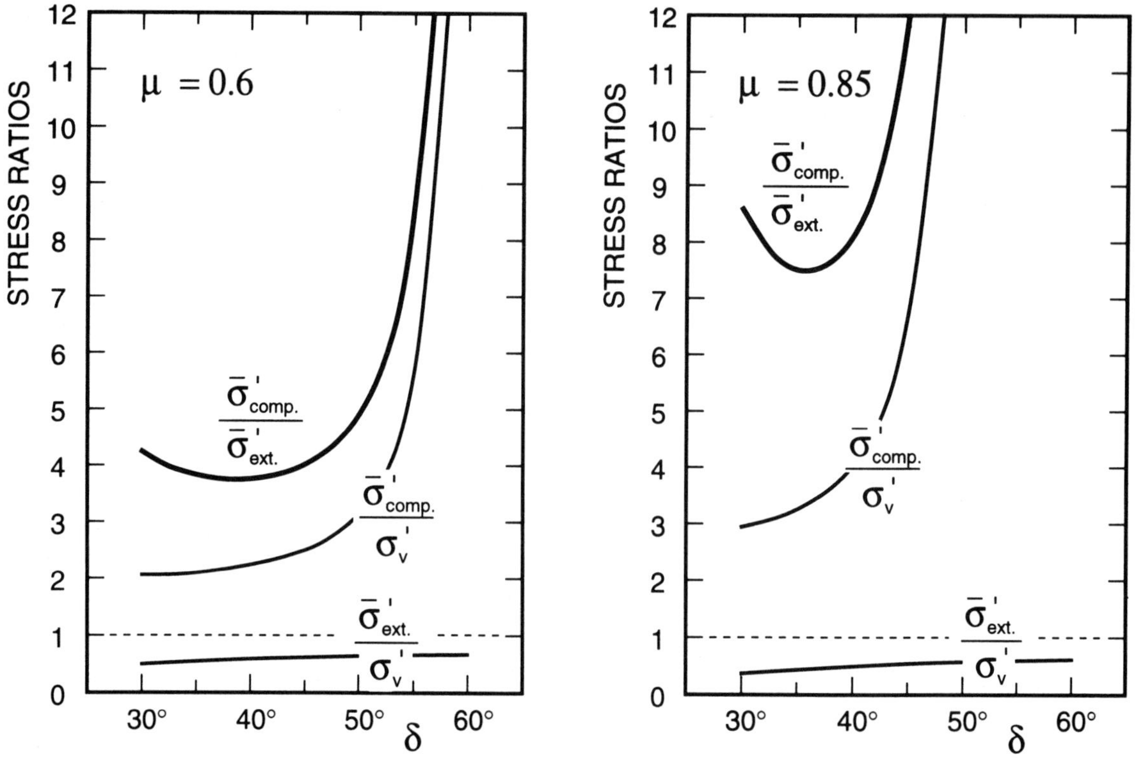

Fig. 9. Mean stress ratios as defined in text plotted against dip angle for $30° < \delta < 60°$, and for $\mu_s = 0.6$ and 0.85, for existing faults at the point of frictional shear failure.

effective mean stress to the least compressive stress at frictional failure for a cohesionless fault oriented at a reactivation angle, θ_r. For a compressional regime, where $\sigma_v = \sigma_3$ and $\delta = \theta_r$, the expression can be written:

$$\frac{\bar{\sigma}_{comp}'}{\sigma_v'} = \frac{2 + \mu_s(\cot\delta - \tan\delta)}{2(1 - \mu_s\tan\delta)} \tag{7}$$

while for an extensional regime with $\sigma_v = \sigma_1$ and $\delta = (\theta_r - 90°)$, it becomes:

$$\frac{\bar{\sigma}_{ext}'}{\sigma_v'} = \frac{2 + \mu_s(\tan\delta - \cot\delta)}{2(1 + \mu_s\tan\delta)} \tag{8}$$

These two equations give the ratio of effective mean stress to effective vertical stress for faults with dip, δ, at the point of frictional failure in compressional and extensional regimes. They can be combined to yield $\bar{\sigma}_{comp}'/\bar{\sigma}_{ext}'$, the ratio of the levels of effective mean stress for a fault with dip, δ, at frictional failure in compression and extension at the same level of fluid pressure. Values from these expressions are plotted for the dip range 30–60° for $\mu_s = 0.6$ and 0.85 (Fig. 9). Clearly, the increase in mean stress during inversion can be quite dramatic; for example, for $\delta = 45°$, $\bar{\sigma}_{comp}'/\bar{\sigma}_{ext}'$ has values of 12.3 and 4.0 for $\mu_s = 0.85$ and 0.6, respectively. Note, however, that these plots presuppose the same level of fluid pressure to be maintained in compressional as in extensional regimes, whereas what is in fact likely to happen is a significant boosting of fluid pressure as mean stress progressively increases.

Combined effect on fluid pressure level

Both the higher fluid storage capacity of extensional stress regimes and the increase in mean stress accompanying the transition from extension to compression are likely to contribute to increased fluid pressures during inversion. The onset of compression will lead to closure of the subvertical microcracks and extension fractures formed during the extensional phase (Fig. 7), inhibiting vertical migration. Coupled with the increasing mean stress, this reduction in void space will either force fluids out of the rock mass, or boost the level of fluid pressure if drainage is inhibited by low permeability caps such as shale horizons. The magnitude of the pressure increase will depend on the permeability characteristics of the rock mass and the length of time over which the transition from an extensional to a compressional regime takes place, shorter transition intervals being more likely to lead to substantial increases in fluid pressure. In some respects, the conclusions reached here are similar to those arrived at by Berry (1973) in his consideration of the development of high fluid pressures in the California Coast ranges.

A good example of the structural features diagnostic of boosted fluid pressures during inversion is provided by a reverse-reactivated fault at Ogof Gynfor on the north coast of Anglesey, Wales (Sibson 1981). The fault, with a finite reverse slip in excess of 20 m, dips moderately to the north and disrupts a sequence of cleaved black shale overlying a quartzite breccia-conglomerate. Two incrementally developed sets of quartz veins document a history of inversion reactivation. Prominent subvertical extension veins striking parallel to the fault were apparently developed during an early phase of extensional normal faulting, but are cut across by a sparsely developed second set of subhorizontal extension veins associated with the reverse reactivation of the fault. The vein-sets demonstrate that overpressuring ($P_f > \sigma_3$) accompanied both extensional normal faulting and compressional reverse reactivation, with fluid pressures intermittently boosted to supralithostatic values ($P_f > \sigma_3 = \sigma_v$) during the phase of reverse reactivation.

Fluid redistribution during fault reactivation

The boosting of fluid pressures during inversion allows existing faults that are not favourably oriented to become reactivated under compression, and provides a plausible mechanism for the selective reactivation that is observed. Any overpressures that develop within a basin undergoing inversion are likely to be very heterogeneous because of varying fluid storage and permeability characteristics. In this connection, it may be significant that Tertiary inversion within the Wessex Basin, for example, tends to be concentrated within the sub-basins which accommodated the thickest accumulations of Late Jurassic–Early Cretaceous sediments (Simpson *et al.* 1989), where fluid content is likely to have been greatest.

Fault-valve activity

Fault-valve behaviour becomes possible wherever active faulting occurs in overpressured portions of the crust and ruptures transect suprahydrostatic vertical gradients in fluid pressure (Sibson 1981). Valving action depends on the ability of faults to behave as impermeable seals through the interseismic period from the presence of clay-rich or cataclastic gouge and/or hydrothermal cementation, but to become highly permeable channelways for fluid discharge immediately postfailure as a consequence of the intrinsic roughness of natural rupture surfaces (Power *et al.* 1987). Breaching of low-permeability seals to overpressured zones by fault rupture thus leads to upwards discharge of fluids along the rupture zone and local reversion towards a hydrostatic gradient (Fig. 10). While valving leading to minor postfailure discharge may occur in any tectonic setting where overpressuring has developed, there are mechanical reasons why extreme fault-valve action involving substantial postfailure discharge tends to be associated with steep reverse faults (Sibson 1990a).

Ordinary versus extreme valving action Progressively increasing fluid pressure may trigger movement on existing faults in accordance with the reactivation criterion (1), thereby promoting valve action. It is apparent from the 2-D analysis of reactivation (Fig. 3) that favourably orientated faults would be the first to be reactivated, and that reactivation of unfavourably orientated faults in areas of increasing fluid pressure depends on the absence of favourably orientated structures. Provided fault orientation lies in the range, $0° < \theta_r < \theta_1 = 2\theta_r^*$, fault reactivation will occur before the condition for hydrofracturing ($P_f > \sigma_3$) can be achieved with fluid pressures then dropping postfailure through valve action. In such circumstances, the volume of fluid discharged is likely to be comparatively small.

For faults that are in the field of severe misorientation, however, the condition $P_f > \sigma_3$ is a prerequisite for reactivation, and hydraulic extension fractures are likely to develop prior to failure. The significance of this is that an array of hydrofractures in the rupture nucleation site provides a reservoir of overpressured fluids with ready access to the fault postfailure. Fluid volumes available for postfailure discharge are thus likely to be much greater in the case of severely misorientated faults; such structures are therefore capable of extreme fault-valve behaviour. In compressional settings, supralithostatic fluid pressures ($P_f > \sigma_3 = \sigma_v$) are required to meet the prefailure condition for steep, severely misorientated reverse faults (Fig. 8), but the prefailure hydrofractures are then subhorizontal and unlikely to breach impermeable sealing horizons themselves. It is this special combination of circumstances that gives steep reverse faults their peculiar ability for extreme fault-valve action, with the discharge of large fluid volumes postfailure (Sibson 1990*a*).

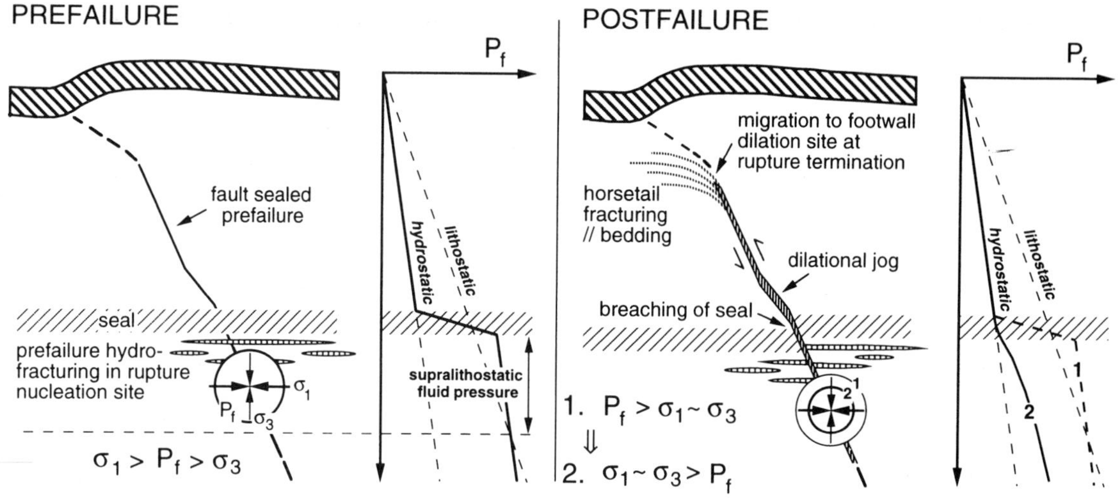

Fig. 10. Cartoon (not to scale) illustrating stress and fluid-pressure states prefailure and postfailure associated with extreme fault-valve action on a steep reverse fault. Only the simplest situation is shown where one impermeable seal separates hydrostatic and overpressured fluid regimes; reality is likely to be more complex.

The stress and fluid pressure cycling associated with extreme valve action on steep reverse faults is illustrated in Figs 10 and 11. In the immediate prefailure state, the condition $\sigma_1 > P_f > \sigma_3$ prevails in the overpressured rupture nucleation site, with formation of gaping subhorizontal hydrofractures. If relief of shear stress along the fault during rupturing is near-complete (and there is textural evidence from vein systems to support this, see Boullier & Robert 1992), the condition immediately postfailure is $P_f > \sigma_1 \approx \sigma_3$. Under these conditions, fluids from the hydrofracture array, and from any grain-scale microfracture porosity developed at low effective stress, are expelled into the rupture zone which becomes the principal avenue for fluid discharge. For total stress release, or if slight 'overshoot' occurs, actual dilation of the rupture zone may take place. Discharge continues until the pressure gradient reverts to hydrostatic or the rupture zone self-seals through hydrothermal precipitation ($\sigma_1 \approx \sigma_3 > P_f$). Differential stress and fluid pressure then start to rebuild towards the next failure episode ($\sigma_1 > \sigma_3 > P_f$). Valving activity thus depends on the competition between creation and destruction of permeability in fault zones (Sibson 1992), for which textural evidence is widespread (e.g. Roberts 1991; Hippler 1993).

Analogy with mesothermal Au-quartz lodes

Mesothermal gold-quartz vein systems developed in sub- to low-greenschist metamorphic environments provide evidence that steep reverse faults of comparatively low displacement can serve as conduits for focused flow of aqueous fluids through extreme fault-valve action (Sibson *et al.* 1988; Cox *et al.* 1991), and demonstrate the massive scale of flow that can occur. Typical vein assemblages comprise a mixture of flat-lying extensional veins in mutual cross-cutting relationships with steep fault-veins hosted by reverse faults. Hydrothermal textures of both vein-sets record histories of incremental deposition (Boullier & Robert 1992), their mutual cross-cutting relationships suggesting that the veins within the two sets formed at different stages of a repeating cycle. The flat-lying extension veins demonstrate that the condition $P_f > \sigma_3 = \sigma_v$ was repeatedly attained and are interpreted as representing prefailure hydrofractures, while the fault-veins are inferred to have developed incrementally during episodes of postfailure discharge (see Figs 10 & 11).

As an example, consider the Mother Lode vein system of early Cretaceous age in the western Sierra Nevada foothills of California (Knopf 1929). The principal quartz veins are hosted on individual reverse faults within the Melones fault zone. In places, continuous veins averaging over a metre in thickness can be traced for kilometres along strike and have been mined to depths in excess of 1 km. Per kilometre of fault strike-length, the volume of fault-hosted quartz therefore commonly exceeds $10^6\,m^3$. Established reverse displacements on the hosting faults range up to 100 m or so, with ribbon vein textures recording hundreds of episodes of hydrothermal deposition. Given the low solubility of quartz (Fyfe *et al.* 1978), something like $10^9 - 10^{10}\,m^3$ of aqueous fluid (more than the

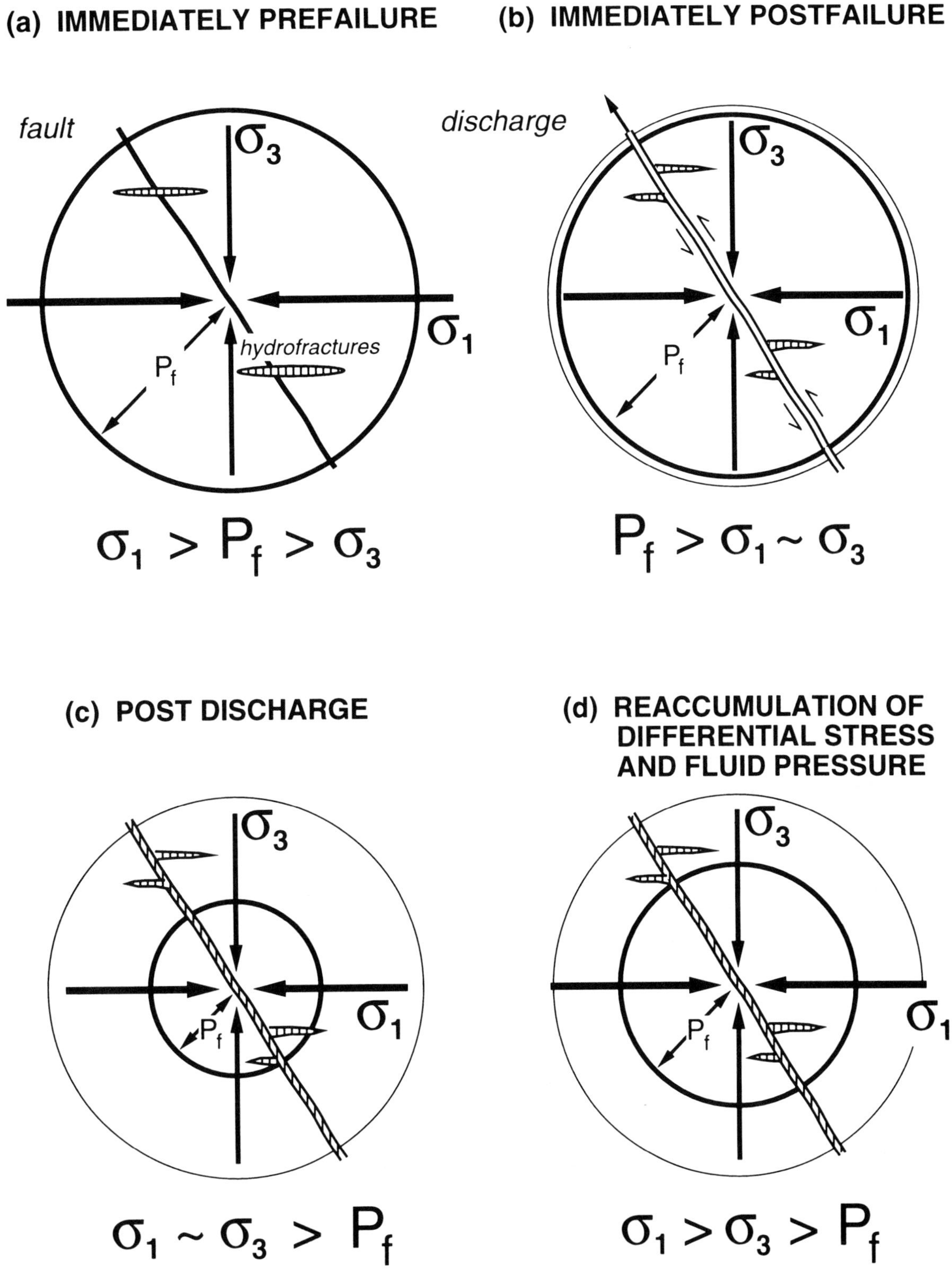

Fig. 11. Cycle of stress and fluid-pressure states accompanying extreme valve action on a steep reverse fault. The radius of the bold circle is a measure of the fluid pressure level in relation to the principal stresses.

fluid volume of a supergiant oil-field!) would have to be flushed through the fault per kilometre strike-length to deposit the hydrothermal infilling. And the Mother Lode vein system can be traced along strike for at least 200 km!

Extent of vertical migration through valving action There are historical instances of surface hydrocarbon discharge, potentially attributable to valving action on steep reverse faults, following shallow crustal earthquakes in areas such as the Western Transverse Ranges of California where overpressuring is widespread (Hamilton *et al.* 1969; Sibson 1990*b*). In general, however, only the larger earthquake ruptures extend to the ground surface. Moreover, in the case of inversion tectonics, normal growth faults developed during the extensional phase may not continue through the upper levels of the stratigraphic succession. This helps to explain the tendency of some steep faults reactivated in compression to shallow into optimally oriented low-angle thrusts at higher levels (e.g. Badley *et al.* 1989).

Such geometrical heterogeneities may play a role in the termination of reverse ruptures, as an upwardly propagating reverse rupture will induce dilation in the footwall near the rupture tip (Pollard & Segall 1987). Termination of upwardly propagating ruptures into 'horsetail' extensional fracturing in the footwall would also be facilitated by bedding anisotropy. Other rheological factors may also play a role. The expectation, therefore, is that individual reverse ruptures are likely to terminate at dilational sites involving horsetail fracturing in the footwall, and it is to these sites that fluids discharged by valve action from overpressured zones at depth may migrate (Fig. 10). Subsequent migration into hangingwall anticlinal structures would depend on the permeability of the fault at high levels (see Knipe 1992).

As a possible example of an area where these processes are likely to be active today, consider the Ventura Avenue anticline in the Western Transverse Ranges of California. Oil is produced from a thick Plio–Pleistocene turbidite sequence in the tight core of the anticline which is disrupted by reverse faults and is strongly overpressured (λ_v *c.*0.8 at 3.8 km depth) (Yeats 1983). The anticline is also flanked by the Red Mountain Fault, a pure reverse structure which dips 60–65° to at least 6 km depth and remains microseismically active (Yeats *et al.* 1987). It seems probable that fault-assisted migration into the upper levels of the structure is still ongoing as the anticline has evolved as an oil-hosting structure within only the last 0.2 Ma!

Diagnostics of fault-valve action Recognition of former fault-valve action (especially extreme valving action) on reverse reactivated faults is of some importance given its potential role in fluid migration during inversion tectonics. While no single item is fully diagnostic by itself, the following characteristics would be expected on or in the vicinity of a fault that has undergone extreme valving action (Fig. 10): (1) evidence of reverse reactivation on a fault dipping steeper than 50–60°; (2) the presence of subhorizontal extension veins (parallel to wallrock bedding?) adjoining faults in formerly overpressured sites of rupture nucleation; (3) traces of hydrothermal and hydrocarbon material along the fault zone, especially in sites such as dilational fault jogs; (4) the presence of 'horsetail' fracturing with hydrothermal/hydrocarbon infilling at former rupture termination sites in the footwall; and possibly (5) the presence of multiple generations of cement in fault-bounded reservoirs, as described by Burley *et al.* (1989) from sandstones of the Tartan Reservoir in the North Sea, and attributed to episodic flushing of hot fluids up fault conduits.

The previously described fault at Ogof Gynfor on the north coast of Anglesey displays some of these structural characteristics. Several generations of lensing quartz veins with surfaces covered with slickenfibres developed along the fault during the phase of reverse reactivation and may result from intermittent valving discharge. Structurally analogous assemblages of gypsum veins occur west of Watchet on the Bristol Channel coast of West Somerset in the vicinity of E–W striking faults that have undergone reverse reactivation juxtaposing crumpled Liassic strata in the hangingwall against Triassic red beds in the footwall (Whittaker 1972). Multiple generations of predominantly flat-lying fibrous gypsum veins are concentrated in the footwall and the fault is coated with several sets of gypsum slickenfibres. The concentration of veining in the footwall, with some local semblance of horsetail geometry with respect to the fault, suggests that the veining may have developed in a footwall dilational site which acted as a migratory sink for fluids transported by valving action during the reverse reactivation.

Discussion

A circumstantial case has been presented that extreme fluid overpressures are likely to develop during the transition from an extensional to a compressional stress regime accompanying basin inversion, and that it is the development of localized overpressuring that allows selective reactivation of faults that would otherwise be unfavourably orientated for reactivation. In such circumstances, extreme valve action on severely misorientated faults may allow episodic vertical migration of large fluid volumes along the fault zone. The association of active high-angle reverse faulting with overpressured basins as in the Western Transverse Ranges of California, coupled with historical records of post-seismic hydrocarbon discharge in such areas (Hamilton *et al.* 1969), lends support to the hypothesis. Judging by the hydrothermal vein systems associated with steep reverse faults of comparatively low displacement in other settings, the fluid volumes transported in this manner may be very large. The case is based on a 2-D analysis of the mechanics of coaxial inversion where pure reverse faulting follows earlier normal slip. Frictional lock-up may also, however, occur in more general situations of oblique slip under triaxial stress, but the fields of severe misorientation where extreme fault-valve action is likely to operate are then critically influenced by the value of the intermediate principal stress, σ_2, whose value is not easily determinable.

A range of other mechanisms may also contribute to the generation of fluid overpressures in sedimentary basins. They include the smectite to illite transition and other diagenetic processes, aquathermal pressuring, disequilibrium compaction, chemical osmosis and the generation of hydrocarbons, especially gas, from kerogen (e.g. Shi & Wang 1986; Buhrig 1989; Williamson & Smyth 1992). No account has been taken of these additional mechanisms in this analysis, but it may prove interesting to integrate these processes, and thermal maturation analyses, with the stress effects considered in this paper to evaluate the optimal conditions under which overpressuring during basin inversion may lead to fault reactivation and hydrocarbon migration. What does seem clear is that the length of time for the transition to take place from an extensional to a compressional stress regime is likely to be a crucial factor.

Direct evidence for the role of faults as migratory conduits for hydrocarbons is accumulating through detailed studies of fault rock assemblages. Hippler (1993) recognizes evidence for hydrocarbon migration through cataclastic material along extensional faults in the Orcadian Basin, while Roberts (1991) describes an association between bitumen and ferroan calcite in the gouge zone of an alpine thrust. Additional evidence comes from comparative studies of incrementally developed cements in reservoir sandstones from the Tartan Field in the North Sea and those developed within adjacent faults (Burley *et al.* 1989).

Other indirect evidence of faults as migratory paths is also coming to hand. In the Sable Basin off Nova Scotia, it appears that overpressured gas has migrated into hydrostatically pressured reservoirs at higher levels only where large listric growth faults are available as conduits (Williamson & Smyth 1992). Likewise, Barnard & Bastow (1991) have emphasized the role of faults in the North Sea basins as conduits for migration into the higher Palaeocene and Eocene reservoirs. Butler & Pullan (1990) note that the second phase of hydrocarbon migration within the Weald Basin of southern England took place during the early to mid-Tertiary accompanying the development of inversion structures, and that there is a clear association between major reactivated faults cutting the entire section, the occurrence of multiple hydrocarbon plays at both deep and shallow levels, and the presence of oil and gas shows along the faults. However, they note that no overpressuring has been encountered within the basin during exploration, suggesting that faulting has been an efficient pressure release mechanism.

In order to evaluate the primary hypothesis developed here – that major episodes of vertical migration are likely to be associated with compressional reactivation of steep faults during inversion – it will clearly be necessary to assess very carefully the relative timing of hydrocarbon migration and reverse reactivation of fault structures in areas that have undergone tectonic inversion.

Many of the ideas expressed here arose through discussions with colleagues at Imperial College, the University of California at Santa Barbara and the University of Otago. I thank referees Tim Buddin and Dick Nieuwland for helpful comments, and Alastair Beach especially, for introducing me to the thought-provoking exposures of fault-related veining along the Somerset coast.

References

ANDERSON, E. M. 1951. *The Dynamics of Faulting and Dyke Formation with Application to Britain*. 2nd edition. Oliver and Boyd, Edinburgh.

BADLEY, M. E., PRICE, J. D. & BACKSHALL, L. C. 1989. Inversion, reactivated faults and related structures: seismic examples from the southern North Sea. *In*: COOPER, M. A. & WILLIAMS, G. D. (eds) *Inversion Tectonics*. Geological Society, London, Special Publication, **44**, 201–219.

BALLY, A. W. (ed.) 1983. *Seismic expression of structural styles*. American Association of Petroleum Geologists, Studies in Geology, **15**, vol. 1–3.

BARNARD, P. C. & BASTOW, M. A. 1991. Hydrocarbon generation, migration, alteration, entrapment and mixing in the Central and Northern North

Sea. *In*: ENGLAND, W. A. & FLEET, A. J. (eds) *Petroleum Migration*. Geological Society, London, Special Publication, **59**, 167–190.

BERRY, F. 1973. High fluid potentials in California Coast Ranges and their tectonic significance. *American Association of Petroleum Geologists Bulletin*, **57**, 1219–1249.

BOULLIER, A.-M. & ROBERT, F. 1992. Paloseismic events recorded in Archean gold-quartz vein networks. *Journal of Structural Geology*, **14**, 161–179.

BUCHANAN, P. & MCCLAY, K. R. 1992. Experiments on basin inversion above reactivated domino faults. *Marine and Petroleum Geology*, **9**, 486–500.

BUHRIG, C. 1989. Geopressured Jurassic reservoirs in the Viking Graben: modelling and geological significance. *Marine and Petroleum Geology*, **6**, 31–48.

BURLEY, S. D., MULLIS, J. & MATTER, A. 1989. Timing diagenesis in the Tartan Reservoir (UK North Sea): constraints from combined cathodoluminescence microscopy and fluid inclusion studies. *Marine & Petrolum Geology*, **6**, 98–120.

BUTLER, M. & PULLAN, C. P. 1990. Tertiary structures and hydrocarbon entrapment in the Weald Basin of southern England. *In*: HARDMAN, R. F. P. & BROOKS, J. (eds) *Tectonic Events Responsible for Britain's Oil and Gas Reserves*. Geological Society, London, Special Publication, **55**, 371–391.

BYERLEE, J. D. 1978. Friction of rocks. *Pure & Applied Geophysics*, **116**, 615–626.

CHADWICK, R. A. 1986. Extension tectonics in the Wessex Basin, southern England. *Journal of the Geological Society*, London, **143**, 465–486.

COX, S. F., WALL, V. J., ETHERIDGE, M. A. & POTTER, T. F. 1991. Deformational and metamorphic processes in the formation of mesothermal vein-hosted gold deposits – examples from the Lachlan Fold Belt in central Victoria, Australia. *Ore Geology Reviews*, **6**, 391–423.

DOSER, D. I. & SMITH, R. B. 1989. An assessment of source parameters of earthquakes in the cordillera of the western United States. *Bulletin of the Seismological Society of America*, **79**, 1383–1409.

DU ROUCHET, J. 1980. Stress fields, a key to oil migration. *American Association of Petroleum Geologists Bulletin*, **65**, 74–85.

ETHERIDGE, M. A. 1983. Differential stress magnitudes during regional deformation and metamorphism: upper bound imposed by tensile fracturing. *Geology*, **11**, 231–234.

FRASER, A. J. & GAWTHORPE, R. L. 1990. Tectono-stratigraphic development and hydrocarbon habitat of the Carboniferous in northern England. *In*: HARDMAN, R. F. P. & BROOKS, J. (eds) *Tectonic Events Responsible for Britain's Oil and Gas Reserves*. Geological Society, London, Special Publication, **55**, 49–86.

FYFE, W. S., PRICE, N. J. & THOMPSON, A. B. 1978. *Fluids in the Earth's Crust*. Elsevier, Amsterdam.

HAMILTON, R. M., YERKES, R. F., BROWN, R. D., BURFORD, R. O & DE NOYER, J. M. 1969. Seismicity and associated effects, Santa Barbara region. *U.S. Geological Survey Professional Paper*, **679**, 47–72.

HAYWARD, A. B. & GRAHAM, R. H. 1989. Some geometrical characteristics of inversion. *In*: COOPER, M. A. & WILLIAMS, G. D. (eds) *Inversion Tectonics*. Geological Society, London, Special Publication, **44**, 17–39.

HIPPLER, S. J. 1993. Deformation microstructures and diagenesis in sandstone adjacent to an extensional fault: implications for the flow and entrapment of hydrocarbons. *American Association of Petroleum Geologists Bulletin*, **77**, 625–637.

HUNT, J. M. 1990. Generation and migration of petroleum from abnormally pressured fluid compartments. *American Association of Petroleum Geologists Bulletin*, **74**, 1–12.

JACKSON, J. A. 1987. Active normal faulting and crustal extension. *In*: COWARD, M. P., DEWEY, J. F. & HANCOCK, P. L. (eds) *Continental Extensional Tectonics*. Geological Society, London, Special Publication, **28**, 3–17.

KING, P. R. & THRASHER, G. P. 1993. Post-Eocene development of the Taranaki Basin, New Zealand: convergent overprint of a passive margin. *In*: WATKINS, J. S., ZHIQIANG, F. & MCMILLEN, K. G. (eds) *Geology and Geophysics of Continental Margins*. American Association of Petroleum Geologists Memoir, **53**, 93–118.

KNIPE, R. J. 1992. Faulting processes and fault seal. *In*: LARSEN, R. M. (ed) *Structural and Tectonic Modelling and it's Application to Petroleum Geology*. NPF Special Publication, **1**, 325–343.

KNOPF, A. 1929. *The Mother Lode System of California*. U.S. Geological Survey Professional Paper, **157**, 88pp.

KOOPMAN, A., SPEKSNIJDER, A. & HORSFIELD, W. T. 1987. Sandbox model studies of inversion tectonics. *Tectonophysics*, **137**, 379–388.

MCCLAY, K. R. 1989. Analogue models of inversion tectonics. *In*: COOPER, M. A. & WILLIAMS, G. D. (eds) *Inversion Tectonics*. Geological Society, London, Special Publication, **44**, 41–59.

PALCIAUSKAS, V. V. & DOMENICO, P. A. 1980. Microfracture development in compacting sediments: relation to hydrocarbon-maturation kinetics. *American Association of Petroleum Geologists Bulletin*, **64**, 927–937.

POLLARD, D. D. & SEGALL, P. 1987. Theoretical displacements and stresses near fractures in rocks: with application to faults, joints, veins, dikes, and solution surfaces. *In*: ATKINSON, B. K. (ed.) *Fracture Mechanics of Rock*. Academic Press, New York, 277–349.

POWER, W. L., TULLIS, T. E., BROWN, S., BOITNOTT, G. N. & SCHOLZ, C. H. 1987. Roughness of natural fault surfaces. *Geophysical Research Letters*, **14**, 29–32.

ROBERTS, G. 1991. Structural controls on fluid migration through the Rencurel thrust zone, Vercors, French Sub-Alpine Chains. *In*: ENGLAND, W. A. & FLEET, A. J. (eds) *Petroleum Migration*. Geological Society, London, Special Publication, **59**, 245–262.

ROBERTS, S. & JACKSON, J. A. 1991. Active normal

faulting in central Greece: an overview. *In*: ROBERTS, A. M., YIELDING, G. & FREEMAN, B. (eds) *The Geometry of Normal Faults*. Geological Society, London, Special Publication, **56**, 125–142.

SECOR, D. T. 1965. Role of fluid pressure in jointing. *American Journal of Science*, **263**, 633–646.

SHI, Y. & WANG, C-Y. 1986. Pore pressure generation in sedimentary basins: overloading versus aquathermal. *Journal of Geophysical Research*, **91**, 2153–2162.

SASSI, W., COLLETTA, B., BALE, P. & PAQUERAU, T. 1993. Modelling of structural complexity in sedimentary basins: the role of pre-existing faults in thrust tectonics. *Tectonophysics*, **226**, 97–112.

SIBSON, R. H. 1981. Fluid flow accompanying faulting: field evidence and models. *In*: SIMPSON, D. W. & RICHARDS, P. G. (eds) *Earthquake Prediction: an International Review*. American Geophysical Union, Maurice Ewing Series, **4**, 593–603.

—— 1985. A note on fault reactivation. *Journal of Structural Geology*, **7**, 751–754.

—— 1990*a*. Conditions for fault-valve behaviour. *In*: KNIPE, R. J. & RUTTER, E. H. (eds) *Deformation Mechanisms, Rheology and Tectonics*. Geological Society, London, Special Publication, **54**, 15–28.

—— 1990*b*. Rupture nucleation on unfavourably oriented faults. *Bulletin of the Seismological Society of America*, **80**, 1580–1604.

—— 1992. Implications of faults-valve behaviour for rupture nucleation and recurrence. *Tectonophysics*, **211**, 283–293.

—— 1994. Crustal stress, faulting, and fluid flow. *In*: PARNELL, J. (ed.) *Geofluids: Origin, Migration and Evolution of Fluids in Sedimentary Basins*. Geological Society, London, Special Publication, **78**, 69–84.

——, ROBERT, F. & POULSEN, K. H. 1988. High-angle reverse faults, fluid pressure cycling and mesothermal gold-quartz deposits. *Geology*, **16**, 551–555.

SIMPSON, I. R., GRAVESTOCK, M., HAM, D., LEACH, H. & THOMPSON, S. D. 1989. Notes and cross-sections illustrating inversion tectonics in the Wessex Basin. *In*: COOPER, M. A. & WILLIAMS, G. D. (eds) *Inversion Tectonics*. Geological Society, London, Special Publication, **44**, 123––129.

STESKY, R., BRACE, W., RILEY, D. & ROBIN, P-Y. 1974. Friction in faulted rock at high temperature and pressure. *Tectonophysics*, **23**, 177–203.

STONELY, R. 1982. The structural development of the Wessex Basin. *Journal of the Geological Society*, London, **139**, 545–554.

WHITTAKER, A. 1972. The Watchet fault – a post-Liassic reverse fault. *Bulletin of the Geological Survey of Great Britain*, **41**, 75–80.

WILLIAMS, G. D., POWELL, C. M. & COOPER, M. A. 1989. Geometry and kinematics of inversion tectonics. *In*: COOPER, M. A. & WILLIAMS, G. D. (eds) *Inversion Tectonics*. Geological Society, London, Special Publication, **44**, 3–15.

WILLIAMSON, M. A. & SMYTH, C. 1992. Timing of gas and overpressure generation in the Sable Basin offshore Nova Scotia: implications for gas migration dynamics. *Bulletin of Canadian Petroleum Geology*, **40**, 151–169.

YEATS, R. S. 1983. Large-scale Quaternary detachments in Ventura Basin, southern California. *Journal of Geophysical Research*, **88**, 569–583.

——, LEE, W. H. K. & YERKES, R. F. 1987. Geology and seismicity of the eastern Red Mountain fault, Ventura County. *U.S. Geological Survey Professional Paper*, **1339**, 161–167.

ZIEGLER, P. A. 1987. Late Cretaceous and Cenozoic intra-plate compressional deformations in the Alpine foreland – a geodynamic model. *Tectonophysics*, **137**, 389–420.

—— 1989. Geodynamic model for Alpine intra-plate compressional deformation in Western and Central Europe. *In*: COOPER, M. A. & WILLIAMS, G. D. (eds) *Inversion Tectonics*. Geological Society, London, Special Publication, **44**, 63–85.

Influence of salt on fault geometry: examples from the UK salt basins

S. A. STEWART[1], M. J. HARVEY[2], S. C. OTTO[3] & P. J. WESTON[1]

[1]*Amerada Hess Ltd, 33 Grosvenor Place, London SW1X 7HY, UK*

[2]*Department of Geology, Imperial College, London SW7 2BP, UK*

[3]*Petroconsultants, 266 Upper Richmond Road, London SW15 6TQ, UK*

Abstract: The influence of salt layers on cover faulting during thick-skinned extensional faulting is examined using seismic examples from the salt basins offshore UK. The ratio of salt layer thickness to basement fault displacement is a key geometric parameter governing spatial offset between cover and basement fault segments. The influence of geological factors such as stratigraphic variation in salt thickness, basement fault zone geometry and basement fault growth through time on this parameter are individually examined. 3D models are constructed to illustrate the spectra of possible fault geometries which result from variations in these factors. Lateral stratigraphic variations within individual 'salt' layers are also considered, as are complications introduced by the addition of further salt layers into cover stratigraphy. Since diapiric intrusions occur after cover fault geometries are established and are localized by such faulting, an understanding of basement-cover fault relationships also illuminates salt diapir-basement fault relationships. An example from the Central North Sea diapir province is subject to such a genetic analysis, relating the diapir and its location to an underlying basement fault. The influence of salt on cover fault reactivation during basin inversion is discussed, focusing on the southern North Sea, where inhibition of reverse fault propagation from basement to cover is noted. It is suggested that analysis of local basement-cover relationships should form a first step in any attempt at a genetic classification of salt bodies.

Structural mapping in the sedimentary basins offshore UK is complicated in many areas by the presence of salt at one or more stratigraphic levels. Particular problems surround the interpretation of fault systems which interact with salt layers. For example, opinions have varied as to whether or not fault systems on the western margin of the southern North Sea link directly from strata above the salt to faults below the Zechstein salt (Christian 1969; Glennie & Boegner 1981; Walker & Cooper 1987; Arthur 1993). The issue is complicated because fault systems in the 'cover' above salt may be kinematically linked with 'basement' faults below salt, but spatial relationships may be difficult to identify due to offset between cover and basement via detachment in the salt layer (Nalpas & Brun 1993; Jackson & Vendeville 1994). For the purposes of this paper, 'cover' and 'basement' are mechanically distinct layers, which may be separated by a viscous (salt) layer (Fig. 1). Whilst it is true that in exclusively thin-skinned structural situations there need be no relationship between fault systems above and below salt (e.g. Crans *et al.* 1980; Duval *et al.*1992), significant basement faults have governed the location, geometry and evolution of the salt-bearing basins offshore UK. The purpose of this paper is to examine the geometries of the upwards continuations of such basement faults through salt layers into the cover. By considering fault systems in their entirety in the vertical sense, and the spatial variations in geometry in relation to geological circumstance within the UK basins, we hope to illustrate the impact of salt layers on the trajectories of thick-skinned extensional faults which post-date salt deposition.

Review

Simple analog models demonstrate that salt layers have an important impact on the manner in which basement faults propagate upwards into cover (Fig. 1). It is easy to envisage a continuous series of models between those examples illustrated in Figs 1a and 1b with varying viscous layer thickness. Although a single fault strand does not always pass directly from basement to cover, the cover and basement fault elements are kinematically linked and could be considered to represent a single fault surface, composed of brittle and ductile portions (Fig. 1d). Furthermore, since salt layers in sedimentary basins are not infinite in lateral extent, one might expect line length balance between pre-rift beds above and below the salt (Nalpas & Brun 1993; Jackson & Vendeville 1994). A case can therefore be made for considering basement fault – detachment – cover faults as single fault surfaces on

From Holdsworth, R.E. & Turner, J.P. (compilers) *Extensional Tectonics: Regional-scale Processes*. The Geological Society,
Key Issues in Earth Sciences, **2 (1)**, 281–308.
1476-315X/02/$15.00 © The Geological Society of London 2002
First published in: "Alsop, G.I., Blundell, D.J. & Davison, I. (eds) *Salt Tectonics*, The Geological Society, London, Special Publications **100**, 175–202".

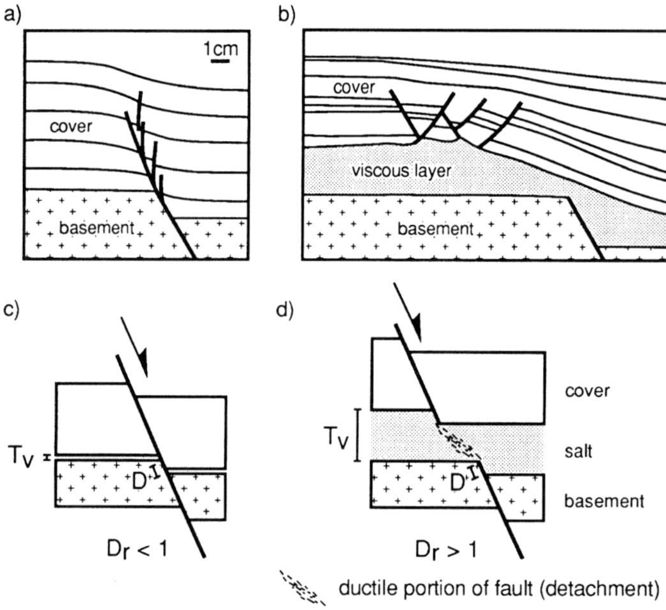

Fig. 1. Analog models illustrating the impact of a viscous layer on the upward propagation of a fault from 'basement' to 'cover', and the relationship of basement-cover fault linkage to displacement ratio. (a) 'Hard' linked, planar fault which has grown from basement to cover where no intervening detachment is present, after Withjack *et al.* (1990). (b) 'Soft' linked, ramp-flat extensional fault with planar, brittle elements visible in basement and cover, and a flat-lying, ductile detachment occupying the viscous layer in the zone offsetting the cover and basement ramp elements (after Larroque in Oudmayer & de Jaeger 1993). Displacement ratio (D_r) is defined in terms of D (basement fault displacement) and T_v (viscous layer thickness), $D_r = T_v/D$ (Koyi *et al.* 1993). (c) $T_v < D$, $D_r < 1$. Basement fault 'hard' linked to cover in a continuous brittle strand. (d) $T_v > D$, $D_r > 1$. Basement fault 'soft' linked to cover fault via ductile detachment in the salt, giving extensional ramp-flat geometry.

the basis of kinematic linkage, the variation in structural style reflecting the rock rheology.

Ramp-flat trajectories are considered to be characteristic of single fault surfaces passing through layered sequences of strongly contrasting rheology in both thrust systems (Boyer & Elliott 1982) and extensional systems (Peacock & Sanderson 1992; Peacock & Zhang 1994). Just as salt layers form important fault 'flats' in thrust systems (Davis & Engelder 1985), they may perform the same role in extensional systems, where the term 'detachment' is commonly used for the tectonically active portion of the salt layer.

Several geological parameters appear to influence the geometry of extensional ramp-flat systems. A review by Koyi *et al.* (1993) indicated that the magnitude, direction and rate of basement fault displacement are important, as are the thicknesses of the overburden and the salt layers. In particular, the magnitude of spatial offset between linked basement and cover fault systems has not yet been quantitatively described in terms of the above parameters. It is noted here that one of the key parameters may turn out to be salt layer top surface slope, which must control localized gravity sliding in the vicinity of a basement fault at the very first stages of fault growth. However, investigating the mechanics of this model is beyond the scope of this paper.

A single, fundamental, parameter can be defined, which encapsulates salt layer thickness and basement fault displacement. The parameter is the displacement ratio D_r which is the thickness of the viscous layer T_v divided by the basement fault displacement D (Koyi *et al.* 1993). D_r is a useful parameter because it embraces several geological controls. T_v can vary spatially for stratigraphic or structural reasons, while D can vary in space and time. One can envisage a special value of D_r, close to 1, at which the magnitude of basement fault displacement exceeds the thickness of the salt layer and a 'hard', direct fault linkage would be established between basement and cover (Fig. 1). This transition is illustrated in analog models presented by Oudmayer & de Jaeger (1993). In this paper, several examples from various basins on the UKCS are examined using seismic data. The aim in each example is to examine the influence of a single geological control, in terms of variation in observed fault system geometry and D_r.

Analogue modelling has shown that the locations of salt diapirs in extensional environments are

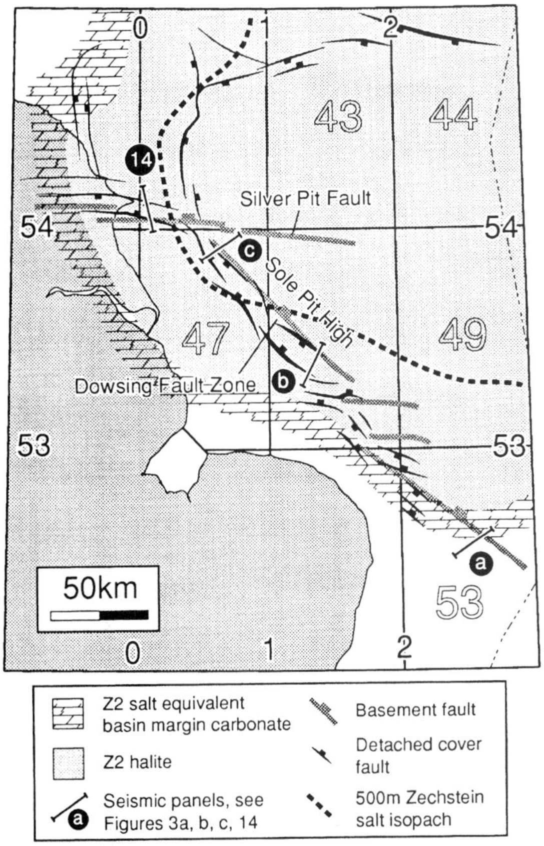

Fig. 2. West margin, southern North Sea Basin, showing location of sections presented in Fig. 3 and the spatial relationship between fault systems and Zechstein facies. Basement fault elements have not been mapped in detail during this study. The detached cover fault elements are only present within the Zechstein salt basin, but their location and orientation is controlled by basement structures.

predetermined by the locations of cover fault systems (Vendeville & Jackson 1992). The modelling also demonstrates that at least 20% extension of the cover (β = 1.2) via normal faulting must occur before diapirs are initiated. The geometry of a given linked basement – cover fault will already have been established at this stage and so the effects of diapirism in the examples discussed here will generally not be emphasized. By taking this approach we hope to illustrate the indirect control exerted by salt on diapir location during thick-skinned extension. The discussions presented here will make the assumption that basement fault movement has been dip-slip. Richard (1991) has shown that the obliquity (strike-slip / dip-slip) of basement fault movement will also affect basement-cover linkage geometry, a result which is relevant to North Sea geology as several tectonic syntheses have noted the importance of oblique slip on North Sea basement structures (e.g. Bartholomew *et al.* 1993).

Stratigraphic variation in salt thickness (T_v) : west margin, southern North Sea

The west margin of the southern North Sea basin is structurally defined by the northeasterly dipping Dowsing Fault Zone (DFZ) which was episodically active as a major basement extensional fault from the middle Triassic to the lower Cretaceous (Cameron *et al.* 1992). Significant quantities of salt are present within the upper Permian Zechstein Supergroup, which accumulated in a flexural basin driven largely by thermal subsidence rather than active faulting. As a result, Zechstein salt isopachs and stratigraphic pinchout are not parallel in trend to the DFZ (Glennie & Boegner 1981; Fig. 2). Serial sections across the DFZ from southeast to northwest illustrate variation in style of the DFZ as it has grown upwards through a salt layer which thickens northwards. Seismic sections presented on Fig. 3 show the spatial variation in fault style. Towards the southeastern end of the DFZ, in the Hewett Shelf area, the Zechstein facies are basin margin carbonates and faults pass without deviation from basement to cover (Badley *et al.* 1989; Fig. 3a). Note the evidence for late (Cretaceous to Tertiary) inversion on these faults while Permian stratigraphic levels remain in net extension.

Proceeding northwestwards, the DFZ passes through the Zechstein salt pinchout in the northwest of Quadrant 53. Mismatch between the cover and basement fault systems in plan view has been noted in the Camelot area, blocks 53/1 and 53/2, where the Zechstein salt (the sum of Z2, Z3 and Z4 halites) is approximately 100 m thick (Holmes 1991). On the southeast margin of the East Midlands Shelf, with just over 100 m of Zechstein salt, there is clearly a degree of detachment occurring between basement and cover faults (Fig. 3b).

Several issues regarding the DFZ are raised on Fig. 3b. The net displacement on the faults at basement level does not balance the displacements seen across the cover faults. The Hewett Shelf section demonstrated that inversion has occurred on the basement DFZ faults. Uplift of the Sole Pit High (Fig. 3b) indicates that inversion also occurred where Zechstein salt is present. It is likely that a significant amount of extensional heave has been lost on most of the basement fault strands. However, reverse displacement has not passed upwards to cover fault segments which had been active during extension. The kinematics of this system during inversion will be examined later. Comparisons of heave across the cover fault system with likely pre-inversion heave across the basement faults (based on the thickness of sediments preserved in the Sole Pit High) suggest that the heave across the cover system may be greater than that which existed at basement level (Arthur 1993;

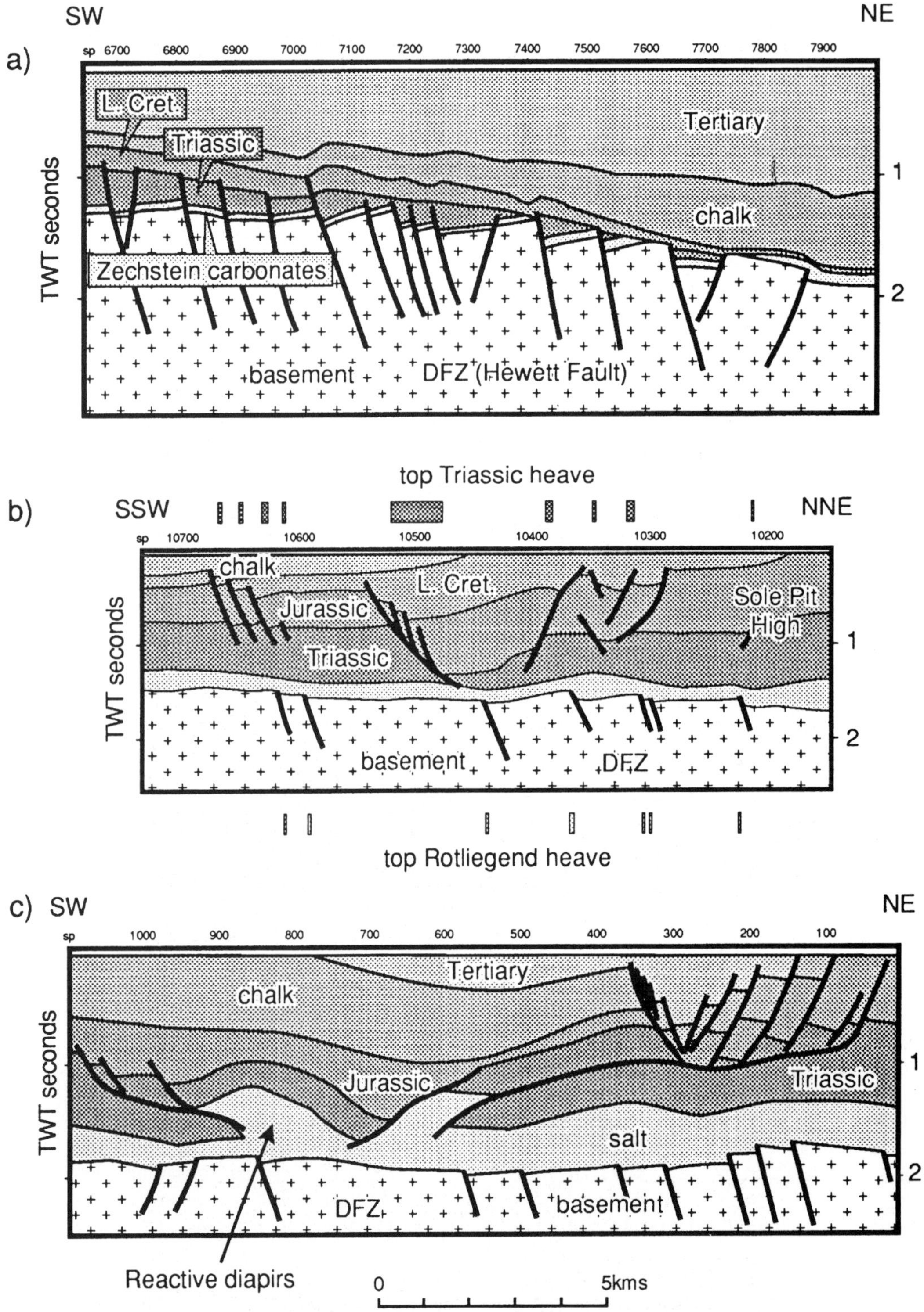

Fig. 3. Dowsing Fault Zone seismic examples and geoseismic interpretations (see Fig. 2 for locations) (a) Hewett Shelf; no Zechstein salt, hard linked faults pass directly from basement to cover. Inversion-related folds in hangingwalls to some faults present at base Chalk and top Chalk. (b) East Midlands Shelf; relatively thin Zechstein. Detachment between basement and cover faults. Approximately 2 km heave discrepancy between basement and cover fault systems. Note westerly facing monoclinal folding at top Jurassic and base Chalk, a ductile cover response to basement inversion. (c) East Midlands Shelf; relatively thick Zechstein. No clear relationship between basement and cover faults. Cover faults intruded by Zechstein diapirs. Significant bedding-parallel detachments on intra-Triassic salt layers.

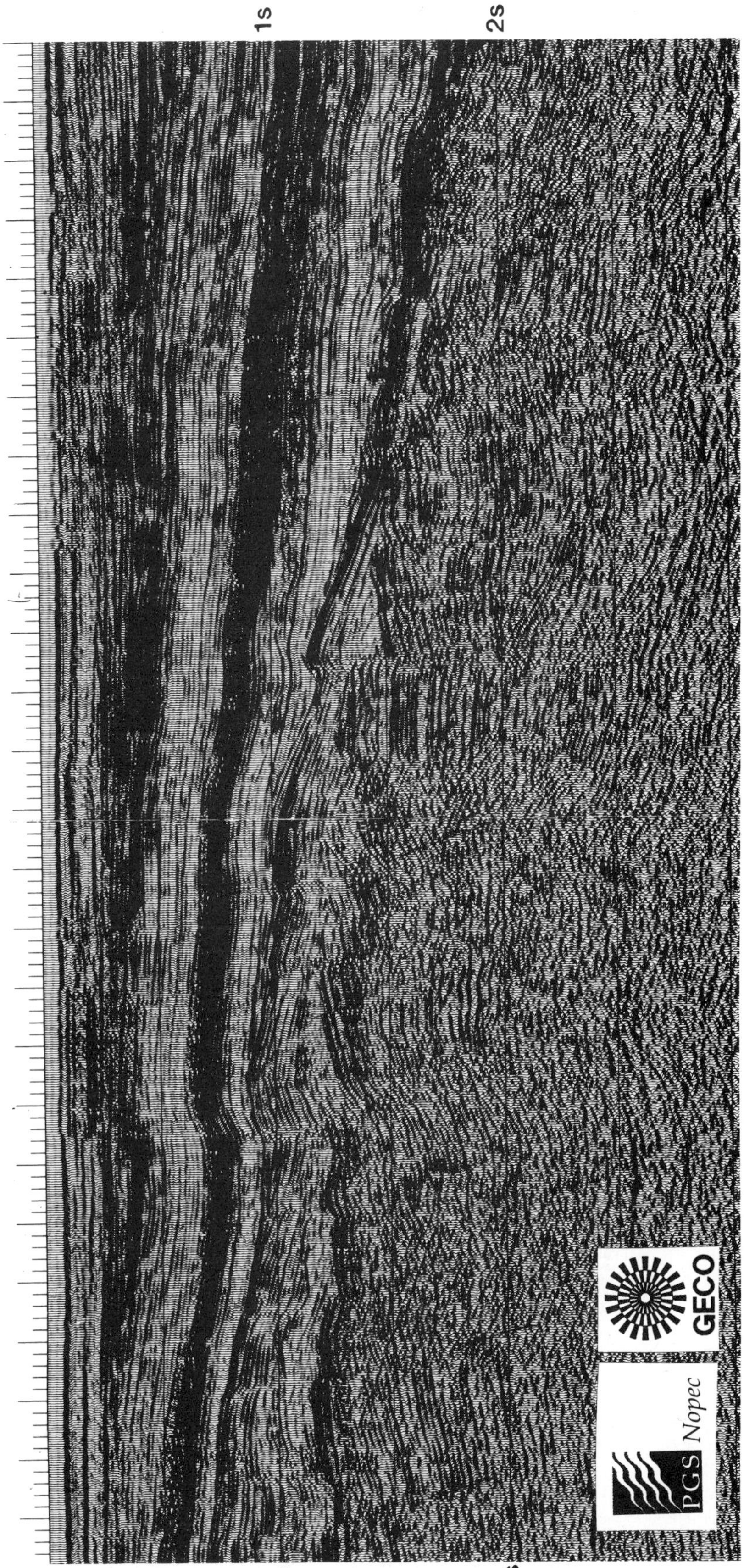

Fig. 3a. Seismic section.

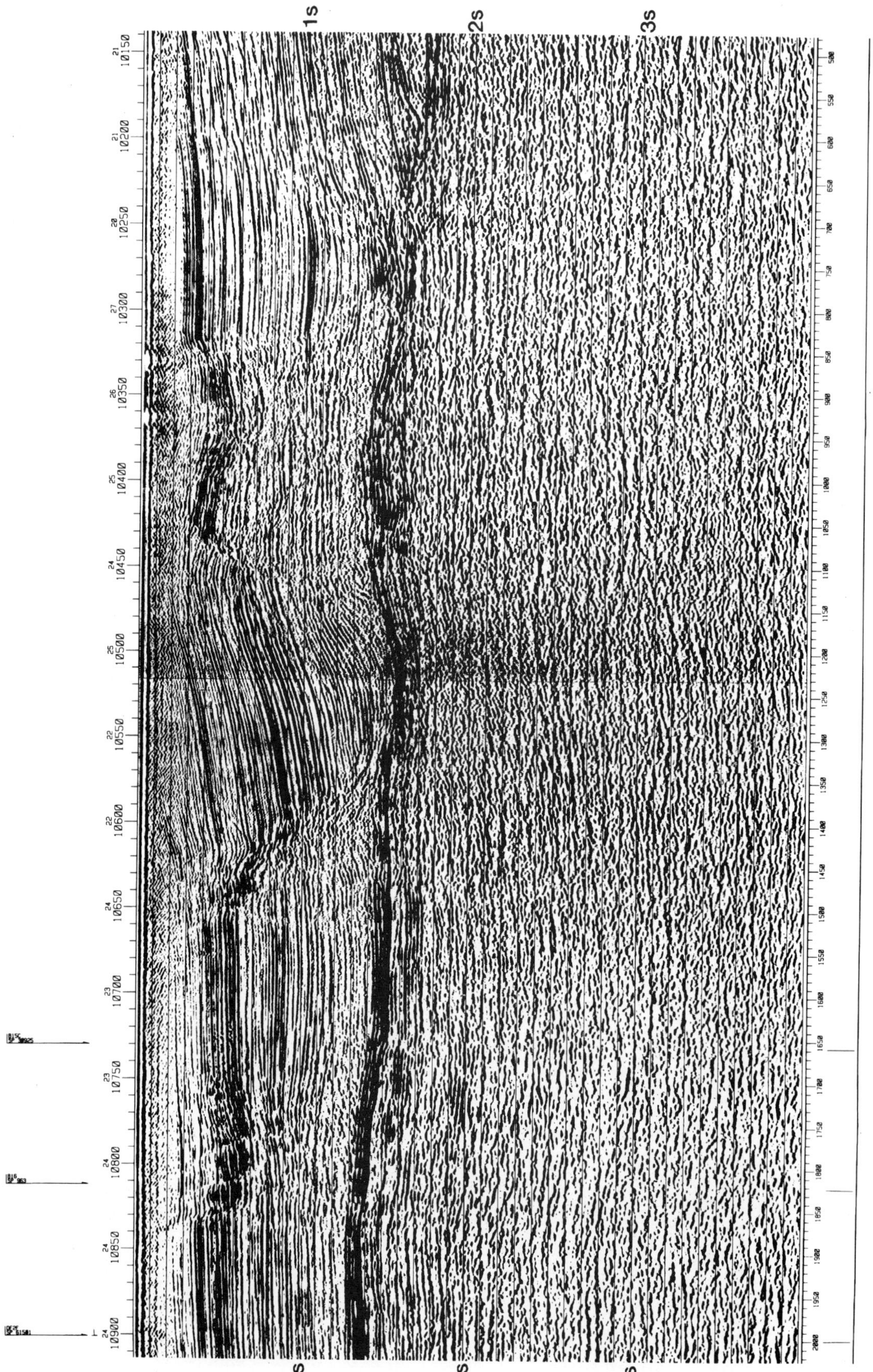

Fig. 3b. Seismic section.

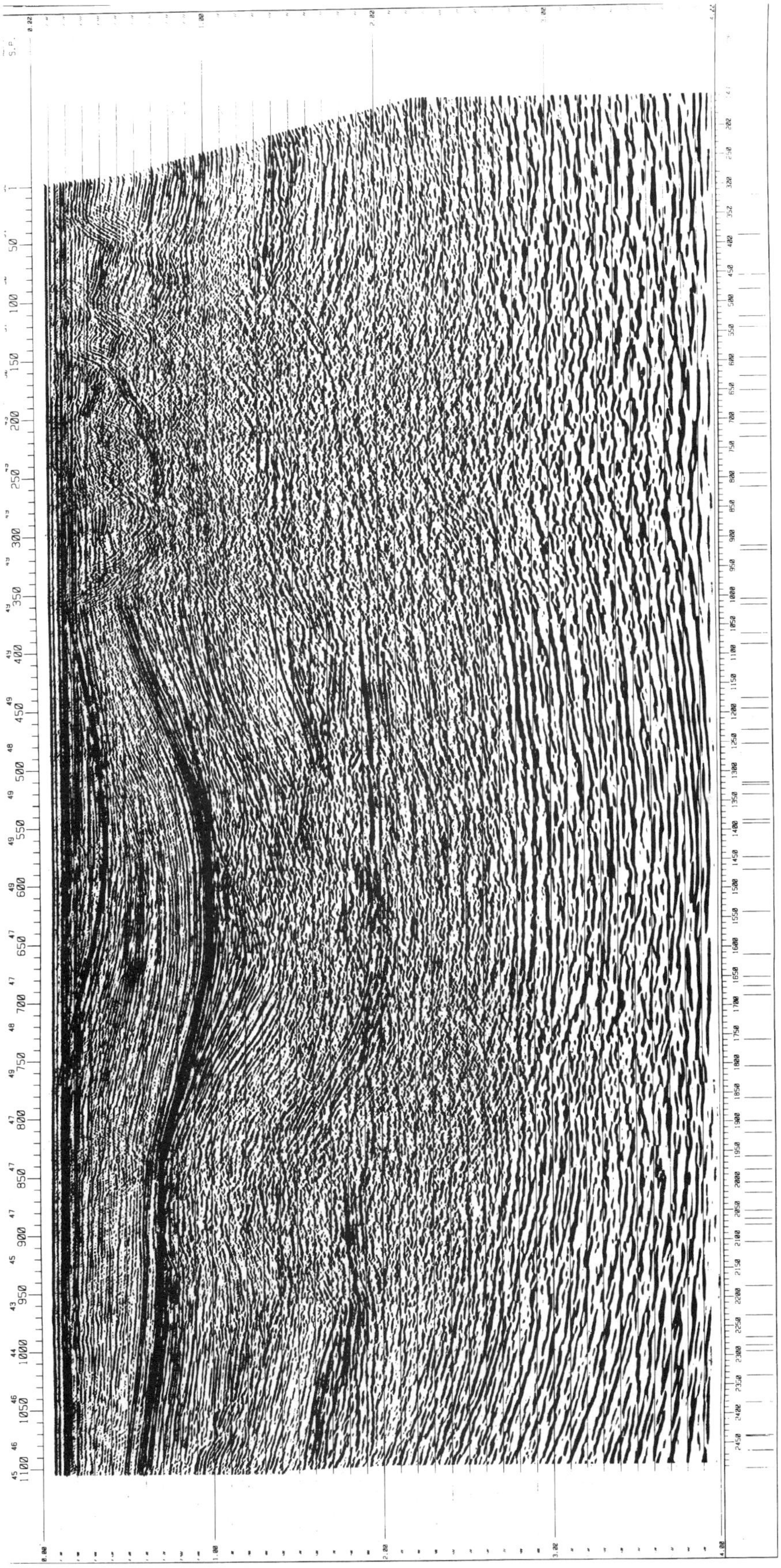

Fig. 3c. Seismic section.

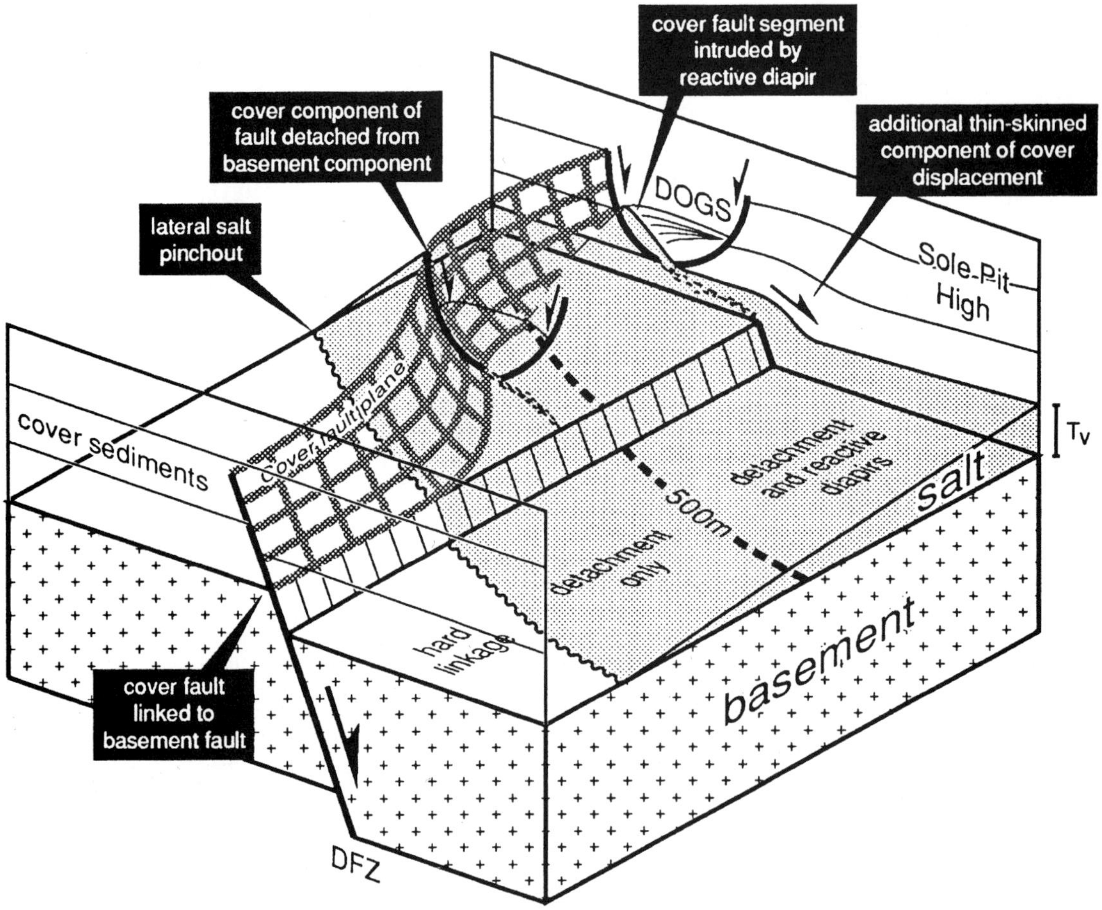

Fig. 4. Block diagram based on the Dowsing Fault Zone summarizing basement-cover relationships along the length of a major fault strand, illustrating the effect of spatial variation in salt layer thickness. Note the onset of diapirism within the 500 m Zechstein salt isopach.

Stewart & Coward 1995). This apparent anomaly can be explained by a component of thin-skinned gliding of the cover section eastwards towards the Sole Pit Trough, balanced presumably by thick-skinned basement extension or thin-skinned cover shortening within the basin. Thin-skinned extension of this nature has been described from the other side of the southern North Sea basin (Allen *et al.* 1994) and the central North Sea (Bishop *et al.* 1995). Given the likelihood that the cover fault system on the east margin of the East Midlands Shelf does not solely represent the upwards continuation of basement faults, reference to the cover fault system as 'Dowsing Fault Zone' could be misleading. Terming the basement faults DFZ and the cover faults Dowsing Graben System (DOGS) has been suggested (Stewart & Coward 1995).

The final example from the DFZ is taken from the northeast corner of the East Midlands Shelf, where the Zechstein Supergroup is approximately 800 m thick and is dominated by the Z2 Stassfurt Halite Formation (Fig. 3c). In this area the spatial relationship between cover and basement faults is far from clear although the DOGS is again present, below the base Cretaceous unconformity. Two important features on this section are: firstly, a major post-Cretaceous thin-skinned collapse graben detaching on upper Triassic evaporites, resulting from uplift of the Sole Pit High; secondly, the DOGS in this area is invaded by reactive diapirs (Vendeville & Jackson 1992), which become less common southeastwards where the salt is thinner.

The DFZ basement – cover relationships are summarized on Fig. 4, which attempts to portray the 3D geometry of a fault system with constant displacement passing through a salt layer which systematically varies in thickness. This illustrates the most basic effect of salt layers on growing extensional fault systems – the importance of the detachment is in proportion to the thickness of the detaching layer (Richard 1991). The examples in Fig. 3 also indicate that as salt increases in thickness along the DFZ, the onset of detachment precedes the onset of diapirism. A tentative conclusion from this observation would be that salt layers can constitute effective detachments though they are not thick enough to support diapirism. In this context it is noted that Price & Cosgrove (1990) demonstrated that the extrinsic mobility of salt in terms of T_v is in proportion to the fourth power of T_v. They calculate that significant lateral salt movement such as that required to feed reactive and active diapirs can only occur if the salt layer is in excess of 500 m thick at temperatures likely to be encountered in sedimentary basins (Price & Cosgrove 1990). The 500 m salt isopach does indeed pass between the locations of the sections presented in Figs 3b and 3c (Figs 3 and 4).

Tectonically-induced variation in salt thickness (T_v) : Forth Approaches Basin, central North Sea

In addition to initial, stratigraphic variations in salt layer thickness, there are several post-depositional processes, independent of basement tectonics, which may affect salt thickness. Examples include thin-skinned extension due to gravity gliding (Vendeville & Jackson 1992), salt redistribution due to buoyancy instability (Geil 1991; Zaleski & Julien 1992) and passive infill of the cores of contractional buckle folds (Koyi 1988). Tectonically-induced salt thickness variation in the North Sea can be found on the platforms either side of the Central Graben, for example the Forth Approaches Basin (Fig. 5).

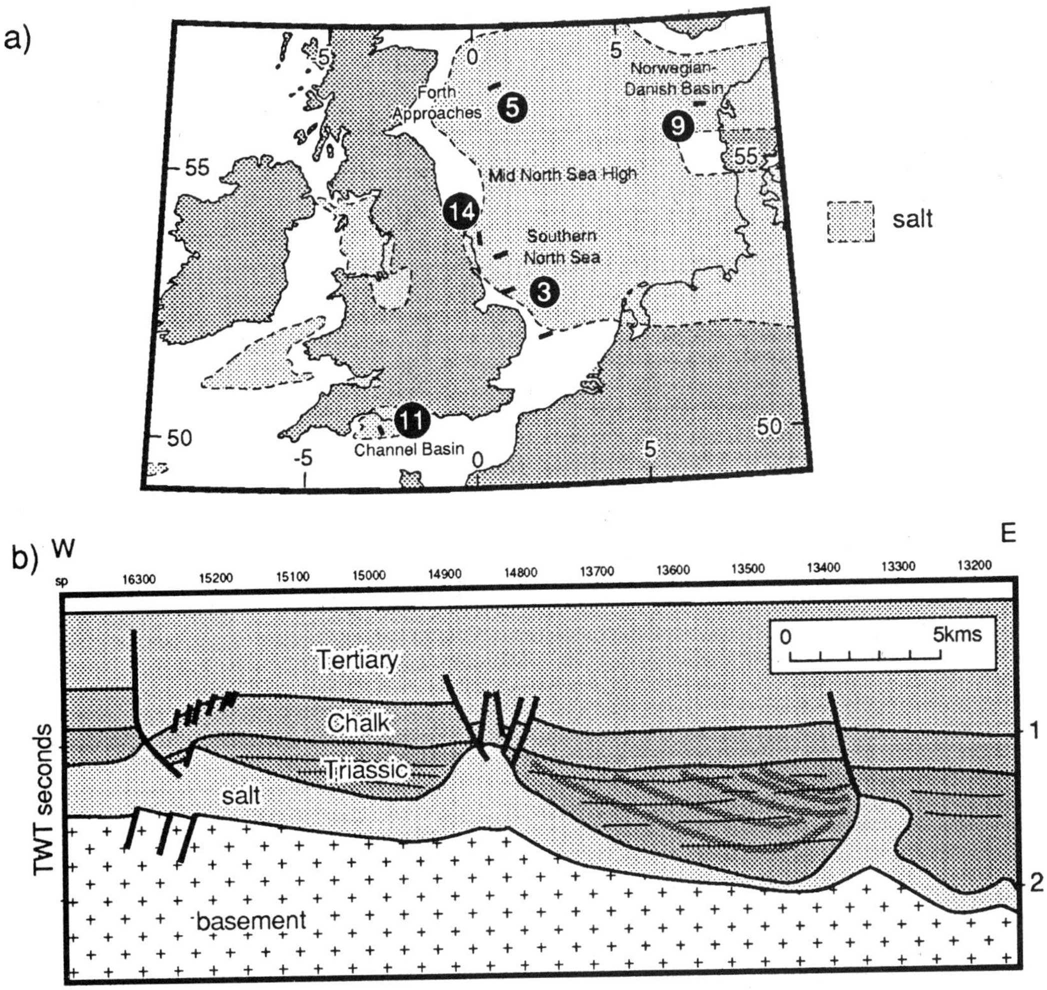

Fig. 5. Forth Approaches Basin seismic and geoseismic. (a) Location map for examples presented in this paper. Figure numbers circled. (b) Variations in salt and Triassic thickness below the base Chalk unconformity (relief at top basement due to velocity pull-up below salt highs). Tertiary thin-skinned extensional faults have been localised above the salt highs. Sub-horizontal reflectors are characteristic of the seismically transparent Triassic Smith Bank shales, regardless of folding recorded by the top Zechstein reflector. The sub-horizontal reflectors often suggest onlap relationships with the top Zechstein. Vestiges of reflectivity within some Triassic pods indicate that the Triassic beds may also be folded and that the sub-horizontal reflectors are multiple energy.

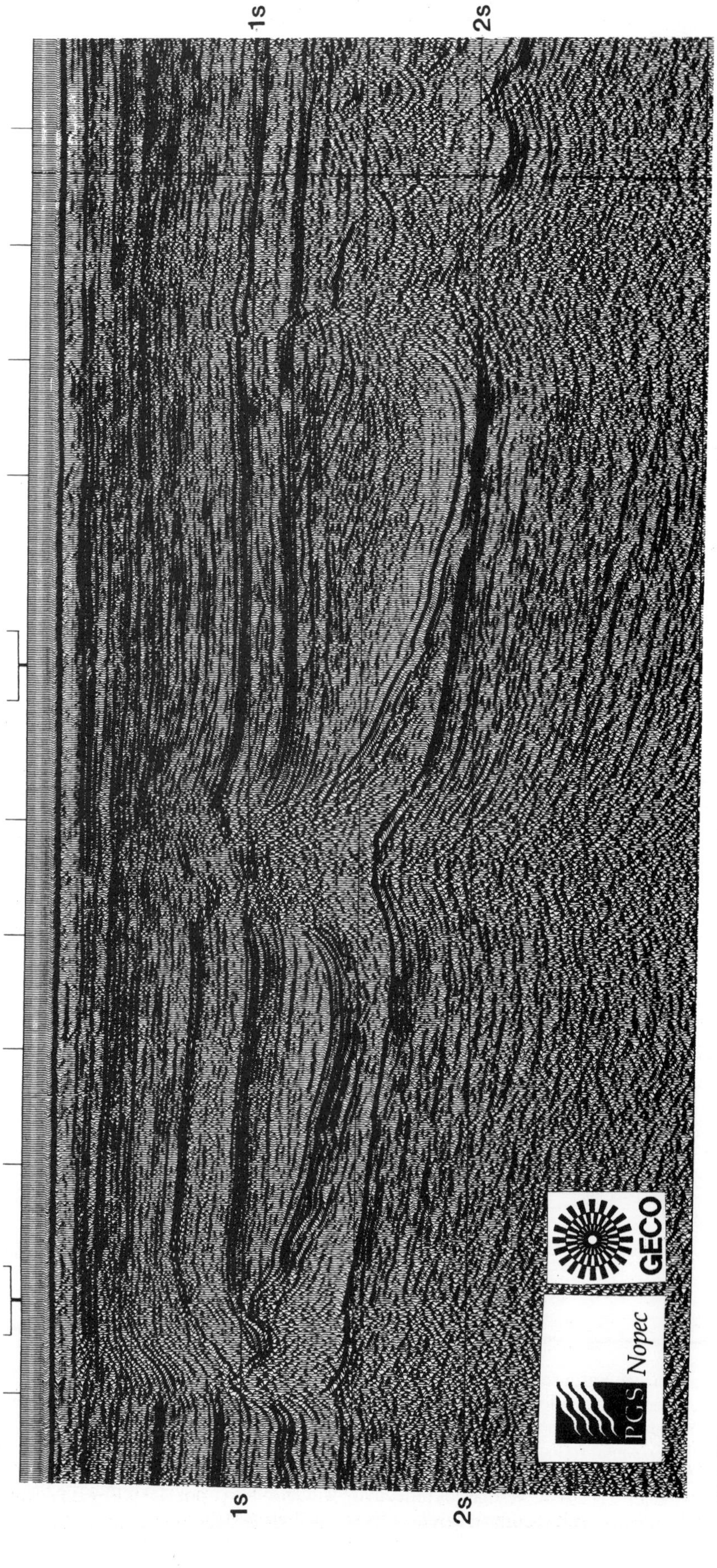

Fig. 5b. Seismic section.

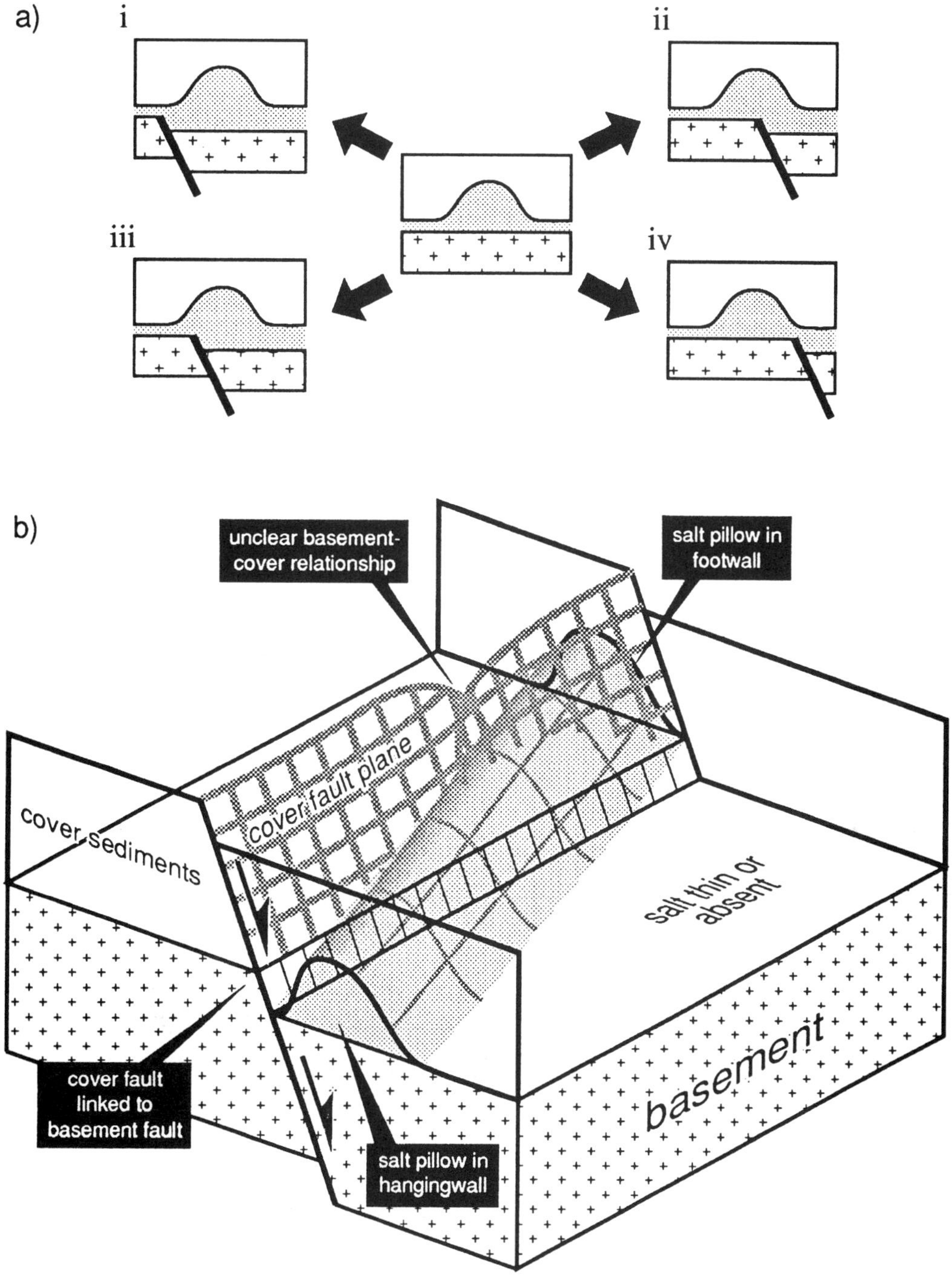

Fig. 6. (a) Spatial variation in T_v (centre) due to salt pillow development prior to basement faulting, after Erratt (1993). The subsequent location of the basement fault will govern whether the pillow is preserved in the hangingwall (i) or footwall (iv) to the fault system. Prediction of basement – cover geometries in the intermediate cases (ii, iii) is not straightforward, as there may be cover structures from the time of pillow growth to inherit and reactivate. (b) 3D variation in D_r along the length of a basement fault strand which has grown slightly oblique in trend to a pre-existing salt pillow. Cover fault is locally detached from basement strand.

Although the interpretation presented in Fig. 5 follows the thin-skinned raft tectonics interpretation of Penge *et al.* (1993), other workers have not yet discounted the possibility of salt redistribution driven by buoyancy instability as an important factor in this area (Hodgson *et al.* 1992; Smith *et al.* 1993). Either way, post-deposittional spatial variations in T_v have been imparted to the Zechstein salt

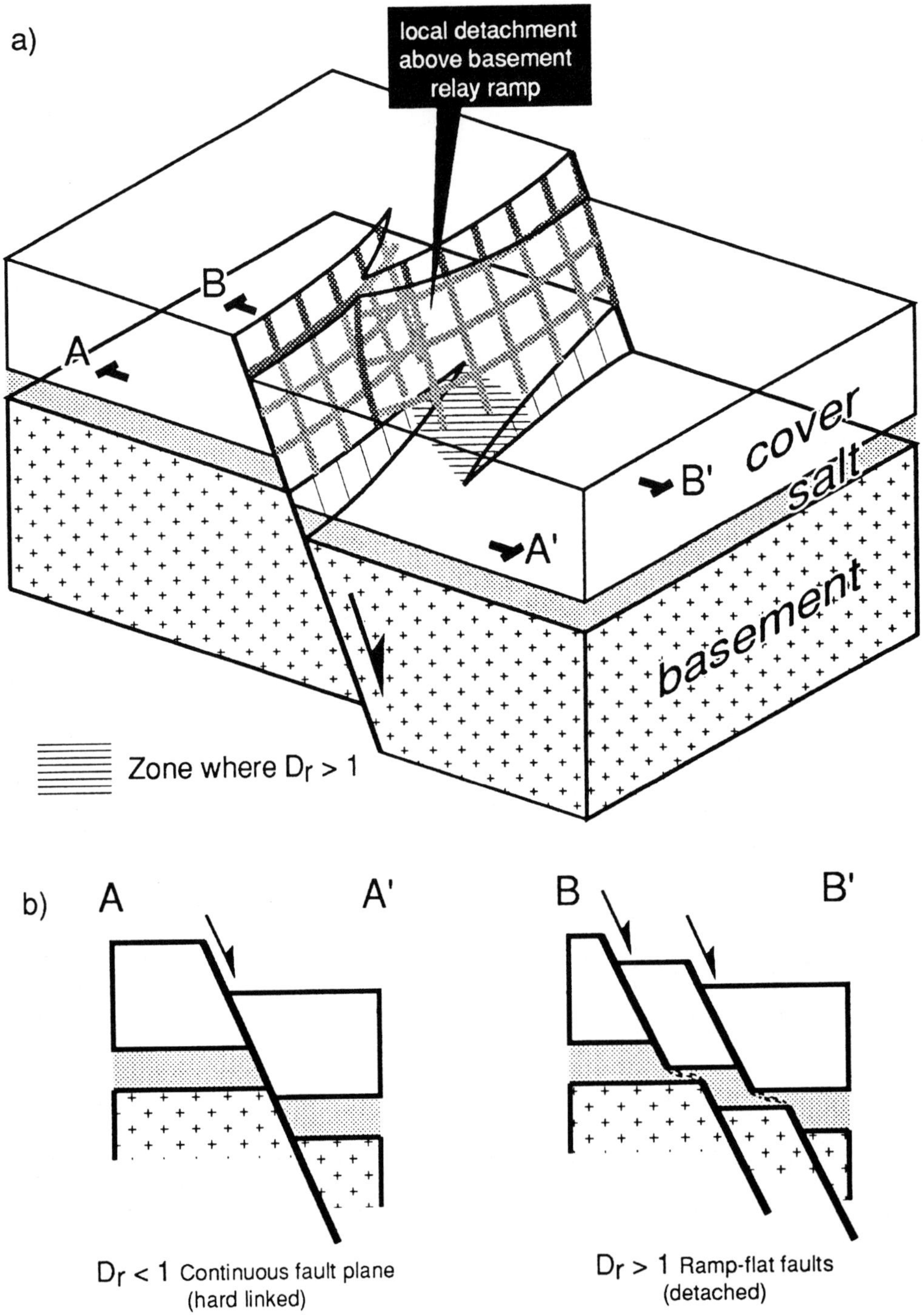

Fig. 7. (a) 3D variation in D_r along the length of individual fault segments within a basement fault zone ($T_v$constant). $D_r < 1$ along most of the length of basement fault strands, but rises above 1 in the relay ramp zone.
(b) Detachment tectonics in the ramp zones of an otherwise hard-linked fault system are predicted - compare sections A–A' and B–B'.

and have preceeded the growth of major basement faults. Were a basement fault to grow now in the Forth Approaches Basin, the spatial location of this notional fault would clearly govern D_r (Erratt 1993; Fig. 6a). The options presented on the 2D cartoons of Fig. 6a can also be considered in terms of variation along the length of a notional basement fault which grows along a trend slightly oblique to a pre-existing salt pillow (Fig. 6b). The salt pillow may be fossilized above the hangingwall or the footwall of

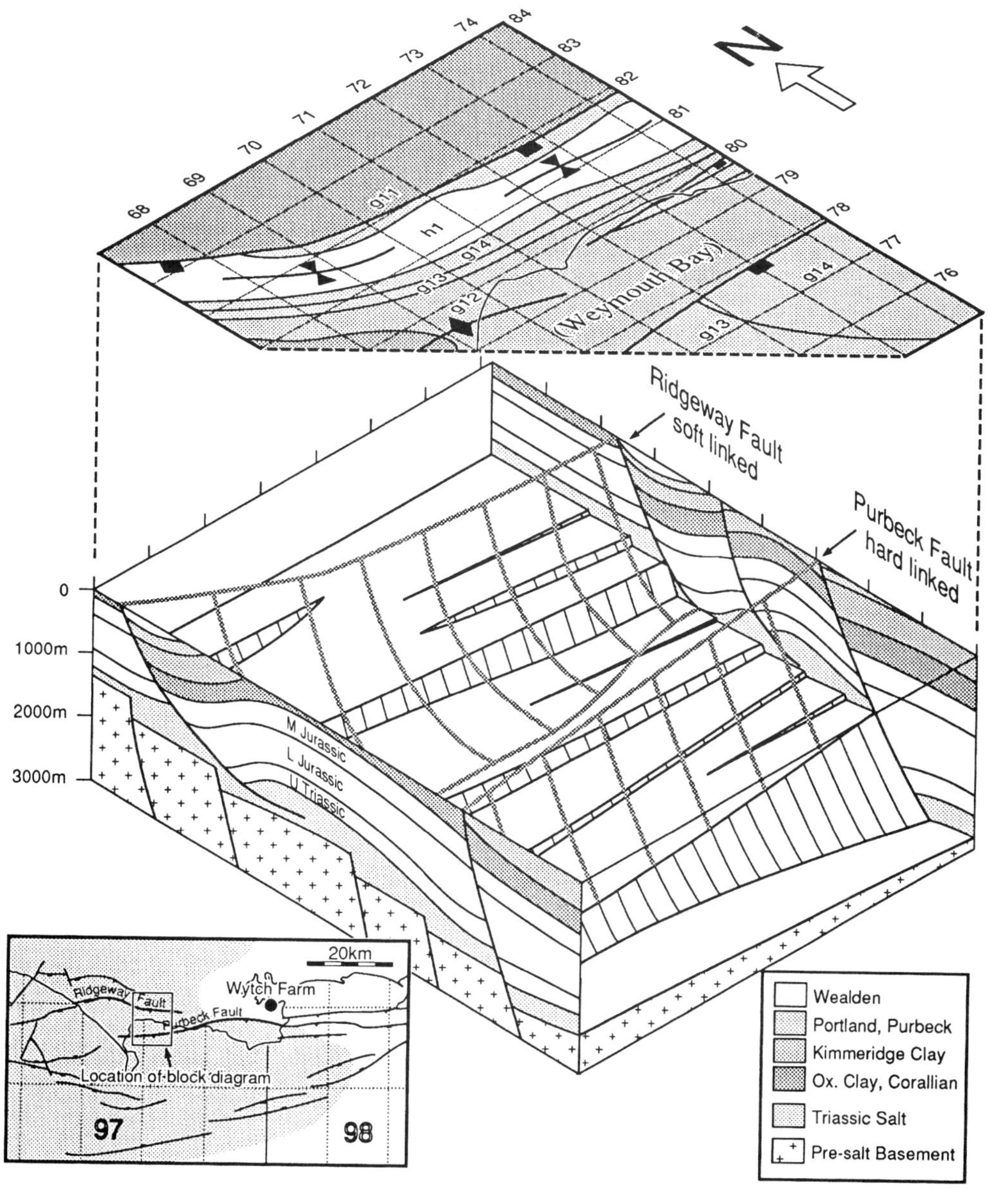

Fig. 8. 3D view of north margin, Channel Basin, showing part of a relay zone in the vicinity of Weymouth. Isometric view; grid on map is Ordnance Survey kilometre grid. Area lies fully within the Triassic salt basin (inset location map), T_v is assumed to be constant over the relay zone. Block diagram is a restoration to Albian (pre-inversion) extensional geometry. Basement fault displacement is partitioned across several minor faults in the relay zone. D decreases westwards on the hard-linked Purbeck Fault ($D_r \approx 1$ on the Purbeck Fault on the western cross section). The Ridgeway Fault is detached in this part of the relay structure.

the basement fault - such geometries mimic those which have been discussed elsewhere in relation to salt movement during and after basement faulting (Roberts *et al.* 1990).

It would appear that variations in T_v of the type observed in the Forth Approaches Basin occurred over a wide area north of the Mid-North Sea High (Fig. 5a) from the Triassic onwards (Hodgson *et al.* 1992; Smith *et al.* 1993), predating the major basement faulting episodes during the Jurassic which formed the Central Graben. Some of the wide variety of salt-related structures on the margins of, and within the Central Graben may be due in part to pre-rifting salt movement (Hodgson *et al.* 1992; Erratt 1993; Gowers *et al.* 1993; Høiland *et al.* 1993). The particular issue of the controls upon diapir location within the Central Graben will be discussed later.

Spatial variation in basement fault displacement (D) : north margin, Channel Basin

In the same way that D_r is sensitive to spatial variation in T_v, spatial variation in basement fault displacement D will also impact D_r and hence fault geometry. For instance, if a basement fault were to vary in displacement along strike while the thickness of the overlying salt layer remained constant, D_r would vary in the same manner as already illustrated in the context of the Dowsing Fault Zone. This effect might become important where basement faults die out, losing their displacement to a spatially offset en echelon strand. Such relay zones maintain the overall displacement of the basement fault zone (Peacock & Sanderson 1994). If the total basement fault zone displacement is such that D_r is less than 1, then D_r will rise and exceed 1 towards the pins of individual fault segments, and local detachment tectonics in the vicinity of basement relay ramps would be predicted (Fig. 7).

Mapping in the Channel Basin has suggested that such localized detachments might be present around relay zones in the major basement fault system on the northern basin margin (Fig. 8). The Purbeck Fault, which defines the northern margin onshore in Dorset, trends from south of the Wytch Farm field in Bournemouth Bay (Block 98/11) 40 km westwards to Weymouth. As the fault dies out towards its western tip, displacement is transferred north to the Ridgeway Fault. In the part of the relay zone detailed in the pre-inversion restoration Fig. 8, the Purbeck Fault ($D_r < 1$) is still hard linked to its basement counterpart, while the Ridgeway Fault is detached. Further west, this situation is reversed. Tertiary basin inversion involved reverse movement on these faults and led to the growth of folds in the relay zone (e.g. Poxwell pericline). The pre-Albian subcrop pattern above the basement relay zone shows, however, that folds related to basin extension are spatially offset from the Tertiary structures (Ridd 1973). The geometries of these folds suggest that they are rollovers related to listric faults soleing on the Dorset Halite (which has been drilled by five adjacent onshore wells). Displacement at top basement level is partitioned between several small faults in the relay ramp, where $D_r > 1$. The 3D geometry of the Purbeck–Ridgeway extensional relay zone is constrained in the east by subcrop below the exposed Albian unconformity (Fig. 8, map and east side panel) and in the west by forward models of rollover geometry adjacent to the Ridgeway Fault (Fig. 8, west side panel). Cover faulting was probably localized above major basement faults in this area, as can be seen elsewhere in the Channel Basin (see below). A sharp swing in strike of the Ridgeway Fault in map view may indicate a second, northwesterly stepping relay ramp in the basement. The degree of Tertiary reversal on the Ridgeway Fault reflects variations in the degree of linkage between basement and cover during inversion, which may have been controlled by local salt welds in the hangingwall block.

Temporal variation in basement fault displacement (D): Danish Basin

For constant T_v, variation in basement fault displacement D due to fault growth may clearly lead to fault systems characterized by $D_r > 1$ early in their history changing to $D_r < 1$ at a later stage (Fig. 9a), indeed, this must be true of all fault strands with $D_r < 1$. It can be seen that cover components of major fault systems which pass through viscous layers may inherently be more numerous than their basement counterparts. A fault system which seems to display this evolutionary sequence is illustrated in Fig. 9. This example is taken from the Norwegian-Danish Basin, an area which has been the subject of several papers discussing relationships between salt and faults in the basement and cover (Hospers *et al.* 1988; Geil 1991; Madirazza 1992; Petersen *et al.* 1992). In Fig. 9 the Triassic package thickens from east to west across the basement fault. Divergent sedimentary packages can be seen in the hangingwalls of detached cover faults above the basement footwall. Balancing movement on the detached cover faults with early displacement on the basement fault as suggested in Fig. 9 accounts for the present-day displacements on the cover and basement fault segments and the thickness variations in Triassic strata across the basement fault zone. Similar geometries can be found on the west margin of the Jæren High (Høiland *et al.* 1993) and can be generated in analog models (Oudmayer & de Jaeger 1993).

Shared detachment: impact on basement-cover fault linkage

The ramp-flat trajectories discussed so far in the context of single fault strands with $D_r > 1$ represent an idealized situation, applicable to single major faults. If on the other hand a basin is characterized by many basement fault strands, which individually have displacement ratios greater than 1, then the zones of detachment in the salt may spatially overlap (Fig. 10). A question then arises as to whether the fault trajectories separate again to proceed independently into the cover as depicted on Fig. 10. An example illustrating this problem can be found in the Channel basin (Fig. 11). In this area the Upper Triassic to Jurassic sediments lie in domino fault blocks defined by fault strands which seem to detach in the Upper Triassic Dorset Halite, while the tops of tilted fault blocks at top basement level can be seen below the evaporite layer (Fig. 11a, b). One can find in this area the opposite situation to that described in the context of temporal D_r variation in that basement faults are more numerous than cover faults (Figs 8 & 11). The depth converted equivalent of this section (Fig. 11c) shows that cover fault segments which appear to follow ramp-flat trajectories in two-way time are planar in depth. The planar faults apparently continue down through the evaporite layer to link directly with basement faults. Of several issues raised by this depth section we first consider the significant discrepancies in displacement between the cover and basement elements of these apparently continuous strands. The same situation was noted on the west margin of the southern North Sea. In that area, the discrepancy was attributed to inversion of the basement faults and basinwards sliding of the thin-skinned cover. In the Channel Basin, one could also argue that the minor basement faults were older than the cover sediments. While none of these suggestions can be ruled out in the Channel Basin, we note that there is approximate balance of basement and cover fault displacements at the scale of individual cover fault blocks (Fig. 11c). The implication could be that the location of cover faults may be controlled by the most significant basement faults, but that the minor basement faults flatten upwards into the salt layer, contributing their displacement to an offset cover fault (Fig. 12). One could alternatively argue that the displacement discrepancy between cover and basement on the main fault strands could be due to salt migration from hangingwall to footwall, indeed, an element of salt redistribution is an inherent feature of the rigid fault block model shown in Fig. 12. If salt redistribution were the only cause of the basement-cover displacement anomaly, however, the displacement across the minor basement faults would not be accounted for.

If a basin were pervaded by many basement faults, each with a flat-lying portion in the same evaporite layer, then these flat detachments might overlap to the extent that they formed a basin-wide décollement. This situation would alias laterally extensive salt detachment characteristic of gravity

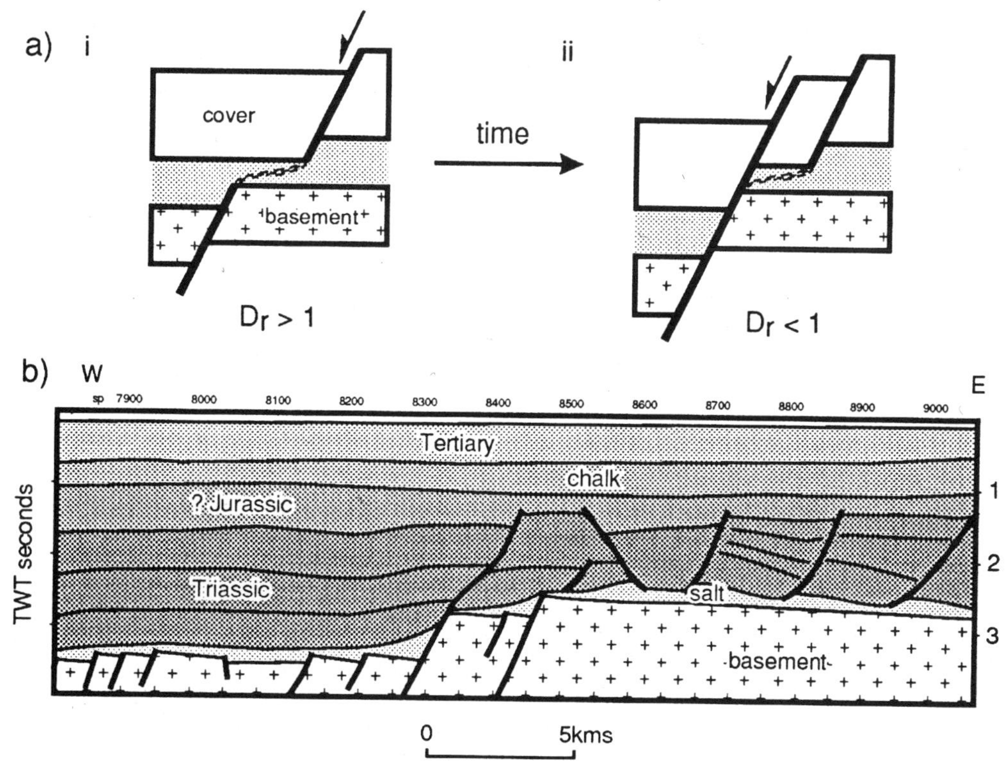

Fig. 9. (a) Increasing basement fault displacement D through time. D_r is initially greater than 1, and ramp-flat linkage may occur (i). As the basement fault grows D_r falls below 1 and hard linkage to a cover fault may occur (ii). The final fault geometry splays upwards from a single basement fault into two or more cover strands. The early spatially offset cover fault segment is 'fossilized' above the basement footwall. (b) Seismic and geoseismic from an example in the Norwegian–Danish Basin. Note thickening of the Triassic section across the basement fault and Triassic growth sequences in the detached faults above the basement footwall suggesting early ramp-flat geometry when $D_r > 1$.

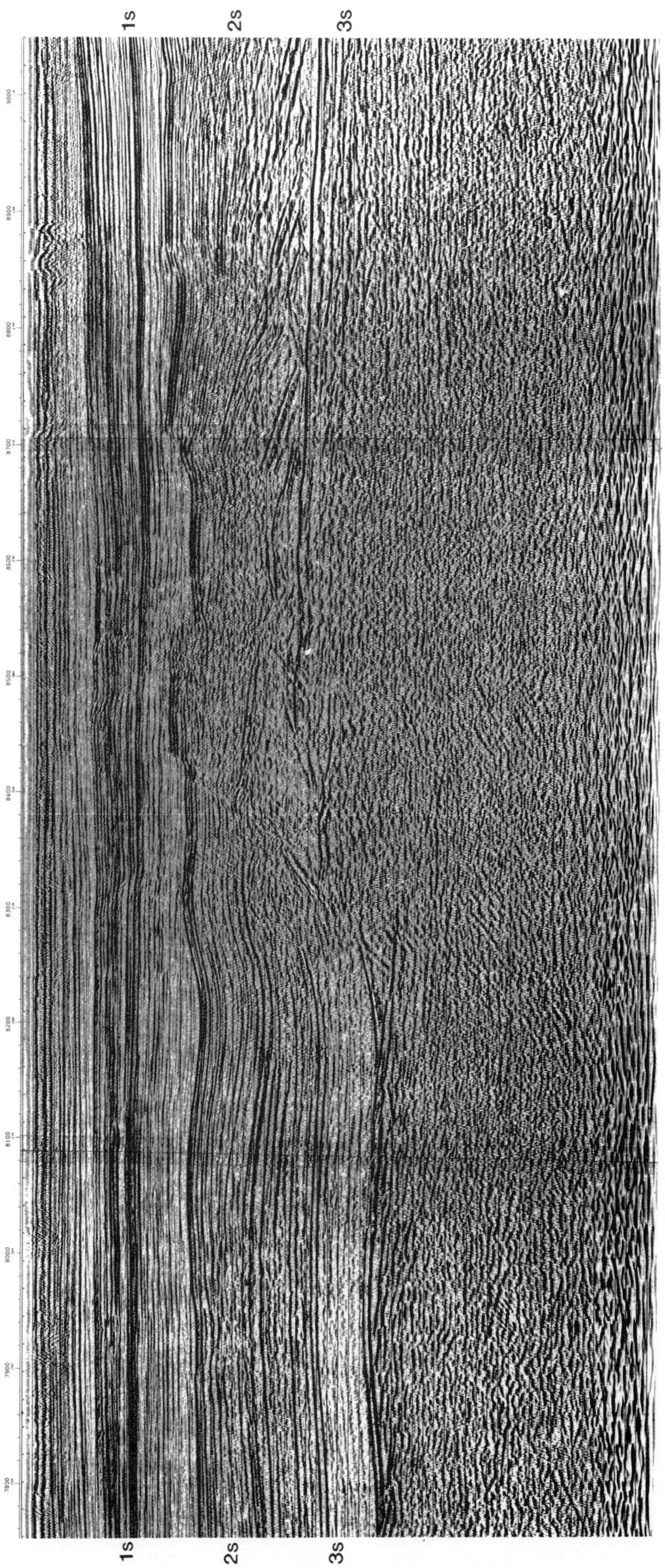

Fig. 9b. Seismic section.

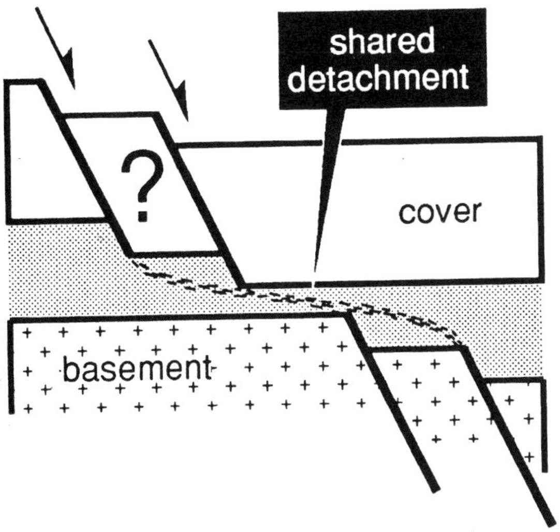

Fig. 10. Closely spaced basement faults which sole out into an overlying detachment. If the detachments overlap, do the individual fault trajectories diverge and proceed separately into the cover ?

gliding, and the fault style imparted to the cover sediments might be expected to be similar. The style of extensional thin-skinned deformation which characterizes gravity gliding is termed raft tectonics following type examples observed offshore Angola (Duval *et al.* 1992). During raft tectonics the cover is extended by normal faults whose location and spacing is largely determined by the material properties of the cover strata (Vendeville & Jackson 1992), although localization of cover faults will occur above pronounced, pre-existing edges (fault scarps) in basement topography (Gaullier *et al.* 1993). During thick-skinned extension it is traction at the base of the cover sequence which will drive cover extension, rather than tension in the cover due to basinward collapse. Of course, both processes could occur simultaneously. An important difference between the two processes is, however, that the former can generate cover raft tectonics in the absence of appreciable regional tilt. This also represents an instance of thick-skinned tectonics where, with the exception of anomalously large faults, no spatial relationship

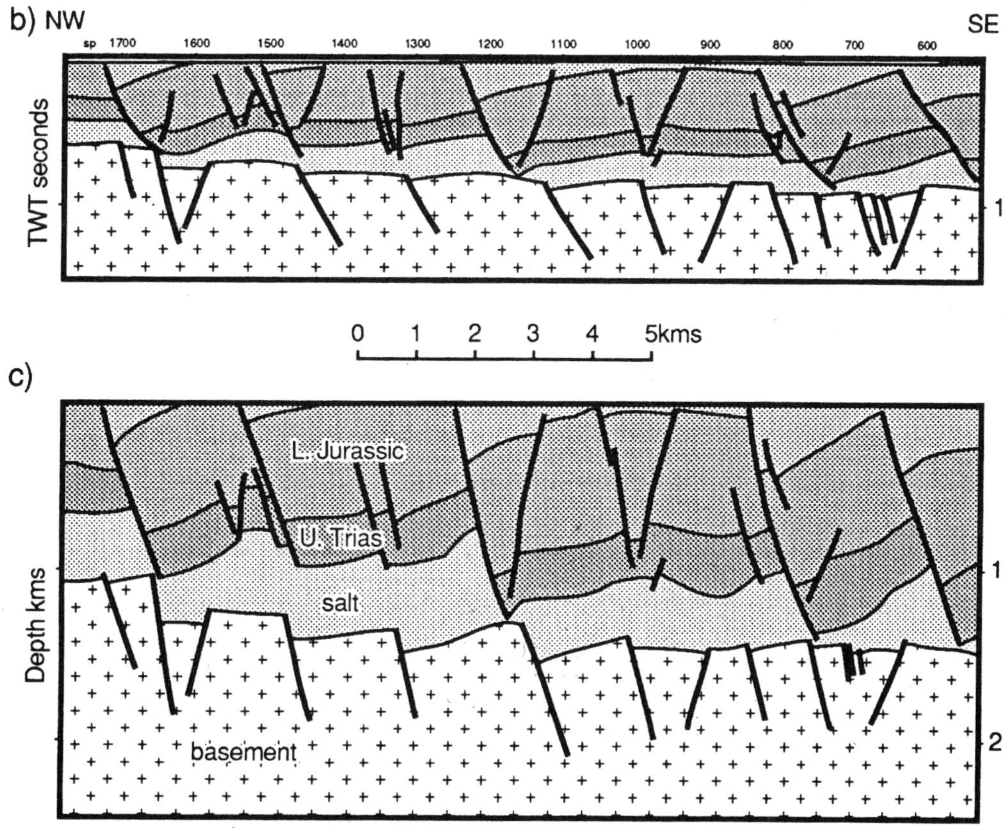

Fig. 11. North-south trending section from the Channel Basin. (a) time section. (b) geoseismic. The main faults are distorted into apparent ramp-flat trajectories by relatively high seismic velocity in the salt layer. (c) Depth converted section, vertical exaggeration = ×3. Main faults appear to be planar and continuous with basement strands. Note discrepancy in displacement between corresponding basement and cover strands.

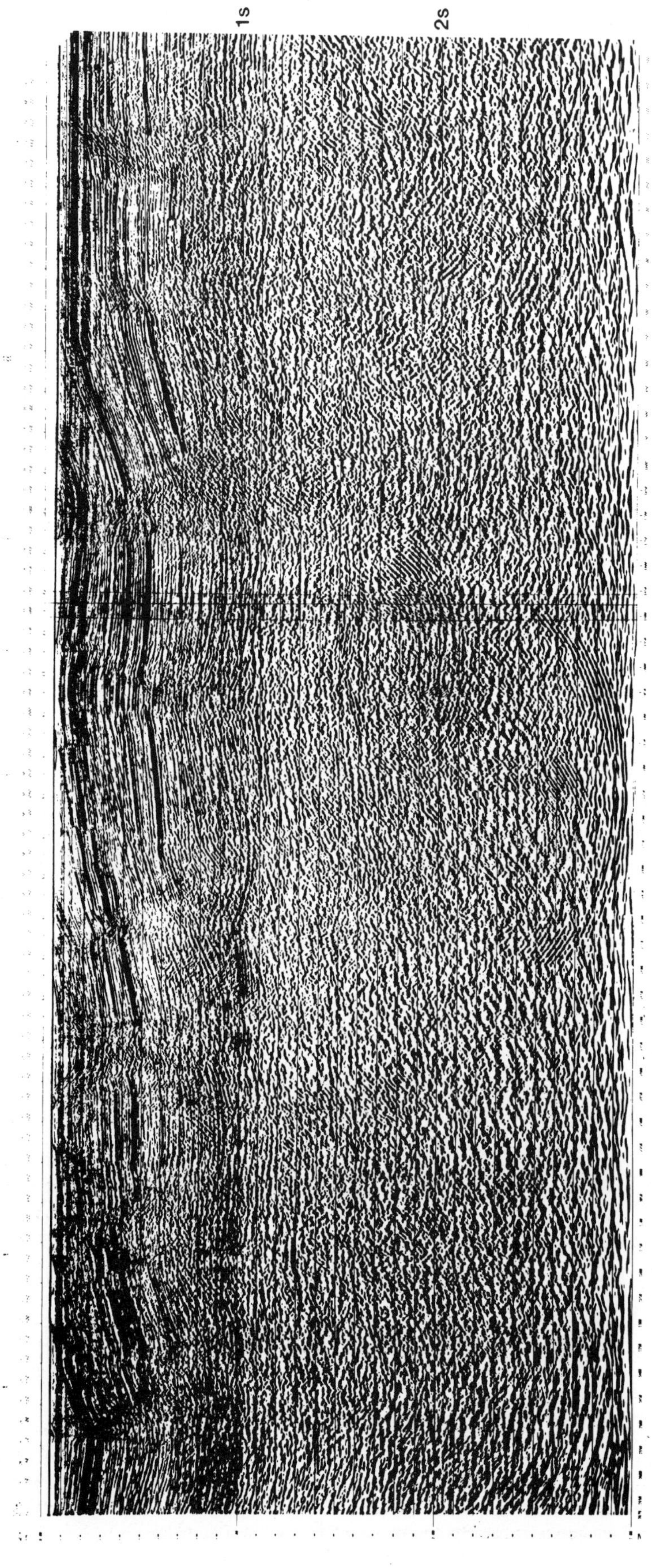

Fig. 11a. Seismic section.

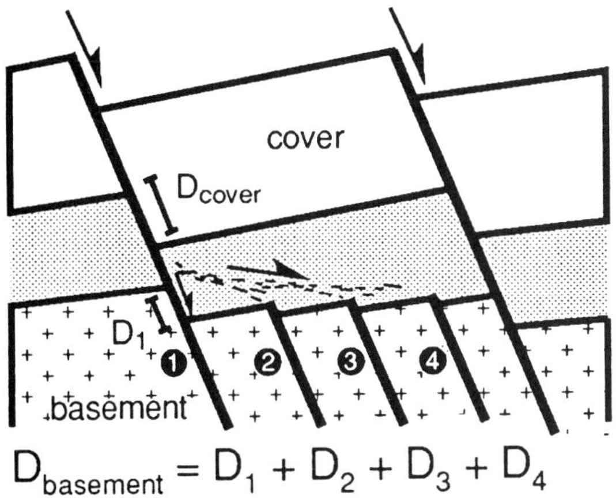

Fig. 12. Channel Basin kinematic model. Several closely spaced basement faults, which could be reactivations of older structures, share a detachment which transfers displacement onto a cover strand localized above the most prominent basement scarp. The rigid fault blocks depicted in this model imply some degree of salt redistribution away from each linked basement-cover strand.

need exist between cover and basement structure. The Whale–Horseshoe Basin system offshore Newfoundland is an example of raft tectonics in a thick-skinned system (Balkwill & Legall 1989; Fig. 4 in Jackson & Vendeville 1994).

Defining effective salt layer thickness (T_{ve})

With the exception of the first example discussed, salt layer thickness T_v has been defined as the distance from the top of the 'basement' to the base of the 'cover'. This approach is perfect for analogue models which contain a single, pure, viscous layer but it is less satisfactory in real sedimentary basins where a salt 'Formation' may contain interbedded anhydrite, dolomite and so on (Taylor 1990). Such interbeds are much stronger than salt and if they are laterally extensive they will reinforce the otherwise weak salt interval. The question arises of how to define an effective T_v value (T_{ve}) which could then be considered in terms of D_r – compare Figs 1 and 13. Choices include using only the thickest salt

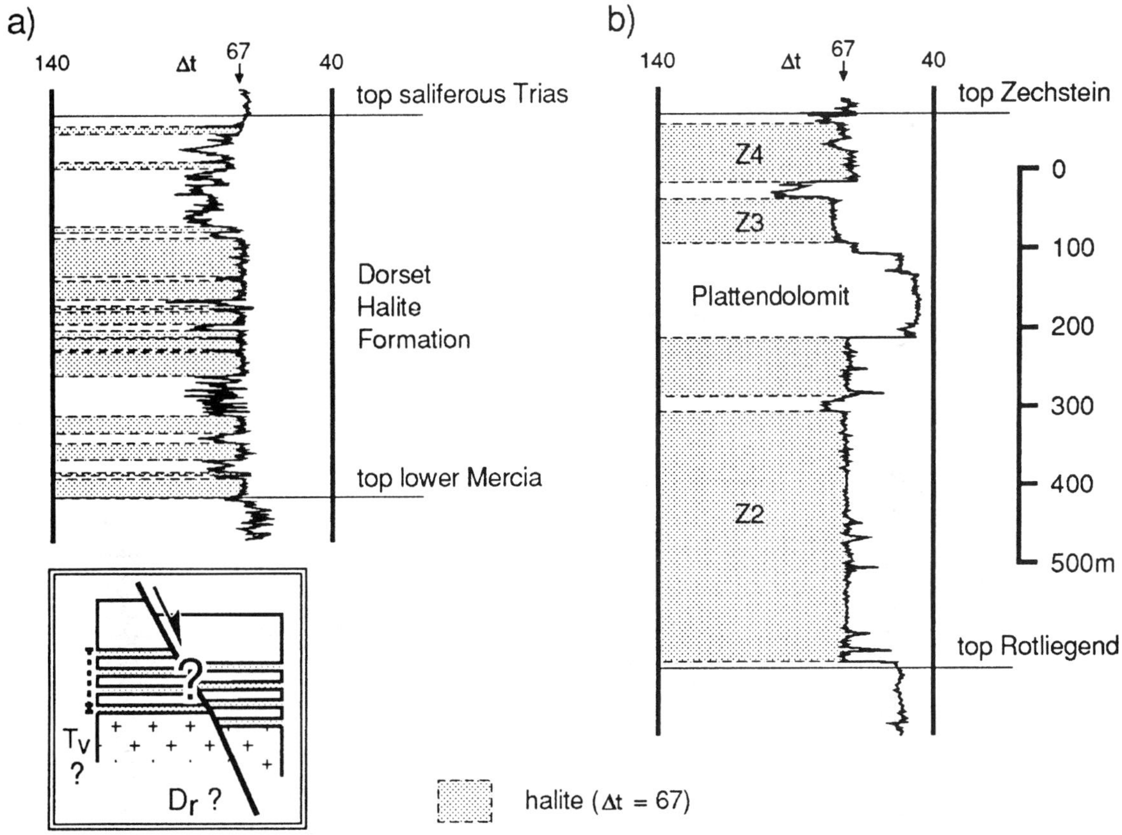

Fig. 13. Comparison of sonic logs of saliferous intervals from released wells in the Channel Basin (a) and the southern North Sea (b). Δt : interval transit time (μs/foot). (a) Channel Basin well, Quadrant 97, showing Dorset Halite Formation (within the Triassic Mercia Mudstone Group). (b) Southern North Sea well, Quadrant 49 (after Taylor, 1990), showing the Permian Zechstein Supergroup. Inset box : if the salt 'layer' contains interbeds of strong, brittle material, what value should be used to represent T_v when calculating D_r and what is the fault trajectory?

layer; summing the salt layers to give the net salt thickness; averaging the salt layer thicknesses, etc. The Channel Basin example (Fig. 11a) has already shown that cover faults may link directly down onto basement faults at anomalously high displacement ratios ($D_r > 1$) if T_{ve} is measured as the entire thickness of the Dorset Halite Formation. A comparison of typical sonic logs through the main saliferous sequences of the Channel Basin and the southern North Sea is shown on Fig. 13. Although the overall thicknesses of the saliferous sections are of the same order, the Channel Basin section contains far more interbeds and the most significant salt layer is an order of magnitude thinner than the thickest halite layer on the southern North Sea log (Z2 Stassfurt Halite Formation). The Channel Basin example indicates that the entire thickness of the saliferous series is an inappropriate choice for T_{ve}, as is the net thickness of salt. Only values close to (but not necessarily equal to) the thickness of the thickest individual salt layer give $D_r < 1$, and account for the observed fault geometry.

Extensional fault trajectories in viscous – brittle multilayers

The examples illustrated in Fig. 13 could represent the same, notional sedimentary formation with a spatially varying stratigraphy. It is easy to imagine a spatially varying stratigraphic section passing from pure halite through an interbedded zone (such as Fig. 13a) to, with increasing thickness of some of the non-salt interbeds, a multilayer sequence containing salt layers separated by significant thicknesses of sediment. It is likely that real examples of fault planes passing through stratigraphies similar to any of these can be found, but the majority of modelling literature focuses on the 'single, pure layer' end of the spectrum (e.g. Koyi 1991; Vendeville & Jackson 1992; Zaleski & Julien 1992; Weijermars *et al.* 1993).

Having already considered examples here of faults passing through stratigraphies similar to those illustrated in Fig. 13, an example of fault trajectories through a cover sequence containing multiple salt layers is presented (Fig. 14). This example is located in the western part of the southern North Sea, where several significant halite layers are present in the middle and upper Triassic, in addition to the Permian Zechstein salt. The thickness of sediments between the lowermost Triassic evaporites and the Zechstein salt in this area is approximately 600 m (Cameron *et al.* 1992). The structure displayed is known as the 'Flamborough Disturbance', part of the Dowsing Fault Zone and its cover counterpart, both of which pass onshore into East Yorkshire (Kirby & Swallow 1987). The style of faulting present on both sides of this symmetrical graben system below the Cretaceous unconformity clearly consists of ramp-flat extensional faults which define domino fault blocks at several stratigraphic levels. Overall, the basement fault system (DFZ) detaches in the Zechstein, then ramps up in synthetic and antithetic systems through the Bacton Group (Lower Triassic) as it does further southeast (Fig. 3b), however the fault surface flattens again in each of the successive evaporite layers in the Triassic, stepping up in ramps which occur progressively further away from the basement fault scarp.

It is likely that the comments pertaining to basement-cover relationships made so far will also apply to each successive ramp-flat-ramp offset in multi-

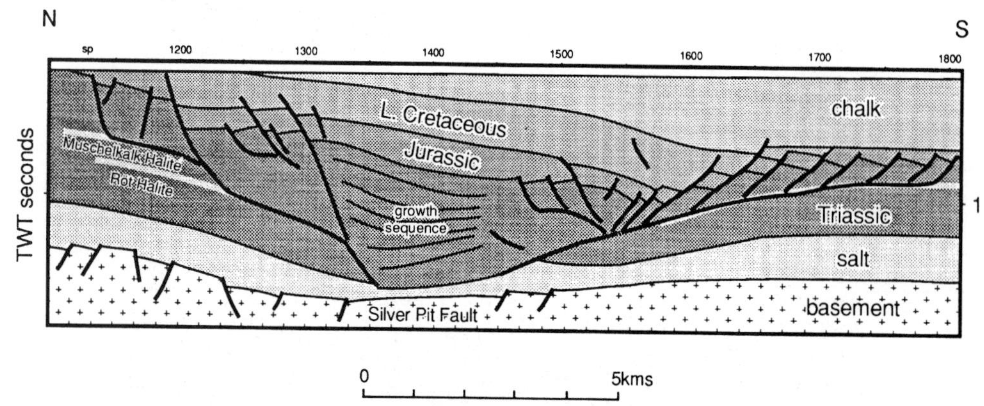

Fig. 14. Seismic profile offshore Flamborough Head, East Yorkshire. Multilevel ramp-flat fault profiles on both sides of the cover graben, detachment occurring on several intra-Triassic salts (Röt Halite, Muschelkalk Halite and Keuper Halite). Note the train of domino fault blocks above the Röt Halite on the south side of the graben, the kink-band style of folding, and southerly facing monoclinal folding at high stratigraphic levels (as seen on the east side of the East Midlands Shelf – Fig. 3b). The ramp-flat geometries in the cover continued to be active while D_r fell below 1 (see text).

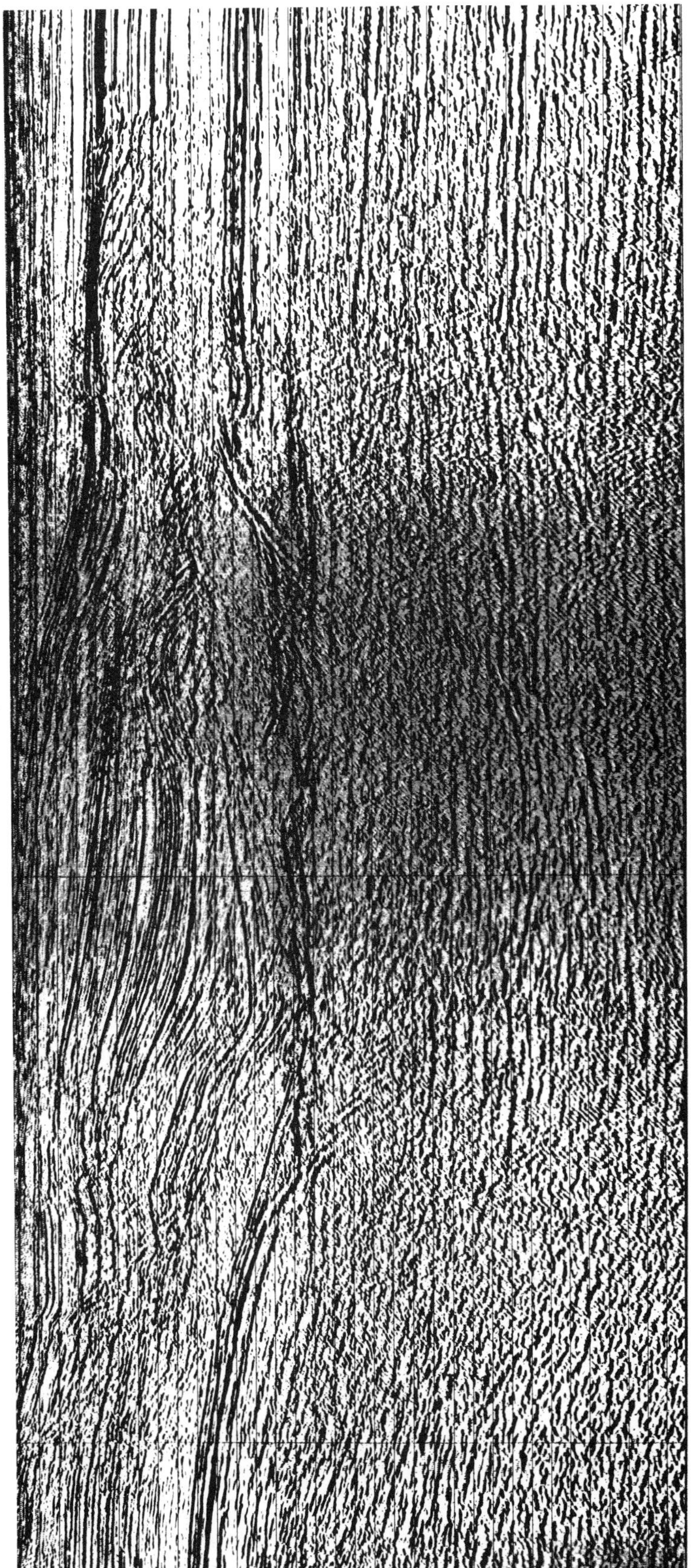

Fig. 14. Seismic section.

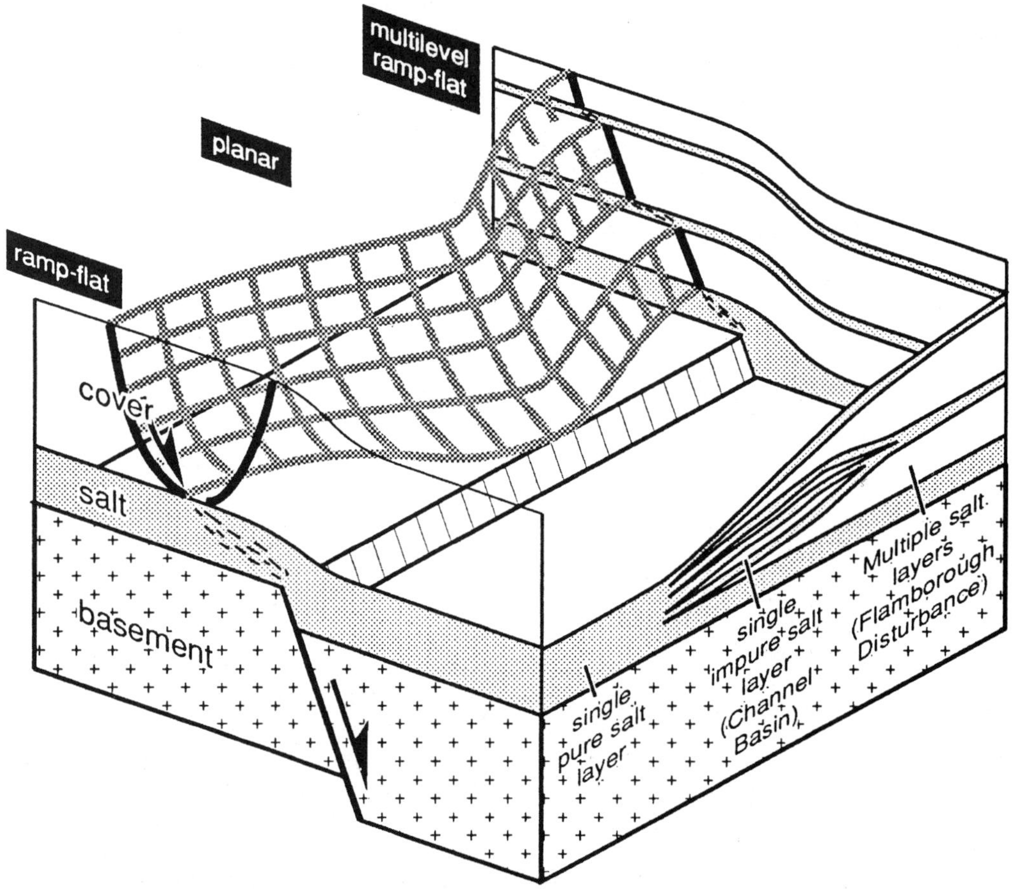

Fig. 15. 3D variation in basement-cover relationship due to intra-salt stratigraphic variation from a single, pure halite layer through interbedded salt and 'shales' to a multilayered salt / cover sequence. Based on analog models (Fig. 1), the Dowsing Fault system (Fig. 3), the Channel Basin (Fig. 11) and the Flamborough Disturbance (Fig. 14).

level stair-stepping fault trajectories. It is noted, however, that the ramp-flat geometries shown in Fig. 14 have persisted in spite of D_r falling below 1 in the Triassic section as the system has grown (cf. Fig. 9). The survival of active ramp-flat surfaces when $D_r < 1$ may reflect a change in the behaviour of the multilayer as it becomes tilted (controlled in this case by Zechstein salt redistribution, Fig. 14), however the influence of multilayer dip is not explored further here. Although the graben system is broadly symmetric, the sub-Chalk Mesozoic sequence clearly thickens northwards, suggesting that the footwall to the system in basement and cover lies on the south side of the graben. This contrasts with the interpretation of Kirby & Swallow (1987) who infer the cover strata south of the graben to be allochthonous.

Ramp-flat geometries within the Triassic can be found in several other parts of the southern North Sea, for example in the crest of the Sole Pit High (Van Hoorn 1987; Walker & Cooper 1987) and on the flanks of major salt-cored fold structures (Owen & Taylor 1983; Jenyon 1988). These examples are entirely gravity driven and arise due to local uplift. The faults in these examples grow downwards towards the Zechstein salt and accommodate local gravity spreading. They are therefore genetically unrelated to the structures seen in association with the cover equivalent of the DFZ. The fault style observed in the Flamborough Disturbance can be considered alongside the patterns seen elsewhere on the DFZ and in the Channel Basin to construct the 3D geometry of a fault growing through a salt-cover stratigraphy which varies from a single to multiple viscous layers (Fig. 15). This figure emphasizes the importance of the purity of the salt layer and selection of appropriate T_v values.

Cover delamination and salt sills in multilayer sequences

Following sufficient cover extension, the Zechstein salt intrudes the cover fault planes as reactive

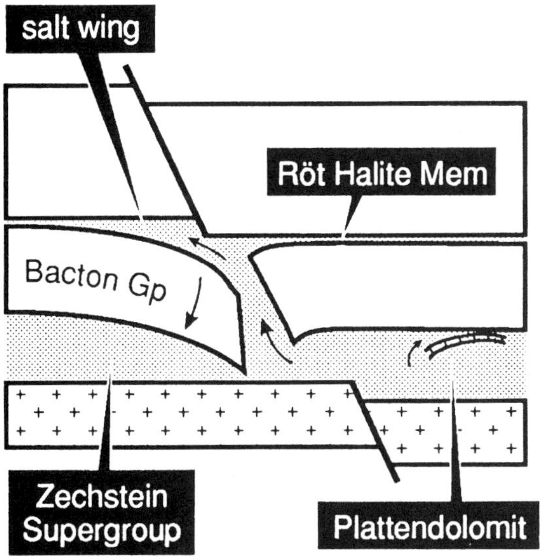

Fig. 16. Model of lower Triassic (Bacton Group) delamination facilitated by diapiric intrusion into a cover fault system. Lateral spreading of Zechstein salt at Röt Halite level is passive, a suction effect driven by unpeeling of Bacton Group into underlying Zechstein salt. Note that if Bacton Group delamination occurred in the hangingwall alone, with the Röt Halite fed by an out of section diapir, the geometry would ape sedimentary growth during Röt Halite times. The process of Bacton Group delamination is a larger-scale example of upper Zechstein carbonate (Z3 Plattendolomit) delamination, which is a common occurrence.

diapirs (Vendeville & Jackson 1992). Once these diapirs have intruded to the top of the Bacton Group, cohesion between the Bacton Group and the surrounding sediments is lost (the Bacton Group is overlain by the Röt Halite Member). With an abundant supply of Zechstein salt adjacent to the Röt Halite layer, the Bacton Group is free to delaminate from the rest of the cover sequence and collapse down, driven by gravity, into underlying Zechstein Supergroup, while mobile Stassfurt Halite feeds up the adjacent conduit to spread laterally into the space created above the Bacton Group (Fig. 16). This process represents localized redistribution driven by buoyancy, facilitated by reactive diapirism. It is a very similar process to that seen on a smaller scale within the Zechstein Supergroup, wherein the Z3 and Z4 carbonates often delaminate on the Z3 and Z4 halite members and sink down into the Z2 Stassfurt Halite, forming allochthonous rafts enveloped by Z2 salt (Jenyon 1989; Fig. 16). Indeed, such rafts can be carried up into the Triassic within reactive diapirs. It is noted that salt plays a passive role in this mechanism for Zechstein sill intrusion.

Zechstein sills may be lopolithic or laccolithic in form and are characterized by divergence of Triassic reflectors which can be misinterpreted as a sedimentary growth sequence. Examples of salt wings associated with lower Triassic delamination are common in the southern North Sea, Danish Central Graben (figure 11 in Sundsbø & Megson 1993) and the northwest German basin (Baldschuhn *et al.* 1991). Sag of the Bacton Group may be more extensive parallel to a cover fault trend than the adjacent reactive diapir conduit, so a given 2D section in the dip direction may contain an anomalously thick Röt Halite swell, apparently unconnected to the Zechstein, though the extra thickness at Röt Halite level is due to movement of Zechstein salt into the section.

Influence of thick-skinned extension on subsequent diapirism

The discussion so far has focused on the geological criteria which control the initial geometry of extensional faults which grow from basement through a viscous layer to the cover. It has been argued that the geometry of such extensional faults is established before diapirism occurs. With continuing extension, salt intrudes the cover fault planes as reactive diapirs, which may be planar in 3D (Jackson & Vendeville 1994). This important period of diapir evolution occurs during active rifting (Late Jurassic in the North Sea) and is a period during which salt in the form of diapirs establishes discordant relationships with cover sediments and attains high, shallow, structural levels. Periods of basin extension via active basement faulting may be followed by periods of thermally driven subsidence (McKenzie 1978). This is the case in the North Sea, where flexural subsidence centred on the Central Graben during the Tertiary has led to the burial of the Jurassic rift system by up to 3 km of post-rift sediment. This second phase of basin subsidence is different from the first in that there is no inherent basement faulting. But since diapirs were already emplaced in association with fault planes at shallow levels, they are ideally placed to passively maintain their altitude relative to the geoid while the surrounding sediments 'downbuild', falling as the basin subsides (Jackson & Vendeville 1994). The conditions of active basement faulting in an area of thick salt, followed by thermal subsidence, are met in the Central Graben, where indeed many impressive salt diapirs can be found, often having 'risen' over 2km to reach juxtaposition with very young Tertiary strata. Some of these diapirs have been drilled and have been found to be capped by condensed sequences which are equivalent to the adjacent stratigraphic sections (Foster & Rattey 1993).

An understanding of the relationship between diapir location and basement structure can be

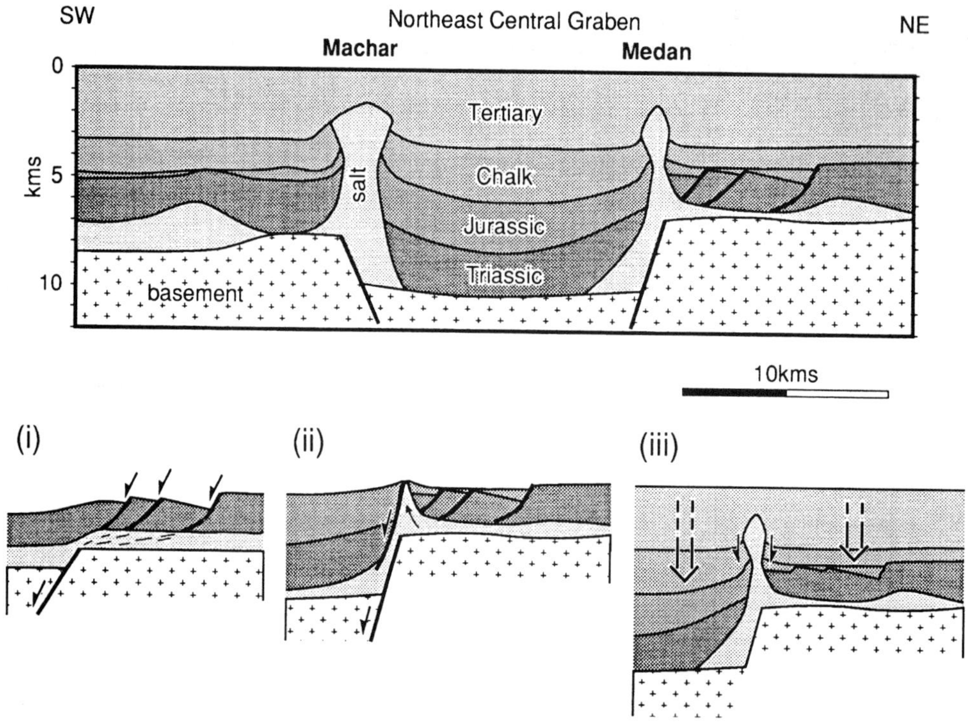

Fig. 17. Cross section across the Machar and Medan diapiric structures, Central North Sea, from Foster & Rattey (1993). The location and evolution of the Medan diapir is explained by a consideration of the basement fault history. (i) Initial basement fault growth ($D_r > 1$) links to cover via detachment in the salt, giving offset cover fault segments with Jurassic growth sequences (of. Fig. 9). (ii) Continued basement fault growth until $D_r < 1$ leads to hard linkage of basement fault to cover, at which point the cover segment is intruded by a reactive diapir. (iii) Cretaceous and Tertiary basin subsidence carries the sedimentary sequence down around the diapir, which passively maintains its altitude relative to the geoid.

approached if one considers the early (pre-diapir intrusion) stages of fault evolution in terms of the geological criteria which have been discussed so far (Fig. 17). Interpretation of mapped spatial relationships is also possible, for example if a train of passive diapirs is spatially offset from a basement trend, they may have evolved from a salt wall which reactively intruded early cover faults offset from their basement roots. Diapirs overlying basement scarps may have intruded faults which attained displacement ratios less than 1, and so on. It should be borne in mind that basement-cover relationships may be modified after active basement faulting by thin-skinned gravity gliding, which would passively move any salt structure present in the cover relative to the basement. Local gravity gliding post-dating basement faulting has been noted on the west margin of the Central Graben (Erratt 1993; Rattey & Hayward 1993) and further west, on the West Central Shelf (Bishop *et al.* 1995).

Influence of salt layers on inversion kinematics

Basin inversion is a common feature of the sedimentary basins offshore UK, particularly those in the south, such as the southern North Sea and Channel Basins (Roberts 1989). There may be many facets to the influence of salt layers on inversion styles, as there are during extension, but a collation of observations made on the west margin of the southern North Sea (Fig. 3) enables some key points to be made. Where no salt is present (Fig. 3a), reverse movements propagate directly into the cover. A key difference is noticed further northwest, where salt is present (Figs 3b, 3c) in that despite reverse movements at basement level (evidenced by monoclinal flexure of the Sole Pit High flanks) there is no reactivation of extensional fault segments in the cover. Given that the cover is certainly pinned to the basement further west at the salt pinchout (with no intervening thin-skinned compressional structures), the basement shortening

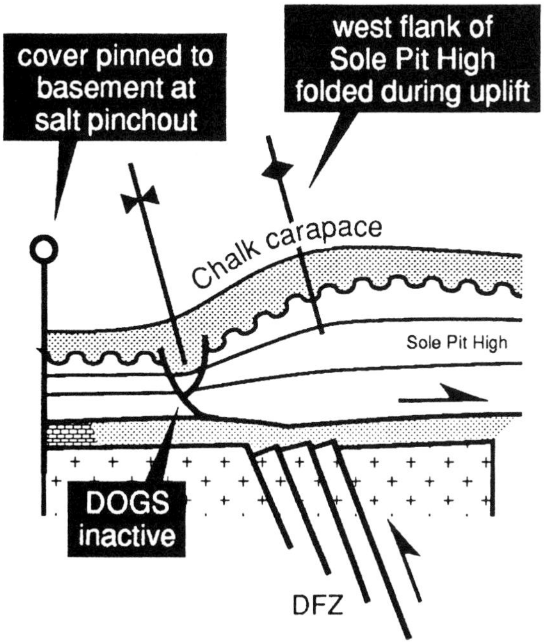

Fig. 18. Inversion kinematics of the Dowsing Fault Zone. Reverse movements on the basement faults are inferred from monoclinal folding of the cover sequence (and are observed along strike). The cover components of the Dowsing Fault system, which were active during extension (DOGS), are not reactivated during basement inversion. This implies basinwards movement of the cover (Sole Pit High) relative to the basement, given that the salt pinches out above the basement footwall. The basement hangingwall block is geometrically similar to a triangle zone.

can only be balanced by basinwards movement of the Sole Pit High relative to the basement hangingwall (Fig. 18). The line length mismatch between cover and basement is balanced by shortening in the cover on the east side of the Sole Pit High, where major salt-cored detached fold structures grew during the Palaeogene, contemporaneous with Sole Pit High inversion as dated by unconformable sediments further south (van Hoorn 1987; Badley *et al.* 1989; Stewart & Coward 1995). The spatial variation in cover fault reactivation is clearly influenced by the Zechstein salt (Fig. 19). This effect has also been noted in analog experiments (figure 7 in Richard 1991). The apparent difficulty in reactivating a stair-stepping fault trajectory during compression may underline an important role of detachments during shortening of multilayers.

Discussion

Several geological criteria which control the geometry of faults which grow from basement to cover through salt layers have been examined using examples from offshore UK. These criteria include spatial variations in salt thickness and basement fault zone geometry. They are summarized by the Displacement ratio D_r which has a key value of 1 marking the transition from soft-linked (detached) basement cover systems to hard-linked (continuous) strands. Examples of spatial coincidence and spatial offset between cover and basement fault elements can be found and accounted for. Although the geological criteria have been individually examined here, any one fault system is likely to be characterized in 3D by simultaneous variation of several criteria, giving complex variations in D_r. Most basin margins are characterized not by a single fault, but by fault systems comprised of several elements in basement and cover. Faults may splay upwards on passing through salt layers, or they may join upwards. In both cases the salt forms a detachment, the difference lying in the number of basement faults contributing displacement to the detachment. One of the most important, if least apparent, criteria is time. The role of salt in the evolution of a fault system changes from the earliest stages, when it contains ductile portions of the fault system, through later stages when it intrudes and deforms the upper portions of the system.

Experimental work has shown that diapir intrusion only occurs after significant amounts of extension on cover fault systems. Given the control of cover systems upon diapir initiation, an understanding of the spatial relationship between cover and basement fault segments should lead to an understanding of the spatial relationship between diapirs and basement faults. Diapir location may additionally reflect cover faulting related partially or wholly to thin-skinned tectonics. Although diapirs can be classified according to their ontogenic progression from reactive through active to passive (Vendeville & Jackson,1992), this does not identify their mode of origin. Many examples of diapirs related to thin-skinned tectonics have been described, as have diapirs related to thick-skinned tectonics (Jackson & Vendeville, 1994). The range of examples discussed here suggests that the problem of salt structure – basement structure relationship is tractable in many places within the UK offshore basins, and a genetic classification of salt structures as an adjunct to the ontogenic classification would be possible.

The authors acknowledge Geco-Prakla (UK) Ltd, PGS Nopec (UK) Ltd, Shelf Exploration Ltd, Western Geophysical, British Gas Exploration and Production, Kerr-McGee Oil (UK) Ltd., OMV (UK) Ltd, and Amerada Hess Ltd for permission to include seismic examples, and reviews by C. Smith, C. Elders and J. Hossack. The views expressed here are those of the authors and not necessarily those of Amerada Hess, British Gas, Kerr McGee or OMV.

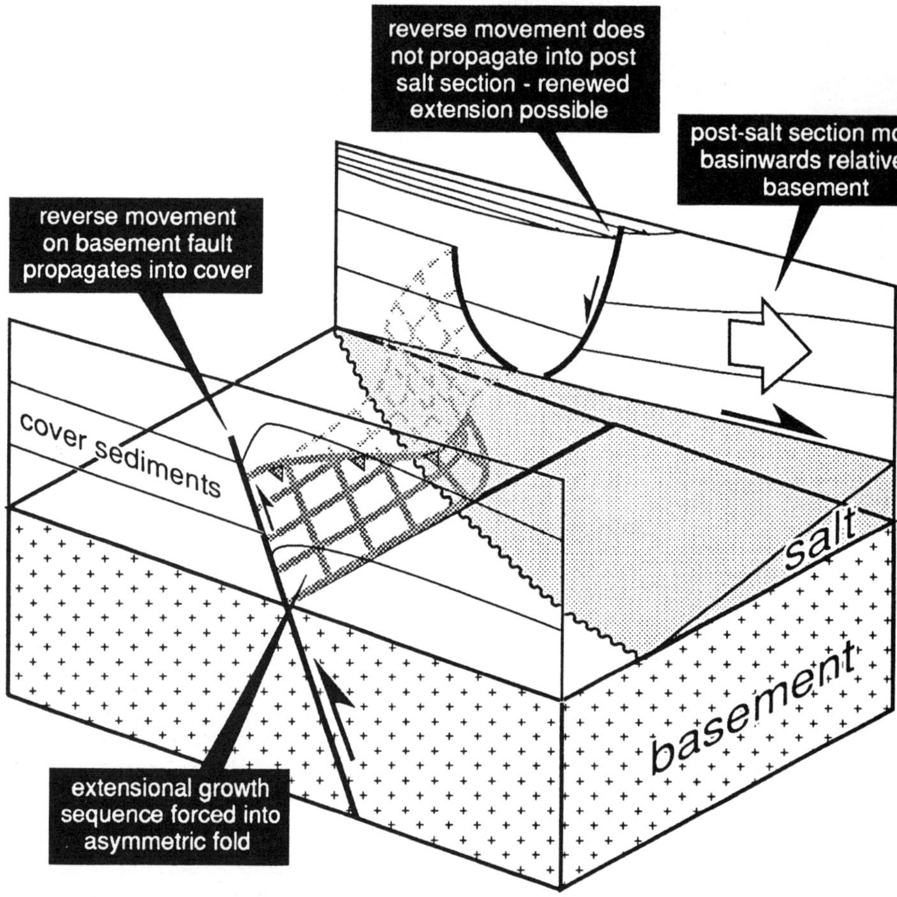

Fig. 19. 3D variation in inversion kinematics based on the Dowsing Fault System (Fig. 3). Without salt, the cover is hard-linked to basement and the cover extensional system is reactivated as a reverse fault, creating an asymmetric, hook-like hangingwall anticline structure. With a significant salt layer lying between basement and cover, reverse movement at basement level does not propagate into the extensional cover system. The basinwards movement and passive uplift of the cover can lead to minor extensional reactivation of the cover fault system. An element of rotation introduced into the cover between the hard-linked shortening zone and the detached zone may contribute to enigmatic basinal thin-skinned cover faulting.

References

ALLEN, M. R., GRIFFITHS, P. A., CRAIG, J., FITCHES, W. R. & WHITTINGTON, R. J. 1994. Halokinetic initiation of Mesozoic tectonics in the southern North Sea: a regional model. *Geological Magazine*, **131**, 559–561.

ARTHUR, T. J. 1993. Mesozoic structural evolution of the UK Southern North Sea: insights from analysis of fault systems. *In*: PARKER, J. R. (ed.) *Pctroleum Geology of Northwest Europe: Proceedings of the 4th Conference*. Geological Society, London, 1269–1279.

BADLEY, M. E., PRICE, J. D. & BACKSHALL, L. C. 1989. Inversion, reactivated faults and related structures: seismic examples from the southern North Sea. *In*: COOPER, M. A. & WILLIAMS, G. D. (eds) *Inversion Tectonics*. Geological Society, London, Special Publication, **44**, 201–219.

BALDSCHUHN, R., BEST, G. & KOCKEL, F. 1991. Inversion tectonics in the north-west German basin. *In*: SPENCER, A. M. (ed.) *Generation, Accumulation and Production of Europe's Hydrocarbons*, European Association of Geoscientists, Special Publication **1**, 149–159.

BALKWILL, H. R. & LEGALL, F. D. 1989. Whale Basin, offshore Newfoundland: extension and salt diapirism. *In*: TANKARD, A. J. & BALKWILL, H. R. (eds) *Extensional Tectonics and Stratigraphy of the North Atlantic Margins*. American Association of Petroleum Geologists, Memoir, **46**, 233–245.

BARTHOLOMEW, I. D., PETERS, J. M. & POWELL, C. M. 1993. Regional structural evolution of the North Sea: oblique slip and the reactivation of basement lineaments. *In*: PARKER, J. R. (ed.) *Petroleum Geology of Northwest Europe: Proceedings of the 4th Conference*. Geological Society, London, 1109–1122.

BISHOP, D. J., BUCHANAN, P. G. & BISHOP, C. J. 1995. Gravity-driven thin-skinned extension above Zechstein Group evaporites in the western central North

Sea: an application of computer-aided section restoration techniques. *Marine and Petroleum Geology*, **12**, 115–135.

BOYER, S. E. & ELLIOTT, D. 1982. Thrust systems. *AAPG Bulletin*, **66**, 1196–1230.

CAMERON, T. D. J., CROSBY, A., BALSON, P. S., JEFFREY, D. H., LOTT, G. K., BULAT, J. & HARRISON, D. J. 1992. *United Kingdom Offshore Regional Report: the Geology of the Southern North Sea*. London, HMSO for the British Geological Survey.

CHRISTIAN, H. E. 1969. Some observations on the initiation of salt structures of the southern British North Sea. *In*: HEPPLE, P. (ed.) *The Exploration for Petroleum in Europe and North Africa*. Institute of Petroleum, London, 231–248.

CRANS, W., MANDL, G. & HAREMBOURNE, J. 1980. On the theory of growth faulting : A geomechanical delta model based on gravity sliding. *Journal of Petroleum Geology*, **2**, 265–307.

DAVIS, D. W. & ENGELDER, T. 1985. The role of salt in fold-and-thrust belts. *Tectonophysics*, **119**, 67–88.

DUVAL, B., CRAMEZ, C. & JACKSON, M. P. A. 1992. Raft tectonics in the Kwanza Basin, Angola. *Marine and Petroleum Geology*, **9**, 389–405.

ERRATT, D. 1993. Relationships between basement faulting, salt withdrawal and Late Jurassic rifting, UK Central North Sea. *In*: PARKER, J. R. (ed.) *Petroleum Geology of Northwest Europe: Proceedings of the 4th Conference*. Geological Society, London, 1211–1219.

FOSTER, P. T. & RATTEY, P. R. 1993. The evolution of a fractured chalk reservoir: Machar oilfield, UK North Sea. *In*: PARKER, J. R. (ed.) *Petroleum Geology of Northwest Europe: Proceedings of the 4th Conference*. Geological Society, London, 1445–1452.

GAULLIER, V., BRUN, J. P., GUÉRIN, G. & LECANU, H. 1993. Raft tectonics: the effects of residual topography below a salt decollement. *Tectonophysics*, **228**, 363–381.

GEIL, K. 1991. The development of salt structures in Denmark and adjacent areas: the role of basin floor dip and differential pressure. *First Break*, **9**, 467–483.

GLENNIE, K. W. & BOEGNER, P. L. E. 1981. Sole Pit inversion tectonics. *In*: ILLING, L. V. & HUDSON, G. V. (eds) *Petroleum Geology of the Continental Shelf of North-West Europe*. Institute of Petroleum, London, 110–120.

GOWERS, M. B., HOLTAR, E. & SWENSSON, E. 1993. The structure of the Norwegian Central Trough (Central Graben area). *In:* PARKER, J. R. (ed) *Petroleum Geology of Northwest Europe: Proceedings of the 4th Conference*. Geological Society, London, 1245–1254.

HODGSON, N. A., FARNSWORTH, J. & FRASER, A. J. 1992. Salt-related tectonics, sedimentation and hydrocarbon plays in the Central Graben, North Sea, UKCS. *In*: HARDMAN, R. F. P. (ed.) *Exploration Britain*. Geological Society, London, Special Publication, **67**, 31–63.

HØILAND, O., KRISTENSEN, J. & MONSEN, T. 1993. Mesozoic evolution of the Jæren High area, Norwegian Central North Sea. *In*: PARKER, J. R. (ed.) *Petroleum Geology of Northwest Europe: Proceedings of the 4th Conference*. Geological Society, London, 1189–1195.

HOLMES, A. J. 1991. The Camelot fields, Blocks 53/1a, 53/2, UK North Sea. *In*: ABBOTTS, I. L. (ed.) *United Kingdom Oil and Gas Fields, 25 years Commemorative Volume*. Geological Society, London, Memoir, **14**, 401–408.

HOSPERS, J., RATHORE, J. S., JIANHUA, F., FINNSTRØM, E. G. & HOLTHE, J. 1988. Salt tectonics in the Norwegian-Danish Basin. *Tectonophysics*, **149**, 35–60.

JACKSON, M. P. A. & VENDEVILLE, B. C. 1994. Regional extension as a geologic trigger for diapirism. *Geological Society of America Bulletin*, **106**, 57–73.

JENYON, M. K. 1988. Overburden deformation related to the pre-piercement development of salt structures in the North Sea. *Journal of the Geological Society, London*, **145**, 445–454.

—— 1989. Plastic flow and contraflow in superposed Zechstein salt sequences. *Journal of Petroleum Geology*, **12**, 477–486.

KIRBY, G. A. & SWALLOW, P. W. 1987. Tectonism and sedimentation in the Flamborough Head region of north-east England. *Proceedings of the Yorkshire Geological Society*, **46**, 301–309.

KOYI, H. 1988. Experimental modeling of role of gravity and lateral shortening in Zagros Mountain Belt. *AAPG Bulletin*, **72**, 1381–1394.

—— 1991. Gravity overturns, extension, and basement fault activation. *Journal of Petroleum Geology*, **14**, 117–142.

——, JENYON, M. K. & PETERSEN, K. 1993. The effect of basement faulting on diapirism. *Journal of Petroleum Geology*, **16**, 285–312.

MADIRAZZA, I. 1992. Comment on 'The development of salt structures in Denmark and adjacent areas: the role of basin floor dip and differential pressure'. *First Break*, **104**, 134.

MCKENZIE, D. 1978. Some remarks on the development of sedimentary basins. *Earth and Planetary Science Letters*, **40**, 25–32.

NALPAS, T. & BRUN, J. P. 1993. Salt flow and diapirism related to extension at crustal scale. *Tectonophysics*, **228**, 349–362.

OUDMAYER, B. C. & DE JAEGER, J. 1993. Fault reactivation and oblique slip in the Southern North Sea. *In*: PARKER, J. R. (ed.) *Petroleum Geology of Northwest Europe: Proceedings of the 4th Conference*. Geological Society, London, 1281–1290.

OWEN, P. F. & TAYLOR, N. G. 1983. A salt pillow structure in the southern North Sea. *In*: BALLY, A. W. (ed.) *Seismic Expression of Structural Styles*. American Association of Petroleum Geologists, Studies in Geology Series, **15**, 2.3.2–7–10.

PEACOCK, D. C. P. & SANDERSON, D. J. 1992. Effects of layering and anisotropy on fault geometry. *Journal of the Geological Society, London*, **149**, 793–802.

—— & —— 1994. Geometry and development of relay ramps in normal fault systems. *AAPG Bulletin*, **78**, 147–165.

—— & ZHANG, X. 1994. Field examples and numerical modelling of oversteps and bends along normal faults in cross-section. *Tectonophysics*, **234**, 147–167.

PENGE, J., TAYLOR, B., HUCKERBY, J. A. & MUNNS, J. W.

1993. Extension and salt tectonics in the East Central Graben. *In*: PARKER, J. R. (ed.) *Petroleum Geology of Northwest Europe: Proceedings of the 4th Conference*. The Geological Society, London, 1197–1209.

PETERSEN, K., CLAUSEN, O. R. & KORSTGÅRD, J. A. 1992. Evolution of a salt-related listric growth fault near the D-1 well, block 5605, Danish North Sea: displacement history and salt kinematics. *Journal of Structural Geology*, **14**, 565–577.

PRICE, N. J. & COSGROVE, J. W. 1990. *Analysis of Geological Structures*. Cambridge, Cambridge University Press.

RATTEY, R. P. & HAYWARD, A. B. 1993. Sequence stratigraphy of a failed rift system: the Middle Jurassic to Early Cretaceous basin evolution of the Central and Northern North Sea. *In*: PARKER, J. R. (ed.) *Petroleum Geology of Northwest Europe: Proceedings of the 4th Conference*. Geological Society, London, 215–249.

RICHARD, P. 1991. Experiments on faulting in a two-layer cover sequence overlying a reactivated basement fault with oblique-slip. *Journal of Structural Geology*, **13**, 459–469.

RIDD, M. F., 1973. The Sutton, Poyntz, Poxwell and Chaldon Herring anticlines, southern England: a reinterpretation. *Proceedings of the Geologists' Association*, **84**, 1–8.

ROBERTS, A. M., PRICE, J. D. & SVAVA OLSEN, T. 1990. Late Jurassic half-graben control on the siting and structure of hydrocarbon accumulations: UK/Norwegian Central Graben. *In*: HARDMAN, R. F. P. & BROOKS, J. (eds) *Tectonic Events Responsible for Britain's Oil and Gas Reserves*. Geological Society, London, Special Publication, **55**, 229–257.

ROBERTS, D. G. 1989. Basin inversion in and around the British Isles. *In*: COOPER, M. A. & WILLIAMS, G. D. (eds) *Inversion Tectonics*. Geological Society, London, Special Publication, **44**, 131–150.

SMITH, R. I., HODGSON, N. & FULTON, M. 1993. Salt control on Triassic reservoir distribution, UKCS Central North Sea. *In*: PARKER, J. R. (ed.) *Petroleum Geology of Northwest Europe: Proceedings of the 4th Conference*. Geological Society, London, 547–557.

STEWART, S. A. & COWARD, M. P. 1995. Synthesis of salt tectonics in the southern North Sea, UK. *Marine and Petroleum Geology*, **12**, 457–476.

SUNDSBØ, G. O. & MEGSON, J. B. 1993. Structural styles in the Danish Central Graben. *In*: PARKER, J. R. (ed.) *Petroleum Geology of Northwest Europe: Proceedings of the 4th Conference*. Geological Society, London,1255–1267.

TAYLOR, J. C. M. 1990. Upper Permian-Zechstein. *In*: GLENNIE, K. W. (ed.) *Introduction to the Petroleum Geology of the North Sea*. Blackwell, Oxford, 153–190.

VAN HOORN, B. 1987. Structural evolution, timing and tectonic style of the Sole Pit inversion. *Tectonophysics*, **137**, 239–284.

VENDEVILLE, B. C. & JACKSON, M. P. A. 1992. The rise of diapirs during thin-skinned extension. *Marine and Petroleum Geology*, **9**, 331–353.

WALKER, I. M. & COOPER, W. G. 1987. The structural and stratigraphic evolution of the northeast margin of the Sole Pit Basin. *In*: BROOKS, J. & GLENNIE, K. W. (eds) *Petroleum Geology of North West Europe*. Graham & Trotman, 263–275.

WEIJERMARS, R., JACKSON, M. P. A. & VENDEVILLE, B. 1993. Rheological and tectonic modelling of salt provinces. *Tectonophysics*, **217**, 143–174.

WITHJACK, M. A., OLSON, J. & PETERSON, E. 1990. Experimental models of extensional forced folds. *AAPG Bulletin*, **74**, 1038–1054.

ZALESKI, S. & JULIEN, P. 1992. Numerical simulation of Rayleigh-Taylor instability for single and multiple diapirs. *Tectonophysics*, **206**, 55–69.

The link between sedimentary basin inversion and igneous underplating

JAMES BRODIE & NICKY WHITE

Bullard Laboratories, Madingley Rise, Madingley Road, Cambridge CB3 0EZ, UK

Abstract: There is now general agreement that many sedimentary basins on the northwest continental shelf of Europe underwent permanent exhumation during the Tertiary. The most dramatic indicator of this process is the present-day absence of up to 3 km of anticipated post-rift subsidence in the midlands of Britain and in the East Irish Sea. Any explanation must take into account the fact that the entire shelf has very small, long wavelength, free-air gravity anomalies. This constraint is of fundamental importance and implies either that the crust has been thickened, that phase changes have occurred within the lithosphere, or that low density material has been added to the lithosphere. Tertiary epeirogenic uplift and exhumation is often attributed to horizontal shortening which is assumed to be related in a general sense to Alpine mountain building. However, the removal of 3 km of sediment from a basin, which was originally 100 km wide, requires 20–30 km of shortening. Whilst minor Tertiary shortening is observed all over the continental shelf, nowhere is it sufficient to account for the inferred amount of denudation. More significantly, exhumation is thought to have commenced in the Early Tertiary and dramatically increases from south to north. Shortening is generally younger (mid-Tertiary) and decreases in intensity from south to north. Here we argue that Tertiary uplift and denudation are a consequence of regional igneous underplating. At the beginning of the Tertiary, rifting associated with the initiation of the Iceland plume generated substantial volumes of melt. Petrological arguments and the results from inversion of rare earth element concentrations of MgO-rich igneous rocks suggest that a minimum of 2–5 km of melt were produced beneath a substantial part of the continental shelf. We infer that much of this melt was trapped within the lithosphere, presumably close to the Moho, which would have acted as a density filter. Such underplating will have caused rapid surface uplift whilst maintaining isostatic equilibrium. Simple calculations based on deep seismic reflection data and the high P-wave velocities observed beneath Scotland are consistent with the petrological arguments.

A substantial part of the British Isles suffered rapid exhumation at the beginning of the Tertiary (Ziegler 1982; Johns & Andrews 1985; Roberts 1989; Lewis *et al.* 1992*a*). This conclusion is based primarily on subsidence analyses, on vitrinite reflectance profiles, and on fission track analyses. Exhumation is also manifest in the onshore and offshore Mesozoic outcrop pattern (Fig. 1). Nonetheless, the extent and magnitude of exhumation are disputed and it is often assumed that most of the British Isles remained at or close to sea-level from the end of the Carboniferous until at least the Cretaceous (see, for example, Ziegler 1982; Glennie 1990; Naylor 1992). Here we briefly review the evidence in favour of Tertiary uplift and consequent erosion. We point out that there is a considerable body of data which suggests that rapid and permanent exhumation must have been caused by regional igneous underplating associated in a general way with the opening of the North Atlantic ocean over the Iceland plume.

The clearest evidence in favour of Tertiary uplift and erosion depends upon our understanding of the evolution of extensional sedimentary basins. According to the lithospheric stretching model (McKenzie 1978; Jarvis & McKenzie 1980), syn-rift subsidence is followed by an exponentially decreasing post-rift or thermal subsidence (Fig. 2). The most recent episodes of extension in northwest Europe took place during the Permo–Triassic and in the Late Jurassic (Ziegler 1982; Glennie 1990). Upper Cretaceous and Tertiary extension events have also occurred but they are largely confined to the continental margin farther north and west of the area of interest and so are not considered here (Brooks & Glennie 1987). In the North Sea basin, where both Permo–Triassic and Late Jurassic extension have been documented (Glennie 1990), the predictions of the stretching model and the observations are in agreement (Barton & Wood 1984). Similar agreement has been documented for many extensional basins throughout northwest Europe (Ziegler 1982). In

From Holdsworth, R.E. & Turner, J.P. (compilers) *Extensional Tectonics: Regional-scale Processes*. The Geological Society, Key Issues in Earth Sciences, **2 (1)**, 309–326.
1476-315X/02/$15.00 © The Geological Society of London 2002
First published in: "Buchanan, J.G. & Buchanan, P.G. (eds) *Basin Inversion*, The Geological Society, London, Special Publications **88**, 21–38".

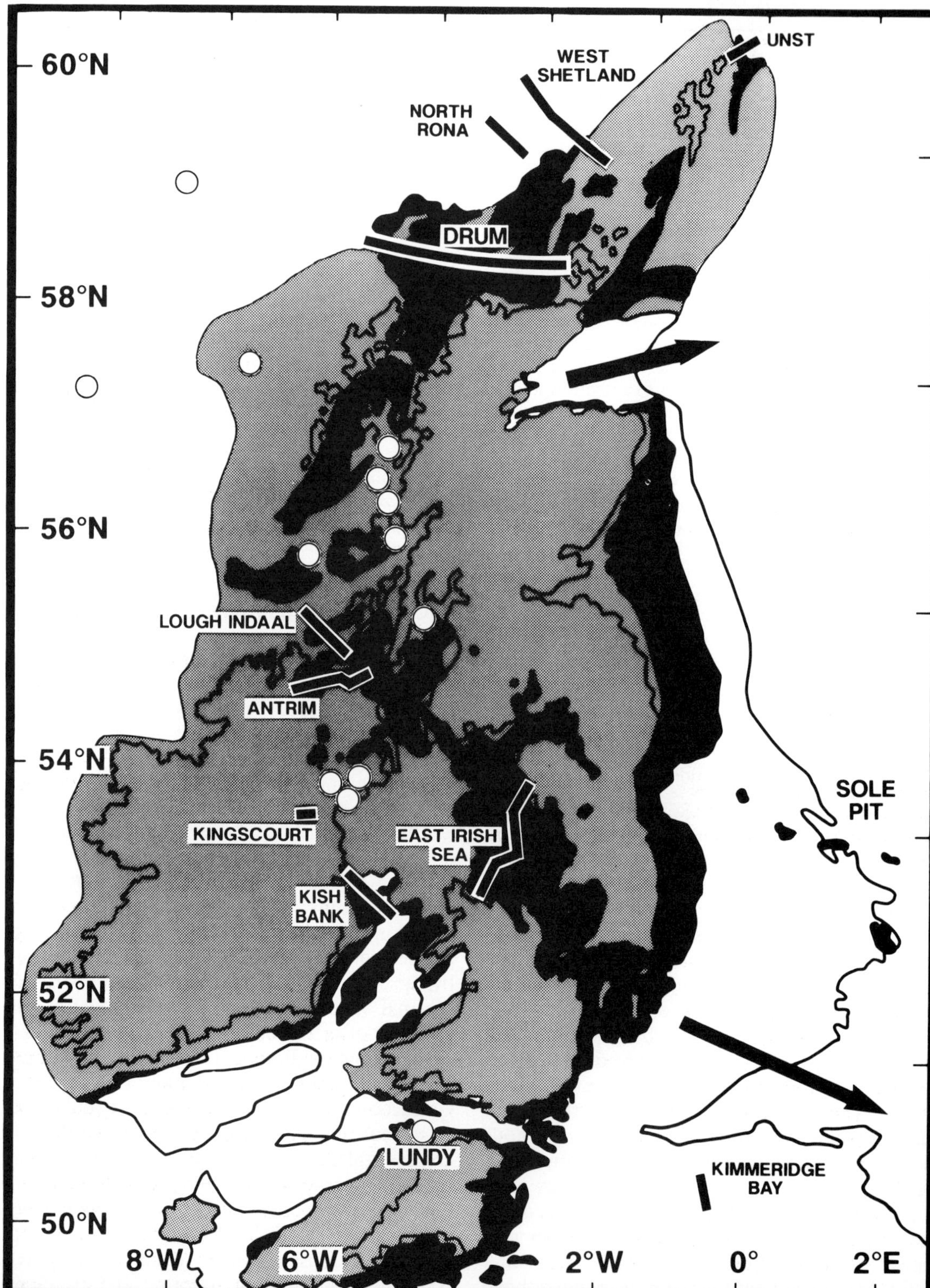

Fig. 1. Onshore–offshore geological map of the British Isles (sub-Pleistocene strata only) showing location of principal sedimentary rocks (redrawn from Woodland 1979). Note that coastlines of southern England and France have been omitted for clarity. Thin black line is boundary between pre-Mesozoic and Cretaceous rocks; thick black line is top of Cretaceous; outcrop pattern of Mesozoic and Cenozoic sedimentary rocks omitted for clarity; black areas show sub-crop of Permian–Triassic rocks, including some Jurassic and Cretaceous rocks in places; shading indicates Palaeozoic and older rocks, each representing basement in this paper. Arrows point in direction of progressively younger Mesozoic and Cenozoic rocks. Circles = major Tertiary igneous centres (note position of Lundy). Thick black lines with white margins and labels = location of profiles shown in Figs 4–6.

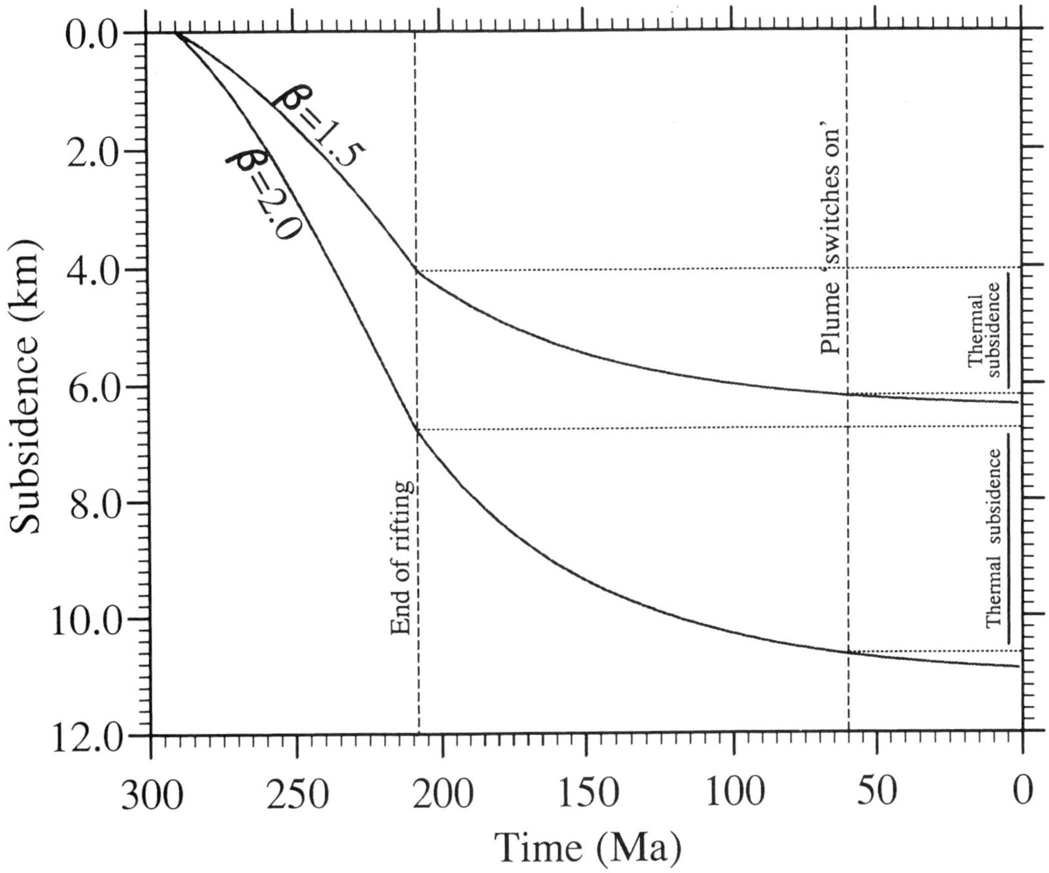

Fig. 2. Theoretical sediment-loaded subsidence curves for stretching factors of β = 1.5, 2. Rifting lasts from 290 to 210 Ma (i.e. Permian–Triassic) assuming a constant strain rate. Sediment density = 2.2 Mg m^{-3}. Vertical dashed lines indicate end of rifting and 60 Ma. Horizontal dotted lines indicate thicknesses of thermal subsidence from end of rifting until 60 Ma for each subsidence curve.

the vicinity of the British Isles, stretching factors for the Permo–Triassic extension are generally small (β = 1.1–1.3), although higher values (*c*. 1.5) have been proposed for basins north of Scotland (Cheadle *et al.* 1987; Enfield & Coward 1987). Late Jurassic extension is often substantial, especially within the Viking and Central Graben of the North Sea where β reaches values of 1.5 (Barton & Wood 1984). In the Porcupine basin west of Ireland, Jurassic stretching factors vary from 1.3–6.0 (Tate *et al.* 1993).

If rifting lasts for 80 Ma, stretching factors of 1.5 can give rise to sediment-loaded post-rift subsidence of *c*. 2 km (Jarvis & McKenzie 1980; Fig. 2). Shorter rifting periods give rise to greater amounts of post-rift subsidence. However, on mainland Britain, in the Irish Sea, and north of Scotland, fault-controlled (i.e. syn-rift) Permo–Triassic sedimentary rocks crop out close to the surface and the predicted post-rift subsidence is absent (Jenner 1981; Evans *et al.* 1982; Enfield & Coward 1987; Kirton & Hitchen 1987; Roberts 1989). Our understanding of extensional sedimentary basin formation suggests that these basins are likely to have subsided thermally after Permo–Triassic rifting. We conclude that post-rift sediments were deposited and have been subsequently removed. This inference is indirectly corroborated by anomalous porosity-depth variation and by high vitrinite reflectance in the Irish Sea, Inner Moray Firth and elsewhere (Roberts 1989; Andrews *et al.* 1990; Hillis *et al.* 1994). There is also excellent evidence that the products of large-scale erosion were deposited during the Palaeocene in rapidly subsiding basins to the north, east and west of the British Isles (Glennie 1990).

Finally, apatite fission track analyses from Mesozoic sedimentary rocks also appear to favour substantial Tertiary denudation (Lewis *et al.* 1992*a*; Green *et al.* 1993). These data can provide estimates of the time at which sedimentary rocks began to cool from maximum palaeotemperatures. Using analyses from the East Midland shelf and from northern England,

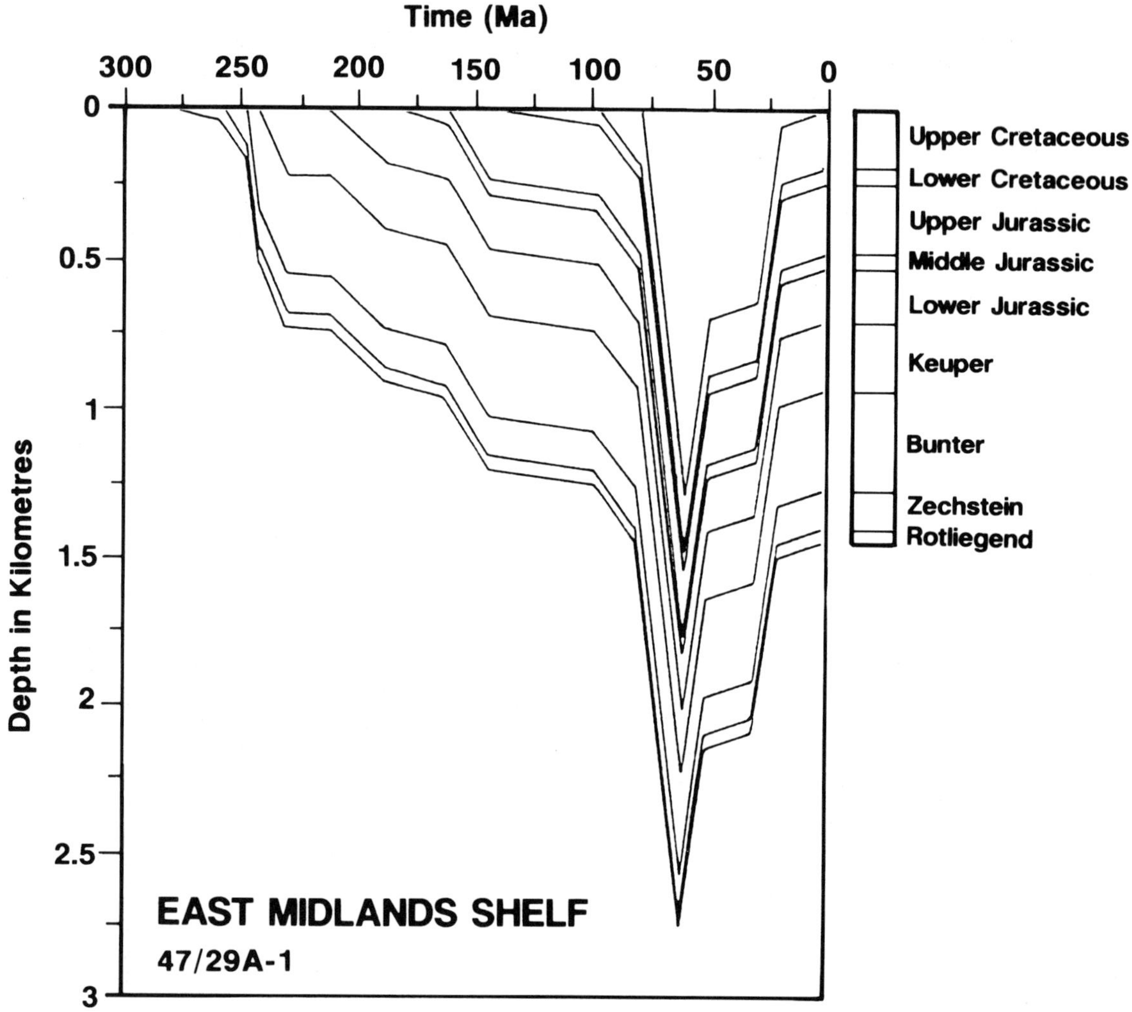

Fig. 3. Reconstructed burial history for post-Carboniferous stratigraphic section in off-shore East Midlands Shelf, based on apatite fission track analyses and vitrinite reflectance data (redrawn from Green *et al.* 1993). See also Roberts (1989) for slightly different interpretation of vertical movements in this area.

Green (1986, 1989), and Bray *et al.* (1992) infer that maximum palaeotemperatures occurred at about 60 Ma and that up to 2 km of section have been removed since that time (Fig. 3). Offshore, in the Irish Sea, Lewis *et al.* (1992*a*) argue that between 2.7 and 3.3 km of overburden was removed. In northern Scotland some fission track results are complicated by local heating effects associated with the Tertiary intrusive centres. Away from these centres, pre-Tertiary fission track ages suggest that a maximum of 1–1.5 km of overburden could have been removed from Scotland (Lewis *et al.* 1992*b*). Inferences based upon fission track analyses must be treated with caution since they rely upon the geothermal gradient remaining relatively constant during the last 60 Ma (Brown 1991). Modelling results are also critically dependent upon important assumptions about the kinetics of fission track growth and annealing (Green *et al.* 1993).

Sedimentary basin geometry

The outcrop pattern of the British Isles attests to exhumation: along the length of Britain, Mesozoic and Tertiary sedimentary rocks offlap to the southeast (Fig. 1). Farther north, eastward-directed offlap is observed in the Moray Firth graben. This general pattern, coupled with our understanding of the evolution of extensional sedimentary basins, can be used to delineate a region within which significant amounts of denudation have occurred (Fig. 1). This region extends from the Unst basin (north east of the Shetland Islands) to the Bristol Channel basin (southwest of Wales) to the west coast of Ireland.

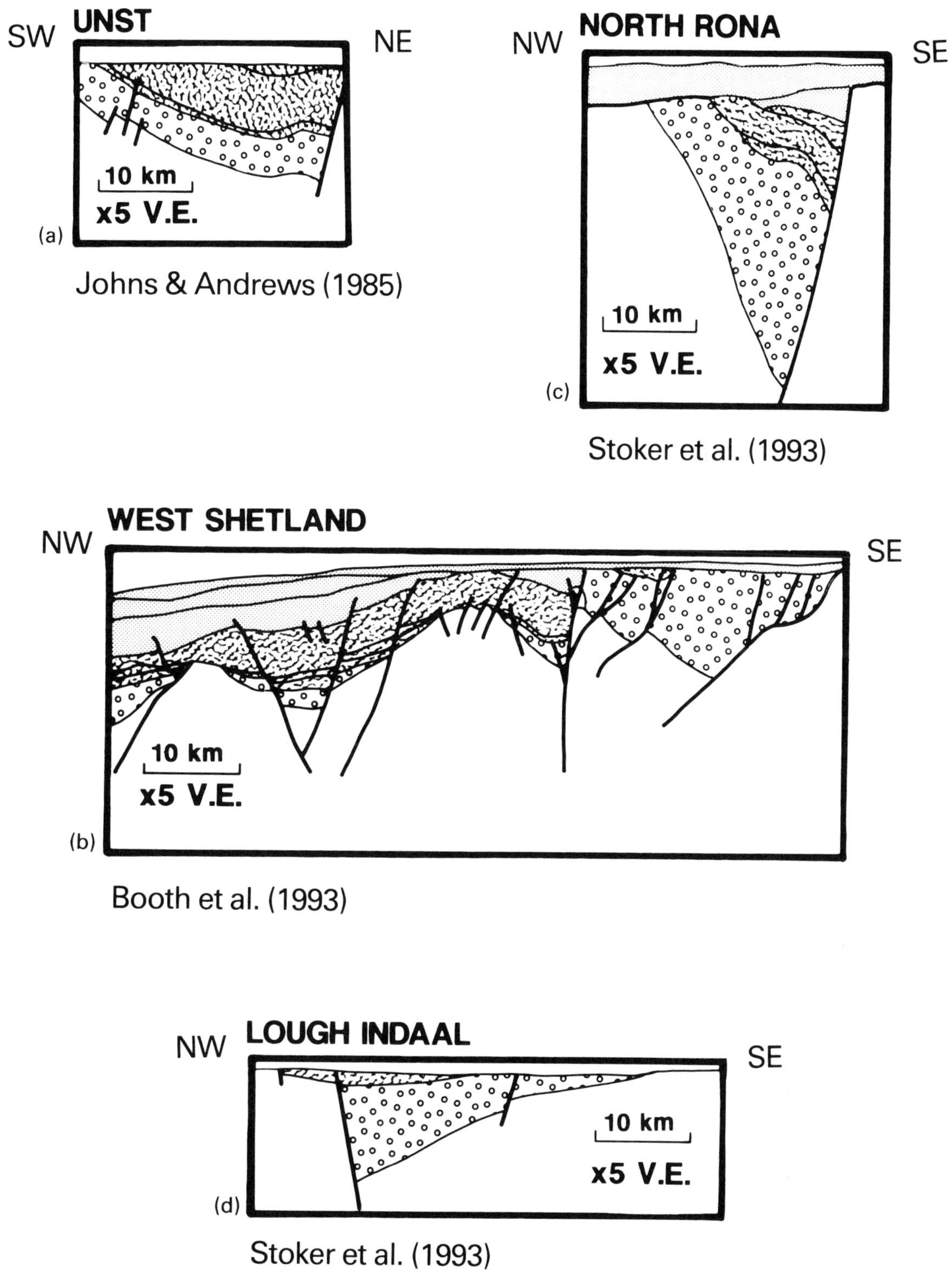

Fig. 4. Depth-converted and interpreted seismic reflection profiles from vicinity of British Isles (see Fig. 1 for locations). Thin solid lines = interpreted strata; thick solid lines = normal faults; scalloped shading = pre-Permian sedimentary rocks and basement; open circle shading = Permian–Triassic sedimentary rocks; irregular shading = Jurassic–Cretaceous sedimentary rocks; fine shading = Tertiary and Quaternary sedimentary rocks. Solid black = Tertiary igneous rocks. Scale, vertical exaggeration (V.E.) and original source indicated in each case. Where appropriate, depth conversion was carried out using a uniform velocity of $3\,km\,s^{-1}$.

continued overleaf

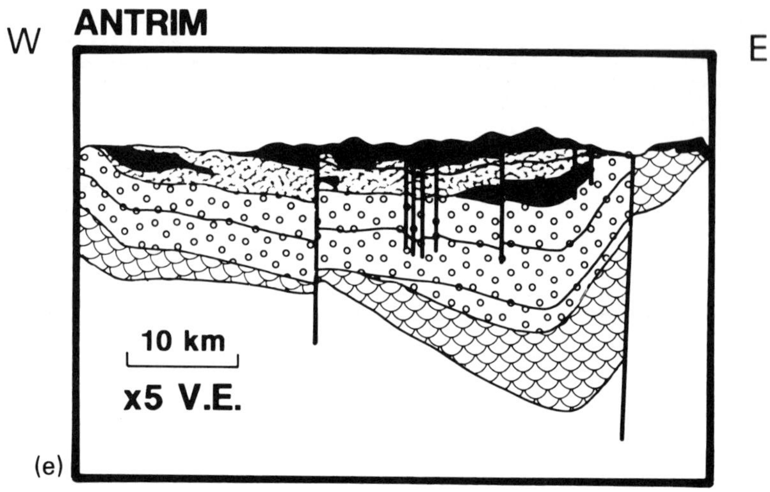

McCann (1988)

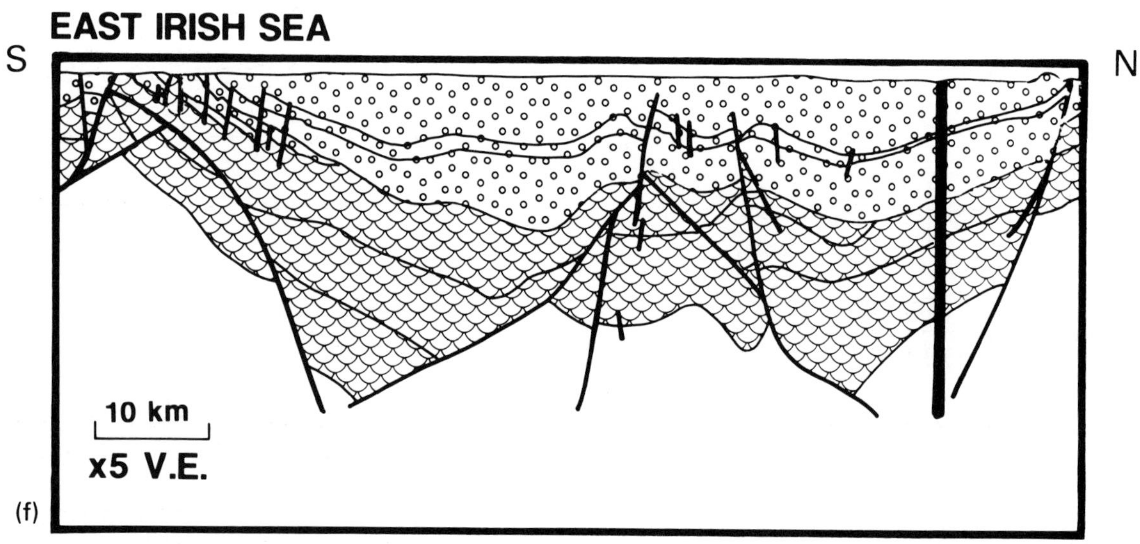

British Geological Survey (1994)

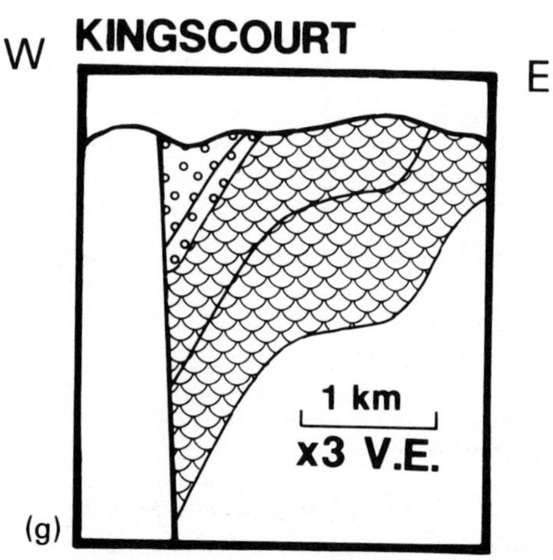

Visscher (1971)

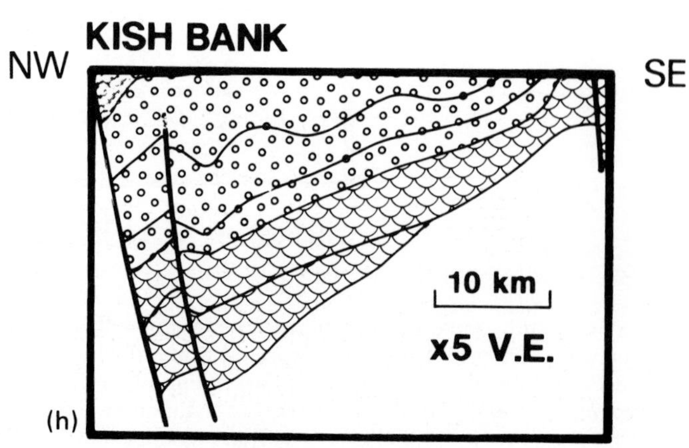

Jenner (1981)

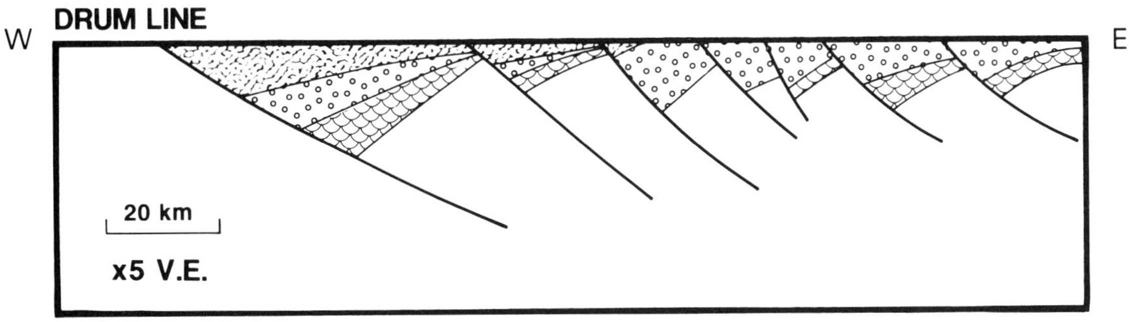

Cheadle et al. (1987)

Fig. 5. Depth-converted interpretation of upper 7 seconds (two-way travel time) of DRUM deep seismic reflection profile. Note absence of thick pile of unfaulted post-rift sedimentary rocks younger than Lower Jurassic. Redrawn from Cheadle *et al.* (1987). See Fig. 1 for location and Fig. 4 for key to shading.

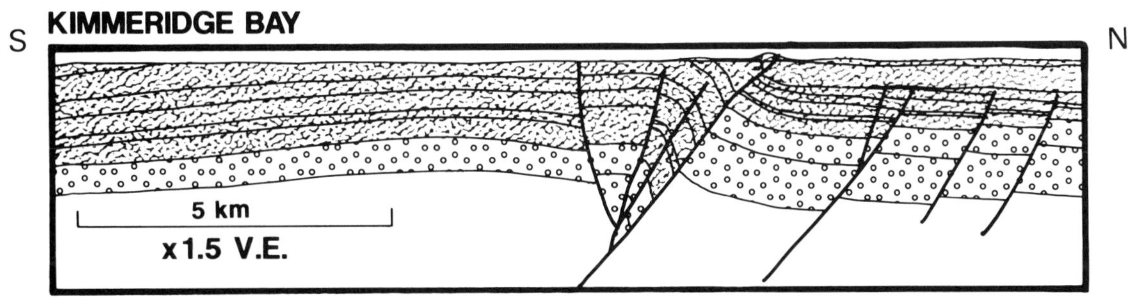

Selley & Stoneley (1987)

Fig. 6. North–south cross-section through Kimmeridge Bay and Wareham oil fields (see Fig. 1 for location and Fig. 4 for key). Redrawn from Selley & Stoneley (1987).

In Fig. 4, a series of simplified cross-sections through selected sedimentary basins located within the exhumed region are shown. In all cases, it is clear that syn-rift sedimentary rocks (i.e. strata which have been rotated and displaced by normal faulting) crop out close to the Earth's surface or beneath a thin cover of late Tertiary and Quaternary sedimentary rocks (e.g. Fig. 4b & c). It is especially clear that mappable syn-rift horizons are truncated close to the surface and that the youngest syn-rift sedimentary rocks occur within the hangingwall, against large normal faults. There is little evidence for regional horizontal shortening. The syn-rift sedimentary rocks are generally 2–5 km thick. Comparison with Fig. 2 suggests that the stretching factor, β, reaches values of about 1.5, which implies that 2 or more kilometres of post-rift sedimentary rocks are missing.

The discrepancy between predicted and observed subsidence is most clearly illustrated by the DRUM seismic reflection profile, which is located 30 km off the north coast of Scotland (Figs 1 & 5). An array of fault-bounded blocks is discernable in the upper crust where individual half graben are characterized by differentially rotated syn-rift sedimentary rocks. As before, unfaulted post-rift sedimentary rocks are absent, despite the fact that estimates of β, determined from the normal faulting geometry and from subsidence analysis, vary between 1.5–2.0 (Enfield & Coward 1987). Whilst acknowledging that the thickness of the Permo–Triassic fill is disputed, we can use these rough estimates of β to infer that 2–4 km of post-rift sedimentary rocks are missing. Cheadle *et al.* (1987) suggested that the absence of post-rift subsidence could be accounted for by stretching crust which was originally 45 km thick by a factor of 1.5–2.0. This scheme is unlikely to account for the absence of post-rift subsidence because it disregards other evidence, cited above, in favour of regional Tertiary exhumation.

Published hypotheses

Despite the fact that most of the observations summarized here are not new, little attention has been paid to the causes of Tertiary epeirogeny. Usually uplift is attributed in a general and rather unspecified way either to flexural and thermal effects associated with the opening of the North Atlantic ocean (Glennie 1990; Lewis *et al.* 1992*a*) or to horizontal shortening associated with the Alpine orogeny (Roberts 1989).

We argue that any explanation must take into account the absence of significant gravity anomalies on wavelengths of greater than 50 km. This important observation means that the entire continental shelf is in approximately local (Airy) isostatic equilibrium. Bearing this important constraint in mind, we will now examine the two main categories of hypotheses used to account for Tertiary epeirogeny.

Flexural and thermal uplift

Admittance studies of the free-air gravity anomalies and load topography in basins like the North Sea (Barton & Wood 1984) together with flexural cantilever modelling of syn-rift topography (Marsden *et al.* 1990) have demonstrated that the elastic thickness of the lithosphere beneath these offshore basins is very small ($\tau_e <$ 5 km). Barton (1992) has shown that the elastic thickness beneath Scotland is equally small. Hence Early Tertiary uplift and denudation of the British Isles cannot be accounted for by lithospheric flexure which, in any case, would generate large amplitude gravity anomalies.

Other models based on transient thermal uplift associated with the opening of the North Atlantic and the initiation of the Iceland plume, would seem to be more plausible (Ziegler 1982). However, the exhumed sedimentary basins around the British Isles have generally subsided very little since the beginning of the Tertiary (Figs 4 & 5). This observation means that thermal uplift would have to have been maintained right through to the present day. The only exceptions occur north of Scotland, where Tertiary subsidence has been important (Fig. 4b & c). The British Isles are now *c.* 1000 km away from the centre of the Iceland plume. Thermal uplift must have decreased by a considerable amount over this distance with a time constant of 40–60 Ma. Once the causes of uplift are removed then sedimentary basins must subside back to the level they would have attained in the absence of an uplift event.

Lithospheric shortening

Permanent, isostatically compensated, uplift can only be generated either by thickening the crust, or by adding material which is less dense than asthenosphere (3.2 Mg m^{-3}), to the lithospheric column. Crustal thickening caused by horizontal shortening associated with Alpine mountain building would initially appear to be a satisfactory way of producing permanent exhumation. There is evidence for slight Tertiary shortening all over the northwest European shelf (Ziegler 1982; Roberts 1989; Evans 1990; Chadwick 1993). The classical 'inversion structures' seen along the southern coast of Britain (e.g. Chadwick 1993) and in the Sole Pit basin of the southern North Sea (Glennie & Boegner 1981) are generally absent further north. An important exception is the Wyville–Thomson ridge in the Faeroe–Shetland basin. This ridge formed by local shortening, which started in the early Eocene and continued until the Miocene (Boldreel & Andersen 1993). However, horizontal shortening reverses the throw across major normal faults, resulting in hangingwall anticlines (Fig. 6). Coupled with erosion, 'structural inversion' produces a complicated Mesozoic outcrop pattern with length scales of tens of kilometres rather than wholesale exhumation with length scales of hundreds of kilometres (Figs 1 & 4). Thus, basins which have undergone even quite modest horizontal shortening should have easily identifiable geometries both in plan view and in cross-section. It is clear that such geometries do not generally occur within the region dominated by Tertiary exhumation (compare Figs 4 & 6).

It is straightforward to quantify the amount of regional shortening that is required to produce a given level of denudation. If a standard column of continental lithosphere is shortened uniformly by a factor, f, then the amount of uplift, U, is given by

$$U = \frac{a\left[(\rho_m - \rho_c)\dfrac{t_c}{a}\left(1 - \dfrac{\alpha T_1}{2}\dfrac{t_c}{a}\right) - \dfrac{\alpha T_1}{2}\rho_m\right]}{\rho_m(1 - \alpha T_1)}(f-1) \quad (1)$$

where the symbols are defined in Table 1. Substitution of the values given in Table 1 yields

$$U = 2.5(f-1) \quad (2)$$

Table 1 *Symbols and values of parameters used in text*

Symbol	Parameter	Value
a	lithospheric thickness	125 km
t_c	crustal thickness	30 km
ρ_w	sea water density	1.0 g cm^{-3}
ρ_c	crust density (at 0° C)	2.8 g cm^{-3}
ρ_m	mantle density (at 0° C)	3.33 g cm^{-3}
ρ_a	asthenosphere density (at 1333° C)	3.2 g cm^{-3}
α	thermal expansion coefficient	3.28×10^{-5} °C^{-1}
T_1	asthenospheric temperature	1333° C

The amplification effect of erosion can result in a maximum amount of denudation, D, which is given by

$$D = \left(\frac{\rho_a}{\rho_a - \rho_s}\right) U \quad (3)$$

where ρ_s is the density of sediment and ρ_a is the density of the asthenosphere. If $\rho_s = 2.4\,\mathrm{Mg\,m^{-3}}$ and $\rho_a = 3.2\,\mathrm{Mg\,m^{-3}}$, horizontal shortening of $f = 1.3$ is required to achieve 3 km of denudation. If the lithospheric mantle does not thicken during shortening then

$$U = t_c\left[1 - \frac{\rho_c}{\rho_m}\left(\frac{2a - \alpha T_1 t_c}{2a(1 - \alpha T_1)}\right)\right](f - 1) \quad (4)$$

Substitution yields

$$U = 4.1(f - 1) \quad (5)$$

In this case, $f = 1.2$ if $D = 3$ km.

These simple calculations show that regional shortening is relatively inefficient at generating substantial amounts of denudation. The level of denudation observed in the vicinity of the British Isles would require a sedimentary basin which was originally 100 km wide to be shortened by 20–30 km. There is no evidence for such large amounts of shortening, either onshore or offshore, on any of the cross-sections shown in Figs 4 & 5. Other simple observations bear out this conclusion. For example, the most significant amounts of Tertiary shortening occur along the south coast of Britain (f c. 1.05; Fig. 6). Despite these signs of horizontal shortening, the level of denudation is modest: most of the Cretaceous sedimentary rocks are still preserved (Glennie 1990). In contrast, the shortened passive continental margin exposed in southeast France has been exhumed to Permo–Triassic and Lower Jurassic stratigraphic levels (i.e. syn-rift). Here, several kilometres of denudation have only been achieved by predictably large amounts of shortening: there is excellent evidence for fault throw reversal, extensive folding, and widespread cleavage development (Lemoine *et al.* 1986).

The only other geologically plausible means of thickening the crust is by igneous underplating (Cox 1980; McKenzie 1984; Brown 1991). We now show that the large volumes of basaltic melt produced at the beginning of the Tertiary must have resulted in rapid uplift and denudation of a considerable portion of the continental shelf.

Tertiary magmatism

The Tertiary igneous rocks of the northwest European continental shelf form part of the North Atlantic Tertiary Igneous Province, which also encompasses the flood basalts of East Greenland and dipping reflector sequences located along the continental margins of northwest Europe and East Greenland (Upton 1988; White & McKenzie 1989). Radiometric dating of the province, combined with biostratigraphic and palaeomagnetic constraints, shows that igneous activity commenced at magnetochron C24R just prior to the initiation of sea-floor spreading in the North Atlantic (Harland *et al.* 1989; Larsen *et al.* 1992). Recent models suggest that rapid generation of the great volumes of igneous material observed in the North Atlantic province resulted from continental rifting in the vicinity of a proto-Iceland plume (e.g. White & McKenzie 1989; Skogseid *et al.* 1992; White 1992; Eldholm & Grue 1994).

The distribution of Tertiary igneous rocks around the British Isles is shown in Fig. 7. Major intrusive centres occur throughout the region, from the west of Ireland to the Erlend centre north of the Shetland Isles. Flood basalts are associated with some of these centres but are more widely distributed in the Rockall Trough, Antrim (Northern Ireland) and in the vicinity of the Faeroe Islands. Minor intrusive bodies and dyke swarms are common both onshore and offshore, as far south as Lundy Island in the Bristol Channel (Thorpe *et al.* 1990), as far east as the western margin of the North Sea basin, and as far west as the Porcupine basin (Tate & Dobson 1988). Further northwest, along the continental margin, the results of wide-angle seismic experiments have indicated the presence of substantial volumes of underplated igneous rock (up to 15 km in thickness) within the lower crust of the Hatton–Rockall margin and the Vøring Plateau (Hinz *et al.* 1987; White *et al.* 1987). In the British Isles, the oldest Tertiary igneous rocks have been dated at c. 63 Ma, with the most intense magmatic activity concentrated at c. 59 Ma (Mussett *et al.* 1988).

Here we are principally concerned with estimating the volume and distribution of melt generated in the vicinity of the British Isles. Recent work by McKenzie & O'Nions (1991) and White *et al.* (1992) suggests that rare earth element (REE) concentrations in MgO-rich igneous rocks ($\geqslant$6 wt% MgO) can be used to constrain the partial melt distribution with depth at the time of melt generation. Estimates of the one-dimensional melt thickness may then be obtained by integration over depth. This inversion technique has been successfully exploited in the ocean basins, where predicted crustal thicknesses are in good agreement with those determined independently by seismic refraction experiments. We shall not describe

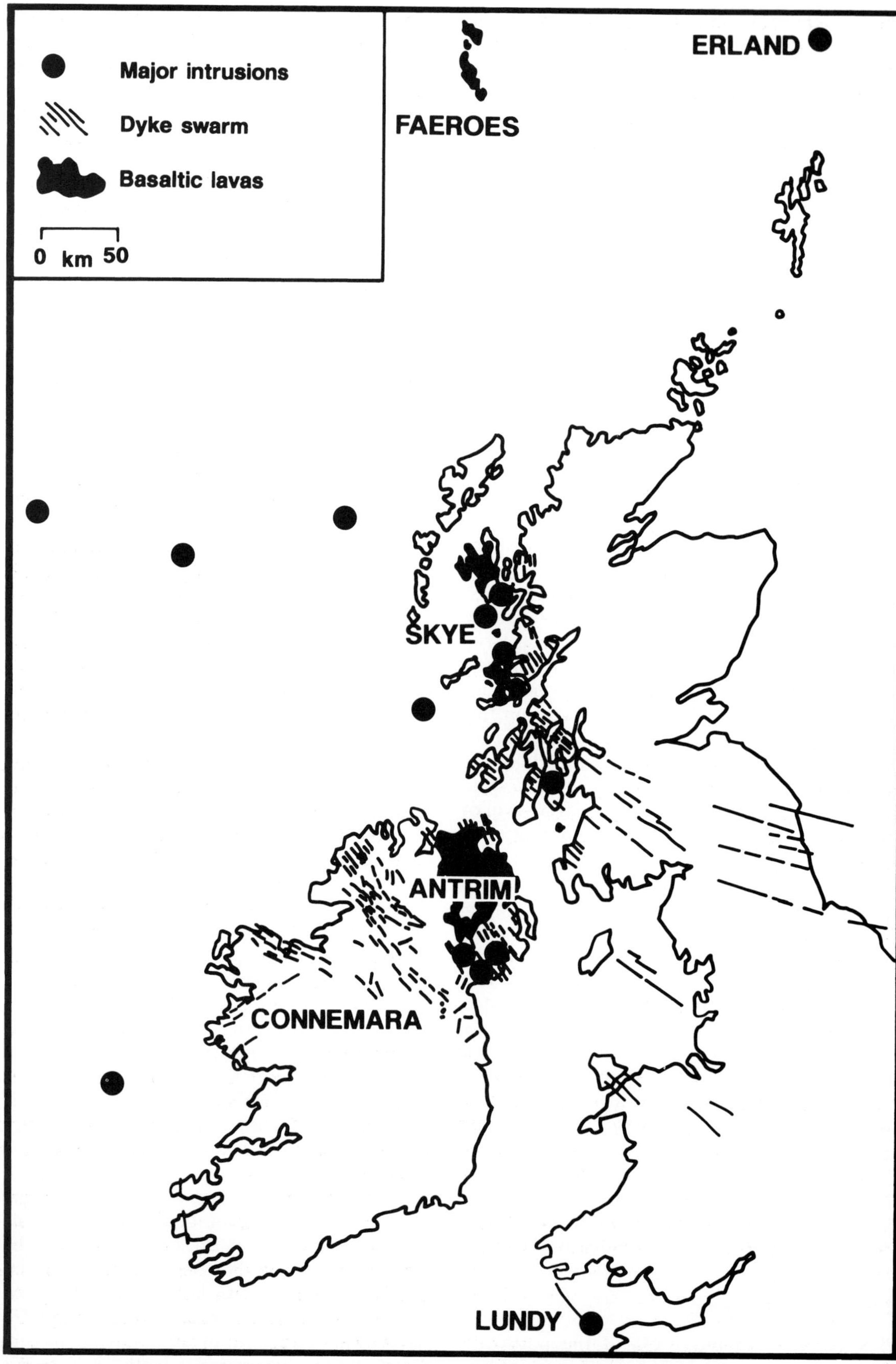

Fig. 7. Early Tertiary magmatism of British Isles, onshore and offshore (redrawn from Mussett *et al.* 1988).

the inverse model here, but refer the interested reader to McKenzie & O'Nions (1991) and to Brodie *et al.* (1994).

There are several important difficulties associated with applying the inversion technique to the British Tertiary Igneous Province. First, there is substantial evidence to suggest that MgO-rich basalts exposed on, for example, the Isle of Skye, have been contaminated with material derived from the continental crust and/or lithospheric mantle (e.g. Dickin 1981; Thompson *et al.* 1982; Thirlwall & Jones 1983; Menzies *et al.* 1987; Thompson & Morrison 1988). The effects of contamination on the REE are difficult to quantify, although selective enrichment in lighter REE is anticipated. This enrichment implies that melt thicknesses calculated using the REE inversion technique are likely to be minimum estimates of the true melt thickness generated prior to contamination (Brodie *et al.* 1994). Kerr (1993) records modest light REE enrichment in crustally contaminated Mull basalts compared to uncontaminated basalts. The degree of enrichment is not sufficient to affect significantly the results of the inversion technique. Similarly, inversion of REE abundances in selected Skye rocks, which are thought to be virtually uncontaminated by lithospheric material (the Beinn Dearg Mhor dykes: Thompson & Morrison 1988), yield partial melt distributions which are similar to those determined for the variably contaminated Skye Main Lava Series. We conclude that the inversion technique is relatively insensitive to the modest amounts of contamination by lithospheric material inferred by previous workers.

The second difficulty is that within a buoyancy-driven flow such as a mantle plume, asthenospheric material is continuously fluxed through the melting region, and in the case of a stationary lithospheric plate, one-dimensional melt thicknesses calculated using the inversion technique are likely to underestimate the true thickness of melt generated (Latin *et al.* 1993). It is not yet obvious how plume fluxes during the Tertiary can be estimated and we have not attempted to correct our predicted melt thicknesses for this effect. These estimates are therefore minima.

Rare earth element abundances have been measured in Tertiary MgO-rich igneous rocks collected from onshore and offshore Ireland and Scotland, and from Lundy Island. We have supplemented our own database with suitable data drawn from the literature. Modelling using the inversion technique suggests that typically 2–5 km of melt were produced after correction for fractional crystallisation of observed melt compositions. Two examples are shown in Fig. 8 for flows sampled from the Skye Main Lava Series and for dykes collected in the west of Ireland. Before correction for fractional crystallisation, the predicted melt thickness typically exceeds *c.* 1.8 km. Calculated melt thicknesses are strongly dependent on composition of the source mantle, which must be specified a priori using the inversion technique. Greater melt thicknesses are predicted from more primitive source compositions (Brodie *et al.* 1994). The cited range of 2–5 km was determined for depleted mantle source (McKenzie & O'Nions 1991), since isotopic constraints suggest that the source of melts in the British province was depleted relative to bulk earth (e.g. Thirlwall & Jones 1983; Kerr 1993). For this reason, and from the arguments outlined in the preceding paragraphs, we suggest that a minimum of *c.* 2–5 km melt was produced beneath a substantial part of the continental shelf encompassing the British Isles.

Although Fig. 7 shows that Tertiary melts are widespread around the British Isles, generally less than 1 km of igneous rock is observed near the surface at any one location (e.g. Upton 1988). The discrepancy between predicted (fractionated) and observed melt thicknesses could arise in several ways. First, melt may have been redistributed by lateral flow prior to solidification. Secondly, erosion could have removed large amounts of igneous material over the past 60 Ma. For example, there is some evidence to suggest that of the order of 1 km of basalt has been eroded from Skye (Anderson & Dunham 1966; Craig 1983). Alternatively, and more probably, much of the predicted melt could have been trapped at depth, as inferred for the Hatton–Rockall margin (White *et al.* 1987). This inference is reasonable since the Moho will act as a density filter for melts which have a density of *c.* 2.8 Mg m^{-3} (Stolper & Walker 1980; Cox 1980).

The addition of material which is less dense than asthenosphere to the lithospheric column must cause uplift, regardless of the precise distribution of added material within that column. Assuming Airy isostasy, the amount of uplift, U, is given by

$$U = (1 - \rho_\chi/\rho_a)\chi \qquad (6)$$

where χ and ρ_χ are the thickness and density of the added material respectively and ρ_a is the density of the asthenosphere (*c.* 3.2 Mg m^{-3}). Thus 5 km of basalt or gabbro with a density of 2.8 Mg m^{-3} causes an initial uplift of 625 m (Fig. 9). As before, the amplification effect of erosion

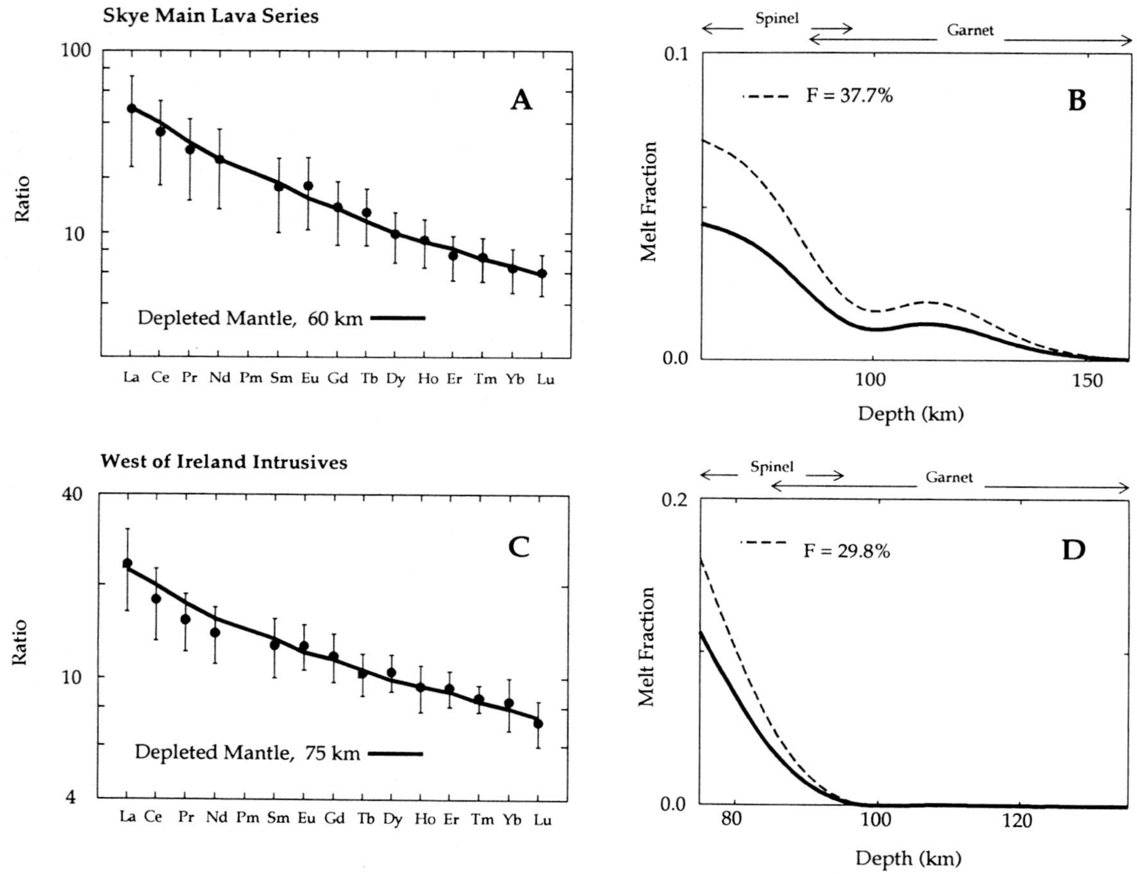

Fig. 8. Inversion of rare earth element concentrations from Isle of Skye, Scotland and from the west of Ireland. (**a**) and (**b**) Skye Main Lava Series. (**A**) Rare earth element concentrations normalised with respect to depleted mantle, circle denotes mean of dataset, bar is 1 standard deviation; thick line shows values calculated from the melt distribution obtained by inversion. (**B**) Corresponding distribution of melt fraction with depth obtained by inversion (thick line); dashed line shows resultant melt distribution when corrected for fractional crystallisation of olivine, based on misfit of modelled Fe and Mg concentrations to observed Fe and Mg concentrations (see McKenzie & O'Nions (1991) for further details). Mantle mineralogy and composition taken from McKenzie & O'Nions (1991): 'spinel' = spinel peridotite; 'garnet' = garnet peridotite. (**C**) and (**D**) Dykes from western Ireland (Brodie, unpublished data). Source composition and depth to top of melting column are indicated on (A) and (C). (B) produce 1.8 km of melt (2.9 km after correction for fraction crystallisation); (D) also produces 1.8 km of melt (2.5 km after correction for fraction crystallization).

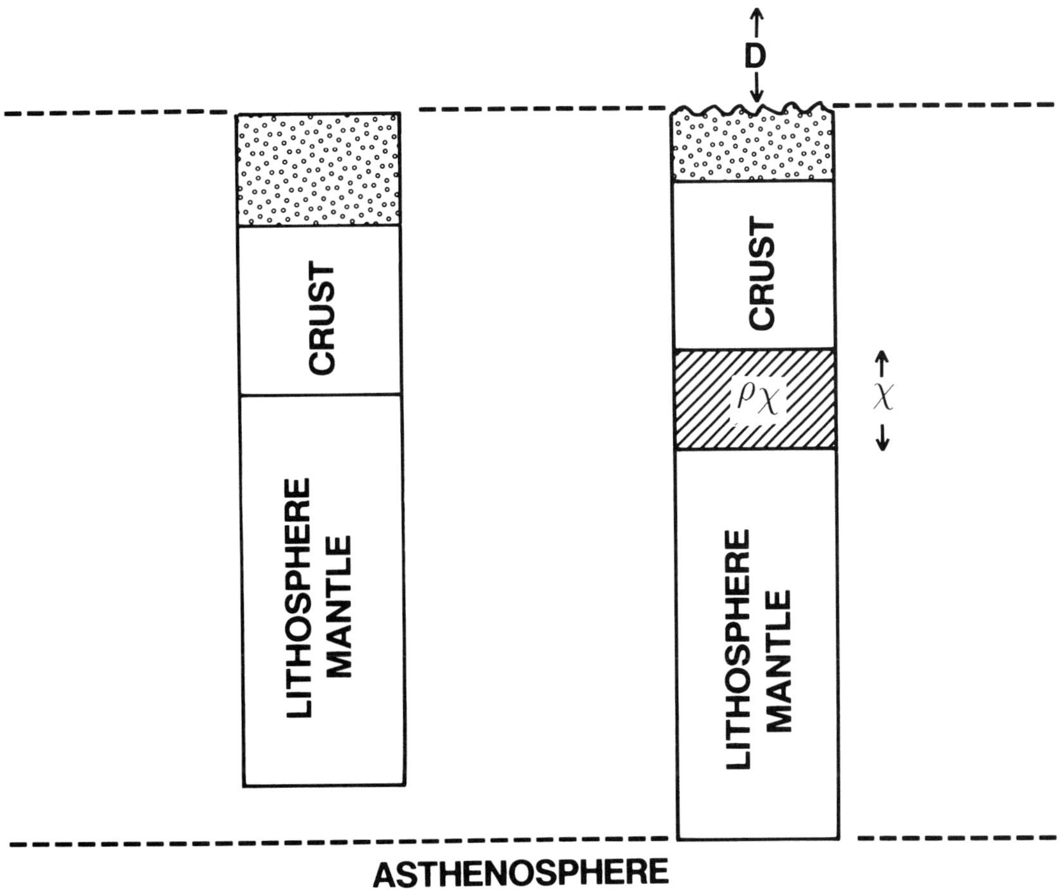

Fig. 9. Cartoons illustrating the effect of adding a thickness of melt, χ, with density ρ_x to the lithosphere. If melt is less dense than asthenosphere (3.2 Mg m^{-3}), surface uplift will result. Erosion can result in a maximum amount of denudation, D, whose value can be determined from Equations (6) and (7). Note that precise distribution of underplating does not affect isostatic balance.

will increase U so that the total amount of material removed, D, is

$$D = \left(\frac{\rho_a - \rho_\chi}{\rho_a - \rho_s} \right) \chi \qquad (7)$$

where ρ_s is the density of sediment. If ρ_s = 2.4 Mg m^{-3} and if U = 625 m, a total of 2.5 km will be removed. Picritic melt, which have densities of *c.* 3.0 Mg m^{-3}, will produce less denudation.

Geophysical evidence for igneous underplating

We have already remarked that about 2 km of post-rift subsidence-related sedimentary rocks are missing from the DRUM deep seismic reflection profile located north of Scotland (Figs 1 & 5). Assuming a crustal velocity of 6 km s^{-1}, the average crustal thickness along the profile excluding syn-rift sedimentary rocks is *c.* 26 km (Cheadle *et al.* 1987). This thickness is in broad agreement with the results of the LISPB wide-angle line, which was positioned north–south along the length of Scotland (Bamford *et al.* 1976; Barton 1992). However, the average stretching factor along DRUM is 1.5–2.0, which implies that the crust should only be 15–20 km thick. If we assume that 5 km of Tertiary magma were generated beneath Skye and trapped within the lower crust then this discrepancy can be explained.

Deep seismic reflection profiles collected around the British Isles (e.g. DRUM profile; McGeary & Warner 1985), indicate that the lower crust is strongly reflective (e.g. Hall *et al.* 1984; Cheadle *et al.* 1987; Freeman *et al.* 1988). It has already been suggested that this reflectivity could be explained by gabbroic intrusions which underplate the lower crust (Warner 1990; Nelson 1991). If so, we suggest that underplating is Early Tertiary in age since this period is the only time since the Late Carboniferous–Early Permian when large quantities of basaltic melt have been generated over such a wide region (e.g. Craig 1983).

Barton's (1992) recent reinterpretation of the LISPB dataset has yielded a detailed cross-section of crustal structure north–south along the length of Scotland. The crust has a relatively uniform thickness (*c.* 30 km) and there is no evidence for crustal roots. Thus any significant topography, such as the Grampian mountains, is likely to be supported by variations in crustal density. Velocities within the upper 20 km of the crust vary from 6.2–6.45 km/s. At 20 km depth, the velocity gradient increases sharply so that the velocity within 5 km of the Moho is 7.1 km/s. Lower crustal igneous rocks are expected to have seismic velocities of *c.* 7.2 km/s (White & McKenzie 1989) and so Barton's (1992) results may be consistent with underplating. However, high velocities are a necessary but not sufficient condition for the existence of igneous underplating. Furthermore, there is no age constraint on timing of formation of lower crustal layering.

The strong correlation between Bouguer gravity anomalies and topography over Scotland show that this region must have an elastic thickness of less than 5 km (Barton 1992). The most significant anomalies are quite clearly associated with granitic and gabbroic bodies within the upper crust. Thus there is little evidence supporting Watson's (1985) argument that Scotland has had a much higher rigidity than the surrounding continental shelf and so was unaffected by Mesozoic extension. Mesozoic fault patterns clearly show that the Scottish massif did not act as a horst block (Fig. 10). With the exception of the northern margin of the Moray Firth, the Scottish coastline is cut at a high angle by numerous faults which bound offshore basins. This observation also applies to the rest of the British Isles landmass.

Discussion & conclusions

Extensional sedimentary basins in the Irish Sea and immediately to the northwest of Scotland formed by lithospheric stretching during the Permo–Triassic and, in some cases, during the Late Jurassic. The general acceptance of the stretching model together with the fact that it is testable has focussed attention on the striking absence of post-rift thermal subsidence in these basins. In many cases, compacted syn-rift sediments occur at the sea-bed, masked only by a thin veneer of Late Tertiary and Quaternary sediment. Subsidence calculations show that 2–4 km of sediment are missing. Knox *et al.* (1981) and Mudge & Bliss (1983) have independently suggested that a period of widespread uplift and erosion must have occurred west of the Moray Firth graben after deposition of Danian chalk and marl since Danian sediments have since been removed from all but the deepest parts of graben. Resedimented chalk deposits, slumping, and debris flows have also been documented in this area (Morton 1982).

Any explanation of these observations must take into account the fact that the entire region is characterized by small gravity anomalies. This fundamental and often overlooked constraint implies that models which appeal to flexural affects arising from the opening of the North Atlantic ocean are incorrect. Furthermore, any thermal effects must decay with time and with increasing distance from the Iceland plume. We argue that the crust has been thickened permanently, either by shortening or by igneous underplating. Basin inversion caused by shortening is observed throughout the region but simple calculations show that the observed shortening cannot account for the magnitude, location, or timing of denudation.

We have shown that it is much easier to explain Tertiary uplift and erosion along a NNE–SSW axis by igneous underplating. Inverse modelling of rare earth element concentrations from MgO-rich igneous rocks collected from the British Isles and continental shelf suggest that at least 2–5 km of melt was generated at the beginning of the Tertiary. Much of this melt has probably been trapped within the lower crust. Simple calculations show that the addition of 5 km of igneous rock can give rise to 2.5 km of exhumation. Since magmatism occurred rapidly and over a wide region, regional epeirogenic uplift would also have been rapid. The consequent tilting about a NNE-SSW trending horizontal axis eventually produced the familiar outcrop pattern of these islands.

Finally, it is often assumed that substantial portions of the British Isles were unaffected by the widespread Permo–Triassic and Upper Jurassic extension documented throughout Western Europe and beyond (Anderton *et al.* 1979; Holland 1981; Ziegler 1982; Glennie 1990). However, the Scottish, Welsh, Cornish, and Irish massifs are completely surrounded by

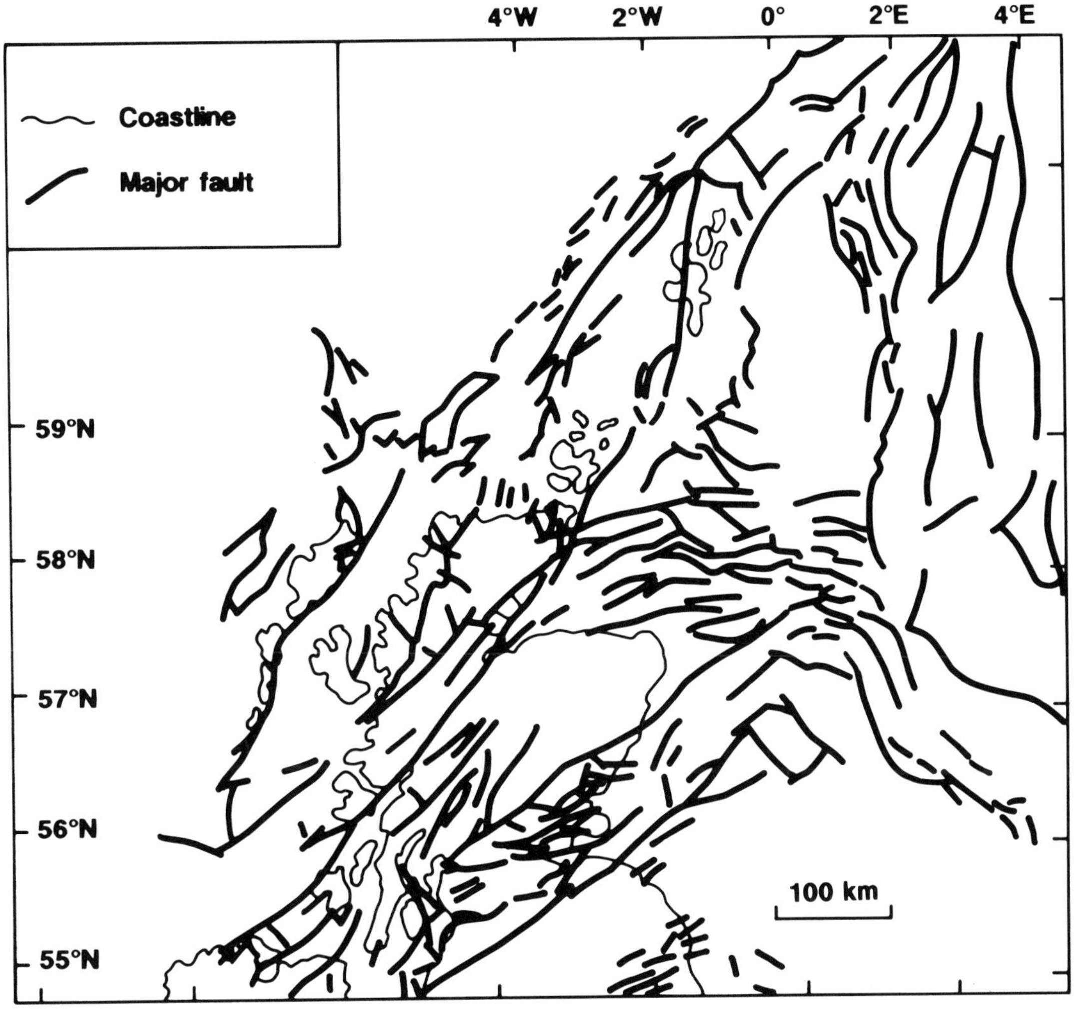

Fig. 10. Map of northern British Isles showing principal post-Palaeozoic normal and strike-slip faulting (onshore and offshore). Redrawn from Cook (1991).

Permo–Triassic and Jurassic outcrop and have dimensions comparable with the thickness ofthe lithosphere. In all cases, significant Mesozoic faults can be traced onshore and coasts are not bounded, except locally, by normal faults (e.g. Fig. 10). More importantly, outliers of Permo–Triassic sedimentary rock (Craig 1983; Holland 1981) together with minor Early Permian volcanic centres (Rock 1983; Penn *et al.* 1983; Mitchell & Mohr 1987) are recorded from the highlands of northern Scotland and from Ireland. We infer from these observations that both landmasses underwent at least modest Permo–Triassic and/or Upper Jurassic stretching. Depending upon location, 1–4 km of Cretaceous post-rift sediments are assumed to have been deposited, followed by extensive denudation in the Tertiary.

JAB gratefully acknowledges receipt of a Shell studentship. We thank Paul Mohr, Paul Lyle and Joanne Wallace for their generous help in the field in Ireland. Major and trace element analyses were carried out on the x-ray fluorescence spectrometer at the Department of Geology and Geophysics, University of Edinburgh. Rare earth element analyses were carried out at the Natural Environment Research Council ICP-MS facility at the Department of Geology, Royal Holloway and Bedford New College, University of London and at CARE, Silwood Park, Ascot. Dan McKenzie kindly provided us with a copy of his rare earth element inversion program. We are grateful to G. Budd, S. Capon, R. England, G. Fitton, B. Hall, H. Iwamori, D. James, K. Jarvis, D. Johnston, A. Kerr, H. Kerr, D. Latin, L. Lonergan, D. Lyness, K. Martyn, R. Newman, G. Sevastopulo, and J. Wills for their help. M. Bamford and R. Sibson provided thorough reviews.

References

ANDERSON, F. W. & DUNHAM, K. C. 1966. The geology of northern Skye. *Memoir of the Geological Survey of Great Britain.*

ANDERTON, R., BRIDGES, P. H., LEEDER, M. R. & SELLWOOD, B. W. 1979. *A dynamic stratigraphy of the British Isles.* George Allen & Unwin, London, 301pp.

ANDREWS, I. J., LONG, D., RICHARDS, P. C., THOMSON, A. R., BROWN, S., CHESHER, J. A. & MCCORMAC, M. 1990. *The geology of the Moray Firth*, British Geological Survey, United Kingdom Offshore Regional Report, HMSO, London.

BAMFORD, D., NUNN, K., PRODEHL, C. & JACOB, B. 1976. LISPB-IV. Crustal structure of Northern Britain. *Geophysical Journal of the Royal Astronomical Society,* **54**, 43–60.

BARTON, P. J. 1992. LISPB revisited: a new look under the Caledonides of northern Britain. *Geophysical Journal International,* **110**, 371–391.

BARTON, P. & WOOD, R. 1984. Tectonic evolution of the North Sea basin: Crustal stretching and subsidence. *Geophysical Journal of the Royal Astronomical Society,* **79**, 987–1022.

BOLDREEL, L. O. & ANDERSEN, M. S. 1993. Late Palaeocene to Miocene compression in the Faeroe-Rocall area. *In*: PARKER, J. R. (ed.) *Petroleum geology of Northwest Europe: Proceedings of the 4th conference.* Geological Society, London, 1025–1034.

BOOTH, J., SWIECICKI, T. & WILCOCKSON, P. 1993. The tectono-stratigraphy of the Solan basin, west of Shetland. *In*: PARKER, J. R. (ed.) *Petroleum geology of Northwest Europe: Proceedings of the 4th conference.* Geological Society, London, 987–998.

BRAY, R. J., GREEN, P. F. & DUDDY, I. R. 1992. Thermal history reconstruction using apatite fission track analysis and vitrinite reflectance: a case study from the UK East Midlands and Southern North Sea. *In*: HARDMAN, R. F. P. (ed.) *Exploration Britain: geological insights for the next decade.* Geological Society, Special Publication, **67**, 3–25.

BRITISH GEOLOGICAL SURVEY 1994. *East Irish Sea (Special Sheet Edition).* 1:250000. Edinburgh, Scotland.

BRODIE, J., LATIN, D. & WHITE, N. 1994. Rare earth element inversion of forward-modelled melt compositions. *Journal of Petrology*, **35**, 1155–1174.

BROOKS, J. & GLENNIE, K. W. (eds) 1987. *Petroleum geology of northwest Europe.* Graham & Trotman, London, 623–632.

BROWN, R. W. 1991. Discussion on thermal and tectonic history of the East Midlands shelf (onshore UK) and surrounding regions assessed by apatite fission track analysis. *Journal of the Geological Society, London,* **148**, 785–787.

CHADWICK, R. A. 1993. Aspects of basin inversion in southern Britain. *Journal of the Geological Society, London,* **150**, 311–322.

CHEADLE, M. J., MCGEARY, S., WARNER, M. R. & MATTHEWS, D. H. 1987. Extensional structures on the western UK continental shelf: a review of evidence from deep seismic profiling. *In*: COWARD, M. P., DEWEY, J. F. & HANCOCK, P. L. (eds) *Continental extensional tectonics.* Geological Society, Special Publication, **28**, 445–465.

COOK, P. J. 1991. *Geology of the United Kingdom, Ireland and the adjacent continental shelf (North Sheet).* British Geological Survey, Natural Environment Research Council.

COX, K. G. 1980. A model for flood basalt volcanism. *Journal of Petrology,* **21**, 629–650.

CRAIG, G. Y. (ed.) 1983. *Geology of Scotland.* Scottish Academic Press, Edinburgh. Second edition.

DICKIN, A. P. 1981. Isotope geochemistry of Tertiary igneous rocks from the Isle of Skye, *Journal of Petrology,* **22**, 155–190.

ELDHOLM, O. & GRUE, K. 1994. North Atlantic volcanic margins: dimensions and production rates. *Journal of Geophysical Research,* **99**, 2955–2968.

ENFIELD, M. A. & COWARD, M. P. 1987. The structure of the West Orkney basin, northern Scotland. *Journal of the Geological Society, London,* **144**, 871–884.

EVANS, C. D. R. 1990. *The geology of the Western English Channel and its western approaches.* British Geological Survey, United Kingdom Offshore Regional Report. HMSO, London.

EVANS, D., CHESHER, J. A., DEEGAN, C. E. & FANNIN, N. G. T. 1982. *The offshore geology of Scotland in relation to the IGS shallow drilling programme, 1970–1978.* Institute of Geological Sciences, Report 81/12. HMSO, London.

FREEMAN, B., KLEMPERER, S. L. & HOBBS, R. W. 1988. The deep structure of northern England and the Iapetus Suture zone from BIRPS deep seismic reflection profiles. *Journal of the Geological Society of London,* **145**, 727–740.

GLENNIE, K. W. (ed.) 1990. *Introduction to the petroleum geology of the North Sea.* Blackwell Scientific Publications, London. Third edition. 402pp.

—— & BOEGNER, P. 1981. Sole Pit inversion tectonics. *In*: ILLING, L. V. & HOBSON, G. D. (eds) *Petroleum geology of the continental shelf of Northwest Europe.* Heyden & Son, London, 110–120.

GREEN, P. F. 1986. On the thermo-tectonic evolution of Northern England: evidence from fission track analysis. *Geological Magazine,* **123**, 493–506.

—— 1989. Thermal and tectonic history of the East Midlands shelf (onshore UK) and surrounding regions assessed by apatite fission track analysis. *Journal of the Geological Society, London,* **146**, 755–773.

——, DUDDY, I. R., BRAY, R. J. & LEWIS, C. L. E. 1993. Elevated palaeotemperatures prior to Early Tertiary cooling throughout the UK region: implications for hydrocarbon generation. *In*: PARKER, J. R. (ed.) *Petroleum geology of Northwest Europe: Proceedings of the 4th conference.* Geological Society, London, 1067–1074.

HALL, J., BREWER, J. A., MATTHEWS, D. H. &

Warner, M. R. 1984. Crustal structure across the Caledonides from the WINCH seismic reflection profile: influences on the evolution of the Midland Valley of Scotland. *Transactions of the Royal Society of Edinburgh: Earth Sciences*, **75**, 97–109.

Harland, W. B., Armstrong, R. C., Craig, L., Smith, A. G. & Smith, D. G. 1989. *A Geologic Time Scale*. Cambridge University Press.

Hillis, R. R., Thomson, K. & Underhill, J. R. 1994. Quantification of Tertiary erosion in the Inner Moray Firth by sonic velocity data from the Chalk and Kimmeridge Clay. *Marine and Petroleum Geology*, **11**, 283–293.

Hinz, K., Mutter, J. C., Zehnder, C. M. & NGT Study Group. 1987. Symmetric conjugation of continent-ocean boundary structures along the Norwegian and East Greenland margins. *Marine and Petroleum Geology*, **3**, 166–187.

Holland, C. H. (ed.) 1981. *A geology of Ireland*. Scottish Academic Press, Edinburgh. 335pp.

Jarvis, G. T. & McKenzie, D. P. 1980. Sedimentary basin formation with finite extension rates. *Earth & Planetary Science Letters*, **48**, 42–52.

Jenner, J. K. 1981. The structure and stratigraphy of the Kish Bank basin. *In*: Illing, L. V. & Hobson, G. D. (eds) *Petroleum geology of Northwest Europe*. Institute of Petroleum, London. 426–431.

Johns, C. R. & Andrews, I. J. 1985. The petroleum geology of the Unst basin, North Sea. *Marine and petroleum geology*, **2**, 361–372.

Kerr, A. C. 1993. The geochemistry and petrogenesis of the Mull and Morvern Tertiary lava succession, Argyll, Scotland. PhD thesis, University of Durham.

Kirton, S. R. & Hitchen, K. 1987. Timing and style of crustal extension N of the Scottish Mainland. *In*: Coward, M. P., Dewey, J. F. & Hancock, P. L. (eds) *Continental extensional tectonics*. Geological Society, Special Publication, **28**, 501–510.

Knox, R. W. O' B., Morton, A. C. & Harland, R. 1981. Stratigraphic relationships of Palaeocene sands in the UK sector of the central North Sea. *In*: Illing, L. V. & Hobson, G. D. (eds) *Petroleum geology of the continental shelf of Northwest Europe*. Heyden & Son, London, 521pp.

Larsen, L. M., Pedersen, A. K., Pedersen, G. K. & Piasecki, S. 1992. Timing and duration of Early Tertiary volcanism in the North Atlantic: new evidence from West Greenland. *In*: Storey, B. C., Alabaster, T. & Pankhurst, R. J. (eds) Magmatism and the Causes of Continental Break-up, Geological Society Special Publication, London, **68**, 321–333.

Latin, D., Norry, M. J. & Tarzey, J. R. E. 1993. Magmatism in the Gregory Rift, East Africa: evidence for melt generation by a plume. *Journal of Petrology*, **34**, 1007–1027.

Lemoine, M., Bas, T., Arnaud-Varneau, A., Arnaud, H., Dumont, T., Gidon, M., Bourbon, M., De Graciansky, P.-C., Rudkiewicz, J.-L., Megard-Galli, J. & Tricart, P. 1986. The continental margin of the Mesozoic Tethys in the western Alps. *Marine and Petroleum Geology*, **3**, 179–199.

Lewis, C. L. E., Green, P. F., Carter, A. & Hurford, A. J. 1992a. Elevated K/T palaeotemperatures throughout Northwest England: three kilometres of Tertiary erosion? *Earth and Planetary Science Letters*, **112**, 131–145.

——, Carter, A. & Hurford, A. J. 1992b. Low-temperature effects of the Skye Tertiary intrusions on Mesozoic sediments in the Sea of Hebrides basin. *In*: Parnell, J. (ed.) *Basins on the North Atlantic Seaboard: petroleum geology, sedimentology, and basin evolution*. Geological Society, Special Publication, **62**, 175–188.

McCann, N. 1988. An assessment of the sub-surface geology between Magilligan Point and Fair Head, northern Ireland. *Irish Journal of Earth Sciences*, **9**, 71–78.

McGeary, S. & Warner, M. R. 1985. Seismic profiling of the continental lithosphere. *Nature*, **317**, 795–797.

McKenzie, D. 1978. Some remarks on the development of sedimentary basins. *Earth & Planetary Science Letters*, **40**, 25–32.

—— 1984. A possible mechanism for epeirogenic uplift. *Nature*, **307**, 616–618.

—— & O'Nions, R. K. 1991. Partial melt distributions from inversion of rare earth element concentrations. *Journal of Petrology*, **32**, 1021–1091.

Marsden, G., Yielding, G., Roberts, A. M. & Kusznir, N. J. 1990. Application of a flexural cantilever simple-shear/pure-shear model of continental lithosphere extension to the formation of the northern North Sea basin. *In*: Blundell, D. J. & Gibbs, A. D. (eds) *Tectonic evolution of the North Sea rifts*. Oxford Science Publications, 240–261.

Menzies, M. A., Halliday, A. N., Palacz, Z., Hunter, R. H., Upton, B. G. J., Aspen, P. & Hawkesworth, C. J. 1987. Evidence from mantle xenoliths for an enriched lithospheric keel under the Outer Hebrides. *Nature*, **325**, 44–47.

Mitchell, J. G. & Mohr, P. 1987. Carboniferous dikes of West Connacht, Ireland. *Transactions of the Royal Society of Edinburgh: Earth Sciences*, **78**, 133–151.

Morton, A. C. 1982. Lower Tertiary sand development in the Viking Graben, North Sea. *American Association of Petroleum Geologists Bulletin*, **66**, 1542–1559.

Mudge, D. C. & Bliss, G. M. 1983. Stratigraphy and sedimentation of the Palaeocene sands in the Northern North Sea. *In*: Brooks, J. (ed.) *Petroleum geochemistry and exploration of Europe*. Geological Society, Special Publication, **12**, 95–111.

Mussett, A. E., Dagley, P. & Skelhorn, R. R. 1988. Time and duration of activity in the British Tertiary Igneous Province. *Geological Society, Special Publication*, **39**, 337–348.

Naylor, D. 1992. The post-Variscan history of Ireland. *In*: Parnell, J. (ed.) *Basins on the North Atlantic Seaboard: petroleum geology, sedimentology, and basin evolution*. Geological Society, Special Publication, **62**, 255–275.

NELSON, K. D. 1991. A unified view of craton evolution motivated by recent deep seismic reflection and refraction results. *Geophysical Journal International*, **105**, 25–35.

PENN, I. E., HOLLIDAY, D. W., KIRBY, G. A., KUBALA, M., SOBEY, R. A., MITCHELL, W. I., HARRISON, R. K. & BECKINSALE, R. D. 1983. The Larne No. 2 borehole: discovery of a new Permian volcanic centre. *Scottish Journal of Geology*, **19**, 333–346.

ROCK, N. M. S. 1983. *The Permo–Carboniferous camptonite–monchiquite dyke-suite of the Scottish Highlands and Islands: distribution, field and petrological aspects*. Institute of geological sciences, Report 82/14, HMSO, London.

ROBERTS, D. G. 1989. Basin inversion in and around the British Isles. *In*: COOPER, M. A. & WILLIAMS, G. D. (eds) *Inversion tectonics*. Geological Society, Special Publication, **44**, 131–150.

SELLEY, R. C. & STONELEY, R. 1987. Petroleum habitat in south Dorset. *In*: BROOKS, J. & GLENNIE, K. W. (eds) *Petroleum geology of northwest Europe*. Graham & Trotman, London, 623–632.

SKOGSEID, J., PEDERSEN, T., ELDHOLM, O. & LARSEN, B. T. 1992. *Tectonism and magmatism during NE Atlantic continental break-up: the Vøring Margin*. Geological Society, Special Publication, **68**, 305–320.

STOKER, M. S. HITCHEN, K. & GRAHAM, C. C. 1993. *United Kingdom offshore regional report: The geology of the Hebrides and West Shetland shelves and adjacent deep-water areas*. London: HMSO for the British Geological Survey.

STOLPER, E. & WALKER, D. 1980. Melt density and the average composition of basalt. *Contributions to Mineralogy and Petrology*, **74**, 7–12.

TATE, M. P. & DOBSON, M. R. 1988. Syn- and post-rift igneous activity in the Porcupine Seabight Basin and adjacent continental margin W of Ireland. *In*: MORTON, A. C. & PARSON, L. M. (eds) *Early Tertiary volcanism and the opening of the N. E. Atlantic*. Geological Society, Special Publication, **39**, 309–334.

——, White, N. & CONROY, J.-J. 1993. Lithospheric extension and magmatism in the Porcupine basin west of Ireland. *Journal of Geophysical Research*, **98**, 905–13,923.

THIRLWALL, M. F. & JONES, N. W. 1983. Isotope geochemistry and contamination mechanisms of Tertiary lavas from Skye, NW Scotland. *In*: HAWKESWORTH, C. J. & NORRY, M. J. (eds) *Continental basalts and mantle nodules*. Shiva. 186–208.

THOMPSON, R. N. & MORRISON, M. A. 1988. Asthenospheric and lower lithospheric mantle contributions to continental extensional magmatism: an example from the British Tertiary Province. *Chemical Geology*, **68**, 1–15.

——, DICKIN, A. P., GIBSON, I. L. & MORRISON, M. A. 1982. Elemental fingerprints of isotopic contamination of Hebridean Palaeocene mantle-derived magmas by Archaean Sial. *Contributions to Mineralogy and Petrology*, **79**, 159–168.

THORPE, R. S., TINDLE, A. G. & GLEDHILL, A. 1990. The petrology and origin of the Tertiary Lundy Granite. *Journal of Petrology*, **31**, 1379–1406.

UPTON, B. G. J. 1988. History of Tertiary igneous activity on the North Atlantic borders. *Geological Society, Special Publication*, **39**, 429–453.

VISSCHER, H. 1971. The Permian ans Triassic of the Kingscourt outlier, Ireland. *Geological Survey of Ireland Special Paper* **1**.

WARNER, M. 1990. Basalts, water, or shear zones in the lower crust? *Tectonophysics*, **173**, 163–174.

WATSON, J. 1985. Northern Scotland as an Atlantic-North Sea divide. *Journal of the Geological Society, London*, **142**, 221–243.

WHITE, R. S. 1992. Crustal structure and magmatism of North Atlantic continental margins. *Journal of the Geological Society, London*, **149**, 841–854.

—— & MCKENZIE, D. 1989. Magmatism at rift zones: The generation of volcanic continental margins and flood basalts. *Journal of Geophysical Research*, **94**, 7685–7729.

——, SPENCE, G. D., FOWLER, S. R., MCKENZIE, D. P., WESTBROOK, G. K. & BOWEN, A. N. 1987. Magmatism at rifted continental margins. *Nature*, **330**, 439–444.

——, MCKENZIE, D. & O'NIONS, R. K. 1992. Oceanic crustal thickness from seismic measurements and rare earth element inversions. *Journal of Geophysical Research*, **97**, 19,683–19,715.

WOODLAND, A. W. 1979. *Sub-Pleistocene geology of the British Isles and the adjacent continental shelf*. Institute of Geological Sciences, Second edition.

ZIEGLER, P. A. 1982. *Geological atlas of Western and Central Europe*. Shell Internationale Maatschappij B. V., The Hague.

Transient regional uplift in the Early Tertiary of the northern North Sea and the development of the Iceland Plume

P. A. NADIN, N. J. KUSZNIR & J. TOTH

Department of Earth Sciences, University of Liverpool, PO Box 147, Liverpool L69 3BX, UK

Abstract: Forward and reverse 2D flexural modelling of syn-rift and post-rift stratigraphy, has been used to investigate departures from McKenzie post-rift subsidence for the Cretaceous-Tertiary following Late Jurassic rifting in the northern North Sea Basin. The analysis shows uplift in the Palaeocene of 375–525 m relative to calibrated McKenzie subsidence curves, followed by a rapid decrease in uplift in the Early Eocene of 160–310 m. Both Palaeocene uplift and Eocene subsidence increase northwards. The timing, distribution and magnitude of regional Palaeocene uplift is consistent with an origin relating to the development of a mantle plume between 65 and 55 Ma. Transient regional Palaeocene uplift in the northern North Sea Basin was produced by long wavelength dynamic uplift generated by the flow field of a mantle plume. Heating of the asthenosphere or lithosphere beneath the North Sea by hot plume material, or magmatic underplating are not believed to have contributed to Palaeocene uplift in the northern North Sea Basin. Rapid Early Eocene subsidence was generated by a decrease in dynamic uplift caused by a reduction in plume activity. The plume responsible for North Sea Palaeocene uplift may have been the early Iceland plume or a distinct British Tertiary Igneous Province plume related to the same mantle convection event.

Keywords: North Sea, Palaeocene, uplift, mantle plumes.

The subsidence histories of the Viking, Central and Witch Ground Grabens of the Mesozoic North Sea rift system (Fig. 1) were generated by two major episodes of rifting in the Triassic (*c.* 250 Ma BP) and Jurassic (*c.* 170–150 Ma BP), each followed by a period of post-rift thermal subsidence. While the total basement subsidence in these basins can be accounted for by Triassic and Jurassic rifting (Giltner 1987; Marsden *et al.* 1990), previous workers using backstripping to analyse basement subsidence history have drawn attention to Early Tertiary departures from post-rift subsidence following Jurassic rifting predicted by the rift basin model of McKenzie (1978). This Early Tertiary subsidence anomaly in the northern North Sea coincides in time with the development of the Icelandic plume and may be related to regional uplift generated by plume dynamics.

Consensus does not exist, however, as to what form this subsidence anomaly takes. Bertram & Milton (1989), Milton *et al.* (1990), and Barr (1991) propose a Palaeocene regional uplift event superimposed on McKenzie post-rift subsidence following Late Jurassic rifting, with an accelerated phase of regional subsidence in the Eocene giving a return to normal post-rift subsidence (Fig. 2). This conclusion is also supported by arguments based on sequence stratigraphy (Jones & Milton 1994). In contrast Joy (1992), White & Latin (1993) and Hall & White (1994) propose a rapid acceleration of subsidence rate in the Early Tertiary relative to that of Cretaceous post-rift subsidence.

Recognition of anomalous Palaeocene and Eocene subsidence patterns are not only confined to the North Sea, but occur across many basins along the NE Atlantic margin (e.g. Joppen & White 1990) and on the western side of the North Atlantic (Keen *et al.* 1990). Apatite fission track (AFT) analysis and vitrinite reflectance (VR) studies across NW and central Britain have also shown the existence of a regional post-Cretaceous uplift event (e.g. Green *et al.* 1993). These AFT and VR studies suggest that Tertiary uplift increased to the W and NW. The resulting broad regional uptilting of onshore Britain to the W and NW in the Tertiary, is consistent with the present-day dip of Mesozoic and Cenozoic stratigraphy down to the SE (Brodie & White 1994).

Analysis of Early Tertiary post-rift subsidence in the northern North Sea Basin

The majority of previous subsidence studies in the northern North Sea Basin performed 1D backstripping on stratigraphy from wells assuming Airy isostasy (e.g. White & Latin 1993; Hall & White 1994). The Airy backstripping technique as it is usually implemented (e.g. Sclater & Christie 1980) requires accurate estimates of palaeobathymetry at each point in time in order to reconstruct basement subsidence history. The use of inaccurate palaeobathymetries leads to complex and misleading basement subsidence histories. Reliable estimates of palaeobathymetry are often not available, as in the case of the Cretaceous of the northern North Sea (see Bertram & Milton 1989 for discussion). An alternative approach (as used by Milton *et al.* 1990, Barr 1991, and Young 1992), is to use a limited number of accurate palaeobathymetry estimates to calibrate the McKenzie (1978) post-rift thermal subsidence model through determination of the β-stretching factor. Another source of error in Airy backstripping is the assumption that sediment loading is compensated using Airy (local) isostasy. This can lead to substantial errors, especially if the thickness of post-rift sediments vary laterally, at length scales of 10–25 km as they do between the fault-block crests and graben centres in the northern North Sea Basin (Roberts *et al.* 1993).

From Holdsworth, R.E. & Turner, J.P. (compiler) *Extensional Tectonics: Regional-scale Processes*. The Geological Society, Key Issues in Earth Sciences, **2 (1)**, 327–332.
1476-315X/02/$15.00 © The Geological Society of London 2002
First published in: "*Journal of the Geological Society*, **152**, 953–958".

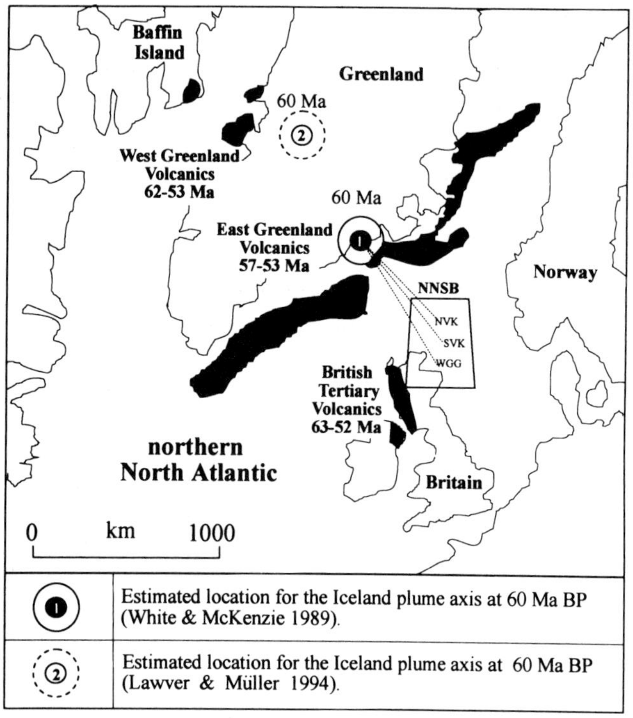

Fig. 1. Map reconstruction of the northern North Atlantic at 55 Ma (adapted from White & McKenzie 1989) showing the location of regional stratigraphic profiles in the Witch Ground Graben (WGG), South Viking Graben (SVK) and North Viking Graben (NVK) of the northern North Sea Basin modelled by Nadin & Kusznir (1995*a*, *b*), together with the distribution of Tertiary igneous activity and estimated locations for the Iceland plume during the Early–Mid-Palaeocene.

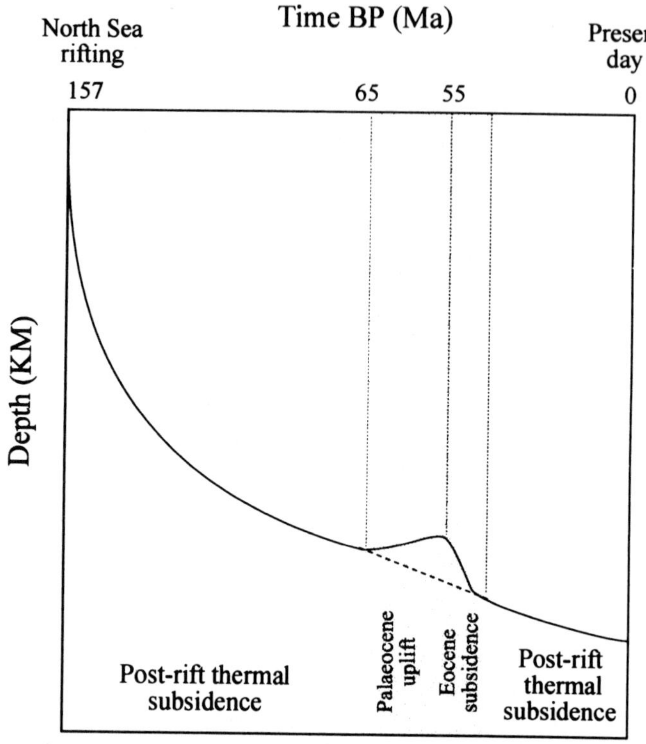

Fig. 2. Schematic Cretaceous and Tertiary post-rift subsidence curve for the northern North Sea Basin, in which transient Palaeocene uplift is superimposed on McKenzie (1978) post-rift thermal subsidence (adapted from Bertram & Milton 1989).

In order to overcome the limitations of Airy backstripping discussed above, Nadin & Kusznir (1995, 1996) carried out subsidence analysis using only accurate palaeobathymetry indicators and flexural isostasy on regional seismic cross-sections in the Witch Ground and Viking Grabens of the northern North Sea. Their objective was to determine departures in Cretaceous and Tertiary subsidence history from calibrated McKenzie post-rift thermal subsidence. Both forward and reverse modelling techniques were used. A detailed description of modelling methodologies, sensitivities and the geological data to which the modelling was applied is given in Nadin & Kusznir (1995, 1996).

A full description of the reverse post-rift modelling methodology, consisting of flexural backstripping, decompaction and reverse thermal subsidence modelling, can be found in Kusznir *et al.* (1995). The β-stretching factors defining the thermal subsidence component of the reverse post-rift models were calibrated by Nadin & Kusznir using the zero palaeobathymetries indicated by the erosion surfaces of Late Jurassic fault-blocks. Since the objective of their study was to determine anomalous Tertiary subsidence, Tertiary palaeobathymetry was not used to calibrate β. Reverse post-rift modelling shows that a regional Palaeocene uplift event must be superimposed on Cretaceous–Tertiary McKenzie post-rift subsidence following Late Jurassic rifting in order to satisfy the observed zero bathymetry indicated by Late Palaeocene coals (Andrews *et al.* 1990). The regional Palaeocene uplift predicted by reverse modelling for the Witch Ground Graben profile (Fig. 3a) relative to calibrated McKenzie post-rift subsidence and the Haq *et al.* (1987) long-term sea-level curve, together with a typical point subsidence history, are shown in Fig. 3b & c.

Nadin & Kusznir (1995) also investigated northern North Sea subsidence histories by independent 2D forward syn-rift and post-rift modelling using the flexural cantilever model of continental rift basin formation (Kusznir & Ziegler 1992). Forward syn-rift modelling provides an independent estimate of the β, derived entirely from information held within the syn-rift and pre-rift stratigraphy. To model the observed post-rift stratigraphic thicknesses and Late Palaeocene zero bathymetry, the forward model required regional Palaeocene uplift followed by rapid decrease in uplift in the Early Eocene. The transient regional Palaeocene uplift event and subsidence for a point in the Witch Ground Graben predicted by forward modelling are shown in Fig. 3d & e.

The subsidence histories for the Witch Ground Graben (Fig. 3c & e) derived from forward and reverse modelling show substantial Cretaceous palaeobathymetries (*c.* 500–1200 m) followed by rapid shallowing during Palaeocene uplift. Palaeocene regional uplift and its reduction in the Eocene are superimposed on exponential McKenzie (1978)-type post-rift thermal subsidence, and gives a marked 'kick' in the basement subsidence curve, similar to that reported by other workers (e.g. Bertram & Milton 1989; Milton *et al.* 1990; Barr 1991). The substantial palaeobathymetries predicted by forward and reverse modelling for the

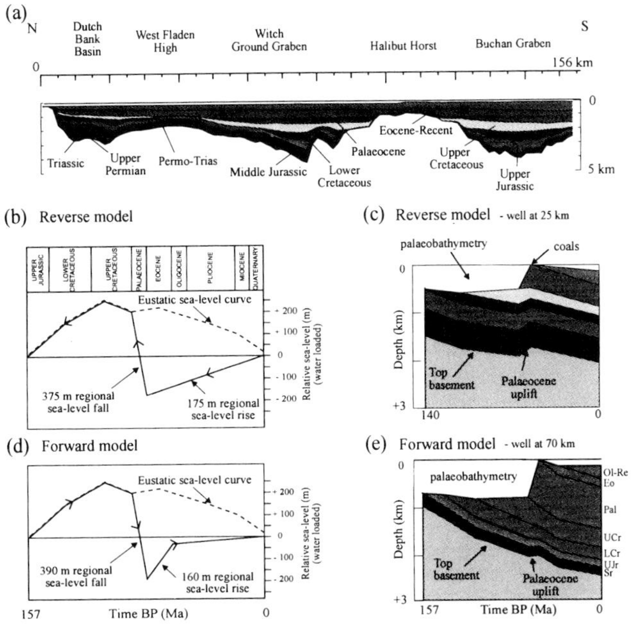

Fig. 3. (**a**) Present-day depth section for the Witch Ground Graben (Andrews *et al.* 1990). (**b**) Departures from calibrated McKenzie post-rift subsidence relative to the Haq *et al.* (1987) long-term sea-level curve predicted by reverse modelling for the Witch Ground Graben. (**c**) Typical subsidence history for a point in the Witch Ground Graben predicted by reverse post-rift modelling. (**d**) Departures from calibrated McKenzie post-rift subsidence relative to the Haq long-term sea-level curve predicted by forward modelling. (**e**) Typical subsidence history predicted by forward syn-rift and post-rift modelling. Global eustasy (dashed line) and transient regional uplift (solid line) in the Palaeocene and Eocene. Sr, Upper Jurassic syn-rift.

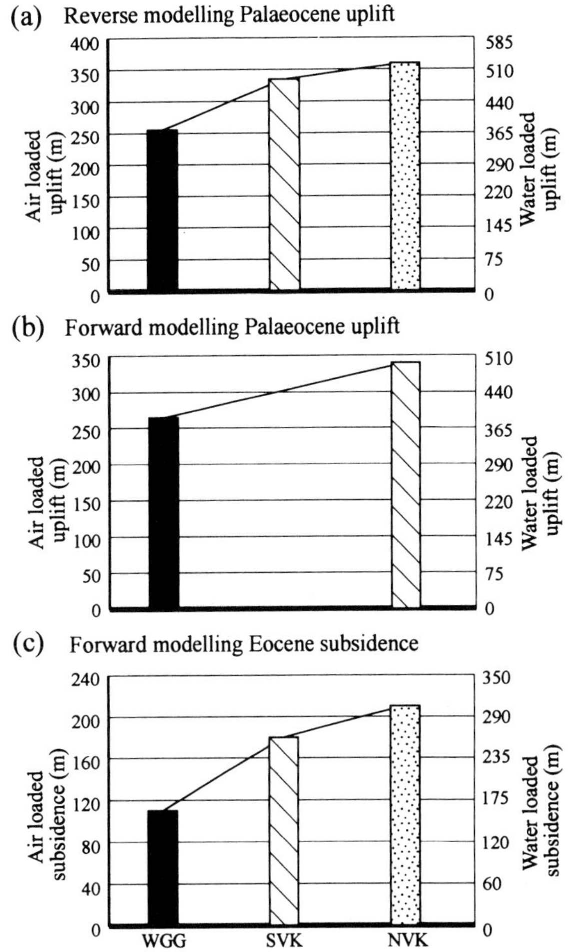

Fig. 4. Estimates of regional Palaeocene uplift and Eocene subsidence for the northern North Sea Basin from: (**a**) reverse post-rift modelling and (**b** & **c**) forward syn-rift and post-rift modelling (WGG, Witch Ground Graben; SVK, South Viking Graben; NVK, North Viking Graben).

basin axis in the Cretaceous are consistent with the conclusions of Hancock (1984) based on palaeontological evidence and Bertram & Milton (1989) and Barr (1991) based on stratigraphic evidence. These large water depths conflict with the assumption, made in much Airy backstripping analysis of northern North Sea subsidence history, that Cretaceous palaeobathymetries were zero or small (e.g. Joy 1992; White & Latin 1993; Hall & White 1994). The Cretaceous and Tertiary subsidence history of the northern North Sea Basin predicted by Nadin & Kusznir (1995, 1996) differs from that predicted by Joy (1992), White & Latin (1993) and Hall & White (1994). This difference is attributed primarily to the use of inaccurate palaeobathymetry estimates for the Cretaceous within Airy backstripping analysis carried out by these latter workers.

Forward and reverse 2D modelling of Cretaceous and Tertiary–Recent post-rift subsidence and stratigraphy in the Witch Ground and Viking Grabens of the northern North Sea Basin by Nadin & Kusznir predict a departure from the McKenzie post-rift subsidence in the Early Tertiary consisting of regional Palaeocene uplift of 375–525 m, followed by a rapid Early Eocene decrease in uplift of 160–310 m. Forward and reverse modelling used independent methods yet gave similar estimates. Results are summarized in Fig. 4. Both Palaeocene uplift and Eocene subsidence increase in magnitude northwards. Forward and reverse modelling show that eustasy, sediment supply and post-rift thermal subsidence alone can not fully account for observed stratigraphic thicknesses in the northern North Sea Basin.

Errors in the determination of Palaeocene uplift from modelling are estimated to be ±60 m, ignoring overpressuring and uncertainty in the long-term global sea-level curve (Nadin & Kusznir 1995, 1996). If long-term sea level was kept constant then the magnitude of the Palaeocene regional uplift was reduced to 175 m for the Witch Ground Graben profile. The preferred T_e of 4 km used in reverse modelling was constrained by forward modelling. For $T_e = 25$ km uplift was decreased by 15 m to 360 m for the Witch Ground Graben, while using a $T_e = 0$ km (Airy isostasy) reduced the uplift estimate to 290 m. The accuracy of the β-factor was estimated to be ±0.05 for the preferred reverse model. Errors associated with the assumption of an instantaneous Late Jurassic rift event are not significant for post-rift thermal subsidence in the Early Tertiary.

The accommodation space for Eocene–Recent sedimentation was largely provided by this starved Late Cretaceous palaeobathymetry, made temporarily unavailable by the transient Palaeocene uplift event. No additional mechanism, such as Tertiary extension or middle lithosphere basalt–

eclogite phase transitions (Hall & White 1994), is required to generate the Tertiary stratigraphy of the northern North Sea Basin. Seismic reflection data does not support Tertiary extension in the northern North Sea of the magnitude proposed ($\beta = 1.05$–1.20) by Hall & White (1994), where Tertiary extension is restricted to small intra-sediment faults associated with slope failure and compaction (e.g. Bertram & Milton 1989).

Regional Palaeocene uplift, Eocene subsidence and the development of a mantle plume

Several potential mechanisms may be examined for the origin of the regional Palaeocene uplift event which effected the northern North Sea Basin. These mechanisms include eustasy (already discounted), Early Tertiary compressional inversion, Early Tertiary extension, intra-plate stress changes, opening of the North Atlantic Ocean, and plume activity. These mechanisms are discussed in Nadin & Kusznir (1995) and, with the exception of the plume mechanism, are discounted. Plume activity, most probably related to the development of the Iceland plume, is felt to be the most likely mechanism for Palaeocene uplift in the northern North Sea.

The timing of Palaeocene uplift within the northern North Sea Basin, and the simultaneous uplift and erosion of the Scottish mainland and northern England (Rochow 1981; Bertram & Milton 1989; Milton *et al.* 1990; Green *et al.* 1993), coincides with intense volcanism within the British Tertiary Igneous Province, in E and W Greenland, and along the North Atlantic continental margin between Greenland and NW Europe (White & McKenzie 1989). This volcanism commenced at *c.* 64–62 Ma and continued to *c.* 55 Ma (Mussett *et al.* 1988; White & McKenzie 1989). This volcanism was associated with the development of the Iceland plume in the Early Tertiary and the initiation of sea-floor spreading in the northern North Atlantic. Figure 1 shows an end-Palaeocene plate reconstruction with two different locations for the Iceland plume axis. White & McKenzie (1989) suggest the plume initiated beneath E Greenland at a distance of *c.* 700–900 km NW of the northern North Sea Basin. Lawver & Müller (1994) in contrast suggested that the plume tracked down from the NW and that in the Palaeocene the plume would have been situated beneath W Greenland at a greater distance from the northern North Sea Basin. At the time of initiation of North Atlantic sea-floor spreading in the Early Eocene (*c.* 55 Ma), the Iceland plume is likely to have underlain where the Atlantic continental margins and the Faeroes-Iceland-Greenland Ridge intersect, in order that the thick ocean crust of this aseismic ridge (Bott 1988) may have been generated by the interaction of the plume with sea-floor spreading. The volcanism of the British Tertiary Igneous Province, which commenced before the volcanism of E Greenland, may represent another plume distinct to that of Iceland, although related to the same mantle convection event responsible for the Iceland plume.

Courtney & White (1986) and White & McKenzie (1989) show that a rising plume of hot mantle generates uplift at the surface of the lithosphere by three distinct mechanisms:

(a) dynamic long wavelength regional uplift due to the fluid-flow field of ascending plume material;
(b) localized uplift due to thermal buoyancy from lithosphere heated by underlying hot plume material;
(c) localized and permanent uplift due to magmatic underplating.

Regional Palaeocene uplift observed in the northern North Sea Basin may be explained by the initiation or approach of a mantle plume. It is proposed that uplift in the northern North Sea is due to long wavelength dynamic uplift due to the fluid-flow field of the plume (Fig. 5a & c). The rapidity of regional Eocene subsidence precludes a thermal mechanism involving heating of lithosphere material under the northern North Sea Basin. Magmatic underplating is not a possible mechanism for regional Palaeocene uplift in the northern North Sea Basin since this mechanism would have produced permanent uplift which is not observed. The northern North Sea Basin may have been too far from the axis of the plume for heating of the underlying asthenosphere or for magmatic underplating to have occurred (Fig. 5). In contrast the Scottish mainland, northern England, and the N Irish Sea appears to have experienced permanent uplift which suggests these areas to the west underwent magmatic underplating (Brodie & White 1994).

Regional Eocene subsidence following plume related uplift (Fig. 5b) could be caused by a decrease in the dynamic uplift from the fluid-flow field as the strength of plume activity diminished. The decrease in regional uplift for the northern North Sea in the Eocene is rapid and precludes a mechanism involving cooling of asthenosphere or lithosphere (heated earlier by plume activity) which would occur too slowly with a time constant of 65 Ma or more. The rapid Eocene subsidence may have been caused by a decrease in the dynamic uplift of the plume fluid-flow field, coinciding with the initiation of sea-floor spreading in the Norwegian–Greenland Sea at around magnetic anomaly Chron 23 times (*c.* 55 Ma). If the Iceland plume was located in the vicinity of E Greenland, as suggested by White & McKenzie (1989) and supported by the formation of the Greenland–Iceland–Faeroes Ridge (Bott 1988), sea-floor spreading may have focused and tapped the rising plume flow onto the sea-floor spreading axis. Plume outward flow within the asthenosphere may have been reduced in the direction perpendicular to the newly formed ridge system, causing a reduction in dynamic support to the NW and SE, i.e. in the northern North Sea Basin. This could have led to rapid Eocene subsidence in the northern North Sea Basin and a return to normal post-rift subsidence in the Mid–Late Eocene (Figs 2 & 3).

Dynamic uplift from a mantle plume

Courtney & White (1986) analysed heat flow, geoid and bathymetric anomalies for the Cape Verde Rise in the North Atlantic. Using an axi-symmetric convection model they concluded that dynamic uplift generated by an ascending mantle plume is of greater importance than uplift caused by heating and expansion of the lithosphere. A dynamic mechanism is able to generate uplift without a corresponding increase in heat flow as is required by observations. In addition, the geoid anomaly data indicates that bathymetry is supported by deep processes within the mantle.

The plume model of Courtney & White assumes a uniform viscosity of 2–4×10^{20} Pa s. Furthermore, it is a

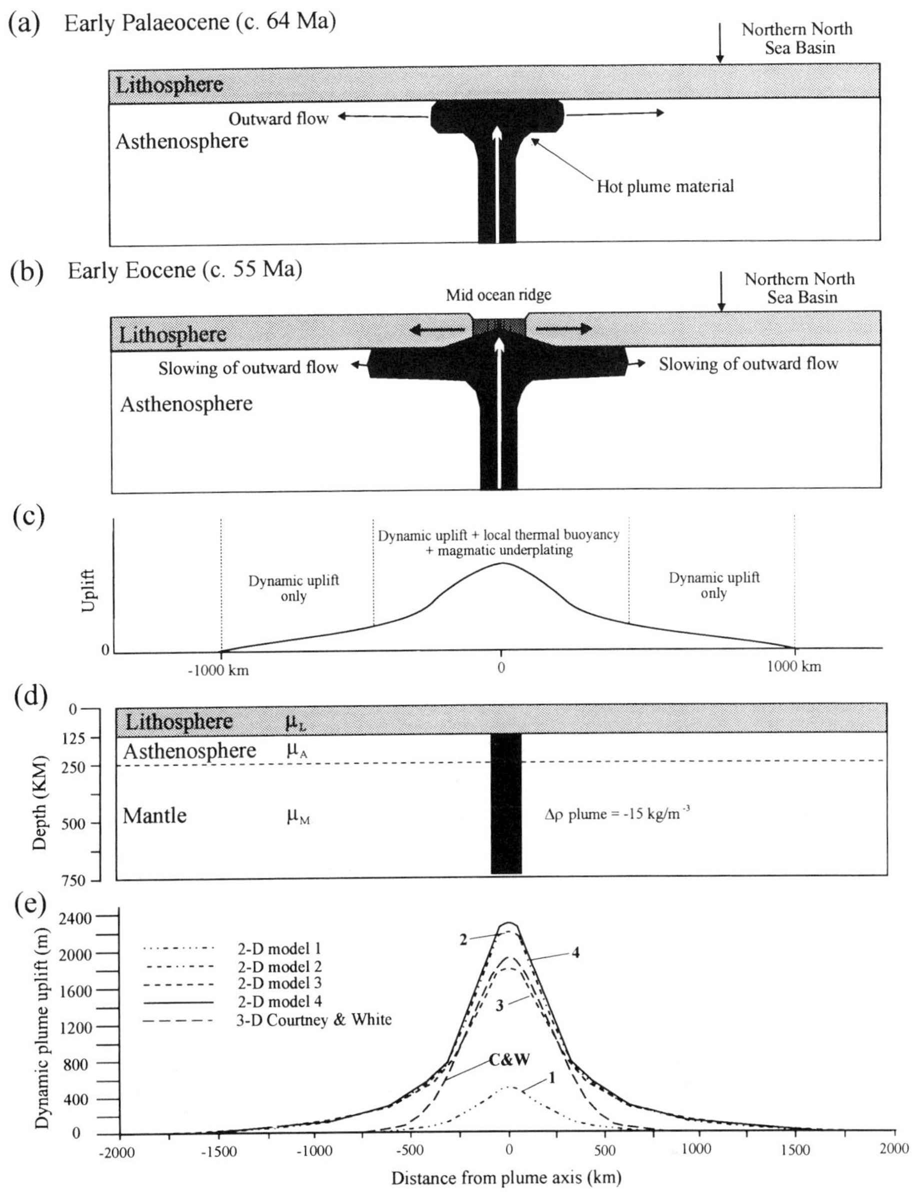

Fig. 5. (**a**) Development of the Iceland plume leads to regional Palaeocene uplift in the northern North Sea Basin. (**b**) Decrease in plume activity leads to rapid regional Eocene subsidence. (**c**) Schematic uplift arising from a mantle plume in the early stages of its formation. Note far field uplift due to lateral fluid-flow in asthenosphere not yet heated by plume material. (**d**) Simplified 2D model of mantle plume which has just risen to the top of the asthenosphere but which has not yet introduced hot material into laterally adjacent asthenosphere. (**e**) Vertical uplift predicted by 2D models 1–4 and 3D model of Courtney & White (1986). Increasing lithosphere viscosity increases surface uplift.

steady state model and has hot asthenosphere material beneath uplifted lithosphere away from the plume axis. A simplified 2D fluid-flow model has been used to investigate the sensitivity of surface uplift to the viscosity of the lithosphere for a plume which is rising through the mantle but has not yet laterally spread hot material within the asthenosphere (Fig. 5d). The fluid-flow model assumes an incompressible Newtonian fluid with flow equations solved using the finite element method. Densities are referenced to a standard density depth profile. A density contrast of $-15\,\mathrm{kg\,m^{-3}}$, estimated from the convection models of Courtney & White (1986) and Watson & McKenzie (1991) has been used for hot plume material and drives the ascending flow. The plume is given a width of 150 km and extends from the base of the lithosphere to a depth of 700 km (Fig. 5d). Surface uplift is computed from the vertical stress on the upper boundary of the model which is constrained for zero vertical flow. Sides and base of the model are free boundaries.

Profiles of vertical surface uplift across the 2D plume model are shown in Fig. 5e. Lithosphere, asthenosphere and mantle viscosities are listed in Table 1. All models show uplift which is maximum over the centre of the rising plume body and which decays to zero away from the plume. Uplift occurs not only above the rising hot plume material but also away to the side where no hot buoyant plume material exists. The magnitude of the uplift is very sensitive to the

Table 1. *Table of viscosities for lithosphere, asthenosphere and mantle used in the plume uplift models* 1–4 (*Fig.* 5)

Model	μ_L (Pa s)	μ_A (Pa s)	μ_M (Pa s)
1	10^{21}	10^{21}	10^{21}
2	10^{23}	10^{21}	10^{21}
3	10^{23}	10^{20}	10^{21}
4	10^{23}	10^{20}	10^{20}
Courtney & White 1986	$2–4 \times 10^{20}$	$2–4 \times 10^{20}$	$2–4 \times 10^{20}$

Plume $\mu = 10^{20}$ Pa s.

viscosity of the lithosphere. Increasing the viscosity of the lithosphere increases the amplitude of the uplift. Model 1 with low lithosphere viscosity of 10^{21} Pa s shows surface uplift of approximately 500 m, while model 2 with a higher lithosphere viscosity of 10^{23} Pa s shows uplift of almost 2500 m. Reducing the viscosity in the asthenosphere and mantle gives a small reduction in uplift (compare models 2 and 3). Model 4 has a similar uniform viscosity to the model of Courtney & White (1986). For the same viscosities the uplift profile of the 2D model (model 4) is wider than the uplift profile of the axi-symmetric Courtney & White (1986) model. As would be expected the uplift of the 3D model attenuates more rapidly with distance away from the plume axis than the 2D model. The results of the 2D model are of limited applicability and should be regarded as being only semi-quantitative, however, they do show that dynamic uplift can occur even when material is not underlain by hot mantle and that this uplift is very sensitive to the strength of the lithosphere.

References

Andrews, I. J., Long, D., Richards, P. C., Thomson, A. R., Brown, S., Chesher, J. A. & McCormac, M. 1990. *The geology of the Moray Firth. United Kingdom offshore regional report.* HMSO, London, for the British Geological Survey.

Barr, D. 1991. Subsidence and sedimentation in semi-starved half-graben: a model based on North Sea data. *In*: Roberts, A. M., Yielding, G. & Freeman, B. (eds) *The geometry of Normal Faults.* Geological Society, London, Special Publications, **56,** 17–28.

Bertram, G. T. & Milton, N. J. 1989. Reconstructing basin evolution from sedimentary thickness: the importance of palaeobathymetrical control, with reference to the North Sea. *Basin Research*, **1,** 247–257.

Bott, M. H. P. 1988. Consequences of the Icelandic hot-spot. *In*: Morton, A. C. & Parson, L. M. (eds) *Early Tertiary Volcanism and opening of the NE Atlantic.* Geological Society, London, Special Publications, **39,** 15–23.

Brodie, J. & White, N. J. 1994. Sedimentary basin inversion caused by igneous underplating: Northwest European continental shelf. *Geology*, **22,** 147–150.

Courtney, R. C. & White, R. S. 1986. Anomalous heat flow and geoid across the Cape Verde rise: evidence for dynamic support from a thermal plume in the mantle. *Geophysical Journal of the Royal Astronomical Society*, **87,** 815–867.

Giltner, J. P. 1987. Application of extensional models to the Northern Viking Graben. *Norsk Geologisk Tidsskrift*, **67,** 339–352.

Green, P. F., Duddy, I. R., Bray, R. J. & Lewis, C. L. E. 1993. Elevated palaeotemperatures prior to Early Tertiary cooling throughout the UK region: Implications for hydrocarbon generation. *In*: Parker, J. R. (ed.) *Petroleum Geology of NW Europe*: *Proceedings of the 4th Conference* Geological Society, London, 1067–1074.

Hall, B. D. & White, N. J. 1994. Origin of anomalous Tertiary subsidence adjacent to North Atlantic continental margins. *Marine and Petroleum Geology*, **11,** 702–714.

Hancock, J. M. 1984. Cretaceous. *In*: Glennie, K. W. (ed.) *Introduction to the petroleum geology of the North Sea*, (1st edition). Blackwell Scientific Publications, Oxford, 133–150.

Haq, B. U., Hardenbol, J. & Vail, P. R. 1987. Chronology of fluctuating sea-levels since the Triassic. *Science*, **235,** 1156–1167.

Jones, R. W. & Milton, N. J. 1994. Sequence development during uplift: Palaeogene stratigraphy and relative sea-level history of the Outer Moray Firth, UK North Sea. *Marine and Petroleum Geology*, **11,** 157–165.

Joppen, M. & White, R. S. 1990. The structure and subsidence of Rockall Trough from two-ship seismic experiments. *Journal of Geophysical Research*, **95,** 19 821–19 837.

Joy, A. M. 1992. Right place, wrong time: anomalous post-rift subsidence in sedimentary basins around the North Atlantic Ocean. *In*: Storey, B. C., Alabaster, T. & Pankhurst, R. J. (eds) *Magmatism and the causes of Continental Break-up*. Geological Society, London, Special Publications **68,** 387–393.

Keen, C. E., Loncarevic, B. D., Reid, I., Woodside, J., Haworth, R. T. & Williams, H. 1990. Tectonic and geophysical overview. *In*: Keen, M. & Williams, G. L. (eds) *Geology of the Continental Margin of Easte Canada.* Geological Survey of Canada, Geology of Canada, **2,** 31–85.

Kusznir, N. J. & Ziegler, P. A. 1992. The mechanism of continental extension and sedimentary basin formation: a simple-shear/pure-shear flexural cantilever model. *Tectonophysics*, **215,** 117–131.

——, Roberts, A. M. & Morley, C. 1995. Forward and Reverse Modelling of Rift Basin Formation. *In*: Lambiase, J. (ed.) *The Hydrocarbon Habitat of Rift Basins*. Geological Society, London, Special Publications, **88,** 3–56.

Lawver, L. A. & Müller, R. D. 1994. Iceland hotspot track. *Geology*, **22,** 311–314.

Marsden, G., Yielding, G., Roberts, A. M. & Kusznir, N. J. 1990. Application of a flexural cantilever simple-shear/pure-shear model of continental lithosphere extension to the formation of the northern North Sea Basin. *In*: Blundell, J. & Gibbs, A. D. (eds) *Tectonic Evolution of the North Sea Rifts*. Oxford University Press, 241–261.

McKenzie, D. P. 1978. Some remarks on the development of sedimentary basins. *Earth Planetary Science Letters*, **40,** 25–32.

Milton, N.J., Bertram, G. T. & Vann, I. R. 1990. Early Palaeogene tectonics and sedimentation. *In*: Hardman, R. F. P. & Brooks, J. (eds) *Tectonic Events Responsible for Britain's Oil and Gas Reserves*. Geological Society, London, Special Publications, **55,** 339–351.

Mussett, A. E., Dagley, P. & Skelhorn, R. R. 1988. Time and duration of activity in the British Tertiary Igneous Province. *In*: Morton, A. C. & Parson, L. M. (eds) *Early Tertiary Volcanism and Opening of the NE Atlantic.* Geological Society, London, Special Publications, **39,** 337–348.

Nadin, P. A. & Kusznir, N. J. 1995. Palaeocene uplift and Eocene subsidence in the northern North Sea from 2D forward and reverse stratigraphic modelling, *Journal of the Geological Society, London,* **152,** 833–848.

—— & Kusznir, N. J. 1996. Forward and reverse stratigraphic modelling of Cretaceous–Tertiary post-rift subsidence and Paleogene uplift in the Outer Moray Firth Basin, Central North Sea. *In*: Knox, R. W. O'B., Dunay, R. E. & Corfield, R. (eds) *Correlation of the Early Paleogene of NW Europe*. Geological Society, London, Special Publications, **101,** 43–62.

Roberts, A. M., Yielding, G., Kusznir, N. J., Walker, I. & Dorn-Lopez, D. 1993. Mesozoic extension in the North Sea: constraints from flexural backstripping, forward modelling and fault populations. *In*: Parker, J. R. (ed.) *Petroleum Geology of NW Europe*: *Proceedings of the 4th Conference* Geological Society, London, 1123–1136.

Rochow, K. A. 1981. Seismic stratigraphy of the North Sea 'Palaeocene' deposits. *In*: Illing, L.V. & Hobson, G. D. (eds) *Petroleum Geology of the Continental Shelf of NW Europe.* Heyden & Son, London, 255–266.

Sclater, J. G. & Christie, P. A. F. 1980. Continental stretching: an explanation of the post-mid-Cretaceous subsidence in the North Sea Basin. *Journal of Geophysical Research*, **85,** 3711–3739.

Watson, S. & McKenzie, D.P. 1991. Melt generation by plumes: a study of Hawaiin volcanism. *Journal of Petrology*, **32,** 501–537.

White, N. J. & Latin, D. M. 1993. Subsidence analyses from the North Sea 'triple junction'. *Journal of the Geological Society, London*, **150,** 473–488.

White, R. S. & McKenzie, D. P. 1989. Magmatism at rift zones: The generation of volcanic continental margins and flood basalts. *Journal of Geophysical Research*, **94,** 7685–7729.

Young, R. 1992. Restoration of a regional profile across the Magnus Field in the northern North Sea. *In*: Larson, R. M., Brekke, H., Larson, B. T. & Talleraas, E. (eds) *Structural and Tectonic Modelling and its Application to Petroleum Geology*. Norges Petroleums forening Special Publication, **1,** 221–229.

Received 24 September 1994; revised typescript accepted 12 June 1995.

SPECIAL

Linking onshore and offshore structures: Mesozoic extension in the Scottish Highlands

A. M. ROBERTS[1] & R. E. HOLDSWORTH[2]

[1]*Badley Earth Sciences Ltd, North Beck House, North Beck Lane, Hundleby, Spilsby, Lincs PE23 5NB, UK (e-mail: alan@badleys.co.uk)*
[2]*Reactivation Research Group, Department of Geological Sciences, University of Durham, Durham DH1 3LE, UK*

Geological relationships in the offshore basins flanking northern Scotland suggest that a regional Caledonian control on the structure of these basins is far from proven, but conversely Mesozoic faults of the offshore set may extend onshore. The Strathconnon fault, long held to be late Caledonian, may be Mesozoic in age, forming the principal kinematic link between the east and west coast basins of northern Scotland. The numerous outcrops of Precambrian basement within the Minch/Hebrides basin nearly all lie in the uplifted footwalls of Mesozoic faults. Similar fault controls on the distribution of onshore outcrop within the Precambrian may exist.

Keywords: Scotland, Precambrian, Caledonian Orogeny, Mesozoic, basins, reactivation.

Both the onshore and offshore regions of northern Scotland are classic areas for the study of structural geology and regional tectonics. The former for its Precambrian and Caledonian orogenic ('basement') structures, the latter for its Mesozoic extensional basins. With some notable exceptions (e.g. Snyder 1990; Underhill & Brodie 1993; Butler & Hutton 1994) there have, however, been few serious attempts to integrate the structural evolution of the onshore and offshore areas, and the prevalent view remains that these are two distinct structural provinces, with old orogenic structures onshore and younger basins offshore. In this paper we advance the hypothesis that some offshore Mesozoic structures extend onshore and that some structures in basement (and traditionally regarded as Caledonian) may be younger, or at least exhibit a younger component of movement.

While the existence of onshore Mesozoic structures has received little previous attention there has been much speculation about the control exerted by Caledonian and other basement structures on the development of offshore Mesozoic basins (e.g. Snyder 1990; Bartholomew *et al.* 1993). It has been proposed, for example, that many Caledonian brittle and ductile thrusts and steeply-dipping strike-slip faults (e.g. the Highland Boundary/Great Glen Fault trends) control the location and geometry of Mesozoic offshore structures. Reactivation of basement structures is a widely recognized phenomenon in the continental lithosphere (e.g. Holdsworth *et al.* 1997), but many advocates of offshore regional reactivation hypotheses present little or no hard evidence to support such models. In the absence of supporting data we see no reason to accept such reactivation hypotheses *carte blanche* and this paper illustrates that it may be instructive to look at alternative ideas, i.e. that the younger faults represent important onshore kinematic links between spatially separated offshore fault sets and that they may form independently of basement structure. What follows is a brief appraisal of relevant examples from the offshore basins bordering northern Scotland, at the conclusion of which the possible influence of Mesozoic faults onshore will be discussed.

The Moray Firth Basin. For a long time, following the work of McQuillin *et al.* (1982), the Moray Firth basin was taken to be a *prima facie* case of a Mesozoic basin which had formed by reactivation of a Caledonian fault, in this case dextral slip on the Great Glen Fault (Fig. 1). Subsequently, however, Underhill (1991) has demonstrated that the Great Glen Fault was dormant during the formation of the Moray Firth and that instead, the basin opened as a result of extensional slip on the Helmsdale Fault (Fig. 1), a structure exposed onshore with syn-depositional Jurassic sediments in its hanging wall.

Kinematically the Great Glen story was very convenient, because by transferring displacement along the length of the Great Glen Fault into the basins of the Scottish west coast region it allowed the problem of locating the onshore accommodation of the Moray Firth extension to be circumvented. Opening the basin by extension on the Helmsdale Fault does not allow this solution, yet the large extension on the Helmsdale Fault must be kinematically linked to other structures, otherwise there is a problem in conserving regional strain at the western margin of the basin. The simplest explanation would be if the displacement were transferred to an (as yet unspecified) en echelon structure (or structures) onshore.

Offshore north coast, Scotland. Following the advent of deep seismic reflection profiling offshore UK, perhaps the most cited example of evidence for Caledonian control on offshore structures has been the correlation of onshore brittle and ductile thrusts (e.g. Barr *et al.* 1986) with structures imaged on seismic data off the north coast of Scotland (MOIST, Brewer & Smythe 1984; DRUM, Snyder 1990) (Fig. 1). The recognition of west-dipping half-graben (West Orkney basin) formed by the apparent reactivation of older east-dipping thrusts gave rise to elaborate theories about Devonian post-orogenic collapse basins (e.g. Enfield & Coward 1987). More recent work, however, has demonstrated that the sediment fill of the West Orkney basin is predominantly Permo/Triassic in age, not Devonian (e.g. Stoker *et al.* 1993, figs 33 & 34) and that onshore extensions of offshore faults demonstrably cut across all basement structures (Holdsworth 1989). This evidence does not preclude the offshore structures being reactivated Caledonian thrusts, but there remains only a geometric correlation of bright seismic events to the mapped onshore geology. The case for such a correlation is weakened still further when it is realized that bright seismic reflections from fault planes are not diagnostic of basement-involved structures, but that any fault exhibiting a significant velocity contrast will produce a bright fault-plane-reflection (e.g. Yielding *et al.* 1991). None of these observations preclude reactivation in the West Orkney basin, but they reduce the evidence for reactivation to no more than spatial proximity.

From Holdsworth, R.E. & Turner, J.P. (compiler) *Extensional Tectonics: Regional-scale Processes*. The Geological Society, Key Issues in Earth Sciences, **2 (1)**, 333–336.
1476-315X/02/$15.00 © The Geological Society of London 2002
First published in: "*Journal of the Geological Society*, **156**, 1061–1064".

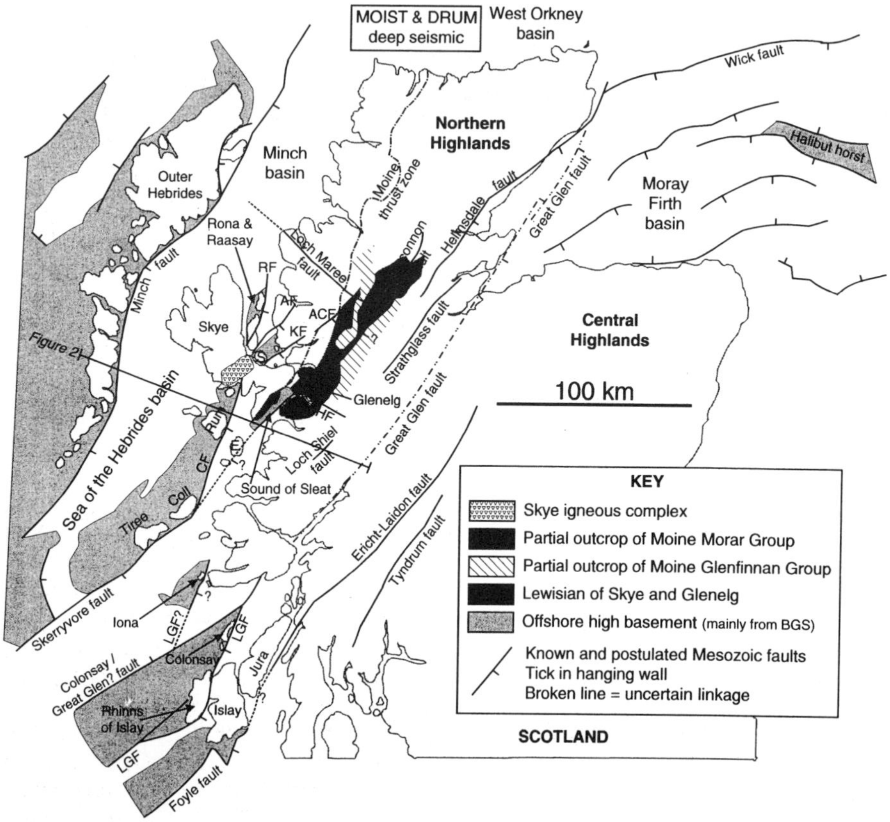

Fig. 1. Location map for the Scottish Highlands and adjacent offshore basins. The main known and postulated Mesozoic faults are shown, highlighting the possibility of structural continuity across the Highlands. The following abbreviations have been used: CF, Camasunary fault; RF, Raasay fault; AF, Applecross fault; ACF, Achnashellach fault; KF, Kishorn fault; KHF, Kinlochourn fault; LGF Loch Gruinart fault; S, Scalpay; E, Eigg.

Offshore and onshore basement highs in the North Sea and Hebrides regions. Within the offshore area north and east of Scotland many high-standing horsts and basement highs have been recognized (e.g. The Halibut horst, Fig. 1; the Shetland Platform 'granite' at 60°N; the Tern Eider ridge, Roberts *et al.* 1995). Such basement highs are commonly interpreted either as areas of long-lived basement elevation, 'shielded' from the surrounding basins by their inherited basement structure, or as 'pop-up' flower structures on strike-slip faults reactivating older Caledonian structures. In fact there is no need to invoke any special basement control on these structures, because they are all situated in the footwalls of major Mesozoic extensional faults (e.g. Roberts & Yielding 1991; Roberts *et al.* 1995). Far from being shielded from Mesozoic tectonism, they have participated fully in the formation of the extensional basins, having been elevated to their current structurally high positions by syn-extension footwall uplift. The case for invoking reactivation is unnecessary.

On a regional scale comparable footwall highs of elevated basement comprise the Outer Hebrides (Fig. 1). It has long been known that the Minch/Sea of the Hebrides area west of Scotland is part of a Mesozoic sedimentary basin (e.g. Morton 1992 and references therein). Seismic data have shown the structure of the basin to be essentially a west-dipping half-graben in the hanging wall of the Minch fault (e.g. Butler & Hutton 1994). In contrast to the North Sea, there are several large outcrops of Precambrian basement (Lewisian, Torridonian) exposed within the basin above sea-level. The outcrops in question form all or part of the islands of the Outer Hebrides, Coll, Tiree, Iona, Skye, Rhum, Raasay, Rona and Scalpay (Fig. 1).

The probable reason for this contrast lies in two separate geological processes. First consider the structural setting of all the Precambrian outliers. The largest and easiest to place in its structural context is the Lewisian outlier forming the Outer Hebrides. Regional cross-sections through the Minch Basin (e.g. Morton 1992; Butler & Hutton 1994; Fig. 2) show that the Outer Hebrides sit in the footwall of the large-displacement Minch fault, the Mesozoic master fault to the Minch/Sea of the Hebrides basin. The Outer Hebrides can therefore be viewed as a classic footwall high, elevated by footwall uplift (Fig. 2) and elongate in outcrop along the Minch fault (Fig. 1).

Next consider two other major Lewisian outliers, the islands of Coll and Tiree (Fig. 1). Morton (1992, fig. 1) showed that these islands sit in the footwall of another large Mesozoic fault, the Skerryvore fault (see also Fyfe *et al.* 1993 & BGS 1:250 000 Tiree map sheet). These two islands also therefore comprise part of an elevated footwall high, in the footwall of the next major fault eastwards from the Minch fault (Fig. 1). Moving northwards, the Skerryvore fault is mapped into the Camasunary fault (Morton 1992; Fyfe *et al.* 1993), which crops out on Skye and passes immediately east of the island of Rum (Figs 1 & 2). Outliers of Precambrian are also exposed in the footwall of the Camasunary fault, although the elevated Precambrian rocks are the younger Torridonian sequence (Morton 1992; Fyfe *et al.* 1993; Butler & Hutton 1994).

Butler & Hutton (1994, fig. 6) suggested that the single strand of the Camasunary fault in SW Skye partitions its displacement northwards into three discrete faults, the Raasay fault, the Applecross fault and the Kishorn fault. The Raasay fault exposes the Torridonian and Lewisian of Raasay and Rona in its footwall, and in the footwall of the Kishorn fault lies the Torridonian of Scalpay (Fig. 1).

By now the structural setting of the Precambrian outliers within the Minch basin is becoming clear; they lie consistently within the footwalls of Mesozoic faults, active during the

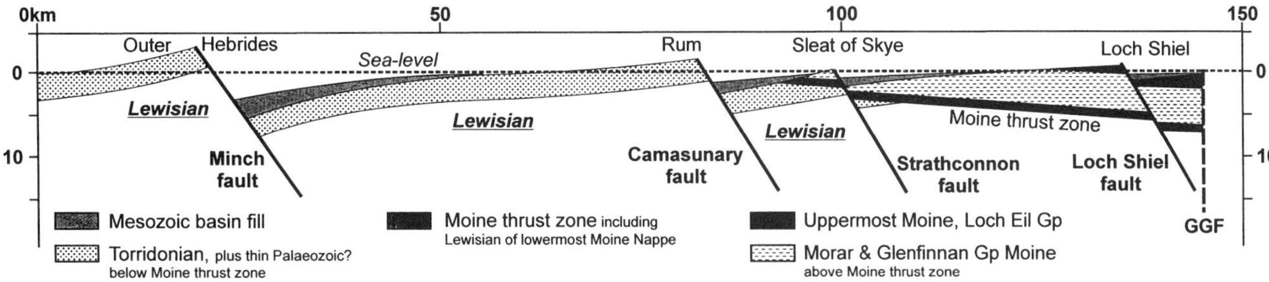

Fig. 2. Schematic cross-section (at true vertical scale) from the Outer Hebrides to the Great Glen Fault (GGF) during the Jurassic, showing how varying components of the Precambrian/Caledonian basement came to be elevated in the footwalls of Mesozoic faults. The present-day erosion level is approximately 1–2 km below the sea-level datum of the section. Stratigraphic thicknesses are not intended to be accurate. See Fig. 1 for location.

formation of the basin. North of the Great Glen fault one remaining 'island outcrop' of Precambrian remains to be explained, the Lewisian and associated Precambrian cover rocks of Iona (Fig. 1). Potts *et al.* (1995) have explained the structures observed on the island of Iona as the result of late Caledonian ductile faulting. Regardless of the Caledonian extensional history in the Sound of Iona, we believe that the geological observation of Precambrian basement exposed on Iona suggests an analogy with Coll, Tiree, Raasay etc. and that there is likely to have been Mesozoic movement and footwall uplift on the Iona fault. Two lines of circumstantial evidence lend support to this idea. First, there is known to have been post-Caledonian fault movement in this area because the Assapol fault on the Ross of Mull has Palaeocene lavas in its downthrown hanging wall (Holdsworth *et al.* 1987). Second, Potts *et al.* (1995) suggested a southwards correlation (with which we agree) of the Iona fault (across the Great Glen Fault) with the Loch Gruinart fault (Fig. 1). The Loch Gruinart fault south of Islay contains Permo-Triassic and Lower Jurassic rocks in its hanging wall (Fyfe *et al.* 1993) and thus must have accommodated Mesozoic displacement. Where the Loch Gruinart fault traverses the island of Islay it exposes the Precambrian basement rocks of the Rhinns complex in its footwall and between Islay and the Great Glen fault the Precambrian of the island of Colonsay is likewise exposed. Fitches & Maltman (1984) documented the tectonics of Islay and Colonsay and described the Loch Gruinart fault as 'post-Permian'.

In summary, we suggest that from the Rhinns of Islay northwards to the Outer Hebrides, wherever Precambrian rocks crop out within the Minch/Sea of the Hebrides Basin, the localized outcrop can be explained by footwall uplift adjacent to the Mesozoic faults which formed the basin.

Having outlined the structural setting of the Precambrian outliers, their presence at the present surface, rather than below a Cretaceous–Recent post-rift sequence (cf. the North Sea), must be explained. Brodie & White (1994) have done so by postulating that the basins west of northern Scotland and the adjacent mainland area were underplated by basaltic melt during early Tertiary opening of the north Atlantic, giving rise to a permanent uplift. This allows the uplifted Precambrian footwalls to be exposed at or above present-day sea-level. Thus the distribution of Precambrian basement in the Hebrides results from fault-controlled footwall uplift during the Mesozoic, followed by early Tertiary regional uplift as a result of underplating.

The continuation of Mesozoic extension onshore: the Strathconnon fault. Having demonstrated that the present day outcrop of offshore Precambrian is controlled by Mesozoic faults, can we extrapolate this relationship (or the faults themselves) onshore into the northern Highlands? The Applecross and Kishorn faults (Fig. 1) are known to extend a short distance onshore and at the east coast, the Helmsdale fault passes onshore. Between these structures consider now the Strathconnon fault (Fig. 1) (see BGS 1:250 000 Great Glen sheet), which exhibits the following features. (1) It shares a common NNE–SSW strike with both the Minch/Hebrides faults and the Helmsdale Fault. (2) It has the outcrop geometry to accommodate relay overlap between the Helmsdale Fault and the faults in the Minch. (3) Assuming that like the Minch, Camasunary and Helmsdale Faults it dips to the east at depth then in the Sound of Sleat and nearby onshore, the Strathconnon fault exposes the Lewisian of Skye and Glenelg (within the Moine Nappe) in its footwall (Fig. 2, Fyfe *et al.* 1993, Great Glen sheet), presenting an analogous geometry to the Minch fault, Skerryvore fault, etc. (4) Again assuming an easterly dip, the Strathconnon Fault onshore is responsible along much of its length for juxtaposing the deeper Morar Group of the Moine in its footwall against the shallower Glenfinnan Group in its hanging wall, i.e. the Morar Group may be uplifted in the footwall. (5) If the trace of the fault in the Sound of Sleat is extrapolated offshore beneath the Tertiary lavas of Eigg it projects directly into the mapped trace of the Skerryvore Fault at Coll (Fig. 1). Arising from point 5, we suggest that the Skerryvore fault splays northwards into the Camasunary Fault and the Strathconnon Fault, perhaps explaining why Lewisian is exposed adjacent to the Skerryvore segment (Coll, Tiree), where the summed displacement is large, but younger Torridonian is exposed adjacent to the Camasunary segment (Rum, Skye), where the partitioned displacement is smaller (Fig. 2).

If it is accepted that the issue of conserving strain at the western margin of the Moray Firth requires transfer of displacement onshore into the northern Highlands and that the extension from the Minch basin may also extend onshore, then we suggest that the Strathconnon fault is the most likely structure in the northern Highlands to be the Mesozoic kinematic link between the east and west coasts of Scotland. Of course not all of the extension present in the offshore basins can be relayed across the mainland, otherwise the 'mainland' too would be predominantly below sea-level. We suggest, however, that a component of extension may be relayed between the offshore basins. Following from this suggestion it is possible, although untested, that the shorter Loch Shiel (Fig. 2), Strathglass and Achnashellach faults (Fig. 1) may be part of the Mesozoic fault set, while the Loch Maree fault (Fig. 1), mapped into the Minch as the Mid-Minch High, has the appropriate orientation (NW–SE) for an extension-parallel transfer fault, as too does the Kinlochourn Fault (Fig. 1).

South of the Great Glen this hypothesis may be developed further. Fyfe *et al.* (1993) have previously suggested that the Ericht–Laidon and Tyndrum Faults (Fig. 1) in the central Highlands may respectively connect across the Malin Sea to the Foyle and Tow Valley Faults in Northern Ireland. As the Foyle and Tow Valley faults are mapped offshore as part of the Mesozoic basin complex (Malin 1:250 000 sheet) this would imply a Mesozoic history for the Ericht–Laidon and Tyndrum faults, comparable with our interpretation of the Strathconnon fault.

Re-examining the orthodoxy. We believe that by integrating onshore and offshore geological relationships enough evidence can be put forward to make the case for a component of Mesozoic extension having traversed the Scottish Highlands on regionally recognized NNE–SSW-trending extensional faults. These faults are traditionally held to be Caledonian, but perhaps it is time now to challenge this assumption. While the Great Glen Fault is demonstrably a Caledonian structure (Stewart *et al.* 1999) with only minor post-Palaeozoic reactivation (Underhill 1991), and the Helmsdale Fault is clearly a Mesozoic fault which extends onshore, what of the other faults? Could they be Mesozoic-only in age, or have they demonstrably reactivated older structures? If the onshore faults are predominantly Mesozoic, then the whole assumption of regional Caledonian control on the development of offshore basins becomes increasingly uncertain.

We do not attempt to provide the definitive answer to these questions here, but rather raise the issue that by attempting to integrate onshore and offshore structural relationships perhaps we can look at an area of classic structural geology in a new light. Our discussion also highlights the need for further studies to test this hypothesis. Such studies might include dating fault movements from a combination of offset igneous features, Ar–Ar dating, and palaeomagnetic data (e.g. Torsvik *et al.* 1992), together with structural studies assessing whether or not reactivation has occurred (e.g. Stewart *et al.* 1999).

The authors would like to thank M. Badley and T. Harris for discussions on new ways to look at old rocks. J. Turner is thanked for his review and N. Rogers for his editorial comments. The template for Fig. 2 was prepared using N. Kusznir's STRETCH software.

References

BARR, D., HOLDSWORTH, R.E. & ROBERTS, A.M. 1986. Caledonian ductile thrusting in a Precambrian metamorphic complex: The Moine of northwestern Scotland. *Geological Society of America Bulletin*, **97**, 754–764.

BARTHOLOMEW, M.J., PETERS, J.M. & POWELL, C.M. 1993. Regional structural evolution of the North Sea: oblique slip and the reactivation of basement lineaments. *In*: PARKER, J.R. (eds) *Petroleum Geology of Northwest Europe, Proceedings of the 4th Conference*. The Geological Society, London, 1109–1122.

BREWER, J.A. & SMYTHE, D.K. 1984. MOIST and the continuity of crustal reflector geometry along the Caledonian-Appalachian orogen. *Journal of the Geological Society, London*, **141**, 105–120.

BRODIE, J.A. & WHITE, N.J. 1994. Sedimentary basin inversion caused by igneous underplating: Northwest European continental shelf. *Geology*, **22**, 147–150.

BUTLER, R.W.H. & HUTTON, D.H.W. 1994. Basin structure and Tertiary magmatism on Skye, NW Scotland. *Journal of the Geological Society, London*, **151**, 931–944.

ENFIELD, M.A. & COWARD, M.P. 1987. The structure of the West Orkney Basin, northern Scotland. *Journal of the Geological Society, London*, **144**, 871–844.

FITCHES, W.R. & MALTMAN, A.J. 1984. Tectonic development and stratigraphy at the western margin of the Caledonides: Islay and Colonsay, Scotland. *Transactions of the Royal Society of Edinburgh: Earth Sciences*, **75**, 365–382.

FYFE, J.A., LONG, D. & EVANS, D. 1993. *The geology of the Malin–Hebrides area*. United Kingdom offshore regional reports. British Geological Survey.

HOLDSWORTH, R.E. 1989. Late brittle deformation in a Caledonian ductile thrust wedge: new evidence for gravitational collapse in the Moine Thrust sheet, Sutherland, Scotland. *Tectonophysics*, **170**, 17–28.

——, BUTLER, C.A. & ROBERTS, A.M. 1997. The recognition of reactivation during continental deformation. *Journal of the Geological Society, London*, **154**, 73–78.

——, HARRIS, A.L. & ROBERTS, A.M. 1987. The stratigraphy, structure and regional significance of the Moine Rocks of Mull, Argyllshire, W. Scotland. *Geological Journal*, **22**, 83–107.

MCQUILLIN, R., DONATO, J.A. & TULSTRUP, J. 1982. Development of basins in the Inner Moray Firth and the North Sea by crustal extension and dextral displacement of the Great Glen Fault. *Earth and Planetary Science Letters*, **60**, 127–139.

MORTON, N. 1992. Dynamic stratigraphy of the Triassic and Jurassic of the Hebrides basin, NW Scotland. *In*: PARNELL, J. (eds) *Basins on the Atlantic Seaboard: Petroleum geology, Sedimentology and Basin Evolution*. Geological Society, London, Special Publications, **62**, 97–110.

POTTS, G.J., HUNTER, R.H., HARRIS, A.L. & FRASER, F.M. 1995. Late-orogenic extensional tectonics at the NW margin of the Caledonides in Scotland. *Journal of the Geological Society, London*, **152**, 907–910.

ROBERTS, A.M. & YIELDING, G. 1991. Deformation around basin-margin faults in the North Sea/Norwegian rift. *In*: ROBERTS, A.M., YIELDING, G. & FREEMAN, B. (eds) *The Geometry of Normal Faults*. Geological Society, London, Special Publications, **56**, 61–78.

——, ——, KUSZNIR, N.J., WALKER, I. & DORN-LOPEZ, D. 1995. Quantitative analysis of Triassic extension in the Northern Viking Graben. *Journal of the Geological Society, London*, **152**, 15–26.

SNYDER, D.B. 1990. The Moine Thrust in the BIRPS data set. *Journal of the Geological Society, London*, **147**, 81–86.

STEWART, M.A., STRACHAN, R.A. & HOLDSWORTH, R.E. 1999. The structure and early kinematic history of the Great Glen Fault Zone, Scotland. *Tectonics*, **18**, 326–342.

STOKER, M.S., HITCHEN, K. & GRAHAM, C.C. 1993. *The geology of the Hebrides and West Shetland shelves, and adjacent deep-water areas*. United Kingdom offshore regional report. British Geological Survey.

TORSVIK, T.H., STURT, B.A., SWENSSON, E., ANDERSEN, T.B. & DEWEY, J.F. 1992. Palaeomagnetic dating of fault rocks: evidence for Permian and Mesozoic movements and brittle deformation along the extensional Dalsfjord Fault, western Norway. *Geophysical Journal International*, **109**, 565–580.

UNDERHILL, J.R. 1991. Implications of Mesozoic-Recent basin development in the western Inner Moray Firth, UK. *Marine and Petroleum Geology*, **8**, 359–370.

—— & BRODIE, J.A. 1993. Structural geology of Easter Ross, Scotland: implications for movement on the Great Glen fault zone. *Journal of the Geological Society, London*, **150**, 515–527.

YIELDING, G., BADLEY, M.E. & FREEMAN, B. 1991. Seismic reflections from normal faults in the northern North sea. *In*: ROBERTS, A.M., YIELDING, G. & FREEMAN, B. (eds) *The Geometry of Normal Faults*. Geological Society, London, Special Publications, **56**, 79–89.

Received 29 April 1999; revised typescript accepted 3 June 1999.
Scientific editing by Nick Rogers.

Index